IRRADIATION TECHNOLOGY

IRRADIATION TECHNOLOGY

Proceedings of an International Topical Meeting
Grenoble, France
September 28-30, 1982

Edited by

Peter von der Hardt

and

Heinz Röttger

C.E.C., Joint Research Centre, Petten Establishment

D. Reidel Publishing Company

A MEMBER OF THE KLUWER ACADEMIC PUBLISHERS GROUP

Dordrecht / Boston / Lancaster

for the Commission of the European Communities

Library of Congress Cataloging in Publication Data
Main entry under title:

Irradiation technology.

Proceedings of the International Topical Meeting on Irradiation Technology sponsored by the European Nuclear Society and others.
1. Irradiation–Congresses. 2. Materials testing reactors–Congresses. 3. Materials–Effect of radiation on–Congresses. 4. Nuclear fuel elements–Effect of radiation on–Congresses. I. Hardt, Peter von der, 1934- II. Röttger, Heinz, 1927- III. International Topical Meeting on Irradiation Technology (1982 : Grenoble, France) IV. European Nuclear Society.
TK9151.4.I77 1983 620.1'1228 83-3042
ISBN 90-277-1568-8

Publication arrangements by
Commission of the European Communities
Directorate-General Information Market and Innovation, Luxembourg

Lay-out: Reproduction service J. R. C. Petten

EUR 8429

Published by D. Reidel Publishing Company
P.O. Box 17, 3300 AA Dordrecht, Holland

Sold and distributed in the U.S.A. and Canada
by Kluwer Boston Inc.,
190 Old Derby Street, Hingham, MA 02043, U.S.A.

In all other countries, sold and distributed
by Kluwer Academic Publishers Group,
P.O. Box 322, 3300 AH Dordrecht, Holland

Printed in The Netherlands

LIST OF CONTENTS

INTRODUCTION XIII

CONFERENCE ORGANISATION XIV

OPENING OF THE MEETING XV

SESSION I-IV: LIGHT WATER REACTORS

FACILITIES AND TECHNIQUES FOR INSTRUMENTED FUEL IRRADIATIONS IN THE NRX REACTOR AT CHALK RIVER
P.J. Fehrenbach 3

ETUDE EXPERIMENTALE ET THEORIQUE DU COMPORTEMENT DE LA GAINE DE L'ELEMENT COMBUSTIBLE
J. Novák, J. Šmíd 13

OPERATIONAL TRANSIENTS ON LWR FUEL IN BR2 REACTOR
H. Nackaerts 23

LA BOUCLE "ISABELLE 4": BOUCLE D'ESSAI PWR DECHARGEABLE ET RECHARGEABLE DANS LA PISCINE DU REACTEUR OSIRIS
M. Lucot 35

THE DIDO WATER LOOP AT HARWELL AND ITS USE IN PARTICULAR FOR PWR CRUD-DEPOSITION EXPERIMENTS
E.J. Mitchell, J. Monahan 49

TECHNIQUES FOR IRRADIATION TESTING OF NON-FUEL MATERIALS IN LIGHT WATER POWER REACTORS
M. Gärtner, D. Alter, P. Dewes 77

L'INSTRUMENTATION DES EXPERIENCES DE RAMPES POUR LA FILIERE A EAU PRESSURISEE
A. Jegat, P. Vulliez, C. Friboulet, P. Bourbousson 93

ERSTE BETRIEBSERFAHRUNGEN MIT EINER BESTRAHLUNGSEINRICHTUNG ZUM TESTEN VON BRENNSTÄBEN UNTER DRUCKWASSERREAKTORBEDINGUNGEN
A. Reymann, J. Ahlf, D. Bellmann, K.H. Blom, H.L. The 103

BATI MULTIPLE IMMERGE POUR LE CONTROLE DE CRAYONS PWR IRRADIES DANS UNE BOUCLE RECHARGEABLE EN PISCINE OSIRIS
G. Farny 117

ANLAGE ZUR UNTERSUCHUNG DES RISSWACHSTUMSVERHALTENS VON DRUCKBEHÄLTERSTAHL UNTER SIEDEWASSERREAKTOR-BEDINGUNGEN
D. Anders, J. Ahlf 133

PREDICTION OF POWER AND ANALYSIS OF RESULTS FROM THERMOHYDRAULIC AND NEUTRONIC MEASUREMENTS IN A PRESSURISED-WATER-LOOP EXPERIMENT IN THE DIDO REACTOR AT HARWELL
T.D.A. Kennedy 143

IRRADIATION EXPERIMENTS AT THE NUCLEAR RESEARCH INSTITUTE, ŘEŽ
O. Erben, Z. Frič, B. Chamrád, A. Kříž, J. Pína,
J. Šalamon, J. Zběhlík 161

CAPABILITIES OF THE POWER BURST FACILITY
W.A. Spencer, A.M. Jensen, R.K. McCardell 175

INPILE TEST RIGS FOR RIA EXPERIMENTS IN THE NSRR
T. Fujishiro, S. Tanzawa, S. Saito 191

DESIGN SYNOPSIS FOR THE NEW ACCIDENT RELATED STUDIES LOOP AT CRNL
W.K. Snelson 201

POWER BURST FACILITY SEVERE FUEL DAMAGE TEST PROGRAM
R.K. McCardell, P.E. MacDonald 213

RIA FUEL BEHAVIOR RESEARCH IN THE NSRR
T. Hoshi 231

FUEL ROD TESTING FACILITIES IN HFR PETTEN
J. Markgraf, D. Perry, M. Leipold 241

SESSION V:
LIGHT WATER REACTORS INSTRUMENTATION

INSTRUMENTATION FOR AUTOMATIC DIAMETER AND EDDY-CURRENT MEASUREMENT OF IRRADIATED LWR FUEL RODS IN THE HFR POOL
D. Klage, J. Markgraf, D. Perry 255

ETUDE DES DEFORMATIONS IMPOSEES PAR L'UO_2 A LA GAINE D'UN ELEMENT COMBUSTIBLE PWR AU COURS D'UN SAUT DE PUISSANCE
R. Schley, P. Salle, R. Atabek 267

THE IN-PILE MEASUREMENT OF THERMAL CONDUCTIVITY OF UO_2 VIBRATORY COMPACT
Tadashi Kubota, Yukio Ikawa, Isao Komine 279

SESSION VI:
FAST BREEDER REACTORS

EXPERIMENTS IN RAPSODIE WITH THERMOCOUPLES INSTRUMENTED PINS FOR MEASUREMENTS OF THE INTERNAL FUEL PELLET TEMPERATURE
F. Marin 291

REACTEUR RAPIDE EXPERIMENTAL RAPSODIE MOYENS DE DETECTION ET LOCALISATION DES RUPTURES DE GAINE
J.C. Cauvin, J. Kermarrec, J.P. Pagès 299

DEVELOPMENT OF THE UN-INSTRUMENTED IRRADIATION SUBASSEMBLY UNIS-B FOR THE EXPERIMENTAL FAST REACTOR JOYO
K. Ogata, H. Hyodo, K. Okada, Y. Ishida 309

TRANSIENT OVERPOWER EXPERIMENT ON LMFBR FUEL PINS IN THE PETTEN HIGH FLUX REACTOR (HFR-TOP)
G. van der Koogh et al. 319

PAPRICA - CODE DE GESTION ET DE CALCUL DU COEUR DU RÉACTEUR RAPSODIE
L. Goutagny, G. Gache, G. Lecardinal 335

DIE KNK II ALS BESTRAHLUNGSREAKTOR
W. Kathol 345

TRANSIENT EXPERIMENTS ON LMFBR FUEL PINS IN THE REACTOR BR2 WITH THE VERSATILE SODIUM LOOP "VIC"
J. DeKeyser 355

PROGRAMME DE PREQUALIFICATION EN PILE D'UN COMBUSTIBLE A OXYDE MIXTE FABRIQUE PAR UN NOUVEAU PROCEDE (GSP)
G. Marinucci, A. Nobili, M. Lanchi, R. Caracchini
G. Dupont, J. Galtier 369

LE PROGRAMME D'IRRADIATION GOCAR
M. Campana 379

EBR-II TRANSIENT OPERATION AND TEST CAPABILITIES
B.R. Seidel, D.C. Cutforth, G.L. Lentz, J.D.B. Lambert 391

EXPERIENCES PAHR A MELUSINE
D. Rousseau, P. Dereymez, H. Guyon, E. Junod,
M. Ploujoux, F. Tournebize, H. Backs 405

SESSION VII:
INSTRUMENTATION

LES MESURES DND FAITES DANS LE CADRE DES IRRADIATIONS DE LA FILIERE A NEUTRONS RAPIDES
P. Vulliez, M. Pluchery 417

THE DELAYED NEUTRON DETECTION METHOD FOR CLADDING TUBE MONITORING AND EXPERIMENTAL RESULTS
S. Jacobi 429

EXPERIENCE ACQUISE DANS L'UTILISATION DE L'ABSORBANT NEUTRONIQUE ^{3}HE POUR L'EXECUTION DE TRANSITOIRES DE PUISSANCE DE FISSION
A. Falla 439

MESURES DES TEMPERATURES D'IRRADIATION ET ETUDE DES SURCHAUFFES LOCALES DE GAINE EN ACIER AUSTENITIQUE, PAR DIFFRACTION X
P. Rousset, R. Cadalbert 451

DEVELOPPEMENT D'EQUIPEMENTS ET D'EXPERIENCES HORS PILE COMME SUPPORT D'IRRADIATIONS EN REGIME TRANSITOIRE
M. Coquerelle, P. Knappik, C. Perrier, R. Schreiber 461

MEASUREMENT AND CALCULATION TECHNIQUES UTILIZED FOR THE NEUTRONIC CHARACTERIZATION OF BR2 IRRADIATION DEVICES OF THE VARIABLE NEUTRON SCREEN TYPE. APPLICATION TO THE "VIC" SODIUM LOOP.
Ch. de Raedt, N. Maene, R. Menil, J. Debrue 475

SESSION VIII: GAS COOLED REACTORS AND FAST BREEDER REACTORS

A CALORIMETRIC DEVICE FOR IRRADIATION OF FBR FUEL PINS
F. Moons 489

IRRADIATION FACILITY FOR TEMPERATURE TRANSIENT AND CYCLING EXPERIMENTS OF HTR FUEL ELEMENTS
R. Conrad, R. Keller 499

EQUIPMENT FOR IRRADIATION EXPERIMENTS IN THE COOLANT CHEMISTRY OF THE AGR
H.A. Prior, A. Harper 509

CHARACTERIZATION OF HTR FUEL ELEMENT SAMPLES BY ACTIVATION TESTING
A.W. Mehner, J.J. Abassin, P. Sciers 519

IN-PILE HIGH TEMPERATURE IRRADIATION TESTING TECHNIQUES FOR THE MATERIALS AND FUELS OF HTGR IN JMTR
Isao Tanaka, Hidetake Takahashi, Hiroharu Itami,
Haruhiko Ito, Noriyoshi Tsuyuzaki 529

SESSION IX: MISCELLANEOUS

DIE EXPERIMENTELLE NUTZUNG DES DEUTSCHEN FORSCHUNGSREAKTORS FR2 - EIN RÜCKBLICK
W. Steiger, G. Wittek 544

BESTRAHLUNGSTESTS MIT NIEDRIGANGEREICHERTEN BRENNSTOFFEN UND BRENNELEMENTEN FÜR FORSCHUNGSREAKTOREN
W. Krug, F. Schmidt, J. Seferiadis, G. Thamm 549

AN AUTOMATIC CONTROLLED, HEAVY WATER COOLED FACILITY FOR IRRADIATION OF SILICON CRYSTALS IN THE DR 3 REACTOR AT RISØ NATIONAL LABORATORY, DENMARK
K. Hansen, K. Stendal, K. Andresen, K. Heydorn 567

QUALITY ASSURANCE AND RELIABILITY ANALYSIS IN THE REALISATION OF IRRADIATION TESTING DEVICES
R.S. Matfield 579

POSTER SESSION

OPERATING EXPERIENCE WITH CAPSULES FOR RPV STEEL IRRADIATIONS WITH LARGE SPECIMEN VOLUME
J. Ahlf, D. Bellmann, H. Martens 593

DETERMINATION TOMOGRAPHIQUE DE LA DISTRIBUTION DES EMETTEURS GAMMA DANS UN CRAYON COMBUSTIBLE EN IRRADIATION (PROGRAMME ISARD)
G. Ducros, F. Michel 605

NICROSIL-NISIL THERMOCOUPLES IN FUEL ELEMENT IRRADIATION EXPERIMENTS
J.M. Hartog 615

DIANE. A SPALLATION NEUTRON SOURCE OF FUNDAMENTAL RESEARCH IN WEST GERMANY
G.S. Bauer, H. Stechemesser, G. Thamm 623

A SMALL, MULTIPURPOSE, LOW PRESSURE WATER LOOP AT BR2 REACTOR
F. Moons 647

AN ULTRASONIC GAUGE FOR MEASURING OF M.T.R. FUEL ELEMENT COOLING GAP DIMENSIONS
K. van Otterdijk 657

MEASUREMENTS OF NUCLEAR HEATING IN THE PETTEN HIGH FLUX REACTOR
A.J. Nolten 669

FACILITIES FOR THE IRRADIATION OF 235-U FOR THE PRODUCTION OF 99 Mo AT THE HFR PETTEN
J. Konrad 677

EVALUATION OF THE PRODUCED RADIOACTIVITY FROM RAPID IRRADIATION IN A NEUTRON FLUX DENSITY PROFILE
J.O. Schmidt 685

CAPSULE TRANSPORT EVALUATION OF A RAPID He-TRANSFER SYSTEM WITH A ^{6}LiD-CONVERTER
A. Salahi, F. Grass, F. Bensch, G. Zugarek, J.O. Schmidt 697

EXPERIMENTAL FACILITIES FOR PEC REACTOR DESIGN CENTRAL CHANNEL TEST LOOP: CPC-1 - THERMAL SHOCKS LOOP: CEDI
C. Calvaresi, L.F. Moreschi 709

PUISSANCE DEPOSEE DANS UN DISPOSITIF COMPLEXE PAR LES GAMMAS DE REACTEUR D'ESSAIS. EXPERIENCES ET CALCULS RELALISES A SILOE
G. Cosoli, R. Caracchini, H. Petitcolas, A. Besson, A. Bevilacqua, S. Bloch 719

OUT-OF-PILE EXPERIMENTS IN SUPPORT OF THE HFR-TOP IRRADIATION
J.M. Plaum, C. Croon, I. Monster, H. Manuel, J.E. de Vries 725

PIRAMID, A LOOP FOR EUROPEAN PAHR IRRADIATIONS ACCORDING TO A MOL INTEGRATED DEVICE
C. Joly, M. Soenen, W. Detavernier, Ch. de Raedt, L. Fouarge, H. Backs 739

CLOSURE OF THE MEETING 749

PROGRAMME SUMMARY 750

LIST OF CHAIRMAN 750

LIST OF PARTICIPANTS 751

AUTHOR INDEX 755

INTRODUCTION

The International Topical Meeting on Irradiation Technology took place two weeks after the Conference on the Neutron and its Applications (Cambridge, U.K.) marking the 50th anniversary of the discovery of the neutron.

The application of neutrons from research reactors for materials testing requires a large variety of irradiation devices (vehicles) and their out-of-pile control and recording equipment. The in-pile sections are sophisticated in design and assembly, expensive, and consumable. There have been only a few international conferences on irradiation testing, the last one being limited to fast breeder reactor work (Jackson, Wyoming, September 1973).

In 1982, however, two conferences picked up similar topics
- the ANS Conference on Fast, Thermal, and Fusion Reactor Experiments (Salt Lake City, Utah, April 1982),
- the Grenoble meeting (these proceedings).

Overlapping has been avoided by putting the accent of the U.S. conference on fast reactor work, and on thermal reactor experiments at Grenoble. Put together, both conferences ligned up more than 130 papers in this very specialised field, demonstrating a high level of technological development as opposed to a decreasing number of large materials testing reactors available.

The editors wish to acknowledge the flawless organisation of the meeting by CEN de Grenoble and the personal commitment of CEN staff to its success. Special thanks go to F. Merchie and J.F. Veyrat of the Service des Piles.

P. von der Hardt
H. Röttger

P. von der Hardt and H. Röttger (eds.), Irradiation Technology, xiii.

CONFERENCE ORGANISATION

Local Committee

General chairman	F. Merchie	CEN Grenoble
Technical Secretary	J.F. Veyrat	CEN Grenoble
Organization	Mme M. Gros d'Aillon	CEN Grenoble
	Mme C. Boujet	CEN Grenoble
	Mlle C. Lauziere	CEN Grenoble

Programme Committee

J. Ahlf	GKSS Geesthacht
J.M. Baugnet	SCK/CEN Mol
E. Bojarsky	KfK Karlsruhe
A. Calza-Bini	ENEA Casaccia Rome
F. Cherruau	CEN Saclay
P. von der Hardt	JRC Petten
K.J. Henry	AERE Harwell
W. Lehrheuer	KFA Jülich
J.A. Leth	Risø National Laboratory
F. Merchie, chairman	CEN Grenoble
J.P. Pagès	CEN Cadarache
H. Röttger, secretary	JRC Petten
R. Swanenburg de Veye	ECN Petten
J.F. Veyrat	CEN Grenoble

Co-sponsors

European Nuclear Society

Commissariat à l'Energie Atomique (CEA), CEN Grenoble

Irradiation Devices Working Group of the
Commission of the European Communities

Joint Research Centre (JRC), Petten Establishment
Commission of the European Communities

CONFERENCE INTERNATIONALE SUR LA TECHNOLOGIE DES IRRADIATIONS

SÉANCE D'OUVERTURE

(F. MERCHIE / CEN de GRENOBLE)

Placée sous le patronage de la Société Nucléaire Européenne, du Centre Commun de Recherche EURATOM, Etablissement de Petten, du Groupe de Travail sur les Dispositifs d'Irradiation de la Commission des Communautés Européennes et du Centre d'Etudes Nucléaires de Grenoble, la Conférence sur la Technologie des Irradiations de Grenoble est ouverte.

Avant d'entamer la première session qui sera consacrée aux Réacteurs à Eau Ordinaire, je voudrais expliquer brièvement l'histoire et l'esprit de cette Conférence sur la Technologie des Irradiations. Comme vous le savez, en Europe, la plupart des pays ont établi des programmes nationaux de développement nucléaire, et parallèlement, ils ont entrepris au sein d'EURATOM, des actions communes complémentaires sur des thèmes variés de recherche et de développement.

C'est ainsi que, sous l'égide d'EURATOM, des groupes de travail se sont constitués dans le domaine des réacteurs de recherche et d'essais, en dosimétrie, en neutronographie et qu'un groupe de travail en matière de technologie des irradiations se réunit périodiquement pour préparer et organiser tous les deux ans environ, dans un des pays de la Communauté Européenne, une réunion plénière où sont présentés et discutés pendant 2 à 3 jours les derniers développements des boucles et dispositifs d'irradiation.

A l'issue de la dernière réunion plénière qui s'est tenue à GEESTHACHT en octobre 1980, le Groupe de Travail a pensé que le moment était venu d'élargir, au-delà des frontières de la Communauté Européenne, la participation à la prochaine réunion qu'il organiserait à GRENOBLE en 1982.

Avec une organisation très légère et des moyens modestes mis en commun par le Centre d'Etudes Nucléaires de Grenoble et par EURATOM PETTEN que je remercie en la personne de M. VON DER HARDT, la Conférence Internationale sur la technologie des irradiations était lancée et je suis très heureux de constater que la première partie de notre pari est gagnée puisque 150 spécialistes représentant 16 pays différents participeront à nos travaux.

Il nous reste à gagner la deuxième partie, c'est-à-dire à faire de ces 3 jours un véritable lieu d'échanges et de communications dont sortira, je l'espère, une nouvelle source de progrès dans l'évolution des techniques d'irradiation et d'instrumentation.

En effet, chacun de nous sait bien qu'à l'heure actuelle, une expérience d'irradiation est devenue une affaire assez complexe dont le succès dépend à la fois des conditions générales délivrées par le réacteur d'essai (par exemple le niveau et la qualité des flux de neutrons, etc.), des conditions particulières offertes par la boucle ou le dispositif lui-même et enfin, et de plus en plus, de la nature, de la qualité et de la diversité des informations et des mesures fournies par l'instrumentation classique ou spécifique.

Beaucoup reste à faire pour connaître et comprendre le comportement des matériaux et des combustibles sous rayonnement. Nos travaux y apporteront, j'en suis convaincu, une large contribution.

P. von der Hardt and H. Röttger (eds.), Irradiation Technology, xv.

SESSION I-IV
LIGHT WATER REACTORS

FACILITIES AND TECHNIQUES FOR INSTRUMENTED FUEL IRRADIATIONS IN THE NRX REACTOR AT CHALK RIVER

P.J. Fehrenbach

Atomic Energy of Canada Limited
Chalk River Nuclear Laboratories
Chalk River, Ontario, K0J 1J0

The requirement for more detailed data on specific aspects of fuel performance, and the desire to verify fuel performance model predictions, has led to the recent development of a number of specialized irradiation facilities and techniques for use in the experimental loops of the NRX reactor at Chalk River. These have been developed to support fuel studies in both normal operating conditions and off-normal or "accident" conditions.

The experimental loops in NRX provide a nuclear environment in which to test experimental fuel under a variety of coolant conditions, ranging from pressurized water to superheated steam. They have been described recently by Nishimura (1) so this presentation will concentrate on the more specialized facilities we have developed to measure the fuel response to changes in operating conditions. These are discussed in relation to their associated programs.

IN-REACTOR MEASUREMENT OF FUEL ROD DIAMETER CHANGES

The objectives of this program are to provide quantitative data for the verification of CANDU fuel performance code predictions of fuel clad stress and strain during normal operating conditions, and to expand our detailed knowledge of fuel element response to power changes. We are also interested in determining the effect of various design and operating parameters, such as ramp rate and fuel density (2,3) outside the range normally experienced on the dimensional response of Zircaloy-clad UO_2 fuel rods. To achieve these objectives, we required the ability to change fuel

P. von der Hardt and H. Röttger (eds.), Irradiation Technology, 3–12.

rod power at controlled rates independently of the experimental reactor power, and to simultaneously measure in reactor, the resulting dimensional changes of the fuel rod.

Element power is controlled by use of a ^{3}He power cycling system. A coiled tubular shroud filled with ^{3}He gas surrounds the central portion of the in-reactor test section of the X-6 loop in NRX. Because ^{3}He has a very high capture cross section for thermal neutrons, the thermal neutron flux and fuel element power inside the shroud can be controlled by varying the pressure of the ^{3}He gas in the coil. The gas pressure is controlled by nitrogen pressure acting through a bellows on the ^{3}He volume. Rates of power change are controlled by micrometer needle valves in the nitrogen circuit. The normal pressure range is between 0.15 and 1.18 MPa (6 and 150 psig) which allows changes in element power of up to 45% at rates between 5×10^{-4}% and 25% maximum power per minute (3×10^{-4} to 15 $kW.m^{-1}.min^{-1}$). Within these limits, any desired power history can be obtained.

In order to monitor the changes in flux at the location of the fuel elements and elsewhere along the test section, the X-6 pressure tube has been instrumented with 14 self-powered flux detectors (4). Ten of these are vanadium detectors for monitoring the steady state flux. Within the region of the ^{3}He coil there are two each of the faster responding platinum and cobalt detectors for monitoring the dynamic changes in flux. Figure 1 shows the effect of the ^{3}He coil on the axial profile in X-6 measured by these detectors as a function of ^{3}He pressure. During most irradiation experiments the fuel element stays within the relatively flat portion of the flux distribution at the center of the test section.

The in-reactor diameter measuring rig (IRDMR) (Figure 2) is a reusable rig into which a single fuel element is placed. After insertion of the rig into the reactor, the element and measuring head are located at the nominal midplane of the reactor core. Diameter measurements are taken by moving the fuel element axially past the stationary measuring head. Axial element motion is provided through the rotation of a lead screw assembly by a remotely controlled stepping motor drive. The measuring head itself contains two pairs of cantilever beams in contact with the 15 mm diameter fuel element. Each beam contains a number of integral lead strain gauges which are used to measure beam deflection as they follow the contours in the element profile.

The upper and lower end caps of the fuel element both contain calibration steps at which the strain gauge signals are recorded before and after each series of diameter measurements. This internal calibration technique eliminates the problem of long term irradiation-induced drift of the strain gauges and permits accurate diameter comparisons throughout the irradiation experiment.

Each diameter scan produces two orthogonal sets of element diameter measurements at 1 mm increments along as much of the element length as required. A typical set of diameter profiles, obtained from a fuel element operating at 58 kW/m, is shown in Figure 3. Calibration steps on the element end caps, end cap to sheath welds and ridges at pellet interface locations are clearly visible. Each diameter profile is analyzed for ridge diameters and ridge heights.

Typical IRDMR results from a standard density (10.6 Mg/m^3) UO_2 fuel element clad in 0.42 mm thick Zircaloy 15.2 mm in diameter are shown in Figure 4. The element was irradiated at a steady low power of about 25 kW/m to a burnup of 94 MW.h/kg U before insertion into IRDMR. During the IRDMR irradiation, the element was reconditioned at about 30 kW/m and then ramped at a steady slow rate, during which diameter scans were taken, to 55 kW/m (3). The data for diameter change vs. power were similar to companion elements which had no previous irradiation.

The technique is now relatively standard at CRNL and is being used to assess the effects of changes in fuel design or operational parameters on fuel dimensional responses to irradiation, such as fuel densification and pellet-clad interaction (2,3,5).

FUEL BEHAVIOUR DURING LOCA TRANSIENTS (X-2)

This program consists of a series of generic tests to verify particular aspects of transient fuel behaviour codes and demonstrate that no unmodelled phenomena are significantly affecting fuel behaviour during high temperature transients. Specific objectives include comparison of clad strains and temperatures measured during coolant depressurization (blowdown) with those predicted by the transient fuel code ELOCA, investigation of fuel element failure mechanisms during blowdown transients, and measurement of the amount and species of fission products released by the fuel during a transient.

The program was first described at Petten in 1979 by MacDonald (6) and more recently by Langman (7). The test program is divided into three phases, beginning with thermalhydraulic tests on stainless steel clad elements to establish loop and blowdown parameters required to achieve the desired clad temperatures during blowdown. Because of the complexity of the loop piping, it has proven difficult to predict local thermalhydraulic conditions and fuel clad temperature rises for a given set of blowdown conditions. This has necessitated a more extensive initial phase of thermal-hydraulic tests than originally proposed. We have now completed this phase and are beginning the code verification tests which will use single, Zr-clad UO_2 fuel elements each subjected to a single, high temperature transient. Following this, a third phase will investigate inter-element effects using trefoil assemblies of three elements.

The X-2 blowdown facility is a pressurized light water loop with a vertical in-reactor test section 3 m long and 38 mm inside diameter. It can be isolated from the main loop and blown down (depressurized) through a variable orifice into a shielded tank. High pressure injection of cold water into the bottom of the test section rewets the fuel after a test.

The single fuel elements for the initial two phases of the program are 19 mm diameter and 500 mm long and fuelled with sintered, high density (10.6 Mg/m^3) UO_2 pellets inside Zircaloy-4 or stainless steel cladding. The single elements (Figure 5) are instrumented with four thermocouples in the fuel, one at the centre and three at the pellet periphery, and up to six thermocouples semi-buried and laser welded into the 0.8 mm thick cladding. The element assembly also contains eddy current pressure transducers for measurement of coolant pressure and internal element gas pressure (not shown) during blowdown, an eddy current displacement transducer to detect the sudden length increase at dryout, a number of coolant and flow tube thermocouples, and several self-powered flux detectors.

Fuel thermocouples are 1.3 to 1.6 mm diameter Mo or Mo-Re clad, with BeO insulation and W5%Re/W26% Re thermoelements. These have performed very satisfactorily at lower temperatures, surviving for up to 3 months at steady temperatures of 1675°C. At higher temperatures (up to 2000°C) experienced in central fuel locations the lifetime was considerably shorter, ranging from several hours to several weeks.

Fuel cladding thermocouples are clad in either 304 L stainless steel or Zircaloy-4 and have flattened tips which are welded directly to the fuel cladding. The 304 L clad, MgO insulated, Cr/Al thermocouples have survived as many as 24 blowdown at temperatures up to 975°C. Two types of grounded Zircaloy-clad thermocouples have been acquired for use with Zircaloy-clad fuel elements. One type, fabricated to CRNL specifications, have a 1 m long x 1 mm diameter Zircaloy-clad section, containing MgO insulation and W/Re thermoelements, with a 135 mm long flattened tip. The other type are full length 7.6 m long, 1 mm diameter Zircaloy clad thermocouples with MgO insulation and Cr/Al thermoelements. The flattened tip section of these is only about 5 mm long. Although neither type has been used yet in blowdowns, they have worked well in reactor for up to 4 months at about 300°C.

In their original state, the balanced-bridge type eddy current coolant pressure transducers exhibited measurement errors up to 100% of full scale output (FSO) when subjected to rapid temperature transients, such as those experienced during a blowdown test. However, as a result of a development effort at CRNL in cooperation with the supplier, electronic and design modifications have been made which reduce this error to less than 10% FSO (1 MPa) in transients up to 40 C/s. Figure 6 shows the response of a modified sensor to such a transient. The differential type internal element pressure transducers exhibited errors up to 30% in their unmodified state. Similar modifications to these sensors reduced this error by a factor of 10 also, to less than 3% FSO (0.3 MPa) during a blowdown transient. One of these modified sensors has now been exposed to 24 in-reactor blowdown tests and functioned satisfactorily throughout. The balanced bridge eddy current transducers used to measure length changes have also exhibited some instability during blowdown tests and a number of these are being electronically modified for future in-reactor tests.

A sweep gas system (8) will be used to determine the amount and species of short-lived gaseous fission products, released from the fuel to internal element voidage during a blowdown transient. During operation, bottled He-2% H_2 carrier gas is introduced into the fuel element at an operating pressure of about 3 MPa, and flows at about 3 mL/s sweeping gaseous fission products out of the fuel element past a gamma spectrometer for measurement. At this flow rate, residence time of the sweep gas in the element is about one second and transit time from element to spectrometer is about 10 minutes. Isotopic data recorded at the spectrometer are converted to concentrations and releases.

The fission product species measured directly at the spectrometer are generally the short-lived xenons and kryptons. Iodine or bromine are only observed when water or steam is inside the element, as would accompany a defect. During normal operation, iodine release to the fuel clad gap is deduced from the decay of I-133 and I-135 to Xe-133 and Xe-135 during reactor shutdowns (8).

One advantage of this sweep gas technique is that it includes all factors in the release mechanisms including fuel power, temperature, temperature gradients and fuel cracking, and provides a relatively direct measure of the short lived fission product activity available for immediate release if the fuel sheath were to defect under a given set of conditions.

Although the sweep gas system has not yet been used during a blowdown test, Hastings et al. have recently reported results obtained during lower temperature transients (8). In these tests, the instrumented fuel element was irradiated at about 55 kW/m and cycled several times from fog cooling into dryout for periods of up to 40 minutes. These dryout cycles resulted in measured temperature increases of 350 to 425 C in the fuel clad and 575 C in the fuel 1 mm from the fuel surface. Fuel and cladding temperatures for one of these transients are shown in Figure 7 together with the corresponding measured rate of activity release. It can be seen that the activity release rate due to the sudden decrease in temperature is greater than the rate resulting from steady operation at the higher fuel temperatures. This is attributed to direct release from fuel cracking in addition to diffusional release (8).

As mentioned, we have completed a number of in-reactor blowdown irradiations in the thermalhydraulic phase of the X-2 blowdown program. In each of these, instrumented fuel assemblies similar to those in Figure 5 but clad in 304 stainless steel were irradiated at linear powers between 45 and 60 kW/m. During the experiments the coolant was blown down from either the top or bottom of the loop, or from both ends simultaneously, from a steady state pressure of 10 MPa. For all tests, cold water injection was initiated when the test section pressure dropped below 2.9 MPa. These thermalhydraulic tests investigated the effects on element temperature of (a) rate of coolant depressurization, (b) time between the blowdown and the reactor shutdown, (c) element power, and (d) fuel to coolant ratio in the test section.

A typical set of results from a recent blowdown test in which the fuel continued to operate at a linear power of 60 kW/m for 21 seconds after initiation of blowdown is shown in Figure 8. Maximum sheath and fuel peripheral temperatures were about 1032 and 1500°C respectively and the element operated in post-dryout for about 30 seconds without any evidence of intermittent quenching. The modified eddy current internal element pressure transducer worked well throughout the test and the output shown in Figure 8 is typical of the other 23 blowdowns.

The ability to achieve steady high (850°C) clad temperatures for up to 30 seconds of post-dryout operation during coolant blowdowns in the X-2 loop has now been demonstrated. Instrumentation to measure most fuel and coolant parameters of interest during the blowdowns is available and is also demonstrated. The program of code verification tests using instrumented Zircaloy-clad fuel elements will begin this year.

SUMMARY

The facilities and techniques developed at CRNL in recent years for instrumented fuel irradiations have enabled us to measure directly in-reactor the response of fuel elements to changes in operating parameters during both normal operating and accident conditions. The data are proving useful in developing and verifying fuel performance codes and in expanding our understanding of fuel performance. Although these facilities were originally developed in support of CANDU fuel programs, they could be applied to fuel testing in support of other reactor programs.

ACKNOWLEDGMENTS

Part of the developments described were funded by CANDEV, the Ontario Hydro-Atomic Energy of Canada Limited Common Development Program.

REFERENCES

1) D.T. Nishimura, "In-Reactor Experimental Facilities at the Chalk River Nuclear Laboratories" presented at ANS Topical Meeting on Fast, Fusion and Thermal Reactor Experiments, Salt Lake City, Utah, USA, 1982 April.

2) P.J. Fehrenbach, P.A. Morel and R.D. Sage, "In-Reactor Measurement of Cladding Strain: Fuel Density and Relocation Effects", Nuclear Technology 56 (1982) 112.

3) P.J. Fehrenbach, I.J. Hastings, P.A. Morel, R.D. Sage and A.D. Smith, "Dimensional Response of CANDU Fuel to Power Changes", presented at the IAEA Specialists Meeting on Power Ramping and Cycling Behaviour of Water Reactor Fuel, Petten, The Netherlands, 1982 September, Atomic Energy of Canada Limited, Report AECL-7837, 1982 August.

4) C.J. Allan and P.J. Fehrenbach, "Operating Characteristics of Vanadium, Platinum and Cobalt Self-Powered Flux Detectors on the X-6 Pressure Tube of the NRX reactor", Atomic Energy of Canada Limited, Report AECL-6957, 1980 September.

5) I.J. Hastings, P.J. Fehrenbach and R.R. Hosbons, "Irradiation Induced Densification in UO_2 and $(Th,U)O_2$" presented at the American Ceramic Society Fall Meeting, San Francisco, 1980 October.

6) R.D. MacDonald and M.J.F. Notley, "In-Reactor Blowdown Tests of Single Instrumented Fuel Elements at Chalk River", Proceedings of the International Colloquium on Irradiation Tests for Reactor Safety Program, Petten, The Netherlands, 1979 June.

7) V.J. Langman, R.D. MacDonald and P.J. Fehrenbach, "Recent CANDU Transient Fuel Behaviour Data from Research Reactor Irradiations" presented at the CSNI Specialists Meeting on Safety Aspects of Fuel Behaviour in Off-Normal and Accident Conditions, Helsinki, Finland, 1980 September, Atomic Energy of Canada Limited, Report AECL-7415, 1982 May.

8) I.J. Hastings, C.E.L. Hunt, J.J. Lipsett and R.G. Gray, "Transient Fission Product Release During Dryout in Operating UO_2 Fuel" presented at the International Meeting on Thermal Nuclear Reactor Safety, Chicago, Ill., 1982 August, Atomic Energy of Canada Limited, Report AECL-7832, 1982 August.

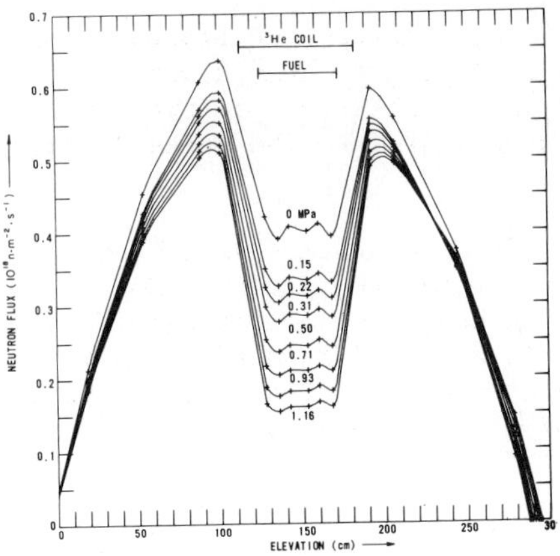

FIGURE 1 Axial flux profile along the X-6 test section as a function of ^{3}He pressure.

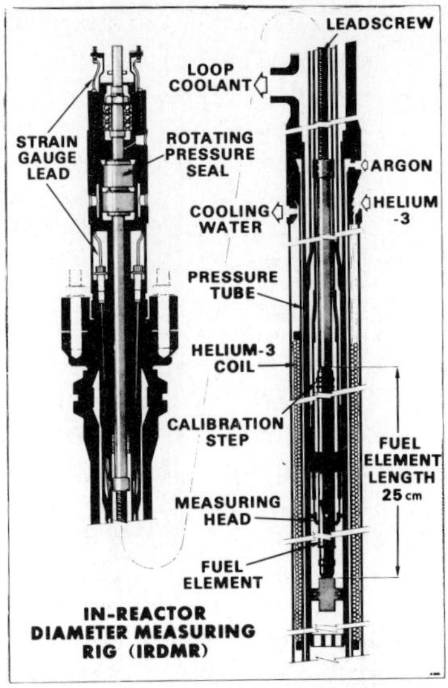

FIGURE 2 Schematic diagram of the In-Reactor Diameter Measuring Rig (IRDMR).

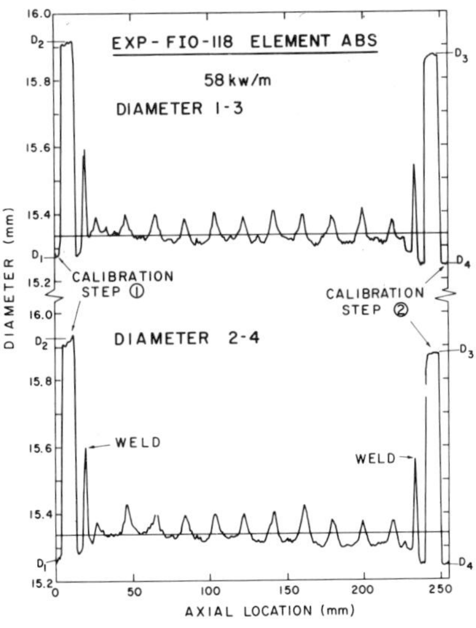

FIGURE 3 Typical set of element diameter profiles from IRDMR showing calibration steps, sheath to end cap welds, and ridges at pellet interface locations.

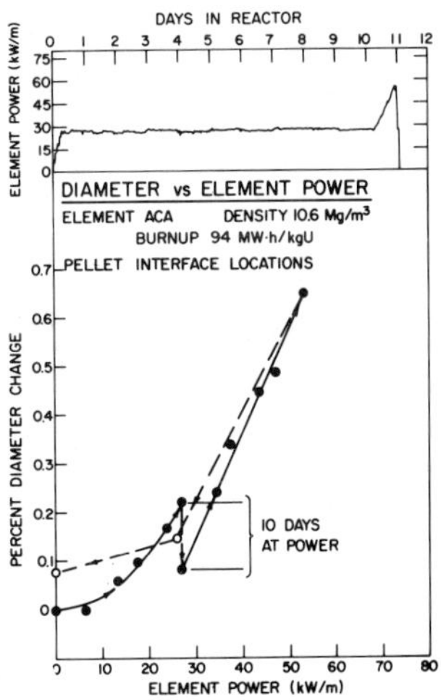

FIGURE 4 Element power and percent change in diameter during power ramping of element ACA in IRDMR (3).

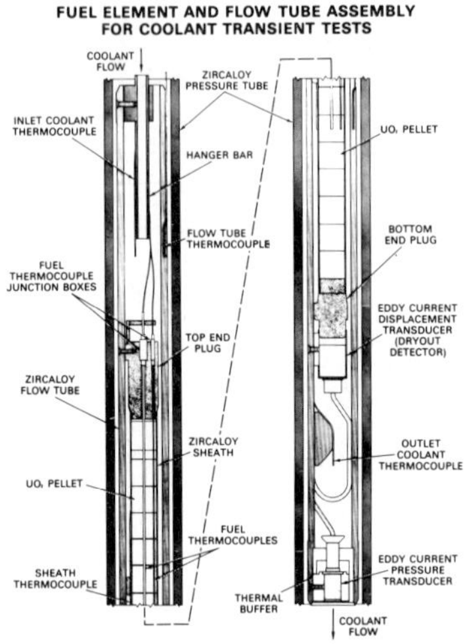

FIGURE 5 Fuel element and flow tube assembly for coolant transient tests.

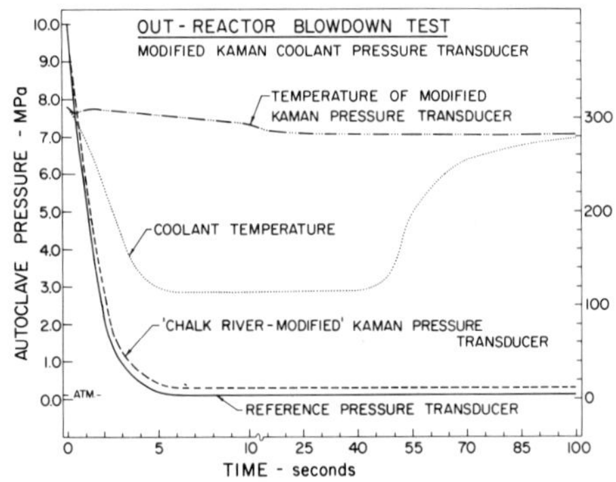

FIGURE 6 Response of modified eddy current pressure transducer to an out-reactor coolant blowdown transient.

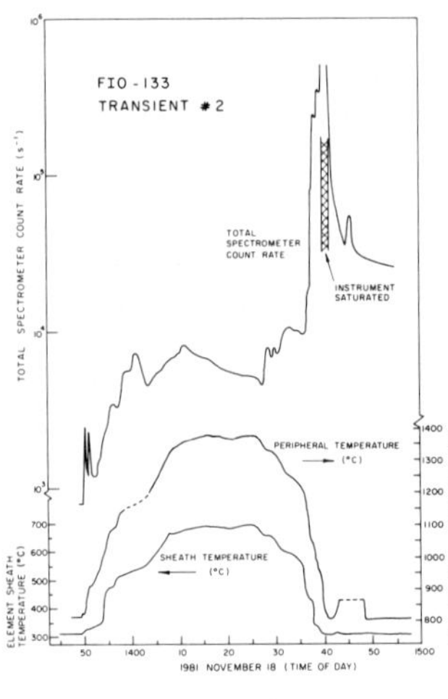

FIGURE 7 In-reactor fission gas activity release rate during a low temperature transient (8).

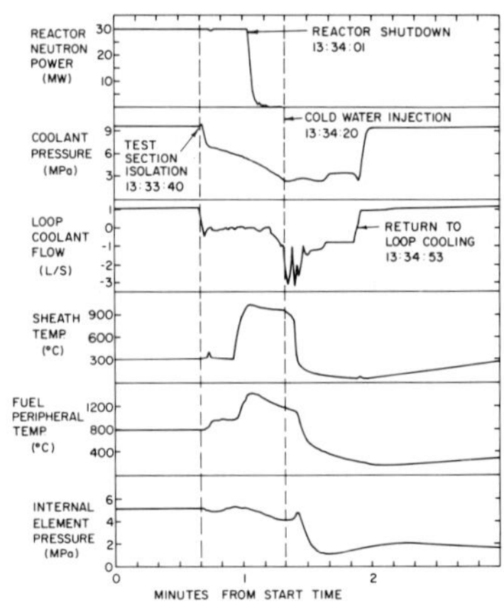

FIGURE 8 Fuel and coolant parameters measured during a coolant blowdown test at 60 kW/m.

ETUDE EXPERIMENTALE ET THEORIQUE DU COMPORTEMENT DE LA GAINE DE L'ELEMENT COMBUSTIBLE

J. Novák, J. Šmíd

Institut de la Recherche Nucléaire

Řež, Tchécoslovaquie

INTRODUCTION

L'expérience actuelle avec les éléments combustibles en régime stationnaire d'exploitation des réacteurs VVER-440 est entièrement satisfaisante. C'est pourquoi l'analyse des résultats d'exploitation ne permet pas l'extrapolation dans la direction des paramètres d'exploitation élevées. Pour pouvoir apprécier les réserves pour l'augmentation du temps d'exploitation et de la puissance et eventuellement l'exploitation des réacteurs en régime nonstationnaire il est nécessaire d'obtenir et utiliser les informations caracterisant les tendences des gaines Zr1Nb à la rupture par corrosion sous contrainte.

APPRECIATION PREALABLE DES RESULTATS PUBLIES

L'appréciation montre que les réserves existent. Parce que pour les alliages Zircaloy nous disposons d'un ensemble plus complet des données caractérisant leurs propriétés et le comportement des éléments combustibles avec les gaines fabriquées de ces alliages, il semble utile de comparer les propriétés décisives des alliages Zr1Nb et Zircaloy. Les différences principales concernant leur composition chimique ressortent du tableau 1.

En considérant les problèmes mentionnés les informations importantes sont contenues dans /1/. Une comparaison des para-

P. von der Hardt and H. Röttger (eds.), Irradiation Technology, 13–21.

Tableau 1. Composition chimique typique /% massique/ des alliages de zirconium pour les gaines des éléments combustibles

	Nb	Sn	Fe	Cr	Ni	O
Zr1Nb	1	-	-	-	-	0,05
Zircaloy-4	-	1,5	0,2	0,1	-	0,12
Zircaloy-2	-	1,5	0,15	0,1	0,05	0,12

mètres critiques /saut de puissance, puissance finale/ pour la perte d'étanchéité des éléments combustibles expérimentaux avec la gaine Zr1Nb et Zr2,5Nb ou Zircaloy semble être en première vue en désaccord avec les résultats des expériences en laboratoire comparant la rupture par corrosion sous contrainte des échantillons annulaires pour les fluences neutroniques différentes. L'analyse du méchanisme d'endommagement /2,3/ mène à la conclusion qu'il y a une liaison étroite entre le fluage, la rupture pendant le fluage et la rupture par corrosion sous contrainte. En plus, il est peut-être plus saisissant de caracteriser le mécanisme de rupture comme "la rupture pendant le fluage modifiée par les effects chimiques du milieu". Pour cette raison l'appréciation des alliages doit sortir d'études de comportement sous déformation, de corrosion sous contrainte et des relations entre ces phénomènes. Quelques comparaisons des propriétés mécaniques des alliages Zr1Nb et Zircaloy irradiés ont été effectuées déjà antérieurement /4,5/; on peut expliquer le comportement des échantillons annulaires en fonction de la fluence des neutrons rapides /1/ par le durcissement plus rapide des alliages Zr1Nb et Zr2,5Nb /4/. Les expériences avec les échantillons annulaires n'ont pas permis une comparaison du comportement des alliages pour les fluences de l'ordre 10^{21} n/cm^2 /E> 1 MeV/, pour lesquelles l'allongement uniforme de l'alliage Zircaloy présente un minimum /6/ et pour lesquelles la saturation des effects d'irradiation aux propriétés mécaniques de l'alliage Zr1Nb commence /7,8/. Pour les fluences élevées l'alliage Zr1Nb peut être equivalent aux alliages Zircaloy.

PROGRAMME EXPERIMENTAL

Pendant l'augmentation de puissance des éléments combustibles en réacteurs de puissance on peut négliger le fluage induit par l'irradiation. C'est pourquoi il est convenable d'effectuer les expériences avec les échantillons de gaine après l'irradiation hors pile. Le programme expérimental comporte les mesures des propriétés mécaniques et les essais de la corrosion sous contrainte en présence du iode avant et après l'irradiation dans le dispositif placé dans le réacteur expérimental VVR-S. Les variantes principales des expériences sont les essais d'éclatement et la déformation avec la vitesse contrôlée dans la direction axiale des tubes. Il est désirable d'exprimer les résultats en forme des lois constitutives et d'une fonction d'endommagement permettant la prédiction de la déformation et de l'endommagement dans les conditions nonstationnaires.

TECHNOLOGIE DES IRRADIATIONS

Une condition nécessaire pour la réalisation du programme est de disposer d'une technologie d'irradiation qui a été éprouvé par des expériences déjà réalisées. On peut trouver une description de celle-ci ailleurs /9/. L'équipement fondamental forment les fours CHOUCA de provenience française en deux variantes /type T avec un volume utile Ø 26 x 300 mm, type MT avec un volume utile Ø56 x 420 mm/. Les fours CHOUCA sont combinés avec les porte-échantillons construits et fabriqués dans l'Institut de la Recherche Nucléaire à Řež d'après les exigences des expériences particulières. Les dispositifs assurant le contrôle des expériences et la regulation ainsi que l'assurance de la technologie et des échantillons expérimentaux en cas de diverses pannes possibles et l'identification et l'enrégistrement des paramètres expérimentaux et d'exploitation sont d'une part des produits de l'industrie tchécoslovaque /un système contenant les régulateurs ZPA type NOTRON et les unités d'alimentation ČKD type NOCONTA/, d'autre part des produits importés /le système complet de contrôle et regulation "Baie electrique EUROPE V 6", la centrale de mesure numérique Schlumberger-Solartron/. La technologie des irradiations s'est bien éprouvée dans les programmes expérimentaux realisés antérieurement: dans le programme de caractère technologique pour la classification des défauts des tubes Zr1Nb pour les éléments combustibles VVER du point de vue de leur admissibilité /10/, dans le programme de dévéloppement des alliages de zirconium résistants à la corrosion par l'eau et par la vapeur surchauffée et dans l'étude des propriétés des aciers pour les cuves du reacteur, l'attestation des propriétés de l'acier de production tchécoslovaque y compris /9/. Les deux programmes concernant le zirconium ont été réalisés en coopération tchécoslovaquo-soviétique.

L'ETAT DES TRAVEAUX

Nous disposons des résultats d'étude du fluage stationnaire des alliages Zr1Nb en laboratoire de l'Institut de la Métallurgie Physique de l'Académie Tchécoslovaque des Sciences à Brno /11-13/. La confrontation et combinaison avec les résultats publiés des mesures des propriétés mécaniques d'alliage Zr1Nb en laboratoire /14,15/ et avec les connaissances de caractère fondamental obtenues en étudiant le fluage du zirconium pur et des alliages de zirconium différents dans l'institut mentionné /16/ sont en cours. On veut comparer les résultats avec les propriétés de Zircaloy. Les données pour Zircaloy ont été exprimées en plusieurs types des lois constitutives. Pour la description des changements de la déformation plastique dans les conditions nonstationnaires il faut utiliser les lois constitutives avec les variables internes.

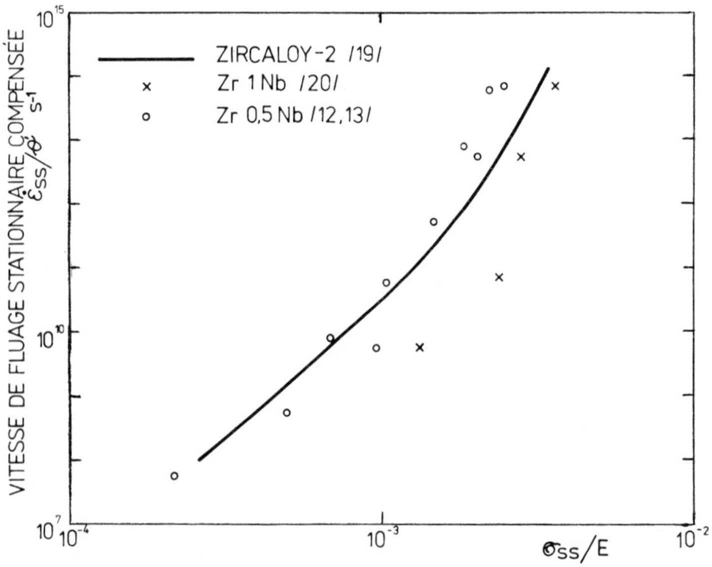

Figure 1. La vitesse de fluage stationnaire compensée par le facteur de type Arrhenius en fonction du rapport de la contrainte et du module d'élasticité.

Les études détaillées /14,16/ montrent que dans le cas du zirconium pur ou Zr1Nb il ne suffit pas d'utiliser une variable de type contrainte interne. Même une variante proche des lois constitutives avec deux variables internes /17/ ne sera pas convenable pour le zirconium /16/. Pour surmonter les contradictions constatées en étudiant le fluage nonstationnaire de l'alliage Fe3,5Si et du zirconium /16/ une conception de Miller et Sherby /18,19/ semble d'être favorable. Elle a été appliquée pour l'essentiel avec succès pour une simulation numérique des chagements de déformation des alliages Fe3Si, Zircaloy etc. Dans les comparaisons des propriétés mécaniques des alliages Zr1Nb et Zircaloy nous utilisons par conséquent les relations de ce type.

Actuellement, il est possible de publier une comparaison de la vitesse de fluage stationnaire des alliages Zircaloy-2 /19/ et Zr1Nb. Les valeurs de vitesse de fluage stationnaire pour l'alliage Zr1Nb ont été dérivées par l'interpolation des données pour les alliages Zr0,5Nb, Zr2,5Nb et Zr4,5Nb traités thermiquement à 823 K/8 heures /20/. On a choisi les valeurs de la contrainte σ_{ss} pour les températures 623 K et 673 K et pour les vitesses de déformation 10^{-9} s^{-1} et 10^{-6} s^{-1}. La figure 1 permet une comparaison de la vitesse de déformation compensée par le facteur de type Arrhenius $\dot{\varepsilon}_{ss}/\theta'$ en fonction du rapport de la contrainte et du module d'élasticité σ_{ss}/E pour les alliages Zircaloy-2 et Zr1Nb. En tous les deux cas, on a utilisé le même facteur θ' dépendant de la température et le même module E /19/. On voit que les valeurs de la contrainte pour l'alliage Zr1Nb sont plus élevées. Une comparaison avec les mesures directes faites avec l'alliage Zr1Nb /14/ n'est possible que pour la température 773 K et la contrainte 60 MPa; pour les contraintes plus élevées et les températures plus basses on n'a pas atteint l'état stationnaire. La vitesse de fluage stationnaire de l'alliage Zr1Nb traité thermiquement à 863 K/20 heures est égale à la vitesse de fluage de l'alliage Zr0,5Nb, qui contient plus de Nb que correspond à la solubilité du Nb dans la phase alpha /3,0.10^{-7} s^{-1}/. Les données pour l'alliage Zr0,5Nb permettent alors d'éliminer pratiquement l'influence de la phase dispersée dans l'alliage Zr1Nb. Elles sont aussi comprises dans la figure 1. Les différences entre les alliages Zr0,5Nb et Zircaloy-2 ne semblent pas être importantes. Il est intéressant de comparer aussi l'expression analytique de la vitesse de fluage stationnaire pour les deux matériaux.

Pour le Zircaloy-2 /19/:

$$\dot{\varepsilon}_{ss} = B\,\vartheta' \left\{\sinh\left[A\sqrt{(\sigma_{ss}/E)^2 - (\sigma_y/E)^2}\right]\right\}^n$$

$$\vartheta' = \exp(-Q^*/RT) \qquad \text{pour } T \geq T_t$$

$$\vartheta' = \exp\left\{\left[-Q^*/RT_t\right]\left[\ln(T_t/T) + 1\right]\right\} \qquad \text{pour } T \leq T_t$$

Pour Zr0,5Nb /12,13/:

$$\dot{\varepsilon}_{ss} = C\,\sigma_{ss}^p \,.\, \exp(-Q/RT) \qquad \text{pour } \sigma_{ss} \leq \sigma_1$$

$$\dot{\varepsilon}_{ss} = D.\exp(E\,\sigma_{ss}).\exp(-Q/RT) \qquad \text{pour } \sigma_1 < \sigma_{ss} < \sigma_2$$

Dans le domaine des vitesses de déformation compensées basses $\sigma_y \rightarrow 0$ et sinh x $\rightarrow$ x; les valeurs des constantes sont proches /Q^* = 60 kcal/mol, n = 4,5, Q = 68,9 $\pm$6,4 kcal/mol, après la correction tenant compte de la dépendence du module elastique de la température Q_{corr} = 65,0 $\pm$6,4 kcal/mol, p = 4,6 $\pm$0,4/. Dans le domaine des vitesses de déformation compensées plus élevées les deux formes passent l'une à l'autre pour $\vartheta' = \exp(-Q^*/RT)$, sinh x $\rightarrow$ 0,5.exp x, $\sigma_y \rightarrow 0$, $E \neq E(T)$.

Après la constatation de l'influence importante de l'état structural des alliages ZrNb au fluage stationnaire on a comparé les autres données sur le fluage de Zr1Nb /21/ et Zr2,5Nb /22/ en forme de tubes. Le fluage stationnaire à 673 K et $\sigma_\vartheta = 2\,\sigma_z$ = = 100 MPa des tubes Zr1Nb pour les éléments combustibles VVER et des tubes Zr2,5Nb pour les éléments combustibles du Whiteshell Reactor 1 est pratiquement equivalent malgré la différence en composition chimique des alliages. Pour la réalisation du programme expérimental suivant il en résulte la nécessité du traitement thermique convenable des échantillons expérimentaux et de la caractérisation détaillée de leur propriétés décisives.

APPLICATION

Les données sur les propriétés décisives de l'alliage Zr1Nb doivent être utilisées pour la prediction des déformations et de l'endommagement des éléments combustibles. Pour ce but on est en train de dévélopper un système des programmes de calcul pour l'analyse de l'interaction combustible - gaine. Un modèle monodimensionel contenant l'approximation des fissures homogénéisées /23,24/ et une description simple des processus de transport de masse dans le combustible /25/ forment une base. Pour la solution du problème monodimensionel on utilise les moyens mathématiques suivants:

1/ pour l'intégration du système des équations différentielles ordinaires du premier ordre représentant le problème de fluage pour une configuration des fissures homogénéisées donnée on utilise les méthodes incrémentales standardes pas à pas dans le temps;

2/ pour la détermination de la configuration des fissures homogénéisées dans certain intervalle du temps on résout deux équations intégrales de Fredholm du deuxième genre dépendantes de deux paramètres /la position du front des fissures radiales et transversales/ par la méthode des approximations successives pourque la condition de relaxation de la contrainte dans le volume nonfissuré soit satisfaite.

Parce que les éléments combustibles VVER standards contiennent les pastilles creuses - on sait que cette configuration contribue à la fiabilité des éléments combustibles - un des problèmes à résoudre est l'appréciation de l'influence de la configuration des pastilles au comportement des éléments combustibles.

CONCLUSION

Les informations accessibles montrent que les propriétés de l'alliage Zr1Nb et des éléments combustibles VVER satisfont les exigences de l'exploitation des réacteurs et qu'ils ont certaines réserves pour l'augmentation du temps d'exploitation et de la puissance. On a formulé et commencé a réaliser un programme de recherche comprenant l'étude expérimentale et théorique du comportement de la gaine des éléments combustibles afin de pouvoir apprécier ces réserves. L'irradiation d'une série des échantillons des tubes dans le dispositif d'irradiation avec le four CHOUCA en réacteur VVR-S - l'usage de la technologie d'irradiation relativement simple - et les essais du matériau après l'irradiation forment une part du programme expérimental. Les mesures du fluage de l'alliage Zr1Nb en laboratoire présentent les premiers résultats obtenus. Après l'élimination de l'influence de la phase dispersée la vitesse de fluage stationnaire est proche à la vitesse de fluage de l'alliage Zircaloy-2. Pour le programme expérimental il est important de choisir le traitement thermique des échantillons expérimentaux et de caractériser leur propriétés décisives.

L'application des résultats ne serait pas possible sans l'ensemble des informations publiées concernant des propriétés des alliages Zr1Nb et Zircaloy et sans les informations permettant la description du comportement de l'élément combustible par un modèle physico-mathématique.

REFERENCES

/1/ D.G. Hardy et al.: Zirconium-niobium alloys as fuel cladding for water cooled reactors. IAEA-SM-233/32, Prague, ČSSR, 6 to 10 November 1978.

/2/ P.J. Pankaskie: Mechanistic consideration pertinent to the PCI fuel failure phenomenon. Specialist's Meeting on Pellet-Cladding Interaction in Water Reactors. Risø, Denmark, 22-26 Sep. 1980. IWGFPT/8.

/3/ J. Novák: Sur le rôle de l'instabilité plastique pendant la corrosion sous contrainte de la gaine des éléments combustibles fabriquée des alliages de zirconium. Le 4ème séminaire Fluage et rupture pendant le fluage, Valtice, ČSSR, Mai 1982.

/4/ C. E. Ells: Deformation of Irradiated Zirconium-Niobium Alloys. ASTM STP 551, 1974, p. 311.

/5/ B. A. Chaedle et al.: Plastic Instability in Irradiated Zr-Sn and Zr-Nb Alloys. ASTM STP 551, 1974, p. 370.

/6/ G. Lestiboudois et al., J. Nucl. Mater. 87 /1979/, 236.

/7/ V.A. Tsykanov et al.: Mechanical Properties and Corrosion Behaviour of Zirconium Alloys under Reactor Conditions. Conf. on Physical Metallurgy of Reactor Materials, Alushta, USSR, 29 May - 1 June 1978.

/8/ V.A. Tsykanov et al.: Hardening and embrittlement of the zirconium cladding alloys under neutron irradiation. Rep. NIIAR - 44/403, 1979.

/9/ B. Chamrád, Jaderná energie 26 /1980/, 368.

/10/ S. Košler et al.: Properties of defected Zr1Nb alloy cladding under LWR conditions. IAEA-SM-233/23, Prague, ČSSR, 6 to 10 November 1978.

/11/ M. Pahutová et al., Kovové mater. 13 /1975/, 417.
/12/ M. Pahutová et al., Kovové mater. 14 /1976/, 252.
/13/ M. Pahutová, J. Čadek, J. Nucl. Mater. 61 /1976/, 285.
/14/ W.R. Thorpe, I.O. Smith, J. Nucl. Mater. 75 /1978/, 209.
/15/ W.R. Thorpe, I.O. Smith, J. Nucl. Mater. 78 /1978/, 49.
/16/ J. Čadek et al.: Dislocationdynamical and structural aspects of high temperature creep in metals and alloys. Studie ČSAV no. 5, Academia, Prague 1979.
/17/ J.L. Chaboche, Bulletin de l'Académie Polonaise des Sciences - Série des Sciences Techniques, vol. 25, no. 1, 1977.
/18/ A.K. Miller: Progress in modeling of zircaloy nonelastic deformation using a unified phenomenological model. ASTM STP 633, 1977, p. 523.
/19/ O.D. Sherby, A.K. Miller: Development of the materials code, MATMOD /Constitutive equations for Zircaloy/. Rep. EPRI-NP-567 /1977/.
/20/ M. Pahutová: travail non publié.
/21/ A.R. Zavialov et al.: Effect of Zirconium Alloy Chemical Composition and Microstructure on Creep Strength. Conf. as /7/.
/22/ Erl Kohn: In-Reactor Creep of Zr2,5Nb Fuel Cladding. ASTM STP 633, 1977, p. 402.
/23/ J. Kofroň et al., internal rep. ÚJV 6099-M /1982/.
/24/ J. Šmíd, On the numerical calculations of stresses and displacements in a fuel pellet with fuzzy cracks. ZAMM, Band 63, Heft 4/5.
/25/ J. Novák, internal rep. ÚJV 6081-M /1981/.

OPERATIONAL TRANSIENTS ON LWR FUEL IN BR2 REACTOR

H. NACKAERTS

S.C.K./C.E.N.

B - 2400 MOL Belgium

ABSTRACT

In the framework of the international programme TRIBULATION, a series of fast operational transients have been executed on preirradiated (40.000 MWd/t), LWR fuel rods in the BR2 reactor.

For this purpose the fuel rods are mounted in a reloadable pressurized water capsule (PWC) which is surrounded by a He^3 absorber screen.

Varying the He^3 pressure allows power transients to about 2 times the initial power.

The pressure in the absorber screen can be increased or decreased very quickly so that fast transients of several percents per second can be obtained.

The main characteristics are :

- stagnant water pressure of the PWC capsule	: up to 150 bars
- maximum linear power	: 600 W/cm
- fuel rod diameter	: 9,5 mm
- active fissile length	: 1 m
- fuel enrichment	: down to natural Uranium
- power ramp rates	: up to 15 %/s.

P. von der Hardt and H. Röttger (eds.), Irradiation Technology, 23–33.

THE TRIBULATION PROGRAMME

The TRIBULATION programme was set up to study the fuel behaviour at high burnup in case an earlier accidental transient has occured in the power plant. /1,2/

The programme was proposed to the international nuclear community in October 1979 by SCK/CEN and BelgoNucléaire. Since then some twenty organisations (fuel vendors, utilities as well as national laboratories) have decided to participate in the programme.

TRIBULATION includes 48 fuel pins, divided in subgroups of typically 4 pins (figure 1), with various design or fabrication parameters, which are irradiated in BR3 at typical PWR conditions, up to 60 to 70 GWd/t_M peak pellet burnup and destructively examined in order to analyse their inpile behaviour.

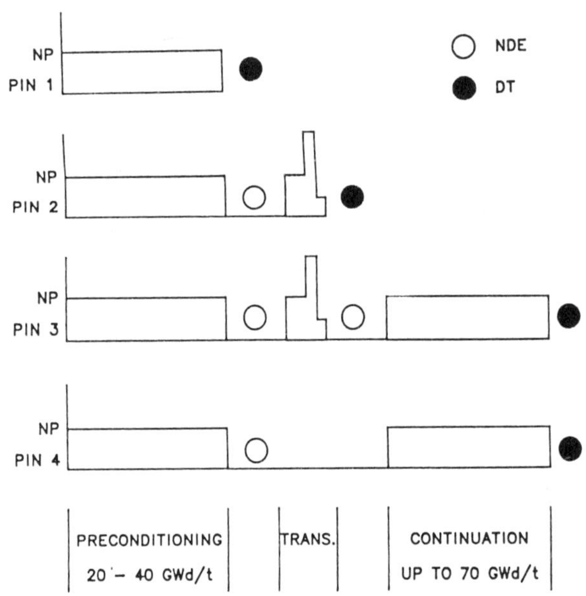

Figure 1 : Irradiation Diagram of Fuel pin Subgroup

After an intermediary burnup of approximately 40 GWd/t_M, 23 selected fuel pins are subjected to fast transients in BR2, so simulating accidental events which could occur during the operation of a PWR.

The main characteristic of TRIBULATION is that after the transients, the fuel pins are being reirradiated up to 60-70 GWd/t_M after which they are destructively examined.

The 48 TRIBULATION pins are of the 17x17 PWR design but modified to the BR3 core height (1000 mm active length).

Some typical pin characteristics are shown on Table I.

PELLET	
NATURE	SINTERED UO2
OUTER DIAMETER	8.0 mm
L/D	1.0 – 1.5 – 1.64
DISHED OR ANNULAR	BOTH
IMMERSION DENSITY	94 – 95 % TD
GRAIN SIZE	5 – 15 µm
CLADDING	
NATURE	Zry 4 CW SR
OUTER DIAMETER	9.5 mm
INNER DIAMETER	8.2 mm
THICKNESS	5.9 – 6.6 % OD
FUEL PIN	
ACTIVE LENGTH	1000 mm
TOTAL LENGTH	1136 mm
INTERNAL PRESSURE	1 to 32 bar

Table 1 : TRIBULATION Typical fuel pin Characteristics

2. TRANSIENTS IN BR2

After the preconditioning irradiation in BR3, some pins of the TRIBULATION programme are subjected to a transient in the BR2 material testing reactor. Testing in BR2 of fuel pins that were preirradiated in BR3 is going on since 1974 in the framework of the Belgian Fuel Development Programme.

The BR2 reactor is a light water cooled, berylium moderated material test reactor, the typical characterisitics of which are given in Table 2. The TRIBULATION transients are performed in one of the 84 mm diameter channels in the reflector region of the reactor.

Water cooled, Be moderated MTR	
Coolant pressure	12 bar
Coolant temperature	45 #C
Normal operating power	80 MW
Fuel type	Conc Shell
U235 enrichment	90 %
Active core height	760 mm
Number of irradiation positions	
84 mm diameter	64
50 mm diameter	10
203 mm diameter	5
Max. therm. neutron flux	1E15 n/cm2s
Max fast neutron flux	7E14 n/cm2s
Gamma heating rate	
in fuel channels	5–16 W/g
in reflector channels	1–6 W/g

Table 2 : BR2 Test Reactor, Typical Characteristics

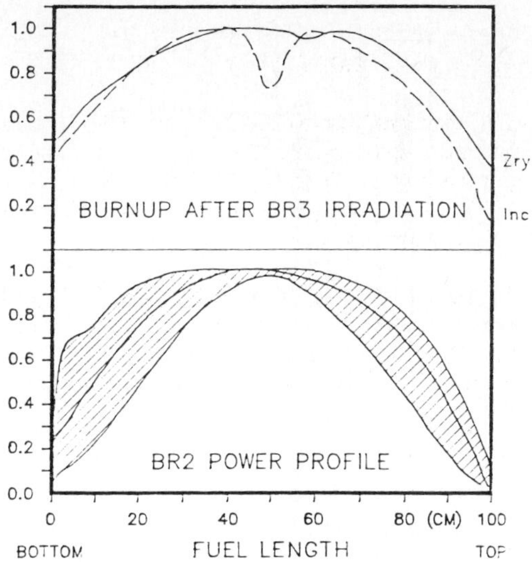

Figure 2 : Axial form factors of BR3 and BR2

The BR2 axial power profile approximates fairly well the axial form factor in BR3 (figure 2), which to a large amount suppresses the difficulties frequently encountered in interpreting results and representativity of reirradiation in MTR's of power pins or rodlets.

This feature is quite unique to the BR2-BR3 tandem, two reactors that are both located at the SCK/CEN at Mol - Belgium.

The transients consist in a preconditioning period at 250-300 W/cm for at least 2 days, followed by a rapid power increase to 320-460 W/cm.

After a hold time of 1 or 10 minutes, the power is rapidly decreased to well below its preconditioning level.

3. EXPERIMENTAL ASSEMBLY

The fast power transients on the TRIBULATION fuel pins are carried out in an irradiation device called PWC (Pressurised Water Capsule) which simulates the coolant pressure and temperature encountered in large PWR's (figure 3).

The fuel pin is centered in a 26 mm diameter thimble tube filled with demineralised water pressurised at 140 bars. The nuclear heat produced by the fuel pin is radially dissipated through the stagnant water annulus.

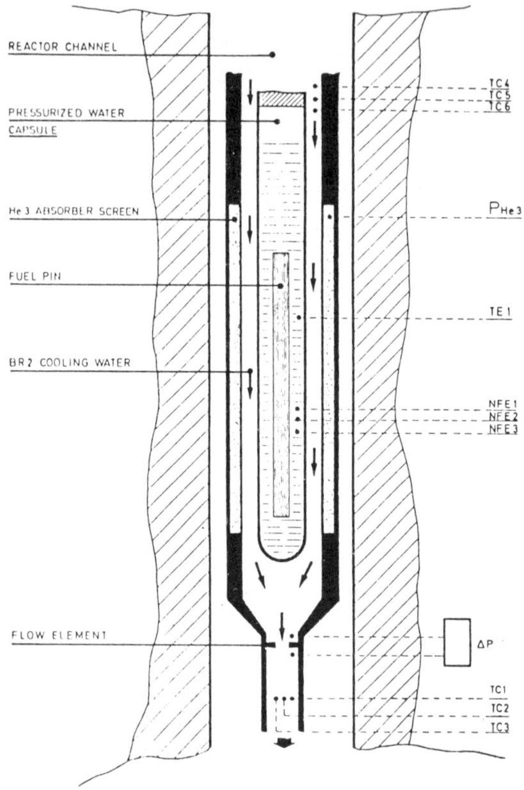

Figure 3 :

Experimental Assembly in BR2

A small water flow is maintained through the inpile section for evacuating radiolytic gases and detecting cladding rupture by the presence of fission gases. Self powered neutron detectors (two cobalt and one vanadium) are installed to record the fuel pin power variations.

The rig is easely reloadable in the BR2 hot cells with BR3 preirradiated fuel pins.

The pressurised water capsule containing the TRIBULATION fuel pin is surrounded by a device called CCD (Control and Calibration Device).
CCD measures the dissipated power of the fuel pin. It also controls the fission power by means of an He^3 gas gap surrounding the fuel pin.

The total dissipated power determination is done by measuring in and outlet temperature and mass flow of the cooling water that flows through the annular space between PWC and CCD.
After correction of the total dissipated power for gamma-heating of the structural materials of PWC and CCD, and after introduction

of the axial neutron flux profile of the corresponding BR2 irradiation channel, the peak pellet linear heat generation rate can be calculated.

The fuel pin fission power can be modified by changing the pressure in the CCD-He^3 gas gap. The fuel pin power is about doubled by reducing the He^3 pressure from 40 to 1 bar (figure 4)

The outpile control panel for CCD is called VNS 2 (Variable Neutron Screen), the flow sheet of which is shown on figure 5.

During steady state conditions and slow power transients, the He^3 pressure is controlled via the secondary side of a bellow pressuriser which is pressurised with nitrogen. A small He^3 gas flow is maintained through the annular gap for purifying the He^3 from tritium produced by neutron capture.

Fast power transients are executed by a fast depressurisation followed by a fast pressurisation of the He^3 screen.

During the fast depressurisation phase, the He^3 is evacuated to the dump tank. The depressurisation rate is controlled by the position of the calibrated needle valve in the depressurisation line.

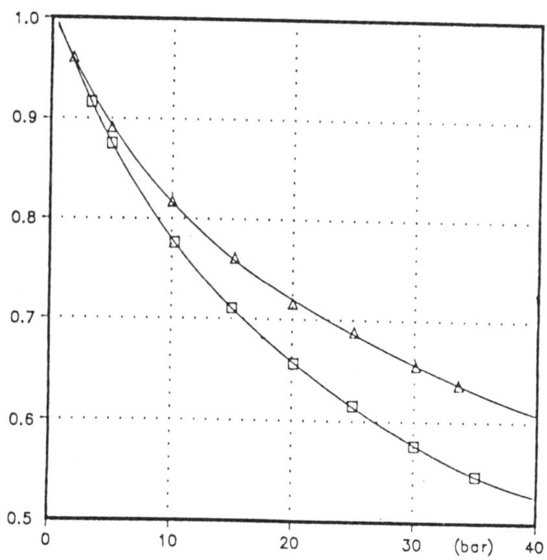

Figure 4 : CCD He3 Screen

Effect of He3 pressure on LHGR

TRIBULATION fuel pins

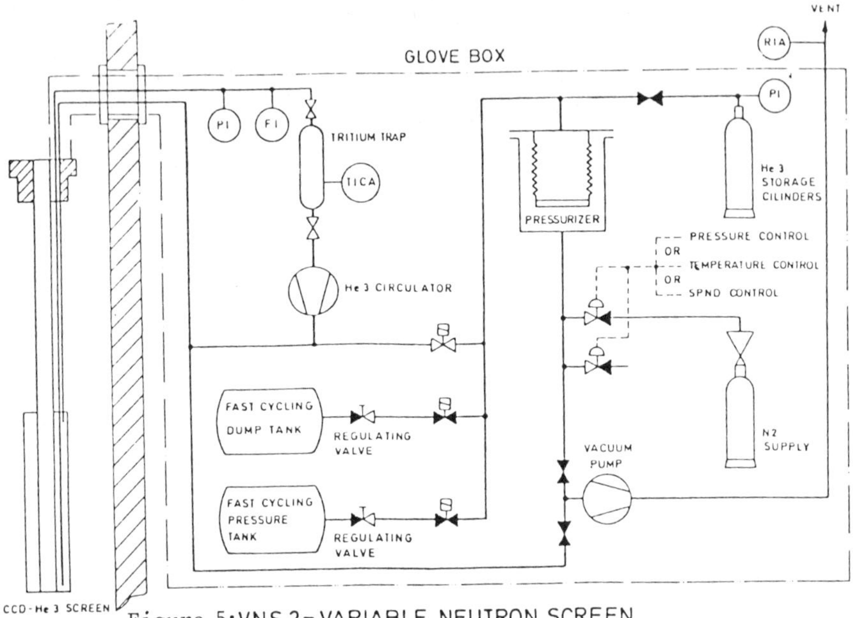

Figure 5:VNS 2-VARIABLE NEUTRON SCREEN

The PWC outpile control equipment accepts 3 rigs simultaneously. Coolant pressure and temperature, self-powered neutron detector signals and bleed water activity are continuosly monitored.

Figure 6 shows a simplified flow sheet of the PWC installation. As the fuel pin is surrounded by stagnant water, the pin cladding temperature is close to the coolant saturation temperature. Changing the cladding temperature can thus be achieved by changing the stagnant water pressure.

Many out-of-pile tests with electrically heated rods have been performed at the SCK/CEN at similar thermal-hydraulic conditions. These tests serve as a data base for cladding temperatures and burn-out limits at different operating pressure and linear heat generation rates.

During irradiation the data are recorded by a computerised data-acquisition system. The equipments used are a computer type HP 2113 E, a digital voltmeter type HP 2402 A, scanners type HP 2912 A, a 20 megabyte disk type HP 7906.

Thermocouple signals, cooling water mass flow and collectron signals are measured at pre-set intervals. At each measurement the average and standard deviation of 10 consecutive measurements are taken in order to increase the accuracy of the final power calculation.

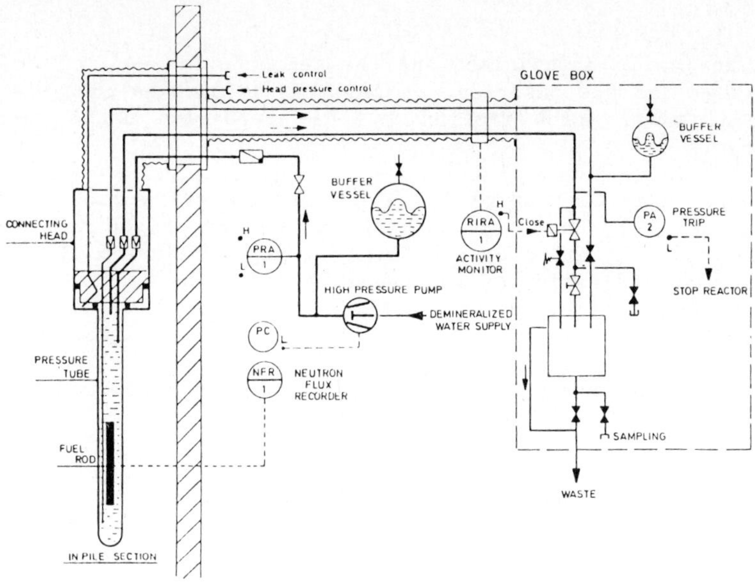

Figure 6 : PWC (Pressurized Water Capsule)

4. TRANSIENT CONTROL

The TRIBULATION transient consists of a fast power increase, a hold time at high power and a fast power decrease to well below its preconditioning power level.

The different power levels are controlled by the gas pressure in the He^3 screen and by the position of the control rod in the neighbouring reactor channel.

During reactor start-up of each TRIBULATION campaign, at 60 percent reactor power, the He^3 pressure in the CCD screen is changed from minimum to maximum, hereby recording the evolution of the fuel pin power.

By extrapolation to 100 percent reactor power the required He^3 pressures at the preconditioning period can be determined. The required power increase during the transient is obtained by controlling the opening time of the depressurisation valve of the VNS 2 installation. During this period He^3 is evacuated to the dump tank.

Formerly made calibration curves of the residual He^3 pressure versus opening time of the depressurisation valve allow an easy control of the maximum fuel pin power.

After a 1 or 10 minutes hold time at maximum power, the He^3

screen is fastly pressurised and the neighbouring control rod is inserted to its low position, so reducing as much as possible the fuel pin power, which remains in this condition until reactor shut-down.

The actual transient shape is determined by calculation of the fuel pin power by an energy balance at the end of the hold time. The evolution of the cobalt collectron signals during the power ramps are an indication of the power increase and decrease rates. Two transient examples plots are given on figures 7 and 8. In the case of figure 8, the reactor was srammed after the transient.

The average positive ramp rates were between 6 and 10 percents per second.

5. PRESENT STATUS AND FUTURE PROGRAMMES

At the present, three PWC rigs are operational and have proven to give satisfaction. Loading and unloading of the fuel pins goes very smoothly. The peak pellet linear heat generation rate is calculated with a precision of about ± 10 percent.

To date nine TRIBULATION transients have been executed without technical difficulties. Another fourteen transients are planned for the rest of the programme. Later on the facility will be used for the international operational transient programme on PWR fuel containing burnable poison /3/ (Gadolinium) for which negociations are underway with several interested parties.

REFERENCES

/1/ F. MOTTE, H. BAIRIOT, E. de MEULEMEESTER
"TRIBULATION High Burnup Behaviour of fuel subjected to transients".
Specialists Meeting on Safety Aspects of Fuel Behaviour on Off-Normal and Accident Conditions,
ESPOO (Helsinki), September 1-4, 1980.

/2/ D. HAAS
"TRIBULATION Objectives and Present Status".
ANS Topical Meeting - LWR Extended Burnup - Fuel Performance & Utilisation, Williamsburg, Va., April 4-8, 1982.

/3/ H. BAIRIOT, D. HAAS, F. MOTTE, H. NACKAERTS
"Gadolinium International Programme Preliminary Proposal"
BN Report 232.40/100/n/001, March 1982.

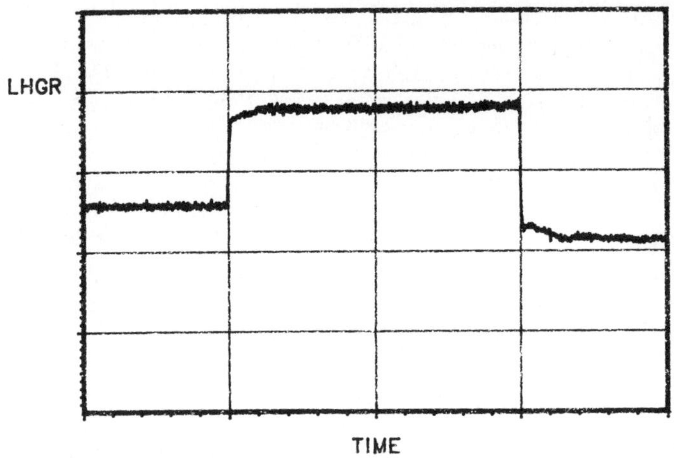

Figure 7 : Typical fast transient, performed with He^3 pressure changes

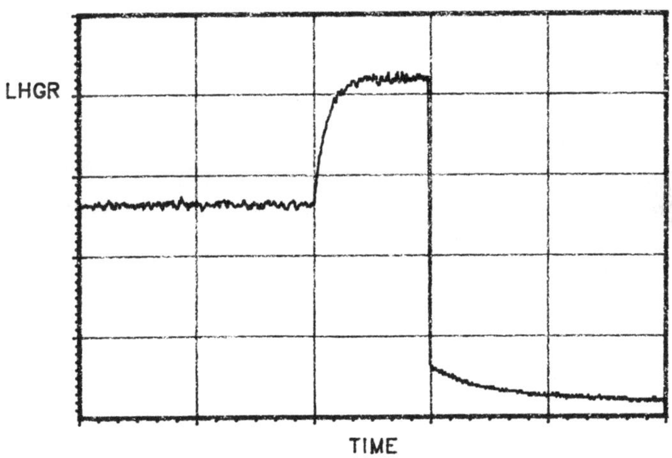

Figure 8 : Typical fast transient, followed by reactor scram

LA BOUCLE "ISABELLE 4" : BOUCLE D'ESSAI PWR DECHARGEABLE ET RECHARGEABLE DANS LA PISCINE DU REACTEUR OSIRIS

Michel LUCOT

Commissariat à l'Energie Atomique - Services des Piles

C.E.N./Saclay - 91191 - GIF-sur-YVETTE Cedex (France)

GENERALITES

Pour effectuer les irradiations de combustibles nucléaires dans des conditions PWR, les Services des Piles de Saclay disposent de deux boucles d'essai : l'une, la boucle IRENE, en service depuis 1970, permet de réaliser les irradiations dans des conditions thermohydrauliques et chimiques, représentatives de celles d'un réacteur de puissance PWR ; elle a été modifiée en 1979 pour permettre d'effectuer des irradiations avec température de saturation à la sortie de la grappe combustible ; l'autre, la boucle ISABELLE, en service depuis 1977, qui utilise la technique des trompes à eau pour la circulation du fluide de refroidissement, permet de réaliser les irradiations dans des conditions normales, anormales, transitoires ou accidentelles.

Ces deux boucles, de caractéristiques légèrement différentes, ont une conception mécanique qui implique la récupération des crayons combustibles en cellules chaudes ; les crayons sont ensuite envoyés vers un laboratoire spécialisé pour subir tous les examens post-irradiation.

Devant la demande croissante des expérimentateurs de pouvoir détecter et suivre l'évolution des dommages subis par les gaines des combustibles en cours d'expérience, les Services des Piles de Saclay ont étudié et construisent actuellement, d'une part un bâti multiple pour le contrôle des gaines, qui sera implanté dans la piscine du réacteur OSIRIS, et d'autre part un nouveau dispositif d'irradiation déchargeable et rechargeable dans la piscine du réacteur. Dans cette boucle, les crayons combustibles seront amovibles et pourront, à n'importe quel moment de l'irradiation, être récupérés pour être examinés sur le bâti multiple (examen visuel, métrologie, courants de Foucault), ainsi que sur les autres dispositifs de

P. von der Hardt and H. Röttger (eds.), Irradiation Technology, 35–47.

contrôles non destructifs en piscine du réacteur : banc de neutronographie et banc de gammamétrie. Après avoir subi les examens de contrôle, les crayons pourront être remis en irradiation.

Le bâti multiple fait l'objet d'une communication présentée par les Services des Piles de Saclay à cette Conférence. Le but de la présente communication est de présenter l'installation qui permet le chargement et le déchargement des crayons combustibles en piscine du réacteur OSIRIS. Cette installation comprend une boucle d'irradiation et le banc de chargement-déchargement associé, immergé dans la piscine.

LA BOUCLE D'IRRADIATION

Conception Générale

La nouvelle boucle déchargeable et rechargeable en piscine est dérivée de la boucle ISABELLE existante. Elle peut recevoir jusqu'à quatre crayons combustibles neufs ou pré-irradiés, dont la longueur peut atteindre 1,80 mètre et permet de réaliser l'irradiation dans des conditions thermohydrauliques et chimiques, représentatives de celles d'un réacteur de puissance PWR. Les irradiations sont prévues pour être effectuées dans la piscine du réacteur OSIRIS où la boucle est positionnée sur un support mobile ; moyennant quelques adaptations la boucle pourrait être implantée dans le coeur même du réacteur.

L'ensemble de l'installation est constitué de trois parties principales :
- la partie en pile,
- la partie hors pile,
- les liaisons entre la partie en pile et la partie hors pile.

La partie en pile (figure 1). Elle comprend le tube de force et le porte-échantillons :
- le tube de force est constitué de 2 tubes concentriques en acier inoxydable dans la partie haute, prolongés en partie basse, dans la zone sous flux neutronique, par 2 tubes en zircaloy. L'espace annulaire entre les 2 tubes assure l'isolement thermique du canal d'essai avec l'eau de la piscine,
- le porte-échantillons comporte, en partie basse, un panier amovible qui contient les crayons combustibles à tester. La circulation du fluide de refroidissement le long des crayons est obtenue à l'aide de trompes à eau (ensemble injecteurs-tuyères) disposées en couronne en partie haute du canal d'essai. Le bilan thermique au niveau du combustible s'effectue à partir du venturi disposé en aval des crayons et de plusieurs thermocouples placés à l'entrée et à la sortie du canal d'essai.

La partie hors pile (figure 2). Les circuits hors pile sont disposés sur les aires expérimentales, dans l'enceinte étanche OSIRIS. Ils assurent la commande et le contrôle de l'ensemble de l'installation. Les principaux paramètres sont mesurés, d'une part par l'appareillage propre à l'instal-

BOUCLE ISABELLE . Partie en pile

ISABELLE LOOP . in_pile part

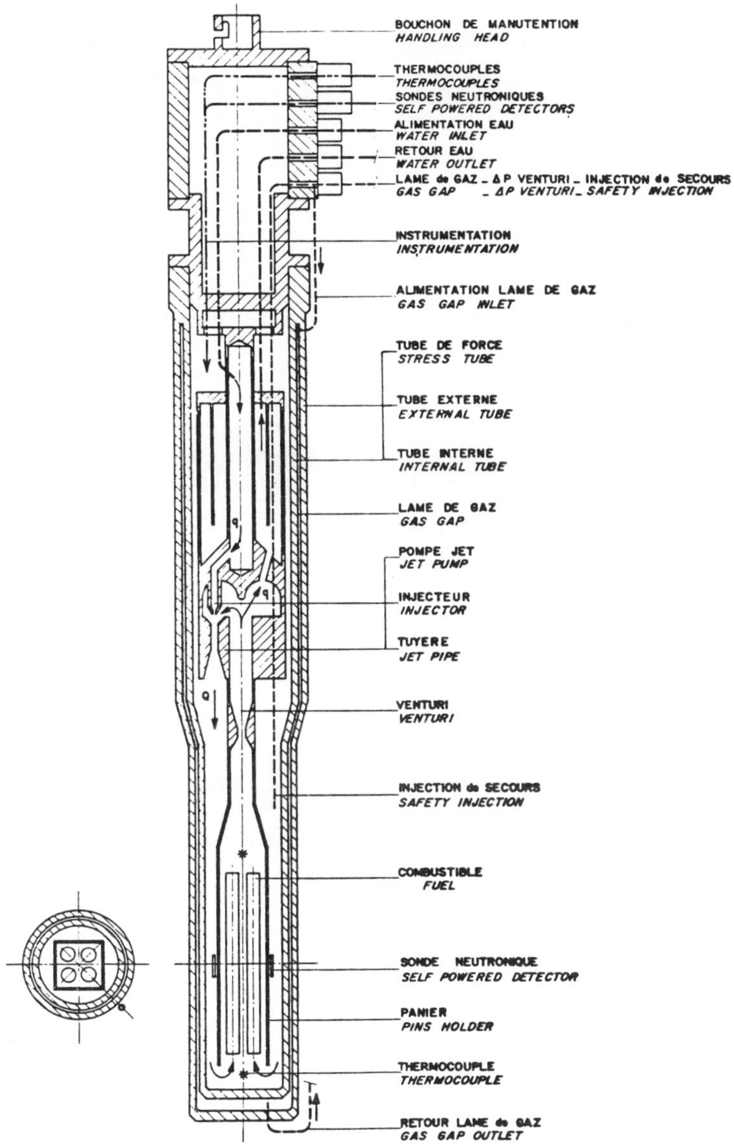

This illustration is printed from the best available original

Figure: 1

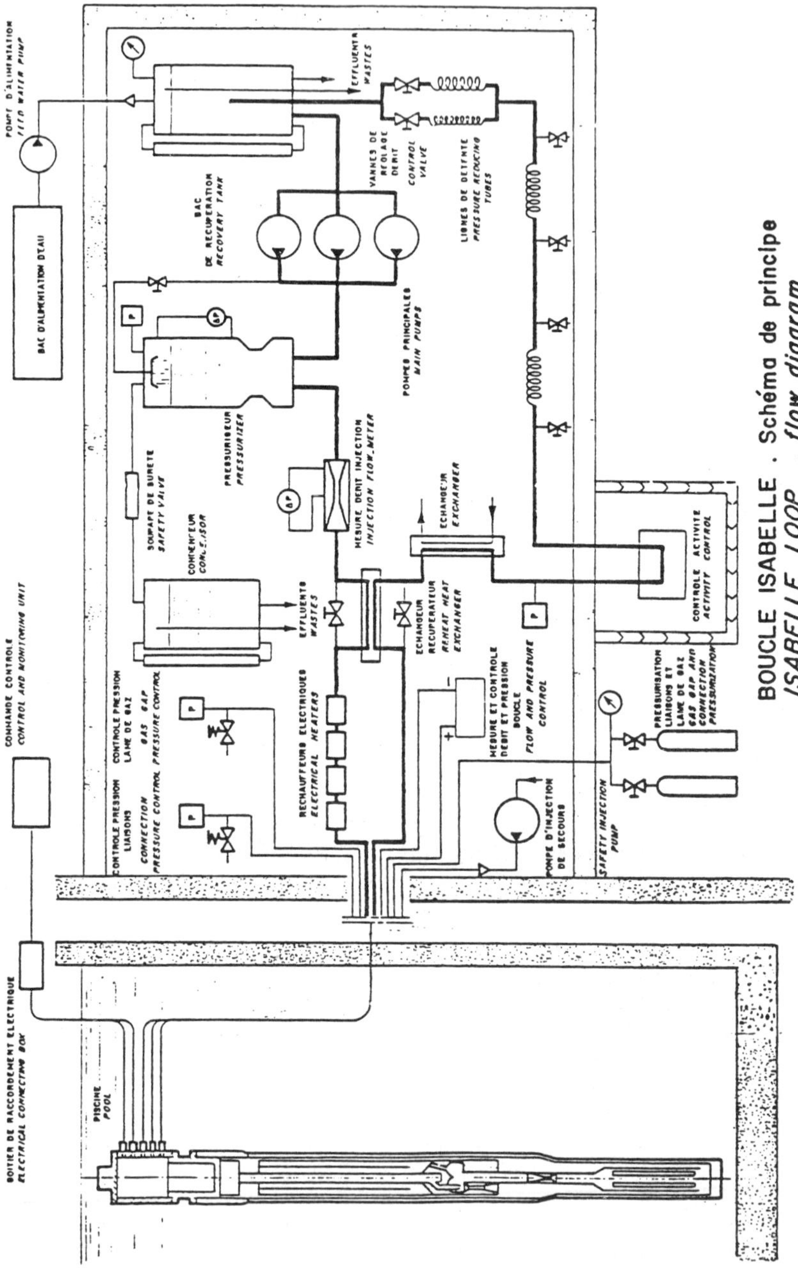

This illustration is printed from the best available original

Figure: 2

lation ISABELLE, et d'autre part par l'un des calculateurs "MITRA-15" du réacteur OSIRIS. L'installation est classique en ce qui concerne l'appareillage de fonctionnement qui comporte essentiellement :

- trois pompes de circulation montées en parallèle (2 pompes en service, 1 en secours automatique),
- un pressuriseur double phase à chauffage électrique,
- un réchauffeur électrique qui permet de réguler la température de l'eau,
- un échangeur-récupérateur pour optimiser l'équilibre thermique de l'installation,
- un échangeur pour évacuation de la puissance produite,
- un venturi pour contrôle du débit,
- deux vannes de réglage du débit,
- une installation de détection de rupture de gaine,
- un volume tampon,
- une pompe de réinjection pour remplir la boucle au démarrage et compenser les fuites d'eau éventuelles et les prélèvements en fonctionnement,
- une pompe de réinjection pour le refroidissement de secours.

Les liaisons entre la partie en pile et la partie hors pile. Le raccordement de la boucle se fait dans la partie supérieure du dispositif. Les liaisons pneumatiques et hydrauliques cheminent sous eau dans des tuyaux métalliques flexibles pressurisés. Pour traverser le mur de la piscine, on utilise un passage étanche situé à -3,70 m au-dessous du niveau libre de l'eau. Ce système assure une protection efficace en cas de transport d'activité par le fluide caloporteur. Les liaisons électriques cheminent en partie sous eau, dans des tuyaux métalliques flexibles pressurisés. Le boîtier de raccordement est situé hors d'eau, au niveau de la structure d'accrochage placée à quelques dizaines de centimètres au-dessus de la surface libre de la piscine.

L'ensemble des liaisons est suffisamment souple pour permettre, sans déconnexion et réacteur à pleine puissance, le transfert de la partie en pile de sa position d'irradiation au banc de chargement-déchargement.

Panier Amovible (figure 3)

En partie basse du porte-échantillons on trouve deux paniers :

a) Un panier "extérieur" en acier inoxydable, ce panier est solidaire des autres éléments du porte-échantillons et supporte l'instrumentation (sondes neutroniques et thermocouples). En partie haute, deux tiges transversales interdisent l'envol des crayons combustibles.
b) Un panier "intérieur" en zircaloy comprenant les 4 crayons disposés au pas de 12,6mm, les tiges support et les grilles d'espacement ; ce panier est amovible, il coulisse à l'intérieur du panier précédent ; en partie basse le panier est renforcé et comporte un verrou "quart de tour" pour l'accrochage sur le panier extérieur. Un pion détrompeur n'autorise qu'une position du panier intérieur dans le panier extérieur. En partie supérieure les 4 crayons combustibles peuvent être saisis individuellement avec une pince spéciale quand le panier intérieur est

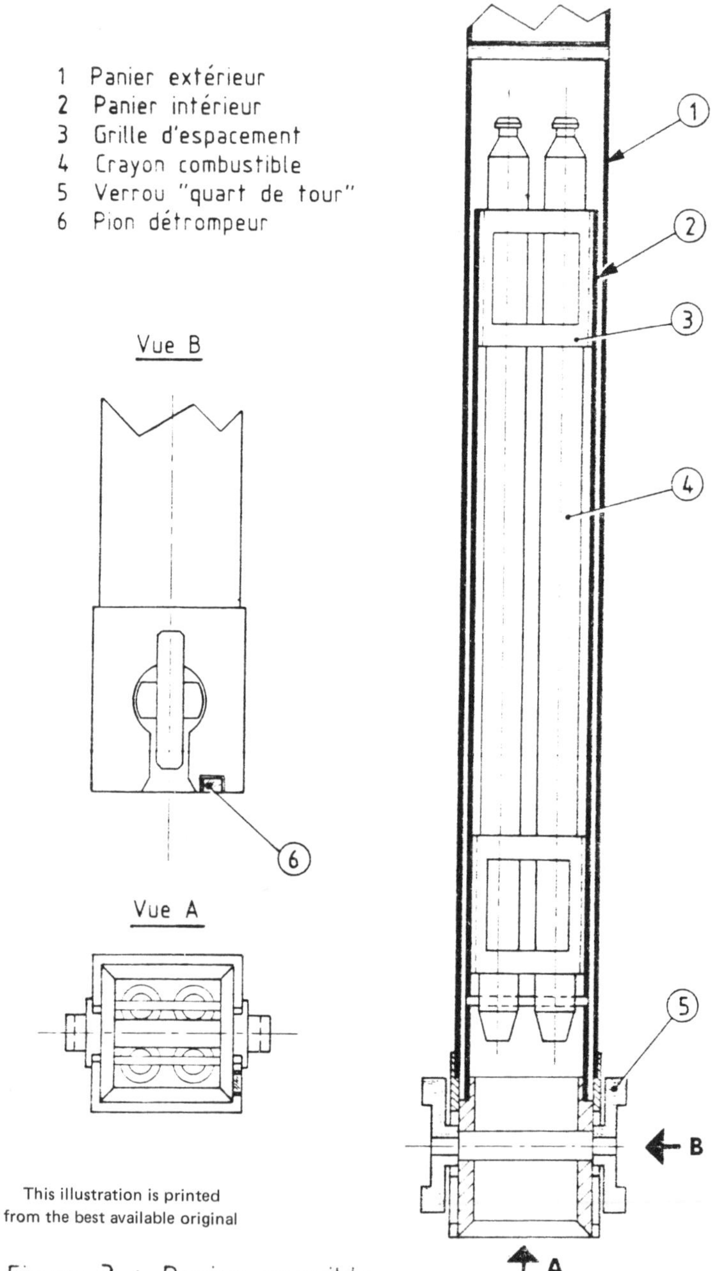

Figure 3 : Panier amovible

sur le banc de chargement-déchargement.

Lorsque le porte-échantillons est en place dans le tube de force, le verrou ne peut pas tourner, le déverrouillage des deux paniers en cours d'irradiation est ainsi condamné.

Caractéristiques Principales

- Géométrie du canal d'essai :
  - . diamètre : 46 mm
  - . hauteur : 2000mm
- Nombre de crayons combustibles : (assemblés en grappe de géométrie 17 x 17)
  - . nombre : 1 à 4
  - . longueur maximale : 1800mm
- Puissances mises en jeu :
  - . évacuable par la boucle : 90 kW
  - . maximale dissipée par les crayons : 500 W/cm
- Caloporteur : eau légère
  - . pression nominale : 155 bar
  - . température maximale : 300°C
  - . débit dans le canal d'essai : 400 g/s
- Instrumentation partie en pile :
  - . thermocouples Chromel-Alumel : 22
  - . sondes neutroniques argent : 8
- Circuits annexes :
  - . conditionnement de l'eau d'alimentation,
  - . laboratoire équipé pour le suivi de la chimie de l'eau,
  - . installation de détection de rupture de gaine.

LE BANC DE CHARGEMENT - DECHARGEMENT

Conception Générale (figure 4)

L'ensemble du banc est constitué par des structures en acier inoxydable, il est fixé en suspension sous une poutre du plancher +0,60 mètre, dans un angle de la piscine du réacteur. Il est en appui à plusieurs niveaux sur la peau de la piscine ; un support en fond de piscine maintient et supporte en partie le poids des structures.

Le banc comprend principalement :

un support de boucle. Ce support est hors d'eau à quelques centimètres du niveau libre de la piscine et sensiblement dans l'axe vertical du banc. La tête de boucle vient sur ce support pour y être fixée, le reste de la boucle étant immergé,

un ascenseur de charge et décharge du tube de force. Il a un mouvement motorisé "montée-descente" et supporte un chariot à mouvement manuel "droite-gauche",

un ascenseur de charge et décharge du panier amovible. Il a lui aussi un mouvement motorisé "montée-descente" et supporte un chariot à mouvement manuel "droite-gauche",

This illustration is printed from the best available original

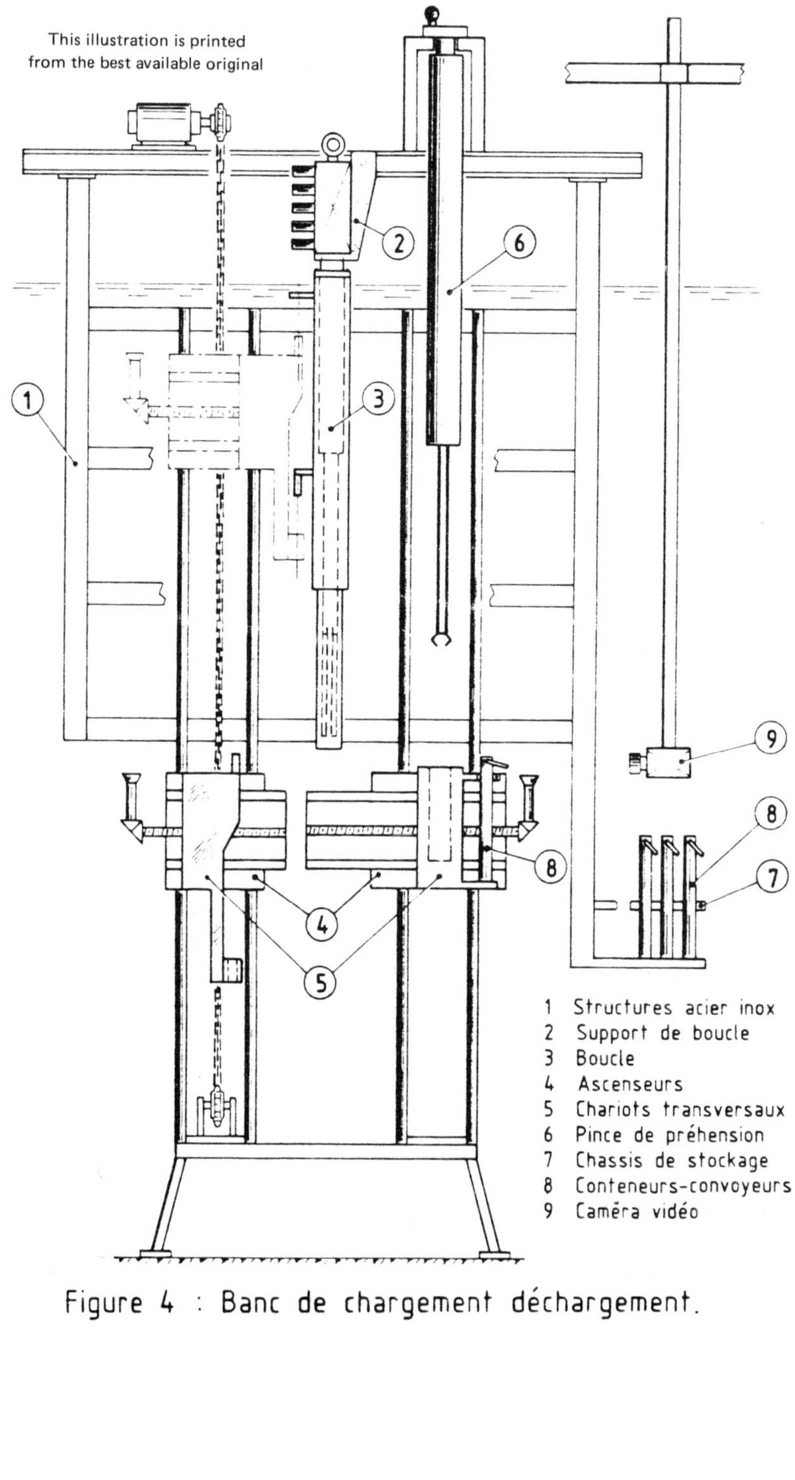

Figure 4 : Banc de chargement déchargement.

une pince de préhension pour l'extraction et l'insertion des crayons combustibles dans le panier. Elle a quatre positions possibles au pas de 12,6mm ; elle peut tourner et descendre de quelques centimètres,

un châssis de stockage des conteneurs convoyeurs. Il supporte les conteneurs-convoyeurs en attente de chargement (ou de déchargement) ou en attente d'expédition vers les bancs de contrôle (gammamétrie, neutronographie, bâti multiple),

quatre conteneurs-convoyeurs de crayons combustibles. Ils sont, soit sur le châssis de stockage (4 emplacements), soit sur l'ascenseur à proximité du panier amovible (1 emplacement), soit sur l'un des bancs de contrôle (3 fois 1 emplacement),

une caméra vidéo. Elle est immergée en suspension au bout d'une perche qui se déplace sur un rail ; elle permet de suivre les opérations de chargement-déchargement ainsi que le contrôle de l'identité des crayons dans le panier.

Les mouvements motorisés des ascenseurs se font avec des moteurs à courant continu, deux sens de marche, vitesse variable et électro-frein. Les mouvements à commande manuelle se font à l'aide d'une perche.

Caractéristiques Principales

- encombrement hors tout :
  - . hauteur : 11.600 mm
  - . largeur : 1.000 mm
- poids de l'ensemble : 1.500 kg
- mouvement des ascenseurs :
  - . montée-descente : par chaîne
  - . droite-gauche : pignon-crémaillère
- course des ascenseurs :

		gauche	droite
. montée-descente	:	4100mm	3300mm
. droite-gauche	:	180mm	230mm

- vitesse des ascenseurs :
  - . montée-descente
  - . droite-gauche : variable de 0 à 15 mm/s
- longueur maximale des crayons pouvant être manipulés : 1800mm

PROCEDURE D'UN DECHARGEMENT - CHARGEMENT

a) Opérations préliminaires (figure 4).

- Les ascenseurs sont mis en position basse fin de course, chariots hors axe du banc, leur position est suivie sur un tableau synoptique au cours de chaque déplacement.

- La boucle est transférée de sa position d'irradiation sur le banc de défournement. Elle est à l'arrêt, mais elle n'est pas déconnectée de l'installation hors pile.

b) Décharge du tube de force (figure 4) :
- Remontée de l'ascenseur concerné hors axe du banc, translation du chariot vers le tube de force, fixation du tube de force sur le chariot, décrochage des vis de fixation tube de force/tête de boucle.
- Descente de l'ascenseur avec le tube de force au niveau bas fin de course, translation vers la gauche en position de stockage.

c) Décharge du panier amovible (figure 5) :
- Translation du chariot support de panier dans l'axe du porte-échantillons, remontée de l'ascenseur jusqu'au panier, déverrouillage du panier et du porte-échantillons, verrouillage du panier sur le chariot.
- Descente de l'ascenseur concerné avec le panier au niveau bas fin de course, translation du chariot dans l'axe de la pince de préhension des crayons, remontée de l'ascenseur pour amener le haut des crayons sous la pince.

d) Extraction des crayons (figure 5) :
- Positionnement de la pince ouverte au-dessus du crayon à extraire, descente de la pince en appui sur le crayon, fermeture de la pince, descente de l'ascenseur, le crayon reste suspendu en bout de pince.
- Translation du chariot pour amener le conteneur-convoyeur sous la pince et le crayon, remontée du conteneur et ouverture de la pince, le conteneur est chargé d'un crayon.
- Descente de l'ascenseur avec le conteneur niveau bas fin de course, transfert du conteneur sur le support de stockage. Recharge de l'ascenseur avec un conteneur vide pour l'extraction suivante.

Les mêmes opérations sont répétées pour l'extraction des autres crayons.

e) Examen des crayons (figure 6) :

Les conteneurs-convoyeurs, avec chacun un crayon, sont dirigés individuellement vers les différents bancs d'examens non destructifs, tous situés à quelques mètres dans la piscine : neutronographie, gammamétrie, bâti multiple. Après ces examens, les conteneurs-convoyeurs reviennent avec leur crayon sur le support de stockage.

f) Réinsertion des crayons dans le panier :

Les opérations effectuées pour l'extraction des crayons sont réalisées en sens inverse. A la fin des opérations, un contrôle rigoureux de l'emplacement de chaque crayon dans le panier est effectué à l'aide de la caméra-vidéo.

g) Recharge du panier dans le porte-échantillons :

Mêmes opérations, en sens inverse, que celles effectuées lors du déchargement.

h) Recharge du tube de force :

Mêmes opérations, en sens inverse, que celles effectuées lors du déchargement.

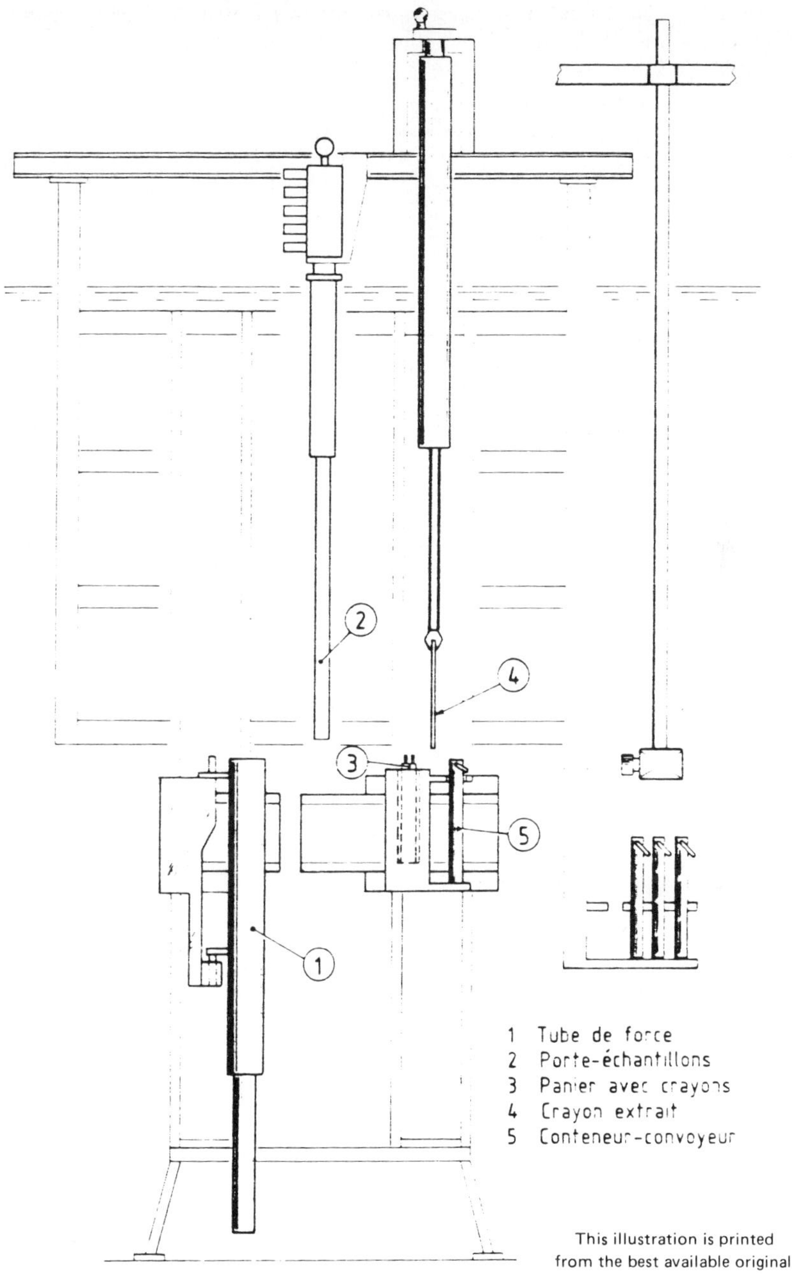

Figure 5 : Extraction d'un crayon.

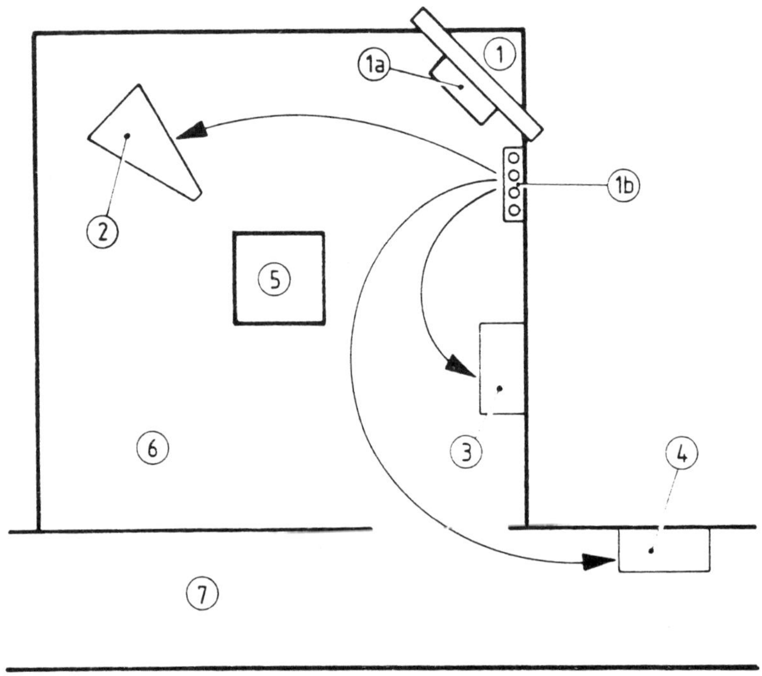

1 Banc de chargement déchargement
  1a - Boucle ISABELLE
  1b - Conteneurs-convoyeurs
2 Neutronographie
3 Gammamétrie
4 Bâti-multiple
  (métrologie , visualisation , courants de Foucault)
5 Cœur réacteur
6 Piscine
7 Canal

Figure 6 : Examen des crayons

This illustration is printed from the best available original

i) Remise en irradiation :

La boucle est d'abord remise en fonctionnement hors irradiation. Après contrôle des paramètres et des sécurités, la boucle est alors remise en irradiation.

CONCLUSION

La nouvelle boucle d'essai PWR, déchargeable et rechargeable dans la piscine du réacteur OSIRIS, ainsi que le bâti multiple, viennent compléter les installations d'essai et de contrôle des Services des Piles de Saclay.

Ces installations permettent désormais à un expérimentateur de suivre l'évolution des caractéristiques et l'état de santé des crayons combustibles qu'il veut tester, les examens pouvant avoir lieu plusieurs fois au cours de l'irradiation ; les conditions d'irradiation peuvent d'ailleurs être modifiées à chaque période d'irradiation en fonction des résultats obtenus lors des campagnes d'analyse.

THE DIDO WATER LOOP AT HARWELL AND ITS USE IN PARTICULAR FOR PWR CRUD-DEPOSITION EXPERIMENTS

E.J. Mitchell & J. Monahan

Atomic Energy Research Establishment

Harwell, Oxfordshire, England

ABSTRACT

A major facility in the DIDO reactor at Harwell is a high-pressure water loop which can operate under typical PWR thermohydraulic and water-chemistry conditions.

The first part of this paper describes the basic DIDO Water Loop and the various supporting facilities. The construction and layout is discussed and the main parameters listed.

The second part concerns a PWR experiment to investigate crud deposition on both a single, irradiated, fuel rod and a parallel, out-of-pile, fuel rod simulator. Details of the irradiation rig, out-of-pile test section, and other special equipment required for the experiment are presented.

A special feature was the necessity to develop a system for generation and injection into the loop water of crud precursors, to form a nickel-iron spinel comparable to that found in a typical PWR circuit. This system is described in the third part of this paper.

P. von der Hardt and H. Röttger (eds.), Irradiation Technology, 49–75.

CONTENTS

PART 1. THE DIDO WATER LOOP

1. Introduction
2. The DWL Circuit
3. General Construction
4. Loop General Data

Fig. 1.1 DWL Simplified Flow Diagram

PART 2. A CRUD-DEPOSITION EXPERIMENT IN SUPPORT OF THE PWR SYSTEM

1. Introduction
2. Technical Requirements
3. Modified DWL Circuit
4. Special Equipment
   4.1 Irradiation Rig
   4.2 Out-of-Pile Test Section
   4.3 Main Loop Heaters
   4.4 Supplementary Cooler
   4.5 Fuel-Rod-Handling Equipment
   4.6 Crud-Injection Requirement

Fig. 2.1 DIDO Water Loop. Simplified Flow Diagram for Crud Deposition Experiment
Fig. 2.2 General Arrangement of Irradiation Rig
Fig. 2.3 Derivation of Flow Channel
Fig. 2.4 Cross-Section through Flow-Divider
Fig. 2.5 General Arrangement of OOPTS

PART 3. DEVELOPMENT OF A SYSTEM FOR GENERATION AND INJECTION OF CRUD PRECURSORS

1. Introduction
   1.1 General
   1.2 Corrosion-Product Composition
   1.3 Selection of Technique
2. Principle of the Chosen Technique
3. Preparation of Crud-Forming Species
4. Development
   4.1 Injection Equipment
   4.2 Plant Commissioning
   4.3 Loop tests and Results

References

Fig. 3.1 Crud Injection Circuit

PART 1 DIDO WATER LOOP

by E.J. Mitchell

1. INTRODUCTION

The DIDO Water Loop (DWL) is a high-pressure, high-temperature, circuit for carrying out irradiation experiments in the DIDO materials-testing reactor at the UKAEA research centre at Harwell, Oxfordshire, England. Although originally designed and built to operate in either pressurised-water (PW) or boiling-water (BW) modes, it is now modified to operate only in the PW mode. Nevertheless, the loop can be operated with some saturated boiling in the irradiated test section and exit-steam quality up to about two per cent.

The all-stainless-steel construction of the DWL with purified feedwater and extensive injection systems for saturated gas, solute, or crud species allied to an extensive sampling system, both from peripheral and isokinetic sources, makes it a unique apparatus in the radiation-water-chemistry field. It is especially suitable for chemistry experiments in conjunction with irradiation of fuel rods in high neutron fluxes. The loop-operating pressure of up to 138 bars means that it can operate at representative PWR conditions, and it is for experiments in support of that reactor system that the DWL has been used since 1976. A recent PWR experiment concerning crud deposition on a fuel rod is described in Part 2 of this paper.

2. THE DWL CIRCUIT

A highly simplified version of the DWL flow diagram is shown at Fig. 1.1. In this we can see the primary circuit where pressurised water is circulated by two variable-frequency, canned-rotor, centrifugal pumps through three heater vessels in series. Each vessel contains three stainless steel rod heaters of 20mm diameter and heated length 500mm, with Nichrome heating elements insulated by boron-nitride powder. The heaters are rated at 11 kW each, thus providing total heater power of 99kW.

After leaving the heaters, the coolant passes through a coarse filter and flowmeter, then through the in-pile rig, where it gains heat from an irradiated fuel rod (with supplementary nuclear heating and sometimes electrical heating), to emerge at the rig head either subcooled or as a low-quality two-phase mixture as required by the experiment. From the rig-head outlet the coolant flows through a wire-wound strainer and then through two water test sections. These are small pressure vessels into which, for example, corrosion or deposition coupons can be installed for experiment purposes.

The water then passes through a further flowmeter and enters a subcooling vessel, where its temperature is reduced by mixing with cooled water returning from the main loop cooler, thereby maintaining a suitably low temperature at inlet to the circulators.

Loop pressure is maintained by a large-volume, spherical, electrically heated pressurising vessel. In parallel with this is a cylindrical condenser vessel - a legacy of the original loop design, which catered for BW operation. It is left in circuit for convenience since it incorporates all the water-level instrumentation; also the extra volume contributes to the stable behaviour of the loop.

Water make-up can be carried out either with straight demineralised water or with specially treated water to suit the particular experiment. The addition of hydrogen, ammonia, lithium hydroxide, or boric acid is achieved by injecting pressurised, saturated solutions of these additives into the loop via small, high-pressure, metering pumps either upstream or downstream of the main loop heaters.

Several sampling points are provided on the loop for taking on-line water samples, as indicated on Fig. 1.1. Two of these - at inlet to and outlet from the in-pile rig - are isokinetic probes able to take samples at representative coolant velocities. The sampled coolant is passed through a high-pressure filter, then through a pressure-dropper. At the lower pressure, a HIAC particle counter can also be inserted in the sample line if required.

The other on-line samples are taken peripherally from selected points on the loop. All the sampling lines pass through a communal cooler and then to a sampling cabinet, where filtration and pressure reduction may be carried out as for the isokinetic samples. Finally, the sample can be passed through a gas stripper for analysis of gases by chromatography.

In addition to the on-line samples it is also possible to obtain samples in removable vessels or "bombs" installed in a glove box adjacent to the main sampling cabinet. One sampling point is upstream and the other downstream of the in-pile test section.

3. GENERAL CONSTRUCTION

The primary-circuit pipework is $1^1/_2$in. nominal bore, schedule 80, AISI 321 material, and all-welded construction has been aimed for where possible. Where mechanically sealed joints are necessary (e.g. pressure-vessel closures) Grayloc couplings have been used for the larger sizes, whilst for the smaller sampling and instrumentation lines Gyrolok and Cajon couplings are fitted.

Because of the many pressure vessels operating at high pressure and temperature plus the possibility of circuit contamination, most of the primary circuit, together with the cooling and purification circuit, is housed in a specially built, shielded plant room on the reactor ground floor. This room is not normally accessible during loop operation, and is sealed off by a shielded, sliding door.

The sampling cabinet, isokinetic-sampling facility, and a glove box for "take-away" samples are sited on the ground floor just outside the shielded room, together with some of the water-treatment, make-up, and dosing equipment. Above this area, on a sub-floor or gallery, are the many control and instrumentation panels for the loop and the experiment therein. These are compactly arranged so as to give the DWL its own Control Room.

The loop primary instrumentation includes facilities for measurement and/or indication of coolant and surface temperatures, coolant mass flows, pressures, pressuriser water level, heater powers, pump and valve operation, and gaseous-fission-product activity. Appropriate warning and trip functions are incorporated to satisfy operational and safety requirements.

Apart from periodic hand logging of all data, certain key readings are continuously displayed on digital panel meters. Also for those parameters of prime importance to the particular experiment, the sensors are linked to a microprocessor which can be programmed to carry out routine scans, provide periodic print-outs of data and calculations derived therefrom, and store data as required. Subject to certain constraints manual intervention via the keyboard is possible, for example in order to carry out a rapid single-point scan on a parameter of particular interest.

4. LOOP GENERAL DATA (PRIMARY CIRCUIT)

Design Pressure	–	172.4 bars (2,500 lb/in^2)
Operating Pressure	–	Up to 138 bars (2000 lb/in^2)
Loop-Operating Temperature	–	Up to T_{sat} at rig outlet (335°C at 138 bars)
Loop Coolant	–	Demineralised water. Nominal condition:- dissolved oxygen <0.01 mg/L chloride <0.1 mg/L
Coolant Volume	–	Total (including pressuriser/condenser) 430 L
	–	Circulating volume 175 L
Circuit Materials	–	Stainless steel AISI 321 or 316
Main Pipe Size	–	$1^1/_2$ in. nominal bore, schedule 80
Circulating Pumps (2 off)	–	1.3 kg/s against 8 bars head. Canned-rotor, glandless, and totally enclosed. Variable-frequency control.
Main Heaters	–	9 off stainless steel rod heaters 11 kW each 99 kW total power 66 kW switched manually 33 kW infinitely variable
Flow-Control Valves	–	Air-operated, double-glanded with drained interspace
Pressuriser Heater	–	48 kW infinitely variable
Loop Cooler	–	130 kW capacity

Experiment Facility

Position	-	Mk V hollow fuel element (50 mm-diameter rig)
Available core depth	-	0.6 m
Peak thermal-neutron flux (unperturbed)	-	$1.85 \times 10^{14} n\ cm^{-2}s^{-1}$
Peak fast-neutron flux (>1 MeV)	-	$1 \times 10^{14} n\ cm^{-2}s^{-1}$

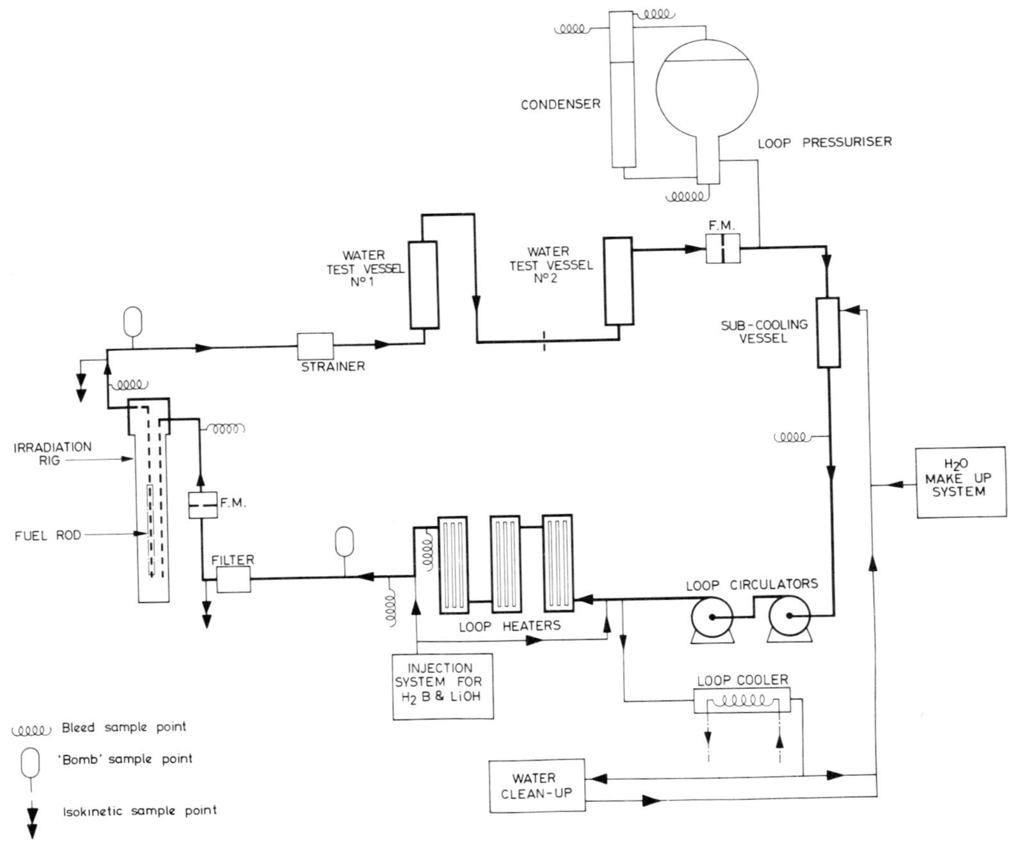

FIG. 1·1 DIDO WATER LOOP

SIMPLIFIED FLOW DIAGRAM

PART 2 A CRUD DEPOSITION EXPERIMENT IN SUPPORT OF THE PWR SYSTEM

by E.J. Mitchell

1. INTRODUCTION

In June 1979, following a design study carried out by Research Reactors Division at Harwell, an irradiation experiment in the DWL was commissioned. The objectives of the experiment were:

(a) To irradiate and to study crud deposition on a fuel rod in a rig serviced by the DWL.

(b) To compare crud deposition in (a) with that on a parallel, out-of-flux, electrically heated rod operating under similar thermohydraulic and water-chemistry conditions.

The experiment programme started in July 1981 and was completed in June 1982 during which time seven reactor operating periods were utilised.

2. TECHNICAL REQUIREMENTS

The general technical requirements of the experiment were as follows.

2.1 Fuel-Rod Irradiation

A single PWR-type Zircaloy-clad fuel rod of length about 1 metre and outside diameter 9.5 mm to be irradiated inside a Zr-Nb pressure vessel. The flow channel cross section to be shaped to simulate a sub-channel of a PWR fuel rod bundle. Thermohydraulic conditions, coolant chemistry and crud composition and concentration to comply with customers specified range.

2.2 Out-of-Pile Test Section

In parallel with the in-pile rig an electrically heated out-of-pile test section is to operate under similar thermohydraulic and water-chemistry conditions. This section to have a Zircaloy-clad heater to simulate the fuel rod.

2.3 Loop Heaters

The nine heater elements to be replaced by Zircaloy-clad heaters each having a nominal power of 13 kW. Also, trace-heating to be installed on the piping between the loop heater and the test sections to balance heat losses and meet the thermohydraulic conditions specified.

2.4 Supplementary Loop Cooler

A supplementary loop cooler to reject the excess heat input by the two test sections (beyond the capacity of the existing loop cooler) to be installed. The primary-side heat-transfer surfaces to be of austenitic stainless steel.

2.5 Fuel-Rod Handling Equipment

Suitable equipment to be designed and developed for handling irradiated fuel rods in a shielded cell and carrying out photography and crud scraping.

2.6 Crud-Injection System

A system for injection of ferrous and nickel species into the loop to be provided in order to achieve the required nickel-ferrite crud composition and concentration in the loop water.

3. MODIFIED DWL CIRCUIT

Figure 2.1 shows the simplified DWL flow diagram modified to suit the crud-experiment requirements. The major addition is the out-of-pile test section, plus its flow-control valve, shown in parallel with the irradiation rig.

The supplementary cooler is shown on a bypass line between the two water test sections, and the crud-injection system is the other major addition to the diagram.

4. SPECIAL EQUIPMENT

4.1 Irradiation Rig

To comply with the Harwell safety practice of double containment of fissile material the irradiation rig is enclosed in a primary pressure vessel subject to the full system pressure and temperature, outside which is a secondary containment vessel or "thimble". Although the thimble is subject only to low pressure and moderate temperature both vessels are designed for the full system pressure so that, should the primary vessel fail, safe containment is provided by the thimble.

Both vessels are made from zirconium/$2^1/_2$% niobium alloy, which gives a combination of high strength and neutron economy. The interspace between the vessels is utilised to control heat loss to the reactor coolant by filling with either helium or neon.

The rig had to fit inside a hollow fuel element in DIDO, so the in-core section was limited to a diameter of 50 mm. Also, because of the double-containment requirements, the actual rig envelope (i.e. the pressure-vessel bore) is only 36.5mm diameter, which dictates a very compact design, as shown in the rig general arrangement at Fig. 2.2.

Pressurised water enters at the rig head and flows downwards in the annulus between the pressure-vessel bore and the outer wall of a hollow flow-divider. The coolant flow reverses at the bottom of the pressure vessel and rises over a single central fuel rod located within the flow-divider section, to emerge at the rig head.

This scalloped and heated hollow flow-divider is a unique feature of the rig. In order to reproduce as far as possible the correct PWR hydraulic conditions, the fuel-rod flow channel is scalloped to simulate the profiles of adjacent fuel rods in a typical 17 x 17 PWR-rod bundle. Fig. 2.3 shows how the flow-channel shape is derived.

A cross-section through the flow-divider is shown at Fig. 2.4. It is made in two parts with an annular gap between them, which fills with stagnant water during operation and acts as a heat-transfer barrier between the down and up flows of coolant. The outer section is a plain stainless steel tube attached mechanically to the main inner section, which is about 1.2m long and is made of Zircaloy 2. In order to provide the scalloped flow channel, the material was hot-extruded over a precisely shaped mandrel of nitriding steel. The Zircaloy billet was copper-coated to provide lubrication during extrusion, and afterwards all traces of copper were removed with nitric acid. The cleaned-up extrusion (which finished up slightly bowed) then had to be straightened and machined to final size on outside diameter, after which a four-start helical groove of semicircular cross-section was machined along most of the length. The purpose of the groove was to accommodate two stainless-steel-sheathed, MgO-insulated heating elements non-inductively wound from bottom to top of the extrusion. The heaters were wound together in this way so that, even if one failed, uniform heat over the flow-divider length would be provided by the other.

These "trim" heaters were designed to provide 8 kW of electrical power to control conditions in the test section and to compensate for predicted changes in fissile power during the reactor cycle. To ensure good thermal contact between the heaters and the Zircaloy flow-divider a plasma-arc metal-spraying technique was used to deposit a layer of stainless steel about 2mm thick over the assembled heater cables, as shown in Fig. 2.4.

The above procedures may appear to be quite straight-forward as described, but considerable development work was involved, and a full paper could be written on this alone! In the first instance several trials were necessary before acceptable Zircaloy extrusions were produced. This was followed by the difficult straightening and machining process and then by a further series of tests to optimise the metal-spraying procedure. As insurance against possible failure of the extrusion process, alternative manufacturing methods were also investigated in parallel.

The stagnant-water annulus between the flow-divider sections is utilised for upward passage of the four "cold tails" of the flow-divider heaters and for instrumentation comprising six self-powered neutron detectors (SPNDs), four sets of fast-flux monitors, and three thermocouples.

The SPNDs are arranged in three diametrically opposed pairs at different rig heights and provide continuous measurement of thermal-neutron flux. The fast-flux monitors consist of iron wires inside Zircaloy capsules, five of which are stacked vertically inside a thin-walled stainless steel carrier tube. The four tubes were arranged to permit sequential removal throughout the experiment at reactor shutdowns.

The thermocouples are for measurement of water temperatures at inlet to the rig and inlet to and exit from the fuel rod. They can only be expected to provide approximate temperatures, so they are backed up by more-accurate platinum-resistance thermometers in the inlet and outlet pipework at the rig head.

The fuel rods for the experiment were manufactured by the Springfields Nuclear-Power Development Laboratories (SNPDL) of the UKAEA. The fuel rod is of conventional PWR design with a stack of UO_2 pellets inside a Zircaloy-4 sheath, but the fuelled length is only 75cm.

A dummy fuel rod having exactly the same external construction was also made by SNPDL for assessment of gamma-heating from the rig. The fuel pellets were replaced by molybdenum of equal weight.

4.2 Out-of-Pile Test Section

The general arrangement of the OOPTS is shown at Fig. 2.5. The outer shell consists of a stainless steel tube with Grayloc end closures and side branches (also terminating in Grayloc couplings) for the coolant inlet and outlet.

Unlike the in-pile rig with its closed-end vessel, there is no need for re-entrant flow in the OOPTS, so the coolant flows straight through over the heater rod. However, in order to reproduce the hydraulic conditions of the rig, the heated, scalloped Zircaloy flow channel is also used in the OOPTS.

The OOPTS is installed in the limited depth of the pipe bridge near the reactor top, so it has to be fitted horizontally. Referring to Fig. 2.5, the coolant enters the left-hand branch and is deflected by a shroud tube to avoid transverse buffeting on the heater rod. The flow then reverses to pass over the heater rod within the scalloped flow channel and emerge via the right-hand branch.

The OOPTS flow channel is of identical cross-section to that of the in-pile rig, but is somewhat longer, and the heater rod therein is also longer than the in-pile fuel rod. This is because additional heat must be added to compensate for the lack of gamma-heating in the OOPTS since we have the same coolant inlet conditions and require the same outlet conditions for each test section for a similar heat flux and identical channel geometry.

At the entry end of the OOPTS is a Grayloc hub closure to which is welded an extension tube terminating in a Gyrolok coupling. The high-flux heater rod engages in and is sealed by this coupling. At the exit end is another Grayloc hub, into which the four heater leads from the two trim heaters are sealed.

The high-flux heater rod simulates the fuel rod during normal operation. It is a uniform-flux, single-ended heater of exactly the same diameter as the fuel rod. However, the heated length is 110cm compared with a the fuelled length of 75cm, and maximum power is 35 kW compared with fissile power of <20 kW. As explained previously, this is to compensate for the lack of gamma-heating in the OOPTS. The heater has an inner stainless steel sheath over which is swaged a Zircaloy-4 tube welded to Zircaloy-2 end fittings. Inside the sheath there are eight Nichrome resistance elements insulated by boron-nitride pellets.

Instrumentation on the OOPTS consists of resistance thermometers in the inlet and outlet pipework and three thermocouples on the vessel surface to monitor its temperature and initiate a heater trip in the event of overheating.

4.3 Main Loop Heaters

The normal loop heaters have been mentioned in Part 1 Section 2.1. However, for this experiment it was thought that stainless steel heaters might contribute to the crud species in the loop, thereby making crud control difficult, so Zircaloy heaters were specified.

The heater has a thick-walled (3mm) Zircaloy-2 sheath enclosing eight Nichrome resistance wires insulated by compacted boron-nitride powder. The heated length is 500mm, above which a cold zone also 500mm long terminates in a raised-diameter feature which engages in a 1 in. Gyrolok coupling to effect a pressure seal. The cold zone was deliberately lengthened from the stainless steel design (340mm) to reduce the temperature and hence possible differential-expansion problems at the Gyrolok seal where the Zircaloy heater engages with the stainless steel coupling.

As a further precaution, tests were carried out on a typical joint subjected to temperatures up to 250°C under a helium pressure of 155 bars (2500 lb/in^2). These tests showed the joint to be satisfactory despite the difference in thermal-expansion coefficients.

The heaters were designed for a maximum power of 15 kW each (compared with 11 kW for the standard stainless steel heaters), thereby providing a maximum heating capability of 135 kW. Control was arranged so that the first two vessels were manually adjustable over the full range, with the third (hottest) vessel under fully automatic control.

4.4 Supplementary Cooler

This cooler is situated on a bypass line between the two water test sections. A fixed orifice is installed in the main line to give a flow through the cooler of about 10% of the total. The cooler consists of a high pressure stainless-steel coil wound on a guide former, housed inside a low-pressure stainless steel shell, and is capable of removing about 70 kW of heat.

4.5 Fuel-Rod-Handling Equipment

A particular requirement of the experiment was the need to photograph the irradiated fuel rods and scrape any accumulated crud from specified positions. Special equipment was therefore designed and made for these purposes.

The equipment is designed to accept the fuel rod and lock it in position, and then raise, lower and rotate it as required for photography or crud-scraping, within the shielded viewing cell of the DIDO reactor. Special tools were developed for attachment to standard remote-handling tongs operated through the tong ports of the viewing cell.

Separate crud-scraping tools were made for axial and circumferential action, and each was connected to a suction tube by means of which the removed crud was sucked into the hollow body of the scraping tool and collected on a fine filter for subsequent removal and analysis.

4.6 Crud-Injection Requirement

Good chemistry control on the DWL results in very low crud levels, and to reproduce the nickel-ferrite species of a typical PWR circuit we had to build a special system for generating and injecting crud precursors into the loop. The invention and development of such a system is described in Part 3 of this paper.

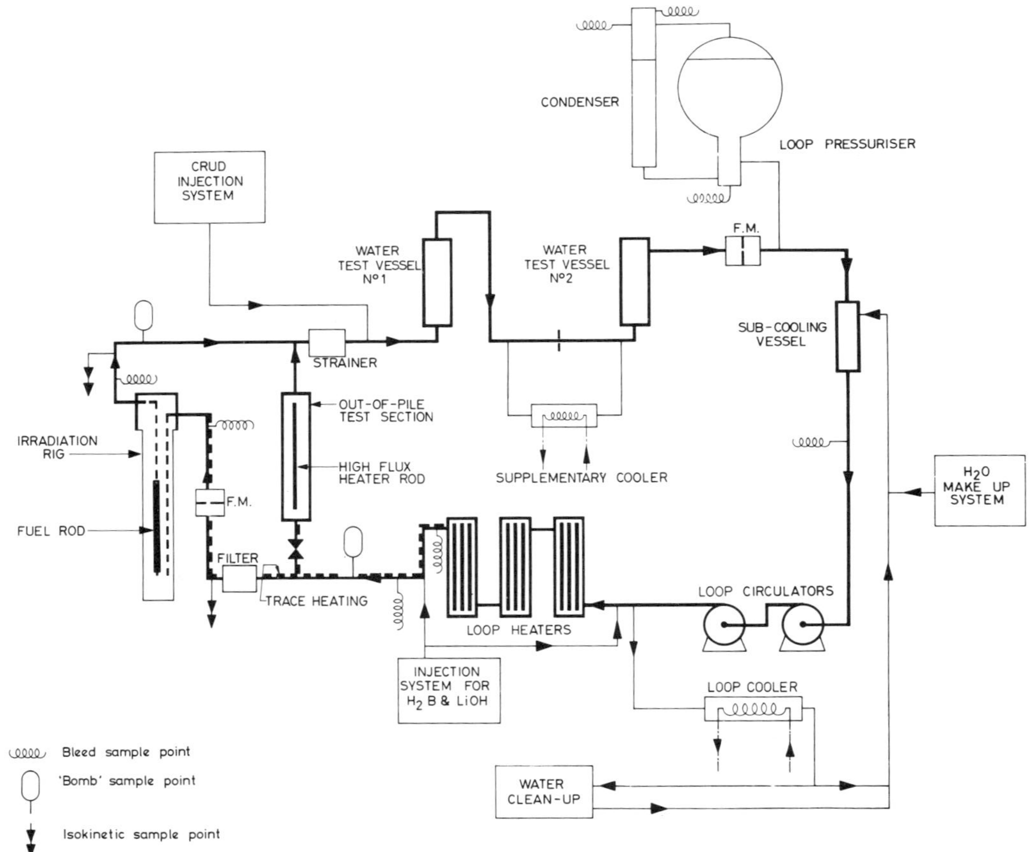

FIG 2·1 DIDO WATER LOOP
SIMPLIFIED FLOW DIAGRAM FOR CRUD DEPOSITION EXPERIMENT

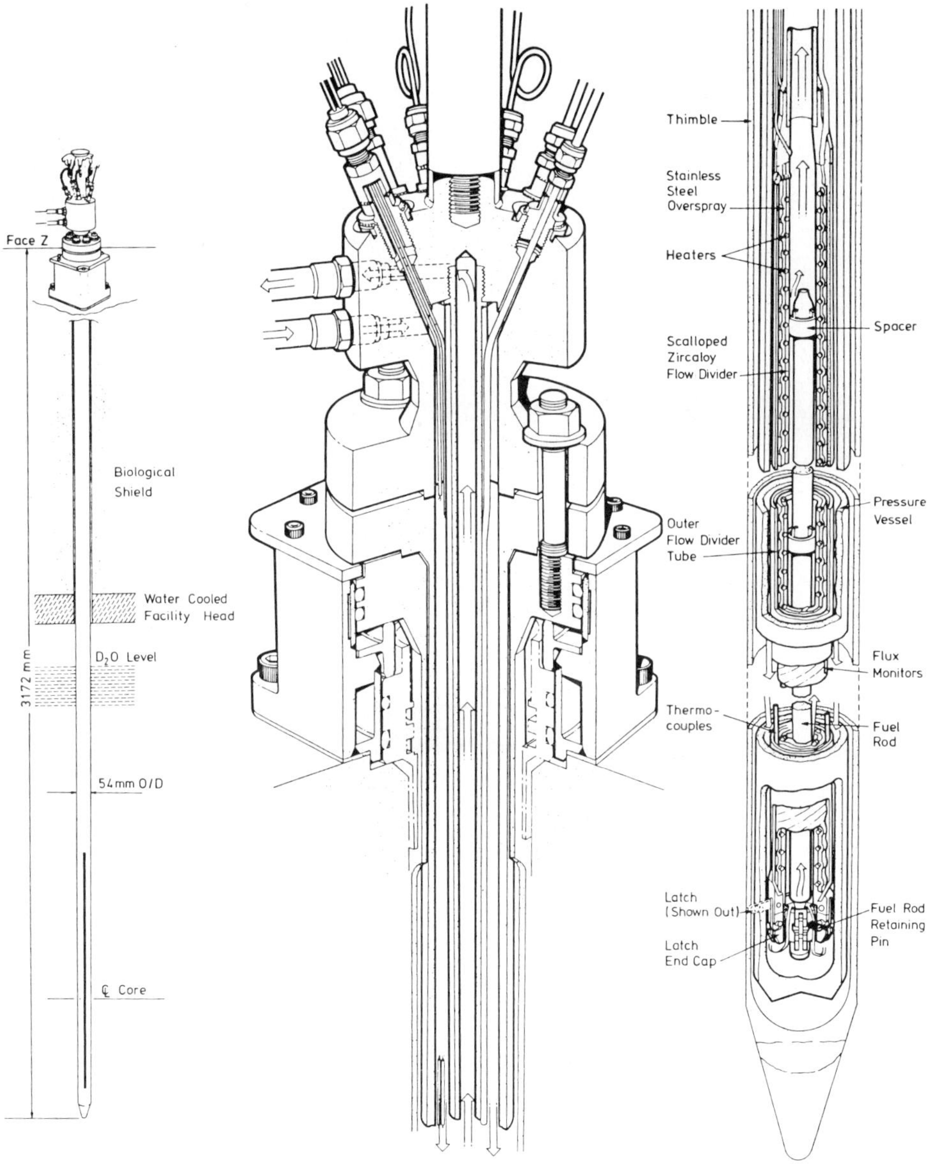

FIG. 2·2 IRRADIATION RIG FOR CRUD DEPOSITION EXPERIMENT

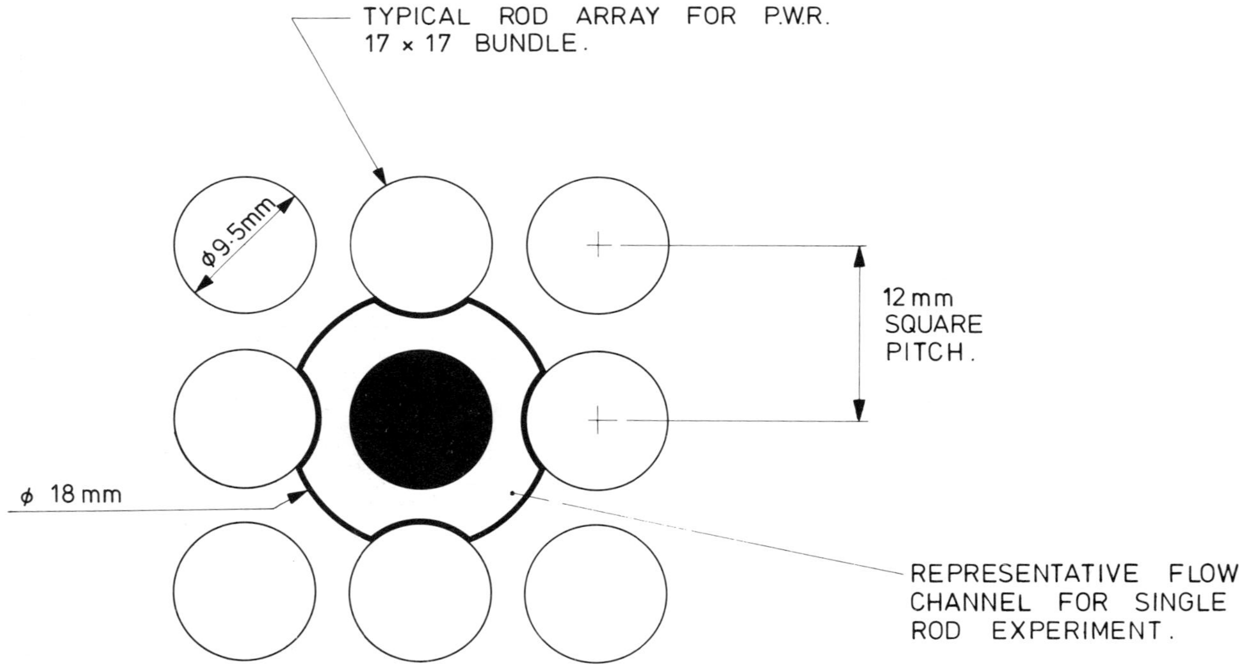

FIG. 2·3 DERIVATION OF FLOW CHANNEL FOR CRUD DEPOSITION EXPERIMENT IN D.W.L.

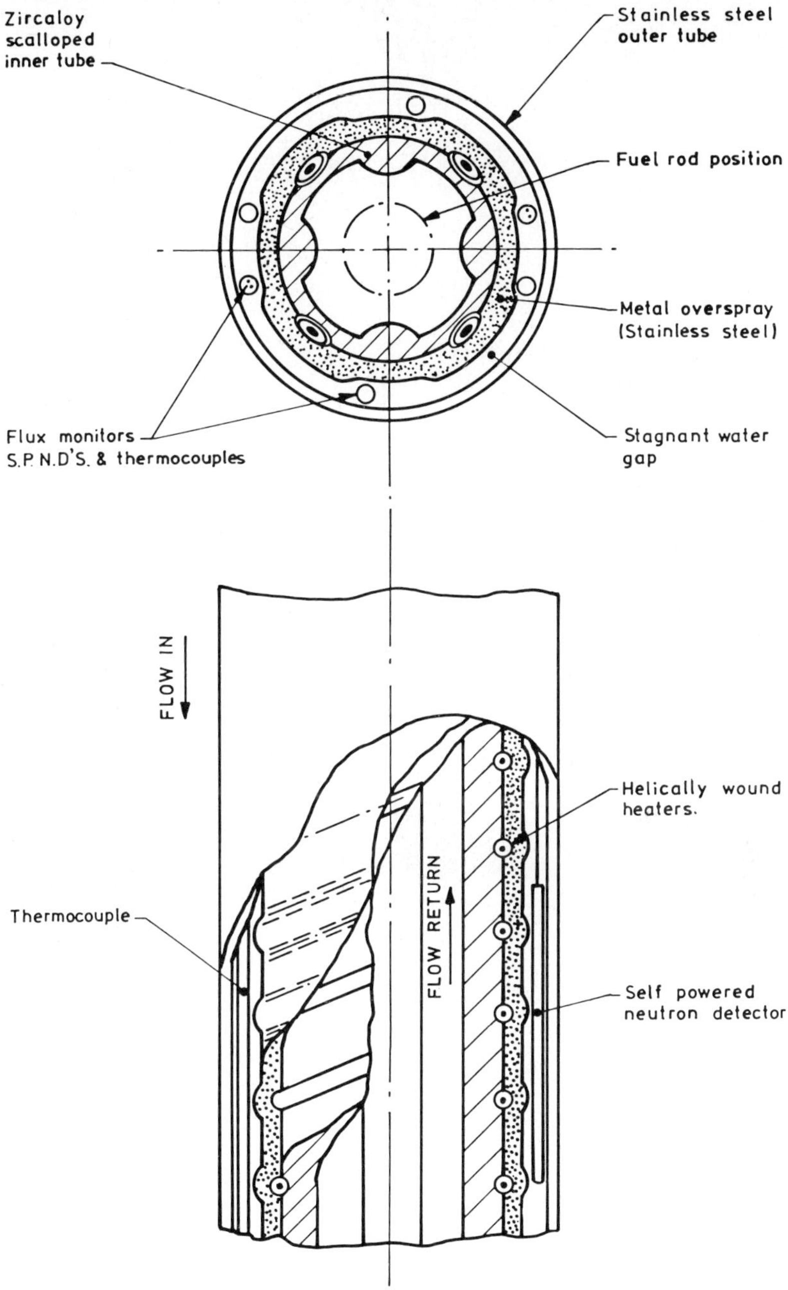

Fig. 2·4 SECTIONS THROUGH FLOW DIVIDER

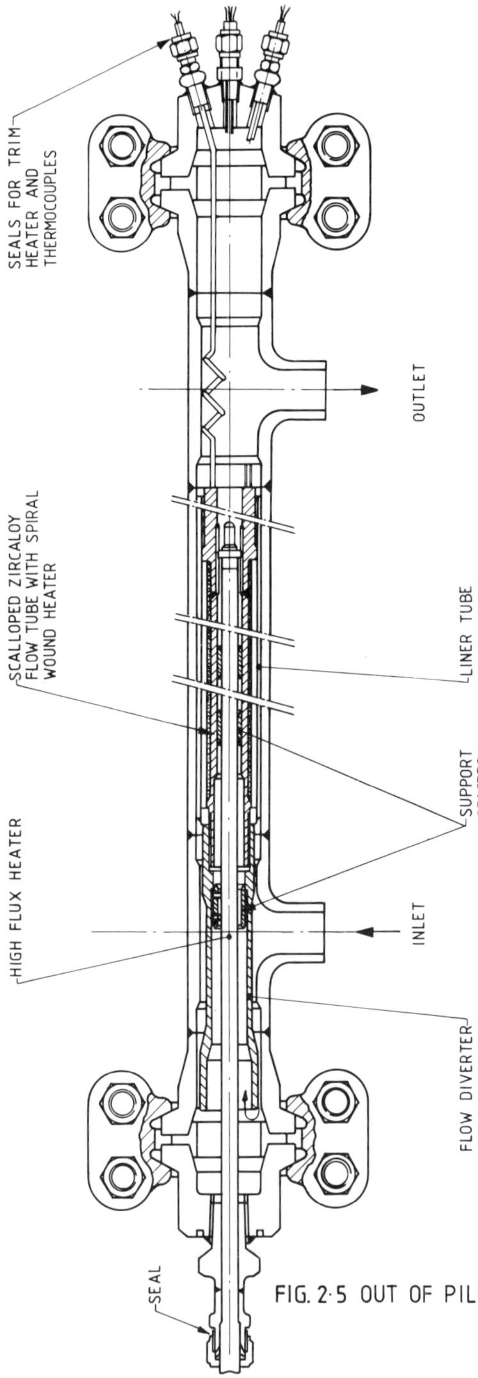

FIG. 2·5 OUT OF PILE TEST SECTION

PART 3 DEVELOPMENT OF A SYSTEM FOR GENERATION AND INJECTION OF CRUD PRECURSORS

by J. Monahan

1. INTRODUCTION

1.1 General

The all-stainless-steel DIDO Water Loop can simulate the water chemistry of a PWR, but its crud-content level is normally less than one ppb, so for some experiments (e.g. deposition or filtration studies) it may be necessary to raise the crud concentration to a level more typical of that in an operating reactor. This can be achieved by adding either preformed crud or crud-forming species.

1.2 Corrosion-Product Composition

The primary-circuit materials of most current PWRs (excluding the Zircaloy cladding of the fuel pins) are about 90% Inconel 600 and 10% stainless steel(1), and the nickel-to-iron ratio in crud deposits can range from about 0.2 to 0.35. The principal constituent of the crud is an inverse spinel with lattice parameters intermediate between those of nickel ferrite and magnetite. According to Sandler(2) the chemical formula is $Ni_xFe_{3-x}O_4$ with x between 0.47 and 0.75. Other minor constituents in the crud are chromium, manganese and cobalt(3).

1.3 Selection of Technique

Increasing the crud concentration in a loop can be achieved by adding either a preformed crud or species which are subsequently converted to crud in the loop. The crud may be introduced batchwise or by continuous injection, but a batchwise process is likely to lead to marked oscillation of crud concentration, so a process suitable for continuous injection is preferred.

(a) Preformed Crud Moroto et al(4) have described several methods of preparing magnetite, any of which could be modified to produce a nickel-substituted spinel, and Dolle(5) has injected a suspension of crud into a high-pressure water loop. However, it was considered that this technique might be difficult to manage, especially control of the particle-size distribution, and its behaviour might differ considerably from that of crud formed in the loop. For these reasons this approach was discarded.

(b) <u>Crud-Forming Species</u> Early in the corrosion process in a PWR primary circuit it is reasonable to suppose that divalent iron and nickel go into solution and then react chemically to form a nickel-substituted magnetite. If divalent iron and nickel are added to an operating loop then presumably they too will be transformed into a similar spinel. Obviously it is important that the iron and nickel do not contain impurities which might affect the system chemistry . This limits the available range of counter ions.

Although iron and nickel hydroxides are readily precipitated from aqueous solutions of their salts by adding any alkali, the precipitates are gelatinous and difficult to wash free of impurities. For iron the washing procedure must be carried out in an air-free environment to avoid oxidation of the ferrous ion, and it is difficult to devise a scheme capable of continuous operation. Furthermore the addition of gelatinous hydroxides to an operating loop is unlikely to produce crud typical of a PWR system because of the solid-state reactions involved.

The addition of divalent iron and nickel in true solution avoids the above disadvantages, and was therefore the method chosen.

2. PRINCIPLE OF THE CHOSEN TECHNIQUE

The solubility of ferrous hydroxide in pure water at 25°C is generally accepted to be in the region of 1.5 mg.L^{-1}[6] or 1.5×10^{-5}M[7]. According to Arden[8] its solubility product is 2.4×10^{-14}, from which it follows that if water is kept free of oxygen and buffered to pH 7.0 the molar concentration of ferrous ion in a saturated solution will be given by:

$$\left[Fe^{2+}\right] = \frac{2.4 \times 10^{-14}}{\left[OH^{-}\right]^2} = 2.4$$

The solubility of nickel hydroxide is 130 mg. L^{-1}.

The coolant in a PWR primary circuit contains boric acid and lithium hydroxide, and at room temperature has a pH of less than 7.0. It is also deoxygenated and contains dissolved hydrogen. It is therefore possible to dissolve ferrous and nickel salts in such a solution and to remove unwanted counter ions by ion exchange. When the resultant solution is injected into a high-temperature-loop circuit containing water of typical

PWR-primary-circuit composition it is rapidly diluted and raised to the operating temperature; the dissociation constant of boric acid is greatly reduced(9,10) and its buffering capacity almost disappears. As a result the solubilities of iron and nickel species are reduced and spinel formation occurs. Since the injected solution has the same concentration of boric acid and lithium hydroxide as the loop water the only change in the water composition is an increase in the iron and nickel concentrations.

3. PREPARATION OF CRUD-FORMING SPECIES

A suitable salt (e.g. ferrous sulphate or nickel sulphate) is dissolved in a solution identical to the loop coolant, typically containing up to 0.09 molar boric acid (1000 $mg.kg^{-1}B$) and up to 3×10^{-4} M lithium hydroxide (2 $mg.kg^{-1}$ Li). An anion exchanger previously conditioned with boric acid is used to remove unwanted anions. The system must be kept under an inert-gas atmosphere to exclude oxygen. For the production of nickel ferrites two such systems are required, the eluants from the exchangers being mixed in the correct proportions and added directly to the circuit. If necessary the technique can be extended to include other metals. In practice it has been found better to add all the lithium to the nickel solution, and under these conditions as little as 200 $mg.kg^{-1}$ of boron is sufficient to keep the ferrous ion in true solution even if the final mixture contains up to 3 $mg.kg^{-1}$ of lithium. The upper limit of alkali in the final mixture has not been established. The method is simple to use and easy to control, and the iron concentration can be varied from its solubility in boric acid down to zero.

4. DEVELOPMENT

4.1 Injection equipment

A simplified circuit diagram of the equipment is shown in Fig. 3.1. The crud-forming species are prepared by dissolving ferrous and nickel sulphates in separate tanks containing solutions of deoxygenated boric acid. The nickel solution also contains sufficient lithium hydroxide to give the correct lithium concentration in the mixing vessel. Deoxygenation is achieved by gas-exchanging with helium, which is gently bubbled through the solutions and vented through a water lute to avoid back-diffusion of air. Unwanted anions (e.g. sulphate) are removed in the process vessels on anion exchangers previously conditioned with an oxygen-free boric-acid solution containing 1000 $mg.kg^{-1}$ of boron. The eluants from the exchangers are pumped into a mixing vessel in the correct proportions by metering pumps and then injected. The make-up tanks are replenished manually; otherwise the system is automatic and controlled by a microprocessor. Injection rates of up to 100 cm^3 per minute are attainable, and typically a mixture containing 45 $mg.kg^{-1}$ of iron and 16 $mg.kg^{-1}$ of nickel with 2 $mg.kg^{-1}$ of lithium and 1000 $mg.kg^{-1}$ of boron is injected.

4.2 Plant Commissioning

The equipment described was used to prepare and inject crud precursors into a water test facility operating at 300°C and 2000 lb/in^2g. The test facility consisted of a stainless steel tube 8ft long and 0.085 in. i.d. coiled to fit into a tube furnace, a cooling coil immersed in water, a crud-filter holder, and finally a length of 0.007 in. i.d. stainless steel capillary tube having sufficient resistance to sustain the required operating pressure. The flow was 16 cm^3 min^{-1}.

The injected solution contained 500 $mg.kg^{-1}$ boron as boric acid, 2 $mg.kg^{-1}$ lithium as lithium hydroxide, and ferrous and nickel ions. In some runs the mixture was saturated with hydrogen gas at atmospheric pressure to give a solution containing 17 $cm^3.kg^{-1}$, calculated from Henry's Law[(11)].

Initial attempts to achieve the specified crud level were unsuccessful because of the large absorption capacity of the system, especially the ion exchangers. It should be noted that in this case the injected mixture was not diluted, whereas in the DWL tests described later the injected solution was diluted a thousand-fold on entering the loop. Eventually, it was decided to operate at a level of a few $mg.kg^{-1}$ of crud precursors, which although about 100 times greater than the DWL level was only about a tenth of the concentration of the solution to be injected during the DWL experiment.

The main purpose of the test was to commission the equipment and check that it produced crud. Samples were collected on 0.45 micron millipore filter papers. Three samples were characterised by both chemical and X-ray-diffraction analysis. The first was prepared in the absence of hydrogen, the second in water saturated with hydrogen at atmospheric pressure at 20°C, and the third from a feed reduced in nickel. All three samples showed the presence of a spinel of the type $Ni_xFe_{3-x}O_4$ and a small amount of nickel metal.

4.3 Loop Tests & Results

After the commissioning trials, the crud-generating and injecting equipment was evaluated on DWL operating at 300°C and 2000 lb/in^2g. The loop coolant contained lithium hydroxide at a nominal level of 1-2 $mg.kg^{-1}$ of lithium and boric acid at a concentration of up to 1000 $mg.kg^{-1}$ of boron.

The equipment was used to generate and inject crud precursors continuously for several periods of about 25 days each. The target was to produce in the loop crud with a nickel-to-iron ratio of 0.35 (i.e. a nickel-substituted spinel). Injection rates were typically 50 $ml.min^{-1}$, and the loop flow was 48 $kg.min^{-1}$. Samples of crud from the loop water were collected on 0.45-micron filter papers at about 60°C through 80ft-long uncooled titanium tubes connected to isokinetic sampling probes, and the nickel-to-iron ratio in the crud was determined by chemical analysis. The residence time in the loop between the injection and sample points was about 2 minutes. Crud levels were also deduced from measurements made with a HIAC particle-size counter. Crud levels of 20 to 50 ppb were achieved, and the nickel-to-iron ratio was about 0.3.

REFERENCES

1. Solomon, Y. An overview of water chemistry for pressurised water nuclear reactors. BNES 1st Water Chemistry Conference, Bournemouth, Oct. 1977.

2. Sandler, Y.L. Structure and solubility of PWR primary corrosion products. WACE Meeting, Houston, Texas. March 1978.

3. Roesmer, J., Rootham, M.W. Estimation of activity inventories in primary circuits of pressurised water reactors. BNES 1st Water Chemistry Conference, Bournemouth, 1977.

4. Maroto, A.J.G., Blesa, M.A., Regazzoni, A. and Urrutia, G.A. The synthesis of corrosion products for simulation studies. BNES 2nd Water Chemistry Conference, Bournemouth 1980. Paper 35.

5. Dolle, L., Pasisot, W., Kahan, H. Radionuclides transport phenomena in a pressurised water in-pile loop. BNES 2nd Water Chemistry Conference, Bournemouth, 1980. Paper 38.

6. Handbook of Chemistry and Physics 60th Edition, Page B87.

7. Seidell, A. Solubilities of inorganic and organic compounds, 4th Ed. 1958.

8. Arden, T.V. J. Chem. Soc. 1950, p.882.

9. Mesmer, R.E., Sweeton, F.H. and Baes, C.F. Jr. Acidity measurements at elevated temperatures VI. Boric Acid Equilibria. Inorg. Chem. 11, 537 (1972).

10. Mesmer, R.E., Baes, C.F. Jr., Sweeton, F.H. Boric Acid Equilibria and pH in PWR Coolants. Proc. 32nd Ann. Mtg. Int. Water Conf. Engineering Society of Western Pennsylvania (1971).

11. Handbook of Chemistry and Physics 37th Edition, p. 1608.

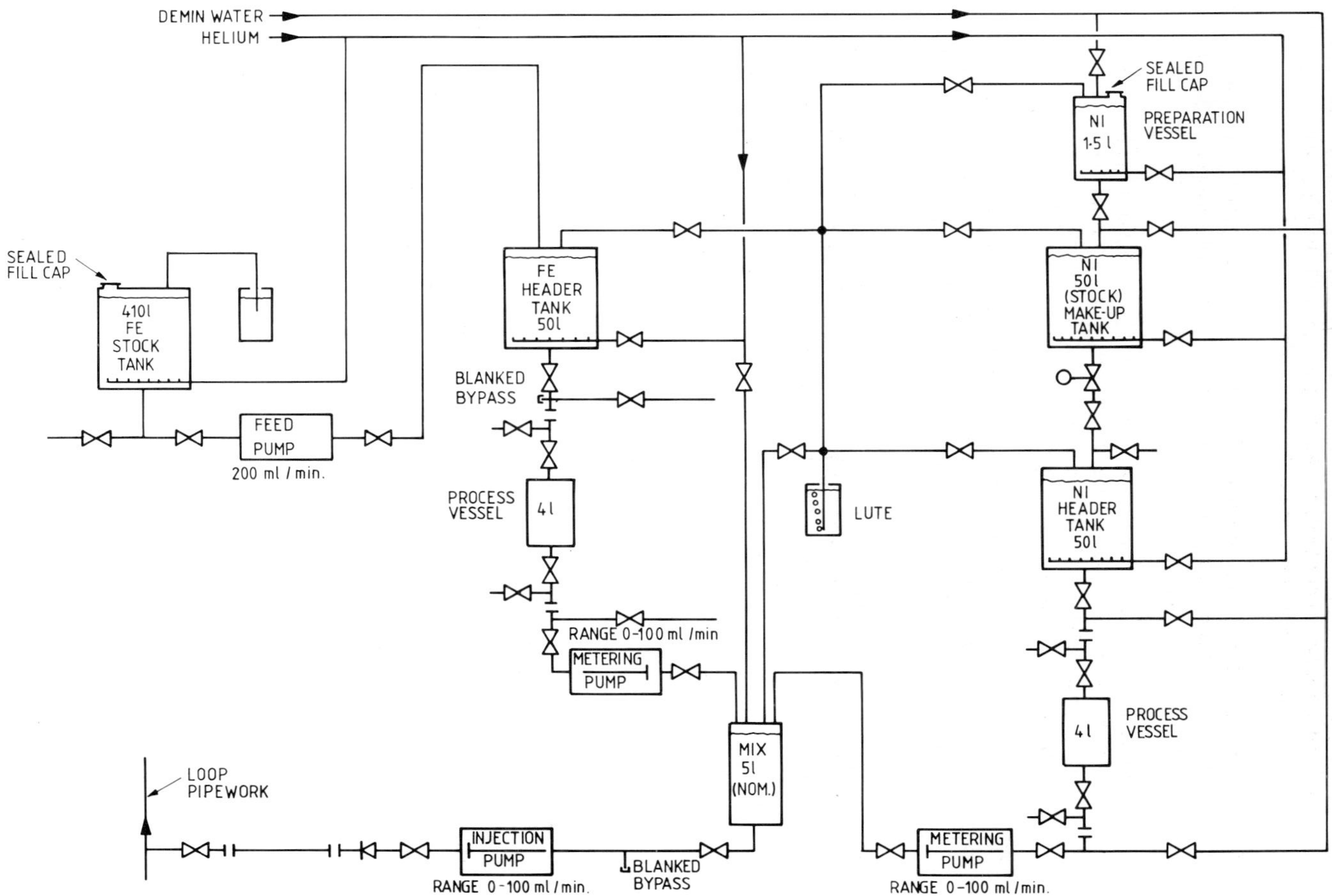

FIG. 3·1 CRUD INJECTION CIRCUIT (SIMPLIFIED)

TECHNIQUES FOR IRRADIATION TESTING OF NON-FUEL MATERIALS IN LIGHT WATER POWER REACTORS

M. Gärtner, D. Alter, P. Dewes

Kraftwerk Union Aktiengesellschaft

8520 Erlangen, Germany

ABSTRACT

Power reactors offer many experimental possibilities to investigate the in-pile behaviour, especially the irradiation induced effects on mechanical properties of in-core materials without influencing the operational duties. Such investigations improve the knowledge of proper material use for fuel assembly and in-core component design. For instance, information on in-pile creep and growth of zirconium alloys as well as on SCC susceptibility of stainless steel under the original power reactor conditions has to be gained. Furthermore, microstructural and dimensional changes and gas release of absorber and neutron source materials are being investigated. Additionally, in-pile relaxation of spacer spring material is being explored by in-reactor experiments.

For such investigations special techniques have been developed. Sample examinations are performed with preference in the spent fuel pool.

This paper presents on the basis of programmes performed by KWU the techniques used for structural material testing in power reactors.

1. GENERAL ASPECTS

Surveillance of standard core components and assemblies as well as testing and evaluation of prototypes are quite common practice in today's developmental strategies. Nevertheless, there are in addition many needs for detailed investigation of the in-reactor behaviour of

P. von der Hardt and H. Röttger (eds.), Irradiation Technology, 77–92.

current and also new improved materials. Especially with respect to the fuel assemblies, larger in-reactor residence times are needed as a consequence of the extended burnup targets.

Power reactors offer many experimental possibilities to investigate the in-pile behaviour of core materials under realistic operational conditions without interfering with reactor operational requirements. Contrary to research reactors nuclear power plants allow the irradiation of large numbers of material specimens and fuel assembly components of original dimensions under actual neutron flux and spectrum, coolant temperature and pressure up to neutron exposures exceeding $10^{22}cm^{-2}$ in due time. However, the large quantity of specimens and hence the good statistical confirmation of the test results have to be reconciled with a minimized experimental risk for the power reactor operation.

The main goals - from KRAFTWERK UNION (KWU)'s point of view - for present and future material irradiation in power reactors are listed in Tab. 1. Information on in-pile creep and growth as well as on mechanical short-term strength of zirconium alloys exposed to high neutron fluences will be gained with respect to the requirements for increased fuel rod burnup. Furthermore, the knowledge of stress relaxation of spacer grid springs and material for the core structure is important in this sense. The absorber and neutron source materials have to be qualified for the desired high neutron exposures. Additionally, the absorber clad shall be improved concerning susceptibility to stress corrosion cracking under reactor conditions.

KRAFTWERK UNION is using nuclear power reactors for core material investigations since 1971. Since that time material test rods without nuclear fuel are being irradiated in pressurized water reactors (PWR) /1, 2, 3/ and later in boiling water reactors (BWR), too. For in-pile testing of structural materials for fuel assembly and core components many techniques have been developed and applied in close cooperation with a well-experienced staff of fuel service engineers.

2. POWER REACTOR AS TEST BED FOR MATERIAL INVESTIGATIONS

Fuel assemblies in nuclear power plants represent a very suitable test bed for in-pile material testing. In PWR fuel assemblies (Fig. 1) material test rods can be loaded either into empty control rod guide tubes or into fuel rod positions, whereas in BWR assemblies (Fig. 2) only fuel rod positions are available for test rod irradiation. These PWR and BWR irradiation positions are characterized by high flexibility in choosing typical irradiation conditions. By taking advantage of the various possible sample positions in the

fuel assembly and assembly positions in the reactor core, temperatures as well as neutron fluxes and spectra can be varied within the frame of typical irradiation conditions of the power reactor (Fig. 3). Positions of realistic coolant flow and chemistry are available for test rod irradiation.

Loading and unloading of the test specimens in fuel rod positions can be easily carried out by removing the lower end fitting (PWR) or the upper tie plate (BWR) of a fuel assembly. Furthermore, special PWR pathfinder assemblies designed for in-reactor component testing have an aperture of the top plate. This aperture allows to rapidly exchange test rods inserted in fuel rod positions also for measurements to be performed during the short refueling periods of the power plant - without turning the assembly upside down. A special flow restrictor keeps the aperture closed when the fuel assembly is loaded and operated in the reactor.

Test rods inserted into the control rod guide tubes ($\emptyset_i = 12.9$ mm) of a PWR fuel assembly are fixed by a proper hold down device, a modified flow restrictor (Fig. 4 + 5). The guide tube can serve for protecting the surrounding fuel rods in case of test specimen defect. The same effect can be realized in a BWR assembly by inserting a test rod of suitable diameter (Fig. 6) into a 7 x 7 or 8 x 8 cladding tube ($\emptyset_i = 12.6$ mm or $\emptyset_i = 10.8$ mm) and loading this carrier tube with the test rod inserted into a fuel rod position of the corresponding fuel assembly.

3. MATERIAL TEST ROD DESIGN

The spacer grid design or the inner diameter of the control rod guide tube predetermines the dimensions of the material test specimen to be loaded into a fuel assembly.

The test rods to be inserted on fuel rod positions exhibit the original outer dimensions of the fuel rod, whereas the outer dimensions of the test rods to be loaded into control rod guide tubes can be chosen rather independently from the fuel assembly design. The necessary guidance and protection against vibration of the test rods in the guide tubes are provided by a proper upper and lower end plug design which considers an axial support at the guide tube top and a diametrical guide e. g. at the lower contractures of the control rod guide tubes.

Normally the test rods assembled from short test samples form segmented rods. This gives the chance for axial variation of material or test conditions. The axial neutron flux and temperature gradients of the fuel assembly can be taken into account for the investigations and the specimen load can be varied i. e. by differ-

ent stressing of tube samples due to internal gas overpressure or a mandrel swelling under neutron irradiation. Furthermore, a full-length segmented material test rod can be disassembled into the short samples for possibly facilitating transport to the Hot Cells and further examinations. For the Hot Cell examinations the samples can already be properly prepared during manufacturing. Also the experimental risk is minimized by the small volume of the individual specimen in case of sample defects, however, the value of the experiment is increased due to the large number of specimens and the experimental variety within a segmented test rod.

In case of a segmented rod the individual samples axially screw-connected to each other have to be protected against disconnecting during irradiation, e. g. by spot welding of the screwed connections.

4. MEASUREMENTS

The irradiation effect on the material investigated is determined by comparing initial and post-irradiation test sample properties. Therefore, the basis for an accurate evaluation of the tests is the accurate sample characterization prior to and after power reactor insertion and the careful determination of the in-pile conditions.

Precharacterization of the Test Samples

The test specimen characterization prior, during and after fabrication of the specimens and test rods may include dimensional measurements, eddy current testing, determination of mechanical (strength, ductility) and structural properties (grain size, porosity, texture, etc.) and chemical analysis.

Intermediate Non-Destructive Sample Inspection

During irradiation, i. e. at reactor refueling shut-downs, and after termination of the irradiation the material test rods and samples are examined non-destructively in the spent fuel pool of the power plant. For the examinations during irradiation the test rods are only temporarily unloaded from the fuel assembly. However, length measurements of tubing for growth investigations are performed without unloading the test rods from the fuel assembly. Such examinations generally include visual inspection and dimensional measurements as well as possibly eddy current integrity testing and oxide/deposit layer thickness measurements.

Destructive Post-Irradiation Examinations

At the final destructive examination in the Hot Cell laboratory metallographic and ceramographic examinations resp., dosimetry,

chemical analyses and measurements of the mechanical properties are performed for characterization of the material after irradiation.

Determination of the Irradiation Conditions

Of high importance for the experiment is to exactly determine the thermal and neutron physical conditions of the material test rods during in-reactor testing. The range of test rod temperature is given by the coolant inlet and outlet temperature of the reactor core. The effective temperatures for individual rod sections are calculated by taking into account nuclear heating of the material specimens, coolant flow and linear heat generation rate of the surrounding fuel rods.

The neutron flux conditions at the material test rod position are determined by in-core detector measurements and additional aeroball measurements in PWR. The ionization chambers of the travelling in-core probe of the BWR as well as the (n , ß)-detectors and the aeroball device of the PWR are located in so-called measuring lances of the reactor core.

Furthermore, basing on the power history of the nuclear power plant 2-dimensional (in PWR) and 3-dimensional (in BWR) neutron flux calculations give information about the time dependence of the neutron flux conditions.

These calculated neutron flux values are verified after irradiation by evaluating the neutron fluence monitors positioned in the test rod specimens (Fe, Nb, Ni, Co wire sections) and by taking gamma-scans of selected test rod components.

5. MATERIAL TESTING PERFORMED BY KWU IN POWER REACTORS

A survey of the objectives and actual activities of KWU programmes for material testing in power reactors was already given in Tab. 1. In the following the experimental goals together with the essential details of the techniques practiced are outlined.

Length Growth of Stress-Free Tubing

In order to investigate the radiation induced effects related to original dimensions, full-length components are inserted into the fuel assembly for irradiation. So, for length growth measurements of the control rod guide tubing, tubes of fuel rod clad dimensions, but representative material conditions are irradiated in fuel rod positions. The tubes are assembled on both ends with special end plugs facilitating tube handling during measuring campaigns, but pressure compensated by coolant water flooding.

Stress Influence on Length Growth

The influence of various stress conditions (ratio of longitudinal and axial stress) on length growth of fuel rods under neutron irradiation is investigated by the irradiation of material test rods without fuel adjacent to standard fuel rods. The test rods allow to investigate the pure irradiation induced growth of cladding tubes filled with Zircaloy pellets and prepressurized as standard fuel rods in comparison to the unstressed, coolant flooded tubing. The former cladding tubes can creep down and interact with the pellets to simulate stress conditions in fuel rod tubing /1/.

Creep and Burst Strength of Tubing

The irradiation induced creep of fuel rod cladding is evaluated from the diameter change of tube samples (Fig. 7) axially connected to each other to form a full-length segmented material test rod in PWR. One type of the samples for these in-pile creep experiments is pressurized internally with helium. The pressures ranging from 1 to 122 bar result in tensile and compressive hoop stress of up to 60 and 110 MPa resp. in the tubing. Other samples are filled with helium to compensate the compressive hoop stress resulting from the outside coolant pressure and are irradiated for later burst tests to be performed in the Hot Cells. Cladding tubes of different manufacturing routines, various heat treatments and hence different material structure have been used for sample fabrication.

Spacer Spring Strength

A large variety of material samples is irradiated in capsules of a material test rod for stress relaxation and strength measurements (Fig. 8). Spacer grid springs prestrained to various levels will give information on the relaxation of the forces holding the fuel rods in the spacer grid during the operation period of the fuel assembly (Fig. 9). Special spring casings allow force measurements during sample fabrication and Hot Cell examination by use of the same measuring equipment.

Core Component Strength

Furthermore, helical springs differently prestrained are inserted in this material test rod. The measurements on these springs will improve the knowledge on the relaxation behaviour of core structure materials due to fast neutron irradiation (Fig. 8). For the same purpose packages of plate and cylindrical tensile specimens are irradiated in this material test rod. Strength measurements on these samples will show the changes of ductility and ultimate tensile strength of nickel base alloys and stainless steel (Fig. 8).

The springs and tensile specimens are loaded into a Zircaloy container coolant flooded by holes drilled through the wall. A hold down spring positioned at the top of the sample column in the container keeps the material samples in positions. The material test rod is inserted into a control rod guide tube of a PWR fuel assembly.

Charcoal Absorption

In a similar carrier loaded to a fuel rod position of a PWR fuel assembly capsules containing charcoal powder in helium atmosphere are irradiated for later out-of-pile testing of the charcoal. These tests are to investigate the irradiation influence on the absorption behaviour of the charcoal.

Dimensional Behaviour of Neutron Source Constituent

Neutron source components are being irradiated in test capsules forming a segmented test rod. Highly dense alumina in form of solid and hollow pellets are inserted with different diametral clearances into stainless-steel capsules (Fig. 10). Antimony-Beryllium pellets, a neutron source material, are loaded without any parameter variation because of the extremely high toxicity of this material.

The specimen capsules are weld-sealed on both ends by end plugs and filled with helium of 1 bar cold pressure for tightness control.

Dimensional Stability and Gas Release of Absorber Rod Components

Dimensional behaviour and gas release of absorber material are being investigated in a PWR. Segmented material test rods containing the various test samples in short capsules with stainless-steel cladding and 1 bar helium pressure are inserted into empty control rod guide tubes (Fig. 11).

Absorber material B_4C is loaded as pellets and powder. Furthermore, AgInCd bars are irradiated in capsules with different diametrical clearances (Fig. 10). A possible interaction between the test material and the cladding under irradiation can be measured via clad deformation in the spent fuel pool. For comparison an unloaded cladding section is integrated into the B_4C specimen capsule and empty tube sections are included in the material test rods.

The gas release of the B_4C pellets and powder will be determined at later destructive post-irradiation examinations in the Hot Cell laboratory.

SCC of Absorber Rod Cladding

To study stress corrosion cracking (SCC) of absorber rod cladding under BWR and PWR conditions, tube sections are filled with Al_2O_3 or Al_2O_3/B_4C swelling mandrels adapted in diameter and weld--sealed on both ends (Fig. 11). Due to volume swelling under fast neutron irradiation the mandrels stress the cladding tubes at different rates.

The test rods are placed on empty fuel rod positions of a BWR assembly and inserted into carrier tubes (Fig. 6). The carrier tube is perforated so that the coolant conditions are representative for the waterside of the absorber rods in a BWR control assembly blade.

Similar in-pile tests are to be performed in a PWR, where the samples are placed in control rod guide tubes.

6. PARALLEL OUT-OF-PILE TESTING

Many of the irradiation experiments in power reactors described above are being accompanied by corresponding out-of-pile experiments. The comparison between the in-pile and out-of-pile behaviour of the samples allows to separate temperature and irradiation effects and hence to study the physical mechanisms.

7. EXPERIMENTAL RESULTS

A typical result of an in-reactor experiment is the creep deformation of Zircaloy-4 tubes of PWR fuel rod dimensions stressed by the inner helium overpressure and the outer coolant overpressure resp. (Fig. 12). This experiment ran over four reactor cycles.

The material test rods consisting of 14 specimens each (Fig. 7) were loaded into the nuclear power reactor Obrigheim in June 1977. The test rods were three times removed from the reactor and the fuel assembly, measured in the spent fuel pool and reinserted during the normal refueling periods of the power plant within the scheduled reactor plan. No extra shut-down time was necessary for the measuring campaigns. Under representative test conditions the creep behaviour of a Zircaloy tube was determined for fuel assembly design in due time.

However, the expenses for this creep experiment performed in the power reactor were extremely low compared to a similar research reactor experiment, too.

8. FURTHER ASPECTS FOR IRRADIATION OF NON-FUEL TEST RODS IN POWER REACTORS

Beside the effort for specimen fabrication and irradiation of the experiments also licensing, transport arrangement, storage and waste disposal of the specimens and test rods have to be taken into account. However also from this point of view the irradiation of non-fuel test rods in nuclear power reactors offers big advantages compared to tests in research reactors where the inevitable irradiation rigs mostly require deductions and concessions especially concerning number and dimensions of the specimens as well as experimental conditions. Hence also in future the favourable irradiation conditions of power reactors will be used for testing of structural materials and components of the fuel assemblies and core structure.

REFERENCES

/ 1 / Gärtner M.,
"Bestrahlung von Teststäben in Normalbrennelementen des Kernkraftwerkes Obrigheim"
BMFT Forschungsbericht K 79-12, Dez. 1979

/ 2 / Gärtner M., Manzel R.,
"Bestrahlung von Absorberstäben in einem Normalbrennelement des Kernkraftwerkes Obrigheim"
Abschlußbericht zum Förderungsvorhaben BMFT AtT 735-72/ Programmpkt. 2, Juli 1975

/ 3 / Schlemmer F.,
"Hochabbrandversuche im Kernkraftwerk Obrigheim"
Abschlußbericht zum Förderungsvorhaben BMFT AtT 735-72/ Programmpkt. 1, Sept. 1976

	Objective	Problem	Activity/Technique for Irradiation Testing
1	PWR Material Development especially for exposure to high neutron fluence (high fuel assembly burnup)	Length growth of Zry control rod guide tubes	Irradiation of tubes, stress-free and with different stress distributions
		Mechanical properties of fuel rod cladding material	In-pile creep testing of cladding tubes under interior and exterior overpressure resp.; irradiation of tubing samples for later burst tests (in Hot Cells)
		Stress relaxation of spacer and helical springs	Preloaded helical and beam springs fabricated of various metal alloys
		Strength of core baffle screws	Irradiation of flat and cyclindrical tensile samples for later tests (in Hot Cells)
		Dimensional stability of AgInCd absorber bars	Irradiation of AgInCd specimens
2	Design Improvement	Swelling and gas release of absorber materials;	Irradiation of B_4C powder and pellets canned in stainless-steel tubing;
		Dimensional behaviour of neutron source material	Irradiation of neutron source specimens
3	Reduction of Susceptibility of BWR Control Rod Cladding to Stress Corrosion Cracking	Defection of absorber pins within the blade of BWR control elements	Absorber cladding stressed by swelling mandrels

Tab. 1 Survey of Non-Fuel Material Irradiation Testing

Kraftwerk Union

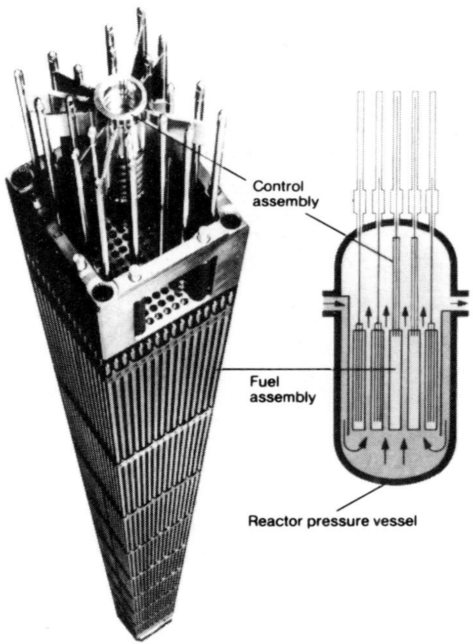

Fuel Assembly for Pressurized Water Reactors (PWR)

Fig. 1

Kraftwerk Union

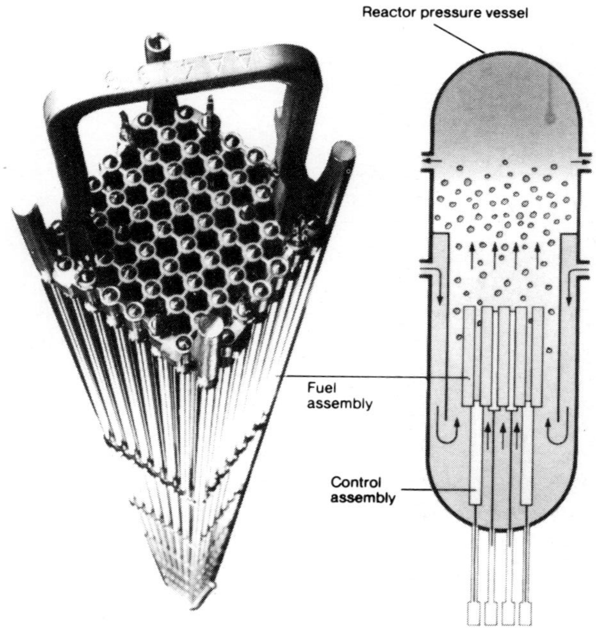

Fuel Assembly for Boiling Water Reactors (BWR)

Fig. 2

Kraftwerk Union

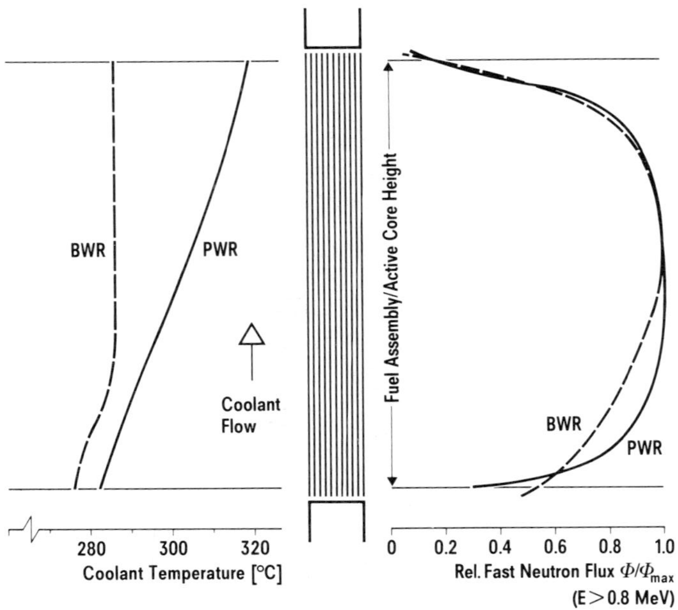

Typical Coolant Temperature and Fast Neutron Flux Distribution in a PWR and BWR Fuel Assembly

Fig. 3

Kraftwerk Union

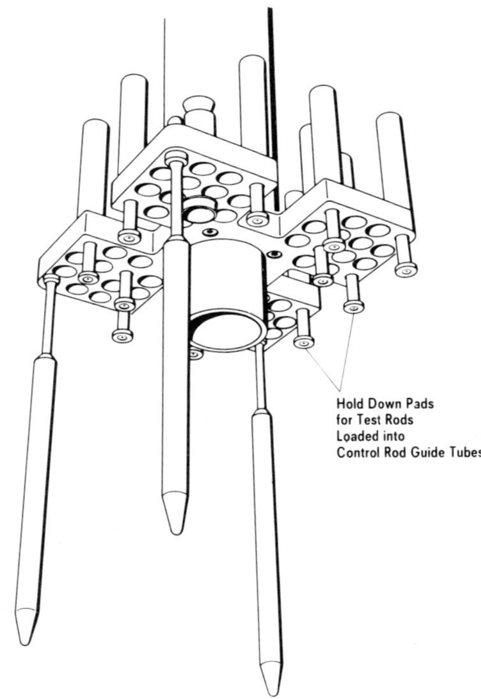

Hold Down Device for Test Rods in PWR Fuel Assembly Guide Tubes

Fig. 4

Kraftwerk Union

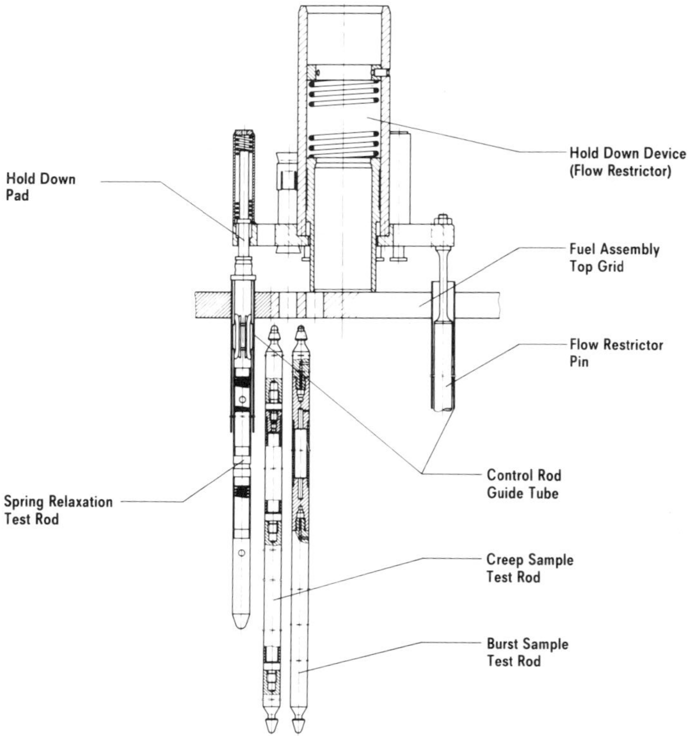

Schematic Diagrame of Material Test Rods in a PWR Fuel Assembly

Fig. 5

Kraftwerk Union

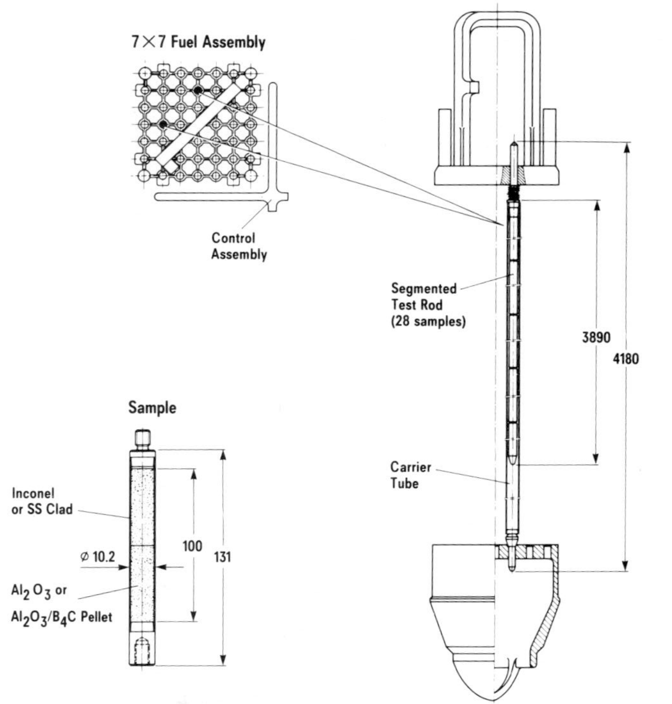

Stress Corrosion Cracking Test in BWR
Test Rod and Specimen Design and Location in the Fuel Assembly

Fig. 6

Kraftwerk Union

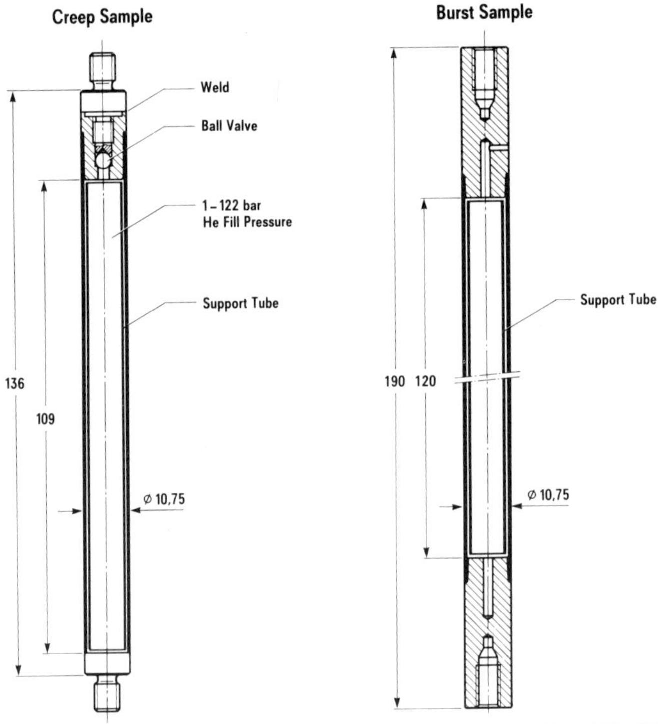

Creep and Burst Test Samples
of Zircaloy-4 Cladding Tubes in PWR

Fig. 7

Kraftwerk Union

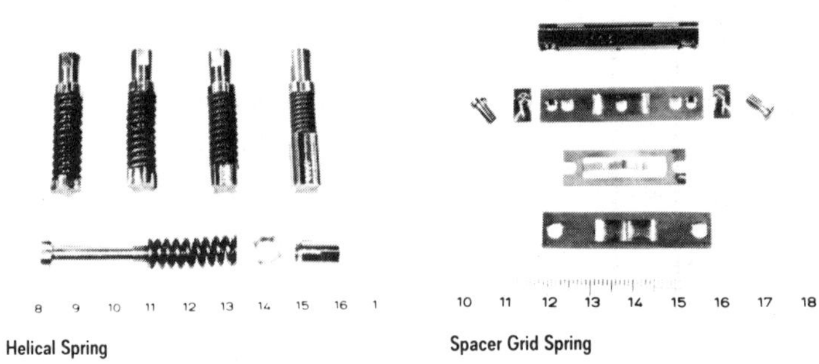

Fig. 8

Spring Samples for Relaxation Testing

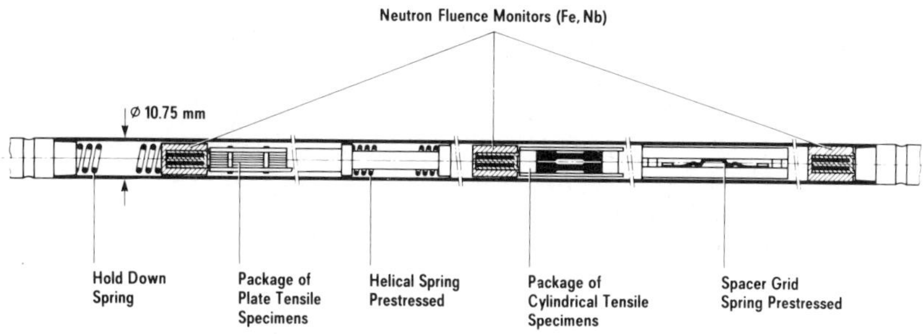

Fig. 9 **Material Specimens for Irradiation Testing in PWR**

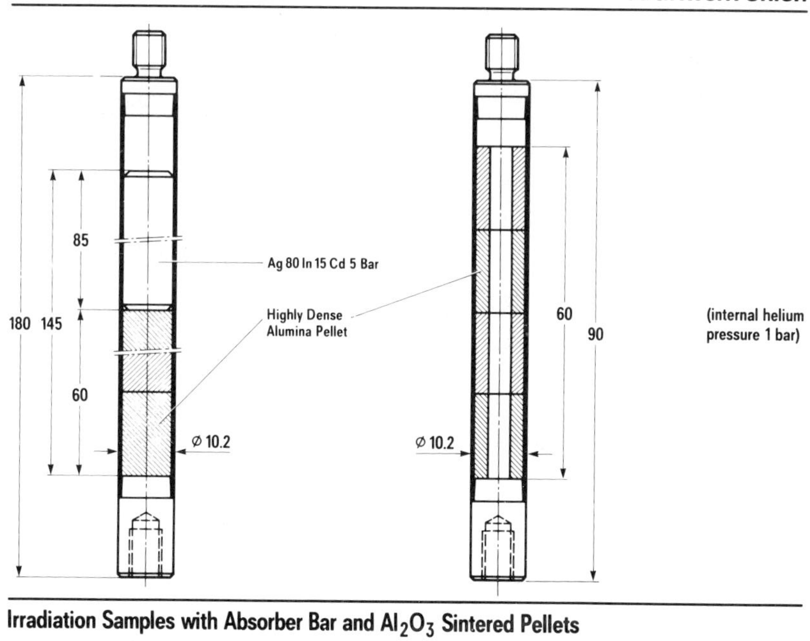

Fig. 10 **Irradiation Samples with Absorber Bar and Al_2O_3 Sintered Pellets**

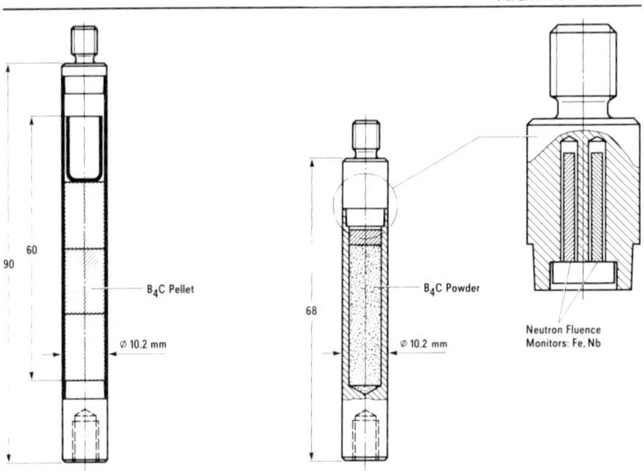

Fig. 11 Irradiation Samples with B_4C Pellets and Powder

E 82 1098 e

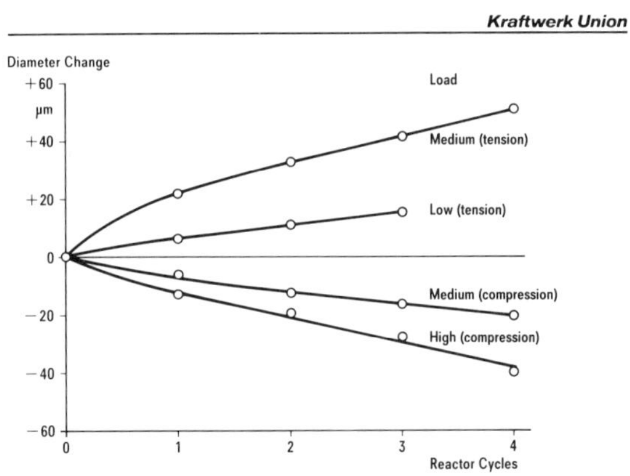

Fig. 12 Diametric Creep Deformation of Zircaloy-4 Cladding Tube Samples under Internal and External Overpressure in the Nuclear Power Plant Obrigheim/Germany

E 82 1091 e

L'INSTRUMENTATION DES EXPERIENCES DE RAMPES POUR LA FILIERE A EAU PRESSURISEE

A. JEGAT, P. VULLIEZ, C. FRIBOULET, P. BOURBOUSSON

CEA - CEN/Grenoble - Service des Piles - SEDTI

1) INTRODUCTION

Le Service des Piles du Centre d'Etudes Nucléaires de Grenoble a développé et utilisé pour les expériences de rampes de puissance sur des aiguilles combustibles de la filière à eau pressurisée, le dispositif AQUILON.

Outre les possibilités habituelles de chargements et de déchargements rapides, de facilité de manutention, le dispositif a l'avantage de pouvoir être instrumenté pour analyser le comportement de l'interaction gaine combustible et pour détecter l'apparition de fissures débouchantes.

2) LE DISPOSITIF D'IRRADIATION

L'installation utilisée est constituée d'une partie en pile placée dans le réflecteur du réacteur SILOE, d'une partie hors pile comprenant un ensemble de conditionnement de l'eau réalisant l'alimentation, la pressurisation et le traitement chimique de l'eau et un ensemble de baies de contrôle commande et d'acquisition de données.

21) Partie en Pile

On distingue
- . la partie basse contenant le crayon
- . la perche ou partie haute comprenant les lignes de liaisons hydrauliques

P. von der Hardt and H. Röttger (eds.), Irradiation Technology, 93–102.

. la boîte à eau à déplacement et le fourreau de refroidissement

La partie basse (fig. 1) est une capsule traversant la grille dans laquelle on peut irradier en flux centré des crayons de grande longueur jusqu'à 1800 mm. La chaleur générée par le crayon est évacuée par vaporisation partielle de l'eau sous saturée qui l'entoure et condensation sur la paroi extérieure refroidie. Des disques annulaires bloquent la convection verticale.

La partie basse est constituée de trois sous-ensembles :
. la partie basse proprement dite où est confinée la pression
. le panier porte coupelles, amovible et instrumenté de thermocouples, "collectrons internes", et capteur de mesure de longueur
. le porte échantillon amovible équipé du crayon fissile et instrumenté de :
  - capteur de bruit acoustique
  - capteur d'émission acoustique
  - d'hydrophone

Il peut recevoir des crayons neufs ou préirradiés.

Le fourreau de refroidissement, monté sur la boîte à déplacement, reçoit la partie sous flux du dispositif et assure son refroidissement. Il est équipé de 6 collectrons au Rhodium dits "collectrons extérieurs".

Les lignes, en partie aériennes et en partie sous marines, assurent la liaison souple entre le dispositif et la partie hors pile. La ligne retour possède des boucles pour analyses DND, gammamétrie en ligne et activité globale.

22) La Partie Hors Pile

Elle est constituée des éléments suivants :
. un étage d'épuration haute pression (résines + filtres)
. un étage de détente de pression possédant deux détendeurs en série
. un étage d'épuration basse pression comprenant :
  - des résines lithiées à lits mélangés
  - du charbon actif
  - des filtres à pouvoir d'arrêt étagé (jusqu'à 0,1 micron)
  - un bidon tampon permettant d'effectuer le changement d'eau de l'installation. Y est associé un poste d'injection d'additifs notamment pour l'ajustement de la concentration en lithine.

. un étage de régulation de niveau constitué notamment d'une cuve à niveau constant qui joue un triple rôle :
  - régulation du niveau d'eau
  - pressurisation du circuit basse pression
  - récupération des gaz de radiolyse et des gaz de fission éventuels.

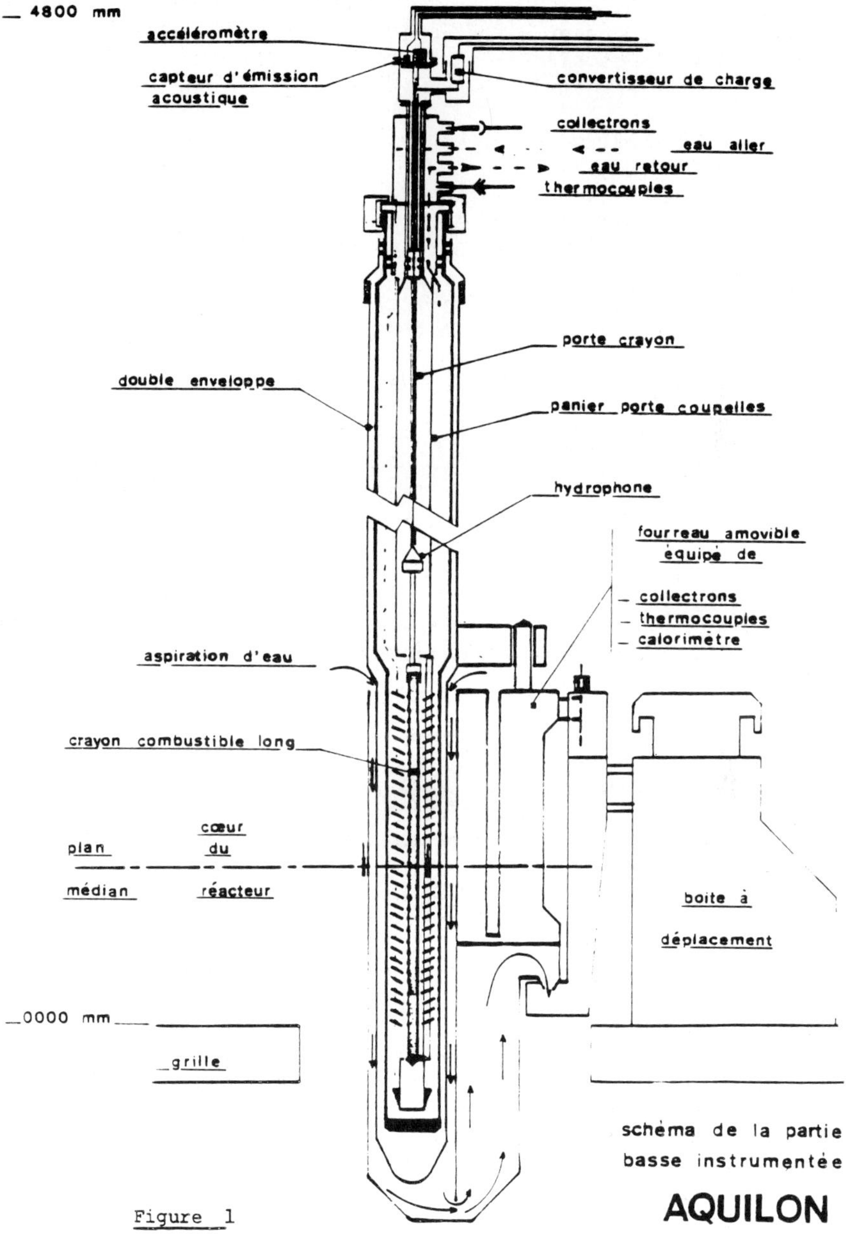

Figure 1

schéma de la partie basse instrumentée

AQUILON

A l'aide de cette cuve, l'eau peut être conditionnée en hydrogène à la concentration désirée.

. un étage d'analyses physico-chimiques de l'eau déminéralisée (oxygène dissous - pH - résistivité - pCl^- - conditionnement en Li7)

. un étage de pressurisation avec pressuriseur - soupape de sureté - capacité de la décharge - capacités d'isolement

23) Analyse des Produits de Fission Eventuels

Dans le cas d'irradiation de crayons rompus, l'eau sortant du dispositif est analysée en continu :

. par DND, qui permet la mesure de la décroissance des neutrons différés.

. par spectrométrie gamma sur l'eau. Une diode Ge-Li mesure l'activité totale de l'eau en continu à travers des collimateurs. Cette mesure permet de tracer le spectre d'énergie de tous les produits de fission présents dans l'eau, et de suivre l'évolution éventuelle d'une rupture.

24) Contrôle Commande

Les baies électriques de contrôle commande assurent les chauffages, les commandes, les contrôles, les régulations, les mesures et la sécurité liés au dispositif. Les mesures sont envoyées sur le calculateur central du réacteur qui gère l'ensemble des données thermiques, neutroniques et hydrauliques.

25) Caractéristiques

Pression nominale de fonctionnement 130 à 155 bars
Température de gaine 320 à 345°C
Puissance surfacique maximale 180 W/cm2
Longueur maximale des crayons centrés 1800 mm
Longueur maximale irradiable 570 mm
Diamètre utile des crayons de 8,6 à 15 mm

3) VARIATIONS DE PUISSANCE

Pour les irradiations de rampes de puissances, la puissance thermique délivrée par l'aiguille combustible doit pouvoir être ajustée à des niveaux choisis et évoluer avec des vitesses prédéterminées. Ces conditions sont obtenues par translation du dispositif dans le réflecteur du réacteur, translation qui a pour effet de faire varier le flux neutronique incident sur l'aiguille, donc la puissance.

31) Détermination de la Puissance

La puissance thermique totale générée par l'aiguille combustible soumise aux flux du réacteur est la somme de deux puissances :
. une puissance induite par le rayonnement gamma qui représente une faible fraction de la puissance totale (5% environ)
. une puissance générée par les fissions provoquées par le flux de neutrons thermiques
La première est déterminée par étalonnage à l'aide de calorimètres. La seconde, la plus importante est déterminée à l'aide de sondes de mesure de flux, ou collectrons (Self Power Neutron Detector = SPND) qui ont la propriété de délivrer un courant électrique proportionnel au flux de neutrons thermiques qui rencontrent l'emetteur.

En répartissant judicieusement ces collectrons à des distances concentriques différentes de l'aiguille combustible, on détermine avec précision la forme du flux thermique incident dans un plan horizontal donné. Connaissant ce flux et les milieux traversés, on calcule à l'aide du code APOLLO la puissance de fission.

Pratiquement, ces collectrons sont placés :
. soit extérieurement au dispositif ; ils sont alors montés sur le fourreau de refroidissement
. soit intérieurement ; ils sont alors montés sur le panier amovible

32) Pilotage du Dispositif

Il s'effectue également à partir des signaux des collectrons et à l'aide de la chaine suivante :
. accélération "apparente" de la réponse du signal de chaque collectron pour rendre compatible le temps de réponse avec les vitesses de variation de puissance, lors de rampes.
. sommation et mise en forme des courants collectrons pour donner directement un signal proportionnel à la puissance de fission
. comparaison par le calculateur du signal résultant avec la valeur de consigne représentative de la puissance à obtenir
. envoi des ordres d'avance ou de recul à l'électronique de commande de la boite à déplacement qui amplifie et transforme le signal reçu en impulsions de commande
. envoi des impulsions au moteur de commande pas à pas de la boite à déplacement

321) Accélération de Réponse Collectron. Le circuit électronique possède un circuit correcteur qui permet de connaitre de façon anticipée la valeur que donnera le collectron à l'équilibre. Il abaisse le temps de réponse apparent à 5 secondes pour 95 % de la valeur.

322) <u>Sommation des Signaux Collectrons</u>.Ce tiroir effectue la somme des signaux "accélérés" fournis par le tiroir précédent et restitue un signal directement représentatif du flux, donc de la puissance. On peut alors l'utiliser pour piloter "en temps réel" le dispositif.

323) <u>Boite à Déplacement</u> (fig. 2). Elle possède un tiroir mobile qui supporte le fourreau du dispositif et qui est déplacé à l'aide d'une commande à distance actionnée par le moteur pas à pas. Le tiroir est équipé d'un système de recopie de position. Le synoptique du système de commande est représenté sur la figure 2.

4) INSTRUMENTATION

Elle vise essentiellement à détecter l'instant précis de la formation de fissures débouchantes ou non.

41) Instrumentation Hors Dispositif en Pile

Une mesure de l'activité globale en ligne par sonde gamma, permet de déceler en conditions thermiques ou neutroniques non stables l'apparition d'une fissure débouchante 2 mn après sa formation. Le matériel utilisé est de haute sensibilité.

42) Instrumentation en Pile (fig. 1)

L'instrumentation est placée le plus près possible de l'aiguille chaque fois que le montage du capteur, ses caractéristiques, le permettent. Pour des raisons d'environnement, les parties sensibles sont montées à l'extérieur de l'enceinte de pression. On y trouve :

. une mesure directe de bruit acoustique par accéléromètre
. une mesure de vibration mécanique transmise par le fluide à un hydrophone
. une mesure d'émission acoustique par un capteur d'émission

421) <u>Accéléromètre</u>. C'est un capteur piézo électrique sensible aux accélérations engendrées par une vibration ou un bruit acoustique. La liaison entre le capteur et l'aiguille, assurée par le porte échantillon lui-même qui transmet le signal, doit être construite solidement. Il travaille selon un axe vertical et pèse 2 grammes. Sa gamme de fréquence est de 1 à 14 000 Hz, sa fréquence de résonance 70 kHz, sa température maximale d'utilisation 250°C.

Le signal émis est envoyé sur un ampli de charge, à haute impédance d'entrée et à faible impédance de sortie, qui assure la

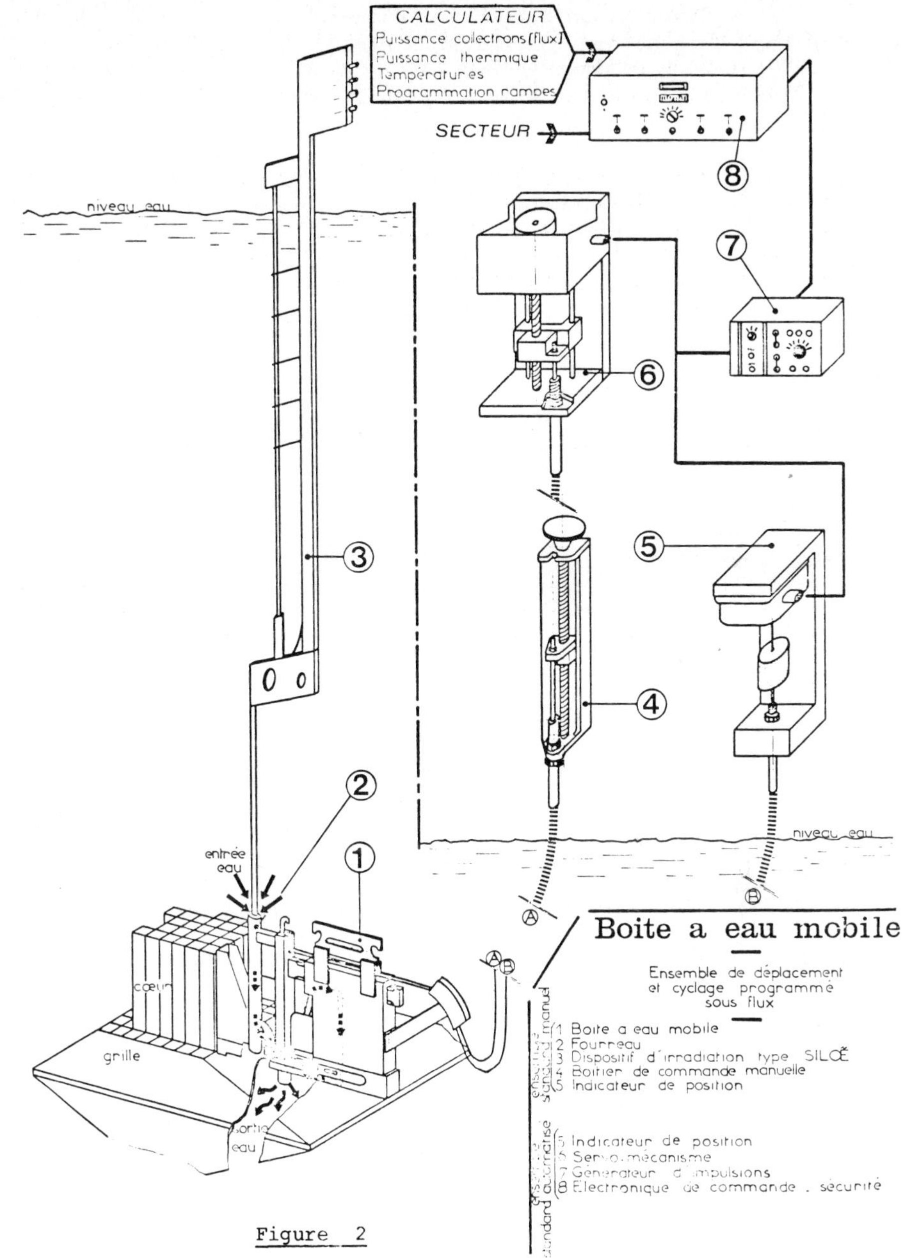
CALCULATEUR
Puissance collectrons [flux]
Puissance thermique
Températures
Programmation rampes
SECTEUR
niveau eau
entrée eau
cœur
grille
sortie eau
Boite a eau mobile
Ensemble de déplacement et cyclage programmé sous flux
ensemble standard manuel
1 Boite a eau mobile
2 Fourneau
3 Dispositif d'irradiation type SILOÉ
4 Boitier de commande manuelle
5 Indicateur de position
standard automatisé
5 Indicateur de position
6 Servo-mécanisme
7 Générateur d'impulsions
8 Electronique de commande . sécurité

Figure 2

transformation charge tension.Le signal résultant est envoyé sur un enregistreur graphique ou sur un enregistreur magnétique pour analyse. Un analyseur "SPECTRAL DYNAMIC" muni de filtres à bande étroite permet d'extraire une partie d'un signal complexe.
La figure 3 montre la forme du signal délivré par ce type de capteur lors d'une rupture de gaine.

422) Hydrophone. C'est un capteur piézo électrique qui fonctionne par compression. Il est sensible aux impulsions dues aux variations de pression engendrées par un bruit et transmises par l'eau qui l'environne. Il permettra donc de détecter des mouvements rapides de gaine, des ébullitions, des éjections d'eau ou de gaz.

Caractéristiques :
Pression maximale d'utilisation 250 bars
Température maximale d'utilisation 350°C
Sensibilité 0,0012 pSI/g
Résistance de sortie 100 MΩ.

L'hydrophone est connecté à un convertisseur de charge placé très près, puis à un amplificateur de charge. Le signal enregistré soit sur enregistreur graphique, soit sur un AMPEX (enregistreur magnétique à vitesse ajustable). L'enregistrement peut être traité par un analyseur de fréquence.

423) Capteur d'émission acoustique. C'est un capteur piézo électrique qui est sensible aux vibrations de surface engendrées par des hétérogénéités de contrainte, donc par des microfissures. Le signal, faible, arrive dans un préamplificateur où il est filtré par bandes de fréquence et amplifié. Un conditionneur met en forme le signal pour en permettre l'analyse. Un analyseur compte dans le temps le nombre d'impulsions dépassant des seuils préfixés et fournit l'information soit sous forme d'une courbe intégrale par sommation des impulsions, soit sous forme d'un comptage d'impulsions. Le résultat est envoyé sur une imprimante.

424) Mesure de longueur de la gaine combustible. L'apparition d'une fissure débouchante provoque une variation de longueur de la gaine de l'aiguille combustible. Un capteur de mesure de cette variation de longueur a été développé et essayé sur une maquette hors flux. Le fonctionnement en est le suivant :

On associe à l'aiguille combustible un cône sur lequel on effectue une mesure de diamètre. Le déplacement axial est donc transformé en une variation diamétrale qui est commodément mesurée à l'aide de jauges de contrainte. L'évolution des jauges placées dans l'environnement du coeur du réacteur est contrôlée périodiquement à l'aide d'un système d'étalonnage manoeuvrable à dis-

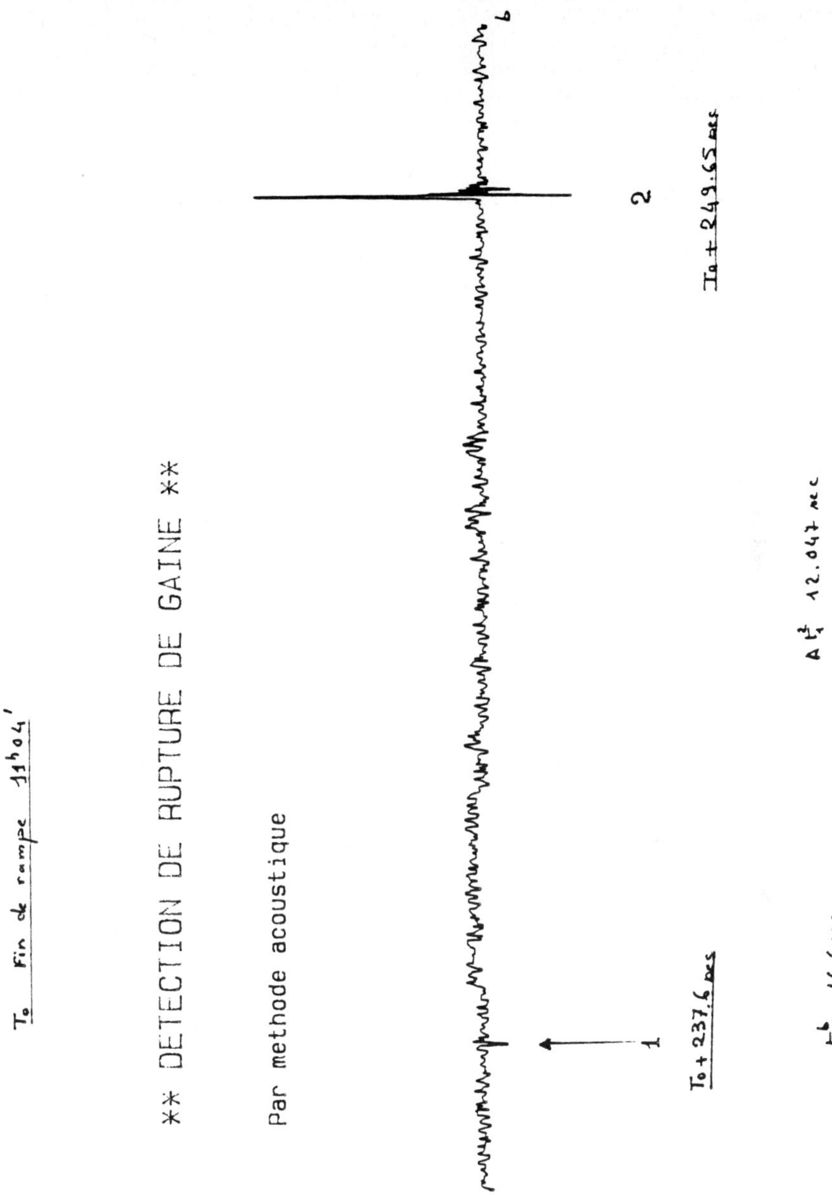

Figure 3

This illustration is printed
from the best available original

tance. La précision de la mesure est directement dépendante des déplacements verticaux du système de prise de diamètre provoqués par les variations de température de l'eau environnante. Cette difficulté à présent surmontée, on peut apprécier des déplacements relatifs avec une précision relative de $\pm$ 0,5 10-6 par degré centigrade de variation de température de l'eau environnante et pour des vitesses inférieures à 20°C/mn.

CONCLUSION

Les possibilités d'irradiation du dispositif AQUILON et les performances du système de pilotage associé nous ont permis d'effectuer de façon automatique un grand nombre de rampes de puissance de caractéristiques différentes.

L'instrumentation complémentaire mise en place nous a permis de déceler l'apparition de fissures débouchantes ou non et de toute façon non détectées par la seule mesure d'activité de l'eau.

La mesure de longueur de l'aiguille venant s'ajouter à cette instrumentation permettra de déterminer l'instant précis d'une perte d'étanchéité de la gaine donc d'affiner la connaissance de l'interaction entre gaine et combustible.

LISTE DES FIGURES

Schéma de la partie basse AQUILON 1
Synoptique du système de commande de la boite 2
Réponse d'un accéléromètre sur une rupture de gaine lors d'une rampe 3

ERSTE BETRIEBSERFAHRUNGEN MIT EINER BESTRAHLUNGSEINRICHTUNG ZUM TESTEN VON BRENNSTÄBEN UNTER DRUCKWASSERREAKTORBEDINGUNGEN

A. Reymann, J. Ahlf, D. Bellmann, K.-H. Blom, H.L. The

GKSS-Forschungszentrum Geesthacht GmbH

2054 Geesthacht, Bundesrepublik Deutschland

ZUSAMMENFASSUNG

Für den Forschungsreaktor FRG-2 wurde eine Bestrahlungseinrichtung entwickelt, mit der LWR-Testbrennstäbe unter Druckwasserreaktorbedingungen (155 bar, 320 °C) bestrahlt werden können. Die abführbare Brennstableistung beträgt 13,5 kW bei einer Stableistung von 450 W/cm. Die Testbrennstäbe können defekt und bereits in einem Leistungsreaktor vorbestrahlt sein.

Es werden die Versuchsapparatur, die experimentellen Methoden sowie die ersten Betriebserfahrungen mit der Bestrahlungseinrichtung am Reaktor FRG-2 beschrieben. Das weitere Versuchsprogramm wird vorgestellt.

ABSTRACT

At the research reactor FRG-2 an irradiation device has been installed for testing fuel rods under PWR-conditions (155 bar, 320 °C). The maximum rod power amounts to 13.5 kW and the linear specific power to 450 W/cm. The maximum length of the test rod is 600 mm. It is possible to test defect and preirradiated fuel rods.

In the paper the test equipment, the experimental techniques and the first operating experience are described. The further irradiation program will be presented.

P. von der Hardt and H. Röttger (eds.), Irradiation Technology, 103–115.

1. EINLEITUNG

Obwohl die Erfahrungsbasis über das Verhalten von LWR-Brennstäben über einen weiten Bereich von Betriebsbedingungen bereits sehr breit ist, gibt es immer noch Fragestellungen, die einer weiteren Bearbeitung bedürfen. Dazu gehört das Spaltproduktfreisetzungsverhalten von defekt gewordenen Brennstäben unter fortgesetztem Leistungsbetrieb.

In Zusammenarbeit mit Partnern aus der Reaktorbauindustrie wurde ein auf diese Fragestellungen bezogenes Untersuchungsprogramm erstellt. Am Forschungsreaktor FRG-2 wurde hierfür eine Bestrahlungseinrichtung in Betrieb genommen, in der LWR-Testbrennstäbe bis zu einer Stableistung von 450 W/cm unter Druckwasserreaktorbedingungen (155 bar, 320 ^{o}C) getestet werden können. Die Anlage wurde bereits in [1] vorgestellt.

2. BESCHREIBUNG DER ANLAGE

Die Inpile-Strecke ist in Abb. 1 dargestellt. Der Testbrennstab, dessen aktive Länge etwa 320 mm beträgt, ist im unteren Teil des Druckrohres untergebracht. An das Druckrohr ist eine dreistufige Spaltrohrmotorpumpe angeflanscht, die Wasser von 320 ^{o}C mit einer Rate von ca. 1 m^3/h umwälzt.

Dabei fließt das Primärkühlwasser im Ringraum zwischen dem Testbrennstab und einem Leitrohr aufwärts zur Pumpe (Aufheizstrecke) und abwärts im Ringraum zwischen dem Leitrohr und dem Druckrohr; das letztere dient als wärmetauschende Oberfläche (Kühlstrecke), die die Wärme über einen Gasspalt und den unteren Sekundärbehälter an das Kühlwasser des Reaktors abgibt.

Das untere Ende des Druckrohres ist mit einer Öffnung versehen, durch die der Testbrennstab in der Heißen Zelle eingesetzt und entnommen werden kann. Die Öffnung wird durch eine Hochdruck-Rohrverschraubung verschlossen.

Im Wasserspalt zwischen Leitrohr und Druckrohr sind Mantelheizleiter installiert. Sie dienen zur Feinregulierung der Betriebstemperatur und zur Aufrechterhaltung der Betriebstemperatur auch während der Reaktorpausen.

Oberhalb der Pumpe ist ein Behälter angeordnet, der zum Ausgleich der Volumenänderungen des Primärwassers dient. In diesem Ausgleichsbehälter befinden sich ein elektrischer Heizer zur Druckhaltung und ein Füllstandsmeßgerät.

Der gesamte Primärbereich ist von einer zweiten Umhüllung umschlossen. Der untere Sekundärbehälter umschließt das Druckrohr und

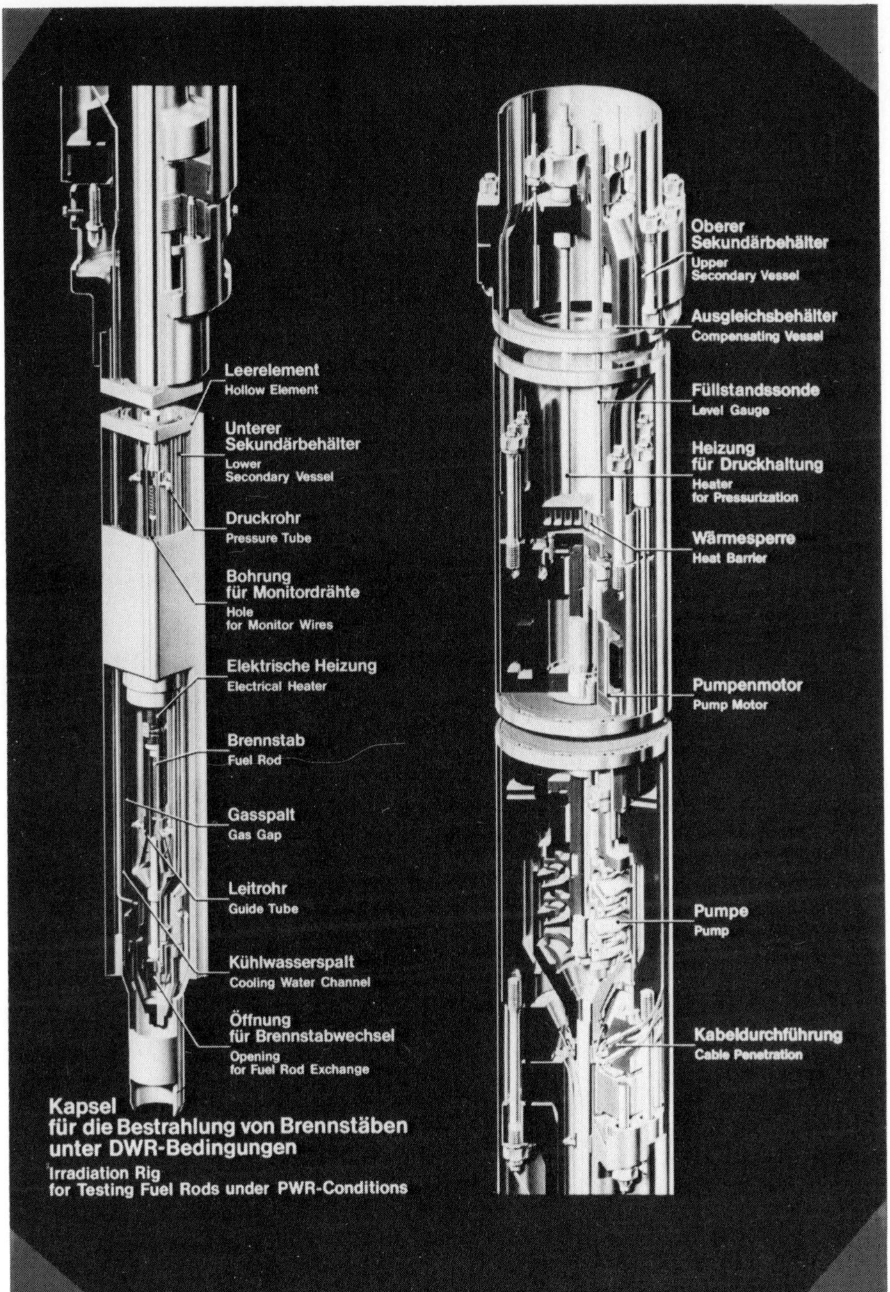

Abb. 1 Bestrahlungskapsel

ist von diesem durch einen etwa 0,5 mm breiten Spalt getrennt, der mit einem Helium-Stickstoff-Gemisch gefüllt werden kann. Über die Gaszusammensetzung wird die Betriebstemperatur im Primärbereich grob eingestellt.

Ein vereinfachtes Fließschema ist in Abb. 2 dargestellt.

Das Frischwasser wird in Vorratsbehältern konditioniert. Der Sauerstoffgehalt wird durch durchperlendes Helium auf die erforderliche niedrige Konzentration herabgesetzt. Bei Bedarf können Borsäure und Lithiumhydroxid hinzugefügt werden; ferner besteht die Möglichkeit, das Wasser mit Wasserstoff zu beladen. Das Frischwasser wird dem Primärsystem mit einer Rate von etwa 10 l/d kontinuierlich zugeführt.

Die Entsorgung des Primärwassers erfolgt über einen Entgasungsbehälter, in welchem die flüchtigen Spaltprodukte mit Stickstoff ausgetrieben werden, und einen Mischbettionenaustauscher. Das Abgas wird über Aktivkohle und Jodfilter geführt.

3. MESSUNG DER SPALTPRODUKTFREISETZUNG

Das Primärwasser wird mit einer Rate von etwa 10 l/d aus der Kapsel abgezogen und an einem Gammaspektrometer vorbeigeführt. Das Gammaspektrometer ist in Abb. 3 dargestellt. Es ist so ausgelegt, daß in einem weiten Konzentrationsbereich gemessen werden kann.

Das Meßvolumen besteht aus einem Teflonschlauch mit einem Innendurchmesser von 0,7 mm und 80 mm Länge, der zur genauen Fixierung in die Nut eines Bleipaßstückes eingelegt ist; es beträgt ca. 30 mm^3. Der Probenträger läßt sich relativ zum Bleikollimator (von 0 bis 150 mm) bewegen, und der gesamte Bleikollimator mit dem Probenträger wiederum relativ zum Detektor (von 0 bis 450 mm) verschieben. Beim ausgefahrenen Zustand des Probenträgers kann die Bleikollimatoröffnung (von 0 bis 80 mm) variiert werden.

Zur Registrierung der Gammastrahlung wird ein Intrinsic Ge-Detektor (Ortec) in "side-looking"-Aufbau mit dem integrierten Vorverstärker und gekühlten FET verwendet. Zur Abschirmung ist der Detektorhals mit dem darin befindlichen Kristall und Vorverstärker in einer Bleiummantelung ausreichender Dicke eingebettet.

Zur Eichung des Spektrometers werden zwei speziell angefertigte ^{152}Eu-Eichquellen verwendet.

Abb. 4 zeigt ein vereinfachtes hard-ware-Schema der verwendeten Elektronik: Für die Aufnahme von Spektraldaten steht ein 200 MHz-ADC des Wilkinson-Typs(Tennelec) zur Verfügung, für die

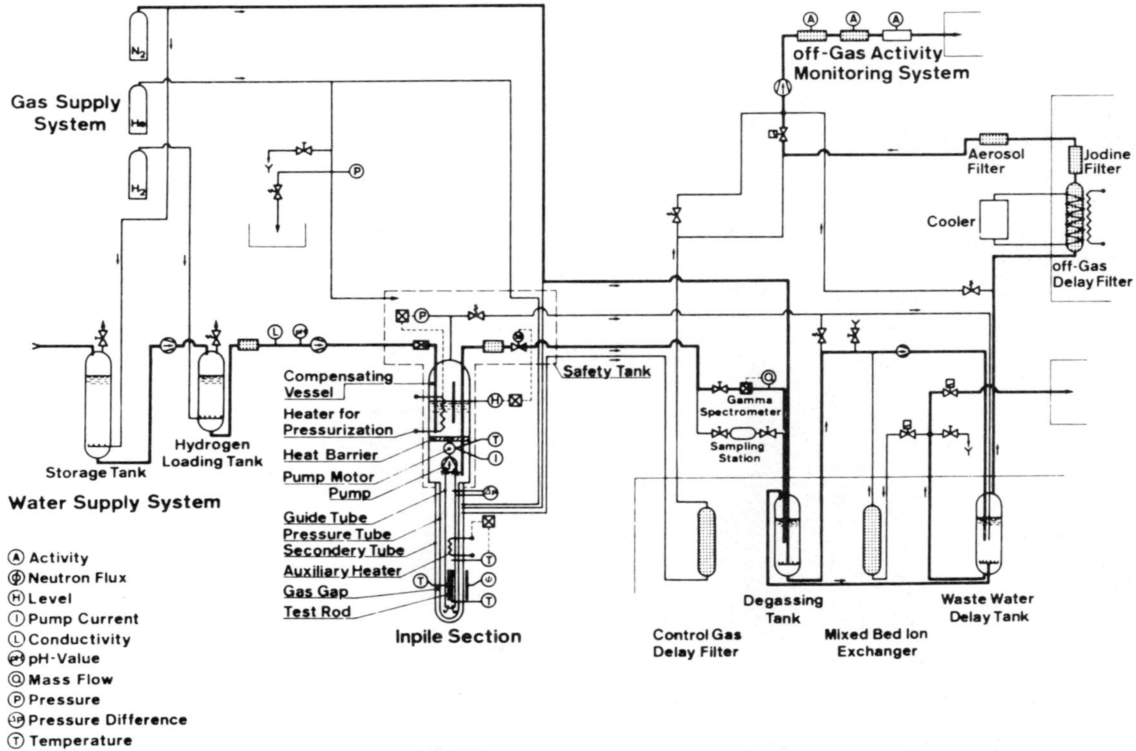

Abb. 2 Vereinfachtes Fließschema

Abb. 3 Gammaspektrometer

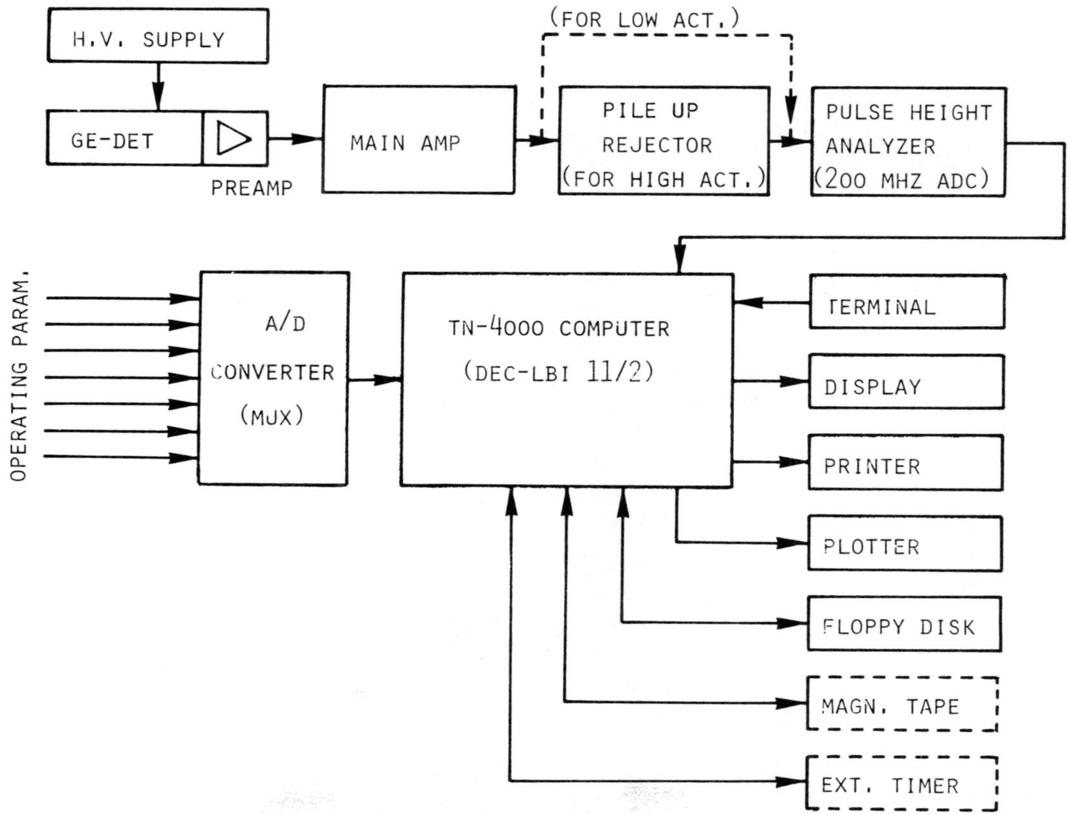

Abb. 4 Vereinfachtes hard-ware-Schema der Datenerfassung

Datenerfassung, -überwachung und -auswertung ein Kleinrechner mit DEC-LSI 11/2-Prozessor (TN 4000 - Tracor Northern) und zur Erfassung der Betriebsparameter (Neutronenfluß, Temperatur u.a.) ein 12 Bit A/D-Wandler mit 35 kHz Konversionsrate (ADAC Corp.) im Multiplex-Betrieb.

Neben den Gammaspektren wird eine Reihe relevanter Betriebsdaten wie u.a. Neutronenfluß und Kühlmitteltemperatur gleichzeitig erfaßt und überwacht; diese werden den gammaspektroskopischen Meßergebnissen eindeutig zugeordnet, um später die Freisetzungsraten berechnen und deuten zu können. Zur Erfassung der Betriebsdaten wird in Verbindung mit dem Kleinrechner ein 12 Bit A/D-Wandler im Multiplex-Betrieb eingesetzt. Andere sekundäre Daten wie Kollimatorbreite und Detektorabstand werden ebenfalls vom A/D-Wandler eingelesen. Für die Spektrenaufnahme wird ein ADC mit 4 K Kanälen verwendet, wobei für jeden Kanal 22 Bits zur Verfügung stehen.

Für den zyklischen Ablauf der Datenerfassung, -auswertung und -überwachung zeigt Abb. 5 ein vereinfachtes Flußdiagramm, bei dem zwei parallel verlaufende Programme - "Data Acquisition/Evaluation Program" und "Control Program" - nach dem "foreground/background"- bzw. "interrupt"-Prinzip die gesamten Meßabläufe regeln. Die Betriebsparameter werden regelmäßig (z.B. alle 5 Minuten) durch das "Control-Program" eingelesen und Mittelwerte gebildet. Bei einer größeren Abweichung von den vorhergehenden Meßwerten wird die laufende Spektralaufnahme beendet, zwischengespeichert und erneut gestartet.

Im "Data Acquisition/Evaluation Program" werden bei fortlaufenden Spektrenaufnahmen jedesmal die Spektraldaten nach Ablauf einer festen Meßzeit zwischengespeichert, um zunächst eine vorläufige Auswertung einiger prominenter Nuklide zu ermöglichen. Dies gestattet eine schnelle Überwachung der aufgenommenen Spektren. Wenn keine Änderung vorliegt, werden die Spektraldaten weiter aufaddiert. Für den Fall, daß sich bei Auswertung des aktuellen Kurzzeitspektrums Abweichungen vom Mittelwert der vorhergehenden ergeben, die einen vorgegebenen Grenzwert übersteigen, wird die Summe der vorangegangenen Spektren abschließend ausgewertet, gespeichert und ein neuer Zyklus begonnen. Eine Auswertung aller Nuklide wird ebenfalls durchgeführt, falls z.B. 10 Spektren bereits aufaddiert sind, oder, wenn die Reaktorbetriebsparameter einer Änderung unterliegen. Dabei wird wieder berücksichtigt, daß sich die Änderung der Betriebsdaten wegen der Laufzeit der Spaltprodukte in der Meßleitung erst nach einer Verzögerungszeit auf die Spektren am Meßort auswirkt.

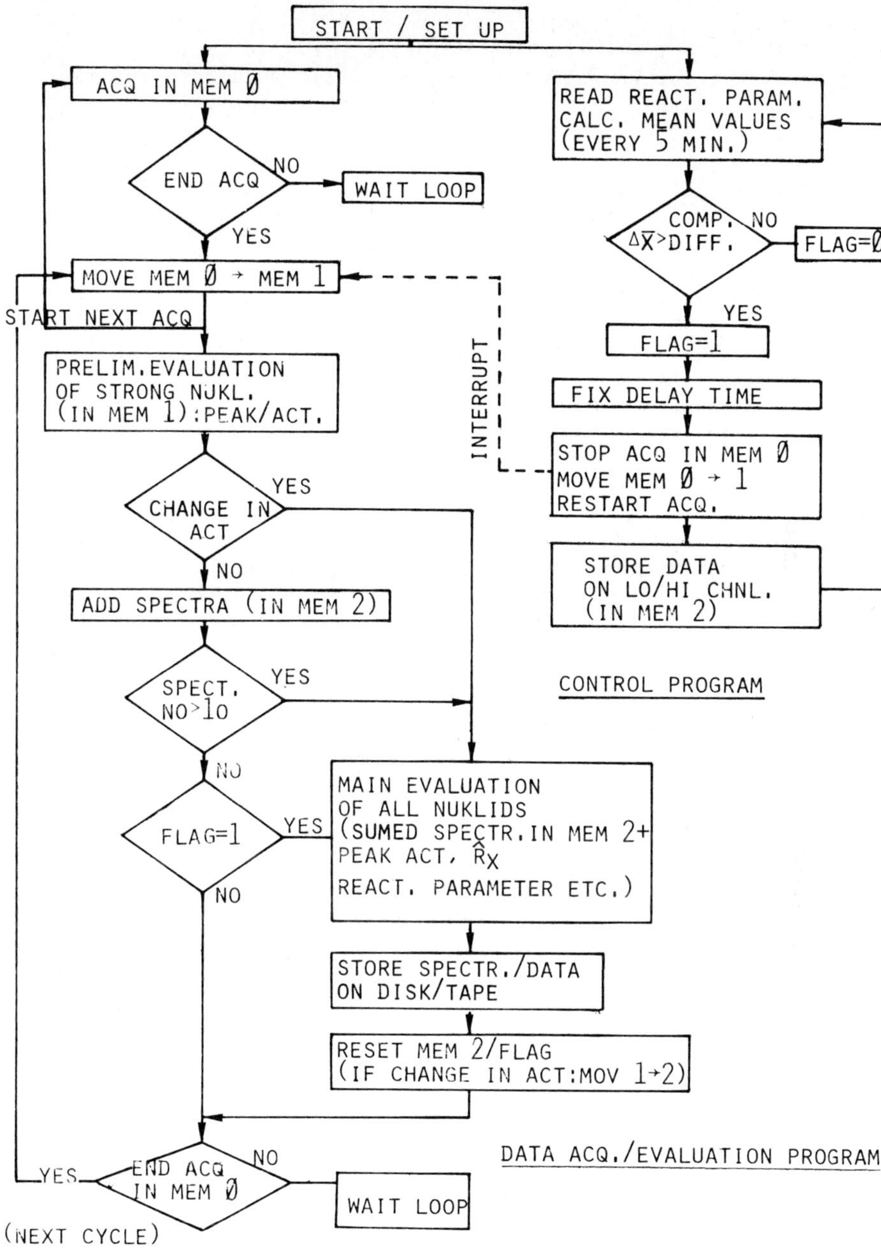

Abb. 5 Vereinfachtes Flußdiagramm der Datenerfassung

4. BETRIEBSERFAHRUNGEN

Im Rahmen der Erprobung der Gesamtanlage wurde die Bestrahlungskapsel zunächst für 2 kurze Zyklen von insgesamt 24 Vollasttagen mit einem intakten, nicht vorbestrahlten Brennstab betrieben. Die Abb. 6 zeigt neben der Reaktorleistung die Kühlmittelaustrittstemperatur und den Systemdruck. Temperatur (320 °C) und Druck (155 bar) konnten auf ± 1 % konstant gehalten werden.

Bei 15 MW Reaktorleistung betrug die Leistung des Testbrennstabes im ersten Zyklus ca. 5,7 kW und im 2. Zyklus ca. 6,9 kW entsprechend einer mittleren Stableistung von 180 W/cm und 216 W/cm. Die Brennstableistung wurde aus der Temperaturdifferenz Kühlwassereintrittstemperatur und -austrittstemperatur und dem Kühlwasserdurchsatz errechnet. Ein Vergleich mit der errechneten Leistung aufgrund der Flußdaten am Brennstab wird noch durchgeführt. Die Temperaturdifferenz von 4 bis 5 K über den Brennstab wurde durch bereits im Primärbereich der Kapsel gegeneinander geschaltete Thermoelemente im Kühlwassereintritt und -austritt erfaßt. Zur Verdoppelung des Signals wurden je Meßstelle 2 Thermoelemente hintereinander geschaltet. Die Kalibrierung erfolgte mit einem elektrisch beheizten Teststab in der Kapsel. Der Kühlmitteldurchsatz wird über eine Meßblende erfaßt.

Die Abb. 7 zeigt die Kühlmittelaustrittstemperatur als Funktion der in der Kapsel freigesetzten Leistung, die sich aus der Brennstableistung und der elektrischen Leistung der Zusatzheizung zusammensetzt. Als Parameter ist die Regelgaszusammensetzung angegeben. Das für die Kapselauslegung benutzte Rechenmodell wurde an die ersten Meßergebnisse angepaßt. Mit der im 1. Zyklus erreichten Brennstableistung von 5,7 kW kann die Betriebstemperatur von 320 °C mit 100 % Stickstoff als Regelgas noch ohne elektrische Zusatzheizung erreicht werden. Mit 50 % He-Anteil läßt sich die Betriebstemperatur durch Zuschaltung von ca. 9 kW elektrischer Leistung aufrechterhalten. Bei abgeschaltetem Reaktor kann die Betriebstemperatur mit 100 % N_2 und ca. 10 kW elektrischer Leistung noch auf etwa 320 °C gehalten werden. Durch eine Vergrößerung des Regelgasspaltes um z.B. 0,1 mm (bei weiteren Kapseln) lassen sich die Temperaturkurven um ca. 30 K anheben.

Die Umwälzpumpe - eine Spaltrohrmotorpumpe -, deren Bau einen erheblichen Entwicklungs- und Testaufwand erforderte, zeigte anfangs unruhiges Laufverhalten mit zeitweise stark schwankender Stromaufnahme. Nach dieser Einlaufzeit lief die Pumpe jedoch zufriedenstellend. Die Radiallager bestehen aus einer Aluminiumoxid-Lagerbuchse und einer Chromoxid-Laufbuchse. Die Axiallager bestehen aus Wolframkarbid-Gleitringen.

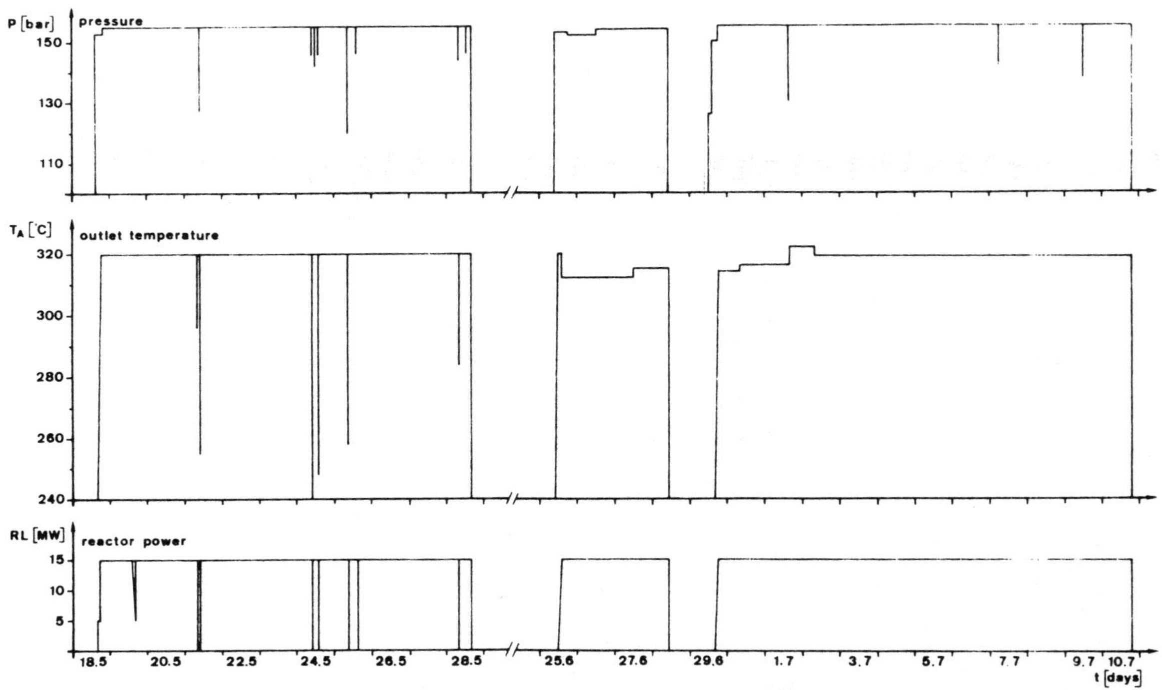

Abb. 6 Bestrahlungsbedingungen

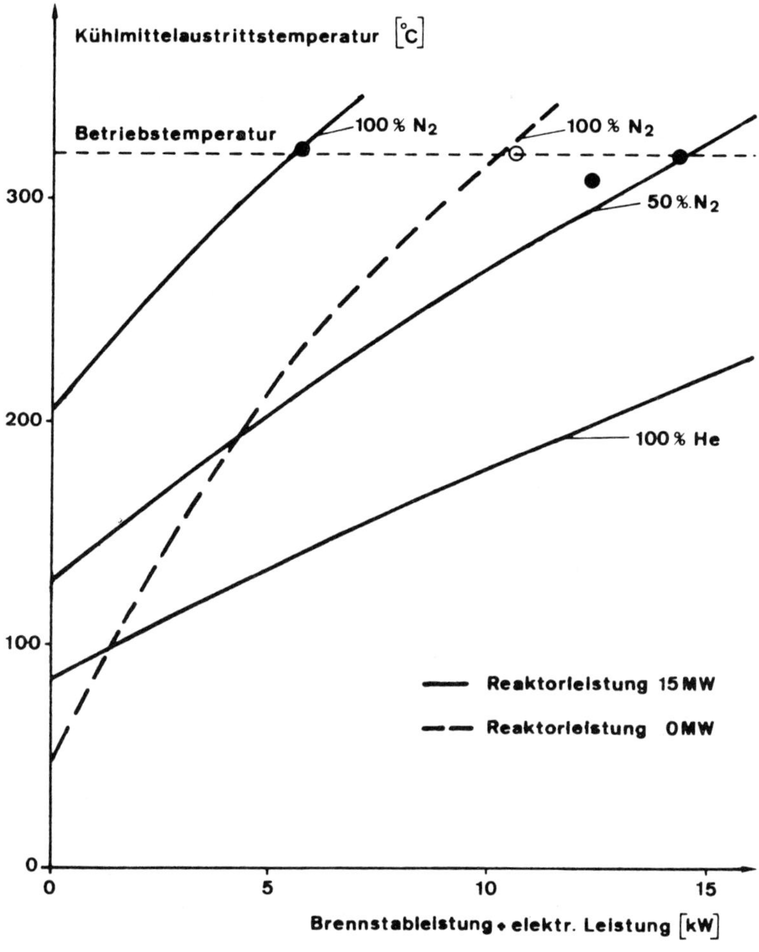

Abb. 7 Kühlmittelaustrittstemperatur als Funktion der Gesamtleistung

Die Füllstandsmessung zur Konstanthaltung des Füllstandes im Ausgleichsbehälter der Kapsel erfolgte über einen elektrischen Heizstab mit innen liegenden Thermoelementen in verschiedenen Höhen. Die Heizstabtemperatur wird über ein im Dampfraum liegendes Thermoelement auf ca. 400 °C geregelt. Anhand der Temperatur der weiteren Thermoelemente wird über eine Elektronik entschieden, ob sich die Meßstelle im Wasser oder im Dampfraum befindet. Ein Temperaturunterschied von ca. 20 K ist für eine sichere Unterscheidung ausreichend.

Vom On-line-Gammaspektrometer wurden in den beiden ersten Zyklen mit intaktem Testbrennstab keine Spaltprodukte gemessen. Es wurden hauptsächlich Aktivierungsprodukte nachgewiesen, die auf den Abrieb an den Pumpenlagern zurückzuführen sind.

5. WEITERES VERSUCHSPROGRAMM

Noch im Rahmen der Erprobungsphase wurde jetzt ein unbestrahlter, jedoch angebohrter Testbrennstab in die Bestrahlungskapsel eingesetzt. Hiermit soll insbesondere die Meßtechnik zur Erfassung der Spaltproduktfreisetzung ausreichend erprobt werden. Danach werden Versuche an im Leistungsreaktor vorbestrahlten Testbrennstäben durchgeführt. Diese Stäbe werden in der 1. Versuchsphase ebenfalls mit einem Laser angebohrt, Lochdurchmesser ca. 0,1 mm. In einer späteren Phase sind Versuche an Testbrennstäben mit PCI-Defekten vorgesehen.

Die Spaltproduktfreisetzung soll unter folgenden Gesichtspunkten untersucht werden:

- Abhängigkeit von der Stableistung
- bei Lastzunahme und Lastabsenkung mit verschiedenen Leistungsänderungsgeschwindigkeiten
- während des Anfahrens und während und nach Abschaltung des Reaktors.

LITERATUR

[1] J. Ahlf; Irradiation Rig Operating at PWR Conditions for Investigating Defected Fuel Rods, IAEA/IWGFPT/6 (Sept. 1979).

BATI MULTIPLE IMMERGE POUR LE CONTROLE DE CRAYONS PWR IRRADIES DANS UNE BOUCLE RECHARGEABLE EN PISCINE OSIRIS

Gérard FARNY

Commissariat à l'Energie Atomique - Services des Piles

C.E.N. Saclay - 91191 - GIF-sur-YVETTE Cedex (France)

INTRODUCTION

La dernière décennie a vu s'opérer une réorientation dans la vocation de nombreux réacteurs expérimentaux du type MTR. Destinés naguère à des essais à caractère statistique de composants de centrales électrogènes, ces essais se justifient de moins en moins eu égard à la quantité de résultats que procure directement et en vraie grandeur, l'exploitation d'un parc de centrales qui ne cesse de croître. Par contre, ces réacteurs restent irremplaçables pour les essais à caractère analytique, destinés à suivre l'évolution de composants après, voire durant des transitoires successifs.

Ainsi, dans le cadre des études menées par le C.E.A. et l'EdF, relatives à la dynamique de la dégradation du combustible PWR en régimes transitoires, normal et anormal, les Services des Piles de Saclay terminent actuellement la réalisation de trois ensembles importants destinés :

- à irradier simultanément 4 crayons aux conditions PWR,
- extraire ces crayons en piscine après la ou les séquences d'irradiation,
- les examiner sur bâti-multiple immergé, puis les recharger pour effectuer de nouveaux cyclages.

La boucle d'irradiation ISABELLE 4 et son système de déchargement font l'objet d'une communication à cette Conférence. Quant au bâti-multiple, objet de cette présentation, il est destiné à assurer trois types d'examen en complément de la neutronographie et de la gammamétrie utilisées depuis plusieurs années, à savoir :

- examen visuel et enregistrement vidéo de l'aspect du crayon,
- métrologie du crayon,
- contrôle par courants de Foucault de la gaine.

P. von der Hardt and H. Röttger (eds.), Irradiation Technology, 117–131.

C'est bien entendu le dernier point que nous développerons, les deux autres étant immédiats ils ne seront simplement qu'évoqués.

COURANTS DE FOUCAULT

Nous ne reviendrons pas sur l'intérêt et la théorie des contrôles par courants de Foucault pour la détection de défauts dans les gaines de combustibles irradiés, d'autant que cette méthode est largement employée en "cellule chaude" depuis plusieurs années.

Aux anomalies géométriques du crayon vient s'ajouter, dans notre cas, l'influence de la puissance résiduelle due à un temps de refroidissement très court. Cette influence fut déterminante dans le choix du mode de détection et de son support.

Mode de Détection

La puissance résiduelle du crayon implique un refroidissement que la convection naturelle suffit largement à procurer, dans la mesure où elle n'est que peu perturbée. Or, la sensibilité d'une bobine encerclante décroît avec l'augmentation de l'espace gaine-bobine rendue indispensable dans le cas présent. Néanmoins, la réduction de cet espace pourrait s'envisager moyennant l'utilisation d'un circuit de refroidissement particulier et secouru. En outre, dépôt, flambage, gonflement, ovalisation impliquent également une majoration de cet espace entraînant ainsi une baisse de sensibilité. Enfin, les problèmes de gradients thermiques excluent quasiment toute méthode absolue.

Ainsi notre choix s'est-il porté sur l'emploi de sondes différentielles, ponctuelles, dont les lignes de champ n'intéressent qu'un faible volume, ce qui en théorie ne peut qu'accroître la sensibilité du système. De plus, la localisation de défauts en z et θ peut présenter un intérêt dans l'analyse des effets d'anisotropie (gradient de flux, effet d'écran ou d'angle, de structure, présence d'absorbant solide .....).

Caractéristiques des Sondes

Ce sont des capteurs de type basse impédance (limitation d'effet d'antenne), à bobinage long sur ferrite. Ces capteurs sont maintenant fabriqués dans nos laboratoires où de nouvelles améliorations sont encore examinées. En outre, leurs constituants ont été choisis en fonction de leur tenue en milieu ionisant et des fluences supérieures à 10^8 rad ne les affectent pas. Enfin, à l'aide d'un montage particulier, nous nous sommes efforcés d'obtenir une reproductibilité des résultats d'une sonde à l'autre afin de limiter les conséquences d'un claquage en cours d'utilisation.

Positionnement des Sondes

C'est un des points les plus délicats de la mise en œuvre de cette méthode, car ce type de sonde utilisé à haute fréquence en fait un excellent capteur de position.

Avec une bobine encerclante et un bon coefficient de remplissage, l'excentrage de la gaine par rapport à l'axe de la bobine entraîne une variation négligeable du signal. Par contre, une variation de distance sonde-gaine de 1/10 de mm perturbe l'amplitude du signal de plus de 30%.

Afin d'éviter une variation de la distance sonde-gaine lors de l'examen, nous avons opté pour la solution du contact avec un léger ressort de rappel. Néanmoins, les mouvements schématisés figure 1 sont générateurs de signaux parasites pouvant, suivant leur fréquence :

- masquer une signature de petits défauts,
- entraîner une figure comparable à celle d'un défaut.

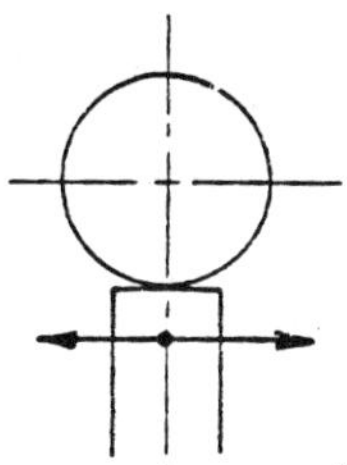

Figure 1

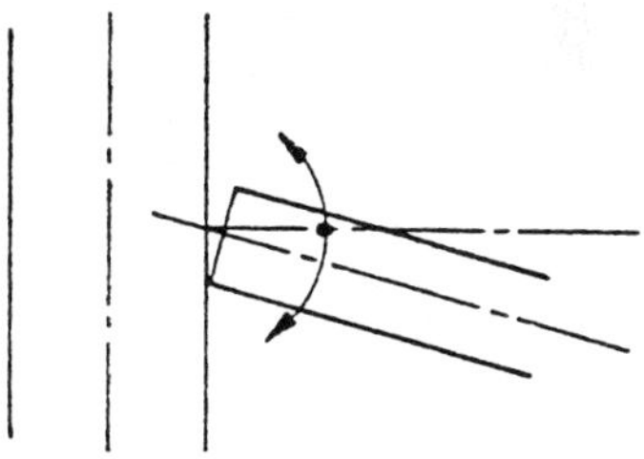

Il est certain que si ces mouvements sont lents en regard de la vitesse de passage du défaut devant la sonde, un filtrage permettra d'extraire le signal recherché. Par contre, des vibrations (dues aux irrégularités de frottement) peuvent avoir des fréquences voisines de celles correspondant au passage du défaut et entraîner des erreurs d'interprétation.

Aussi, une importante partie de notre travail a consisté à mettre au point un système permettant de limiter les mouvements du type A ou B et leurs combinaisons malgré les anomalies géométriques du crayon (flambage, gonflement, ...).

Lors de nos premiers essais effectués sur un appareil à guidage et centrage mécaniques, les résultats obtenus furent décevants puisqu'un trou traversant de 1/10 de mm ne fut pas décelé à cause du bruit de fond. C'est donc vers un système totalement différent que nous nous sommes orientés.

Tête de Mesure

La conception de la tête de mesure repose sur l'utilisation des techniques de guidage et supportage par paliers fluides. Cette tête est constituée d'un ensemble comprenant des patins hydrauliques prenant appui sur la gaine. Ainsi, les frottements sont inexistants, sauf la faible surface de contact de la sonde ($\leqslant 1mm^2$), le refroidissement du crayon est largement assuré par l'échappement de chacun des patins. En outre, cet ensemble a été conçu pour limiter le déplacement latéral de la sonde, notamment lors du passage d'une déformation.

Pour n'apporter aucune contrainte au crayon, la tête suspendue est flottante et possède trois degrés de liberté dans des limites qui ont été déterminées à partir de considérations de flambage maximum.

Quant à l'alimentation hydraulique, elle se doit d'être souple, malgré la pression du fluide véhiculé, tout en ayant une bonne tenue aux rayonnements.

Eu égard aux difficultés rencontrées lors de la mise au point, nous avons opté pour une tête de lecture fixe et un crayon mobile suivant z et θ , l'inverse est néanmoins envisageable.

A titre d'exemple voici quelques comparaisons de signatures de défauts identiques, obtenues avec l'appareil à guidage et centrage mécanique, puis avec le système hydraulique.

a) Sur un trou traversant de 2/10 de mm de diamètre (figure 2).

b) Autre différence obtenue sur une fente transversale interne de 1/10 de mm de profondeur. (figure 3)

Remarque :

Ces examens ayant été effectués à des dates différentes, ne serait-ce qu'en raison des délais de réalisation de la tête flottante, il n'y a pas eu de recalage en phase, la sensibilité de détection étant le premier objectif. Cependant, aux dérives près, excitation et gain sont restés identiques.

Comparaison Sonde-Bobine Encerclante

Bien que l'emploi de bobine encerclante soit quasiment exclu dans notre cas, nous avons effectué quelques comparaisons en fixant une telle bobine sur la tête flottante. Un tronçon de gaine vierge parfaitement cylindrique nous a permis de limiter le jeu pour accroître la sensibilité de la bobine encerclante.

Sur des défauts identiques, nous avons effectué les comparaisons suivantes :

- longitudinal externe 1,5 (figure 4)

a) Sur un trou traversant de 2/10 de mm de diamètre (figure 2).

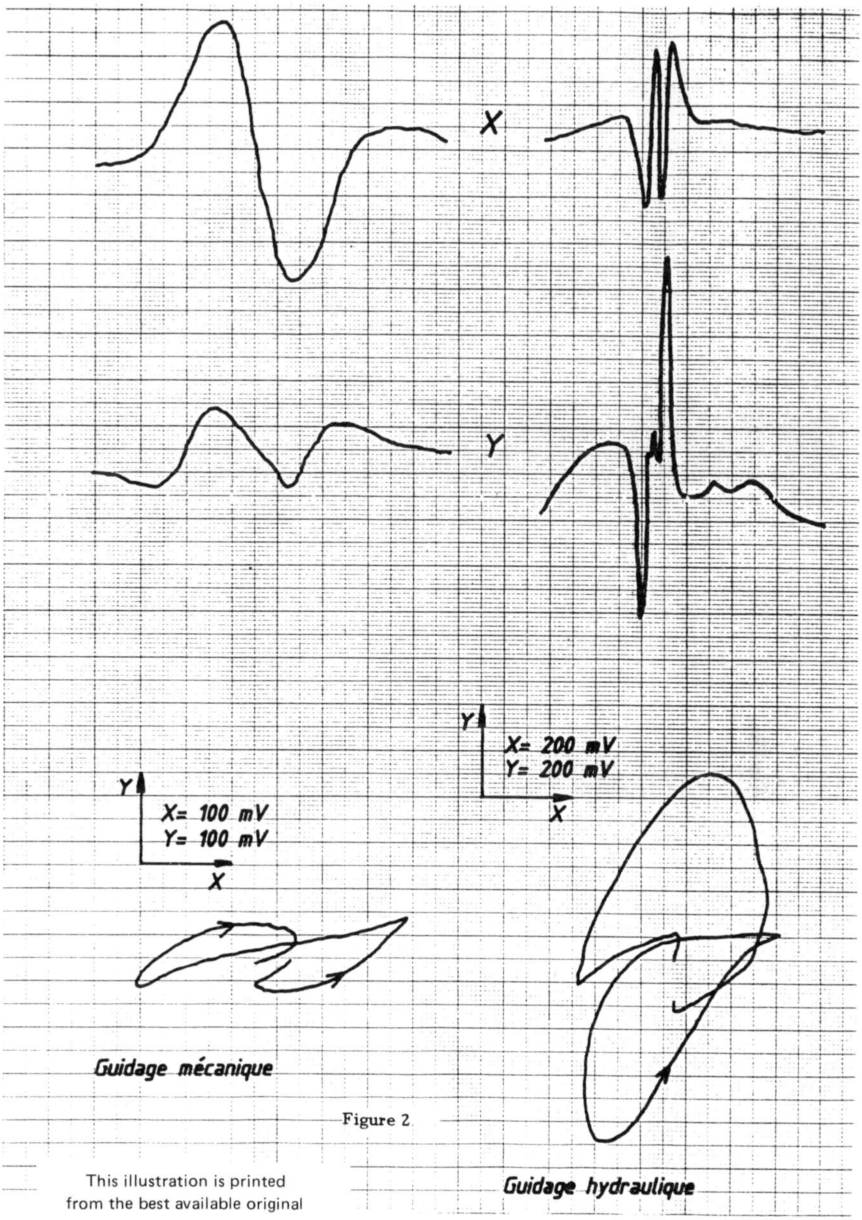

Figure 2

This illustration is printed from the best available original

b) Autre différence obtenue sur une fente transversale interne de 1/10 de mm de profondeur. (figure 3)

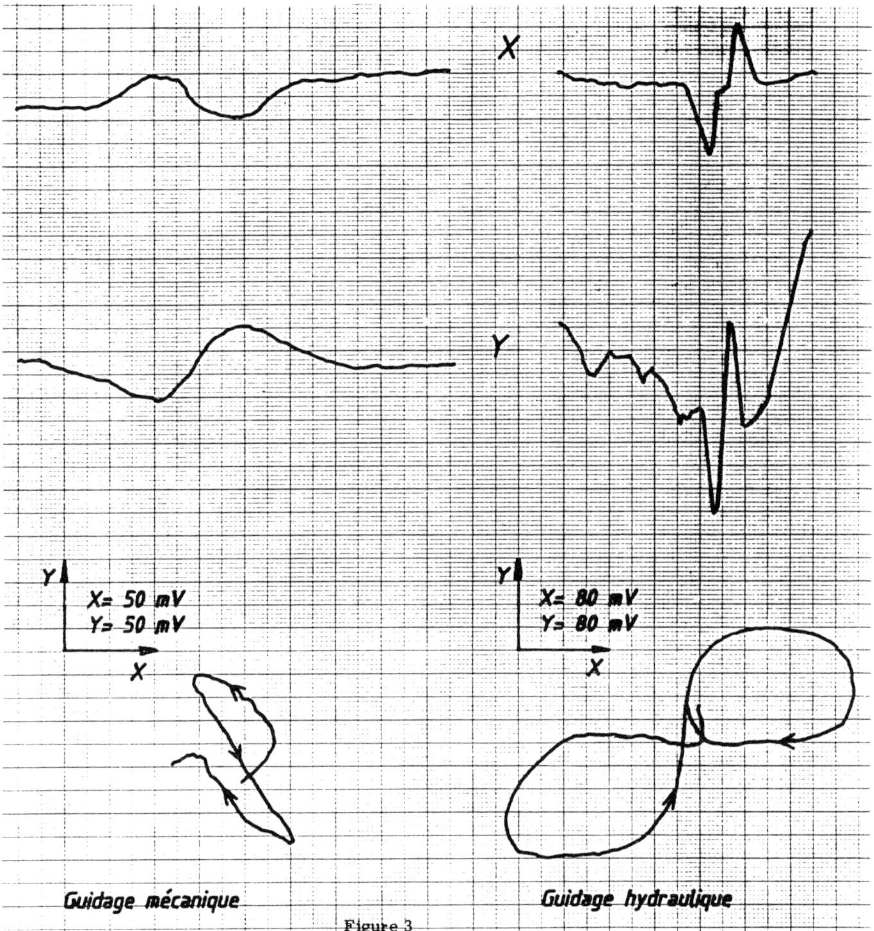

Figure 3

Remarque :

Ces examens ayant été effectués à des dates différentes, ne serait-ce qu'en raison des délais de réalisation de la tête flottante, il n'y a pas eu de recalage en phase, la sensibilité de détection étant le premier objectif. Cependant, aux dérives près, excitation et gain sont restés identiques.

This illustration is printed from the best available original

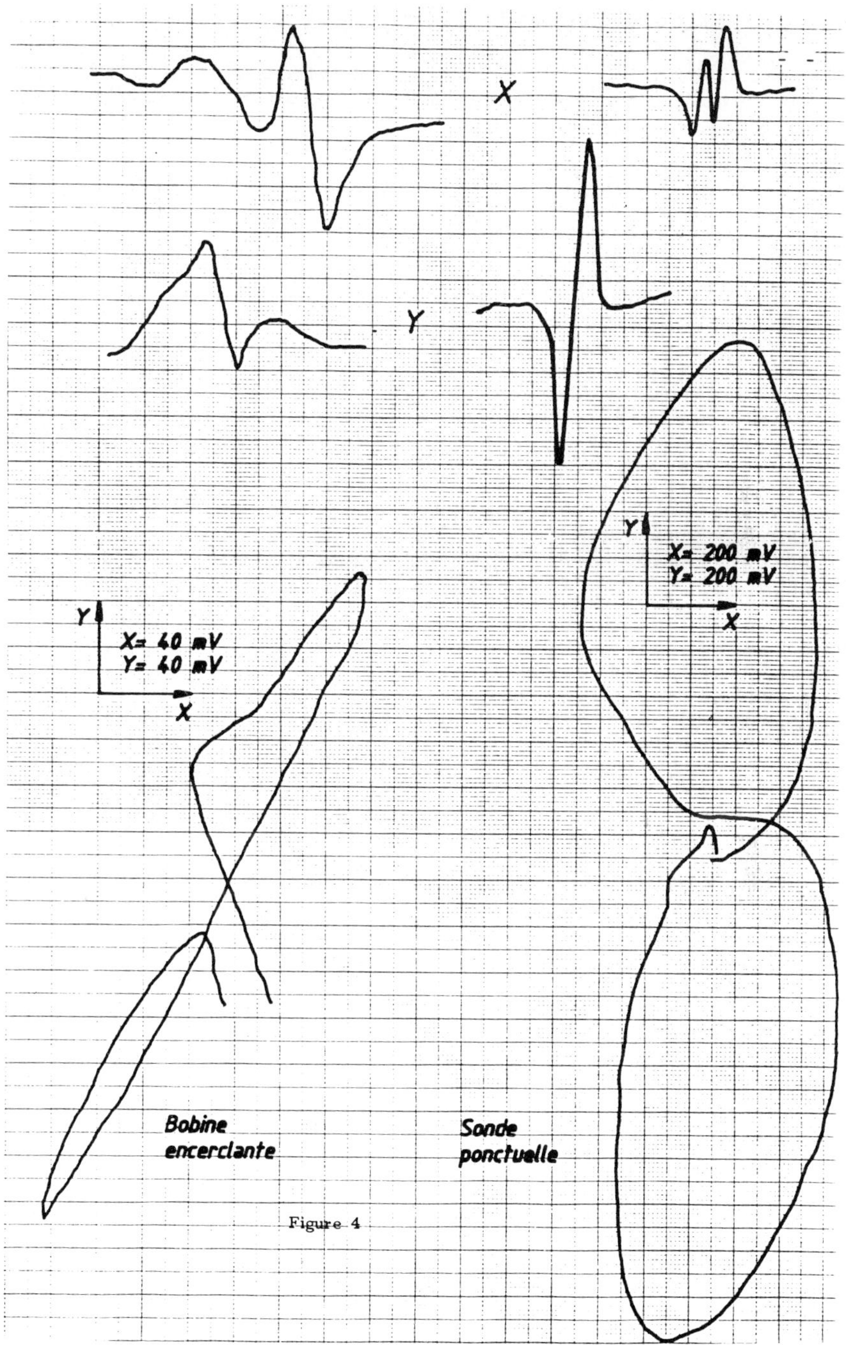

Figure 4

This illustration is printed
from the best available original

- trou débouchant de 1/10 de mm

Sur un trou de 1/10 de mm, qui ne fut pas détecté par la bobine encerclante, on remarquera l'incidence de l'angle θ sur la signature du défaut (figure 5).

Mouvement du Crayon

L'exploration s'effectuant suivant z et θ, un guidage du porte-crayon est indispensable. Pour ne pas générer de vibrations par des frottements au niveau de ce guidage, nous avons également opté pour la mise en œuvre de patins hydrauliques au niveau du porte-crayon, et c'est sur cette installation prototype que nous avons obtenu les résultats présentés.

Quelques Résultats

Nous disposons d'un ensemble de défauts calibrés réalisés par électro-érosion, auxquels il faut ajouter d'autres types de défauts obtenus par fatigue. Pour la détection aux limites de sensibilité, l'emploi de filtres passe-bande s'est avéré nécessaire pour diminuer l'incidence des paramètres non recherchés, comme par exemple les variations d'épaisseur de la gaine. A ce sujet, il est à noter que, sur notre installation, une variation d'épaisseur interne de 1/10 de mm entraîne un signal plus important que celui d'un trou débouchant de 2/10 de mm de Ø avec, bien entendu, une phase différente. C'est dire que les faibles variations d'épaisseur inhérentes à la procédure de fabrication entraînent des variations correspondantes du signal venant se superposer, en l'absence de filtre, à la signature du défaut recherché.

Outre les résultats déjà montrés pour les comparaisons de montage et de capteurs, nous avons retenu les signatures d'un ensemble de défauts, simulant l'évolution d'une fissure interne :

Il s'agit de cinq fentes internes longitudinales de 1 mm de longueur, de 4/100 entre lèvres et de profondeur variable, soit en mm :

0,075 - 0,1 - 0,15 - 0,22 - 0,35 - traversant,

Les signatures données correspondent aux enregistrements du type F (X,Y) (figure 6).

Enfin, la dernière signature présentée (figure 7) provient de l'examen d'un défaut obtenu par fatigue entraînée par une pression pulsée d'eau vers 600 bars. Ces contraintes cyclées ont entraîné une fissuration traduite par une fuite inexistante à une pression inférieure à 60 bars.

Approche d'un Traitement Informatique

Nous ne reviendrons pas sur les étalonnages classiques qui d'ores et déjà permettent de déterminer ou plutôt d'estimer la nature et l'importance du défaut, dans la mesure où il n'y a pas interférence de plusieurs signatures,

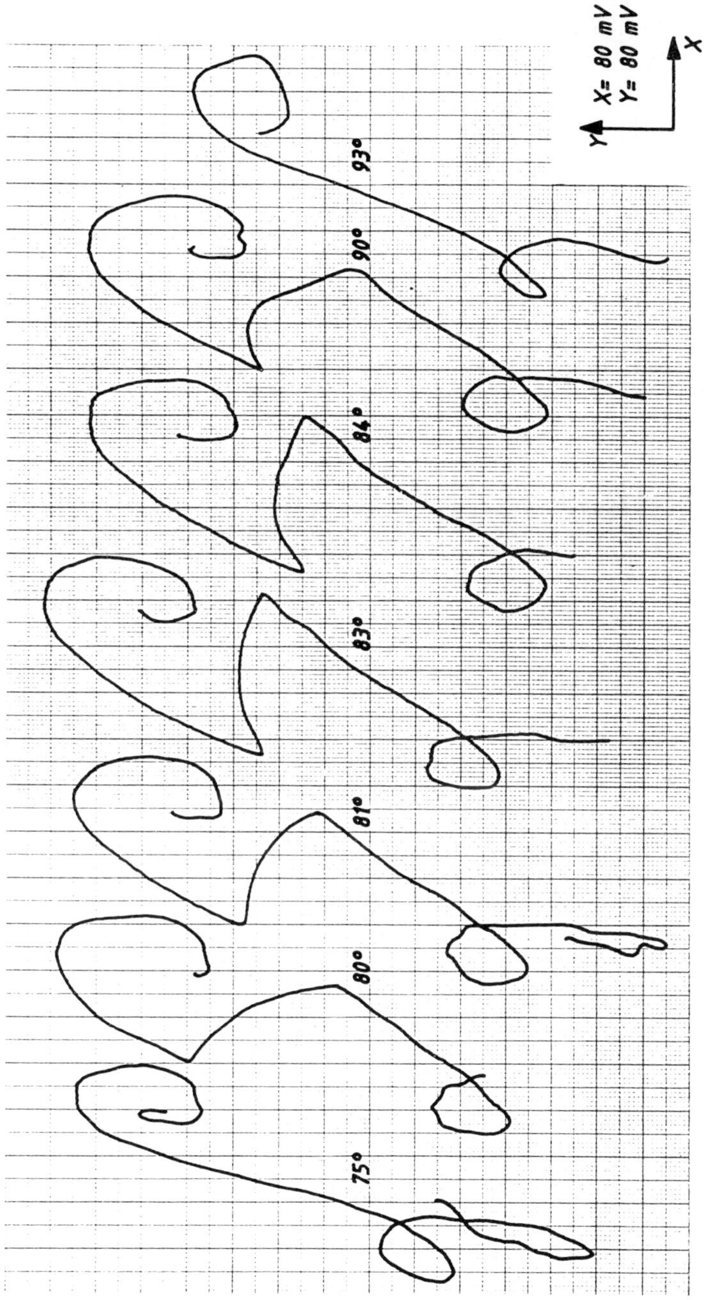

INCIDENCE DE L'ANGLE

Figure 5

This illustration is printed
from the best available original

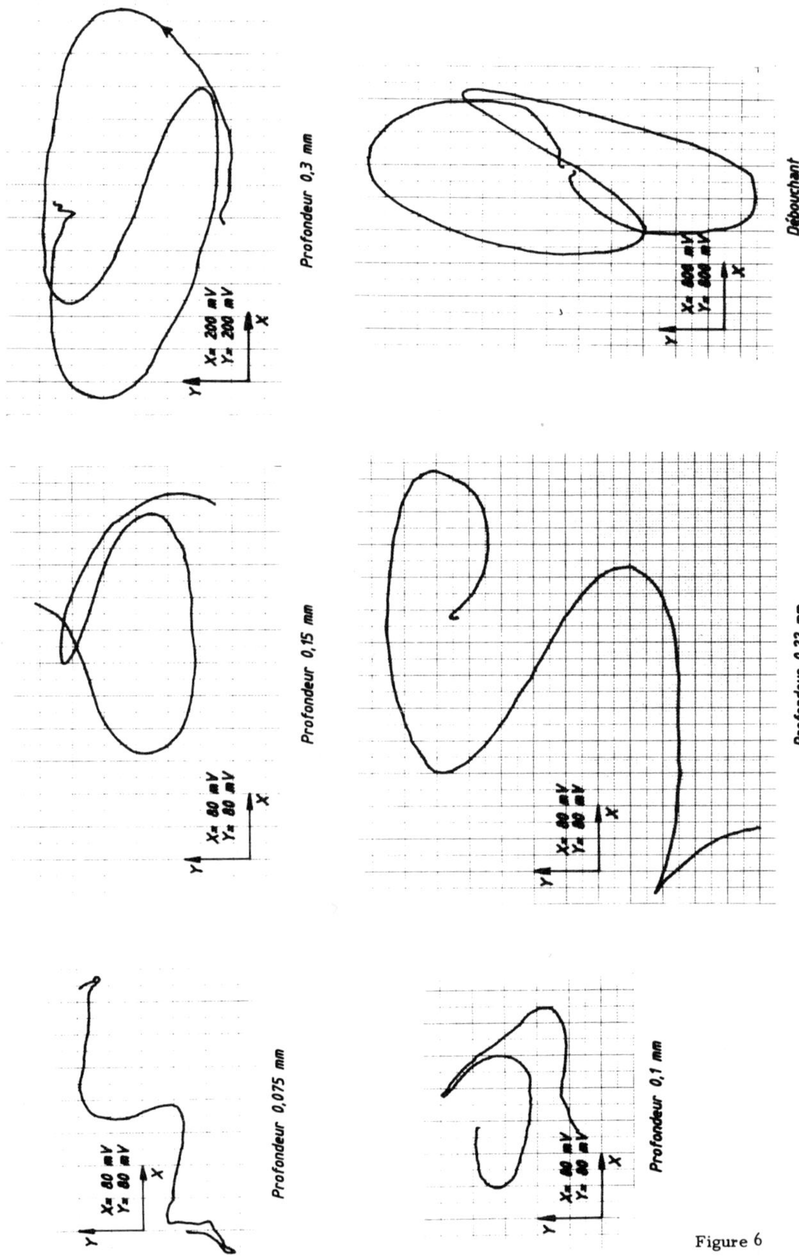

Figure 6

This illustration is printed from the best available original

Enfin, la dernière signature présentée (figure 7) provient de l'examen d'un défaut obtenu par fatigue entraînée par une pression pulsée d'eau vers 600 bars. Ces contraintes cyclées ont entraîné une fissuration traduite par une fuite inexistante à une pression inférieure à 60 bars.

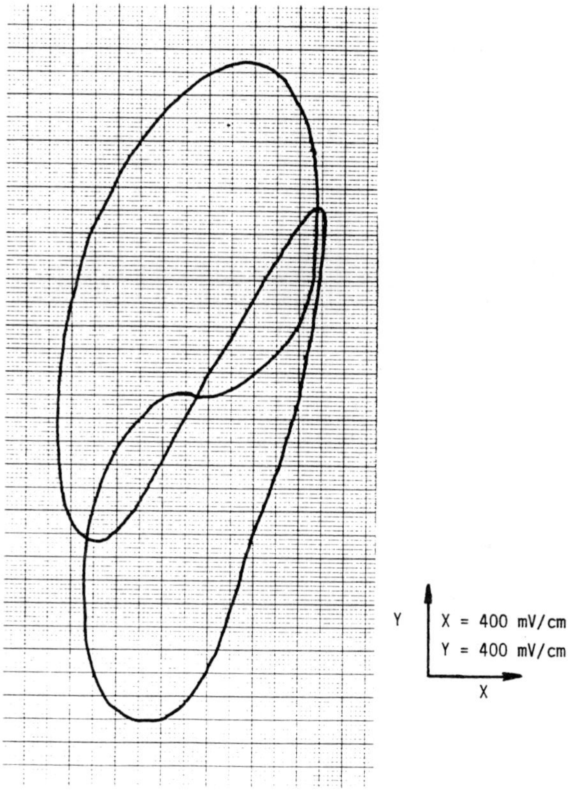

Figure 7

This illustration is printed from the best available original

interférence que l'emploi de sonde ponctuelle ne peut que minimiser.

Quant à l'analyse de la signature proprement dite, image du type de défaut, elle peut découler d'une représentation interne ou d'une représentation externe. La représentation interne implique une corrélation des signaux au système qui le génère alors que la représentation externe consiste en une description directe du signal. Les défauts naturels étant loin d'avoir des formes géométriques simples, une approche par représentation interne ne peut être que complexe, voire impossible. Par contre, une comparaison des signatures de défauts de nature inconnue à celles de défauts de nature connue, pourrait constituer un moyen de reconnaissance théoriquement automatique, aux réserves d'appareillage près d'une part, et d'un nombre raisonnable de signatures possibles d'autre part.

Bibliothèque. Outre les signatures des défauts artificiels calibrés du type de ceux dont on dispose, il est envisageable de les compléter par les signatures de défauts réels obtenus par fatigue ou par irradiation et dont la nature serait précisée a postériori par examens destructifs en Laboratoires Chauds (micrographies, ...). Cette méthode impliquerait la constitution d'une bibliothèque évolutive, en fonction des divers types de défauts décelés lors de la vie d'un crayon en pile. Si cette bibliothèque n'est pas trop importante, une signature de défaut détecté serait comparée systématiquement aux signatures des défauts alors en bibliothèque.

Méthode d'analyse. Toujours dans l'optique de comparer la signature d'un défaut à identifier aux signatures de défauts en bibliothèque, nous nous trouvons amenés à déterminer des coefficients de corrélation entre chacune des fonctions associées aux défauts "étalons" et la fonction relative au défaut inconnu.

Le programme dont nous disposons actuellement utilise différentes méthodes de corrélation, notamment dans le domaine temporel et fréquentiel. Il tient compte des translations possibles lors de l'enregistrement des X_i (t) et Y_i (t), du filtrage éventuel et de la norme de chacune des fonctions. Les entrées peuvent être directes ou via une table à digitaliser.

Les essais ont été effectués sur les signatures des défauts artificiels, l'une supposée inconnue étant comparée à toutes les autres.

A partir des matrices de corrélations obtenues, voici quelques exemples de coefficients trouvés sur des signatures de défauts différents, étant entendu que pour k = 1, les défauts sont théoriquement identiques (figure 8).

Installation Prototype et Schéma
de l'Installation en Cours de Realisation
(figure 9)

Sur la tête de lecture, à la sonde de courants de Foucault sont associés

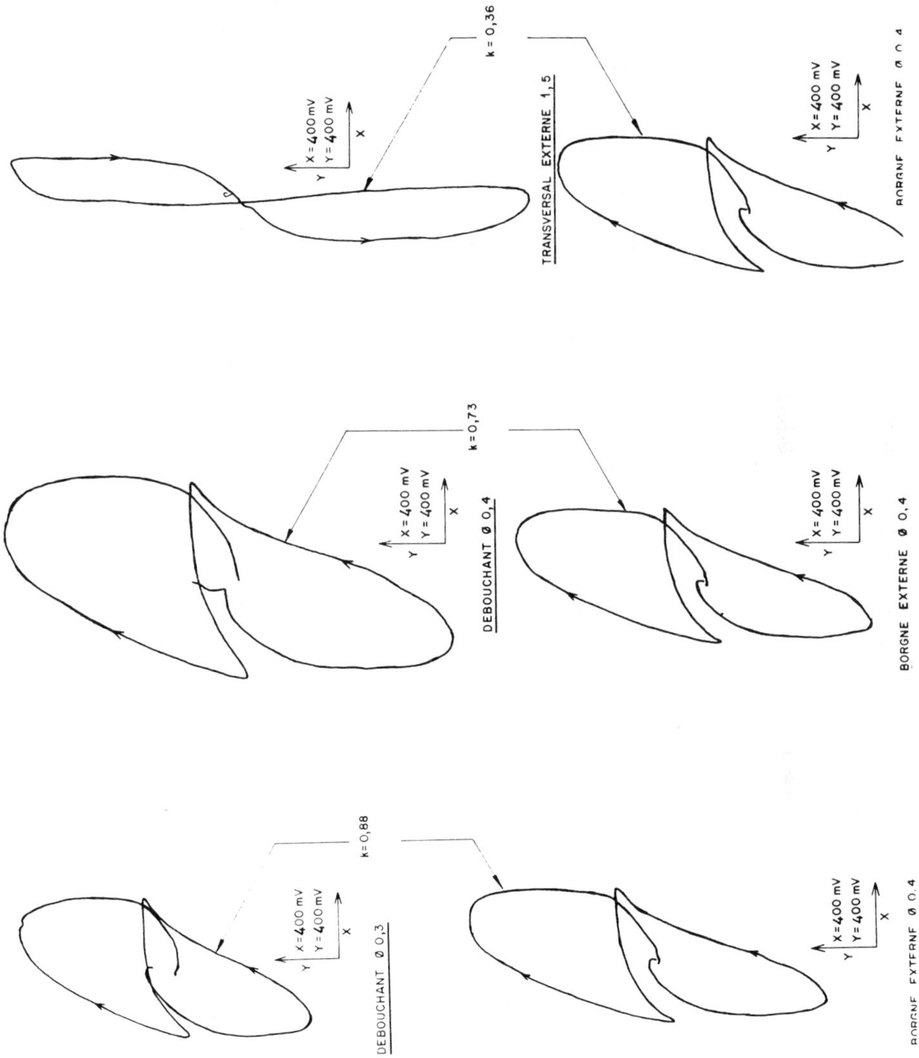

Figure 8

This illustration is printed
from the best available original

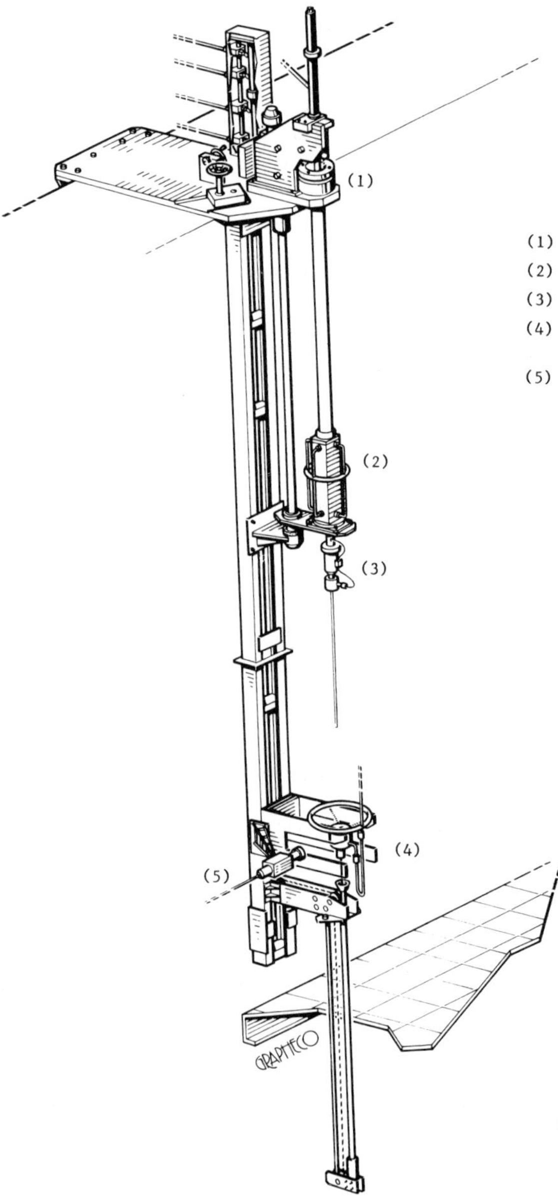

(1) plateau tournant

(2) guidage hydraulique

(3) pince porte-crayon

(4) tête flottante porte-capteurs

(5) caméra

Figure 9

deux palpeurs pour effectuer les mesures de déformation du crayon. La quasi absence de gradient thermique due à la circulation d'eau entraîne une précision de l'ordre de 5 µm sur la mesure de diamètre.

Quant à l'installation définitive, elle comportera en outre l'ensemble vidéo dont la mise en œuvre ne présente pas de difficultés méritant d'être soulignées.

Enfin, signalons que le banc sera couplé à un enregistreur magnétique multipiste et magnétoscope afin d'être à même de réexaminer des données précédemment acquises, pour compléter une interprétation ou de nouveaux résultats.

CONCLUSIONS

L'étude entreprise a permis de montrer que :

- des contrôles de crayons combustibles par courants de Foucault sont possibles en piscine, malgré une puissance résiduelle élevée, sans apporter de contraintes mécaniques à la gaine et interférer sur son refroidissement,
- la sensibilité de détection est élevée malgré certaines contraintes liées à des anomalies géométriques du crayon,
- le mode d'exploration sur coussins d'eau éliminant les frottements, la vitesse d'exploration n'est limitée que par des critères de sécurité et le mode d'acquisition des données,
- la tête de lecture mise au point est utilisable sur n'importe quelle longueur de crayon d'un type donné (17 x 17) ; l'emploi d'une autre tête serait nécessaire pour des crayons du type 15 x 15,
- les sondes ponctuelles, beaucoup plus sensibles que les bobines encerclantes, donnent des signatures de défauts nettement plus caractérisées. En outre, eu égard au faible volume examiné, la discrimination entre deux défauts voisins ne peut que s'accroître,
- l identification "automatique" du type de défaut détecté par l'informatique reste un sujet à approfondir et l'on ne saurait être trop prudent à ce stade. Néanmoins, il semble possible d'envisager la constitution d'une bibliothèque de signatures pour lesquelles les défauts correspondants seraient ultérieurement identifiés par examens destructifs en Laboratoires Chauds. La méthode proposée donne des résultats encourageants et différentes méthodes mathématiques sont en cours d'essais.

Ainsi, dans le cadre des études sur la modélisation du combustible en transitoires normal ou anormal, OSIRIS disposera début 1983 d'un nouveau bâti multiple immergé pour le contrôle des gaines des crayons irradiés dans la boucle ISABELLE 4 crayons défournable et rechargeable en piscine.

ANLAGE ZUR UNTERSUCHUNG DES RISSWACHSTUMSVERHALTENS VON DRUCKBEHÄLTERSTAHL UNTER SIEDEWASSERREAKTOR-BEDINGUNGEN

D. Anders, J. Ahlf

GKSS-Forschungszentrum Geesthacht GmbH

2054 Geesthacht, Bundesrepublik Deutschland

ZUSAMMENFASSUNG

Es ist eine Bestrahlungseinrichtung entwickelt worden, die es gestattet das Rißwachstumsverhalten unter Reaktorbetriebsbedingungen zu untersuchen.

Der Bestrahlungseinsatz besteht im wesentlichen aus einem stabilen Rahmen mit einer hydraulischen Belastungseinrichtung, dem Probenstrang und der Meßinstrumentierung. Die Kraft zur Belastung des Probenstranges wird von einer 100 kN-Materialprüfmaschine geliefert; sie bewegt einen Kolben in dem Druckzylinder, der über eine Rohrleitung mit den Belastungsbälgen des Bestrahlungseinsatzes verbunden ist.

Die Meß- und Regeleinrichtungen sowie ein Tischrechner zur Erfassung und Aufbereitung der Daten sind in der Reaktorwarte untergebracht.

Es ist ein Vorversuch durchgeführt worden, der zur Erprobung der Wegaufnehmer sowie der Ermittlung von Daten über die Temperatur- und Neutronenflußverhältnisse in der Bestrahlungsposition diente.

ABSTRACT

An experimental device is developed to investigate the crack growth behaviour of RPV steel specimens under service conditions. It will be installed in the experimental power station VAK-Kahl (BWR, 16 MWe).

P. von der Hardt and H. Röttger (eds.), Irradiation Technology, 133–142.

The in pile part is composed of a stable frame with a hydraulically actuated load mechanism, the specimen chain and a measuring instrumentation. The specimen chain, fastened between load mechanism and a lower fixing point at the frame, is made up of five compact tensile specimens (CT40) and the associated connecting links. Specimen strain, crack opening and temperature are measured; for neutron dose monitoring activation wires are disposed.

Out of pile, in the reactor hall, the hydraulic loading system is installed. The loading force is generated by a 100 kN-material testing machine; it moves a piston in the control cylinder, which is connected to the loading bellows of the in pile section.

The measuring and control equipment and a desk computer serving for data preparation and reduction is placed in the reactor control room.

1. EINLEITUNG

Reaktordruckbehälter sind im Lauf ihrer Betriebszeit einer großen Anzahl von Belastungsänderungen ausgesetzt. Die Ursachen dafür sind vielfältig; sie reichen von Druckproben vor Inbetriebnahme über Anfahr- und Abschaltvorgänge, Leistungsschwankungen bis zu Pumpenvibrationen o.ä.

Wechselbelastungen können unter bestimmten Voraussetzungen im Lauf der Zeit zu Ermüdungsrissen führen.

Das Vorgehen nach dem Entdecken eines solchen Fehlers ist nach dem ASME Boiler and Pressure Vessel Code [1] geregelt. Die Prozedur beinhaltet eine Ermüdungsbruchanalyse aus deren Resultat die zu ergreifenden Maßnahmen abgeleitet werden. Falls eine Reparatur nicht möglich ist, liefert eine bruchmechanische Betrachtung die Grundlagen für die Abschätzung der verbleibenden Zeit für einen sicheren Weiterbetrieb der Komponente; ggf. kann das Resultat auch zu Anordnungen für eine geänderte Betriebsweise führen.

Für eine derartige Beurteilung ist die Kenntnis des Rißwachstumsverhaltens in dem zu begutachtenden Werkstoff die entscheidende Voraussetzung. Belastbaren Daten darüber kommt also eine erhebliche sicherheitstechnische und wirtschaftliche Bedeutung zu. Aufgrund dieser seit langem bekannten Tatsache sind bereits viele Untersuchungen zu diesem Thema durchgeführt worden. Ein von der IAEA im vergangenen Jahr veranstaltetes Specialists' Meeting [2] hat jedoch gezeigt, daß auf diesem Gebiet noch Fragen offen sind.

Der Rißfortschritt wird bei Wechselbelastung von zahlreichen Parametern beeinflußt; dazu gehören: der zeitliche Verlauf der Belastung, das Umgebungsmedium, die Zusammensetzung des Werkstoffs sowie die Neutronen und γ-Strahlung des Reaktors, die sich über die Änderung der Materialstruktur bzw. über die radiochemische Veränderung des Wassers auswirken kann.

Zwischen diesen - nicht vollständig aufgezählten - Einflußparametern gibt es gegenseitige Beeinflussungen, so daß sie nicht isoliert zu ermitteln sind.

Das Untersuchungsziel - die Ermittlung der Rißwachstumsrate in RDB-Stahl unter betriebsnahen Bedingungen - läßt sich deshalb nur in einem Bestrahlungsexperiment erreichen, bei dem alle Einflußparameter simultan wirken.

Das Experiment wird im Versuchsatomkraftwerk Kahl (SWR, 60 MW_{th}) durchgeführt. Dort liegen relativ günstige Platzverhältnisse vor, die für einen derartigen Versuch erforderlich sind, weil bruchmechanische Untersuchungen erst bei einer Mindestprobengröße aussagefähige Ergebnisse liefern.

2. VERSUCHSBEDINGUNGEN

Der Bestrahlungseinsatz wird am Kernrand innerhalb des Kernmantels untergebracht. Der dort zur Verfügung stehende Raum gestattet die Unterbringung von 5 Proben der Größe CT40. Aus den für den Versuchsbeginn vorgesehenen Spannungsintensitäten K_{max} = 1053 $N/mm^{3/2}$; K_{min} = 733 $N/mm^{3/2}$ ergeben sich mit der Probengröße die zugehörigen Lasten F_{max} = 45,2 kN und F_{min} = 31,6 kN. Es ist ein sinusförmiger Belastungsverlauf mit einer Frequenz von 16,7 mHz = 1 min^{-1} vorgesehen. Mit der Lastwechselzahl von ca. $1{,}4 \cdot 10^5$ ergibt sich eine Versuchsdauer von ca. 3 Monaten. In dieser Zeit wird bei dem in der Bestrahlungsposition herrschenden Fluß eine Neutronendosis von ca. $2 \cdot 10^{19}$ n/cm^2 akkumuliert. Die Probentemperatur soll 288 ^oC betragen.

3. VERSUCHSAUFBAU

Die Versuchsanlage (Abb. 1) läßt sich grob in die Bereiche
- Bestrahlungseinsatz
- Außenanlage
- Meßwerterfassung und Verarbeitung

gliedern.

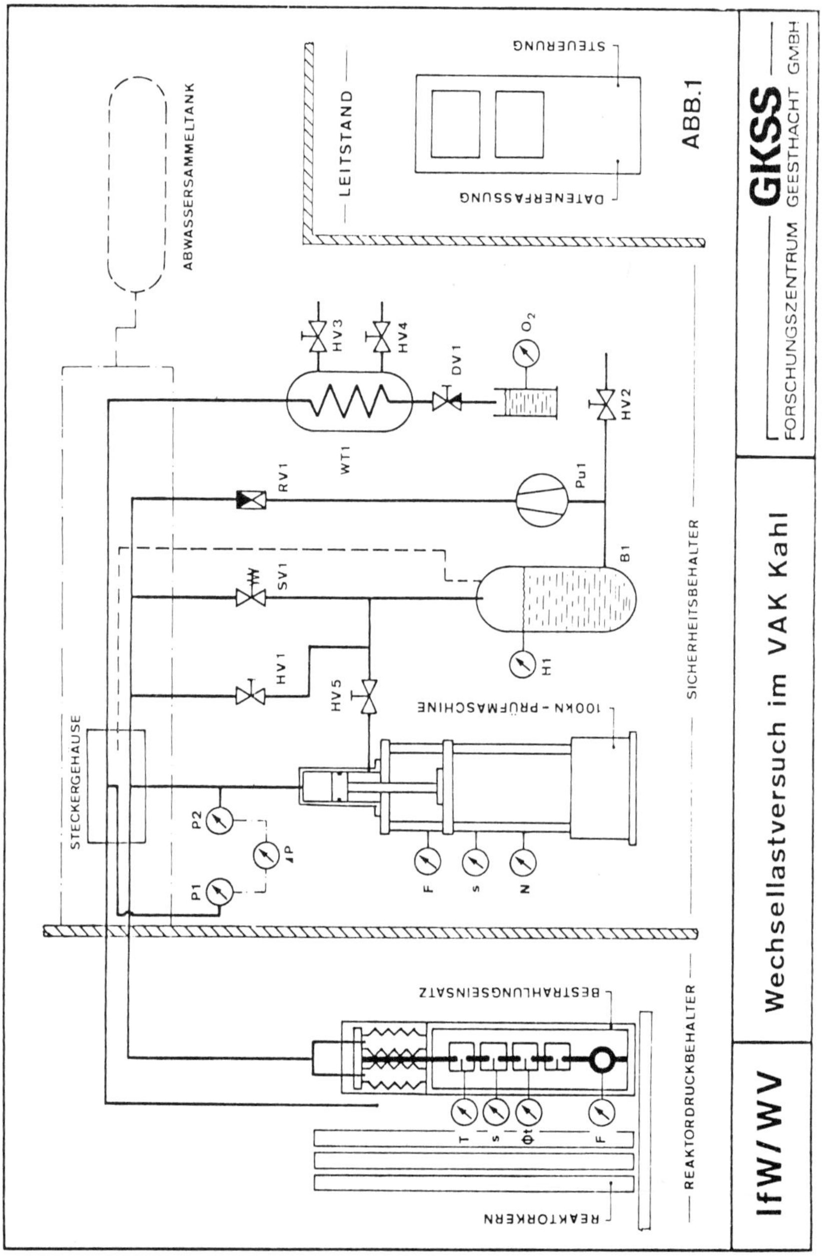
ABWASSERSAMMELTANK
LEITSTAND
STEUERUNG
DATENERFASSUNG
ABB.1
HV3
HV4
DV1
O_2
WT1
RV1
Pu1
HV2
SV1
B1
HV1
HV5
H1
SICHERHEITSBEHALTER
100KN - PRÜFMASCHINE
STECKERGEHAUSE
P1
P2
ΔP
F
s
N
BESTRAHLUNGSEINSATZ
REAKTORDRUCKBEHALTER
REAKTORKERN
T
s
Φt
F
IfW/WV
Wechsellastversuch im VAK Kahl
GKSS
FORSCHUNGSZENTRUM GEESTHACHT GMBH

3.1 Bestrahlungseinsatz

Der Bestrahlungseinsatz (Abb. 2) besteht im wesentlichen aus einem stabilen Rahmen, der hydraulischen Belastungseinrichtung, dem Probenstrang und der zugehörigen Instrumentierung.

Der Belastungsmechanismus wird von zwei Doppelkompensatoren gebildet, die mit einem inneren Überdruck gegenüber dem Reaktorbetriebsmedium beaufschlagt werden. Sie stützen sich auf der oberen Tragplatte ab und wirken über eine Traverse und die beiden Zugstangen auf den Probenstrang, dessen Widerlager die untere Tragplatte bildet.

Der für die Aufbringung der maximalen Zugkraft (ca. 45 kN) erforderliche Überdruck in den Kompensatoren beträgt ca. 50 bar (Absolutdruck ca. 120 bar).

Der Probenstrang besteht aus 5 Kompaktproben der Größe CT40, die mit Verbindungsklauen und Bolzen aneinandergekoppelt sind. Die äußeren Brückenglieder lassen nur einen vorgegebenen Maximalwert (1,5 mm) für die Rißöffnung zu und gewährleisten den Weiterbetrieb des Experimentes bei eventuellem Versagen einer Probe. Zur Messung der Rißöffnung, aus der auf die Rißlänge geschlossen wird, ist jede Probe an der Stirnseite mit zwei Wegaufnehmern (Abb. 3) bestückt. Davon dient einer als Reserve für den Fall des Versagens eines Aufnehmers.

Ein unabhängiger Nachweis darüber, daß die Rißlänge einen bestimmten Wert erreicht hat, ist durch feine Bohrungen möglich, die durch die Rißebene führen. In die Bohrungen sind Mantelmeßleitungen eingeschweißt, deren Isolationswiderstand überwacht wird. Wenn die Bohrung vom Riß erreicht wird, dringt Wasser ein und verschlechtert den Isolationswiderstand, der überwacht wird. Z.Zt. wird noch geprüft, ob sich die Bohrungen (Funkenerosion 0,3 mm Ø) nicht störend auf den Rißfortschritt auswirken können.

Zur Ermittlung der Probentemperatur ist jede Probe mit einem Thermoelement bestückt.

Die Neutronendosis wird mit Aktivierungsdrähten bestimmt, die in den Proben untergebracht sind; sie werden nach Abschluß des Experimentes ausgewertet.

Im unteren Bereich des Probenstranges befindet sich der Kraftaufnehmer. Er besteht aus einem Ring, dessen Verformung ebenfalls mit Wegaufnehmern bestimmt wird. Die Wegaufnehmer (Differentialtransformatorprinzip), die bei kleinen Abmessungen und extremen Umgebungsbedingungen eine hohe Meßgenauigkeit haben müssen, wurden bei GKSS entwickelt.

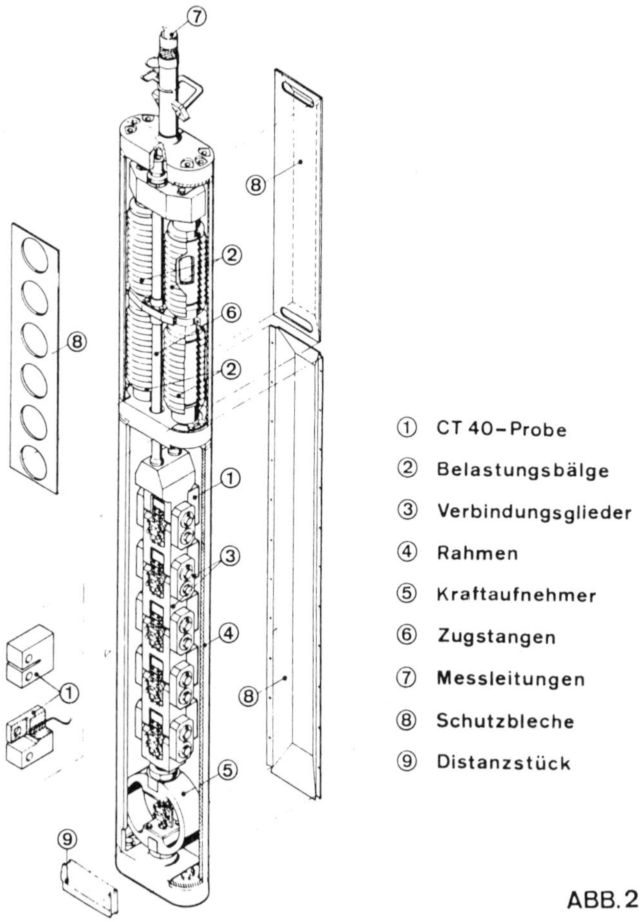

ABB. 2

BESTRAHLUNGSEINSATZ FÜR RISSWACHSTUMSUNTERSUCHUNGEN

IfW / WV

GKSS FORSCHUNGSZENTRUM GEESTHACHT GMBH

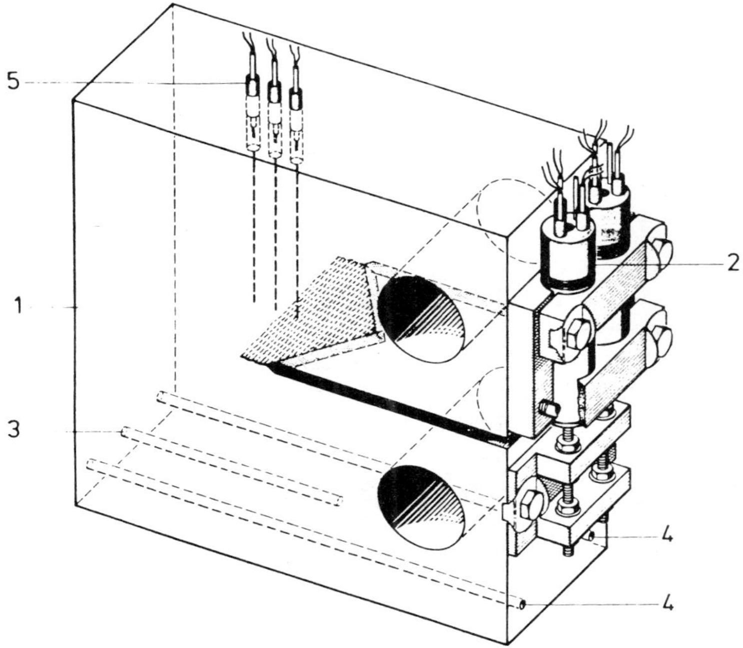

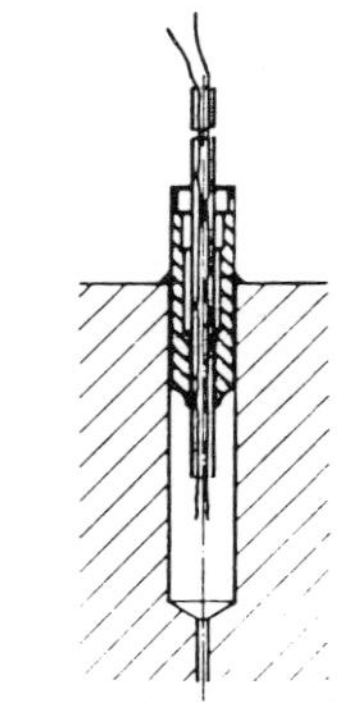

Einzelheit Teil 5

1.......Kompaktprobe CT 40
2.......Wegaufnehmer LVDT
3.......Bohrung für Thermoelement
4.......Bohrung für Aktivierungsdrähte
5.......Rißdetektoren

ABB.3

IfW/WV	Wechsellastversuch im VAK Kahl Instrumentierte Kompaktprobe	GKSS FORSCHUNGSZENTRUM GEESTHACHT GMBH

Die Meßleitungen, es handelt sich um mineralisolierte Metallmantelkabel, werden ebenso wie die Zuleitungsrohre zu den Kompensatoren über eine Flanschverbindung aus einem Stutzen des Druckbehälters herausgeführt.

3.2 Außenanlage

Die erforderliche Kraft zur Belastung des Probenstranges wird von einer 100-kN-Prüfmaschine geliefert, die in dem Reaktorsicherheitsbehälter aufgestellt ist. Sie bewegt einen Kolben in dem auf der oberen Quertraverse montierten Zylinder, der über eine dünne Rohrleitung mit den Belastungsbälgen des Bestrahlungseinsatzes verbunden ist.

Das von einem Sollwertgenerator gelieferte sinusförmige Signal wird mit der im Probenstrang gemessenen Kraft verglichen. Entsprechend der Abweichung wird dann der Druck innerhalb des Belastungssystems durch Verfahren der Traverse gesteuert; er bewegt sich zwischen ca. 105 und 120 bar.

Der Druck innerhalb des Belastungssystems und der Reaktorbetriebsdruck werden gemessen. Die Differenz aus beiden Drücken ist ebenfalls ein Maß für die Kraft; dieses Signal dient als Reserve für den Fall, daß der Kraftaufnehmer im Probenstrang versagt. Ein weiteres bedingt nutzbares Referenzsignal für die Kraft wird von einem Aufnehmer an der Traverse der Prüfmaschine geliefert.

Das Sicherheitsventil begrenzt den Belastungsdruck; es bläst in einen Vorratsbehälter ab. In denselben Behälter wird das Leckwasser aus dem Belastungszylinder geleitet. Es kann von dort mit einer Hochdruckdosierpumpe in das System zurückgespeist werden.

Zur Probennahme für die Kontrolle des Gehalts an gelöstem Sauerstoff im Primärwasser dient eine Leitung, die vom Bestrahlungseinsatz über einen Wärmeaustauscher zu dem Probenventil führt.

3.3 Meßwerterfassung und Datenverarbeitung

Zur Verarbeitung der Versuchsdaten wird ein Kleinrechner (HP 9826) mit Meßstellenumschalter, Drucker und Magnetplattenspeicher eingesetzt.

Die Wegaufnehmer werden stündlich einmal für die Dauer von 400 Sekunden mit einer Frequenz von 2 Hz abgefragt. Die Meßsignale werden mit den Kalibrierwerten der einzelnen Aufnehmer in Wege umgerechnet und zusammen mit den übrigen Daten (Datum, Uhrzeit, Lastwechselzahl, Temperaturen, Systemdrücke) auf Magnetplatten abgespeichert.

Mit Hilfe eines Ausgleichsprogramms, das einen Sinus durch die Meßpunkte legt, werden die Meßdaten der Wegaufnehmer reduziert. Die so ermittelten Funktionswerte werden dann zusammen mit anderen wichtigen Meßwerten (s.o.) zur Datensicherung und Kontrolle auf einem Protokolldrucker ausgegeben.

Eine Reihe von Betriebswerten wird außerdem parallel zum Rechner auf Kompensationsschreibern registriert.

4. STAND DER ARBEITEN

Um Erfahrungen für das Originalexperiment zu sammeln, wurde zunächst ein stark vereinfachter sog. Dummy-Einsatz gebaut, der sich seit November '81 im Reaktor in Kahl befindet. Hauptgründe für diesen vorbereitenden Versuch sind:

- Die Kühlverhältnisse in der Bestrahlungsposition waren unbekannt;
- die Wegaufnehmer mußten einem Test unter Originalbedingungen unterzogen werden;
- das Profil des Neutronenflusses war zu bestimmen und
- es waren Erfahrungen für die Handhabung des Einsatzes zu sammeln.

In dem Dummy-Einsatz sind 2 Proben untergebracht - eine im Flußmaximum und eine im äußeren Bereich. Die Zentralprobe ist mit zehn, die andere mit 2 Thermoelementen bestückt. Bei den z.Zt. herrschenden Betriebsbedingungen wird in der Zentralprobe eine Maximaltemperatur von 268°C und eine Minimaltemperatur von 264,3 ^{o}C gemessen. Die maximale Temperaturüberhöhung gegenüber dem Primärmedium beträgt 5,2 K. Daraus ergibt sich, daß die Betriebsbedingungen des Reaktors, die z.Zt. aufgrund anderer Experimente herabgesetzt sind, für den Originalversuch wieder angehoben werden müssen. Die entsprechenden Genehmigungen sind beantragt.

Ein eingebauter Wegaufnehmer lieferte ca. 8 Monate ein unverändertes Ausgangssignal und wurde dann defekt.

Der Einbau des Dummy-Einsatzes in den Reaktor und die Herausführung der Meßleitungen bereitete keine Schwierigkeiten. Auch die Übertragung der Meßwerte in die Schaltwarte war problemlos.

Der Ausbau dieses Versucheinsatzes wird im Oktober '82 erfolgen.

Mit dem Wechsellastversuch soll nach gründlicher Vorerprobung in einem dafür vorgesehenen Autoklaven im Mai 1983 begonnen werden.

LITERATUR

[1] ASME Boiler and Pressure Vessel Code, Section XI, Rules for Inservice Inspection of Nuclear Power Plant Components

[2] IAEA - Specialists' Meeting on Subcritical Crack Growth (13.-15. Mai 1981), Freiburg, FRG, Proceedings

[3] ASTM E 647-81, Standard Test Method for Constant-Load-Amplitude Fatigue Crack Growth Rates above 10^{-8} m/Cycle.

PREDICTION OF POWER AND ANALYSIS OF RESULTS FROM THERMOHYDRAULIC AND NEUTRONIC MEASUREMENTS IN A PRESSURISED-WATER-LOOP EXPERIMENT IN THE DIDO REACTOR AT HARWELL

T.D.A. KENNEDY

ATOMIC ENERGY RESEARCH ESTABLISHMENT,

HARWELL, OXON, OX11 ORA, ENGLAND

1. INTRODUCTION

The experiment irradiated a fissile fuel pin simulating typical PWR conditions of pressure, heat flux, water flow, and geometry. The pressurised water loop[1] had a parallel out-of-pile channel with electrically heated test section, which ran under very similar flow and heat-flux conditions, to act as a control on radiation effects on the water chemistry. The fuel pin, containing 1.05%-U^{235} enriched fuel, was irradiated for 103.4 days in a mean measured thermal-neutron flux of 1.1×10^{14} n/cm²s. U^{235} burn-up was ≈ 50% at the end of the experiment, and (Pu^{239} + Pu^{241}) build-up was of similar magnitude; any calculation of fissile power from neutron-flux measurements therefore required accurate knowledge of these parameters. Thermohydraulic measurements were used at intervals to calculate the pin power, but because the experiment required some bulk boiling of outlet, these did not give a continuous measure of power. Continuous measuring of fissile power was therefore restricted to neutron-flux measurements. The two independent methods of measuring power gave good agreement at the times that thermohydraulic measurements could be taken. This gave confidence in the use of the self-powered neutron-flux detectors (SPNDs) to provide a method of measuring fissile power.

The burn-up of U^{235} and build-up of Pu isotopes was predicted by solution of the differential equations governing the processes, using information obtained from an actinide analysis of a similar fuel pin irradiated in the same experiment hole in DIDO. Absorption cross-sections were fitted to the relevant equations to obtain agreement with the measured results from the actinide analysis. These cross-sections thus include the appropriate resonance integrals for the neutron spectrum in the correct fuel-element position in DIDO.

P. von der Hardt and H. Röttger (eds.), Irradiation Technology, 143–160.

The effective reaction area for fission was predicted to remain constant within 1% over 300 days irradiation at a mean thermal-neutron flux of 1.1 x 10^{14} n/cm²s - the burn-up of U^{235} being almost exactly balanced by the build-up of Pu^{239} and Pu^{241} - up to the time when 80% of the U^{235} has been burnt up. This prediction is substantiated by the measured fissile power from the experiment fuel pin over the 103 irradiation days that the pin remained in the reactor. It is interesting to note that this effect was not designed into the fuel enrichment, which was designed to give a required power level at the start of the experiment - a welcome inversion of 'Murphy's Law'.

2. THERMOHYDRAULIC MEASUREMENTS

Fig. 1 shows the main features of the DIDO Water Loop (D.W.L.) set up for the experiment. This paper is concerned only with the In-Pile Test Section (I.P.T.S.). The flow through this section is measured by the calibrated orifice-plate flowmeter FI 39. The heat transferred to the water in the I.P.T.S. is calculated from the measured water flow, and inlet and outlet temperatures. The temperatures are measured by calibrated resistance-bulb thermometers, TI22 at inlet and TI2 at outlet. Water pressure is measured at outlet by the transducer PDR 2.

Heat added to I.P.T.S. water, $Q^1_{TS} = \dot{m}\,\bar{C}_p(T_{out} - T_{in})$, - (1)

where $\dot{m}$ = Water mass flow (Kg/s)

$\bar{C}_p$ = Mean specific heat of water (J/Kg K)

T_{in} = Water temperature at inlet to I.P.T.S. (°C)

T_{out} = Water temperature at outlet from I.P.T.S. (°C).

Because there is a significant heat loss from the I.P.T.S., this must be measured independently before the reactor is taken to power, with the loop water at operating temperatures.

Heat loss, $Q^1_L = \dot{m}\,\bar{C}_p\,(T_{in} - T_{out})$ - (2)

The nuclear heating, Q^1_n, is made up of two components; gamma heating, Q^1_γ, and fissile heating, Q^1_f. To establish the γ-heating rate, the experiment was run for one cycle with a non-fissile fuel pin, and the heat added to the water, Q^1_{TS}, found.

$$Q^1_\gamma = Q^1_{TS} + Q^1_L \qquad - (3)$$

Finally, the fissile fuel pin was loaded for the main experiment, and the fissile power, Q_f', was calculated from:

$$Q_f^1 = Q_{TS}^1 + Q_L^1 - Q_\gamma^1 - Q_e^1, \qquad - (4)$$

where Q_e^1 = I.P.T.S. electrical trim-heater power.

The γ-heating rate was measured at 13.0 KW, and the heat loss was found as a function of loop-H_2O to reactor-D_2O temperature difference:

$$*Q_L^1 = 0.038\ (\bar{T}_H - \bar{T}_D), \qquad - (5)$$

where $\bar{T}_H$ = mean temperature of light water in I.P.T.S.

$\bar{T}_D$ = mean temperature of heavy water flowing past I.P.T.S.

3. NEUTRONIC MEASUREMENTS

Because pin power could only be calculated from thermohydraulic measurements with loop water subcooled throughout, these measurements were only made for limited periods, mainly at start-up and shutdown for each reactor-cycle (≈ 25 days). The main chemistry experiment required saturated boiling over at least 25% of the fuel pin, and these conditions preclude an accurate calculation of power from thermohydraulic measurements. Operating fissile power was therefore obtained from neutron-flux measurements made with six SPND monitors spaced at three axial levels around the experiment fuel pin. Fig. 2 shows the axial positioning of these monitors-in the top right inset. The general axial flux shape in the DIDO reactor is known(2), and such a curve is fitted through the mean measured flux at each level. Reference to Fig. 2 shows that the use of 'Simpson's rule' for integrating under the curve gives a close approximation to the integral under the given flux shape. This approach was therefore adopted to obtain mean fluxes throughout the experiment.

The thermal-neutron flux was measured on the outside of the fuel pin, and an allowance was then required for flux depression in the fuel pin. This was computed by the Harwell code 'RAFTER'(3), which gave a depression factor of 0.909.

*This expression is valid for helium in the gas space in the experiment thimble. The thimble gas was changed during the experiment to neon, and the applicable expression is then:

$$Q_L^1 = 0.018\ (\bar{T}_H - \bar{T}_D) \qquad - (6)$$

The fissile power was computed from:

$$Q_f^1 = 0.909\ \bar{\Phi}\left(\sum_{i=1}^{3} n_i\ \sigma_{f,i}\right) \times 3.092 \times 10^{-11} \text{ watts,} \qquad -\ (7).$$

where $\bar{\Phi}$ = mean measured flux (n/cm² s)

n_i = No. of atoms of fissile isotope i

$\sigma_{f,i}$ = Westcott cross-section for fission of isotope i.

The relevant isotopes are U^{235}, Pu^{239} and Pu^{241}. At the start of irradiation, the fuel pin contained only U^{235}. The mass of U^{235} was measured at 3.675g.

$$\text{Then} \quad n_o^{235} = \left[\frac{6.023 \times 10^{23}}{235} \times 3.675\right] = 9.419 \times 10^{21}, \qquad -\ (8)$$

where n_o^{235} = No. of atoms of U^{235} at start of irradiation

σ_f^{235} = 565 barns = 565×10^{-24} cm²

$$Q_f^1 = 0.909\ \bar{\Phi}\ (9.419 \times 10^{21} \times 565 \times 10^{-24}) \times 3.092 \times 10^{-11}$$

$$= 1.496 \times 10^{-10}\ \bar{\Phi}. \qquad -\ (9)$$

Equation (9) holds only at the start of irradiation. As irradiation proceeds, U^{235} is burnt up, and Pu^{239} and Pu^{241} are built up. A similar experimental fuel pin, irradiated in D.W.L. for 98.17 days at a mean perceived thermal-neutron flux of 1.02×10^{14} n/cm² s was analysed by the Harwell Chemistry Division Actinide Analysis Group[4]. The pin contained 1.10% U^{235} at the start of irradiation, and 0.60% U^{235}, 0.35% Pu^{239}, and 0.05% Pu^{241} at the end. For the DIDO neutron-flux spectrum, fission cross sections for these isotopes are[5]: U^{235} - 565 barns, Pu^{239} - 713 barns and Pu^{241} - 1200 barns. With this information, it is possible to normalise total fission reaction areas for each isotope at the end of 98.17 days irradiation to that of U^{235} at the start of irradiation. The numbers of atoms of each isotope are calculated from the following equations:

$$\frac{dn^{235}}{dt} = -\lambda_{f+a}^{235}\ n^{235} \qquad -\ (10)$$

$$\frac{dn^{236}}{dt} = \lambda_a^{235}\ n^{235} - \lambda_a^{236}\ n^{236} \qquad -\ (11)$$

$$\frac{dn^{239}}{dt} = \lambda_a^{238}\, n^{238} - \lambda_{f+a}^{239}\, n^{239} \qquad - (12)$$

$$\frac{dn^{240}}{dt} = \lambda_a^{239}\, n^{239} - \lambda_a^{240}\, n^{240} \qquad - (13)$$

$$\frac{dn^{241}}{dt} = \lambda_a^{240}\, n^{240} - \lambda_{f+a}^{241}\, n^{241} \qquad - (14)$$

$$\frac{dn^{242}}{dt} = \lambda_a^{241}\, n^{241} - \lambda_a^{242}\, n^{242}, \qquad - (15)$$

where $\lambda_a^m = \Phi\sigma_a^m$, $\lambda_{f+a}^m = \Phi\sigma_{f+a}^m$

These equations neglect the intermediate steps of the decay of U^{239}, Np^{239} and Np^{240}, and, instead, assume that Pu^{239} and Pu^{240} are formed directly by neutron absorption in U^{238} and Pu^{239} respectively. This simplification will not cause significant error in the solutions because the half-lives of U^{239}, Np^{239} and Np^{240} are relatively short.

The solutions to (1) and (11) are:

$$n^{235} = n_o^{235} \cdot e^{-\lambda_{f+a}^{235} t} \qquad - (16)$$

$$n^{236} = n_o^{235} \cdot \frac{\lambda_a^{235}}{\lambda_a^{236} - \lambda_{f+a}^{235}} \cdot \left[e^{-\lambda_{f+a}^{235} \cdot t} - e^{-\lambda_a^{236} \cdot t} \right] \qquad - (17)$$

and to (12) to (15) are:

$$n^{239} = \frac{ab}{(c-b)} \left[e^{-bx} - e^{-cx} \right] \qquad - (18)$$

$$n^{240} = \frac{abh}{(c-b)} \left[\frac{e^{-bx} - e^{-gx}}{(g-b)} - \frac{e^{-cx} - e^{-gx}}{(g-c)} \right] \qquad - (19)$$

$$n^{241}\ \frac{gabh}{(c-b)} \left[\frac{e^{-bx} - e^{-kx}}{(g-b)\ (k-b)} - \frac{e^{-gx} - e^{-kx}}{(g-b)\ (k-g)} - \frac{e^{-cx} - e^{-kx}}{(g-c)\ (k-c)} + \frac{e^{-gx} - e^{-kx}}{(g-c)\,(k-g)} \right] \qquad - (20)$$

$$n^{242} = \frac{lgabh}{(c-b)}\left[\frac{e^{-bx} - e^{-mx}}{(g-b)(k-b)(m-b)} - \frac{e^{-kx} - e^{-mx}}{(g-b)(k-b)(m-k)} - \frac{e^{-gx} - e^{-mx}}{(g-b)(k-g)(m-g)} + \frac{e^{-kx} - e^{-mx}}{(g-b)(k-g)(m-k)} - \frac{e^{-cx} - e^{-mx}}{(g-c)(k-c)(m-c)} + \frac{e^{-kx} - e^{-mx}}{(g-c)(k-c)(m-k)} + \frac{e^{-gx} - e^{-mx}}{(g-c)(k-g)(m-g)} - \frac{e^{-kx} - e^{-mx}}{(g-c)(k-g)(m-k)}\right] \quad - (21)$$

where $a = n_o^{238}$, $b = \lambda_a^{238}$, $c = \lambda_{f+a}^{239}$, $g = \lambda_a^{240}$, $h = \lambda_a^{239}$

$k = \lambda_{f+a}^{241}$, $l = \lambda_a^{241}$, $m = \lambda_a^{242}$ and $x = t$

Table I shows the results from an actinide analysis of a 1.1%-U^{235} enriched fuel pin irradiated in E2 fuel-element position for 98.17 days in an earlier experiment in D.W.L.

Time (s)	U^{235}	U^{236}	U^{238}	Pu^{239}	Pu^{240}	Pu^{241}	Pu^{242}
		(No. of atoms x 10^{-21})					
0.0	9.868	0.0	887.2	0.0	0.0	0.0	0.0
8.48×10^6	5.298	0.971	882.9	3.114	0.776	0.292	0.047

TABLE I

The absorption cross-sections were obtained by fitting the coefficients in equations (16) to (21) to the known values for the proportions of U and Pu isotopes shown in Table I:

σ_a^{235} = 153 barns,

σ_a^{238} = 6.86 barns, σ_a^{239} = 702 barns

σ_a^{240} = 1494 barns, σ_a^{241} = 571 barns.

These cross-sections include resonance integrals and consequently exceed the thermal-neutron cross-sections by over 50% for U^{235} and Pu^{241} and up to ≈ 400% for Pu^{240}. The equations used neglected the short-lived U^{239} and Np isotopes, which would lead to some error in their values; they are, in any case, specifically of use for the neutron spectrum existing in DIDO fuel-element positions, and not for general use. The absorption cross-section of U^{236} was taken as 5.2 barns, which is a standard value for thermal-neutron flux. This cross-section is low, and as U^{236} is an end product as far as the fissile isotopes are concerned, it will have no effect on their concentrations.

Table II shows the predicted reaction areas for each fissile isotope in the experimental fuel pin for up to 1000 days of irradiation in a mean thermal-neutron flux of 1.08 x 10^{14} n/cm²s in the E2 fuel-element position in the DIDO reactor. These reaction areas have then been normalised to the U^{235} fission reaction area at the start of irradiation, and the results are plotted for 0-250 days in Fig. 3. The surprising result is that the sum of the combined reaction areas (U^{235}, Pu^{239} and Pu^{241}) is predicted to remain constant within 1% for the first 300 days of irradiation (2.80 x 10^{21} n/cm² neutron fluence).

4. RESULTS

The experiment fuel pin was run with the cooling-water outlet below saturation temperature for three days at the start of its first irradiation cycle in the reactor. The water temperatures at inlet to and outlet from the test section, measured by resistance-bulb thermometers, are shown in Fig. 4. The water flow was measured by a calibrated, variable-orifice flowmeter, accurate to ± 2%. The heat transferred to the water was calculated from this flow and temperature information, and the consequent fissile power was then calculated using the method described above in Section 2. These heat rates are shown in Fig. 5. The rise of 4.4KW in heat to the water results from changing the experiment thimble gas from helium to neon, with a consequent reduction in heat lost to the reactor D_2O.

The SPND neutron-flux readings over the first three days are shown in Fig. 6. The mean flux, derived as described in Section 3 above, is also shown. This mean flux was then used to calculate the fission power, and this is plotted in Fig. 7 together with the fission power obtained from thermohydraulic measurements. The mean power from flux measurements is 4.8% below that from thermohydraulic measurements, so agreement between the two methods is better than the accuracy of the thermohydraulic measurements. The fissile power was monitored continuously for the 103.4 days irradiation by the SPNDs, and at discrete intervals by thermohydraulic measurements. Fig. 8 shows the resulting powers obtained by the two methods. Agreement continued to be better than ± 7½% throughout.

5. CONCLUSIONS

The experiment confirmed the interesting result of the prediction that fissile power from the 1.05%-enriched fuel should remain constant for a constant 'driver' flux - within the limits of accuracy of measurement. The ability to retain a steady power while 80% of the original U^{235} is burnt up, without the use of any additional flux control (by absorber) is an obvious asset. The semi-empirical method used to calculate U^{235} burn-up and Pu build-up was possible since the analysis of a previous experiment fuel pin provided detailed information at a particular position (in time) on the build-up/burn-up curves. It was used in preference to one of the large computer codes available e.g. 'WIMS'(6), since these did not provide a ready solution to this specific problem.

Irradiation time t (seconds) or ( ) in days	No. of atoms of U^{235}, (x 10^{-21})	Total reaction area U^{235}, $(n\sigma_f)^{235}$ (cm²)	No. of atoms of Pu^{239} (x 20^{-21})	Total reaction area Pu^{239} $(n\sigma_f)^{239}$ (cm²)	No. of atoms of Pu^{240} (x 10^{-21}) (cm²)	No. of atoms of Pu^{241} (x 10^{-21}) (cm²)	Total reaction area Pu^{241} $(n\sigma_f)^{241}$	Total fission reaction area $-\Sigma(n\sigma_f)$	$\frac{\Sigma (n\sigma_f)}{(n\sigma_f)^{235}_{t=0}}$
0.0	9.419	5.32	0.0	0.0	0.0	0.0	0.0	5.32	1.0
2.16×10^6 (25)	7.966	4.50	1.210	0.86	0.093	0.0103	0.012	5.37	1.01
4.32×10^6 (50)	6.738	3.81	2.077	1.48	0.300	0.063	0.076	5.37	1.01
6.48×10^6 (75)	5.699	3.22	2.698	1.92	0.545	0.165	0.20	5.34	1.00
8.64×10^6 (100)	4.820	2.72	3.142	2,24	0.794	0.303	0.36	5.32	1.0
10.80×10^6 (125)	4.077	2.30	3.459	2.47	1.019	0.462	0.55	5.32	1.00
12.96×10^6 (150)	3.448	1.95	3.685	2.62	1.224	0.626	0.75	5.32	1.00
17.28×10^6 (200)	2.465	1.39	3.959	2.82	1.514	0.934	1.12	5.33	1.00
21.60×10^6 (250)	1.764	1.00	4.093	2.92	1.708	1.180	1.41	5.33	1.00
43.2×10^6 (500)	0.330	0.19	4.178	2.98	1.957	1.621	1.93	5.10	0.96
86.4×10^6 (1000)	0.012	0.01	4.052	2.89	1.913	1.611	1.93	4.83	0.91

TABLE II

1. Mitchell, E.J. and Monahan, J. - "The DIDO Water Loop at Harwell and its use in particular for PWR Crud Deposition Experiments" - Int. Topical Meeting on Irradiation Technology, Grenoble, 1982.

2. Moore, P.G.F. and Lay, R.P., "Neutron Fluxes in DIDO experimental holes for a MK5/4 core operating at 22MW" - Harwell - A.E.R.E. R8214, 1976.

3. Forgan, R., "RAFTER, a computer program for the evaluation of reactivity absorbed by experimental rigs and the internal thermal flux distribution". - Harwell - A.E.R.E. M1599, 1965.

4. Foy, M.E., Private Communication, Harwell, Chemistry Division Actinide Analysis Group 1981.

5. Nicholson, K.P. Private Communication, Harwell, Materials Physics Division, 1982.

6. Askew, J.R., Fayers, F.J. and Kemshell, P.B. - "A general description of the lattice code WIMS". B.N.E.S. Journal 5, 564, 1966.

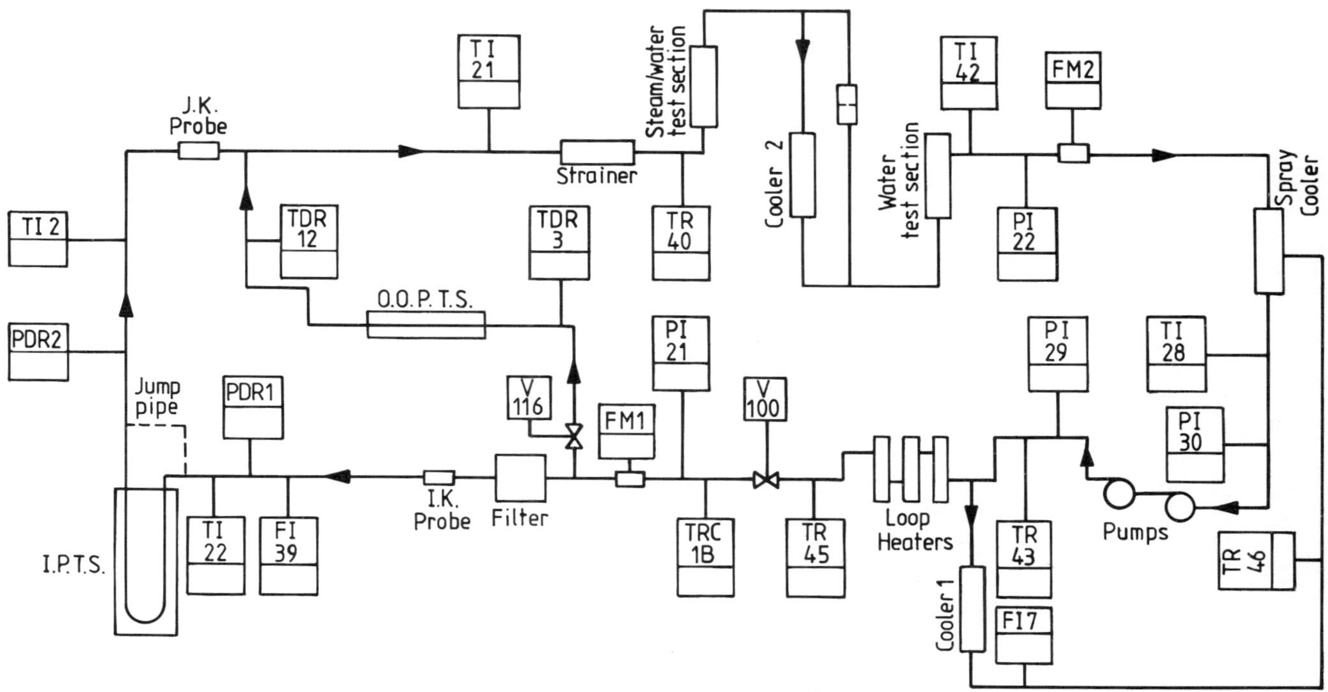

FIG. 1 D.W.L. CIRCUIT LOCATION OF INSTRUMENTS

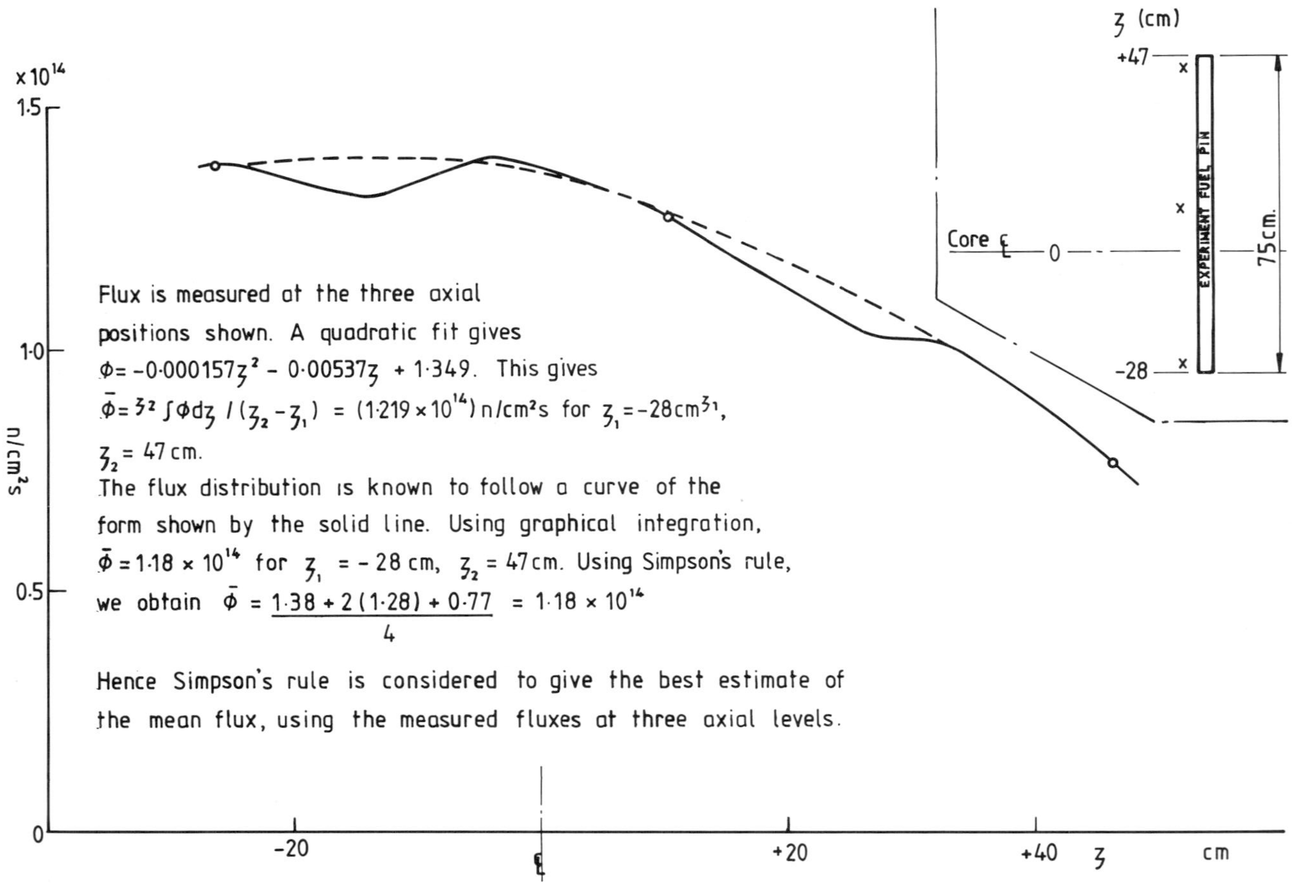

FIG. 2 D.W.L. MEASURED NEUTRON FLUXES AT THREE AXIAL LEVELS-
TAKEN 300 HOURS AFTER REACTOR START UP.

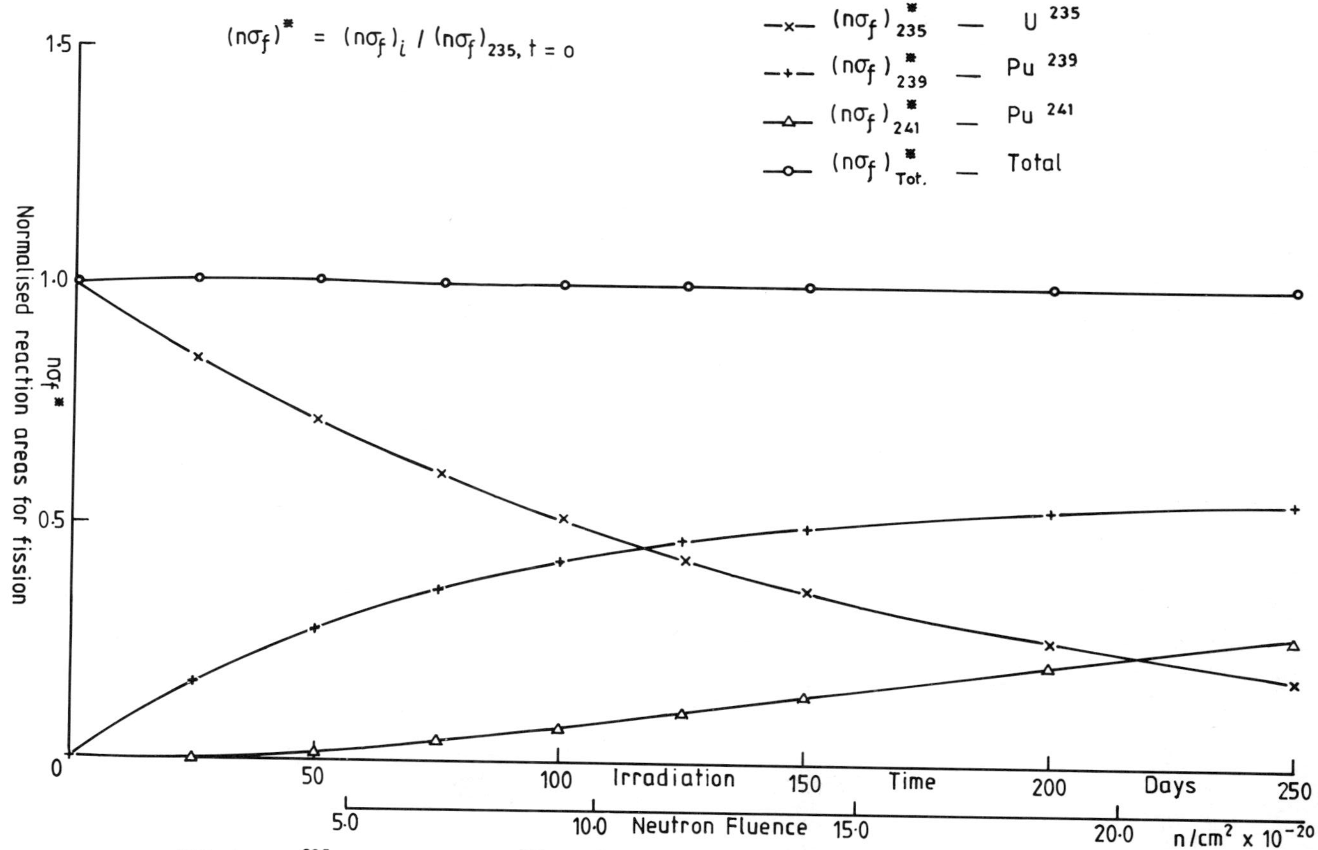

FIG. 3 U^{235} BURN UP & Pu^{239}, Pu^{241} BUILD UP FOR D.W.L. ($\bar{\Phi} = 1.08 \times 10^{14}$ $n/cm^2s.$, 1.05 % U^{235} ENRICHMENT IN EXPERIMENT FUEL PIN)

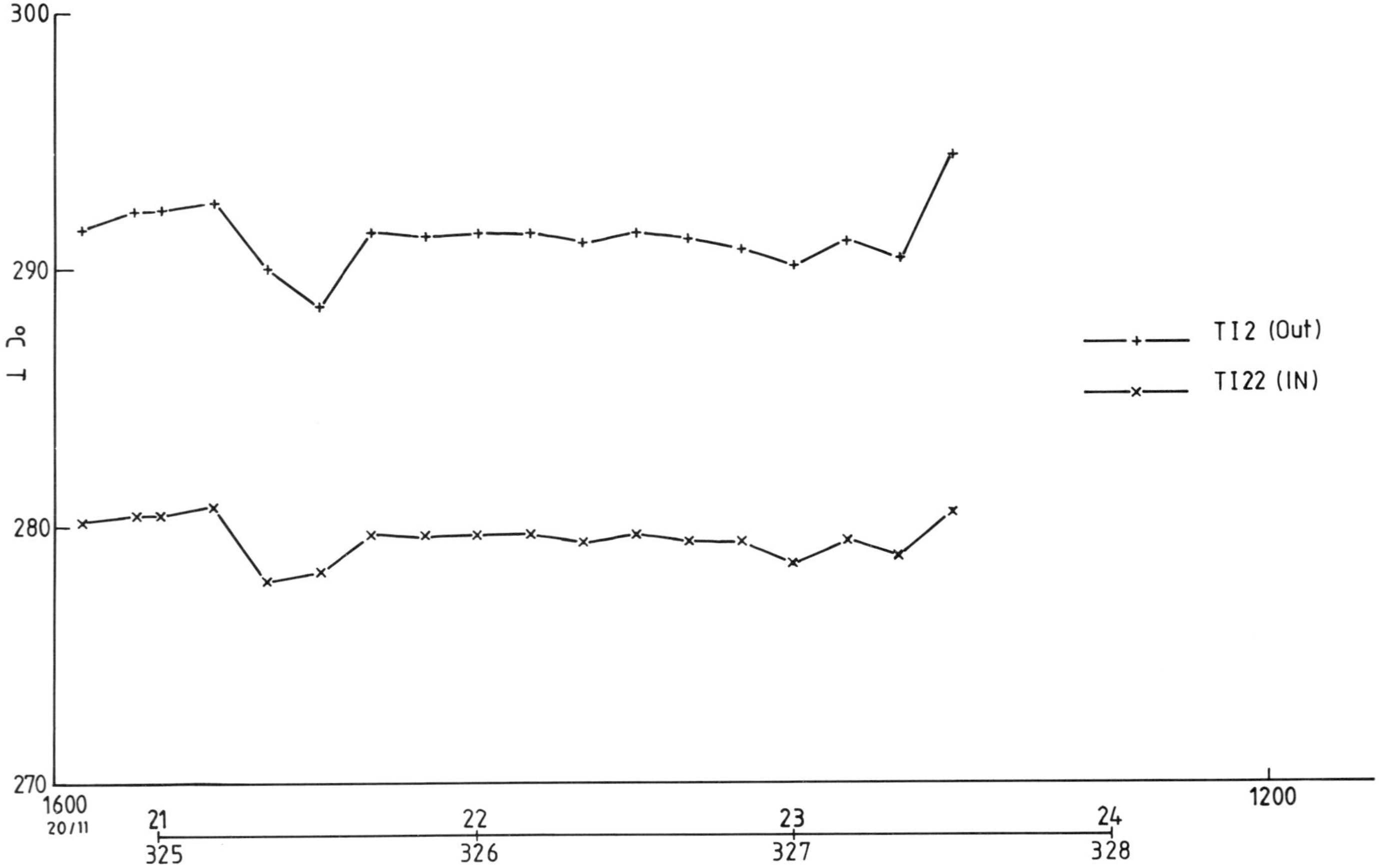

FIG. 4 D.W.L.- I.P.T.S. INLET & OUTLET TEMPERATURES (SUBCOOLED RUN - P = 138 Bars, T_{sat} = 335·4°C) TI 22 (In) & TI2 (Out)

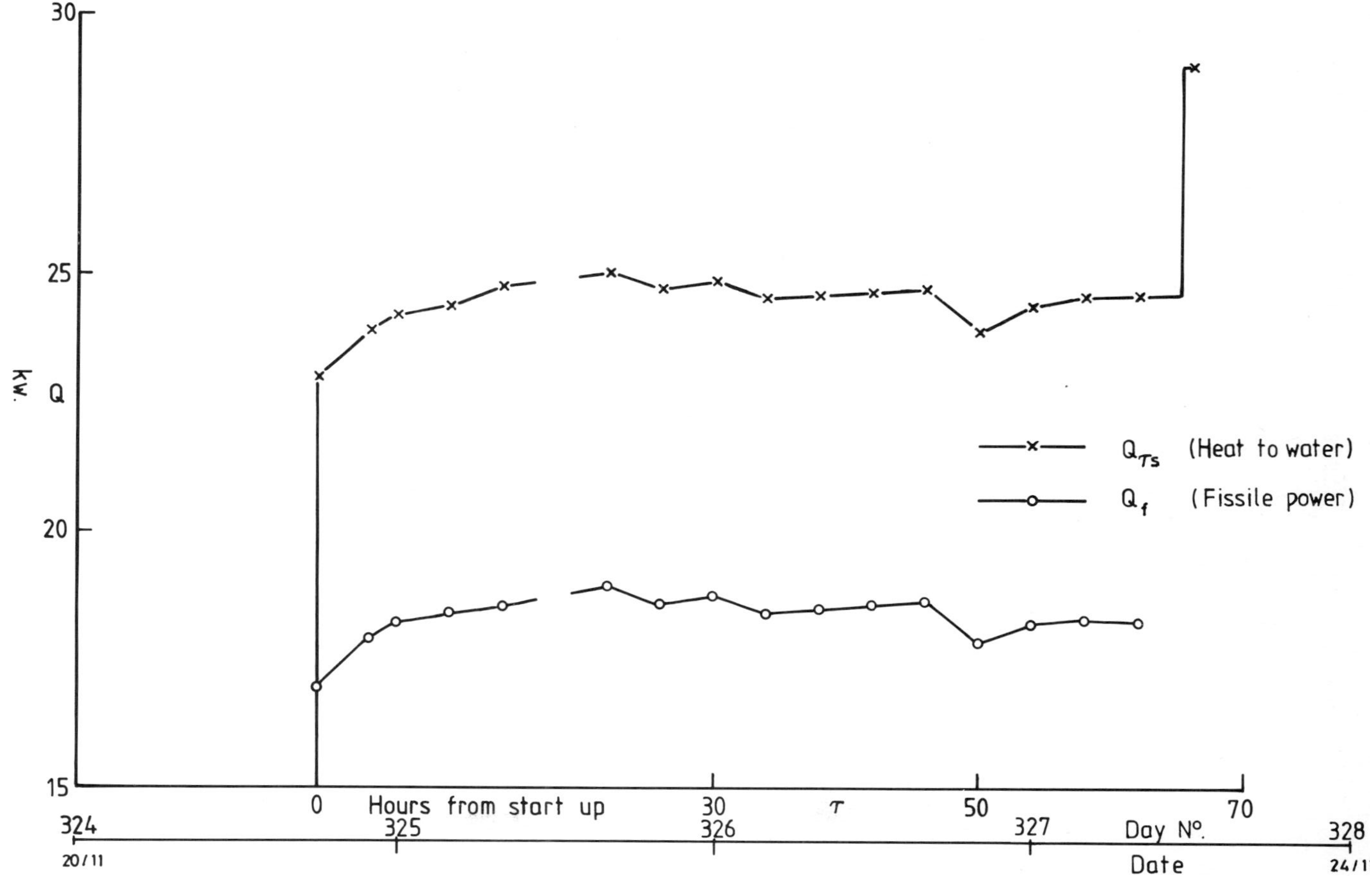

FIG. 5 D.W.L – HEAT IN TO WATER (I.P.T.S. INLET TO OUTLET) AND FISSILE POWER
(ASSUMED $Q_\gamma = 13$ kw, $Q_l = 0{\cdot}038\ (\bar{T}_H - \bar{T}_D)$)

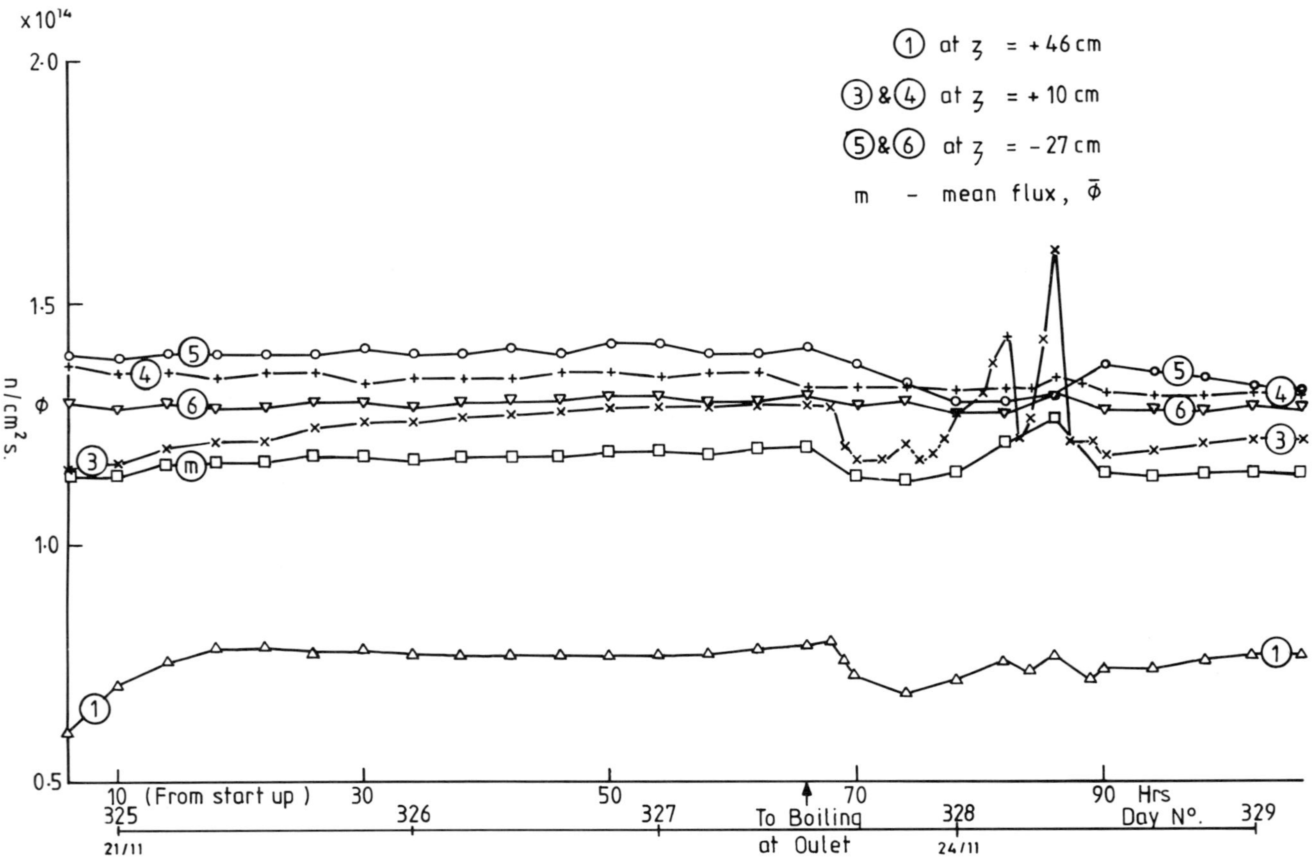

FIG. 6 D.W.L.- MEASURED NUTRON FLUXES

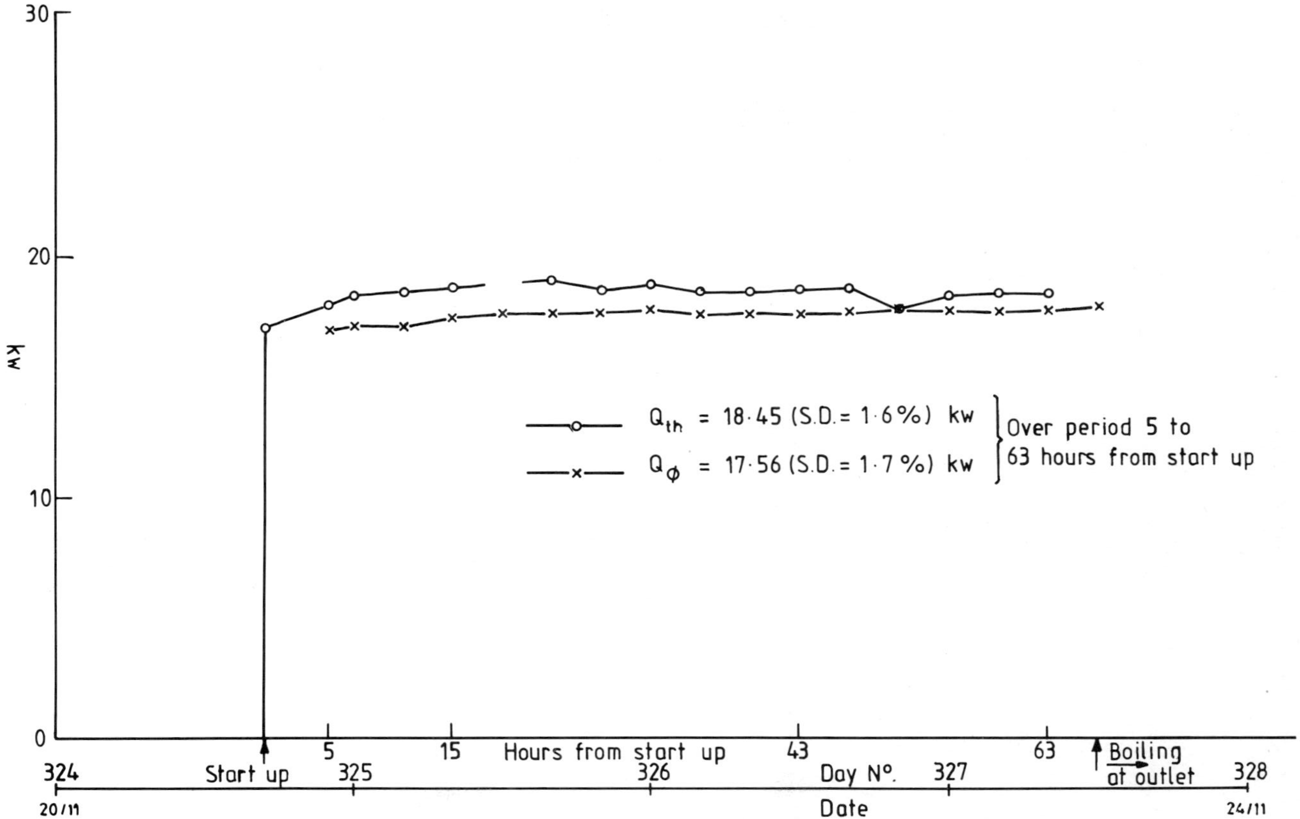

FIG. 7 D.W.L.- FISSILE POWER FROM (1) HEAT BALANCE (Q_{TH}) & (2) NEUTRON FLUX MEASUREMENTS (Q_{ϕ})

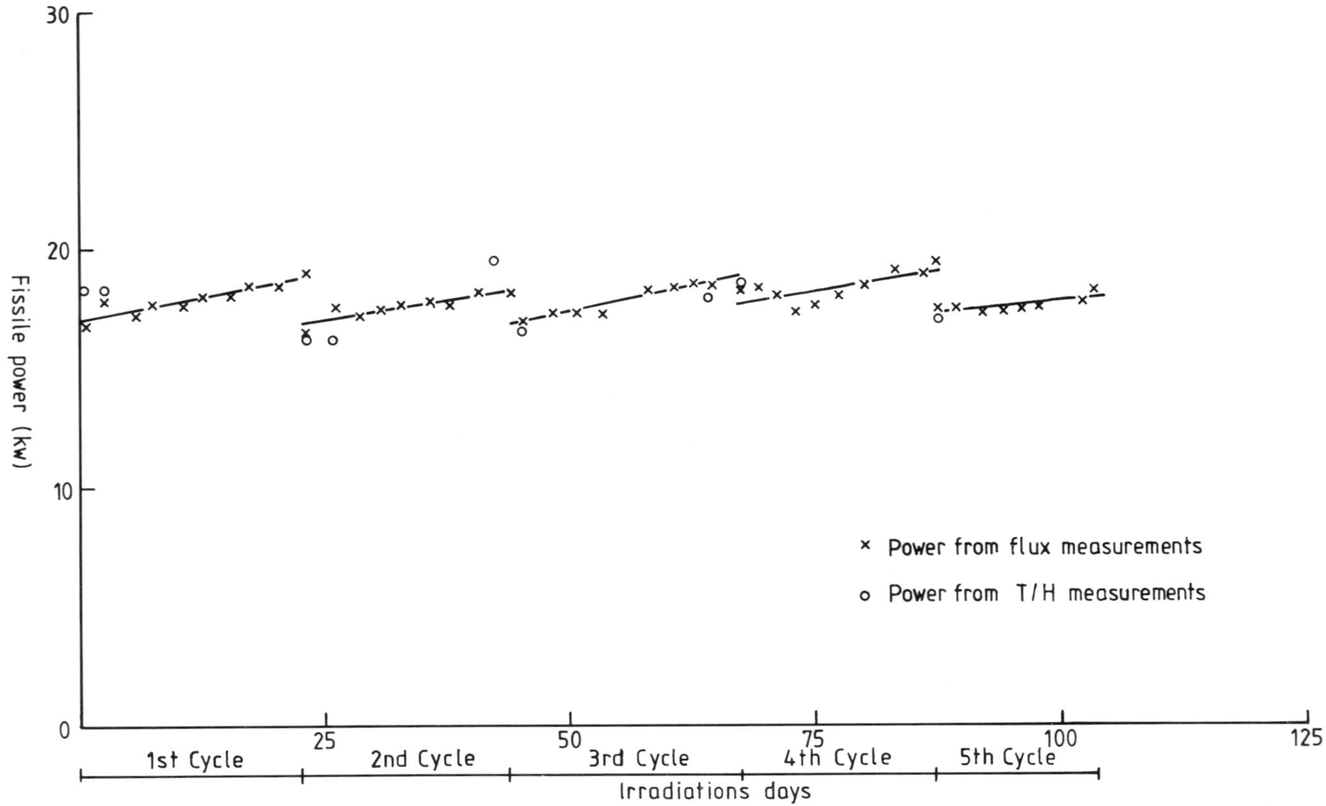

FIG. 8 D.W.L. EXPERIMENT FUEL PIN - COMPARISON OF FISSILE POWER CALCULATED FROM NEUTRON FLUX MEASUREMENTS & FROM THERMAL HYDRAULIC MEASUREMENTS

IRRADIATION EXPERIMENTS
AT THE NUCLEAR RESEARCH INSTITUTE, ŘEŽ

O. Erben, Z. Frič, B. Chamrád, A. Kříž, J. Pína, J. Šalamon, J. Zběhlík

Ústav jaderného výzkumu
250 68 Řež, ČSSR

Abstract

The paper deals with the irradiation facilities used and irradiation experiments carried out in the research reactor WWR-S at the Nuclear Research Institute (NRI), Řež. Two reactor water loops, RVS-2 and RVS-3, are described. Their main parameters, characteristics and fields of use are given.

The experiments carried out in our reactor research loops are closely connected with the development of electromagnetic filters for purification of the WWER cooling water from corrosion products. Three generations of these filters are described: a filter cooled by air, a water-cooled filter and, finally, a filter whose exciting coil is placed in a cryostat cooled by nitrogen and helium.

A description of the irradiation laboratory with irradiation rigs CHOUCA-TR and CHOUCA-MT for testing structural materials is presented and a review of the principal experiments carried out so far is given.

The measurements of irradiation conditions for the irradiation experiments are mentioned.

Introduction

The programme of irradiation experiments at NRI was

P. von der Hardt and H. Röttger (eds.), Irradiation Technology, 161–173.

started shortly after the research reactor WWR-S had been put into operation. The first experiments served mainly for purposes of the construction of the Czechoslovak HWGR power plant and they included, for example, irradiation of structural and fuel materials and models of fuel elements, the development and testing of in-core instrumentation and measurement methods, etc. These experiments were realized in irradiation rigs produced at NRI and mainly in the reactor gas loop, cooled by CO_2 (maximum flowrate of 2 000 kg/hour, maximum pressure of 4 MPa, maximum temperature of 450 °C, maximum output power of 35 kW). At present, most of our irradiation experiments are connected with the programme of construction and safe operation of the WWER nuclear power plants. In these experiments instrumented irradiation capsules, and irradiation loops and rigs are used.

Good results have been achieved in the development of electromagnetic filters for purification of the WWER cooling water from corrosion products.

Reactor WWR-S

The reactor WWR-S is sited in the NRI area. Since the first operation in 1957 with the output of 2 MWth the reactor has been modernized and at present the maximum thermal output is 10 MWth.

The IRT-M Fuel elements (U-Al metallic dispersion, enrichment 80 % U-235) consist of three or four concentric tubes of quadratic form. The lattice pitch is 71.5 mm, active length 580 mm. The core is cooled and moderated by demineralized light water, the reflector consists of beryllium and demineralized light water. The automatic control rod is situated in the beryllium reflector element. The control or safety rod is placed in the centre of the three-tube fuel element /1/.

The reactor is used in many ways. The horizontal channels with a maximum neutron flux density of 10^{13} n/ /m^{2}s at the exit can serve for research in the fields of nuclear and neutron physics, solid state physics, neutronography, etc. A graphite thermal column, which is a part of the reactor, can be also used.

The core average neutron flux density is 10^{18} n/m^{2}s. Two pneumatic rabbits can serve for purposes of activation analysis. Irradiation capsules, rigs and loops can

fuel element assembly
CR control rod
SR safety rod
AR automatic regulator rod

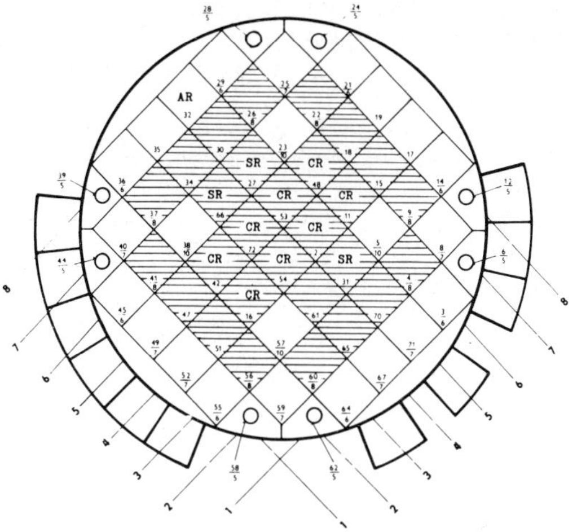

Diagram of a typical configuration of the WWR-S core

be used for irradiation in the reactor core and for production of radioisotopes.

Reactor Water Loop RVS-2

The reactor water loop RVS-2 has been operated since 1962. The loop active channel in the form of a Field tube is placed in the 14/6 channel of the reflector. Maximum working pressure in the loop is 6.5 MPa. After some modifications it was possible to increase the maximum working temperature from the original 90 ^{o}C up to 200 ^{o}C. Working medium (light water with chemical ingredients corresponding to the needs of a given experiments) has a maximum flowrate of 10 000 kg/hour. The loop has served for research on reactor structural materials, WWER coolant behaviour (radiolysis of ammoniac in the presence of boron acid), and for the development and testing of remote metering devices and methods (measurement of conductivity and pH, and measurement of corrosion rate by the polarized resistance method). The loop has recently been modified and at present is used mainly for the development and testing of electromagnetic filters. A typical arrangement of the loop primary circuit in this type of experiments is presented in Fig 1. Besides the primary circuit the loop consists of a cooling circuit, gas branches, and charging, dump and sampling branches.

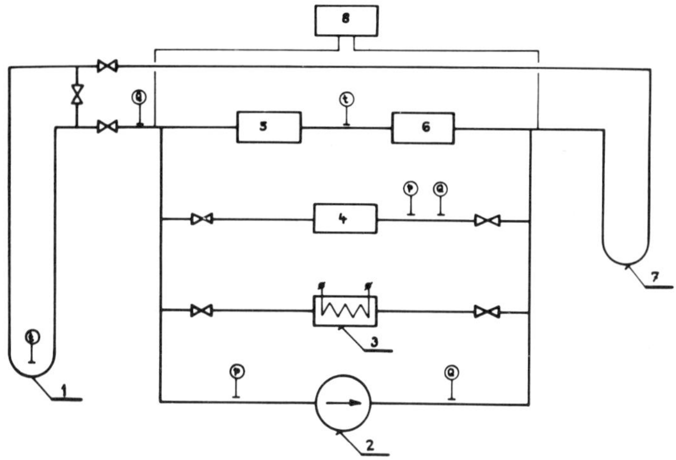

t - temperature measurement, p - pressure measurement, Q - flowrate measurement, 1 - irradiation channel, 2 - primary circulation pump, 3 - tested filter, 4 - measuring devices, 5 - cooler, 6 - comparative channel, 7 - pressurizer

Fig. 1. Primary circuit of the RVS-2 loop

Reactor Water Loop RVS-3

The reactor water loop RVS-3, which is being put into operation this year, models the WWER-1000 primary circuit conditions. Its main parameters are:

maximum working pressure	15.7 MPa,
maximum temperature	350 °C, and
maximum flowrate	10 000 kg/hour.

The primary circuit of this loop (Fig. 2) is composed of the main circuit, hot measuring circuit, cold measuring circuit and filter circuit. All pipes and equipment are made of stainless steel. The loop is so equipped that it is possible during its operation to add any required amount of water or other chemicals into the primary circuit and to tap samples at different locations for further analyses. Besides the primary circuit, the loop consists of auxiliary circuits and branches (cooling, gas and ventilating circuit, etc.). The digital information system CIS 3000 with minicomputer ADT 4100 provides automatic acquisition and recording of measured data, monitoring of pre-set values, and partial on-line evaluation of experiments.

The irradiation channel of the loop consists of a Field tube (outer diam. 57/8 mm, inner diam. 30/3 mm), situated in the 5/10 channel of the WWR-S reactor core. In the centre of this channel, a dry measuring probe can be placed. Two fixed measuring probes are located on the outer side of the channel. One of them contains five self-powered neutron detectors monitoring the thermal flux density distribution, the other probe has five tungsten calorimeters, monitoring the gamma radiation dose rate distribution.

Coolant flowrate is controlled by pump by-pass flowrate. The irradiation channel can be separated from the remaining parts of the loop. In this case, water flow is maintained by natural circulation.

The hot measuring circuit has water with the same parameters as those of water in the main circuit. When working with equipment and apparatuses requiring lower temperatures, the cold measuring circuit with the temperature of water decreased down to 30 °C is used.

The filter circuit serves for the purification of coolant and is also used for tests of filter assemblies and filter materials. Water in this circuit has a minimum temperature of 30 °C; the maximum flowrate of 100 kg/hour is the same as in the hot and cold measuring circuits.

The loop RVS-3 will serve for development and testing of primary circuit components of nuclear power plants, namely electromagnetic filters and devices for control of boron acid concentration, for the investigation of technological processes (radioactive water purification, regeneration of decontamination solutions) and operational problems (behaviour and transport of corrosion and fission products, corrosion of structural and fuel materials, behaviour of fuel elements in off-normal conditions, etc.) of WWER primary circuits, and for the development and testing of detectors and measuring methods.

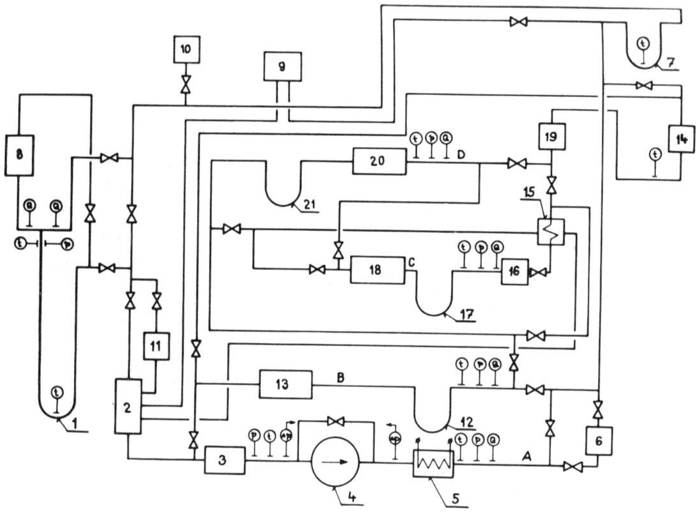

t - temperature measurement, p - pressure measurement, Q - flowrate measurement, p - pressure gradient measurement, A - primary circuit, B - hot measuring circuit, C - cold measuring circuit, D - filter circuit, 1 - irradiation channel, 2 - mixing chamber, 3 - gas separator, 4 - primary circulation pump, 5 - heater, 6 - auxiliary chamber, 7, 12, 17, 21 - comparative channels, 8, 11, 18, 19 - coolers, 9 - pressurizer, 10 - expansion tank, 13, 18 - experimental apparatuses and equipment, 14 - auxiliary cooler, 15 - heat exchanger, 20 - experimental filters, equipment and apparatuses

Fig. 2. RVS-3 water loop primary circuit

Electromagnetic Filters

The development of electromagnetic filters for purification of WWER cooling water from corrosion products is tightly connected with the experiments carried out in our reactor loops. Filters of the first generation had a ball-charged filter section and an air-cooled outer solenoid. The height of the ball charge was 700 mm, water flowrate 600 kg/hour and magnetic field intensity

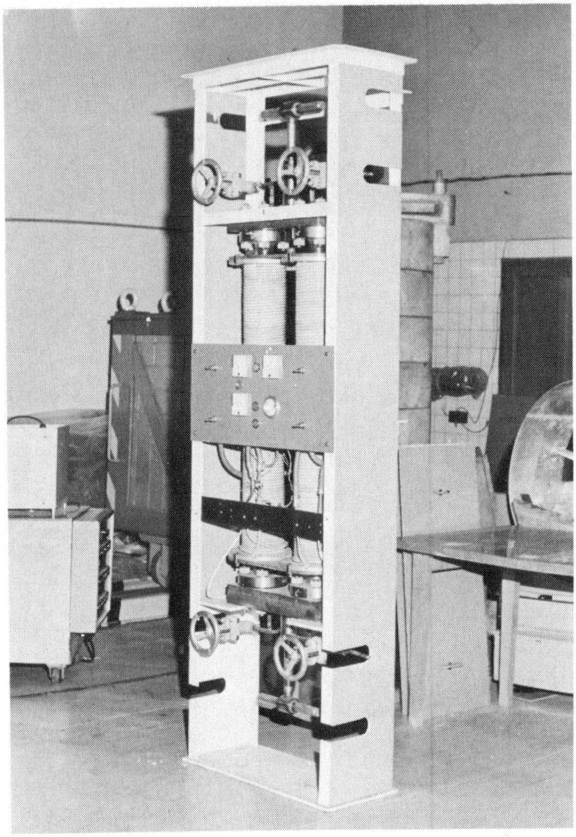

Fig. 3. Filter of the second generation

0.7×10^5 A/m.

The effort to increase the parameters of these filters lead to difficulties with cooling the solenoid winding. That was the reason why the second generation filters (two-chambered with a closed electromagnetic field) were developed (Fig. 3). Their winding is made from a hollow copper conductor with inner flow of cooling water. Filters of this type with a ball-charge height of 1000 mm, water flowrate of 2600 kg/hour, pressure of 18 MPa, temperature of 350 °C and magnetic field intensity of 1.5×10^5 A/m have been prepared and tested in RVS-2 loop.

Tests of these filters under the design conditions will be carried out in the loop RVS-3.

At present, research on the third generation filters is in progress. These filters use supraconductive materials and cryogenic technique with the aim to achieve magnetic fields over 1 tesla in the filter charge area together with a substantially lowered input power. A helium cryostat and supra-conductor NbTi are components of the filter. The first function tests of this filter have been succesfully carried out. Water flowrate was 600 kg/ /hour and a maximum magnetic field of 5 tesla in the filter charge area was achieved.

Irradiation Rigs

The activity of irradiation rig department closely follows the demands of the Czechoslovak programme of construction and operation of nuclear power plants. Attention is paid mainly to the effects of neutrons on the mechanical and physical properties of WWER nuclear power plant structural materials.

Until 1970, the NRI produced irradiation rigs with internal electric heating had been used. In 1970, two irradiation rigs CHOUCA-TR were bought in CEN/Grenoble and another rig, CHOUCA-MT, was bought in 1978. At the same time, the control room of the irradiation experiments was modernized and at present it is possible to operate as many as three irradiation rigs in the WWR-S reactor core. The rigs are controlled by automatic electronic systems EROS-76 and EROS-78, developed at NRI, and by a control system BAI ELECTRIQUE EUROPA, produced in CEN/Grenoble. A block diagram of the control and operation of the irradiation rigs is in Fig. 4. Temperatures of the irradiated specimens are measured by chromel-alumel insulated thermocouples, thermal neutron flux density is monitored by self-powered neutron detectors, and fluences are determined with the aid of a set of activation detectors. Measured quantities (pressure, temperature, thermal neutron flux density, reactor output power) are converted into voltages and recorded in the pre-determined time intervals by data transfer unit SOLARTRON DTV 3240. The evaluation of the irradiation conditions is carried out off-line on HP-9600 E computer.

At present, the predominating irradiation experiments are connected with the development and testing of the WWER pressure vessel steels. These experiments are carried out mostly in CHOUCA-MT rig, where 40 impact CHAPPY-V specimens and approx. 20 specimens for study of hydrogen embrittlement can be placed. The required temperature and fluence of these specimens is 561 K and 5×10^{23} n/m^2 ($E > 1$ MeV), respectively. The irradiated specimens are placed in inert environment (helium) with pressure of 0.15 MPa. The average neutron flux density of 10^{17} n/m^{2}s in the CHOUCA-MT rig leads to temperatures not greater than 533 K. Owing to the use of the control system the deviations of temperature along the 300 mm length of the specimen carrier did not exceed 1.5 K during the three-weeks irradiation cycles.

In Table 1, a summary of main irradiation experiments carried out in the period 1977-1981 is presented. The other experiments performed served for the study of radiation stability of fuels, of evolution of fission products release, and of characteristics of thermometers based on the electrical resistance change of special salts. Steel specimens at temperatures from 393 K to 430 K were irradiated as well.

After irradiation, the carrier with irradiated specimens is clipped off, placed into a container, transported into hot cells and dismounted there. Decontaminated specimens are transferred to hot laboratories for post--irradiation testing.

Measurements of Neutron Spectra and Radiation Heating

The neutron energy spectrum needed for analyses of irradiation experiments is being determined by activation method. The spectra are unfolded by SAND-II code with cross section data taken from ENDF/B-IV. Detailed description of measurements carried out in three irradiation channels and their results is available in multigroup ABBN format in /1/.

The reactor WWR-S is in operation from Monday till Friday. For monitoring of thermal neutron flux density, beta-emission self-powered neutron detectors are used. The dependency of the detectors signals on the height of the irradiation channel 5/10, as measured on Monday, 8 p. m. (u_1 curve), on Tuesday, 8 a. m. (u_2), and on Thursday, 4 p. m. (u_3) is shown in Fig. 5. Zero height is placed in the centre of the core. The research, which has been carried out in the field of properties and application

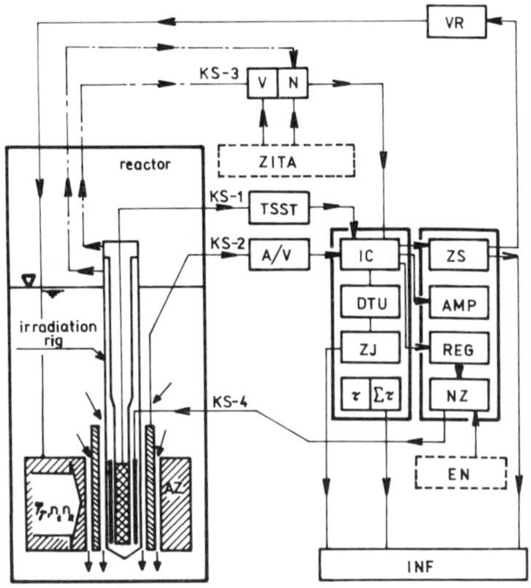

AZ - reactor core, pV - high pressure gas branch, pN - low pressure gas branch, ZITA - source of inert and technological atmospheres, TSST - thermostat of thermocouples cold junctions, A/V - current/voltage convertors, IC - information center, DTV - data transfer unit, ZJ - registration unit, VR - reactor control room, ZS - emergency system, AMP - analog measuring apparatuses, REG - regulators, NZ - power supply, φ_γ , n_t, n_R - gamma radiation, and thermal and fast neutron fluxes, KS-1, KS-2, KS-3, KS-4 - communication systems, τ - time, $\Sigma\tau$ - total time of irradiation, EN - energy entry, INF - output of information

Fig. 4. Block diagram of the control and operation of irradiation rigs

of self-powered neutron detectors, is summarized in /2/.

For the determination of the radiation heating, both the temperature field measurements in models of irradiation rigs and reactor calorimeters data are used. As an example, the measurements of temperature distributions in both vertical and horizontal directions in irradiation rig CHOUCA-MT with carriers for vast programme of irradi-

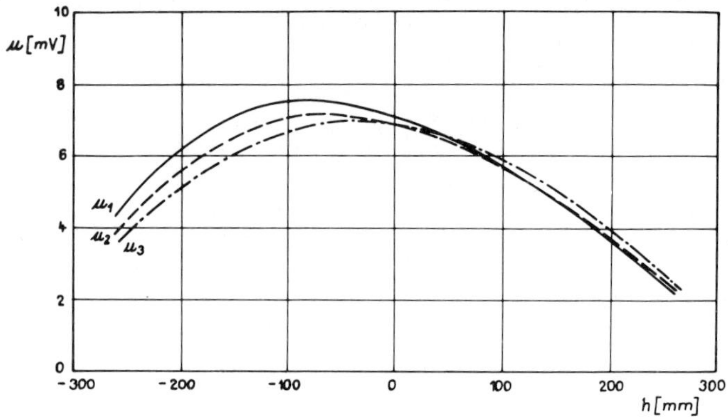

Fig. 5. Signals of rhodium self-powered neutron detectors

ation of structural materials, together with the simultaneous measurement of heat leakage into the upper non-heated part of the rig lead to radiation heating (corresponding to 1 MW thermal output of reactor) 0.3 W/g in steel and 0.24 W/g in aluminium. Simultaneously with the measurements with self-powered neutron detectors (shown in Fig. 5) the distributions of gamma radiation dose rate were measured with tungsten calorimeters. The results for reactor thermal output of 6 MW are presented in Fig. 6. In near future, a programme of detailed measurements of gamma field properties in WWR-S reactor core is going to be realized. For these measurements, measuring probes with graphite, aluminium, steel, tungsten and lead calorimeters are manufactured in ŠKODA works.

Conclusion

With increasing number of operated nuclear power plants in Czechoslovakia, irradiation experiments will serve mainly for the improvement of their safety, reliability and economy. This research programme will include the above described types of experiments, the development of electromagnetic filters, and the development and testing of measuring detectors, chains and measurement methods for the determination of the status and function of reactors of WWER nuclear power stations.

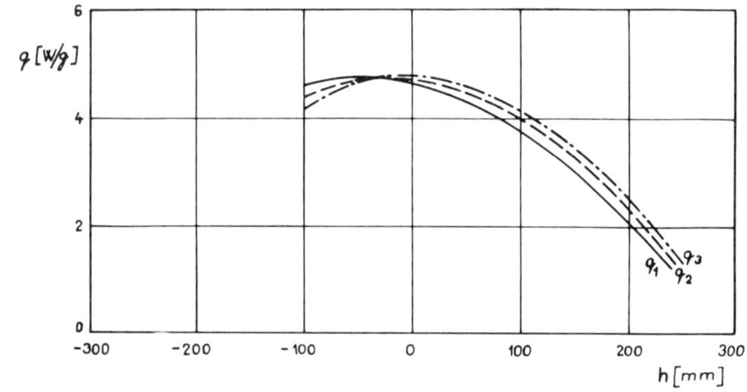

Fig. 6. Gamma radiation dose rate distributions measured with tungsten calorimeters

References

/1/ B. Ošmera et al. "Neutron Spectra Measurements in WWR-S Reactor", 4th Symposium on Reactor Dosimetry, Washington, USA, 22-25 March, 1982.

/2/ O. Erben, J. Šalamon "On Measurement of Thermal Neutron Flux Density and Gamma Radiation", Panel on Research Reactors Modernization in CMEA States, Moscow, USSR, 22-26 June, 1982 (in Russian).

Name of exper.	Temp. [K]	Press. [MPa]	Fluence $[m^{-2}]$	Rig	Start	End
REX-ZIRKON						
-0X/77		0,15			12. 4.77	2. 6.77
-01/77	643 ± 10	12,00	5.10^{23}	CHOUCA--T	11. 7.77	9. 9.77
-02/77		12,00			19. 9.77	21.10.77
-03/77		12,00		(Zr)	28.11.77	30.12.77
-04/78		12,00			6. 2.78	10. 3.78
REX-ZIRKON-I						
-02/78			1.10^{23}		19. 6.78	29. 6.78
-03/78	573 ± 10		1.10^{23}		3. 7.78	6. 7.78
-04/78			5.10^{23}		18. 9.78	4.12.78
-05/78		0,15	5.10^{23}	CHOUCA--T	4. 9.78	4.12.78
-07/78	723 ± 10		1.10^{23}		5. 2.79	23. 3.79
-08/78			1.10^{23}	(Zr)	5. 2.79	23. 3.79
-06/79	573 ± 10		1.10^{23}		14. 4.79	22. 6.79
-09/79			1.10^{23}		3. 9.79	16.11.79
REX-OCEL				CHOUCA-T		
-01/77	561 ± 10	0,15		(Steel	31.10.77	24.11.77
-02/78					3. 1.78	26. 1.78
REX-OCEL-A						
-01/78	561 ± 10	0,15	2.10^{24}	CHOUCA-T	3. 4.78	15. 6.78
-02/78			5.10^{23}	(Steel)	10. 5.78	15. 6.78
REX-OCEL-S				CHOUCA-		
-01/78	561 ± 10	0,15	1.10^{23}	MT (Steel)	15.11.78	5.12.78
REX-OCEL-A						
-03/79	561 ± 10	0,15	5.10^{23}	CHOUCA-	2. 4.79	6. 5.79
-04/79			1.10^{24}	-MT	5. 9.79	16.11.79
-05/79			2.10^{23}	(Steel)	26.11.79	14.12.79
REX-OCEL-S				CHOUCA-		
-02/79	561 ± 10	0,15	2.10^{23}	-MT	22. 5.79	22. 7.79
-03/79			2.10^{23}	(Steel)	26. 9.79	16.11.79
REX-OCEL-A						
-06/80					14. 1.80	1. 2.80
-07/80	561 ± 10	0,15	5.10^{23}		10. 3.80	28. 3.80
-08/80					8. 4.80	25. 4.80
-09/80				CHOUCA-	12. 5.80	30. 5.80
-10/81			2.10^{23}	-MT (Steel)	12. 1.81	27. 2.81
-11/81			$2,5.10^{23}$		9. 3.81	27. 3.81

Table 1. The main experiments carried out in the rigs

CAPABILITIES OF THE POWER BURST FACILITY[a]

W. A. Spencer
A. M. Jensen
R. K. McCardell

EG&G Idaho, Inc.
P. O. Box 1625
Idaho Falls, Idaho 83415
U.S.A.

ABSTRACT

The unique and diverse test capabilities of the Power Burst Facility (PBF) are described in this paper. The PBF test reactor, located at the Idaho National Engineering Laboratory, simulates normal, off-normal, and accident operating conditions of light water reactor fuel rods. An overview description is given, with specific detail on design and operating characteristics of the driver core, experiment test loop, fission product detection system, test train assembly facility, and support equipment which make the testing capability of the PBF so versatile.

INTRODUCTION

The Power Burst Facility, located at the Idaho National Engineering Laboratory in the United States, is extremely versatile. The experimental program for the PBF is structured to provide relevant and pertinent data necessary to (a) develop and assess computer models needed for licensing analysis, (b) confirm the adequacy of specific U.S. Nuclear Regulatory Commission licensing regulations that are designed to ensure plant safety with respect to specific accident conditions, (c) help resolve

---

a. Work supported by the U.S. Nuclear Regulatory Commission, Office of Nuclear Regulatory Research, under DOE Contract No. DE-AC07-76ID01570.

P. von der Hardt and H. Röttger (eds.), Irradiation Technology, 175–189.

unanswered safety issues and provide a data base from which new regulations can be established, and (d) indicate where new or revised regulations may be appropriate.

The PBF reactor and associated experiment test loop has the capabilities to simulate reactivity initiated accidents (RIA), power-cooling-mismatch accidents (PCM), anticipated transients with and without scram (ATWS), loss-of-coolant accidents (LOCA), and severe fuel damage accidents (SFD) using up to 32-rod clusters of test fuel rods. The PBF has an integral experimental program beginning with test requirements, pretest predictions, and safety analysis; test hardware and instrumented test fuel rod design, fabrication, and assembly; data acquisition, reduction, and processing; posttest analysis and postirradiation examination; and final test reporting.

During the PBF experiments, measurements of test rod coolant conditions, cladding and fuel temperatures, internal rod pressure, and rod elongation data are recorded on magnetic tape using the computer-controlled data acquisition and reduction system which can accommodate up to 287 data channels. Further data processing and analysis are performed using a larger dual Control Data Corporation 176 System. A very sophisticated fission product detection system makes it possible to determine the time-dependent release of fission products during accident simulation tests that result in cladding failure. Pre- and posttest computer code calculations are performed for comparison with the measured data and the data obtained from extensive postirradiation examinations performed in the INEL hot cells. In the recent past, the light water reactor fuel behavior research in the PBF has been solely sponsored by the United States Nuclear Regulatory Commission (USNRC). However, the PBF is now available to other sponsors, and a LOCA test in the PBF using CANDU fuel rods is presently being co-sponsored in PBF by Atomic Energy of Canada, Limited, and Ontario Hydro.

An overview description is given with specific detail on design and operating characteristics of the driver core, experiment test loop, fission product detection system, data systems, and test train assembly facility, which make the testing ability of the PBF unique and versatile.

GENERAL

The PBF is designed for remote, steady state and transient nuclear operation. It is housed essentially in two buildings--a reactor building and control building--located appoximately 80 m apart. The reactor building houses the reactor, associated process equipment, experimental facilities, fuel handling facilities, and essential controls and instruments. The control building contains the control room, data acquisition systems, and limited office space.

REACTOR

The PBF reactor is an open vessel type in which a closed-loop coolant system provides forced cooling of the core. This system, which uses light water as coolant, is capable of removing up to 28 MW of heat during steady state operation. The reactor core, or driver core, is a right circular annulus, 1.3 m in diameter and 0.91 m high, with a centrally located, vertical test space, 0.21 m (8.25 inches) in diameter. Figure 1 shows a cut-away view of the major reactor components.

The core is undermoderated and is surrounded by a ring of stainless steel rods that reflect neutrons back into the core. The reactor is fueled with 18.5% enriched ^{235}U contained by approximately 2400 fuel rods. These rods are grouped in closely packed, square-lattice canisters, each containing from 28 to 62 rods. The fuel rods are composed of ternary (20.6% urania, 61.8% zirconia, and 7.6% calcia) ceramic fuel pellets contained in a ceramic (zirconia, calcia) thermal insulator and a stainless steel cladding tube. This design restricts the rate of heat transfer from fuel to coolant during transient heating of the core, thereby providing an inherent shutdown mechanism through Doppler feedback of the ^{238}U. Stainless steel shim rods are interspersed with the fuel rods. Reactivity may be increased by replacing shim rods with additional fuel rods, or may be decreased by replacing fuel rods with additional shim rods. Figure 2 illustrates the driver core cross section.

Two sets of moveable poison rods--control rods and transient rods--control reactivity of the driver core. Boron carbide (76.5 weight % boron) is the poison used in both sets of rods. Eight control rods maintain basic reactivity control. The four transient rods can be used manually or automatically to change reactor power at various rates.

An in-pile tube (IPT) occupies the vertical test space in the center of the core. The spacing of fuel rods in the core provides an intentionally undermoderated core which, combined with the relatively low ^{235}U atom density, results in a high fast neutron flux. The central test space moderates the fast neutron flux and provides a thermal flux peak in the IPT. This increases the power density in the test fuel relative to the driver core. The dimensions of the IPT are an inside diameter of 155 mm and an inside length of 4.47 m. The IPT extends 2.67 m above the core and 0.88 m below the core. A typical PBF neutron spectrum inside the IPT is shown in Figure 3, and the axial power profile for a water-filled IPT is shown in Figure 4.

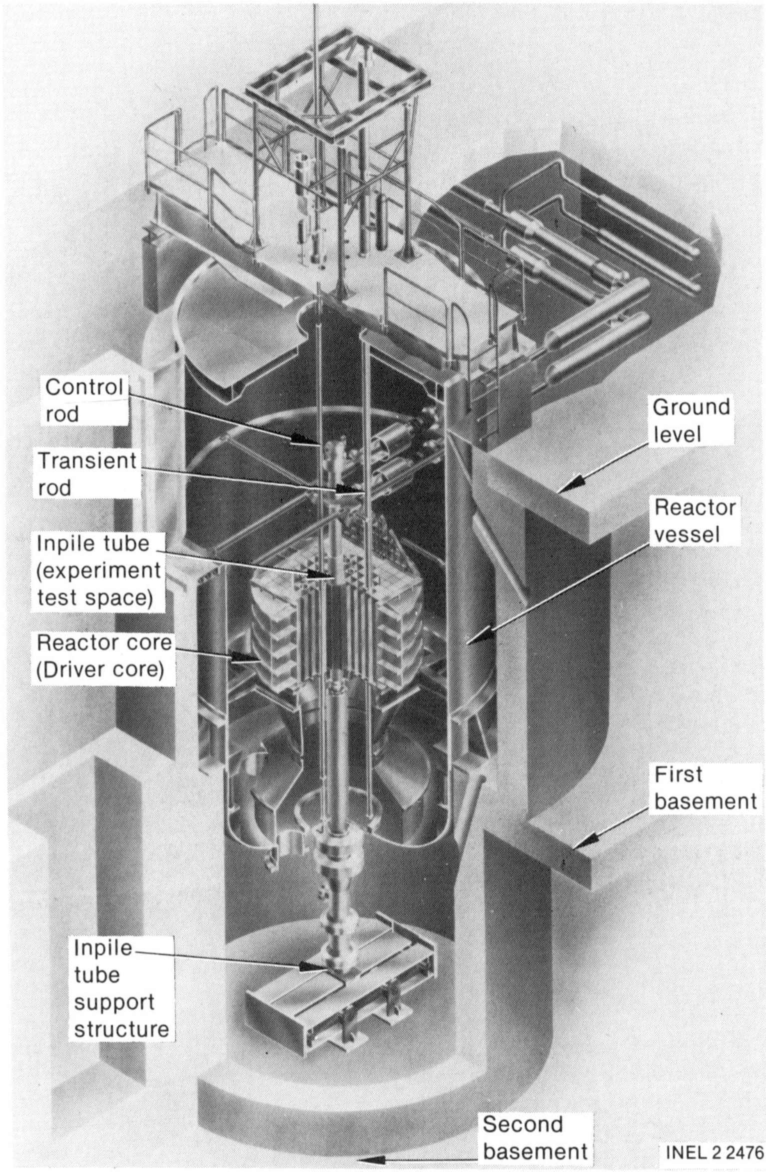

Figure 1. Cut-away view of PBF reactor.

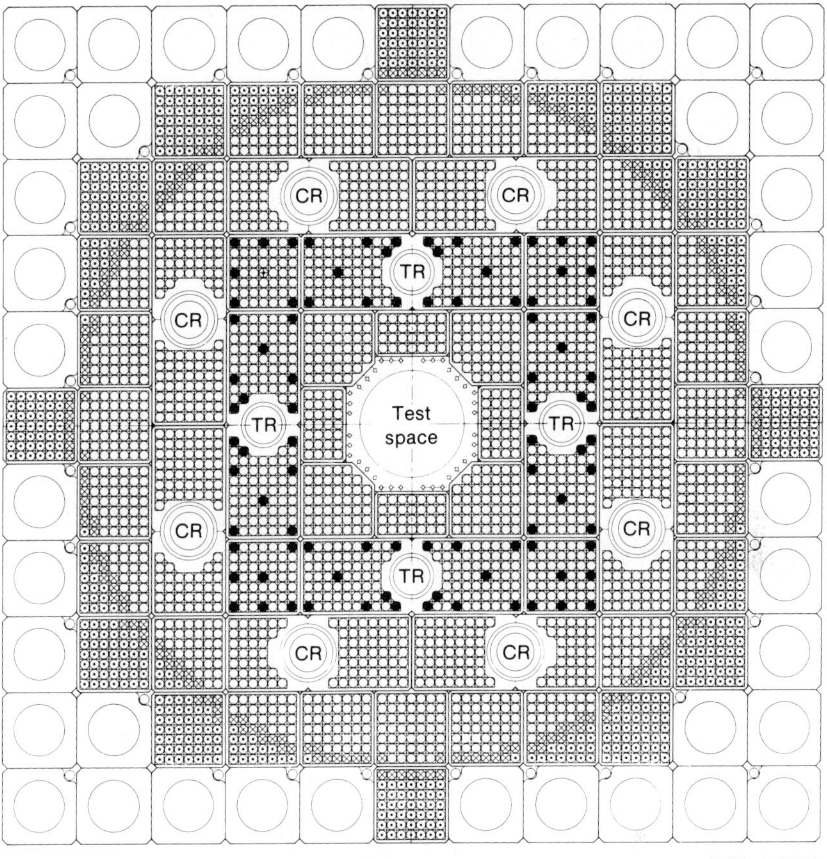

○ Fuel rod
⊗ Stainless steel reflector rod
⊙ Aluminum filler rod
● Stainless steel shim rod
⊕ Source rod
CR Control rod
TR Transient rod

Figure 2. Driver core cross section.

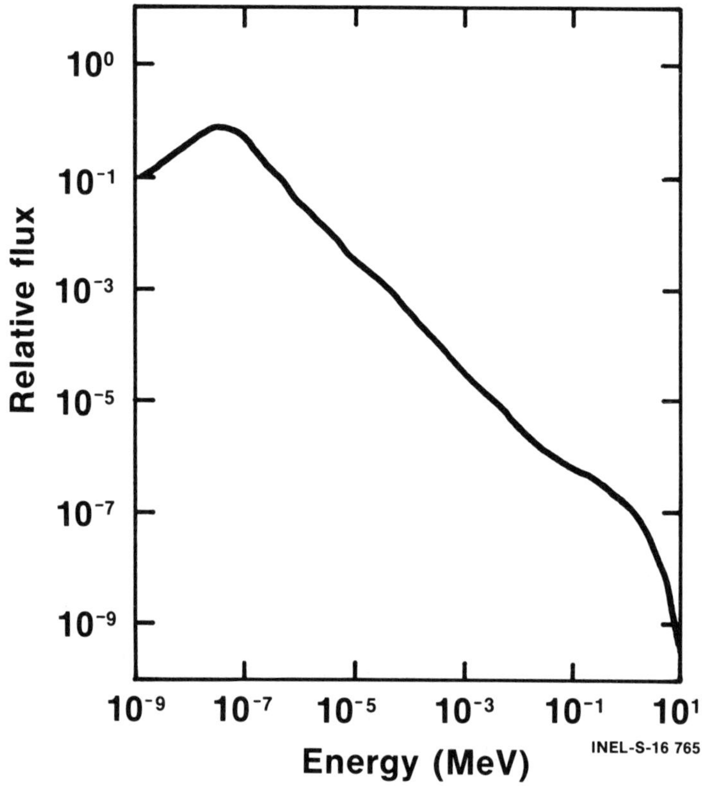

Figure 3. Typical PBF spectrum inside wall of the in-pile tube.

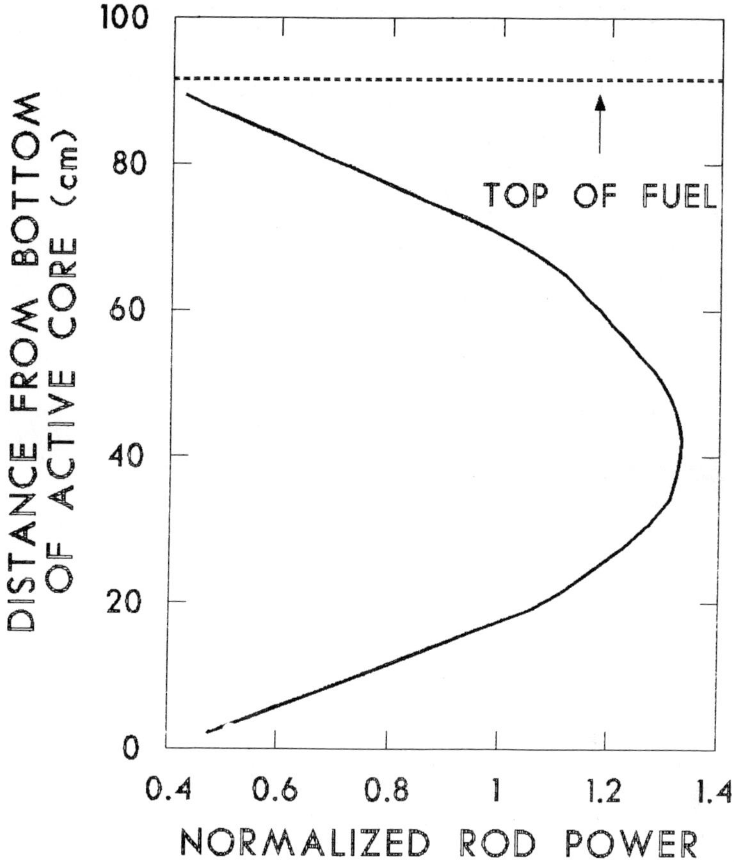

Figure 4. PBF axial power distribution.

This illustration is printed from the best available original

The reactor can be operated in several modes as illustrated in Figure 5: (a) a steady state mode with power levels up to 28 MW, (b) a natural burst mode yielding reactor periods as short as 1.0 ms and peak powers as large as 270,000 MW, (c) a shaped burst mode yielding integrated reactor energies as large as 1350 MJ, or (d) a power shaping mode in which the control system drives reactor power to simulate anticipated transient power test conditions. In the natural burst mode, the transient rods are ejected from the core at 952.5 cm/second; in the servo control mode, they travel at 88.9 cm/second out of the core and 419.1 cm/second into the core.

To illustrate the expected steady state test rod power capabilities of the PBF, calcuated test rod power as a function of burnup is plotted in Figure 6 for single BWR/6 and PWR 15 x 15 test fuel rods for a PBF power level of 27.5 MW. Figures of merit (test rod power in kilowatts per meter per megawatt of core power) are listed in Table I for various test rod array sizes.

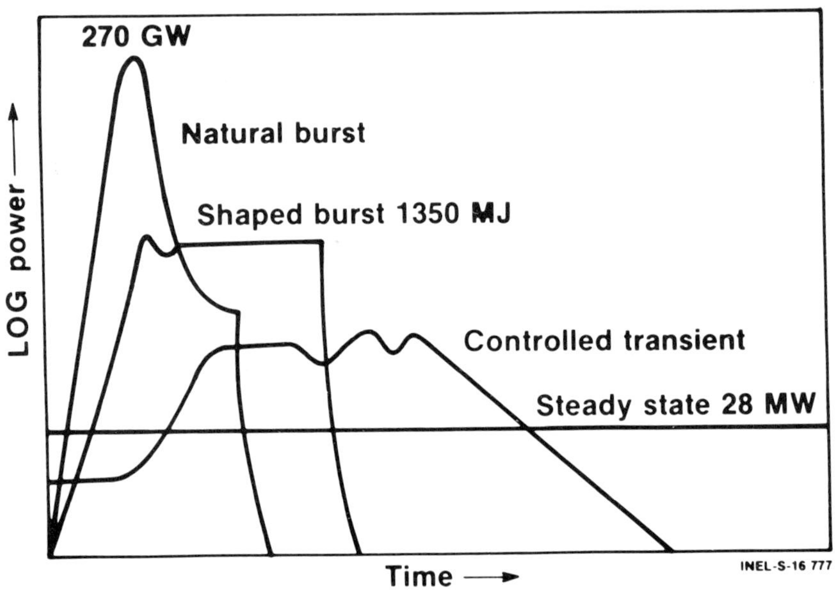

Figure 5. PBF modes of operation.

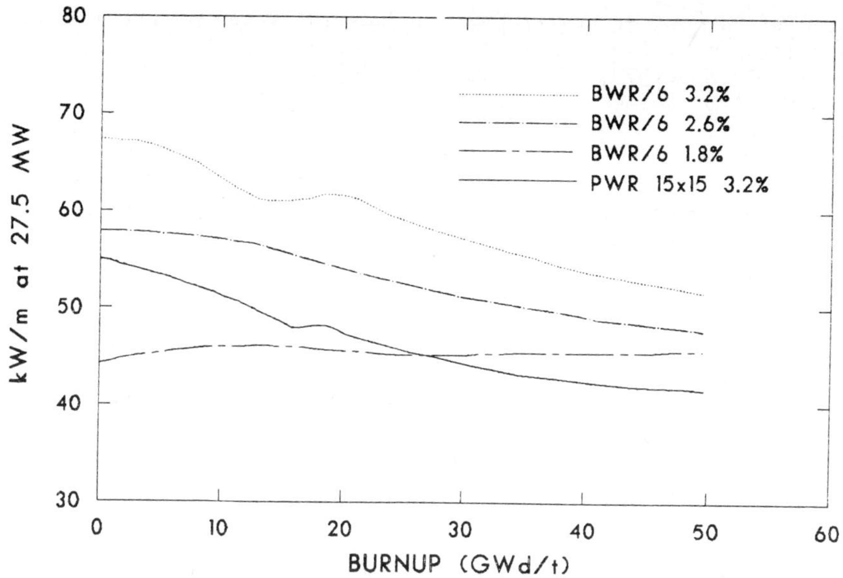

Figure 6. Calculated test rod power as a function of burnup.

EXPERIMENT TEST LOOP

The experiment test loop, illustrated in Figure 7, consists of the in-pile tube, isolation valves, a bypass valve, acoustic filters, rupture disks, heat exchangers, line heaters, a flow control valve, a variable-speed recirculation pump, a pressurizer, instrumented spools, quick-opening blowdown valves, Henry nozzles, a blowdown tank, a reflood system, and a quench tank.

The IPT is a thick-walled (2.54 cm), side-reentrant pressure tube fabricated from Inconel 718. It is designed to contain steady state operating pressures and pressure surges that may result from test fuel failure. A flow tube with a central section (through the reactor core) of zircaloy-2 is positioned within the IPT to direct coolant flow. Coolant enters the top of the IPT above the reactor core and flows down the annulus between the IPT wall and flow tube. At the bottom of the IPT, the coolant reverses direction and flows up through the test fuel, and then exits at the IPT outlet.

The acoustic filters and rupture disks, placed at the IPT inlet and outlet, reduce or control pressure surges which may propagate down the loop piping during test fuel failure.

The recirculation pump is a linear motor pump rated at 223.7 kW and 3.02 m^3/min. The pump motor is driven by a variable-speed motor/generator set.

TABLE I. FIGURE OF MERIT VERSUS BUNDLE SIZE[a]

Number of Rods	Description	Axial Peak FOM (kW/m/MW)
4	12.5% depleted (11,000 to 15,000 MWd/t) 15 x 15 PWR	2.0
4	2.9% depleted (4,000 to 15,000 MWd/t) BWR/6	1.5
4	2-10% fresh BWR/6 2-2.9% depleted (10,000 MWd/t) BWR/6	3.2 1.8
4	2-9.5% zircaloy shroud 15 x 15 PWR	2.7
	2-9.5% stainless steel shroud 15 x 15 PWR	2.4
9	2.9% depleted (5,000 to 20,000 MWd/t) BWR/6	1.4 (bundle average)
9	5.7% depleted (∿5,000 MWd/t) MAPI	1.4 (bundle average)
9	93% center 15 x 15 PWR 35% side 15 x 15 PWR 20% corner 15 x 15 PWR	2.7 2.6 2.6
32	6.2% subcooled 17 x 17 PWR 6.2% 6.9 MPa saturated steam 17 x 17 PWR	0.28 0.17
60	Peripheral rods 3.2% 17 x 17 PWR Center 16 rods 3.2% 17 x 17 PWR Center 16 rods 8.0% 17 x 17 PWR Center 16 rods 20.0% 17 x 17 PWR	0.07 0.08 0.11 0.06 0.12 0.22

a. Figure of merit is test rod power (kilowatts per meter) per megawatt of core power.

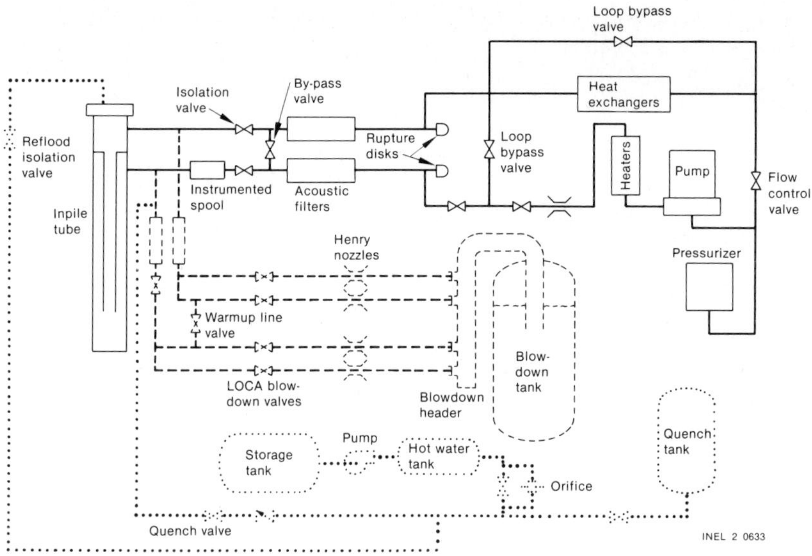

Figure 7. PBF experiment loop schematic.

Instrumented spools are located in the experiment loop inlet piping, cold leg blowdown piping, and hot leg blowdown piping. These spools contain instruments (such as turbine flowmeters, drag disks, gamma densitometers, thermocouples, resistance temperature devices, pressure transducers, and differential-pressure transducers) to accurately quantify test conditions.

Achievable loop conditions are: flow rates from 1.0 to 50.3 L/S; pressure from 0.69 to 15.5 MPa; and temperatures from 340 to 615 K.

Flow within the loop is controlled by changing position of the flow control valve or by varying frequency of the motor/generator set driving the pump motor.

Loop temperature is controlled by use of the loop heat exchangers, loop heat exchanger bypass, line heaters, and driver core heat source. Flow through each heat exchanger can be varied by individual flow control valves to provide fine control of loop coolant temperature. The loop heaters are resistance-type heaters which are clamped to the loop piping.

System pressure is maintained by a pressurizer connected to the loop piping near the pump inlet. Two immersion heaters in the pressurizer maintain pressurizer water at the saturation temperature corresponding to the desired loop operating pressure.

Capability of simulating loss-of-coolant accident conditions in the core test section is provided by the blowdown lines connected to the IPT inlet and outlet lines (cold leg--IPT inlet, and hot leg--IPT outlet). Each line contains an instrumented spool, two quick-opening-and-closing (<100 ms) blowdown valves, and two Henry nozzles which provide the break plane and control the break flow rate and depressurization rate. The blowdown header and tank collect and contain the coolant ejected from the IPT and piping during blowdown and subsequent quench cooling.

Capability of simulating posttest quench cooling, reflood cooling, and postblowdown cooling is provided by the quench and reflood portion of the test loop. The quench tank is pressurized by a nitrogen gas system. Upon opening the reflood valve or the cold leg blowdown valves, water from the quench tank is allowed to enter the IPT. After the quench tank is emptied (about 60 s), coolant is provided from the storage tank for up to 4 hours.

The loop isolation and bypass valves, the blowdown valves, and the quench and reflood valves are controlled and sequenced for the desired test conditions by a computer system that monitors plant status and controls sequenced outputs.

During tests which void the IPT through the computer-programmed blowdown portion of the system, the loop isolation and bypass valves are actuated automatically to isolate the IPT from the experiment loop and divert loop coolant flow through the IPT bypass line.

For the Severe Fuel Damage (SFD) tests, four modifications to the experiment test loop were completed:

1. Seperate test coolant flow lines were added to supply coolant through the experiment. The previously existing coolant flow lines are used to cool the outside of the SFD experiment shroud.

2. An emergency quench system was added to inject water to the outer shroud region in the event of an unplanned experiment test loop loss-of-coolant accident.

3. A shroud pressurization system maintains the pressure inside of the SFD shroud walls at the desired level relative to the test bundle pressure. This is to ensure that the SFD shroud, which is composed of ZrO_2 insulation contained between outer and inner zircaloy walls, does not fail prior to reaching high temperatures.

4. A safety and shutdown modification was added to cause a slow scram if both flowmeters measuring shroud cooling bypass flow indicate a low flow condition. An annunciated alarm is provided if either one of the bypass flowmeters indicates a low flow condition.

One additional modification was completed on the experiment test loop. During preconditioning of the SFD test bundle prior to the transient test, the coolant flowing through the test bundle will enter the experiment test loop through a check valve at the top of the bundle. However, when transient testing begins, a positive pressure will be maintained across the experiment shroud and therefore the check valve will be held shut. Subsequently, all of the coolant flow from the SFD bundle will enter the fission product detection system after preconditioning is completed so that all fission product release can be monitored.

FISSION PRODUCT DETECTION SYSTEM

The fission product detection system as presently installed in the PBF is illustrated in Figure 8. Before exiting the in-pile tube, the steam will be cooled to ensure that temperatures downstream of the in-pile tube do not exceed 980 K. Six effluent sample bombs will be located upstream from a condenser; the sample

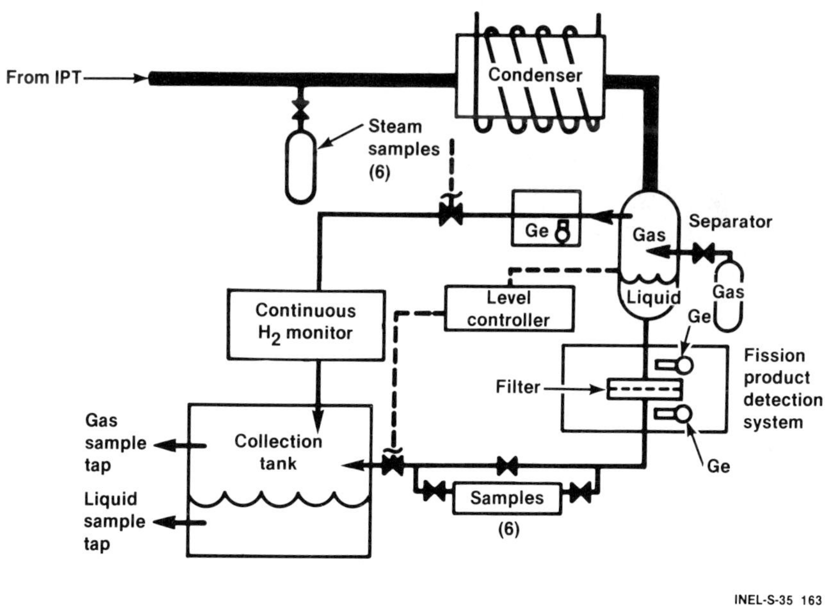

Figure 8. Fission product and H_2 monitoring system.

bombs will be initially at low (ambient) pressure and argon filled. The samples will be remotely operated to provide effluent samples for posttest analysis. The effluent will then pass into a condenser, where the steam will be reduced to liquid at a temperature below 340 K, and enter the separator vessel with a minimum residence time. A continuous purge of the separator gas space with high-purity nitrogen gas will sweep hydrogen, fission gases, and other noncondensibles out of the separator, past a gamma spectrometer and hydrogen monitor, and into the collection tank. The liquid flowing from the separator vessel will be monitored by two gamma spectrometers located upstream and downstream from a particle filter. The liquid will then pass through one of six flow-through grab samples, through a flow control valve, and into the same collection tank as the gaseous effluent. The liquid samples will be remotely operated sequentially such that the sample stream always flows through either a sample bomb or the bypass. There are also provisions for posttest grab samples of the gas and liquid in the collection tank.

DATA SYSTEMS

A Data Acquisition and Reduction System (DARS) records 287 channels of data on magnetic tape in pulse code modulation format. Wideband and narrowband channels are available with appropriate sampling rates.

All narrowband parameters are available on a CRT screen in an on-line, real-time mode. Up to 56 hours of narrowband data can be accumulated on tape reels because of the system's high-density recording. Uncertainty estimates are provided for all data obtained for the experiments.

A second computer system, the Event Sequence and Monitor System, is used to monitor plant parameter status. This system is capable of (a) monitoring 256 discrete input signals in the reactor building with 1-ms resolution capability, (b) monitoring 256 discrete input signals in the control building with 1-ms resolution, and (c) providing 24 controlled discrete output signals to plant equipment with 100-ms control capability.

TEST TRAIN ASSEMBLY FACILITY

The Test Train Assembly Facility (TTAF) was designed to assemble nonirradiated and irradiated experimental test assemblies and to store fuel. The general work areas within the TTAF consist of a fuel rod assembly area, a test train assembly area, a storage canal, a machine shop, and a metal cleaning facility.

Test fuel rods are characterized by dimensional measurements and fuel pellet density measurements in the fuel rod assembly area. Also, pellets are loaded into cladding, end caps installed and welded, void volume measured, and test fuel rod instrumentation installed in the fuel rod assembly area.

Test fuel rods are then loaded into the test hardware (train) and instrumentation is installed on the test train in the test train assembly area.

CONCLUSIONS

The design features of the Power Burst Facility make it a unique and versatile test facility. The reactor system, experiment test loop, and support equipment provide the capability of simulating light water reactor conditions for nuclear test fuels. To date, successful PCM, RIA, LOC, and ATWS test series, simulating light water reactor fuel behavior under these conditions, have been completed. Data regarding fuel behavior obtained from these test series have helped focus on and resolve key safety issues. The PBF is now prepared to begin severe fuel damage testing for the USNRC and is also available to other sponsors. A LOCA test using CANDU fuel rods is presently being co-sponsored in the PBF by Atomic Energy of Canada, Limited, and Ontario Hydro.

The first phase of the PBF Severe Fuel Damage test program is being performed to (a) characterize fuel rod damage resulting from severe cladding oxidation and melting, UO_2 dissolution and melting, and fuel rod fragmentation; (b) measure the release rates, transport, and deposition of fission products; (c) measure the magnitude and timing of hydrogen generation; (d) measure the coolability characteristics of test bundles with various types and degrees of damage; and (e) determine the effects of irradiated fuel and control rods on fission product release.

NOTICE

INPILE TEST RIGS FOR RIA EXPERIMENTS IN THE NSRR

T.Fujishiro, S.Tanzawa and S.Saito

Japan Atomic Energy Research Institute

Tokai-mura, Naka-gun, Ibaraki-ken, Japan

ABSTRACT

This report introduces the major inpile test rigs for the NSRR project conducted in Japan Atomic Energy Research Institute since 1975. The NSRR project is the inpile experimental program to understand the behavior of LWR fuels under reactivity initiated accident (RIA) conditions. LWR fuel rods are irradiated in the pulse reactor, NSRR, to realize a prompt power excursion in the RIA. Different types of inpile capsules and loops have been developed to accomodate the test conditions for the various parametric studies on the RIA fuel behavior. Major design features and test capabilities of these test rigs and the instrumentations for them are described.

1. INTRODUCTION

Inpile experiments to study the LWR fuel behavior under reactivity initiated accident (RIA) conditions are being conducted in the Nuclear Safety Research Reactor (NSRR) project in Japan Atomic Energy Research Institute since 1975. LWR fuels are irradiated in the pulse reactor, NSRR, to realize a prompt power excursion in the RIA.

Fig.1 shows the general arrangement of the NSRR. The reactor is a modified TRIGA Annular Core Pulse Reactor whose features are (1) the large pulsing power capability to heat up the moderately enriched fuel over the melting temperature of UO_2 by nuclear fission, and (2) the large (22 cm in diameter) dry irradiation space located in the center of the reactor core to accomodate a sizable experiment. The core structure is mounted on the bottom of 9 m

P. von der Hardt and H. Röttger (eds.), Irradiation Technology, 191–200.

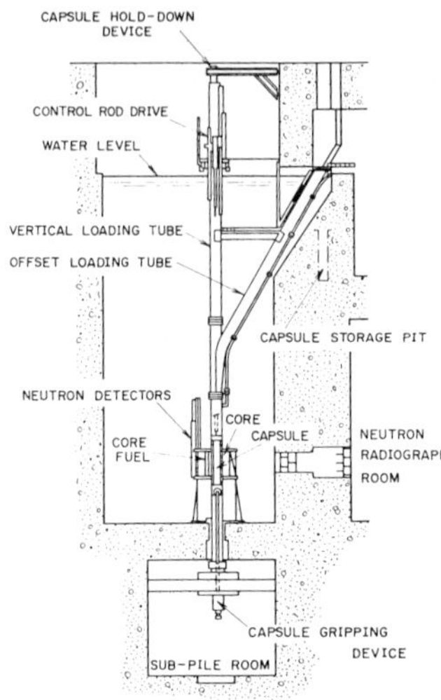

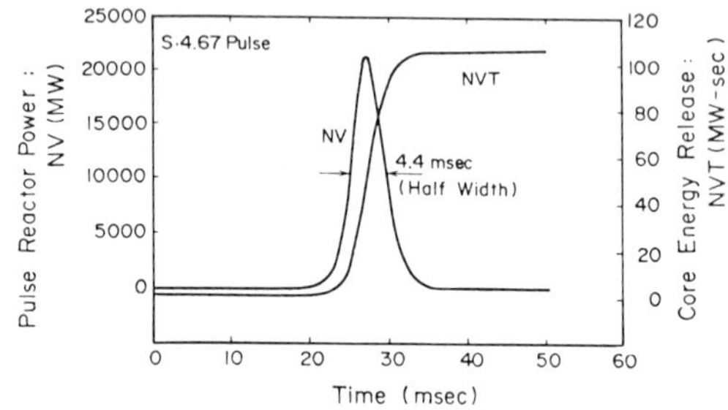

Fig.2 Pulse reactor power and core energy release

Fig.1 General arrangement of the NSRR reactor

deep open top water pool. The center of the core is the experimental cavity to which a test capsule or a loop is inserted through the off-set or vertical loading tubes. Typical transient histories of the reactor power and core energy release are shown in Fig.2.

Various inpile test rigs have been developed to accomodate the test conditions and to obtain good measurements of the transient fuel behaviors. Major test rigs are an atmospheric pressure capsule, a high temperature high pressure (HPHT) capsule, a fuel behavior observation capsule and water loops. Major design features, test capabilities and the state of the arts are described of the test capsules, loops and the instrumentations in the following sections.

2. TEST CAPSULES AND LOOPS

2.1 Atmospheric Pressure and HPHT Capsules

The atmospheric pressure capsule was designed as a handy and versatile irradiation vessel to conduct numbers of test for the parametric study. Fig.3 shows the schematic drawing of the capsule in which a single test fuel rod is installed. The test space is 120 mm in diameter and 800 mm in hight. As the water in the capsule functions as the moderator as well as the coolant, the diameter of the capsule internal was chosen so as to give the maximum energy generation in the test rod for a unit reactor power.

This short sized capsule is inserted to the core through the off-set loading tube. As no shielding plug is needed to this off-set loading tube, the capsule is quickly loaded to the experimental cavity and transfered shortly after the test to the storage pits at the exit of the loading tube. This convenient test procedure enabled us to conduct the fuel failure experiments as many as 500 in 5 years, and to provide a large data base of RIA fuel behaviors on the effects of various fuel design and cooling condition parameters.

The operating pressure of this capsule is 1 atm (0.1 MPa). However, the capsule was designed as a strong pressure vessel made of type 304 stainless steel to sustain pressure pulses and water hammer actions of the coolant slug caused by steam explosion at fuel failure. The design pressures are 13 MPa for the pressure pulses and 61 MPa for water hammer which were estimated from SPERT experimental data. In order to avoid excessive neutron absorption, the capsule wall thickness was cut down as thin as possible by applying elastic-plastic stress analysis in the design. Plastic deformation of 2 % at maximum was allowed in the design, and the margin to the vessel rupture by the shock loads was confirmed by the chemical explosive tests of fully mock-up test capsules.

Based on the experiences with this atmospheric pressure capsule, HPHT capsule was developed to extend the test conditions of the atmospheric capsule to the operating pressure of PWR or BWR. Fig.4 shows the outline of the HPHT capsule. The test vessel is connected to the pressurizer tank by which the system pressure is adjusted. The vapor pressure in the pressurizer is controlled by an electric heater immersed below the water level. The temperature in the test vessel is adjusted by three sheathed heaters in the test vessel. A suppression tank is attached to the top to condense the

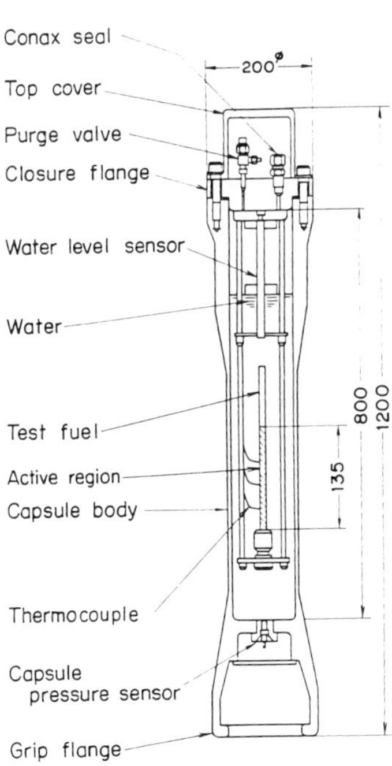

Fig.3 Atmospheric pressure capsule arrangement

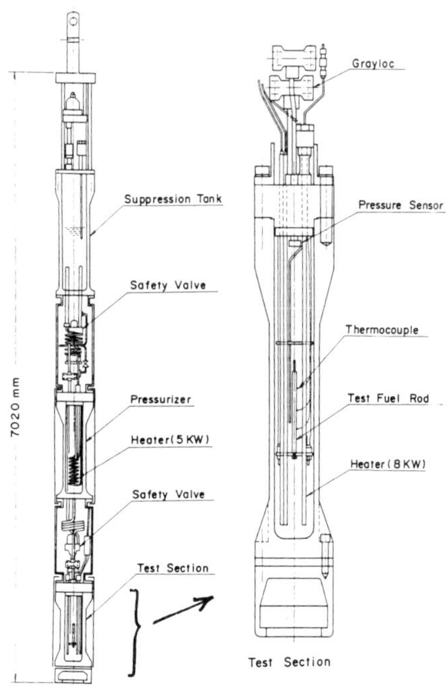

Fig.4 Outline of the HPHT capsule

steam from the safety valves of the pressurizer and the test vessel. This 7 m long HPHT capsule system is assembled on the assembling tower in the reactor room, and inserted to the core through the vertical loading tube. Eight scoping tests in which single test rods were irradiated under PWR and BWR conditions have been completed, and the fuel design parameter tests, in which three different rods are irradiated simultaneously in the same vessel, are now progressing.

2.2 Fuel Behavior Observation Capsule

Inpile high speed cinematography of the fuel rod during an RIA transient was attempted. The ambient pressure capsule was connected to a high speed cinematography system as shown in Fig.5. In the test vessel at the bottom, a periscope head and 3 kw tungsten lamp are installed with the test fuel in the water. The image of the fuel rod is led through the periscope to the high speed camera at the top.

The periscope, 2.6 m in length and 30 mm in outer diameter, is composed of a prism, 2 mirrors and 22 lenses. The prism is set at the bottom of the periscope and used to turn the image of an object at right angles. The light path in the periscope is turned two times at right angles by the mirrors at the middle of the path to protect the film from the radiation streamed through the periscope. To prevent the coloring of the glasses under high neutron and gamma-ray irradiation, all the lenses and the prism were made of non-browning glass. The visible light transmission of the periscope becomes about 30 % in the whole visible range because each non-browning lens, which is 3 mm thick and 14 mm in diameter, has light transmission of about 95 %. The portion of the periscope in the test capsule is covered with a protection tube which has a thick quartz window at the end to protect from the shock pressures and the attack of hot fuel fragments dispersed at fuel failure.

A small sized high speed camara which can be installed inside the 15 cm diameter cavity is set at 2.6 m apart from the midplane of the core. The upper portion of the periscope and the high speed camera are also covered by a containment to prevent the release of the fission products to the cavity in case of the failure of the anti-shock quartz window. The high speed camera and film are protected from neutron and gamma-ray radiation by shielding structure consisting of lead and boron contained paraffin.

Eight tests have been conducted with this capsule system successfully. In Fig.6, some pictures from the motion color film taken at 250 fps are shown. The case shown is a PWR type rod experiment in which the test rod was subjected to about 300 cal/g of UO_2 under atmospheric pressure and stagnant coolant. Wavy vapor film surrounding the red-hot fuel rod was clearly visualized.

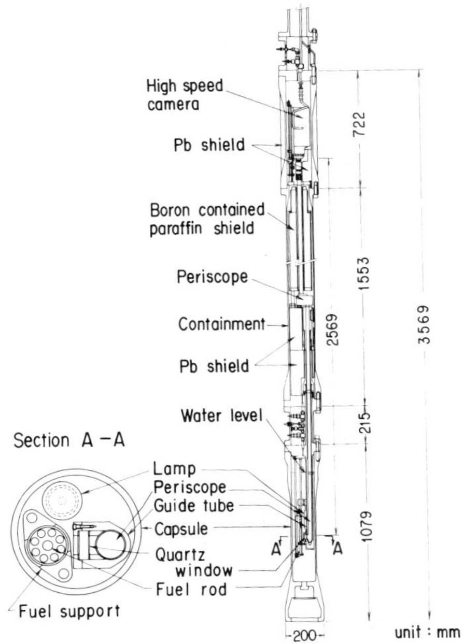

Fig.5 Schematic configuration of fuel behavior observation capsule

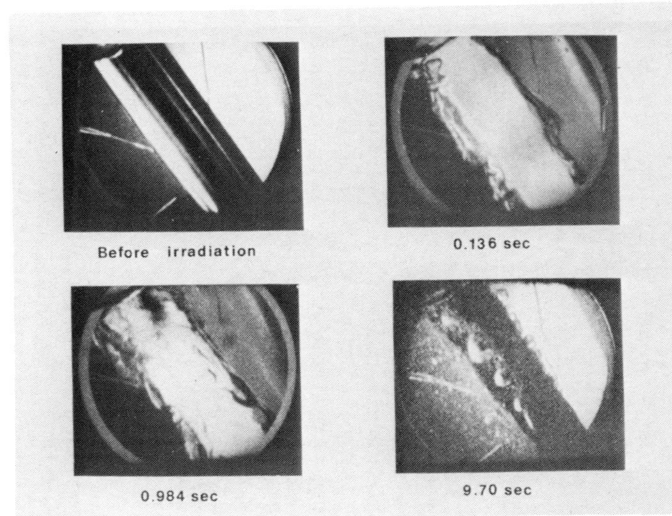

Fig.6 Pictures printed from motion color film taken at 250 fps. The fuel rod was subjected to about 300 cal/g of UO_2.

2.3 Water Loop

Water loop test facilities were developed to examine the effects of coolant flow conditions on RIA fuel behaviors. Two types of inpile water loops, low pressure loop and high pressure high temperature (HPHT) one, have been constructed so far. The low pressure loop was designed to conduct the experiments under comparatively low pressure of 1.5 MPa at maximum to obtain preliminary information on the effects of coolant flow and pressure, and to obtain technical know-how on the loop experiment under relatively easier conditions. The coolant velocity of maximum 7 m/s was provided for a 9 rod cluster of PWR type under the temperature of near saturation. Scoping tests have been completed and the data indicated large influences of coolant flow and pressure on fuel thermal behaviors.

Based on the experiences with this low pressure loop, HPHT loop has recently been installed in the NSRR. Fig.7 shows the flow chart of the HPHT loop. The loop consists of inpile flow tube, which has a test section at its lower end, and the coolant circulating system located at the reactor side. Maximum operating conditions are pressure of 16 MPa, temperature of 347°C and the coolant flow rate of 280 l/min. The inpile tube is designed for the pressure of 25 MPa considering the pressure pulse generation at fuel failure. In the experiment, the fuel rods in the test section are irradiated under a large pulse power operation to simulate the RIA while the system conditions are kept constant.

Fig.8 is a test section arrangement of the HPHT loop for the irradiation of four single test rods. Four test rods are supported in the independent flow shrouds. The coolant flows down in the outer side of the flow shrouds and, turning at the bottom of the test vessel, goes up in the annular coolant space between the fuel rod and the shroud. Each flow shroud has an inlet orifice, inlet and exit flowmeters, pressure sensors and a density meter for void ratio measurement. With this device, four differently designed rods can be irradiated simultaneuosly under the same condition, or the same type of rods under different coolant flow rates by using different inlet orifices.

3. INSTRUMENTATIONS

3.1 Measuring Items

Instrumentation is the another very important element of an experimental facility. The major measuring items for the NSRR experiments are summarized in Table 1. Particular requirements for the RIA test instrumentations are, firstly, to have a fast response capability to follow a very quick transient initiated by a prompt heat up of the fuel rod, and secondly, not to be influenced by a large pulse irradiation of neutron and gamma-rays. In the following, instrumentations for cladding temperature, fuel and cladding

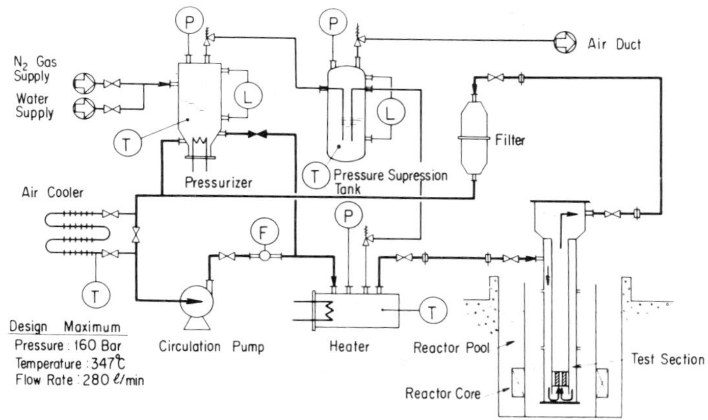

Fig.7 Flow chart of the NSRR HPHT loop

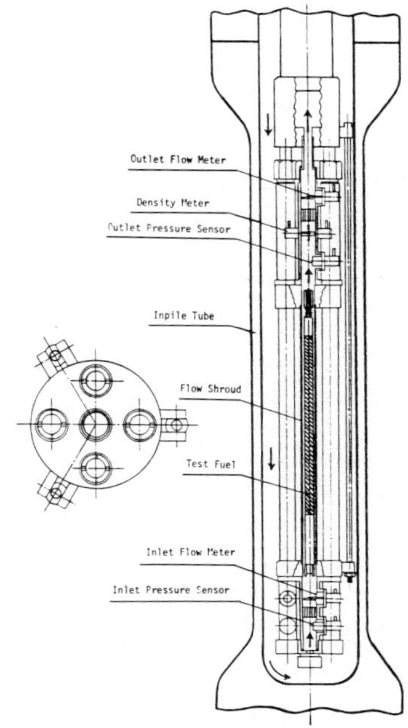

Fig.8 Schematic of test section arrangement of the NSRR HPHT loop

Table 1 Summary of Test Instrumentations

Measuring Item	Instrument	Measuring Range	Frequency Response	Test Hardware*3	Comments
Temperature					
Cladding Surface	pt/pt-13%Rh Bare-wire T/C *1 D = 0.1 - 0.3mm	RT-1800°C	200 Hz	AC,HC,LL,(HL)	*1 :- spot-welded on cladding surface
Fuel Centerline	W-5%Re/W-26%Re Tantalum Sheathed T/C D = 1.6mm	RT-3000°C	5 Hz	AC	*2 :- resonanse frequency
Coolant	C/A Sheathed T/C D = 1.0 mm	RT- 500°C	10 Hz	AC,HC,LL,(HL)	
Pressure					*3 :-
Fuel Rod Internal	Flash Diaphram Type	5-100 MPaFS	170 KHz*2	AC,HC,(HL)	AC : Atmoshperic pressure Capsule
Capsule Internal	Strain Gauge Transducer	1- 20 MPaFS	110 KHz*2	AC,HC,LL,(HL)	HC : High Pressure Capsule
Displacement					
Cladding & Fuel Stack	LVDT	± 10 mm	1 KHz	AC,HC	LL : Low Pressure Loop
Meachanical Energy					
Coolant Ejection Speed	Float Type Senson	0-20 m/s	-	AC	HL : High Pressure Loop
Strain					
Flow Shroud and other Internal Structire	Strain Gauge	100-6000μstrain	10 KHz	AC	
Coolant Flow Rate					
Inlet and Outlet Coolant Flow Rate	Drag-disc Type Transducer	50-200 l/minFS	100 Hz	AC,LL,(HL)	
Void Ratio					
Coolant Density	Vidrating Vane Type	0-100 %	10 Hz	AC,HL	

displacements, coolant flow rate and void ratio are presented among the items listed in Table 1.

3.2 Cladding Temperature

Thin bare wires of Pt/PtRh thermocouples are directly spot-welded on the cladding to obtain good response. Fig.9 shows a cross-section of the cladding and spot-welded thermocouple wires. Each thermocouple wire of 0.1 to 0.3 mm in diameter is welded separately as shown in the photograph to minimize the disturbances to the measuring point. Response time of less than 10 ms and the temperature disturbance of less than 150°C in the range up to 1600°C were attained by this method.

The same method was applied to the high pressure high temperature experiments. Bare wires at the end of a 1.0 mm dia. seathed thermocouple are spot-welded to the cladding as shown in Fig.10. Several kinds of insulation at the thermocouple sheath end have been tested under high pressure and temperature of PWR condition. The final conclusion obtained was that no insulation is necessary for pure water coolant for the duration of our RIA experiments, i.e. one full day under atmospheric pressure during the preparation and around 10 hours under high pressure and temperature during the test.

3.3 Displacements of Fuel Stack and Cladding

LVDTs (Linear Variable Differential Transformer) are used for the axial displacements of fuel stack and cladding. Fig.11 shows the schematic of the LVDT's arrangement at the upper end of the test fuel. The LVDT for high pressure and temperature application has been developed. The simultaneous measurement of fuel stack and

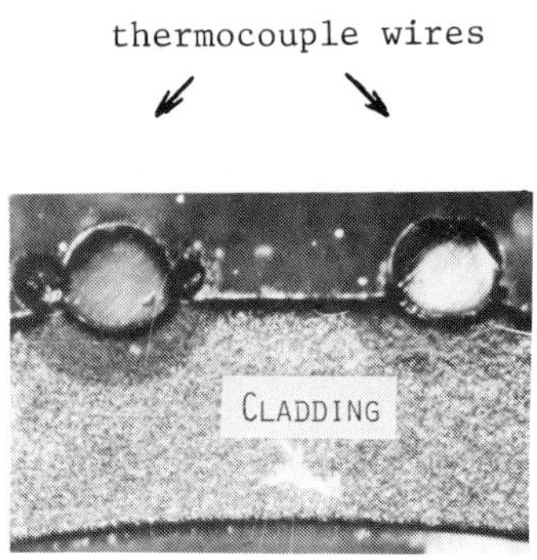

Fig.9 Cross-section of spot-welded thermocouple wires on the cladding surface

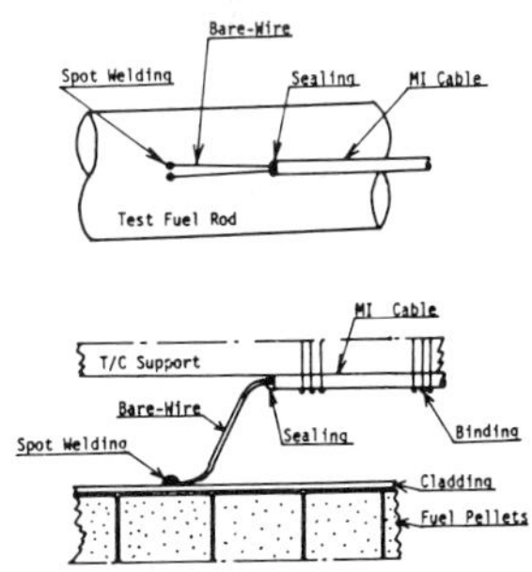

Fig.10 Application of bare-wire thermocouples to the HPHT environment

cladding axial displacements under BWR or PWR conditions are now progressing for the RIA transient with these LVDTs.

3.4 Flow Rate and Void Ratio

Coolant flow rate in the loop tests are measured by drag-disc type flowmeters. Four strain gauges are used in Wheatstone bridge for each transducer in order to cancell out induced noises by the large pulse irradiation and to decrease the drift by the temperature. Good response and simple structure are also the merits of this type. To obtain wider dynamic range, small sized turbine flowmeter changeable with the drag-disc one is under development with the cooperation of Grenoble Nuclear Research Center.

Vibrating vane type density meter is in the development for measuring the transient void ratio at the flow shroud exit. Average density around a flat vibrating vane is detected by the change of vibrating frequency. Fig.12 illustrates an exit measuring spool consisting of a drag-disc flowmeter and a density meter. Good results have been obtained with this test spool in the preliminary experiments under atmospheric pressure.

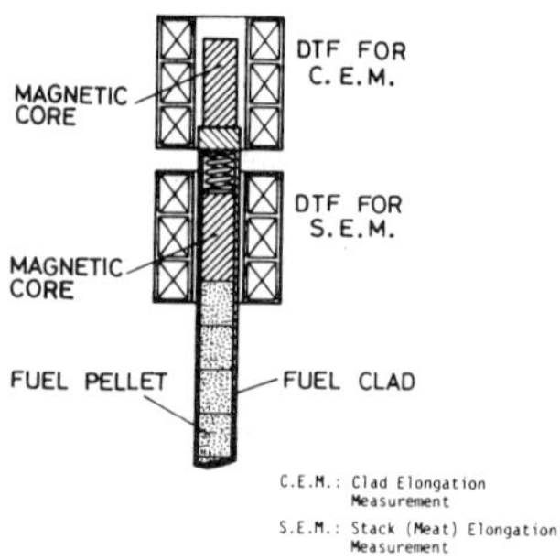

Fig.11 LVDTs' arrangement at upper end of test rod

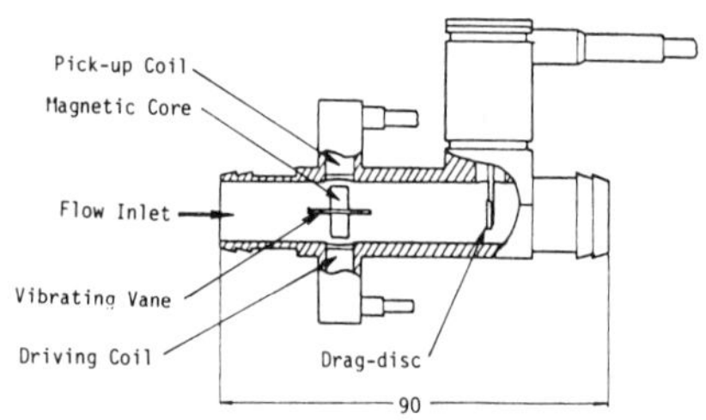

Fig.12 Measuring spool for the coolant density and velocity

3. CONCLUSIONS

Various types of inpile experimental tools have been developed for the NSRR - RIA experiments. Our major efforts are at present focused on the high pressure and temperature experiments and on the preparation of future burn-up fuel experiments in Phase-II program.

We would be very happy if the experimental technique for the RIA study introduced can give some useful suggestions to other inpile experimental fields.

DESIGN SYNOPSIS FOR THE NEW ACCIDENT RELATED STUDIES LOOP AT CRNL

W.K. Snelson

Atomic Energy of Canada Limited
Chalk River Nuclear Laboratories
Chalk River, Ontario Canada K0J 1J0

ABSTRACT

This paper describes a new in-reactor test loop in the NRX research reactor at Chalk River Nuclear Laboratories. Test results from this facility will be used to verify the mathematical models which predict fuel behaviour and fission product releases during postulated LOCA and LORA accidents in CANDU power reactors.

The loop is designed to operate with pressurized water, superheated steam, or fog cooling, with the capability of blowing down from any mode. The main loop and auxiliary systems will be capable of containing the fission product releases caused by severe fuel failures which may result from some of the accident tests.

INTRODUCTION

In April 1981 a proposal to construct a new test loop to further the studies of fuel performance under postulated in-reactor accident conditions (Loss of Coolant and Loss of Regulation Accidents) was approved. The new facility, subsequently designated as X-9, was required to measure releases and assess the transport of fission products during tests involving increasingly severe fuel damage. Experiments were also considered necessary to experimentally verify computer code predictions currently used to analyze fuel behaviour during postulated LOCA and LORA situations. Improved accuracy in these codes may serve to eliminate excessive conservatism in the accident analyses used for CANDU power reactor licensing.

P. von der Hardt and H. Röttger (eds.), Irradiation Technology, 201–211.

TEST FACILITY REQUIREMENTS

For LOCA tests the loop is required to blow down (depressurize) from superheated steam, fog or pressurized water. Steam cooled operation is necessary to achieve fuel sheath temperatures in excess of 1100°C (2000°F), and to accurately predict blowdown heat transfer coefficients used in the fuel modelling codes.

For LORA tests the loop is required to operate at steady power in superheated steam, fog or pressurized water. In these tests fuel sheath dryout will be induced, causing bowing of the fuel element and possible contact with the pressure tube. Results from these tests will provide information regarding the effects of post-dryout fuel behaviour on pressure tube integrity. Again the superheat mode is required to avoid unpredictable changes in heat transfer regimes associated with dryout.

The loop and blowdown system must be capable of containing the fission product releases caused by severe fuel failures which may result from some of the above accident tests.

LOOP SCHEMATIC DESCRIPTION

Figure 1 shows a simplified flowsheet.

In superheated steam mode coolant flows from the pumps through a preheater which raises the temperature to saturation. Steam is generated by the boilers and a superheater provides additional heat to produce the required test section inlet conditions. Maximum steam flow will be about 0.3 kg/s and test section inlet temperature 350°C. The superheated steam flows through the test section to a jet condenser which is connected in parallel with a surge tank. The steam is condensed by a spray flow of water from the cooler circuit and condensate returns to the pump suction line.

In pressurized water mode the coolant flows directly to the test section from the preheater. Components are sized for a maximum flow of 1.5 kg/s and typical test section inlet temperature will be about 280°C. Operating pressure will be in the 10 MPa (1450 psi) range for most tests. Downstream of the test section flow bypasses the jet condenser and returns directly to the pump suction line.

In fog mode superheated steam and pressurized water are combined in mixer #1 to give the required test section inlet conditions. Fog flows are expected to be similar to those for steam, with qualities up to 50%. The fog produced passes through

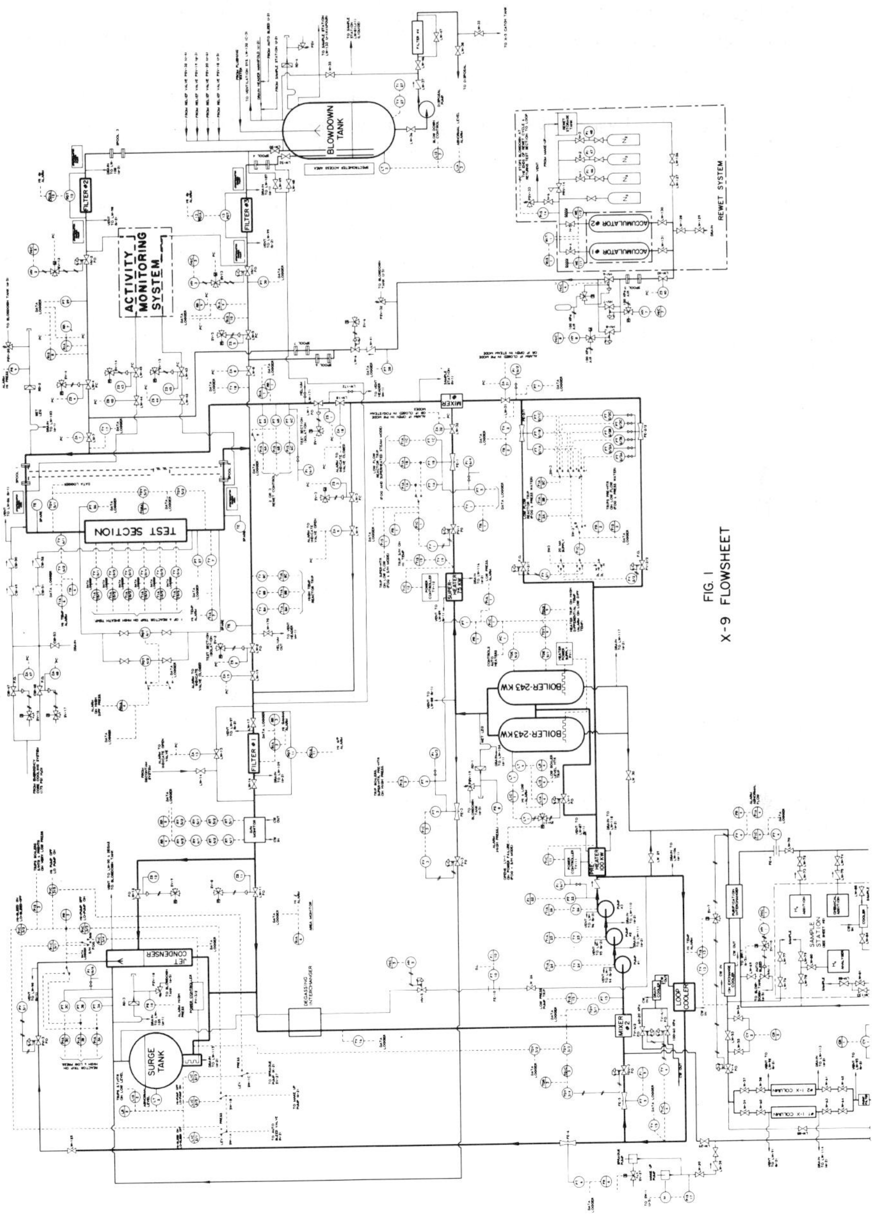

FIG. I
X-9 FLOWSHEET

the test section and flows to the jet condenser and back to the pumps as in the superheated steam mode.

The cooler circuit is connected in parallel with the main loop system and supplies cool water as required to the jet condenser sprays and to mixer #2. Outlet temperature from the cooler is pre-set according to operating mode conditions and is controlled by modulating cooling water flow on the secondary side.

Loop pressure control in the superheat and fog modes is achieved by varying the flow of cool spray water into the jet condenser. Boiler pressure is controlled by modulating the heater power supply. In pressurized water the surge tank and jet condenser ride on the main loop circuit and pressure is controlled by the surge tank heaters.

A purification system containing ion exchange columns is also connected in parallel with the main loop circuit. Connections are made between this system and a glovebox which is used to sample and control loop coolant chemistry.

Flow through the test section can be directed upwards or downwards by changing spool pieces. A bypass line around the test section allows the loop to continue operating while the test section is isolated during blowdown tests.

Separate blowdown lines run from the top and bottom of the test section to a closed blowdown tank partly filled with water for quenching purposes. Blowdown rates are controlled by pre-set valves located in these lines.

A sampling and monitoring system is connected to each blowdown line to detect and follow the release and transport of fission products during the blowdown and rewet phases.

A pressurized rewet system cools the fuel after blowdown. The rewet fluid is contained in accumulators, and automatically-operated valving enables rewetting to occur at various pre-set flow rates. Spool pieces are used to direct the flow to either top or bottom of the test section. The rewet flow discharges to the blowdown tank.

A disposal system is used to empty the blowdown tank once the activity level of the contents is acceptable. The fluid may be pumped away for further treatment or to storage tanks.

Connections are provided for decontaminating the main loop and blowdown circuits. A flushing water spray connection is used for cleaning the insides of the blowdown tank.

MECHANICAL DESIGN

The loop design conditions are 13.8 MPa (2000 psi) at 454°C (850°F) and 16.5 MPa (2400 psi) at 343°C (650°F).

The vessels are designed and constructed to ASME Code Section VIII Division 1, the piping to the Power Piping Code ANSI B31.1, and the test section is designed using the rules of ASME Section III Class 1 as a guide. Where code compliance is not possible the design principles and criteria used are approved by the Canadian Atomic Energy Control Board. In addition the test section is manufactured, inspected and tested to the highest quality assurance level (Z299.1) of the Canadian Standards Association.

MATERIALS

Due to the high design temperature the vessels, piping, and test section are built of stainless steel. Type 347 SS was chosen to avoid possible carbide precipitation effects.

Although Zr-2.5 wt % Nb or Zircaloy would be preferable for the test section the highly accelerated creep rates of these materials at high temperatures make them unsuitable for superheat testing. The inner and outer tubes of the cooling jacket are made of Zr-2.5 wt % Nb and Zr-2 respectively.

EQUIPMENT LOCATION

A layout showing location of the main loop components is indicated in Figure 2. This site is situated inside the NRX reactor building on the southeast side of the reactor. It permits relatively short piping lengths running below ground level to and from the lattice, and easy excavation due to soft soil conditions.

The equipment is contained within an underground concrete enclosure 10.1 m (33 ft) x 9.45 m (31 ft) x 4.6 m (15 ft) deep. For shielding and safety purposes the enclosure is subdivided into five areas separated by concrete walls and covered by concrete slabs. Concrete is used where possible rather than movable lead bricks due to the large cost differential.

Most of the process equipment is grouped in packages and located in the Main Loop Room. The blowdown tank, filters and ion exchange columns may develop high radioactive fields and these components are located in a separate Filter Room surrounded by sufficient concrete and lead to reduce the activity to an acceptable level. These rooms are not accessible during

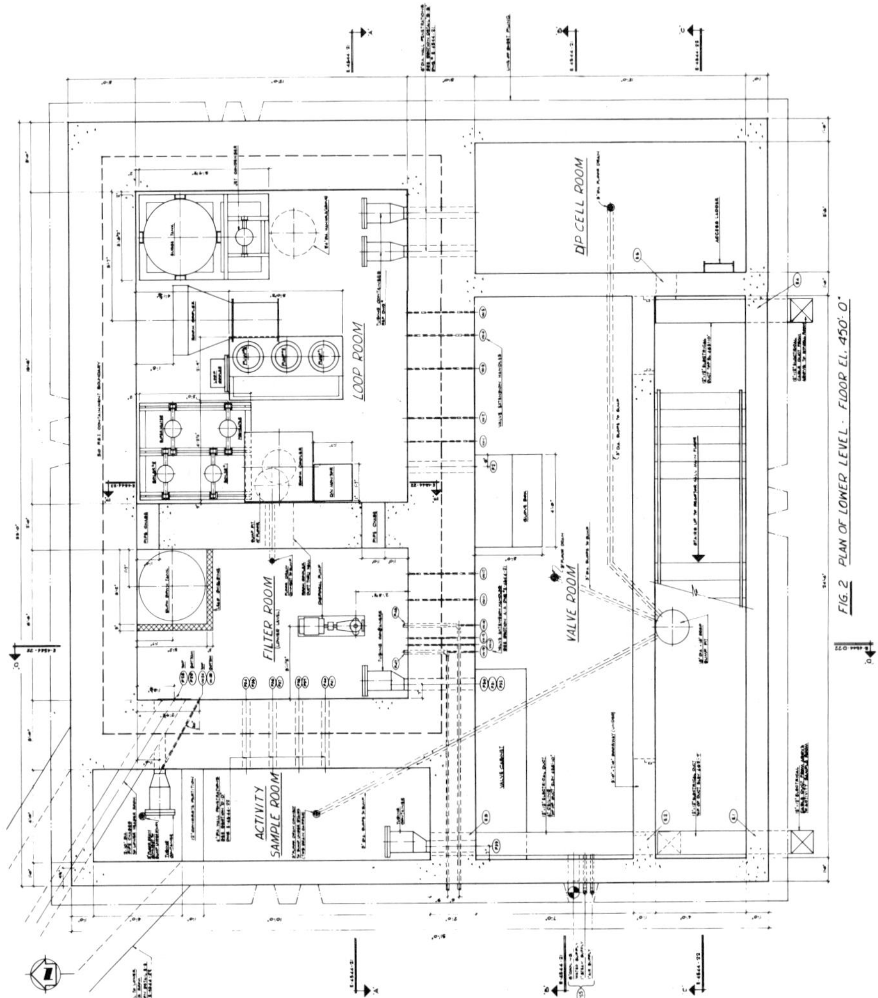

FIG. 2 PLAN OF LOWER LEVEL · FLOOR EL. 450' 0"

This illustration is printed
from the best available original

operation. Manual valves in the main loop circuit are operated remotely from the Valve Room using extension handles passing through the shielding walls. An open stairwell leads down to the Valve Room which also provides access to the glovebox and a valve cabinet containing all the vent, drain and purification system valves. The D/P Cell Room contains all the transmitters required to convert instrumentation signals used for indication and control purposes. The Activity Sample Room is subdivided into two sections separated by a concrete block wall. One area contains the activity monitoring system used for analyzing grab samples. The other provides access to spectrometers located in the Filter Room and in the tunnel containing the interconnecting piping between the reactor and loop room.

CONTAINMENT

The concrete enclosure for the Main Loop Room and Filter Room forms a sealed containment and is designed to withstand an internal pressure of 0.45 MPa (50 psig) to prevent activity release to the reactor hall in the unlikely event of a loop pipe rupture.

The two rooms are interconnected to equalize pressure. The cover slabs are bolted down and leakage is minimized by neoprene gaskets. The pipe sleeves carrying main loop piping along the tunnel to the Lower Header Room of the reactor are also part of the containment. A dousing spray system is provided to ensure that the room pressure does not exceed the design pressure. Dousing is actuated by an increase in room pressure such that even in the worst postulated failure case there is adequate time for the sprays to be effective long before the room pressure approaches 0.45 MPa (50 psig).

The enclosure is designed in accordance with the National Building Code of Canada. The ANSYS engineering analysis system was used to analyze the finite element model of the containment area.

Hot loop piping leading in and out of the tunnel must penetrate the concrete wall of the Filter Room and the concrete shielding outside the Lower Header Room. To avoid overheating of the concrete in these areas the pipes are located inside sleeves, and cooling air is blown down the annulus between the pipe insulation and the sleeve. Penetrations through the containment walls have been specially designed for valve extension handles, and sealed bulkhead fittings are used to accommodate piping, tubing and cabling penetrations.

TEST SECTION

Reference design for the test section is a 3-element single bundle (trefoil) fuel assembly. A typical cross-section is shown in Figure 3.

The fuel assembly consists of the fuel elements and a flow tube used to protect the pressure tube from possible damage due to fuel/pressure tube interaction. This assembly will be used for the code verification test work. For severe fuel damage testing an insulating type flow tube will probably be necessary to prevent any possibility of injury to the pressure tube. A fuel catcher is used to prevent any large fuel particles from reaching the main loop during fuel rupture tests.

The pressure tube is welded directly to the loop inlet and outlet piping and must be cut when necessary to remove it from the reactor. Eliminating mechanical connections avoids any potential thermal transient problems associated with material section changes.

A cooling jacket surrounds the pressure tube similar to other NRX loops to insulate the aluminum calandria tube from high temperatures.

The fuel assembly is installed and removed from the test section through the top closure. This is a specially designed compression seal unit using a Flexitallic gasket, and it is operated remotely from the reactor deck plate when removing fuel to the transfer flask at the end of a test.

PIPING

All main piping runs have been analyzed using UCCPIPE, a proprietary finite element stress analysis computer program.

The use of stainless steel and the high design temperature combine to make pipe routing difficult to accommodate all the expansion loops, hanger locations, and anchor points required.

VALVES

All isolation and control valves in the primary circuit are bellows sealed to prevent activity release due to leakage at the stem or body/bonnet joints. The numerous small valves in the vent and drain system which connects directly to the main loop are also bellows sealed.

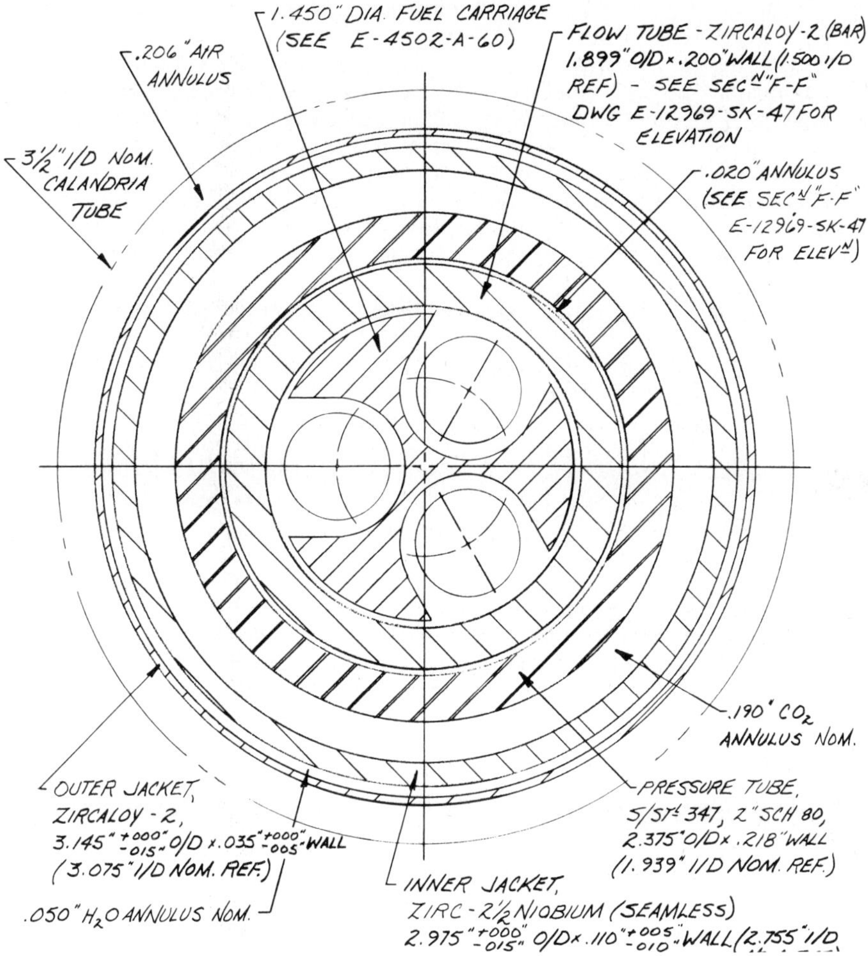

FIG. 3 TEST SECTION

All control valves are installed with R-Con pipe connectors for ease of removal for maintenance purposes. All other valves are butt welded directly into the piping.

PUMPS

The pumps are Westinghouse canned motor type 150-D units which are no longer required in existing test facilities. This design of pump has given continuous reliable service in other CRNL loops for over 20 years with a minimum of maintenance required. It is clearly the type to use in an accident related studies loop where a high level of pumping reliability is mandatory.

VENTILATION

Since the main loop components are located in a tightly sealed containment the ventilation requirement for this area is minimized. However, it was considered desirable to have a small purge flow of air and to maintain a negative pressure in the containment.

23.6 L/s (50 cfm) of filtered air is drawn into the containment and exhausts through a filter train to the reactor exhaust ventilation system. The filter train consists of a heater to reduce humidity, a HEPA filter to remove particulate, two charcoal adsorbers in series to remove gaseous fission products, a fan and a check valve. An identical filter train is located in parallel with the first. Only one train operates at a time, the other providing a redundant system which allows contaminated filters and adsorbers decay time before removal. The filter trains are located in the upper section of the Valve Room such that slabs at reactor hall floor level provide easy access for filter cartridge removal.

For the peripheral rooms 472 L/s (1000 cfm) of air is drawn from the reactor hall into the Valve, D/P Cell, and Activity Sample Rooms through various cracks, ports and doors. The outlet air is collected in an exhaust header which leads to a shielded filter train located on the east side of the loop.

FILTRATION

The main loop filter located downstream of the test section is required to trap any fuel particulate which passes through the fuel assembly catcher. The blowdown line filters are required to fulfill a similar function.

A disposable cartridge type filter has operated well in other loops and similar units have been designed for X-9. The filter beds are about 50 cm (20 in) deep and will contain either graphite particles or a suitable mesh capable of fine filtration.

The cartridges are located in 8 in. (nominal) diameter vessels accessible at the top through R-Con connectors. When a filter has to be removed a shielded filter transport flask is used to transfer the loaded cartridge and covering shield plug from the vessel to disposal.

SHIELDING

Lead and concrete are used to minimize radiation fields around accessible areas of the loop for operating and maintenance personnel.

The concrete walls around the containment and the slabs covering the Main Loop Room are 0.91 m (3 ft) thick. The filters and ion exchange columns are shielded by self-contained lead wells around the sides and by lead caps on top. The blowdown tank is shielded by lead walls around the sides and on top. Piping runs in the Upper and Lower Header Rooms are shielded with as much lead as possible, consistent with the space available.

STATUS

Most of the design work has been done in-house by Plant Design Division staff, and detailed design is now nearly complete.

Construction is also being handled by CRNL forces. Excavation and structural steel installation is finished, the vessels are being built, and loop commissioning is scheduled for late 1983.

TENTATIVE TEST PROGRAM

After initial calibration testing it is planned during the next three years to complete the code verification test program and to do some severe fuel damage tests. These experiments will involve blowing down from increasingly high fuel sheath temperatures up to at least 2000°C.

POWER BURST FACILITY SEVERE FUEL DAMAGE TEST PROGRAM[a]

R. K. McCardell
P. E. MacDonald

EG&G Idaho, Inc.
P.O. Box 1625
Idaho Falls, Idaho 83415
U.S.A.

ABSTRACT

As a result of the Three Mile Island Unit 2 (TMI-2) accident, the United States Nuclear Regulatory Commission (USNRC) has initiated a severe fuel damage research program to investigate fuel rod and core response, and fission product and hydrogen release and transport during degraded core cooling accidents. This paper presents a discussion of the expected benefits of the PBF severe fuel damage tests to the nuclear industry, a description of the first five planned experiments, the results of pretest analysis performed to predict the fuel bundle heatup for the first two experiments, and a discussion of Phase II severe fuel damage experiments. Modifications to the fission product detection system envisioned for the later experiments are also described.

INTRODUCTION

The light water reactor (LWR) accident at the Three Mile Island Nuclear Station-2 (TMI-2) clearly demonstrated that large hydrogen and fission product releases to the containment can occur despite current regulatory procedures and LWR engineered safety

---

a. Work supported by the U.S. Nuclear Regulatory Commission, Office of Nuclear Regulatory Research, under DOE Contract No. DE-AC07-76ID01570.

systems. Accordingly, the United States Nuclear Regulatory Commission (USNRC) is considering fundamental changes in the regulatory policies which have guided LWR design and siting over the last 25 years and has initiated a Severe Fuel Damage (SFD) research program to investigate fuel rod and core response, and the release and transport of fission products and hydrogen during degraded core cooling accidents. A significant portion of the USNRC's experimental program is being performed in the Power Burst Facility (PBF) at the Idaho National Engineering Laoratory. The SFD experimental program in the PBF is designed to obtain the data necessary to understand fuel behavior and coolability under severely degraded conditions; fission product release, deposition, and transport; and hydrogen generation. The data will be obtained from well-controlled, highly instrumented tests that will not only help reconstruct the TMI-2 scenario, but will also investigate other accident scenarios leading to severe damage.

The following sections contain a discussion of the expected benefits of the PBF severe fuel damage tests, a description of the first five planned tests, a discussion of experiment predictions for the first two experiments, and a discussion of the Phase II SFD experiments. Modifications to the PBF fission product detection system envisioned for later experiments are also described.

EXPECTED BENEFITS OF THE PBF SEVERE FUEL DAMAGE TESTS

The PBF severe fuel damage test data will be combined with results from the TMI-2 examination, results from small-scale experiments being performed at other laboratories, and mechanistic computer models to (a) develop more accurate and reliable Probabilistic Risk Assessment (PRA) consequence analysis, (b) help the USNRC set performance requirements for nuclear power plant designs and engineered safety features that will best protect the public from radiation, (c) ensure correct emergency response actions during a severe accident, and (d) provide an in-depth understanding of what happened at TMI-2.

Probabilistic Risk Assessment

Probabilistic risk assessment is the methodology currently used by the USNRC to systematically analyze the probability and consequences of severe accidents over a broad range of system conditions, initiating events, and sequence paths. Improved estimates of risk from PRA studies will be possible only when uncertainties in the complex, coupled physical processes during degraded core cooling accident sequences are better understood and quantified. The planned SFD tests in the PBF will contribute to an improved understanding of fuel-related phenomena during degraded core cooling events and provide data necessary to reduce the uncertainties in severe accident PRAs. Fission product release, transport, and deposition are important examples of the applicability of SFD data in quantifying current uncertainties in PRAs.

Large uncertainties currently exist in PRAs regarding primary system fission product holdup, which is very important for those risk-dominant accident sequences in which the containment is bypassed or fails prior to or during core melt. Four of the nineteen risk-dominant accident sequences in the review of PRAs for Indian Point II, Indian Point III,[1] Zion,[2] Limerick,[3] and Sequoyah[4] involve complete bypass of the containment (interfacing systems loss-of-coolant accident), four involve core melt after containment overpressure failure, and three involve containment failure soon after core melt. Primary system retention is the dominant mechanism for source term reduction in such cases. It has been suggested that a factor of ten decrease in the estimate of fission product release assumed in WASH-1400[5] would eliminate all early fatalities.[6] Therefore, the importance of fission product retention in the primary system is evident.

Reactor Designs and Engineered Safety Features

Commercial power reactors incorporate both passive and active design features to limit the exposure of the public to reactivity. These design features limit the temperatures reached in the core and the fission products released to the environment during conditions varying from normal operation through design basis accidents. The consequences of degraded core cooling accidents can also be mitigated through improvements in design and/or operation, when the physical processes are understood. Data obtained from the SFD program will provide important information regarding hydrogen generation; fission product release, transport, and deposition; damaged fuel characteristics and coolability; the effects of control rods and irradiated fuel; as well as temperatures, pressures, and the timing of various events. This information is required to help set performance requirements for reactor designs and engineered safety features that mitigate severe accidents. Presently under consideration are questions regarding containment size, containment venting, hydrogen ignitors or containment inerting, fission product scrubbers, core liquid level detectors, and damaged core heat removal systems.

Emergency Response

It is clear from the TMI-2 experience that emergency preparedness is extremely important to the public in their perception of the risk of nuclear power. The public demands and the operating utilities have the responsibility to provide timely and accurate information regarding nuclear reactor accidents. Further, decisions concerning evacuation or sheltering of the public require an understanding of the real risk and actual consequences associated with an accident. The information necessary to implement appropriate levels of emergency response will come, in part, from recognition of PRA-determined accident "signatures" (unique features that characterize the type of accident occurring), the output of selected monitoring instruments designed to operate reliably under severe accident conditions, and an understanding of real hazards.

The primary driving force in emergency planning is the radiological hazard to the general public. We clearly need radiological source term information as a function of accident conditions. This information can be provided by a mechanistic computer code if measurements of source terms, transport, and retention are available (under conditions which bound the range of possible severe accident conditions and sequences) for the development and assessment of such a code. Results from the planned SFD tests in the PBF will contribute to the source term data base. With proper instrumentation to measure the state of core damage during an accident, and given the expected response to various accident management actions, source terms can then be used to guide emergency response alternatives in an accident situation.

Understanding TMI-2

A thorough examination of the TMI-2 core is a critical step in understanding the behavior of severly damaged reactors. The PBF severe fuel damage program will aid the efforts to examine and understand the TMI-2 accident in three principal areas, each of which is expected to ultimately benefit the nuclear industry. The first area is the duplication of the main TMI damage features. The SFD program will be conducted using a variety of heatup rates, peak fuel temperatures, and cooldown rates and will therefore produce varying degrees of fuel rod damage. Although the condition of the TMI core will make reconstruction of damage event sequences difficult, the knowledge gained from the SFD program on event sequences, the effect of varying peak temperatures, and fuel rod breakup during quenching will be critical to deducing the actual causes and progression of the TMI damage.

The second area of the SFD program that will aid the TMI-2 examination is in technique development. Many of the postirradiation examination techniques being developed for the SFD program will be directly applicable to TMI debris examinations. The third area of aid is enhanced cost effectiveness of the TMI-2 core examination. The experience gained from examining the damaged fuel from the PBF SFD tests is expected to reduce the costs and accelerate the hot cell examination of the TMI core.

In summary, results from the planned SFD test program will contribute to improved PRAs, formulation of requirements for improved designs for accident mitigation, emergency response actions based on a better understanding of risk, and proper interpretation of the TMI-2 core examination. We believe these contributions will benefit the nuclear industry through risk reduction, improved regulatory decisions, and, therefore, greater public acceptance of nuclear power. A flowchart showing the ordered sequence of these activities is illustrated in Figure 1.

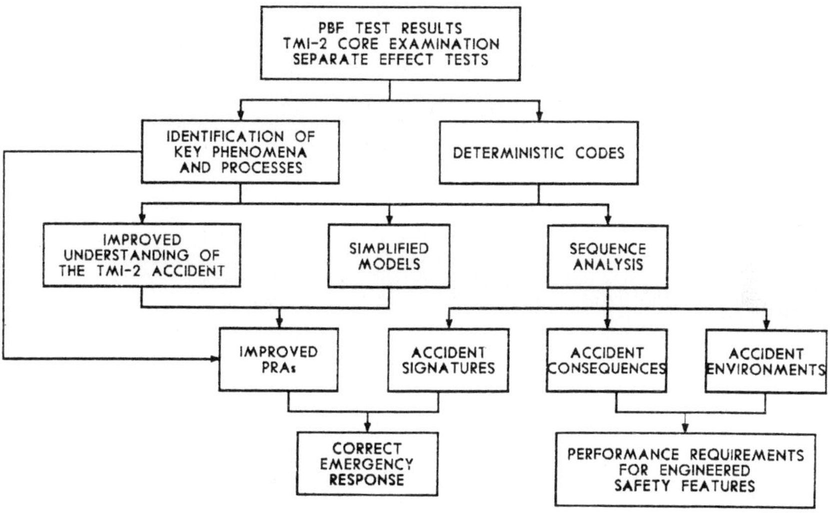

Figure 1. Ordered sequence of severe fuel damage research.

DESCRIPTION OF PHASE I SFD EXPERIMENTS

The primary objectives of Phase I of the PBF severe fuel damage test program are to (a) characterize fuel rod damage resulting from severe cladding oxidation and melting, UO_2 dissolution and relocation, and fuel rod fragmentation; (b) measure the release rates, transport, and deposition of the fission products; (c) measure the magnitude and timing of the hydrogen generation; (d) measure the coolability characteristics of test bundles with various types and degrees of damage; and (e) determine the effects of irradiated fuel rods and control rods.

Five bundle experiments are presently planned for Phase I of the SFD program. Each bundle will be operated at full power for three days and then cooled for about two weeks to build up an appropriate ratio of cesium and iodine. The inlet flow to the bundle will then be reduced or the power will be increased to force the water level to decrease, leaving the upper level of the bundle steam cooled. The upper regions of the bundle will heat up in superheated steam to peak temperatures of up to 2400 K. Parameters to be varied during the five experiments are heatup rate (amount of cladding oxidation), inlet flow (hydrogen-to-oxygen ratio), cooldown rate (amount of fragmentation), test rod burnup, and presence of control material. Unirradiated test fuel rods will be used in the first three experiments. Irradiated test fuel rods will be used in the fourth and fifth tests, with control rods included in the fifth test. The Phase I SFD tests are summarized in Table I.

TABLE I. SFD SERIES I TESTS

Number	Heating Rate	Inlet Flow (g/s)	Test Bundle	Cooling	Working Schedule
SFD-ST	0.5 K/s	13.3	Fresh rods	Quench	10/18/82
SFD-1-1	TMI-2[a]	0.6	Fresh rods	Slow	01/28/83
SFD-1-2	TMI-2[a]	0.6	Fresh rods	Quench	05/13/83
SFD-1-3	TMI-2[a]	0.6	Irraidated rods	Slow, no reflood	12/16/83
SFD-1-4	TMI-2[a]	0.6	Irraidated rods Control rods	Slow, no reflood	03/16/84

a. Characterized by slow heating rate up to 1600 K and rapid heating rate above 1600 K, driven by metal-water reaction.

The SFD test bundle and shroud were designed and fabricated by Battelle Northwest Laboratories. The SFD test bundles consist of 32 zircaloy-clad UO_2 fuel rods arranged in a 6 x 6 array without the four corner rods. The active fuel length in the bundle is about 1 m, and the fuel rods are of typical 17 x 17 PWR design. The test bundle is contained in an insulating shroud consisting of low density zirconia insulation sandwiched between inner and outer zircaloy walls. The low density zirconia insulation reduces the radial heat loss through the shroud wall. This reduces the power required to raise the test rods to the desired surface temperature of 2400 K, and also minimizes radial temperature variations in the bundle. A constant test fuel rod enrichment of 6.2% was selected to balance the radial power distribution within the bundle against the radial heat losses.

The test bundles are installed in the PBF in-pile tube, a thick-walled Inconel cylinder designed for a wide range of test coolant conditions. Figure 2 shows a cross-sectional view of the test hardware in the in-pile tube. Coolant will enter the inlet to the in-pile tube during operation of the PBF loop, flow downward to the lower plenum, reverse direction, and flow up through the region between a cylindrical flow tube and the outer wall of the insulating shroud. This flow will be maintained throughout all the tests to provide continuous cooling to the outer wall of the insulating shroud. The coolant flow in the interior bundle (fuel) region of the test train will be supplied separately by four inlet lines.

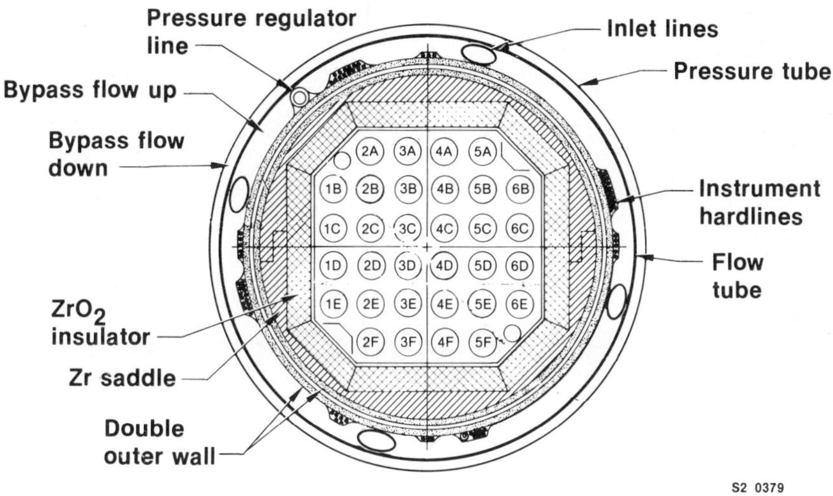

Figure 2. SFD test train in core region.

The coolant flow will be maintained constant at 13.3 g/s and the bundle power will be ramped up to about 80 kW, resulting in boiloff of the coolant, dryout in the upper regions of the fuel rod bundle, and cladding temperatures of about 2400 K during the first experiment, the SFD Scoping Test. Cladding heatup will occur over about 4 hours, allowing considerable time for oxidation of the cladding. The second experiment, SFD-1, will be initiated from a bundle power of 12 kW and a coolant flow rate of 0.6 g/s. After the cladding temperatures reach 1550 K, the power will be increased to 27 kW. The SFD-1 cladding temperatures will rise to 2400 K in about 25 minutes, allowing little time for oxidation of the cladding.

The first two SFD tests will be highly instrumented, with each experiment including 6 fuel rod centerline thermocouples, 25 internal cladding thermocouples, 28 shroud thermocouples, 16 steam probes, 2 flowmeters, 14 pressure sensors, 12 fission chambers, 2 flux wires, 2 temperature profile detectors, and a shroud penetration detector. The internal cladding surface thermocouples will provide data up to about 2100 K; the fuel centerline thermocouples will operate to temperatures of about 2400 K. Following the experiment, the entire test bundle and shroud will be removed for examination. Postirradiation examination techniques planned for the SFD test bundles include gamma scanning of the test train, neutron tomography, photo-visual examinations, hot cell flow tests, flux wire scanning and burnup analysis, spectral gamma scanning, metallographic examinations, scanning electron microscopy, Auger Spectroscopy, and X-ray spectroscopy.

The fission product detection system presently installed in the PBF is illustrated in Figure 3. The entire experiment effluent will be drawn out of the test bundle and into a steam pickup line leading to a condenser. Six effluent sample bombs will be located upstream from the condenser. The sample bombs will be initially at low (ambient) pressure and argon filled, and they will be remotely operated to provide superheated steam, fission product, and hydrogen samples for posttest analysis. The steam will then be reduced to liquid in the condenser at a temperature below 340 K, and enter a separator vessel with a minimum residence time. A continuous purge of the separator gas space with high-purity nitrogen gas will sweep hydrogen, fission gases, and other noncondensibles out of the separator, past a gamma spectrometer and hydrogen monitor, and into the collection tank. The liquid flowing from the separator vessel will be monitored by two

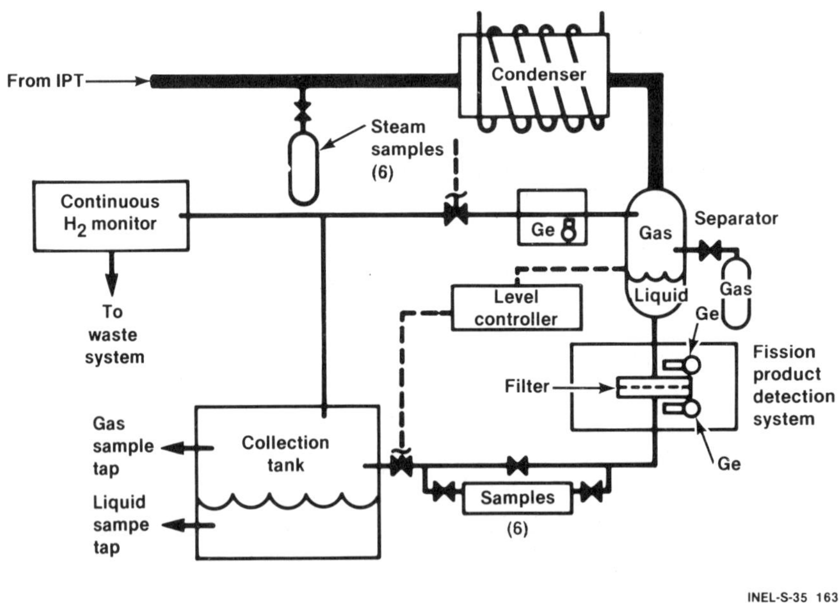

Figure 3. Fission product and H_2 monitoring system.

gamma spectrometers located upstream and downstream of a particle filter. The liquid will then pass through one of six flow-through grab samples, through a flow control valve, and into the same collection tank as the gaseous effluent. The liquid samples will be remotely operated sequentially such that the sample stream always flows through either a sample bomb or the bypass. There are also provisions for posttest grab samples of the gas and liquid in the collection tank.

This system will provide important data on (a) the fission products in the condensed and gaseous effluent, (b) particulate and nonparticulate fission product partition factors, (c) fission product release during quench, (d) fission product leaching from a rubble bed, and (e) integral isotopic fission product release fractions. The fission product detection system will be improved for the fourth and fifth SFD tests, SFD-3 and SFD-4, to focus on fission product transport and retention and provide order-of-magnitude aerosol generation and transport data (including the effects of control rod materials on aerosol generation), and yield general indications of fission product chemical form and deposited materials. The expanded scope will significantly improve the mass balance following the tests.

The modifications include a larger diameter pipe leading from the test fuel rod area to above the in-pile tube head, as shown in Figure 4. Contained within this large pipe will be a small-diameter rod containing deposition coupons. The rod and coupons will be lifted into a shielded cask after each test and transported to a hot cell. Other modifications include moving one of the germanium detectors to monitor the steam line prior to the sample bombs and condenser, and instrumenting the condenser with deposition coupons. The entire condenser will be removed after each test for analysis. The gaseous and condensed effluent collected during each test will also be recirculated past the germanium detectors after cooldown to measure the long-lived isotopes, which were masked by the short-lived isotopes during the initial measurement.

Analysis of the deposition coupons will include gamma spectroscopy to determine the quantity and identity of deposited fission products, scanning electron microscopy and micro-Raman spectroscopy to determine particle size and elemental makeup, and microchemistry to determine the solubility and oxidation state of the deposited materials.

EXPERIMENT PREDICTIONS

Calculations have been performed to evaluate the bundle and shroud response during the SFD Scoping Test and SFD-1 high temperature transients using the TRAC-BD1[7] computer model. Although

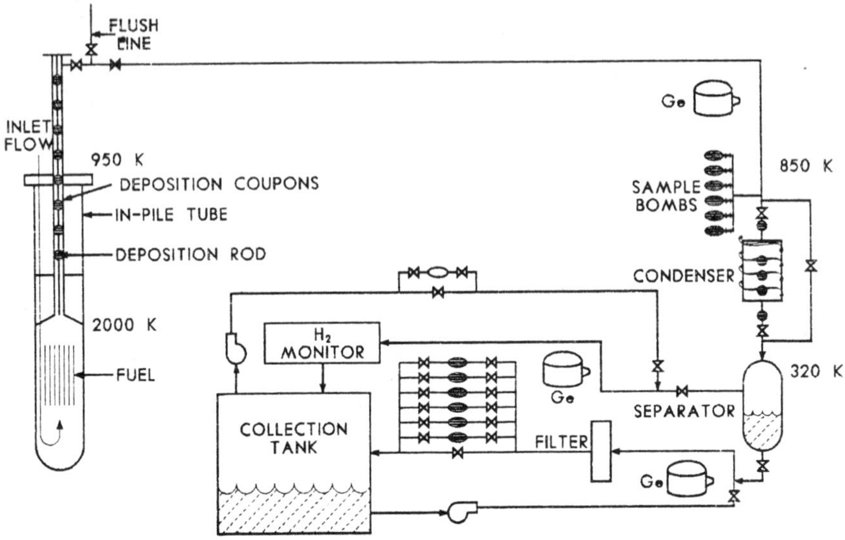

Figure 4. Fission product detection system modifications for Tests SFD-3 and SFD-4.

TRAC-BD1 was designed primarily for analysis of boiling water reactor large-break LOCAs, it can also be applied to many other configurations, including the SFD test train.

The input model developed for use with the TRAC-BD1 computer code included eight equal axial levels and three radial rod groups within the bundle, as shown in Figure 5. The insulating shroud was included in the model, with three regions representing the internal zircaloy wall, the low density zirconia insulation, and the outer zircaloy saddle and wall. The metal-water reaction and heat of fusion of the fuel rod and shroud wall cladding were also included in the calculation.

The SFD Scoping Test is designed to have a very slow cladding heatup rate of about 0.15 K/s. Severe oxidation and considerable fuel rod fragmentation is expected. Shown in Figure 6 is the power history for the test, illustrating the various power ramps and plateaus planned for the 5-hour experiment. The calculated temperature history of the fuel rod surface near the top of the test bundle and the temperature history of the steam exiting the bundle are also shown in Figure 6. The temperature of the rod surface is driven by the power level, which reaches a maximum of 80 kW at 16,000 seconds (4.4 hours). At that time, the rod surface temperature is calculated to be about 2450 K, and the exit

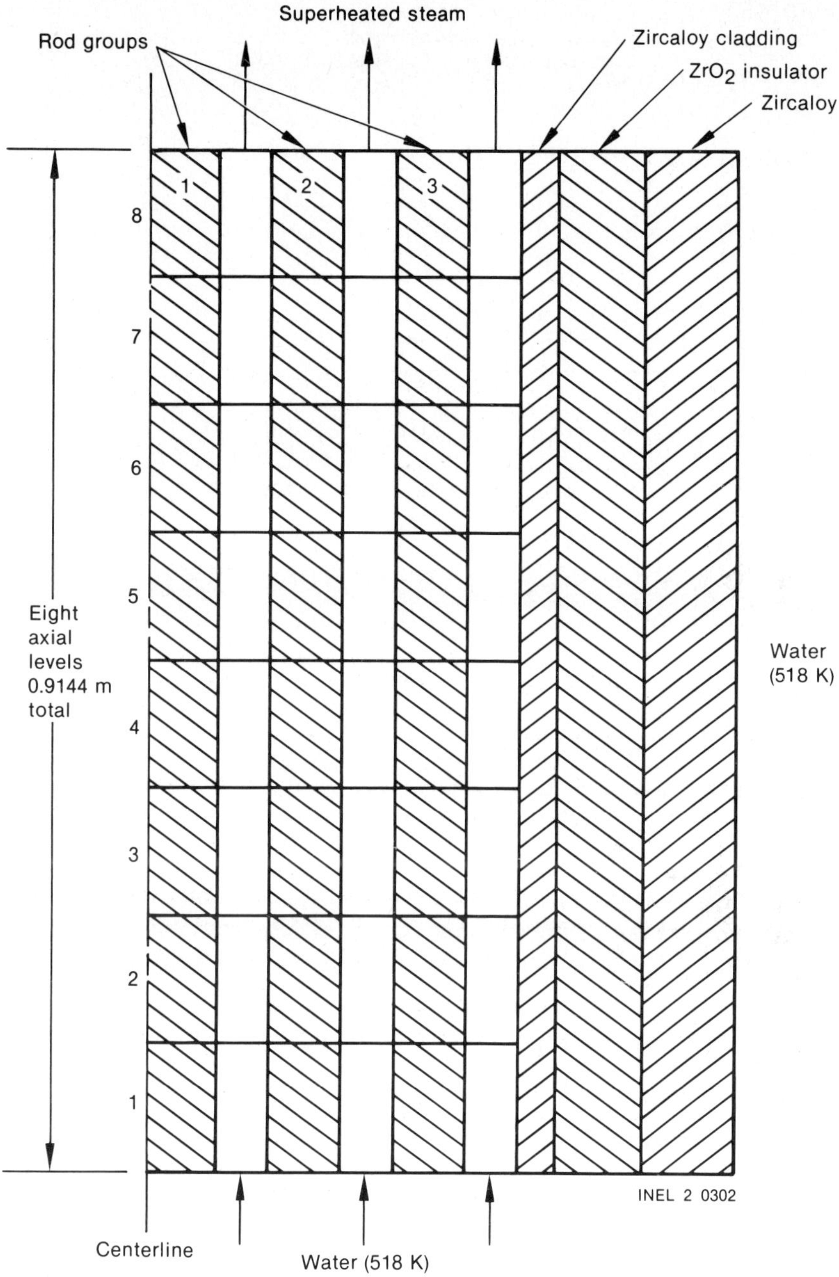

Figure 5. TRAC-BD1 input model.

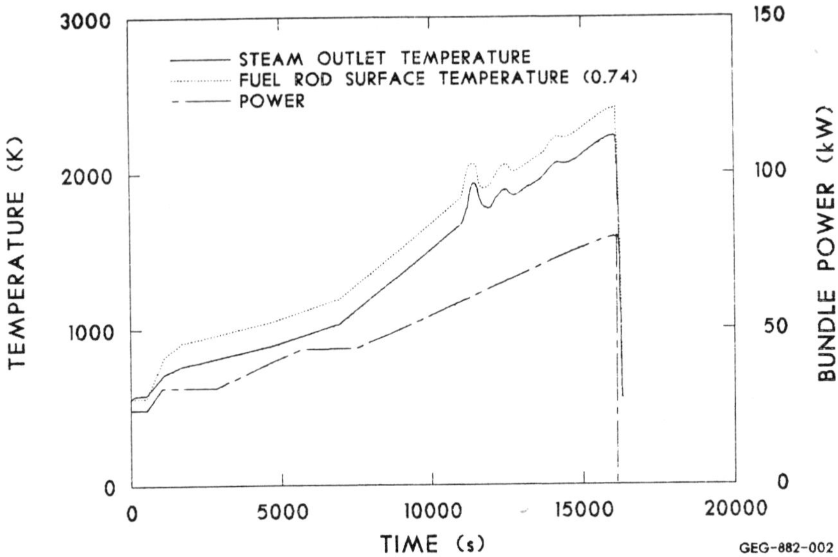

Figure 6. Predicted results for SFD Scoping Test.

steam temperature about 2280 K. The reactor is then scrammed, the flow is increased to 30 g/s, and the bundle is quenched within about 3 minutes.

The test temperatures are strongly affected by the metal-water reaction of the zircaloy cladding and shroud inner wall. The calculated peak surface temperature shows the effect of metal-water reaction at about 12,000 seconds (3.3 hours) into the transient. At this point, the bundle is calculated to reach a cladding temperature in excess of 1850 K and an extremely fast metal-water reaction is predicted. The metal-water reaction has an Arrhenius temperature dependence and increases rapidly with increasing temperature. Above 1850 K, the heat released by the metal-water reaction rapidly increases the bundle temperature, which in turn increases the reaction rate, until all of the cladding in a given axial node has been totally oxidized. The reaction starts at about the 0.82-m elevation and propagates to the top of the bundle. The runaway reaction is then predicted to propagate down to the lower regions of the bundle as the power is increased. In the actual test, a flame front will propagate down the bundle. Below the flame front the cladding will be metallic zircaloy with a thick oxide layer. As the flame front propagates through a given elevation, the metal-water reaction will consume all of the metallic zircaloy. Finite nodalization was used in the calculation and the metal-water reaction was predicted to occur over the whole node, giving rise to the second and third temperature peaks at 12,000 and 14,000 seconds. In the actual case, it is expected that the flame front will spread out in time and there will be no distinct peaks.

The SFD-1 test is designed to have a cladding heatup rate of about 1 K/s to 1600 K, followed by a very rapid heatup rate, until the cladding temperatures reach values of about 2400 K, and a slow cooldown rate. Because of the fast heatup rate, less time is available for cladding oxidation. The cladding surface temperature of the central bundle fuel rods at the axial midplane calculated by the TRAC-BD1 code is shown in Figure 7. The transient begins at a cladding peak temperature of 900 K near the top of the bundle, with a bundle flow rate of 13.3 g/s and a bundle power of 30 kW. The axial midplane cladding temperature is at saturation because the two-phase mixture level is above the midplane. The

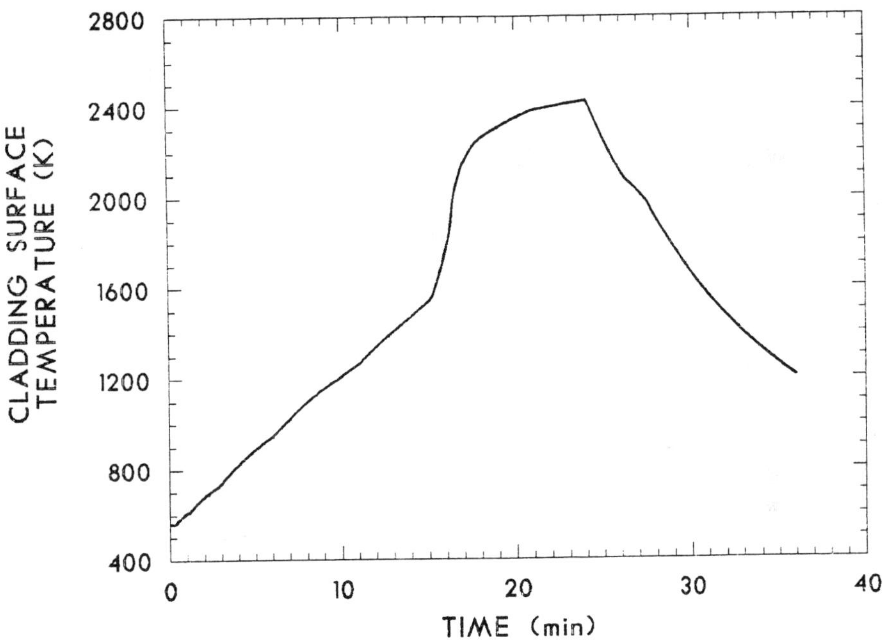

Figure 7. SFD-1 predicted cladding peak temperature history.

bundle flow rate is then reduced to 0.6 g/s in 4 seconds and the bundle power is decreased to 12 kW in 18 seconds. The cladding temperature at the axial midplane is calculated to increase to about 1550 K in 15 minutes at an average rate of about 1 K/s. The bundle power is then increased at 15 minutes to 27 kW within 15 seconds to ensure a final cladding temperature of 2400 K. The calculated cladding temperature increases from 1550 to 2150 K in

2 minutes. This rapid temperature increase is primarily due to the metal-water reaction. Only about 20% of the cladding is oxidized when the cladding temperature reaches 2150 K, and it is assumed that the remaining 80% melts and slumps down. Therefore, when the cladding reaches 2150 K, the metal-water reaction is "turned off" in the calculation. This accounts for the change in slope of the temperature rise just above 2150 K. The calculated cladding temperature continues to increase at a slower rate for about 7 minutes to a temperature of about 2400 K. At the end of the 7-minute period, the transient is terminated by manually scramming the reactor while maintaining the low coolant flow of 0.6 g/s. About 30 minutes are calculated to be required for the SFD-1 cooldown, as compared with the 3-minute quench for the SFD Scoping Test.

The SFD-1 transient is calculated to be steam limited so that the metal-water reaction can proceed from one node to another only when the oxidation is completed or "turned off" for the former node. This is because at any one time there is only enough steam available for 10 to 20% of the cladding to be oxidizing at temperatures above 1850 K. Therefore, because of the steam-limited condition, the calculational node at the axial midplane reaches 2150 K first, then the node just below the axial midplane oxidizes until that node reaches 2150 K, followed by oxidation of the node just above the axial midplane, and so on.

The TRAC-BD1 code was developed for the analysis of design basis accidents and is being stretched to the limit of its applicability when used to analyze beyond design basis events such as severe damage accidents. Although some modifications were required to properly analyze these tests, the number and scope of the modifications were kept to a minimum. Potentially large uncertainties exist in the calculated results at temperatures above 1850 K, particularly above 2150 K. The metal-water reaction rate at cladding temperatures above 1850 K has a significant effect on the transient, yet it is based on a relatively small data base. The assumption that the melted cladding slumps to a cold region and no longer oxidizes is highly uncertain. The fuel bundle geometry is assumed fixed throughout the transient, and TRAC does not take into account any slumping or debris formation in the damaged fuel. Finally, there is no provision for treating the flow and heat transfer characteristics of the hydrogen, which is abundant at temperatures above 1850 K.

However, even in view of these limitations, the analysis is probably adequate for setting the test parameters of bundle power and coolant flow. The analysis performed for the SFD Scoping Test indicated that the test objective of a slow heatup to 2400 K with considerable cladding oxidation and little cladding melting will be achieved using a linear ramp in bundle power to 80 kW with an inlet flow of 13.3 g/s. Similarly, the objective of a fast heatup with little cladding oxidation and considerable cladding melting for SFD-1 is expected to be achieved using a bundle power below 30 kW and reducing the coolant flow rate to 0.6 g/s.

DESCRIPTION OF PHASE II SFD EXPERIMENTS

The primary objective of the Phase II SFD experiments is to measure fission product transport and retention in environments representative of selected segments of commercial reactor primary systems. Control rod material aerosol release, fuel and cladding aerosol release, fission product chemistry in a typical environment, and fission product and aerosol interaction in a typical environment will be measured. Risk-dominant accident sequences in which UO_2 melting temperatures are reached will be simulated during the Phase II tests. The associated high temperatures will produce significant fission product release and fuel and cladding aerosols for interaction studies. Also of importance is the hydrogen release, damaged core coolability characteristics, and melt progression and melt interaction with support structure phenomena which may occur during a degraded core cooling accident.

Modifications to the Phase I test train hardware envisioned for the higher temperature Phase II tests are illustrated in Figure 8. The primary changes are that the zirconium oxide insulator used in the Phase I hardware will be replaced with thorium oxide, and the zirconium saddles in the Phase I hardware will be replaced with zirconium oxide insulation. Thorium oxide was

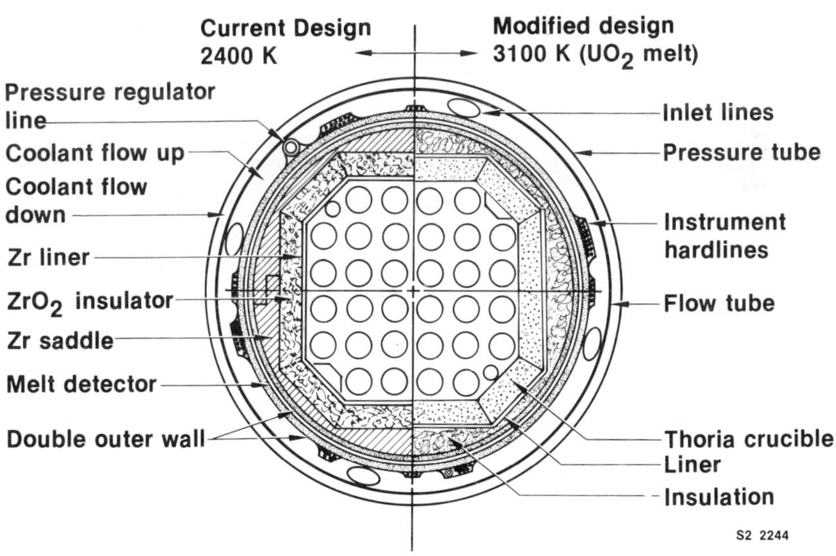

Figure 8. SFD high temperature test train.

chosen for the crucible because it will maintain its form and position at the UO_2 melting temperature, and it is chemically inert at high temperatures and does not react with steam.

The fission product detection system planned for the Phase II SFD experiments is illustrated in Figure 9. The system is intended to provide environments typical of selected segments of a commercial reactor primary coolant system so that fission product retention measurements can be performed. The system will include measurements of the relative timing of volatile fission product and aerosol release, time-dependent measurement of aerosol release in the primary system, specific measurement of selected fission product chemical forms, and discrete characterization of aerosol release and integral fission product and aerosol retention measurements. The modifications include a simulated upper plenum structure and hot leg, a pressurizer volume, and two letdown systems.

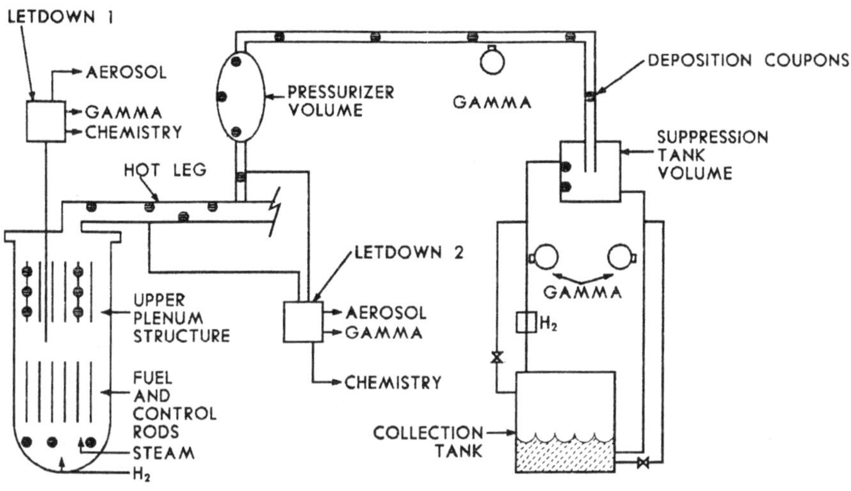

Figure 9. Fission product detection system concept for Phase II SFD tests.

The letdown systems provide an interface between the high pressure/high temperature experiment loop, and lower pressure/lower temperature instrumentation. The letdown system takes a sample of the effluent stream, adds a dilution gas (Ar or N), and expands the mixture into a volume. The fast expansion and dilution inhibit condensation and agglomeration of the sample. The diluated sample can then be routed to instrumentation to measure aerosol characteristics (impactors, diffusion batteries, filters), fission product isotopes (gamma spectrometers), and chemical compounds (gas chromatographs with microwave emission systems, on-line chemistry). The chemistry systems are being designed to measure selected chemical compounds containing I, Cs, and Te. The on-line chemistry system is being adapted from an existing system currently used to perform fast, on-line chemistry on short-lived (minutes) fission products deposited on NaCl aerosols. The system will operate in either a semi-continuous on-line mode or store (via a collection tape) a continuously collected sample for posttest batch analysis.

CONCLUSIONS

The PBF severe fuel damage test data will be combined with results from the TMI-2 examination, results from small-scale experiments being performed at other laboratories, and mechanistic computer models to (a) develop more accurate and reliable Probabilistic Risk Assessment (PRA) consequence analysis, (b) help the USNRC set performance requirements for reactor designs and engineered safety features (c) ensure correct emergency response actions during a severe accident, and (d) provide a comprehensive understanding of what happened at TMI-2.

The objectives of the Phase I SFD experiments are to (a) characterize fuel rod damage resulting from severe cladding oxidation and melting, UO_2 dissolution and eutectic melting, and fuel rod fragmentation; (b) measure the release rates, transport, and deposition of fission products; (c) measure the magnitude and timing of hydrogen generation; (d) measure the coolability characteristics of test bundles with various types and degrees of damage; and (e) determine the effects of irradiated fuel rods and control rods. The PBF fission product detection system will provide important data on fission products in the condensed and gaseous effluent, partition factors for particulate fission products, and integral isotopic fission product release during the first three Phase I SFD experiments. The fission product detection system will be improved for the last two Phase I SFD tests to focus on fission product transport and retention, and provide a complete mass balance of the fission products.

The analyses performed for the SFD Scoping Test indicate that the test objective of a slow heatup to 2400 K with considerable cladding oxidation and little cladding melting will be achieved using a linear ramp in bundle power to 80 kW with an inlet flow of 13.3 g/s. Similarly, the objective of a fast heatup with little cladding oxidation and considerable cladding melting for SFD-1 is expected to be achieved using an initial bundle power of 12 kW and a reduced coolant flow rate of 0.06 g/s. After the cladding temperature reaches about 1550 K, the bundle power will be increased to 27 kW.

The Phase II SFD experiments will reach UO_2 melting temperatures of 3133 K. The main focus of the Phase II experiments is directed toward fission product transport and retention in environments representative of selected segments of commercial reactor primary systems.

NOTICE

This paper was prepared as an account of work sponsored by an agency of the United States Government. Neither the United States Government nor any agency thereof, or any of their employees, makes any warranty, expressed or implied, or assumes any legal liability or responsibility for any third party's use, or the results of such use, of any information, apparatus, product or process disclosed in this paper, or represents that its use by such third party would not infringe privately owned rights. The views expressed in this paper are not necessarily those of the U.S. Nuclear Regulatory Commission.

REFERENCES

1. "Indian Point Probabilistic Safety Study," Copyright 1982, Power Authority of the State of New York, Consolidated Edison Co. of New York, Inc.

2. "Zion Probabilistic Safety Study," Copyright 1981, Commonwealth Edison Co., Chicago, IL.

3. "Probabilistic Risk Assessment, Limerick Generating Station," Philadelphia Electric Co., April 1982.

4. D. D. Carlson et al., "Reactor Safety Study Methodology Applications Program: Sequoyah #1 PWR Power Plant," Sandia National Laboratory, NUREG/CR-1659, February 1981.

5. U.S. Nuclear Regulatory Commission, "Reactor Safety Study--An Assessment of Accident Risks in U.S. Commercial Nuclear Power Plants," WASH-1400, October 1975.

6. S. Levine et al., "Source Terms: Investigation of Uncertainties, Magnitudes, and Recommendations for Research," NUS-3808, ALO-1008, March 1982.

7. F. Aguilar et al., "Development of the TRAC Code for BWR Analysis," <u>Quarterly Technical Progress Report on Water Reactor Safety Programs Sponsored by the Nuclear Regulatory Commission's Division of Reactor Safety Research April-June 1980</u>, NUREG/CR-1400(2), EGG-2048, July 1980.

RIA FUEL BEHAVIOR RESEARCH IN THE NSRR

T. Hoshi

Japan Atomic Energy Research Institute

Tokai-mura, Naka-gun, Ibaraki-ken, Japan

ABSTRACT

An in-reactor experimental research on fuel behavior under reactivity initiated accident(RIA) conditions has been progressed in the Nuclear Safety Research Reactor(NSRR). LWR type fuel rods were irradiated by a pulse burst power in the water environment.

Through the experiments, it is found that fuel failure modes are characterized in followings in the RIA events. Those are;

(1) cladding melting and/or brittle failure,
(2) fuel pellet melting failure,
(3) high temperature cladding burst,
(4) low temperature cladding burst, and
(5) pellet-cladding mechanical interaction (PCMI) failure.

The failure modes change depending on heat generation, burn-up, internal pressure and existense of warer in the fuel rod.

In this report, fuel failure modes observed in the RIA, status of studies on failure mechanism and future work of the NSRR program are introduced.

1. INTRODUCTION

An in-reactor tests on the LWR fuel behavior under the RIA conditions has been conducted in the NSRR, at Japan Atomic Energy Research Institute (JAERI), and a total of 550 tests have been performed since 1975. The major objectives of experimemts are to determine the fuel failure threshold and clarify the mechanism and consequence of fuel failure during the RIA.

In the tests, a number of shorten LWR type fuel rods have

P. von der Hardt and H. Röttger (eds.), Irradiation Technology, 231–240.

been irradiated by a pulse burst power of the NSRR reactor in the capsules.(1) A PWR type fuel rod was selected as the NSRR standard type test rod which was used to produce the base line data on the fuel behavior under typical RIA condition. Then, many parameter tests were conducted to determine the effects of fuel design and cooling environment variations. A schematics of the NSRR standard type test rod and parameters changed in the experiments are shown in Fig.1 and Table 1, respectively.

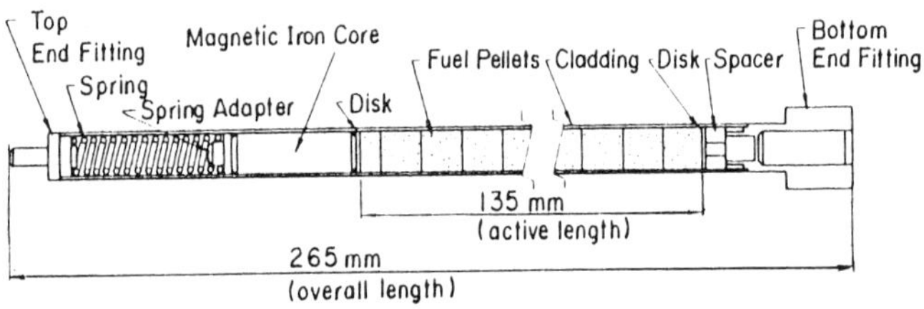

Fig.1 NSRR standard type fuel rod

This illustration is printed from the best available original

	Test	Parameter
Atmospheric Pressure Capsule Tests	Standard Fuel Tests	PWR type fuel rod
	Fuel Design Variation Tests	
	Fuel Enrichment	5 ∿ 20%
	Gap Width	0.05 ∿ 0.195 mm
	Gap Gas Composition	Xe, Ar
	Rod Pressure	-0.1 ∿ 5 MPa
	Clad Thickness	0.40 mm
	Clad Heat Treatment	Cold Worked, Annealed
	Clad Material	Stainless Steel
	PCI-Remedy Clad	Cu, Zr Lining
	Others	
	Defected Fuel Tests	
	Waterlogged Fuel	Water Volume, Flux Peak
	Fretting-Corroded Fuel	Depth of Clad Defect
	Cooling Environment Variation Tests	
	Subcooling	10 ∿ 80°C
	Shrouded Single Rod	Shroud Size
	Clustered Rods	5 Rod Bundle
	Forced Convection	0.5 ∿ 1.8 m/s
	Optical Observation of Fuel Behavior	
High Pressure Capsule Tests		7.2 MPa - 276°C / 16 MPa - 305°C
Loop Tests		1.1 MPa, 100 ∿ 150°C, 3 ∿ 6 m/s

Table 1 Major test parameters and conditions

2. GENERAL FUEL BEHAVIORS AT RIA

2.1 Fuel Behaviors of Standard Type Fuel Rod (2)

In the standard type fuel rod tests, the rods were subjected to energy depositions ranging from about 50 to 550 cal/g.UO_2 in the atomospheric pressure capsule to obtain the general response of un-irradiated fuel rods. Photographs of typical post-test fuel rods are shown in Fig.2 and typical time histories of the cladding temperatures are illustrated in Fig.3.

Onset of departure from nucleate boiling(DNB) occured at an energy deposition of about 140 cal/g.UO_2. Above this energy, the cladding surface temperature reached a maximum over 1000°C immediately after power burst,and remained high for several second. The rod was suddenly quenched when the cladding surface temperature decreased to around 900°C. During film boiling, the cladding surface was oxidized and discolored over active length due to Zircaloy and steam reaction. Rod diametral deformation and bowing were also observed. The extent of oxidation and deformation increased with energy deposition.

Incipient fuel failure occured at about 260 cal/g.UO_2. The rods generally failed with circumferential cracks in the cladding, involving melting of cladding inner surface. The measured cladding surface temperatures of failed rods reached near the melting temperature of Zircaloy (~1840°C). Loss of structural integrity of the rods occured in the energy deposition range above 300 cal/g.UO_2 where conspicuous melting of the cladding and pellets were observed.

In the energy deposition above 380 cal/g.UO_2, fuel rods broke into small particles immediately after or during power burst, inducing the mechanical energy generation.

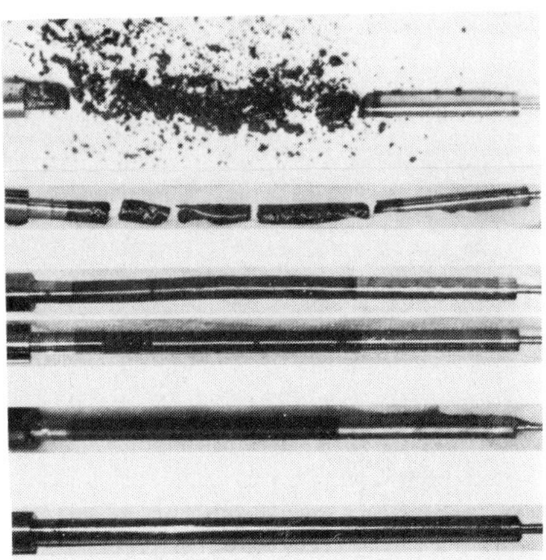

Energy deposition
(cal/g.UO_2)
334
292
237
214
159
102

This illustration is printed from the best available original

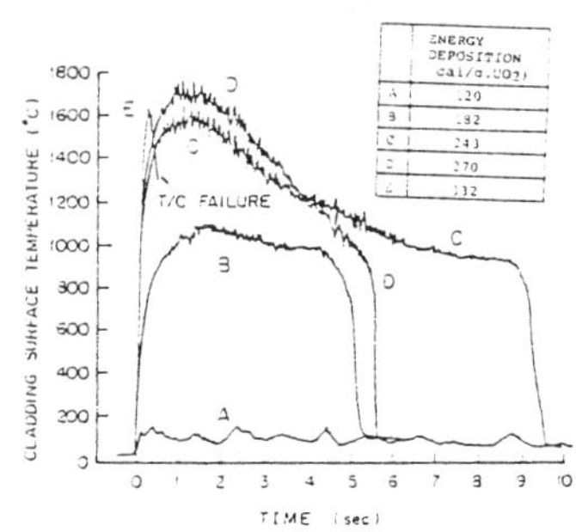

Fig.1 Post-test appearance of standard type rods

Fig.2 Cladding surface temperature responce of standard tests

2.2 Effects of Fuel Design Parameters

1) Design of Fuel Pellets and Cladding

Fuel design parameters such as fuel enrichment, gap gas compositions and cladding wall thickness change the fuel failure threshold energy from that of the standard fuel rod, but don't change the failure mode. The changes of failure threshold energy are illustrated in Fig.4.

2) Effects of Rod Internal Pressure[3]

Among the fuel design parameters, the initial internal fuel rod pressure shows the most evident effects, characterized by the change of fuel failure mode of cladding burst as shown in Fig. 5. This is caused by the overpressure of the internal fuel rod and significant weakening of heated cladding.

The failure of rods initially pressurized to less than 0.6 MPa has been determined to be the same manner as that of the standard fuel rods. It is, therefore, said that the fuel rods with internal pressure of more than 0.6 MPa from the outside coolant pressure shows a failure mode of cladding burst.

This illustration is printed from the best available original

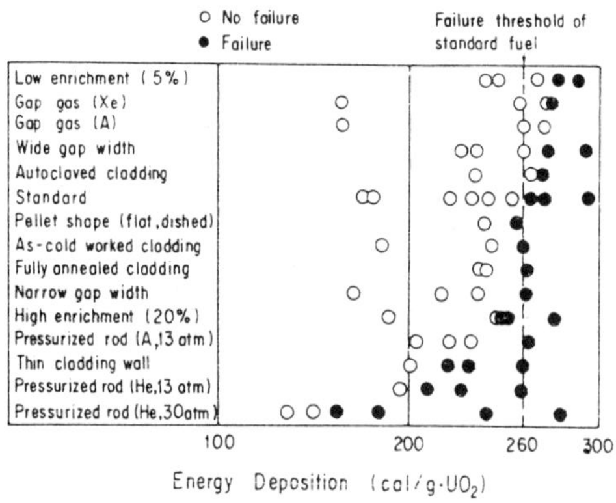

Fig.4 Fuel failure threshold in in fuel design parameter tests

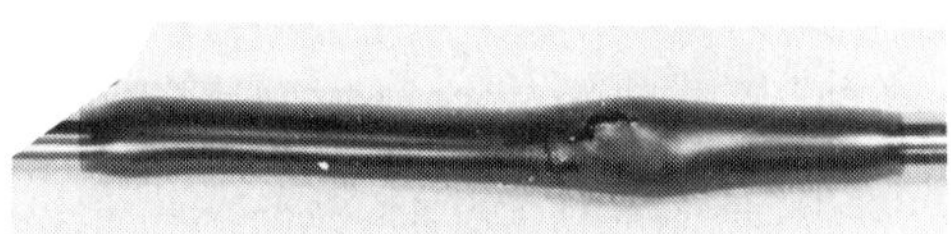

Fig.5 Post-test appearance of pre-pressurized fuel rod

2.3 Waterlogged Fuel Rod[4]

To investigate the fuel behavior of defected fuel rods, tests of waterlogged fuel rods were performed. The waterlogged fuel rods generally failed at energy depositions of about 100 to 130 cal/g.UO_2 without significant cladding temperature rise. Mode of fuel failure was a cladding rupture as shown in Fig.6. Most of fuel pellets were expelled into water from the opening of the cladding, inducing mechanical energy generation.

2.4 Effects of Cooling Environments (5)(6)

Fig.7 illustrates a comparison of typical cladding temperature histories for different coolant subcooling and flow rates. As shown in this figure, decreasing of subcooling increases the cladding temperature and increasing of flow rate decreases the temperature. Low cladding temperature is enhanced by good heat transfer from cladding to coolant and this increases the threshold energy for failure.

Fig.8 shows the appearance of fuel rod irradiated under the high pressure and high temperature condition. Under the high pressure condition, cladding collapse occured but did not cause fuel failure. Fuel failure was caused by melting and embrittlement of the cladding.

2.5 Burn-up Fuel Rod

Fig.9 shows a post-test appearance of burn-up fuel rod of about 32,000 MWd/t which were tested in the SPERT-CDC (9). The rod failed with a long axial crack and no elongation of cladding is observed at the crack. The failure seems to be caused by the swelling of pellets. It is also reported that the failure occured during the power burst and this means that the failure occured at the period when the cladding temperature was still low.

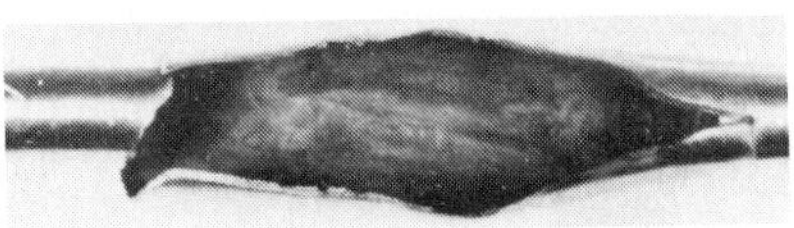

Fig.6 Post-test appearance of waterlogged fuel

Fig.8 Post-test appearance of fuel rod under the high press. and temp.condition

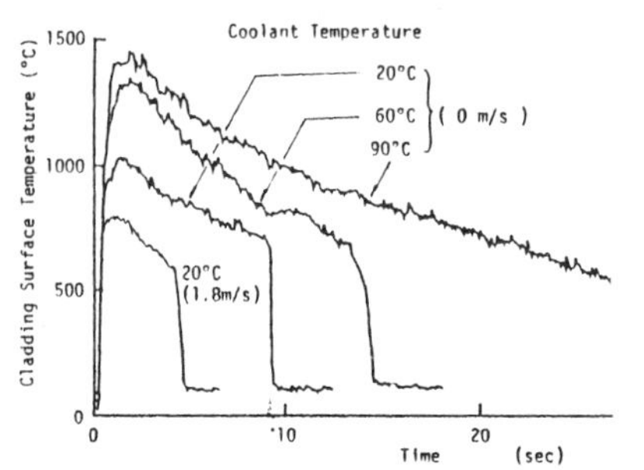

Fig.7 Cladding surface temperature under various cooling conditions (180 cal/g.UO_2)

Fig.9 Post-test appearance of burn-up fuel rod (31,800 MWd/t, SPERT-CDC (9))

These illustrations are printed from the best available originals

3. DISCUSSION ON FUEL FAILURE MODE AND MECHANISM

3.1 Fuel Failure Modes at the RIA Events

As described above, it has been found that the fuel failure modes at the RIA events are characterized in following five modes depending on heat generation, internal pressure and existence of water in the fuel rod and burn-up. Those are;

(1) cladding melting and/or brittle failure,
(2) fuel pellet melting failure,
(3) high temperature cladding burst,
(4) low temperature cladding burst, and
(5) pellet-cladding mechanical interaction (PCMI) failure.

3.2 Cladding Melting and/or Brittle Failure (2)(7)(8)

It is thought that this failure occures for unirradiated or low burn-up fuel rods of which internal pressure do not exceed more than 0.6 MPa from the coolant pressure.

Fig.10 shows crack characteristics of the rod irradiated at 265 cal/g.UO_2 in the atomosphric pressure capsule. Severe oxidation of cladding and decrease of cladding thickness at the location of crack were found. Fig.11 shows the microscopically measured variation of cladding wall and oxide layer of the fuel rod. The oxide thickness is about 80 μm at maximum and the cladding thickness is reduced to about a third of the original at the locations of the crack due to melting of inner surface. Thus, the ratios of oxide thickness to wall thickness increases and the cladding becomes brittle. Brittle failure of the cladding occures during quenching since the strong binding are being made by UO_2-Zry reaction which prevents the slippage between the cladding and pellets.

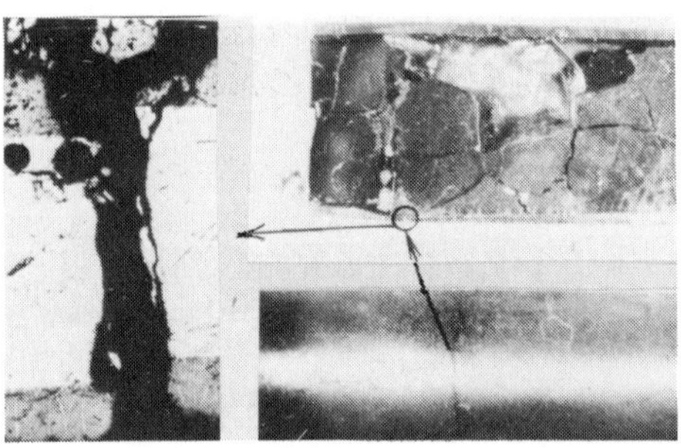

Fig.10 Crack characteristics of the fuel rod irradiated at 264 cal/g.UO_2

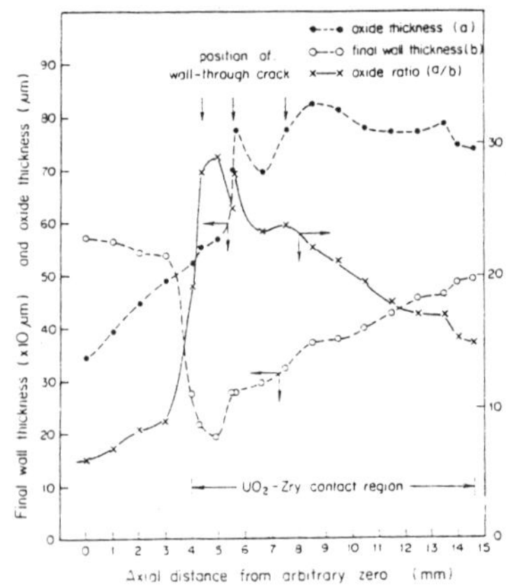

Fig.11 Variation of clad.wall thickness and oxidation after irradiation at 264 cal/g.UO_2

These illustrations are printed from the best available originals

3.3 Fuel Pellet Melting Failure

This failure occurs in the condition when the energy deposition exceed 380 cal/g.UO_2 in the NSRR tests. The energy deposition of 380 cal/g.UO_2 is high enough to cause gross melting of fuel pellets and vaporization at the position of power peak in the rod. In the NSRR test, about 20 MPa of UO_2 pressure is expected at the maximum position where is at the fuel periphery at the stack end of 20 % enriched fuel when the energy deposition (radially average at the stack center) is about 350 cal/g.UO_2. This pressure is thought to be sufficient to burst the cladding since the cladding temperature goes up to about 500°C at the failure.

3.4 High Temperature Cladding Burst (3)

This failure occures in the fresh or low burn-up rods when the rod internal pressure exceeds the outside coolant pressure (more than about 0.6 MPa). The cladding ballooned and ruptured at elevated cladding temperature as shown in Fig.12. The relationships among burst temperature, burst pressure, and ballooning showed a good agreement with out-of-pile experiments.

3.5 Low Temperature Cladding Burst (4)

This failure occures in the waterlogged fuel rods. Fig.13 illustrates the transient response of waterlogged fuel test. The pressure of more than 100 MPa was measured in many cases. This high pressure rise is brought by expansion and/or vaporization of the water inside the rod due to quick heating.

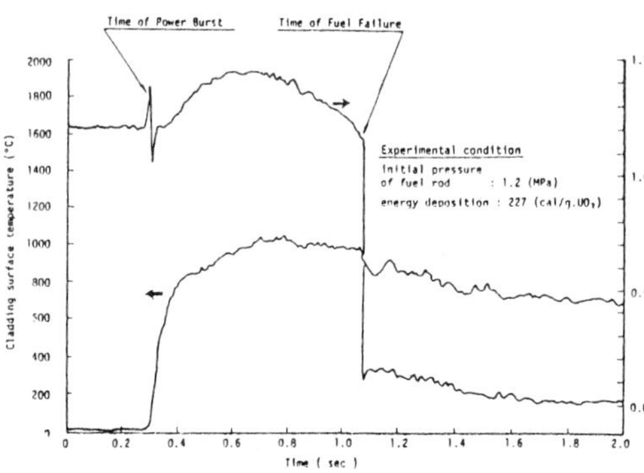

Fig.12 Transient response of clad. temperature and rod pressure in a pre-pressurized fuel test

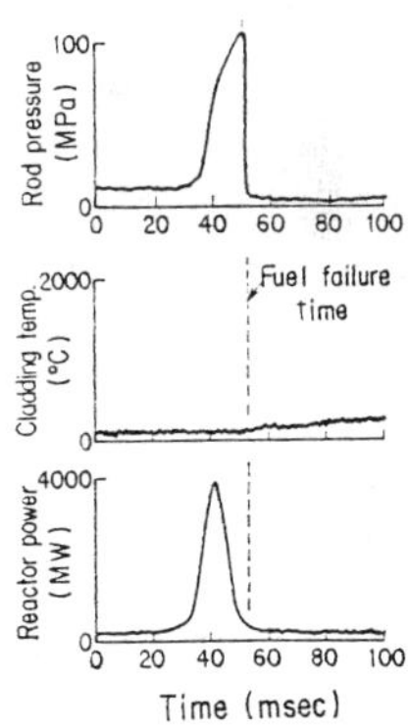

Fig.13 Transient response of a waterlogged fuel test

These illustrations are printed from the best available originals

3.6 Pellet-Cladding Mechanical Interaction (PCMI) Failure

This failure is caused in the burn-up fuel rods and it is characterized by a axial crack. The failure is believed to be caused by the thermal expansion of fuel pellets which stretches the embrittled cladding after contact.

To determine the PCMI failure for the embrittlement cladding rods, tests with the embrittled cladding were performed. Embrittlement of cladding was achieved by oxidation and fresh pellets were inserted in it. Appearance of post-test rod is shown in Fig.14 where the PCMI failure is clearly observed.

Annealing of irradiation effects is not expected because this PCMI failure occurs during the power burst when the cladding temperature is still low at the RIA conditions.

3.7 Summary on Fuel Failure Mechamism

Fuel failure sequences at the RIA conditions are summarized in Fig. 15.

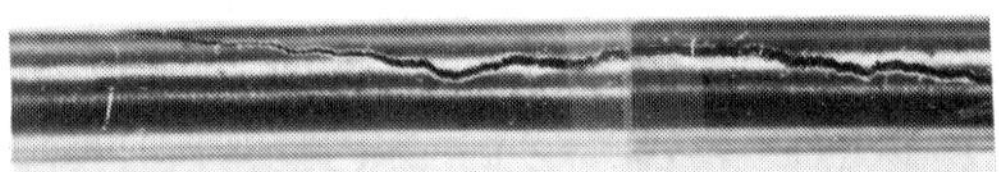

Fig.14 Appearance of post-test rod in embrittled cladding tests

This illustration is printed from the best available original

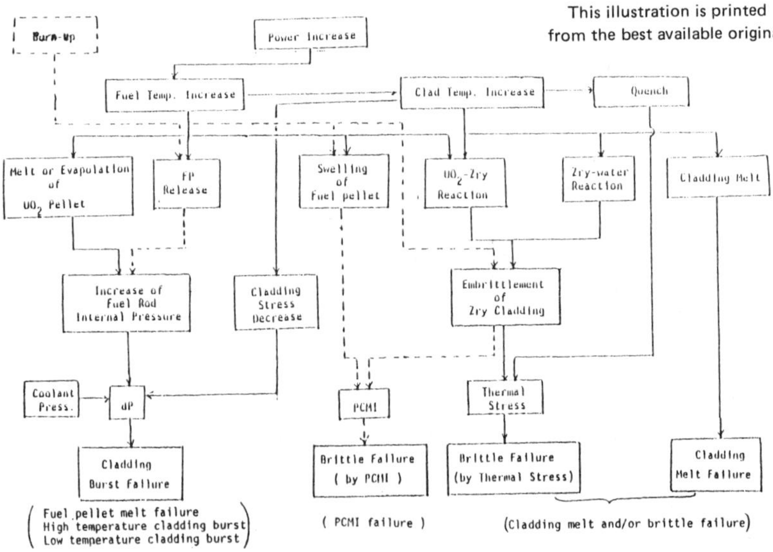

Fig.15 Possible fuel failure mechanism at RIA conditions

4. FUTURE PLANS OF NSRR PROGRAM

4.1 Experiments with Unirradiated Fuel Rods

Experiments to study the effects of fuel design parameters have been completed. Studies on coolant conditions are carried out mostly with the high pressure high temperature capsule and loop, since fuel behaviors and failure threshold are affected by cooling conditions.

Studies on mechanical energy generation are also continued.

4.2 Experiments with Burn-up Fuel Rods

Fuel failure threshold might be decrease largely with increasing of burn-up because cladding becomes brittle due to neuton irradiation. Therefore, determination of failure threshold becomes important.

In the NSRR, burn-up fuel rod tests are being planned and the design of fuel handling system have been started. Start of the experiments with burn-up fuel rods is scheduled to be in the end of 1986 as shown in Table 2.

4.3 Modification of Pulse Characteristics

To investigate the failure threshold for burn-up fuel rods, initial power of the fuel rod before the power burst is very important because the pellet contacts with cladding well at the power operation condition. Small gap between pellets and cladding causes decreasing of failure threshold due to strong PCMI.

Controlled pulse operation as shown in Fig. 16 is considered and a linear heat rate of 200 to 500 W/cm is expected as an initial power of the test rod before the pulse operation.

4.4 Experiments on Other Accident Conditions

(1) A studies on fuel behaviors under the severe core accident (SCD) conditions is being planned. Fuel melt-down behaviors, fuel fragmentation, and coolability of debris will be investigate in the capsule.

(2) Modification of the pulse characteristics can simulate very wide spectrum of accidents. The experiments under an ATWS condition and /or a rapid power ramp are being discussed.

REFERENCES

(1) Fujishro,T., et al.: This proceeding

(2) Hoshi,T., et al.: J. At. Energy Soc. Japan(in Japanese),20(9), 651(1978)

(3) Saito,S., et al.: J. Nucl. Sci. Technol.,19(4), 289(1982)

(4) Ohnishi,N.,et al.: J. At. Energy Soc. Japan(in Japanese),24(4), 289(1982)
(5) Fujishro,T.,et al.: J. Nucl. Sci. Technol.,18(3), 196(1981)
(6) Kobayashi,S., et al.: ibid.,15(6),448(1978)
(7) Shiozawa,S.,et al.: ibid.,24(5),368(1982)
(8) Yanagihara,S.,et al.: ibid.,24(6),469(1982)
(9) Miller,R.W.: ANCR-1280 (1976)

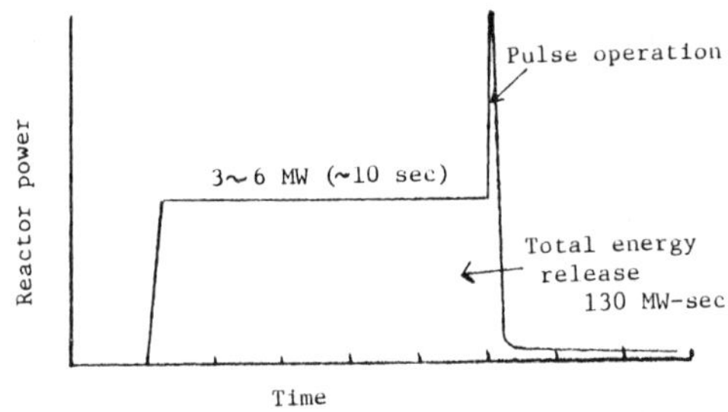

Fig.16 Controlled pulse chracteristics in the NSRR reactor

These illustrations are printed from the best available originals

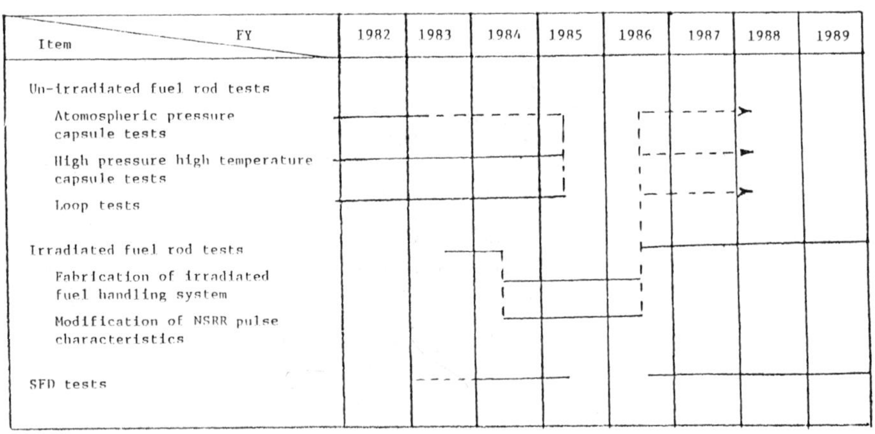

Table 2 Time schedule of NSRR program

FUEL ROD TESTING FACILITIES IN HFR PETTEN

J. Markgraf, D. Perry, M. Leipold

Joint Research Centre (JRC) of the
Commission of the European Communities
Postbus 2
1755 ZG PETTEN (N.H.)/The Netherlands

ABSTRACT

The HFR Petten provides with its Pool Side Facility (PSF), LWR environments simulating, pressurized Boiling Water Fuel irradiation Capsules (BWFC) and non-destructive test equipment in the reactor pool together with well equipped hot cells and a wide spectrum of irradiation test capabilities for LWR fuel rods.
These test capabilities have been successfully employed in more than 160 transient tests, mainly using in power reactor pre-irradiated fuel rods from BWR's and PWR's. The scope of applications covers fuel rods with burn-up levels up to 45 MWd/kg(U), max. linear heat generation rates up to 700 Wcm^{-1} and constant power change rates between 0.05 and $10^3 Wcm^{-1} min^{-1}$.
BWFC irradiation devices with instrumentation for measurement of e.g. fuel rod length, fuel stack length, fuel rod internal pressure, central fuel temperature and fuel rod diameter are available (or in preparation) and are described in the paper together with associated non-destructive test equipment in the HFR pool.

1. INTRODUCTION

HFR's LWR fuel irradiation support has consisted throughout the last 20 years mainly:

- in the 60's: of performance studies for advanced fuel concepts and investigations of effects occuring during steady state operation. Specially designed fuel rods were irradiated in a high pressure loop /1/ operating under steady state PWR conditions and in irradiation capsules in NaK environment /2/.
- Since the late 60's: of investigation of transient effects (power cycling, - ramping and - bumping) on standard LWR fuel rod segments under typical LWR system conditions. More than 160 transient tests - with in power reactor pre-irradiated fuel rods - have been performed in HFR's Pool Side Facility (PSF) /3/ by means of Boiling Water Fuel irradiation Capsules (BWFC).

The object of this contribution is to give a survey of the various BWFC-type irradiation devices, their further development in view of the new test capabilities of the HFR after its reactor vessel exchange (1983) and a general view of the associated non-destructive equipment at HFR.

P. von der Hardt and H. Röttger (eds.), Irradiation Technology, 241–251.

2. HFR's POOL SIDE FACILITY

For transient testing of pre-irradiated segmented fuel rods from power reactors, HFR's PSF is in combination with reloadable BWFC irradiation devices of particular interest because - for standard BWR and PWR fuel-:

- Individual transient scenarios in the range between zero and 700 Wcm^{-1} linear fuel rod power and at e.g. constant power change rates between 0.05 and 1000 $Wcm^{-1}min^{-1}$ can be operated - by controlled displacement of the irradiation rig relative to the HFR core-.
- Operation of tests is independent from reactor operation. Tests can be started and terminated at any time due to easy access to the PSF from the reactor pool.
- Fuel rod failures can be "frozen" in their primary condition. This is due to fast withdrawal of the experiment into a powerfree position after failure detection.
- Capsule loading and discharging is performed in the reactor pool - close to the non-destructive test facilities -, thus fuel rod irradiation behaviour and condition can be checked at regular intervals, without any lengthy transfer and without transportation to hot cells. Furthermore the irradiation test facility performance can be checked at any time.
- All these features together provide a fast, reliable and relatively low cost test facility.

Due to the present geometry of the reactor vessel, irradiation rigs follow a pendulum movement when displaced in the PSF, fig.1. The present form of the PSF restricts fuel rod length to 600 mm. The new reactor vessel will have flat walls at the PSF sides over its entire height, fig.2, thus obviating the fuel rod length limitations. It will have a PSF at both sides of the reactor core, increasing the number of PSF positions from 10 to 22. For long irradiation rigs, displacement devices enabling parallel displacement of the rigs relative to the PSF wall and others providing supplementary vertical displacement will be introduced, fig. 2.

3. BWFC CHARACTERISTICS AND BASIC DESIGN

The characteristic feature of the BWFC-irradiation capsule is the provision of typical LWR system conditions at the fuel rod surface. Both design principles as described by Mr. Reichardt et al. /4/ and Mr. Hebel et al. /5/ are applied in the BWFC designs. The basic concept consists of

- a pressurized water-filled capsule, in which, the power generated by the fuel rod, is removed radially by subcooled boiling, and
- a capsule support, providing forced convection cooling of the capsule. As the fuel rod surface temperature, after beginning of subcooled boiling - above approximately 200 Wcm^{-1} -, depends only on the pressure in the capsule both BWR and PWR operating conditions can be simulated, fig. 3.

At HFR Petten out-of-pile installations for operation and control of up to 8 BWFC capsules at BWR and PWR conditions are available.
The pressurizing primary water system is conceived as a miniature loop enabling monitoring of cladding failures by means of activity monitors. In the cooling system (capsule carrier) for the capsule thermocouples and a flow meter are used to determine fuel rod power by the thermal balance method. Cobalt self powered neutron detectors are provided for supplementary supervision of transient conditions.

For irradiation testing of pre-irradiated fuel rods - since 1973, HFR's main LWR fuel testing activity - reloadable BWFC irradiation devices as illustrated in fig. 4 were developed. Loading and discharging of the irradiation rigs is performed about 3 m under water in the HFR pool, fig. 5. Close to this loading and discharging station, non-destructive test equipment for fuel rod inspection and checks is situated. During all operations the fuel rod is handled vertically. Handling of fuel rods with PCI-defects in the HFR pool has been successfully conducted.

4. BWFC IRRADIATION DEVICES WITH IN-PILE INSTRUMENTATION

Supplementary to the basic type of BWFC irradiation device, instrumentated BWFC irradiation devices for pre-irradiated and "fresh" fuel rods are in use and under preparation, table 1 and fig. 6 to 9. A modular design concept has been adopted for these devices.

Table 1. Survey on instrumented BWFC-irradiation devices.

BWFC-type	Instrumentation, application, status
BWFC-RA (fig. 6)	Linear displacement transducer, length measurement on pre-irradiated fuel rods, used for supplementary failure detection /7/.
BWFC-RT	Thermocouples, used for thermo-hydraulic studies of BWFC capsules, temperature measurement inside the capsule
BWFC-I	Two linear displacement transducers, pressure sensors; fuel rod length measurement, fuel stack measurement and fuel rod pressure measurement on "fresh" fuel rod. Two tests completed /6/.
BWFC-PT (fig.7)	Pressure sensors (3) and high temperature thermocouple; fuel rod pressure and fuel central temperature measurement on "fresh" fuel rod. Test of a new "irradiation resistant" pressure transducer. Two tests in preparation (1982).
BWFC-RD (fig.8)	Strain gauges; diameter measurement during power operation along two 260 mm long axial traces on pre-irradiated fuel rods. Every measuring cycle includes run over calibration standards. In preparation.
BWFC-RP (fig.9)	Pressure sensors (2), valve and tap; pressure measurement on pre-irradiated fuel rods. Attachment of measuring system in hot cells. Capabilities for refill, void volume determination, fission gas analysis etc. In preparation.

Particulars regarding these special irradiation capsules are:

Fig.6: BWFC-RA: As both the transducer coil support and the transmission system of the length changes are situated in the same temperature field, the measurement is partly compensated for temperature effects.

Fig.7: BWFC-PT: Fuel rod and instrumentation are assembled separately. The high temperature thermocouple is installed during final assembly and sealing is performed by HT-soldering in a Zry-S.S. joint. Void volume of the measuring system is less than 1.5 cm^3.

Fig.8: BWFC-RD: The measuring system originated from the CEA/CENG design /8/. Loading of fuel rod is performed from bottom end. The hydraulic displacement system operates with water from the primary system.

Fig.9: BWFC-RP: Assembly, by remote welding, of the pre-irradiated fuel rod and monitoring system is performed in the hot cells. Connection of fuel rod plenum to the measuring system is made by drilling through the end plug. The drill is an integral part of the connection head and after drilling, is subsequently sealed by an end cap with a metal seal.

5. NON-DESTRUCTIVE TEST EQUIPMENT AT HFR

Fig. 5 gives a summary of the non-destructive LWR fuel rod test facilities available, respectively under preparation at HFR.
The neutron radiography device in the reactor pool /9/ is used for inspection of fuel rods up to 1.2 m in length. The dry neutron radiography device /10/ can take fuel rods up to 2.5 m in length. For LWR fuel inspection the direct image method using nitrocellulose film is mainly applied /11/.
Fuel rod cladding condition and geometry is checked by eddy current and diameter measurements. The fuel rods are moved vertically in a combined unit (speed up to 600 mm min^{-1}) through an encircling eddy current coil and a "floating" linear displacement transducer, along axial or helical traces. Data is recorded on analog recorders and floppy discs /12/. Extension of the installation with sensors for oxide layer- and clad metal wall thickness is under preparation.
A gamma scan facility has recently been designed and is under consideration for procurement. Scanning for short lived isotopes and Kr-85 measurement in the plenum region for intermediate, non-destructive determination of fission gas release data will be its main tasks.
The fission gas release measuring station will become available with the introduction of the BWFC-RP irradiation device. (Providing possibilities for refill, void volume determination, fission gas sampling for analysis etc).
In conclusion it should be mentioned that the hot cells at Petten possess similar non-destructive equipment to that installed at the HFR together with supplementary installations for preparation and checks on pre-irradiated fuel rods and for execution of full non-destructive and destructive PIE works on LWR fuel rods.

REFERENCES

/ 1/ N. van der Kley et al., RCN-145, 1971
/ 2/ J. Markgraf, EUR 4639e, part 4, 1972
/ 3/ H. Röttger et al., HFR Characteristics, EUR-5700e, 1981
/ 4/ K. Reichardt et al., Kerntechnik 10, 1968
/ 5/ W. Hebel et al., Kerntechnik 12, 1975
/ 6/ I. Ruyter et al., IAEA Specialists' Meeting Petten, Sept. 1982
/ 7/ J. Markgraf et al., IAEA Specialists' Meeting Petten, Sept. 1982
/ 8/ M. Masson et al., IDWG Geesthacht, Oct. 1980
/ 9/ H.P. Leeflang et al., NR-installation at HFR, EUR-7915, 1981
/10/ H.P. Leeflang et al., Dry NR-installation, EUR 7915, 1981
/11/ J. Markgraf et al., Application of NR, EUR 7915, 1981
/12/ D. Klage et al., contribution to this meeting, session V

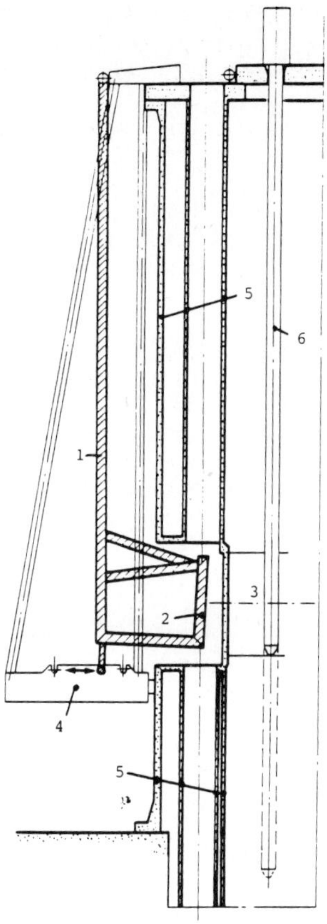

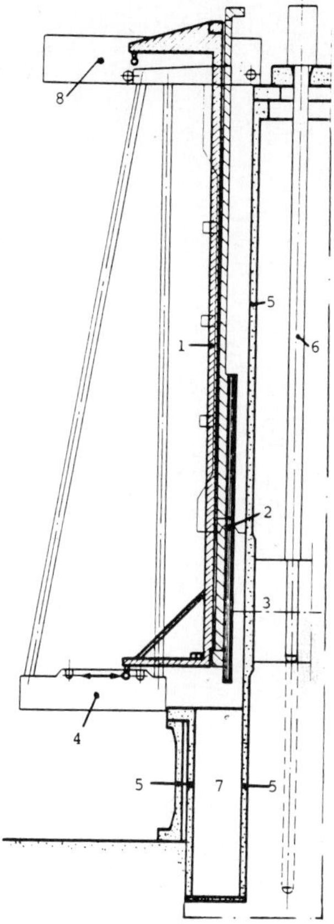

Legend for fig. 1 and 2

1 - Capsule carrier
2 - Irradiation capsule
3 - HFR core
4 - PSF table with trolleys
5 - HFR vessel (outer structure only)
6 - In - tank irradiation device
7 - Extension of PSF beneath HFR core
8 - Upper displacement unit of parallel displacement device

Fig. 1 Present HFR vessel

Fig. 2 New HFR vessel

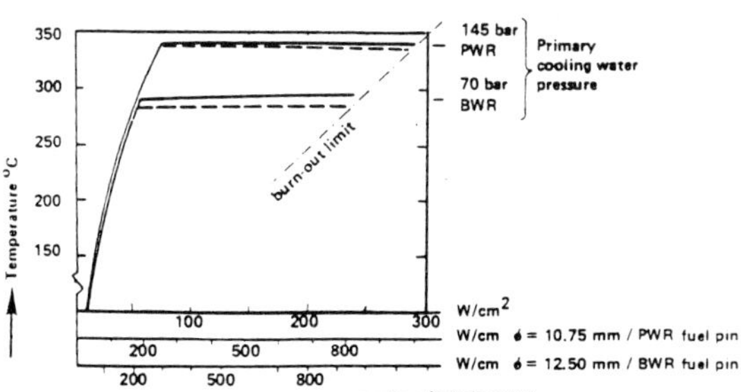

Fig. 3 Fuel rod surface temperature versus fuel rod power in BWFC- devices

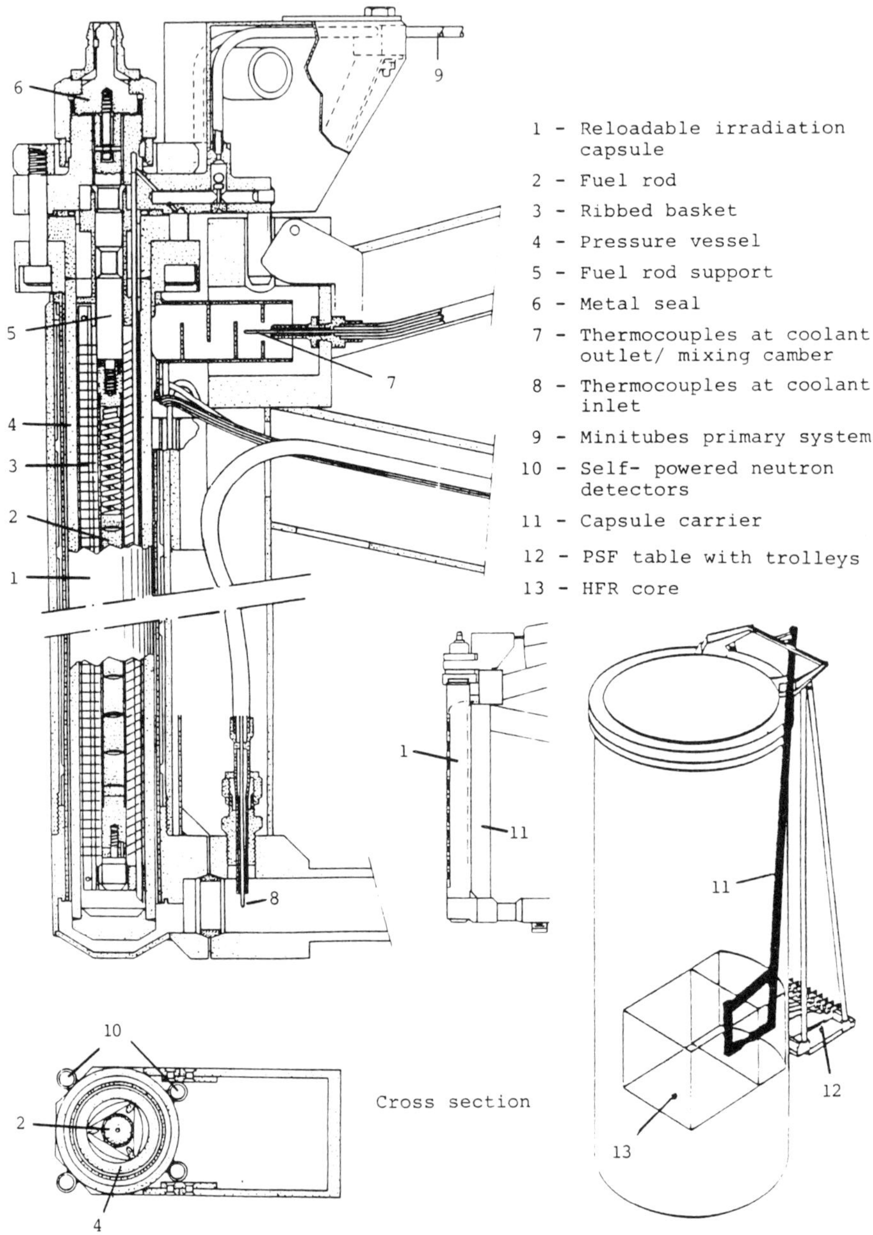

Fig. 4 Schematic drawing of base- type BWFC- irradiation capsule

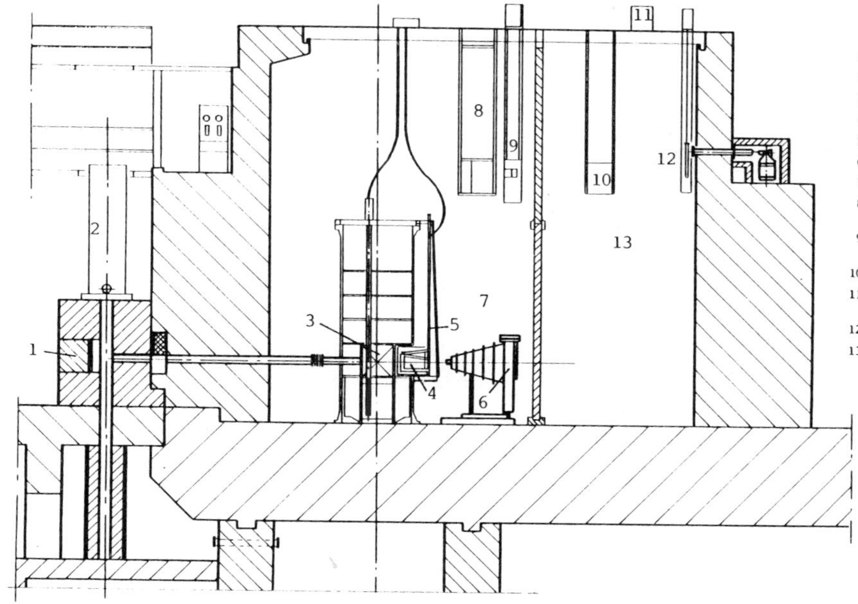

1 Dry neutron radiography installation (DNR)
2 Transport container for DNR
3 HFR core
4 Pool side facility (PSF)
5 PSF irradiation device
6 Neutron radiography camera
7 HFR reactor pool
8 Loading and discharging station for fuel rod irradiation devices
9 Eddy current check and diameter measuring station
10 Fuel rod storage rig
11 Fission gas release measuring station (under consideration)
12 Gamma- scan installation (in preparation)
13 HFR storage pool

Fig. 5 Overall arrangement of LWR- fuel rod test equipment at HFR Petten

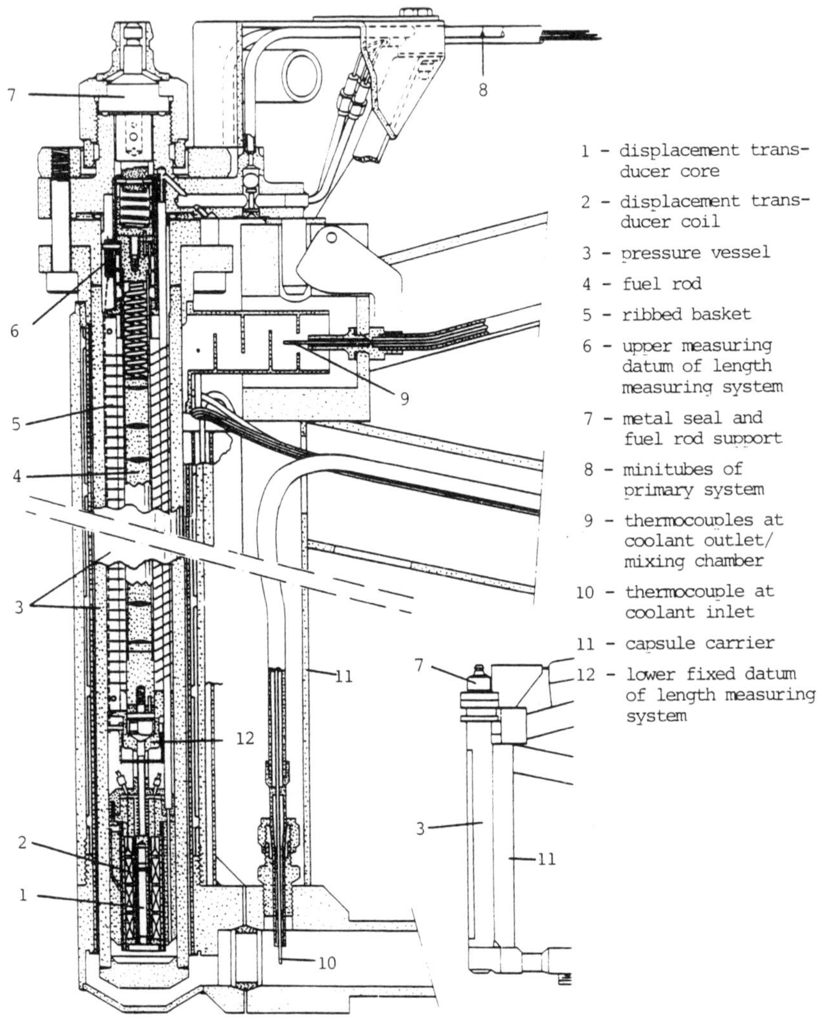

Fig. 6 Schematic drawing of BWFC- RA irradiation capsule with length measuring system (for pre- irradiated fuel rods)

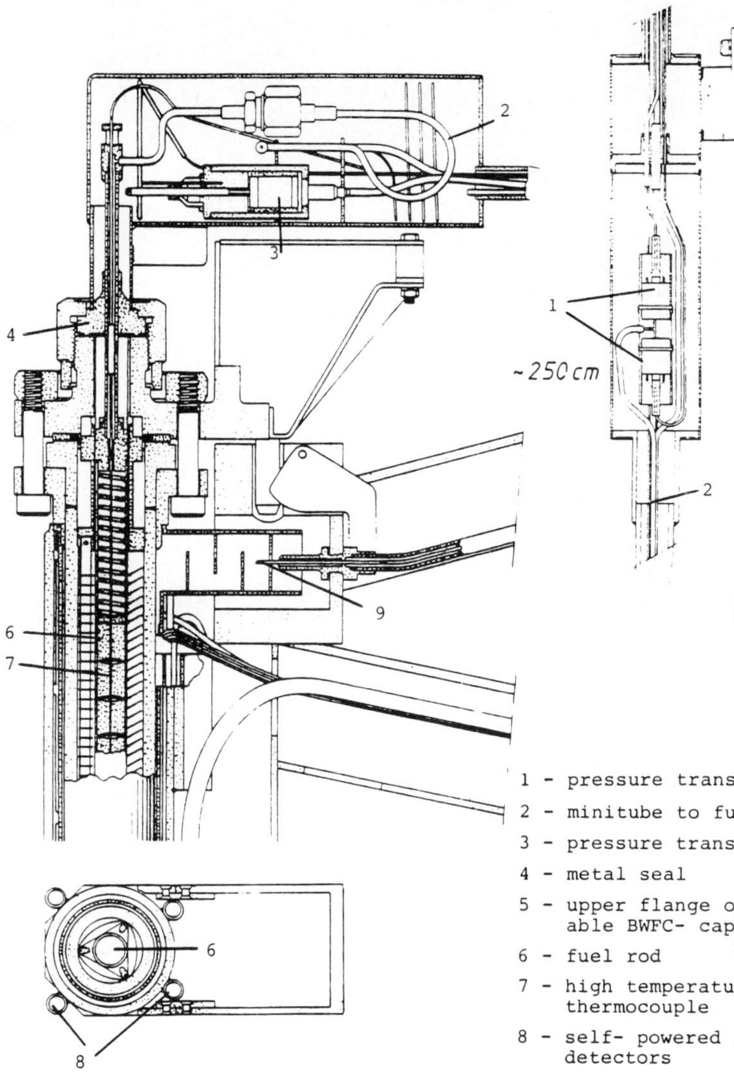

1 - pressure transducers
2 - minitube to fuel rod
3 - pressure transducer
4 - metal seal
5 - upper flange of reloadable BWFC- capsule
6 - fuel rod
7 - high temperature thermocouple
8 - self- powered neutron detectors
9 - thermocouples at coolant channel outlet/ mixing chamber

Fig. 7 Schematic drawing of BWFC- PT irradiation capsule, positioning of in- pile instrumentation for fuel rod pressure and centerline temperature

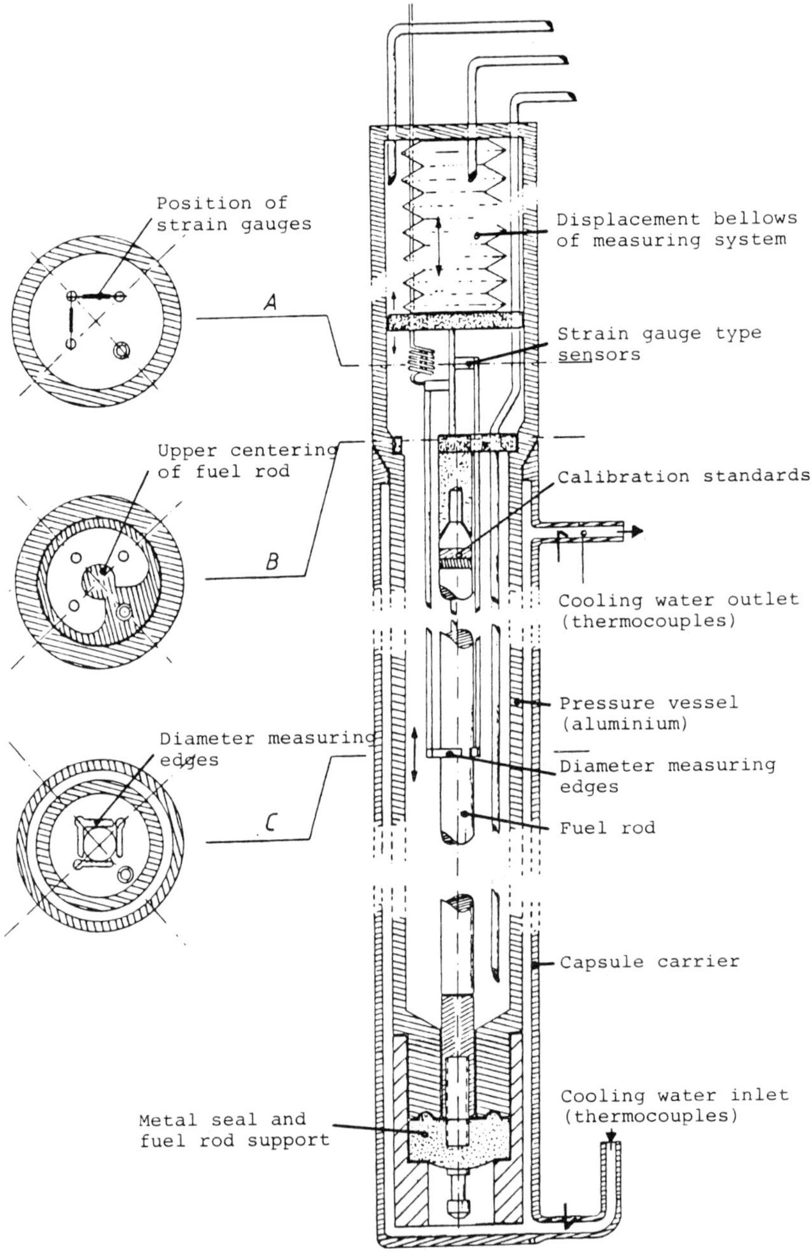

Fig. 8 Schematic drawing of profilometry device for use with pre- irradiated fuel rods

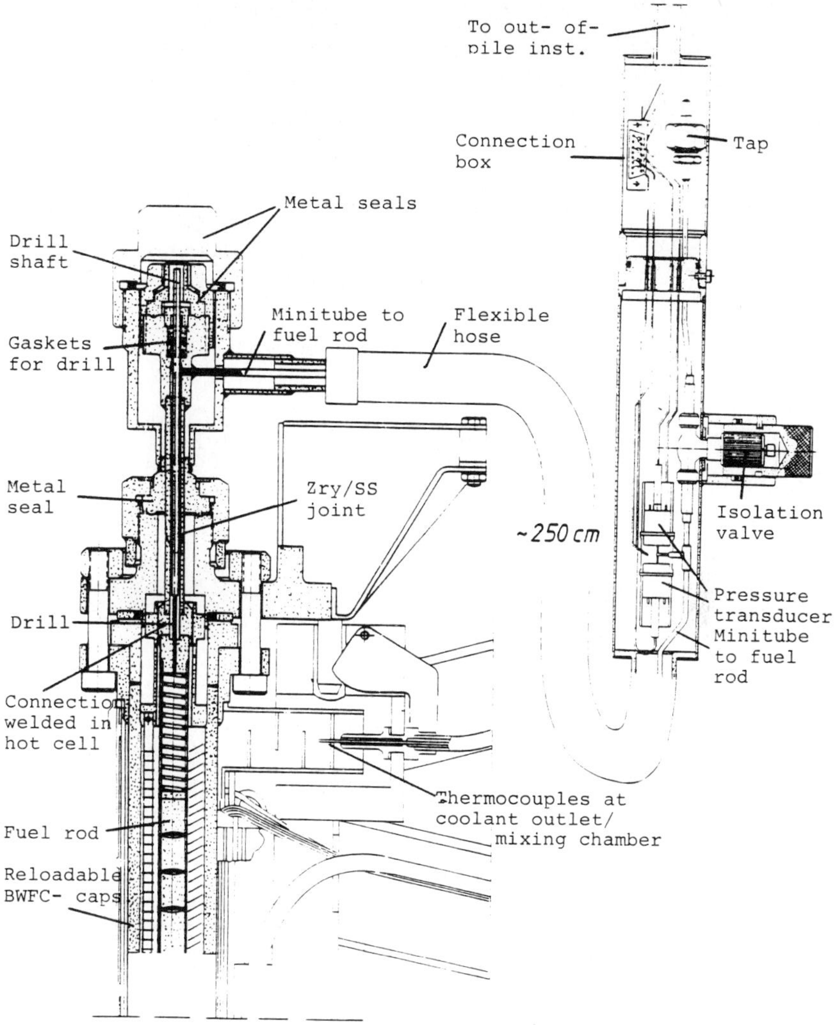

Fig. 9 Schematic drawing of BWFC- RP irradiation device with pressure measuring system for pre- irradiated fuel rods (upper part of capsule only)

SESSION V
LIGHT WATER REACTORS. INSTRUMENTATION.

INSTRUMENTATION FOR AUTOMATIC DIAMETER AND EDDY-CURRENT MEASUREMENT OF IRRADIATED LWR FUEL RODS IN THE HFR POOL

D. Klage, J. Markgraf, D. Perry
Commission of the European Communities
Joint Research Centre Petten Establishment
1755 ZG PETTEN (N.H.)/The Netherlands

ABSTRACT

The programme of tests on Light Water Reactor fuel in the High Flux Reactor (HFR) at Petten, has necessitated the installation of specialised under-water inspection equipment in the HFR pool for eddy current and diameter scanning of the fuel rods before and after irradiation. The inspection measurements provide valuable information on the state and physical dimensions of the fuel cladding after being irradiated under simulated power reactor conditions.
A laboratory computer MINC[R] used with the fuel inspection facility, stores measurement data on floppy disc for further analysis.

The following paper includes a brief description of the test equipment and instrumentation used, together with a comprehensive account of the computerised data collection and treatment system as applied to these particular measurements.

R) Digital Equipment Corporation

1. INTRODUCTION

During the past six years, the HFR at Petten has been an important research tool in irradiation testing of LWR fuel rods pre-irradiated in BWR and PWR power reactors. The irradiation tests have been conducted in the HFR pool side facility (PSF) /1/ using irradiation capsules which simulate typical LWR system conditions. To date, more than 160 power transient tests have been conducted on PWR and BWR fuel rod segments /2,3/.

Equipment for neutron radiographic inspection /4/ and under-water inspection by eddy current and diameter scanning of fuel rods has been installed in the HFR pool.

The main advantages of having inspection and irradiation facilities in close proximity under water are:

- No special transport requirement, thus minimum handling reducing chance of mechanical shocks on fuel rod.
- Fuel rods are always handled vertically also minimizing mechanical disturbances within the rod.
- No special post irradiation cooling of fuel rod required in conducting early inspection scanning.
- Inspection scans can be directly performed before and after irradiation of the fuel rods.
- Fuel rods having cladding perforation can be inspected.
- Fast provision of information on fuel rod condition.

P. von der Hardt and H. Röttger (eds.), Irradiation Technology, 255–266.

Whereas in the past the signals were merely recorded on conventional analogue chart recorders, the addition of a computerised system now facilitates simplification of data acquisition, cataloguing and graphical presentation for analytical study.

2. MECHANICAL EQUIPMENT (Fig. 1)

The eddy current (EC) and diameter scanning equipment installed in the HFR pool comprises the following:

- Support Frame:

A support frame extends down into the pool for a depth of 2 metres below the water surface. It is secured to the inner wall of the pool next to the under-water work table where the irradiation capsules are placed for loading and unloading of the fuel rods.

- Eddy Current and Diameter Measuring Rack:

The rack is mounted vertically and supported by the frame described in the preceding paragraph. A mobile platform fixed above the water level accommodates the scanning drive unit. A fixed platform at the bottom locates two measuring heads side by side. The upper mobile platform permits the drive unit to be displaced horizontally and locked such that it is vertically in-line with the desired measuring head.

- Measuring Heads:

Two types of LWR fuel rods are tested, namely PWR rods of 10.75 mm diameter and BWR rods, 12.5 mm in diameter. To accommodate these differences in diameter, two measuring heads having alternative internal dimensions are provided.

At the top of each head a funnelled guide tube facilitates loading of the fuel rods. Rollers inside the head centralise and guide the fuel rod through an eddy current encircling coil.

Fitted horizontally just above the EC coil, an independently sprung "U" sectioned yoke spans the fuel rod under test. Fitted to one arm of the yoke, perpendicular to the fuel rod, a spring loaded displacement transducer probe contacts the surface of the rod. From the remaining arm of the yoke, a fixed anvil register contacts the rod and because the yoke is self-centralising always follows the rod profile during scanning. The transducer probe and anvil contact points are diameterically opposed to each other, the distance between them representing the fuel rod diameter.

- Drive Unit:

The drive unit comprises a D.C. motor and gearbox with a linear drive system of 500 mm displacement. An electrical potentiometer connected to the drive gearing derives linear position.

A tacho-generator fed from the motor provides a signal for linear scanning speed control.

At the front face of the unit is a register into which the fuel rod carrying tool can be locked. The register can be raised and lowered by the drive system at speeds of up to 650 mm/min. For helical scanning, a gear wheel fixed at the back of the register enables combined rotation and linear motion of the fuel rod, one revolution of the fuel rod relating to 3 mm linear (vertical) motion. A gear change mechanism together with a radial position adjuster and indicator are provided on the drive unit.

- Fuel Rod Handling and Carrying Tool:
Comprises a stainless steel tube with a self locking coupling fitted to the bottom end which can be locked on or off the mating coupling forming the top attachment of the fuel rod. A gear wheel is welded close to the top of the tool and provides a location for carriage and rotation within the register on the drive unit.
- Calibration rods: They comprise:
For E.C. Instrumentation, zircaloy cladding tubes having machined-in calibration defects such as holes, sawcuts and grooves.
For Diameter Instrumentation, stainless steel rods having 50 µm steps which provide a range of calibration diameters within the fuel rod dimensions to be measured.

3. INSTRUMENTATION

The block diagram given in Fig. 2, shows the basic layout of the instrumentation used to control and monitor the EC and diameter scanning system. The following Table 1 gives the instrument specifications and a brief description of their function and purpose:

Table 1

Instrument	Specification	Function
Motor Drive Unit	Model 187 /5/ (JRC Petten design)	Provides: electrical power to motor via speed controller, manual on/off switching, 0-20mA signal derived from potentiometer in drive mechanism, scan position setting and digital display.
Speed Controller	Model 255 /5/ (JRC Petten design)	Enables: setting of scan speed and selection of high or low speed range mode. Motor speeds between 650 to 1350 rpm corresponding to 200 to 650 mm/min linear speed of scan.
Marker Generator	Model 256 /5/ (JRC Petten design)	Provides: automatic pulse of 24 volt DC for every 50 mm displacement step over the 0-500 mm range of scan. Manual switch to generate pulse also available. Enables: automatic and manual drive of linear position marker pens on Kompensograph chart recorders.
Paper Chart Recorders	Siemens Kompensograph III. Two channel units Type M73922. Ranges used: 30 mV DC for diameter scan, 30 V DC for EC scan Chart speed setting: 600 mm/min.	Two recorders provide: paper trace record of EC and diameter measurement signal scans, record of linear position on pin relative to scan, back-up of scan data recorded by computer system.
Eddy Current Test Unit	Multitest EM-3300 (Automation Industries B.V. Sperry Div.) Frequency set at 200 KHz	Provides: supply to EC coil in reactor pool, display of both the phase and amplitude vectors of EC signal on a built-in storage oscilloscope, signals in DC form as horizontal (X) and vertical (Y) components for display on recorders and storage in computer. Note: Settings are such that, in general the X-signal is sensitive to dimensional variation of the rod, whilst the Y signal is sensitive to cladding thickness variation or defects.
Eddy Current Test Coil	Encircling Coil Probe for under-water use (Developed and manufactured by KWU)	Installed in measuring head in HFR pool and together with EM3300 enables: detection of fuel rod wall defects, cracking, porosity, thinning, diameter undulation.
Amplifier for Diameter Measuring Transducer	Type D7M (RDP Electronics) Provides transducer supply of 5 V rms at 5 KHz. ± 10 V at 10 mA filtered output.	This unit drives and conditions the signal from the linear displacement transducer used to measure fuel rod diameter.
Displacement Transducer for Diameter Measurement	Type D5/100 AW for under-water use. (RDP Electronics) Designed for 5 V rms at 5 KHz. Range ± 2.5 mm.	Installed in measuring head in HFR pool and together with amplifier D7M enables: fuel rod diameter measurement scanning. Output signal sensitivity of the order 0.3 mV/10^{-3} mm.
Interface Unit	Special design and manufacture (JRC Petten)	Enables simultaneous start and stop of the drive unit, two paper recorders and computer with only one push button. Each instrument can be enabled or disabled from the interface by individual switches if so desired. Scaling of signals.

4. DATA ACQUISITION

4.1 Considerations

Well-known features of a computerised systems are:

- handling of a large amount of data in a short time,
- presenting the results in a conveniently reduced and clearly arranged manner to the experimentor.

On the basis of fuel rod length (400 mm), scanning step (.2 mm) and number of parameters (4≙ position, X- and Y-eddy current signals, diameter) about 8000 data points (words) are yielded from one axial scan.
Experience /6,7/ has shown that scanning steps between 0.15 to 0.35 mm are sufficient for display and detection of fuel rod profile and failures.
It is desirable to limit the time required for the measuring scans in order to reduce the period during which the operators, wearing pressurized suits, have to wait for loading and discharging operations. Usually one scan takes about 40 seconds.

4.2 Hardware

The characteristics and configuration of the computer system comprising a MINC-11 (Modular Instrument Computer) from Digital Equipment Corporation (DEC) are shown in table 2.

Table 2 MINC configuration

Processor	PDP11/03
Memory	60KB MOS
On-line Mass Storage	1 MB dual Floppy Disc, double density
Standard Interfaces	IEEE-488, 4 x RS-232C
Terminals	VT 105 Video graphing terminal, DECwriter IV
I/O-Modules	16-channel A/D-Converter, programmable clock
Software	MINC BASIC, Versions 1.2 and 2.0
Max. data rate	<130 operations per sec.(version 1.2) <600 operations per sec.(version 2.0)

4.3 Software

Figure 3 shows in a simplified manner the four main functions of the data acquisition routine /8/. A cycle trigger routine, called "SCHMITT", waits for a flag to be set by a manual START/STOP-switch via the clock module.

After setting of this trigger flag the ADC routine - called 'AIN' - scans with a chosen sample frequency ≤ 30 Hz. The input voltages of 4 channels are digitized and stored in a ring buffer (array B%), which forms part of the computer memory (± 6Kbyte workspace) also containing the programme.

As soon as the ring buffer B% is half filled, the content is transfered to the 'virtual' data file 'DATA.DAT' on the system floppy disc (sy0). In the meantime, the A/D converter continues to run by means of the 'CONTINUE' routine and fills the second part of the ring buffer B%. When this is completed, the content is transfered to file 'DATA.DAT' again, etc.
The 'TERMINATE' - routine stops data collection and conversion after 10584 data points have been stored on the system floppy.
In the next step the data from the 'DATA.DAT' file on the system floppy is transfered together with reference data of the experiment and measurement (project nb., date, time, scan frequency, rod orientation, type of measurement etc.) to the data floppy (sy1) into a file 'Cx.DAT' as shown in fig. 4.
One data floppy can store 10 different scans.
Further evaluation of the data is performed by the host computer on the basis of information on the data floppy, fig. 5.
Supplementary to the data acquisition routine, service routines for check and determination of free storage space and reading of data from the data floppy are available.

5. EVALUATION METHODS AND TYPICAL RESULTS

Eddy-current check and diameter measurements provide - together with other non-destructive tests, such as neutron radiography and gamma scanning - the main data source for investigation of the fuel rod and its behaviour e.g. in transient tests.

Analogue display of the measured data during the course of the measurements yields immediate information on the fuel rod condition, e.g. failure positions, and provides a check on the measurement procedure itself. It is therefore maintained in parallel to the digital data recording system described in the preceding chapters.

Figure 5 gives an overview of the digital data handling and evaluation on a PDP11/34 computer system. After filtering, scaling and calibration, the measured data is stored in files on hard disc and thus available simultaneously with data from previous and other relevant measurements.
It is therefore feasible to perform comparative checks (e.g. before v/s after transients), presentation of several measurements in one plot (e.g. gamma-scan and eddy current data, fig.6), evaluation of data referred to local power, pellet mid-position or pellet interface position etc.

On the basis of digital records and analogue recordings the following evaluations are usually being performed:

o Eddy current check:
  - Determination of failure position(s),
  - qualitative evaluation of the dimension sensitive X-signal for ridge height determination;
  - comparison of checks recorded throughout the irradiation test at different stages of the programme.

o Diameter measurement:
  - On the basis of axial traces: Determination of ridge heights; average, maximum and minimum diameter; Comparison of actual measurement with previous measurements; On the basis of helical traces: Supplementary to the afore mentioned evaluation, determination of ovality.

Depending on the requirements of a specific irradiation programme, adaptation of evaluation routines and form of presentation of results are feasible due to the modular lay-out of the computer routines. New measuring techniques (e.g. oxide layer thickness measurement) can also be easily introduced into the system.

6. FUTURE DEVELOPMENTS

Further improvements and extensions are under consideration on the following items:

- The measuring head will be exchanged for a new unit of a modular "cassette"- type, which accommodates the centering devices and individual measuring systems on separate supports, held in a special rack. It also provides as an integral device, a calibration rod and position transducer.
- Introduction of sensors for measurement of oxide layer and clad thickness.
- Introduction of a faster MINC- BASIC, version 2.0.
- On-line data transmission of digital data to host computer.

ACKNOWLEDGEMENTS

The authors wish to thank all staff of JRC Petten, especially Mr. B. Wittebrood and Mr. H. Rother, who are working on this project and contributed on specific subjects.

REFERENCES

/1/ H. Röttger et al., HFR Characteristics, EUR 5700e, 1981
/2/ W. Vogl et al.,IAEA Specialists' Meeting Petten, Sept. 1982
/3/ J. Markgraf et al., IAEA Specialists' Meeting Petten, Sept. 1982
/4/ NR at HFR, contr. to NR-Conference in San Diego, EUR 7915, 1981
/5/ W. Olthoff et al., Technical note P/10/78/23, 1978
/6/ K. Reichardt et al,. 18th Euratom Hot Cell W. G. Meeting, 1978
/7/ A. Okuchi, et al. PIE-Conference, BNS, 1980.
/8/ MINC-11 manuals, books 3, 4, 6 and 7.

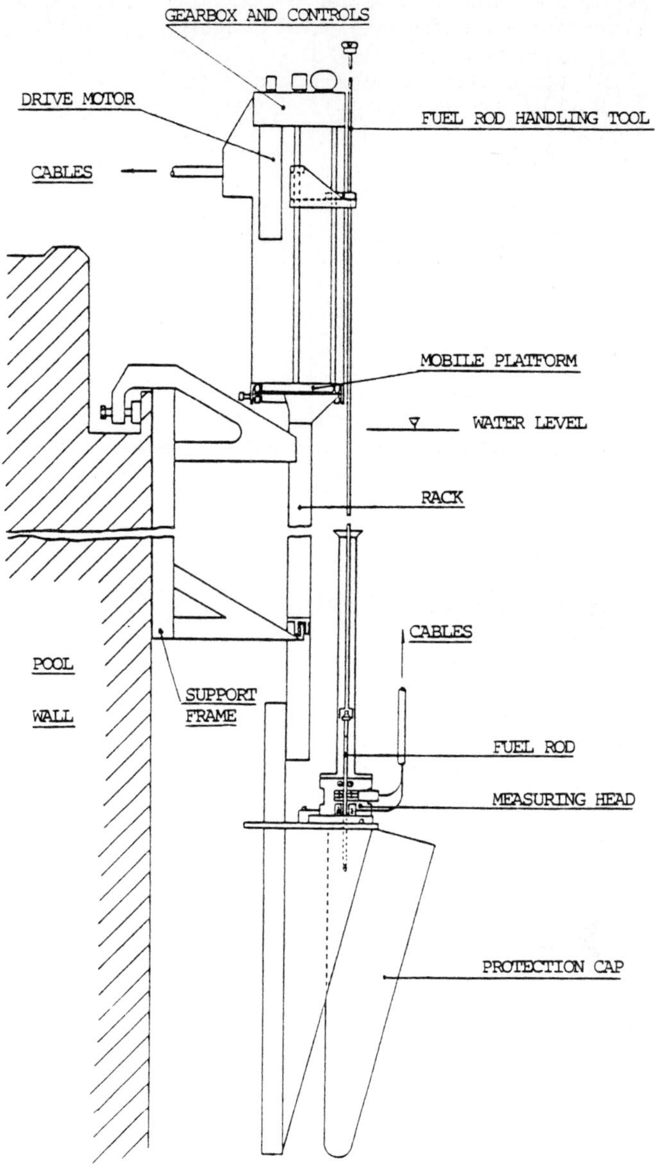

Fig. 1 EDDY CURRENT AND DIAMETER MEASURING DEVICE IN HFR POOL

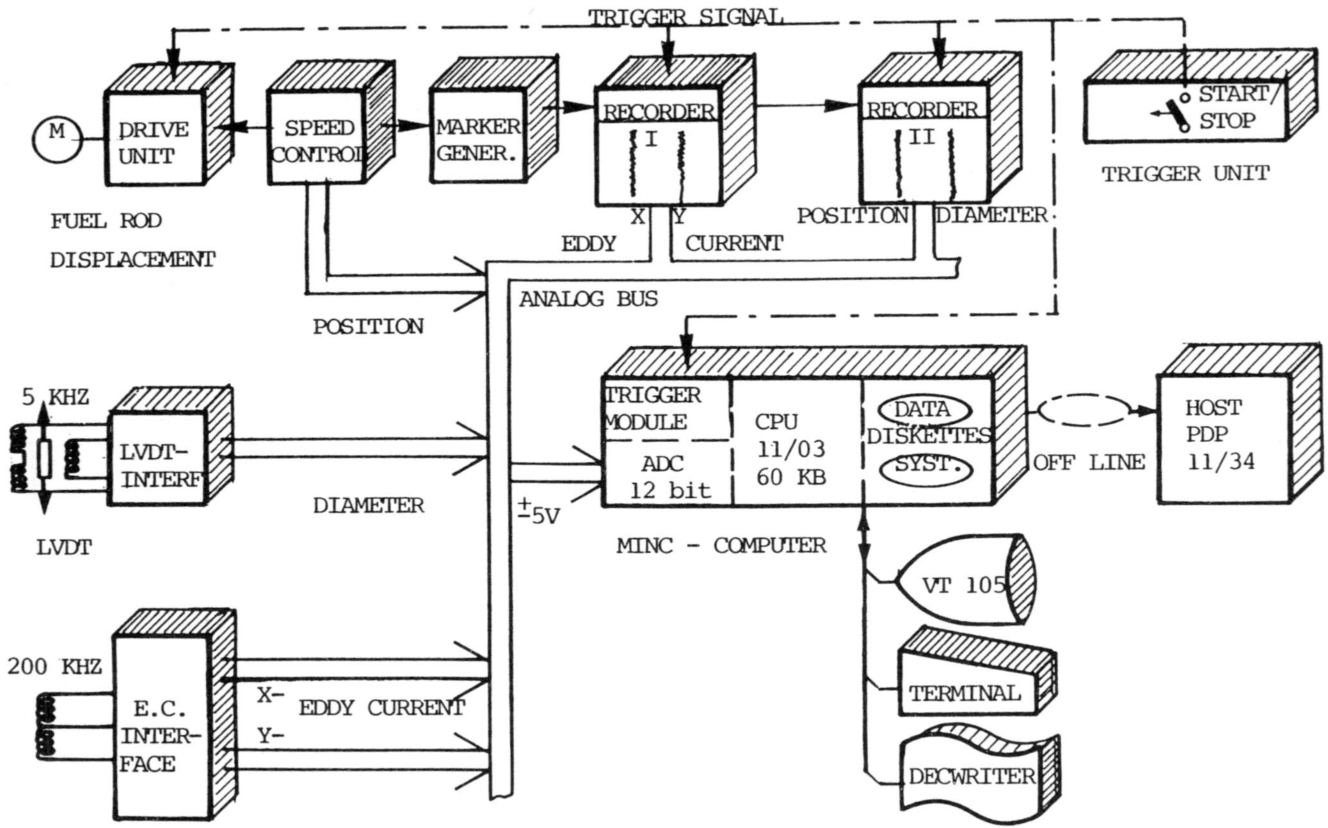

Fig. 2 INSTRUMENTATION FOR AUTOMATIC DIAMETER AND EDDY CURRENT MEASUREMENTS

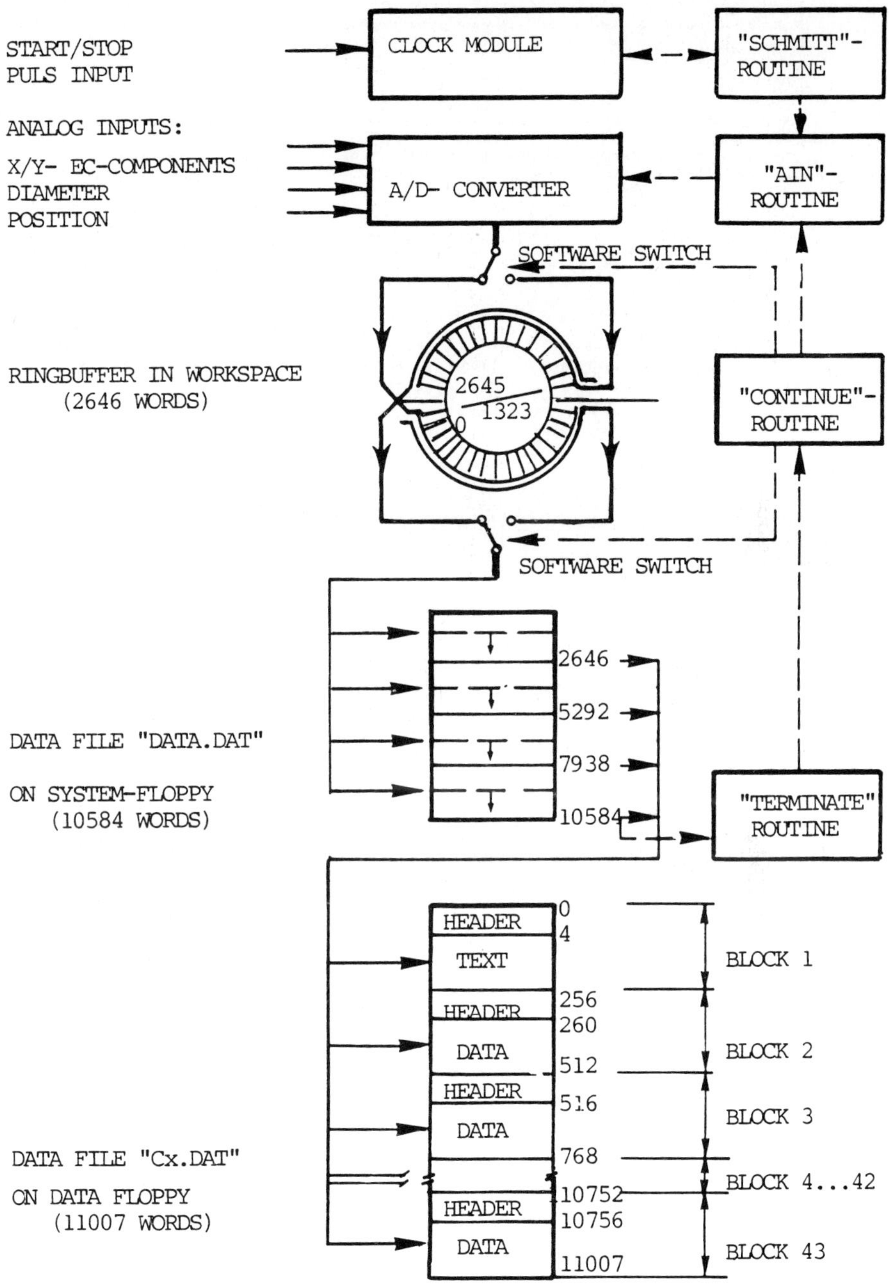

Fig. 3 MAIN FUNCTIONS OF DATA ACQUISITION PROGRAMME

	WORD NR.	CONTENT	MEANING	
BLOCK-HEADER	1	16705	AA	BEGIN TEXT-BLOCK
	2	23130	ZZ	
	3	42	MEASUREMENT-CYCLE NUMBER	
	4	1	BLOCK-NUMBER	
CONSTANTS	5	0-30	SCANFREQUENCY	(Mantisse)
	6	0	" "	(Exponent)
	7	1,2 or 3	PIN ORIENTATION	(Mantisse)
	8	0	" "	(Exponent)
	9	1,2,3 or 4	TYPE OF MEASUREMENT	(Mantisse)
	10	0	" " "	(Exponent)
	11 - 128	INTEGERS	ROOM FOR OTHER CONSTANTS AND VARIABLES	
TEXT	129 - - - - - - - 256	INTEGERS (ONE INTEGER PER ASCI CHARACTER) ASCI-CODE	TEXT OF: Project nr. Expl: E125-75 Date " 27-nov-80 Time " 08:55:00 Scanfrequency " 30 Pin Orient. " 2 Type of " 3 Measurement THE NOT-USED SPACE IS FILLED WITH THE ASCI-CODE OF THE ASCI CHARACTER: ✱	

	WORD NR.	CONTENT	MEANING	
BLOCK-HEADER	1	16962	BB	BEGIN DATA-BLOCK
	2	23130	ZZ	
	3	42	MEASUREMENT-CYCLE NUMBER	
	4	2 - 43	BLOCK-NUMBER	
SCAN-POINT (1)	5	0-20000	MEASURED VALUE: POSITION	(CH.7)
	6	" "	" " : DIAMETER	(CH.8)
	7	" "	" " :EDDY CURR.X	(CH.9)
	8	" "	" " :EDDY CURR.Y	(CH.10)
SCAN-POINT (2)	9	" "	" " : POSITION	(CH.7)
	10	" "	" " : DIAMETER	(CH.8)
	11	" "	" " :EDDY CURR.X	(CH.9)
	12	" "	" " :EDDY CURR.Y	(CH.10)
SCAN-POINT (3-253)	//	//	//	//
SCAN-POINT (254)	253	" "	" " : POSITION	(CH.7)
	254	" "	" " : DIAMETER	(CH.8)
	255	" "	" " :EDDY CURR.X	(CH.9)
	256	" "	" " :EDDY CURR.Y	(CH.10)

Fig. 4 ORGANIZATION OF DATA AND TEXT FILES ON DATA DISKETTE

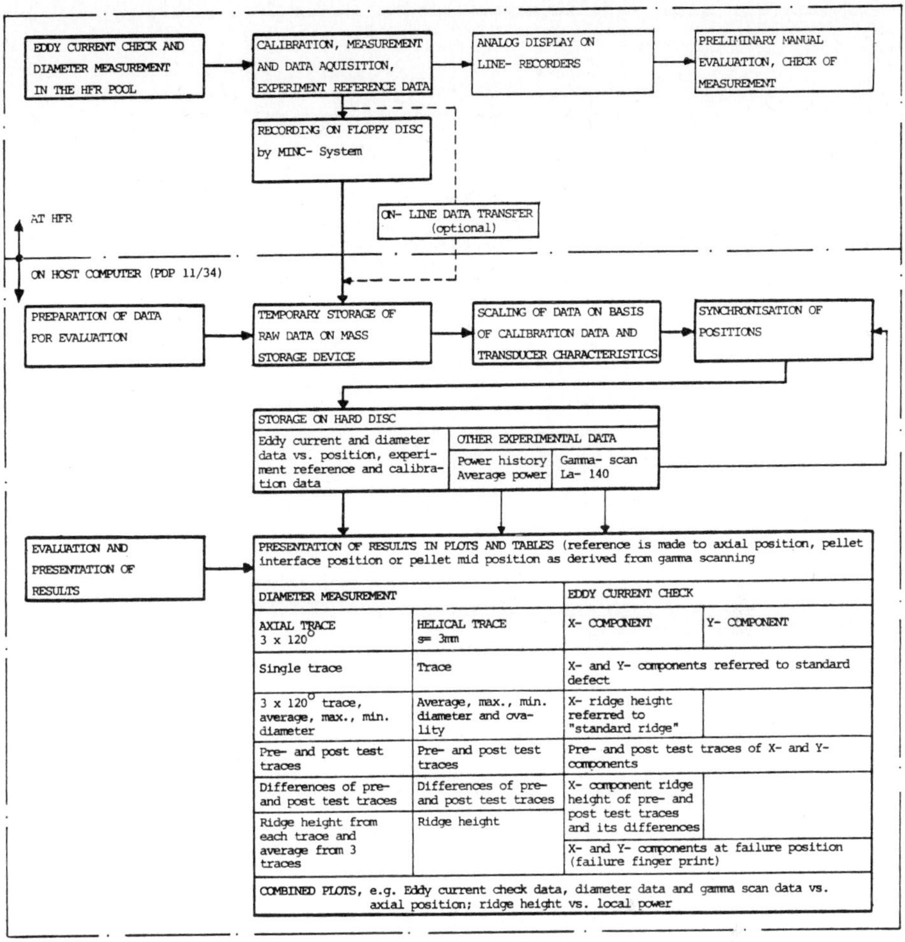

Fig. 5 DATA TREATMENT AND EVALUATION OF EDDY CURRENT AND DIAMETER MEASUREMENTS

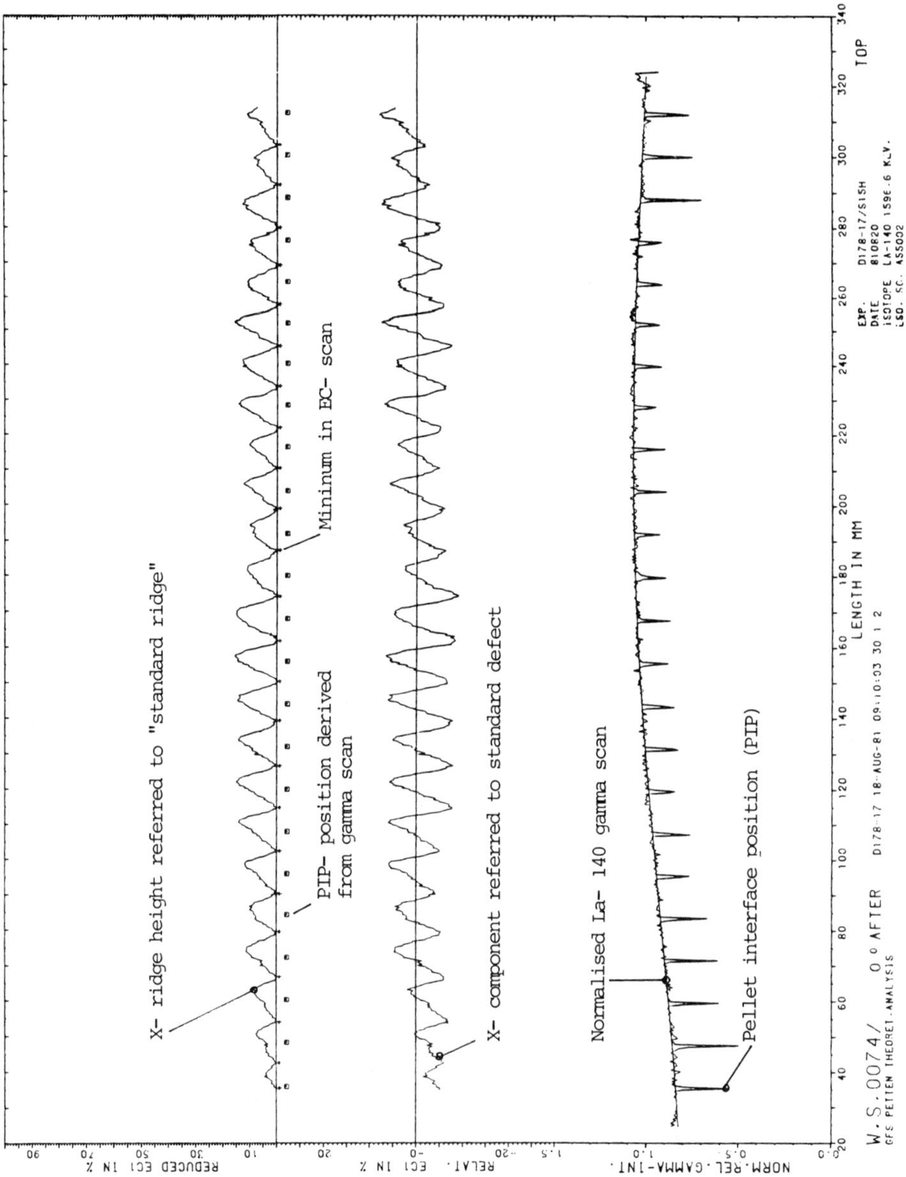

Fig. 6 TYPICAL EXAMPLE OF A COMBINED PLOT

ETUDE DES DEFORMATIONS IMPOSEES PAR L'UO_2 A LA GAINE D'UN ELEMENT COMBUSTIBLE PWR AU COURS D'UN SAUT DE PUISSANCE,

par Robert SCHLEY, Pierre SALLE, Rosemarie ATABEK

(CEN/SACLAY - D.TECH. SECS/SEECRE)

RESUME

L'expérience baptisée ZADIG 01 a consisté à mesurer la déformation imposée par l'UO_2 à la gaine d'un élément combustible de réacteur à eau pressurisée au cours d'un saut de puissance.

Cette expérience a été réalisée en début de vie de l'élément combustible alors qu'il ne s'est pas encore formé de produit de fission susceptible de corroder la gaine en zircaloy.

Le but de l'expérience est d'étalonner, en début de vie, le modèle mathématique d'interaction UO_2 gaine tel qu'il est décrit dans le code VERDON.

L'instrumentation, réalisée sur un crayon combustible court, consistait à équiper la gaine en zircaloy de ce crayon de deux ponts de jauges de contraintes soudables.

L'irradiation a eu lieu dans une boucle à eau pressurisée en périphérie du réacteur OSIRIS implanté à SACLAY, la variation de puissance a été obtenue par déplacement du dispositif dans un module à commande pneumatique.

Le présent texte décrit les méthodes employées pour réaliser l'implantation du crayon combustible, les travaux d'étalonnage effectués pour optimiser les résultats obtenus en pile avec des jauges de contraintes soudables, ainsi que le déroulement de l'expérience et de son interprétation.

Les résultats obtenus montrent que les déformations engendrées sur la gaine du crayon combustible sont mesurables avec le type de jauges de contraintes utilisées et que leurs valeurs sont en bon accord avec celles obtenues avec le modèle mathématique.

P. von der Hardt and H. Röttger (eds.), Irradiation Technology, 267–277.

I - INTRODUCTION

Le présent travail ne vise pas essentiellement à présenter les résultats de l'étude de la déformation imposée par l'UO_2 à la gaine du crayon combustible, mais plutôt les moyens technologiques qui ont permis d'accéder à ces résultats.

C'est donc à la description des méthodes employées pour réaliser l'instrumentation du crayon combustible que nous allons nous attacher.

Cette expérience étant réalisée avec un crayon combustible court et neuf, il est aisé d'implanter une instrumentation conséquente, ce qui n'aurait pas été le cas pour un élément combustible pré-irradié.

Dans cette expérience, nous poursuivions le but d'étalonner le modèle mathématique d'interaction UO_2-gaine tel qu'il est décrit dans le code VERDON du Commissariat à l'Energie Atomique.

Les paramètres essentiels à connaître sont donc déterminés par la mesure des déformations longitudinales et radiales de la gaine.

Pour accéder à la connaissance de ces déformations, nous avons utilisé des jauges de contraintes.

Le dispositif ainsi réalisé a été étalonné en laboratoire puis implanté dans une boucle à eau placée dans le réacteur OSIRIS du Centre d'Etudes Nucléaires de SACLAY.

Dans ce qui suit, nous allons développer les phases essentielles de cette expérience baptisée ZADIG 01.

- Les jauges de contraintes
- L'instrumentation du crayon
- L'étalonnage en laboratoire du dispositif
- Le dispositif d'irradiation
- Le déroulement de l'irradiation
- L'interprétation des résultats
- La critique de l'instrumentation

II - LES JAUGES DE CONTRAINTES

Les conditions de l'expérience constituent une rude épreuve pour l'instrumentation implantée, en effet, au cours du cycle d'irradiation on aura libéré entre le 18.11.81 et le 21.12.81 une énergie correspondant à 1853 MWj, soit 26,47 JEPP dans un réacteur fonctionnant à 70 MW pendant 630 heures, la température de l'eau du réacteur étant de 320°C.

Il convenait donc d'implanter des jauges de contraintes susceptibles de résister à ces conditions.

Le choix peut donc uniquement se fixer sur le seul modèle de jauges répondant à ces critères : des jauges à isolement céramique et structures métalliques soudables.

L'élément actif de ces jauges est un filament d'alliage nickel-chrome.

Pour mesurer les contraintes longitudinales, nous avons adopté un modèle de jauges droites comme l'indique la figure (1a). Les contraintes radiales sont évaluées par une jauge préformée suivant la circonférence du crayon combustible, comme le montre la figure (1b). Les jauges droites sont auto-compensées et constituent en elles-mêmes un demi-pont. Les jauges préformées constituent un quart de pont. Elles sont soumises aux déformations d'origine mécanique que nous souhaitons connaître, mais aussi aux déformations et dérives d'origine thermique qui se comportent en parasites des mesures. Il est donc nécessaire d'y associer des jauges inertes du point de vue mécanique mais évaluant les effets thermiques. Ces jauges inertes sont intégrées dans l'ensemble formé par les jauges droites.

La jauge préformée et sa jauge de compensation sont montées en demi-pont, comme l'indique la figure (1c), cette disposition rend compte uniquement des contraintes mécaniques recherchées. Le demi-pont des jauges droites est représenté par cette figure.

Les jauges subissent un étalonnage sur un support de même nature que l'échantillon à étudier, soit dans notre cas, du zircaloy IV. De cet étalonnage, il résulte la détermination d'un facteur de jauge naturel Kn, lié à la résistance RJ_A de la jauge considérée.

La mesure relative de déformation est exprimée en µmètre par mètre en fonction de la résistance R de la jauge par l'expression :

$$\frac{\Delta R}{R} = K \frac{\Delta L}{L}$$

où ΔR est l'accroissement de résistance de la jauge sous l'effet d'un accroissement de longueur ΔL résultant de la déformation. R et L étant respectivement la résistance et la longueur de la jauge au repos.

Suivant le schéma de la figure(1), au niveau du bras de mesure de l'appareil, on prend en compte :

$$RJ_A + 2\ RL = Kc$$

où RJ_A est la résistance de la jauge active au repos et RL la résistance en ligne. Il en résulte un facteur de jauge calculé Kc suivant :

$$Kc = \frac{Kn \times RJ_A}{RJ_A + 2\ RL}$$

Ce facteur de jauge n'est pas constant mais il décroît linéairement de 1,8% pour 100°C dans le cas des jauges nickel-chrome. Ce fait implique la nécessité d'effectuer un étalonnage au niveau du pont de jauge considéré.

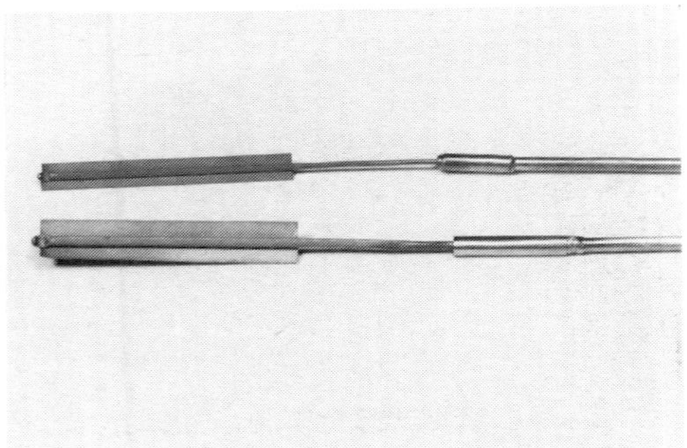

1a-Jauges droites auto-compensées

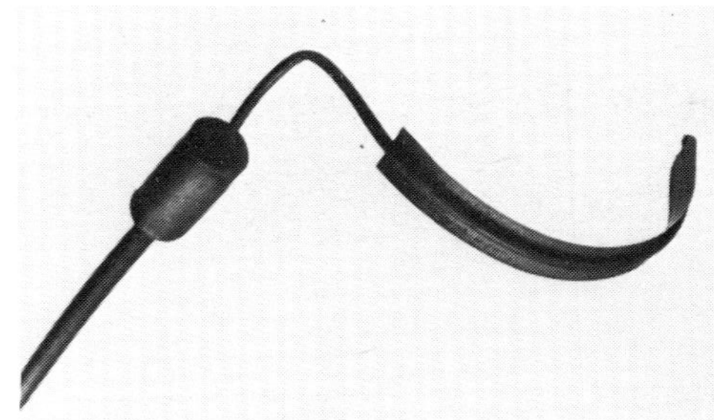

1b-Jauge préformée

Montage en 1/2 pont de jauges droites auto-compensées.

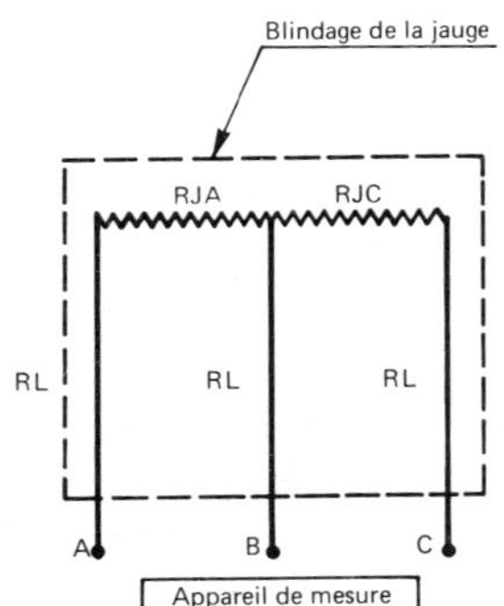

Montage en 1/2 pont d'une jauge 1/4 de pont préformée avec une jauge 1/4 de pont de compensation.

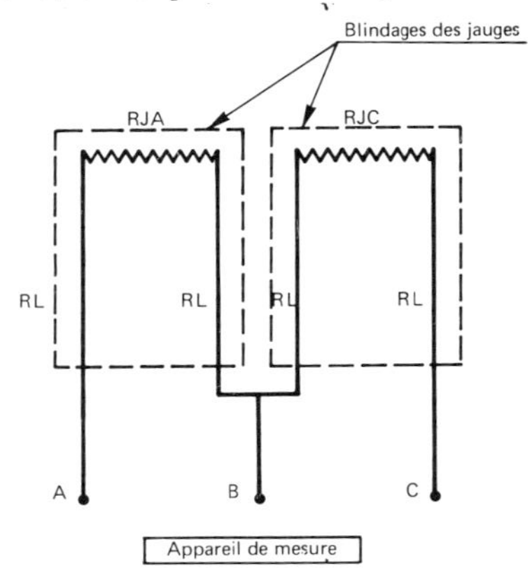

1c-Montage en pont

Figure 1 - Jauges de contraintes soudables

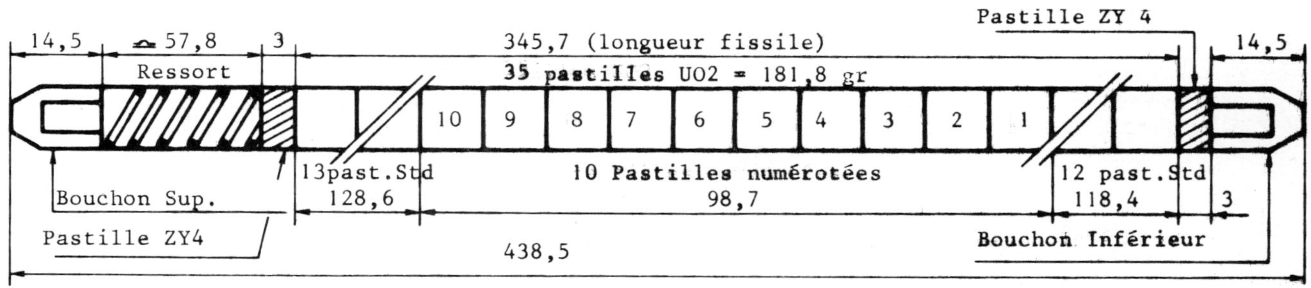

2a-Constitution du crayon

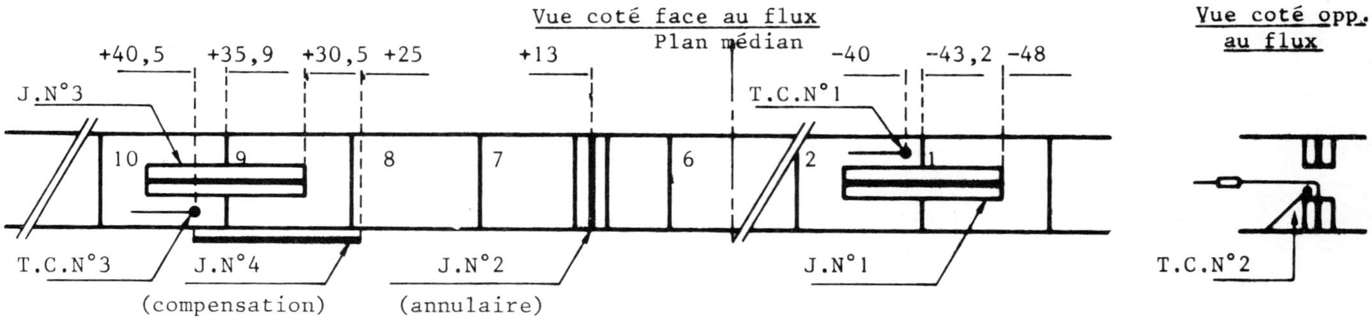

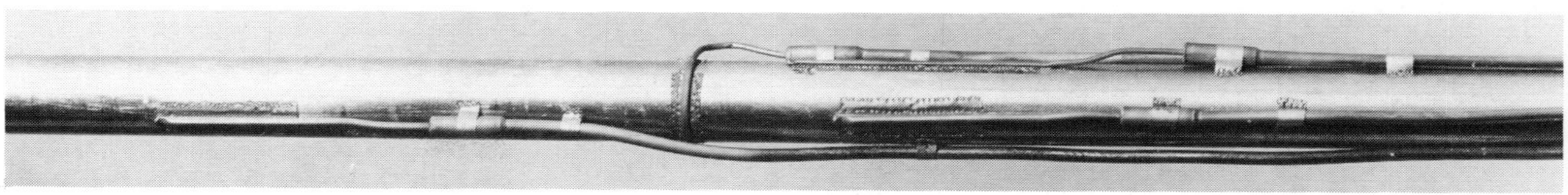

2b-Disposition de l'instrumentation

Figure 2 - Constitution et Instrumentation du crayon expérimental.

III - L'INSTRUMENTATION DU CRAYON COMBUSTIBLE

Le crayon expérimental est constitué par un empilement d'UO_2 enrichi à 7% d'U 235 d'une longueur de 345 mm, soit 35 pastilles. Cet empilement est placé dans une gaine de zircaloy IV recristallisé type CEA d'un diamètre intérieur de 8,2 mm, une épaisseur de 0,6 mm et une longueur de 420 mm. A l'extrémité du crayon, un plénum de 62 mm comporte un ressort d'inconel 600, le tout est rempli d'hélium à une pression de 1 bar absolu figure (2a).

Le jeu UO_2-gaine a été volontairement réduit par l'utilisation de pastilles dont le diamètre extérieur était de 8,15 mm. Ce faible jeu doit favoriser un contact UO_2-gaine dès la première montée en puissance.

La température du crayon est mesurée par 3 thermocouples de diamètre 0,5 mm fixés à la surface de la gaine.

L'instrumentation essentielle est constituée par 4 jauges de contraintes dont deux jauges auto-compensées droites, 1 jauge active circonférentielle et 1 jauge de compensation thermique. Chaque thermocouple est associé à chacune des jauges actives. Le schéma de la figure (2b) de même que la photographie rendent compte de ces dispositions.

L'instrumentation est répartie suivant le plan médian du coeur du réacteur de la manière suivante :

- Jauge auto-compensée droite n° 1 : - 48 mm
- Jauge active circonférentielle n° 2 : + 13 mm
- Jauge de compensation droite n° 4 : + 25 mm
- Jauge auto-compensée droite n° 3 : + 30,5 mm

Le milieu de la partie fissile du crayon est décalé de + 2 mm par rapport au plan médian du coeur.

IV - L'ETALONNAGE EN LABORATOIRE DU DISPOSITIF

Un ensemble expérimental de géométrie extérieure rigoureusement identique à celle du crayon combustible a été réalisé. Il comporte la même gaine que le crayon fermé par des bouchons soudés suivant la spécification adaptée, mais il ne comporte ni pastilles d'UO_2, ni ressort d'inconel. Sur un des bouchons est soudé un tube capillaire pour permettre la pressurisation interne de ce faux crayon.

Dans un premier temps, l'ensemble ne comporte pas de jauges de contraintes, mais des micromètres mécaniques placés exactement au point d'application des futures jauges. On applique successivement une pression P d'origine pneumatique puis hydraulique.

Les déformations sont relevées à l'aide des micromètres. On constate alors que la déformation par allongement précède la déformation par gonflement. Pour P identique, les déformations d'origine pneumatique et hydraulique sont identiques.

On en déduit que l'élévation de température due à l'application de la pression pneumatique n'a pas d'incidence.

A ce stade, on procède au soudage des jauges de contraintes et la même série d'essais est effectuée, les micromètres demeurant en place. Ces instruments indiquent les mêmes valeurs que précédemment, ce qui montre que le soudage des jauges et leur effet de frettage sur la gaine n'ont pas d'effets néfastes sur sa résistance mécanique nominale.

Enfin, la même série d'essais est effectuée avec les jauges seules qui indiquent les mêmes valeurs que précédemment, donc la présence des micromètres n'a pas d'incidence sur le comportement de la gaine.

Après chaque essai, on constate qu'une dérive de zéro apparaît et qu'un recalage de l'appareil de mesure est nécessaire.

L'ensemble expérimental est ensuite placé dans un four tubulaire régulé avec précision et comportant une zone de gradient thermique nul sur la longueur de l'échantillon. Les essais de pressurisation sont effectués et mettent en évidence des dérives du facteur de jauge pour chaque palier de température.

Ce sont ces essais qui permettront, par comparaison avec l'essai en pile, d'effectuer les recalages nécessaires des résultats de l'expérience pour les différentes variations de température résultant du scénario d'essai.

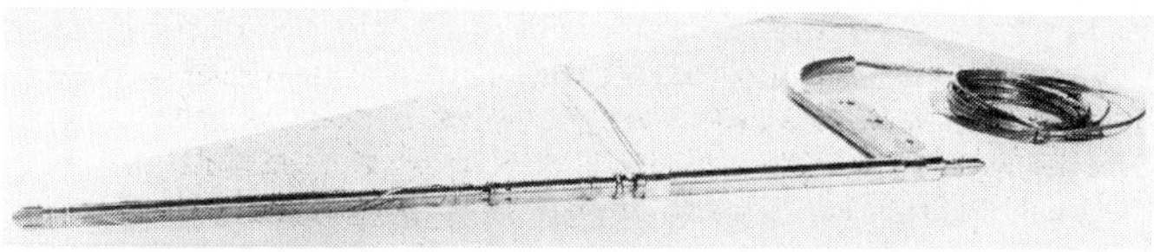

(3a) : Ensemble de la capsule d'irradiation

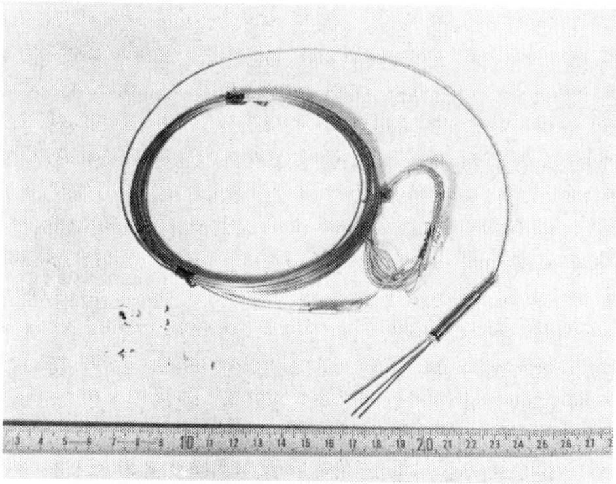

(3b) : Sonde de mesure du bilan thermique

(Figure 3) Eléments du dispositif d'irradiation

V - LE DISPOSITIF D'IRRADIATION

Le dispositif d'irradiation est du type " bouilleur ". Il est rempli d'eau dégazée pressurisée à 150 bars. Son débit de refroidissement est de 1,6 $m^3 \cdot h^{-1}$. La variation de puissance neutronique sur le crayon expérimental est obtenue par déplacement de l'ensemble du dispositif par rapport à la périphérie du coeur du réacteur à l'aide d'un appareil de commande pneumatique. La figure (3a) montre l'ensemble de la capsule d'irradiation.

La puissance dissipée par le crayon est estimée par deux procédés différents.

Un bilan thermique effectué à partir de la mesure du débit dans la capsule et de celle de la différence de température ΔT entre l'entrée et la sortie du dispositif.

La mesure précise de la température de sortie est obtenue à l'aide d'une sonde thermo-électrique spéciale permettant de mesurer la température en plusieurs points de la veine d'eau et d'en délivrer directement la moyenne sous forme d'un signal exprimé en millivolts. La figure (3b) montre cette sonde.

Un bilan neutronique est effectué au moyen de sondes de flux neutronique de type " collectron ". Ces sondes sont placées sur le tube de force de la capsule, au niveau du flux maximum.

Les deux estimations de bilan sont appréciées pour un rendement de fission de 180 MeV.

Les valeurs de la puissance obtenues par bilan thermique excèdent de 15% celles qui résultent du bilan neutronique.

VI - LE DEROULEMENT DE L'IRRADIATION

Les opérations de conditionnement ont duré 3 jours au cours desquels le réacteur OSIRIS a été porté à une puissance de 70 MW. Le dispositif étant maintenu en position arrière, le crayon est alors soumis à une puissance linéique de 36 $W \cdot cm^{-1}$.

A ce stade, le scénario de montée en puissance est le suivant :

Montée à P_1 = 306 $W \cdot cm^{-1}$
Descente en deux paliers à P_1 = 190 $W \cdot cm^{-1}$ puis 131 $W \cdot cm^{-1}$
Montée lente à P_1 = 575 $W \cdot cm^{-1}$ puis descente à 37 $W \cdot cm^{-1}$

La durée totale du scénario est de 3 jours à l'issue desquels le dispositif est défourné.

Nous avons ensuite procédé à la dépressurisation du dispositif et aux examens neutronographiques.

Ces examens révèlent une fissuration de l'UO_2 dans la zone de flux maximal.

L'appréciation de la combustion massique par la moyenne des bilans neutroniques et thermiques donne une valeur moyenne de 178 MWj/tU.

Le suivi en permanence de l'activité γ et de l'activité neutronique de l'eau permet d'accéder à la détection de rupture de gaine.

Le cyclage de puissance décrit dans le scénario cité ci-dessus a permis de mettre en contact les pastilles et la gaine, puis après relaxation de la contrainte par baisse de puissance, celle-ci a été augmentée progressivement à une valeur maximale de 575 W - cm^{-1}.

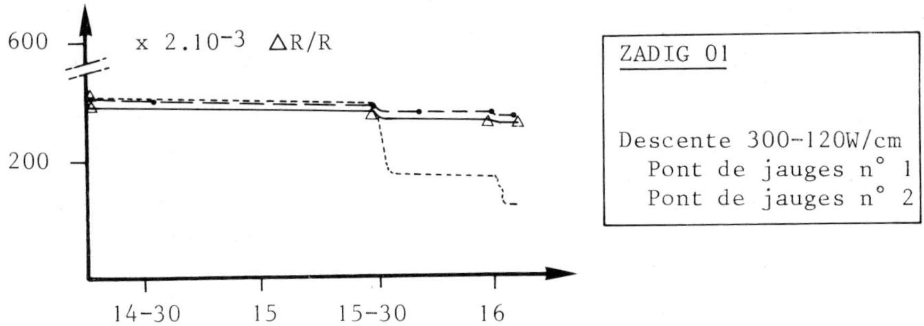

(4a) : Evolution des mesures de $\frac{\Delta R}{R}$ de 300 W - cm^{-1} à 130 W - cm^{-1}

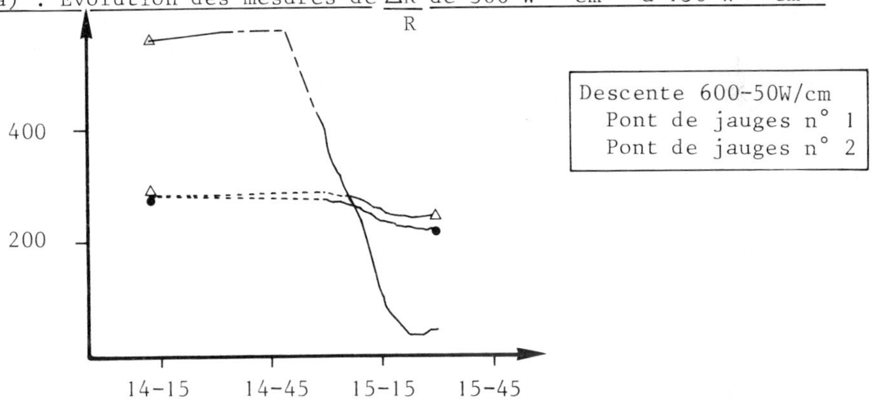

(4b) : Evolution des mesures de $\frac{\Delta R}{R}$ de 575 W - cm^{-1} à 37 W - cm^{-1}

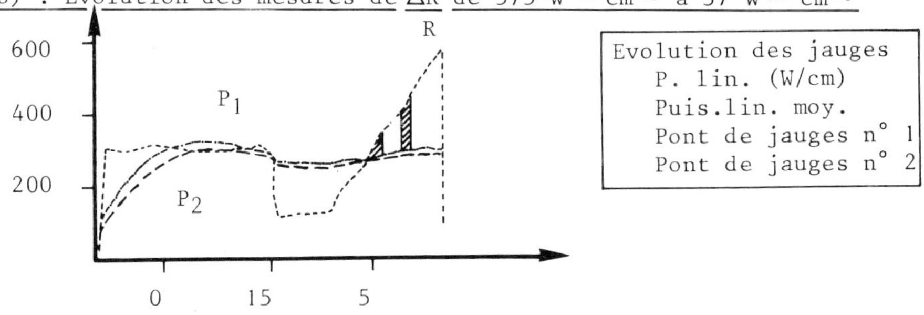

(4c) : Evolution des déformations au cours du scénario d'essais

(Figure 4) Résultats de l'expérience

VII - L'INTERPRETATION DES RESULTATS

La figure (4a) présente l'enregistrement de l'évolution des signaux délivrés par les ponts de jauge 1 et 2 (jauges droites) de 300 W - cm^{-1} à 130 W - cm^{-1}, c'est-à-dire pendant la première partie du scénario.

La figure (4b) indique les mêmes données dans la phase finale du scénario, de 575 W - cm^{-1} à 37 W - cm^{-1}.

Dans les deux cas, les valeurs sont comparées à la courbe d'évolution de puissance.

La figure (4c) rend compte des évolutions des déformations de la gaine du combustible au cours des différents évènements du scénario. Ces valeurs sont comparées aux valeurs calculées et référencées à la valeur de la pression de contact résultant des codes VERDON et RESTA du CEA.

A partir de ces documents, nous constatons les résultats qui suivent et la cohérence de cette expérience.

- L'établissement de la pression de contact UO_2-gaine à 220 W - cm^{-1} et de la plastification à 250 W - cm^{-1} apparaissent sur les courbes par des points d'inflexion.
- La formation des grains balsatiques dans l'UO_2 à 1 750°C semble être mise en évidence.
- Les deux ponts de jauge situés à 50 mm de part et d'autre du plan médian voient sensiblement les mêmes phénomènes, les différences de niveau constatées s'expliquent par l'homogénéisation thermique du dispositif utilisé.
- Au cours de l'expérience, un recalage du zéro des ponts de jauge a été effectué.
- Dans le domaine élastique, les mesures sont sensiblement cohérentes avec les valeurs calculées, par contre dans le domaine plastique, il y a tout lieu de penser que les valeurs mesurées sont correctes alors que le calcul semble imprécis.
- Il convient de noter la reversibilité des phénomènes constatés entre 350 et 400 W - cm^{-1}.

VIII - LA CRITIQUE DE L'INSTRUMENTATION

Pour des raisons fortuites, la jauge circonférentielle n'a pas fonctionné. Nous ne pouvons que le regretter. Au-delà, les jauges de contraintes soudables, si elles nous ont apporté des résultats appréciables ne sont pas exemptes d'incertitudes.

Ainsi, au cours des essais d'étalonnage mentionnés au § IV, des dérives importantes du zéro et du facteur de jauge ont été constatées.

les Dérives du Facteur de Jauge

Elles pourraient provenir de l'association en demi-pont d'une jauge active préformée et d'une jauge de compensation droite. La réalisation de jauges de compensation préformées n'étant pas techniquement possible, il conviendrait, au stade de la fabrication des jauges, de procéder à un appairage en établissant les coefficients de jauges à partir d'un échantillon de gaine d'élément combustible rigoureusement conforme à celui de l'essai.

Il est en effet nécessaire de s'assurer que le coefficient de dilatation thermique est bien identique d'une gaine à l'autre, car les variations de composition ont une incidence importante : une variation de composition de 2 ppm de l'alliage peut entraîner une dérive correspondant à 20 um/m à 100°C et 60 um/m à 300°C.

Les variations de l'état métallurgique (écrouissage, détensionnement, recristallisation) ont des conséquences identiques et le nickel-chrome y est particulièrement sensible.

En fait, les effets de l'irradiation étant de même nature, le choix du filament en nickel-chrome n'est pas forcément le meilleur, d'autant que les filaments d'alliage platine-tungstène permettent de réaliser des jauges dont le coefficient Kn est deux fois supérieur à celui des jauges nickel-chrome. Il en résulte de nombreux avantages pour les jauges platine-tungstène, du point de vue de la sensibilité et de la réduction des dérives d'origine thermique et neutronique.

les Dérives de Zéro

Les jauges soudables sont affectées par le problème de dérive du zéro qui conduit à recaler les appareils de mesure après chaque cycle de déformation. Les origines de ces dérives sont multiples et les fabricants de jauges eux-mêmes ne se les expliquent pas. Une des causes de ces dérives se situe dans les contraintes initiales résultant de la fabrication des jauges et de leur soudage. Un traitement thermique initial des ensembles comportant des jauges soudées permet de s'en affranchir en grande partie, mais l'application de ce traitement n'est pas toujours compatible avec la procédure expérimentale. Des essais sont poursuivis dans notre laboratoire sur cette question.

THE IN-PILE MEASUREMENT OF THERMAL CONDUCTIVITY OF UO_2 VIBRATORY COMPACT

TADASHI KUBOTA, YUKIO IKAWA and ISAO KOMINE
MITSUBISHI ATOMIC POWER INDUSTRIES, INC.

1-297 KITABUKURO, OMIYA, SAITAMA, JAPAN

Abstract

This report describes the measurement of thermal conductivity of UO_2 vibratory compact directly at in-pile by aiming to measure at high temperature near to melting point of UO_2 and also to measure for longer irradiation period. The instrumented capsule made use of a tantalum gas bulb thermometer and null-balance diaphragm type pressure transducer to monitor fuel center temperature. The fuel specimen was doubly contained in two concentric thin walled stainless steel shells. The annulus between these two containers was filled with stacked niobium sleeves, in which three pairs of stainless steel sheathed, Al_2O_3 insulated chromel-alumel thermocouples were positioned 120 degree apart for the purpose of measuring heat generated in fuel. A secondary gas was introduced into the transducer on the upper side of the diaphragm from an out of pile control system.
When the secondary gas pressure exceed the primary gas pressure, an electrical contact was broken and the pressure was recorded. The center temperature was measured as a function of gas pressure in the primary system. The measurement of thermal conductivity was made up to 2400°C. In addition, the measurement was made, although at lower temperature such as below 800°C, continuously for fifteen days.

I Introduction

In the design of nuclear fuel, thermal conductivity is regarded as one of important parameter as determining the limitation of heat flux and temperature in operating with safety. So far, there are only several[1)2)3)] reports on the direct measurement of thermal conductivity during irradiation.

P. von der Hardt and H. Röttger (eds.), Irradiation Technology, 279–287.

Most data on thermal conductivity is derived from the microscopic observation on the irradiated sample and heat flux calculated from neutron flux.

In this report, thermal conductivity of vibratory compacted UO_2 fuel was measured while irradiating in nuclear reactor by using instrumented capsules, which are equipped with thermometer composing of tantalum bulb and null-balance diaphragm type pressure transducer and niobium calorimeter for measuring heat generated in UO_2. Irradiation and in-pile measurement was performed at BRR, BMI.

Table 1

Summary of fuel specimen

Fuel type	UO_2 vibratory compact of fused UO_2 particles
enrichment (U^{235} o/o)	3.5
O/U	2.00
UO_2 density (g/cm^3)	9.54
UO_2 dimension (mm)	20^{mm} ø x 150^{mm} l
The clearance between UO_2 and cladding (mm)	0
Outer diameter of fuel rod (mm)	40
UO_2 spacer	
dimension (mm)	20^{mm} ø x 10^{mm} l (top and bottom)
enrichment U^{235} o/o	Natural
density (g/cm^3)	10.42 (sintered pellet)
Plenum volume of fuel rod (cm^3)	4.19

Table 2

Irradiation condition at peak status

effective neutron flux in UO_2	calculated temperature °C				heat flux at the surface of cladding	fuel burn-up
	at cladding		at UO_2			
	Outer surface	inner surface	surface	center		
$n/cm^2.sec$					W/cm^2	MWD/TUO_2
1.4×10^{13} perturbed 1.0×10^{13}	62	276	412	2643	44.5	720

II EXPERIMENT

1. FUEL SPECIMEN

The fuel specimen is summarized in Table 1. Designed irradiation condition at peak status is shown in Table 2. the fuel rod specimen was set in instrumented capsule after confirming the soundness by helium leakage testing, Xray radiography and auto-clave testing, -280°C. 70 Kg/cm^2, 80 hrs.

2. Capsule

The details of capsule is shown in Fig.1. The capsule makes use of a gas-bulb thermometer and a null-balance diaphragm type pressure transducer to monitor the fuel centerline temperature. The fuel specimen is doubly contained in two concentric thin-walled stainless steel shells. The annulus between these two shells is filled with stacked niobium-1 w/o zirconium sleeves. The bases of each containment shell are sealed with end plugs. A plenum chamber for fission gas is provided above the end plug of the inner shell and below the fuel specimen. The inner shell is closed at the top with a plug welded to a stainless steel tube which connects the transducer and the tantalum pressure bulb. A capillary tube is used in the tantalum bulb so that the volume in the tube is insignificant with respect to the large volume at the base of the bulb. The inside of capsule is shown in Photo 1. The helium gas is contained below the diaphragm in the transducer, as well as in tantalum bulb.

A secondary gas is introduced into the transducer on the upper side of the diaphragm from an out-of-pile control system (Fig.2). When the secondary gas pressure exceeds the primary gas pressure, an electrical contact is broken and the pressure is recorded. The temperature is measured as a function of the gas pressure in primary system. Prior to irradiation, the relation between pressure and temperature is determined at out of pile up to 1400°C and the relation above this temperature was estimated by extrapolation. The reproducibility of measured value during thermal cycling was satisfactory. Once the pressure has been recorded, the secondary pressure is then manually decreased by an out-of-pile control system. The transducer assembly is contained in the outer shell. Three pairs of stainless steel sheathed, Al_2O_3 insulated chromel-alumel thermocouples are positioned 120 degrees apart in the niobium sleeve. Each pair serves as a calorimeter to determine the amount of heat generated in the capsule. Each sheath of thermocouple is sealed by welding at the hot end. The top of each sheath extends above the top of the capsule. The thermocouples, a one mm OD stainless steel gas tube, and the pressure balance gas tube containing the electrical lead pass through a top plug above the transducer assembly. The gas tubes and thermocouples are brazed to this plug forming secondary containment for the system. The top of the one mm OD gas tube extends 15 cm above the capsule and the bottom of the tube ends below the pressure transducer in the upper capsule chamber. The stainless steel gas tube has several purposes. This tube was used to leak test both the inner and outer container shells. The tube was connected to the helium/argon leak detector. After leak testing, an extension tube was brazed to the gas tube. The space between the two containers was evaculated and then back-filled with helium. This fills the internal gas gaps between the niobium and the inner and outer shells with helium. The extension tube is closed with a valve. Periodically during the irradiation, samples of gas was taken from the extension tube and analyzed for radioactive fission gases. This provides a method of leak testing the inner container of the capsule during the irradiation.

3. Irradiation

During the initial reactor startup the temperatures of the fuel specimen was monitored closely using the exterior thermocouples and the gas bulb thermometer. Helium was flown automatically into the secondary side of the transducer so that the pressure in the system rises at a slow uniform rate. the rate was manually controlled. When the pressure on the secondary side balance the pressure on the primary side, the electrical contact in the transducer is broken closing the inlet valve. The pressure was then measured by aircraft-use precision pressure gauge which is capable of reading up to 0.01 kg/cm and an outlet valve was manually opened allowing the gas to flow from the system to an arbitrary pressure which was slightly less than the primary pressure. Upon recording the pressure at the time of system balance, the thermocouple temperatures from the three sets of paired thermocouples was measured. During the time that thermocouple measurements are not made, one pair of thermocouples were hooked to a temperature recorder.

III Result

1. The first capsule was operated for measuring thermal conductivity for longer irradiation period, while center temperature was controlled to comparatively lower value. Fig.3 shows the relation between irradiation time and center temperature. The temperature of cladding inner surface, which was calculated from measured data on the inner surface of niobium sleeve, is shown in the same figure. The center temperature is controlled to about 400 - 500°C. Fig.4 shows the same relation in case that the center temperature is raised to about 800°C. The relation between $\int_{T_s}^{T_c} KdT$ and center temperature is shown in Fig.5.

2. The purpose of the second capsule is to measure the thermal conductivity at higher temperature. The relation between effective thermal conductivity and center temperature is shown in Fig.5. The effective thermal conductivity was calculated by the following formula.

$$\int_{T_s}^{T_c} K(T)\,dT = K'\Delta T = \frac{1}{4\pi} q \frac{I_0(Ka)-1}{\frac{1}{2}Ka\,I_1(Ka)} \quad \ldots (1)$$

where

T_c : center temperature
T_s : surface temperature
K' : effective thermal conductivity
K : thermal conductivity
q : heat generation rate (w/cm^3)
$\frac{I_0(Ka)-1}{\frac{1}{2}Ka\,I_1(Ka)}$: depression factor of neutron flux
ΔT : $T_c - T_s$

Measurement was made during irradiation time of four hours. The maximum temperature measured is 2400°C. The tantalum bulb was broken down after this measurement. The failure of the bulb was quickly detected by the rapid drop-down of primary pressure, which is continuously measured by transducer and then the measurement on this capsule was stopped.

IV Conclusion

1. While the measurement data of thermal conductivity is comparatively scare in case of #2 capsule, which was measured up to such a high temperature as 2400°C, the formulation of thermal conductivity was tried. In consideration of overall microstructural change of vibratory compact with the variation of fuel center temperature during irradiation, it was considered to adequate to formulate the relation between effective thermal conductivity and center temperature. The effective thermal conductivity of UO_2 vibratory compact can be nearly expressed as

$$K' = 3.36 - 1.37 \times 10^{-3}(T - 273) + 2.0 \times 10^{-7}(T - 273)^2. \quad ..(2)$$

2. The effect of irradiation period on the thermal conductivity of UO_2 vibratory compact was not significant at least while the center temperature was controlled below 800°C.
This is consistent with that the microstructure of vibratory compact, such as by sintering and the formation of cracking are not significant at this temprature. Regretfully, the effect at higher temperature was not confirmed by the failure of tantalum bulb in case of #2 capsule at earlier irradiation time.

Acknowledgements

The authors wish to thank Dr. S. Takahashi, Mr. O. Kawaguchi, Dr. M. Mizuta and Dr. K. Maruya for valuable advice in the course of this study.

Reference

1) I. Chen, B. Lustman and J. D. Eichenberg
WAPD - 228 (1060)
2) M. E. Cunningham, J. O. Barner and M.D. Freshley
Trans. ANS (1981) 276
3) A. M. Tachis, W. G. Mothers and J. E. May
AECL - 2521 (1966)

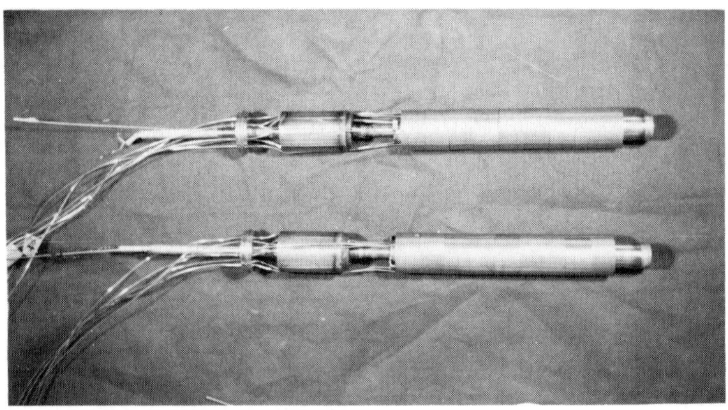

Photo1: The inside of instrumented capsule

1	TOP END PLUG	8	ELECTICAL LEAD WIRE
2	UO_2. COMPACT (E,U)	9	1ϕ S/S PIPE
3	UO_2. PELLET (N,U)	10	INNER CONTAINER SHELL
4	Ta. BULB	11	S/S PELLET
5	THERMOCOUPLE (A,C)	12	BOTTOM END PLUG
6	Nb. SLEEVE	13	Co-Al MONITOR
7	OUTER CONTAINER SHELL	14	PRESSURE TRANSDUCER

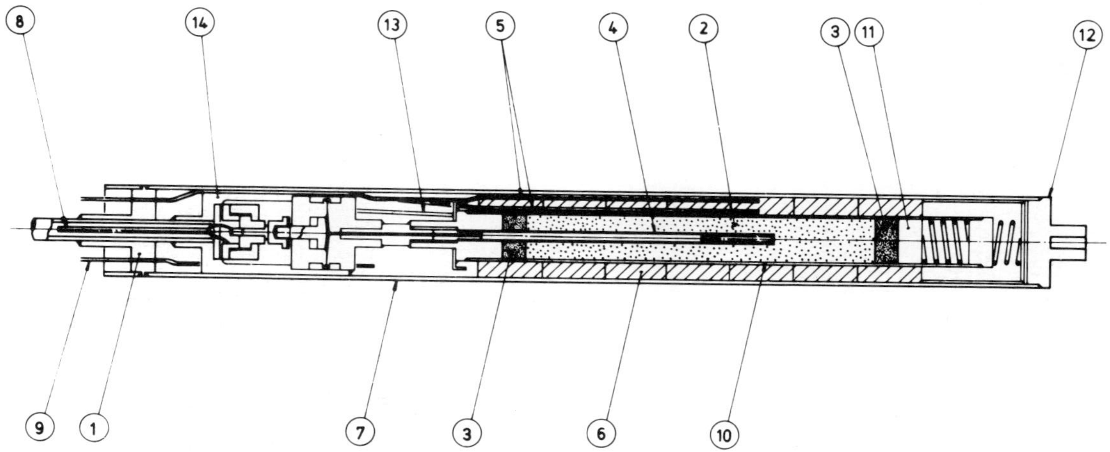

CROSS SECTION OF CAPSULE

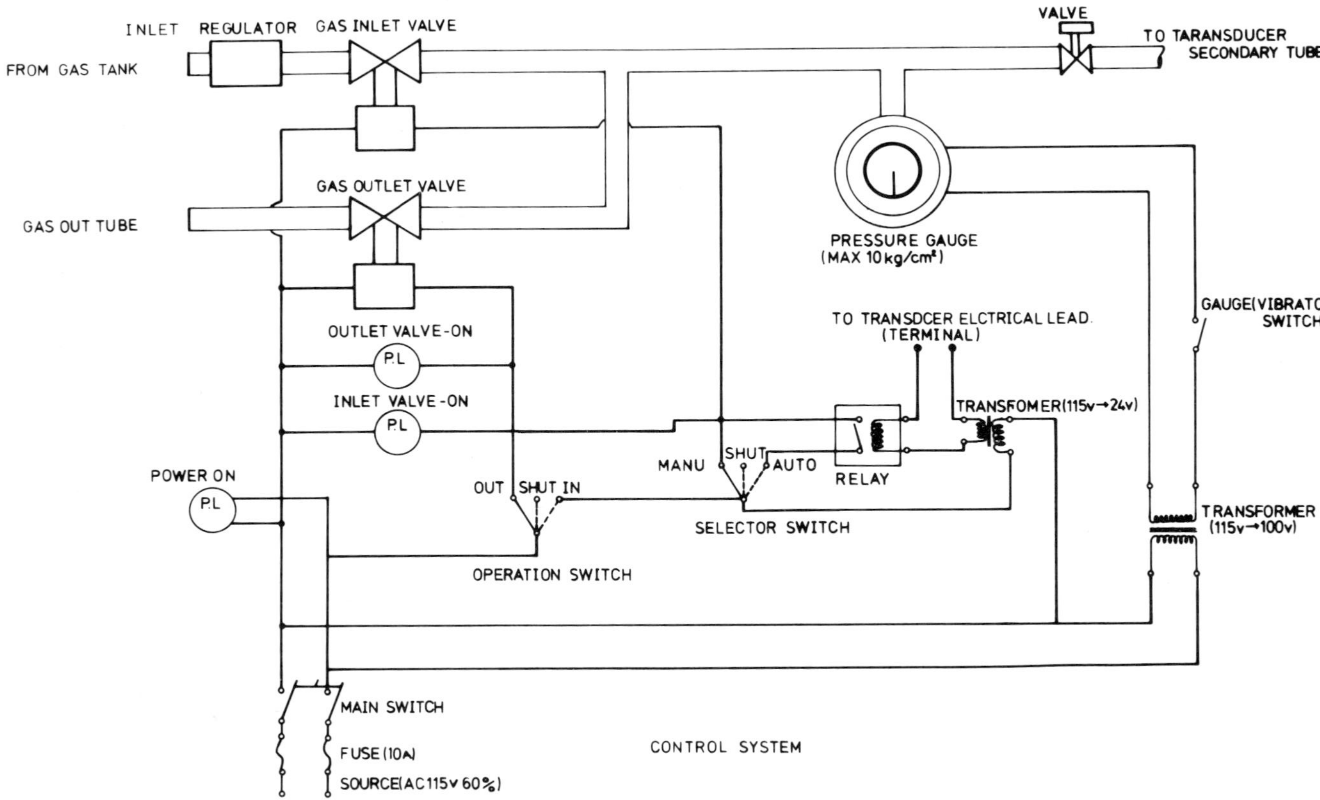

CONTROL SYSTEM

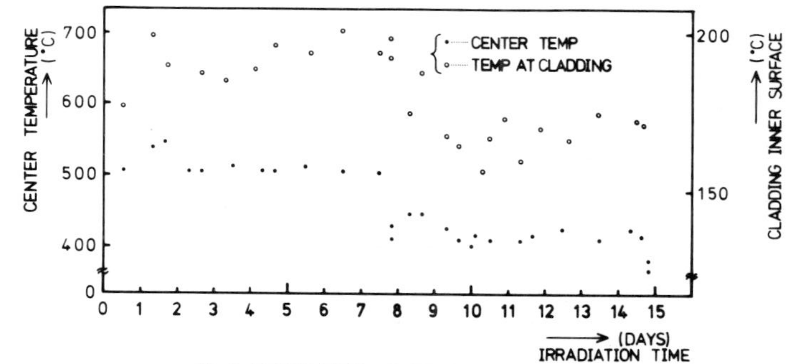

FIG: 3 TEMPERATURE VS IRRADIATION TIME (NO.1 CAPSULE)

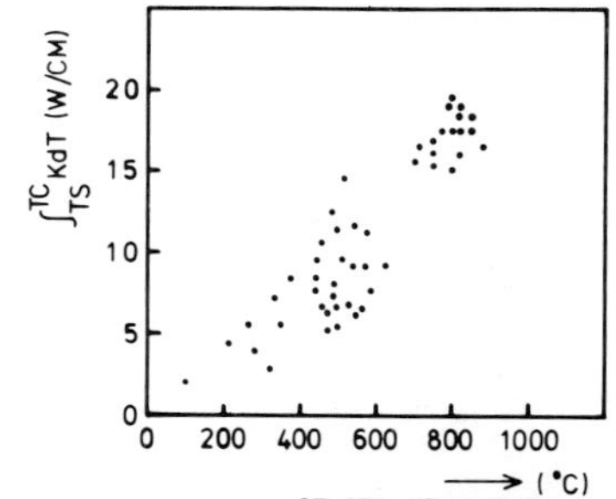

FIG: 5 $\int_{TS}^{TC} KdT$ VS CENTER TEMP (NO. 1 CAPSULE)

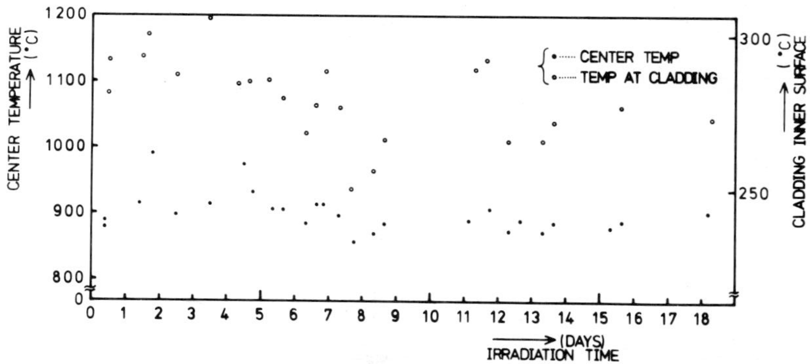

FIG: 4 TEMPERATURE VS IRRADIATION TIME (NO.1 CAPSULE)

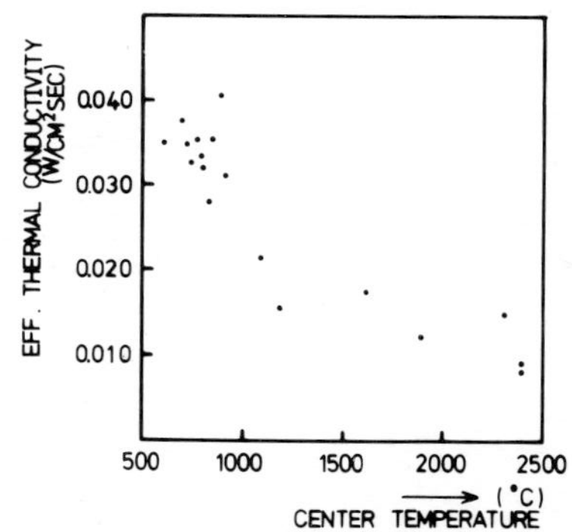

FIG: 6 EFFECTIVE THERMAL CONDUCTIVITY (NO.1 AND NO.2 CAPSULE)

SESSION VI
FAST BREEDER REACTORS

EXPERIMENTS IN RAPSODIE WITH THERMOCOUPLES INSTRUMENTED PINS FOR MEASUREMENTS OF THE INTERNAL FUEL PELLET TEMPERATURE

F. Marin

Centre d'Etudes Nucléaires de Cadarache (DER/SEER)

B.P. n°1 - St Paul Lez Durance - France

FOREWORD

The safe operation of a fast fuel element is notably dependant on its thermal behaviour, particularly in the beginning of its irradiation. As a complement to computer calculations, we use on-line temperature measurements in the centre of fuel pins. The main problemsrelated to such experiments are essentially of three types :

- to perform a good connection of thermocouples between the pin center and the outside of reactor,
- to have at any moment the possibility of cutting this connection in order to allow both the handling of the rig and the use of rotating plugs,
- to have temperature sensors able to run at high temperatures, above 2000°C.

Our purpose is to describe such a rig which was successfully used in Rapsodie for the first time in 1978.

DESCRIPTION OF THE RIG

It is composed of four main parts :

1. The irradiation compartment which can accomodate up to 7 fuel pins of small diameter or 5 of larger diameter like Super-Phenix 1 pins, each one with a thermocouple reaching the pellet center and positionned at mid-length of the fissile zone. This irradiation

P. von der Hardt and H. Röttger (eds.), Irradiation Technology, 291–297.

compartment or inner rig is itself positionned in a receiver fuel sub-assembly. This one is a special Fortissimo fuel sub-assembly in which the three central pins rows have been removed in order to leave a sufficient place for positionning the experiment inner rig. This rig is maintained in its receiver by rotating lugs (fig. 1 and 3).

2. The inner rig is connected to a "measuring tree" in which thermocouples run. This tree is designed for two functions :

   - the connection with the irradiation compartment,
   - the cutting of thermocouples at the end of irradiation by means of a special cutting tool which is driven from the top of the reactor. The cutting of thermocouples is achieved by rotating this tool and so releasing the inner rig in order to discharge it out of the reactor core. This tool acts really as shears. Such a solution was chosen for having a more convenient handling of the whole rig at the end of its irradiation i.e. discharging separately the measuring tree and later the inner rig. But if necessary, for instance in case of failing to cut the thermocouples, it remains possible to discharge both the inner rig and its measuring tree, still connected (fig.2).

3. The measuring tree is itself connected to a tube which drives the cutting tool at the end of irradiation.

4. The whole system, i.e. fuel compartment, measuring tree, driving tube is retractable in a sleeve directly positionned above the receiver fuel sub-assembly in the central channel of the reactor control plug. This retraction allows the movement of the reactor rotating plugs during the dischargement cycles and also the locking of the receiver fuel sub-assembly during the unloading operations of the inner rig.

TEMPERATURE MEASUREMENTS

In the fuel pellets, we have already used W-5% Re/W-26% Re thermocouples, 1.1mm diameter, positionned in renium sleeves of 1.45mm diameter, the compensating wires being made of Hoskins alloy (fig.4).

In the inner rig, we have also positionned CrAl thermocouples at the outlet of fuel pins in order to measure their power. The flow rate was preliminary adjusted, by means of a preliminary hydraulic test, with a nozzle placed at the bottom of the inner rig.

OPERATION OF THE RIG IN RAPSODIE

We have made two experiments in 1978 with three fuel pins and W-Re thermocouples each time. These experiments were fully successful, although the mean life of thermocouples was about 20 hours at temperatures ranging from 1800 to 1900°C. New experiments are ready with 5 Super-Phenix 1 fuel pins and also special boron carbide pins.

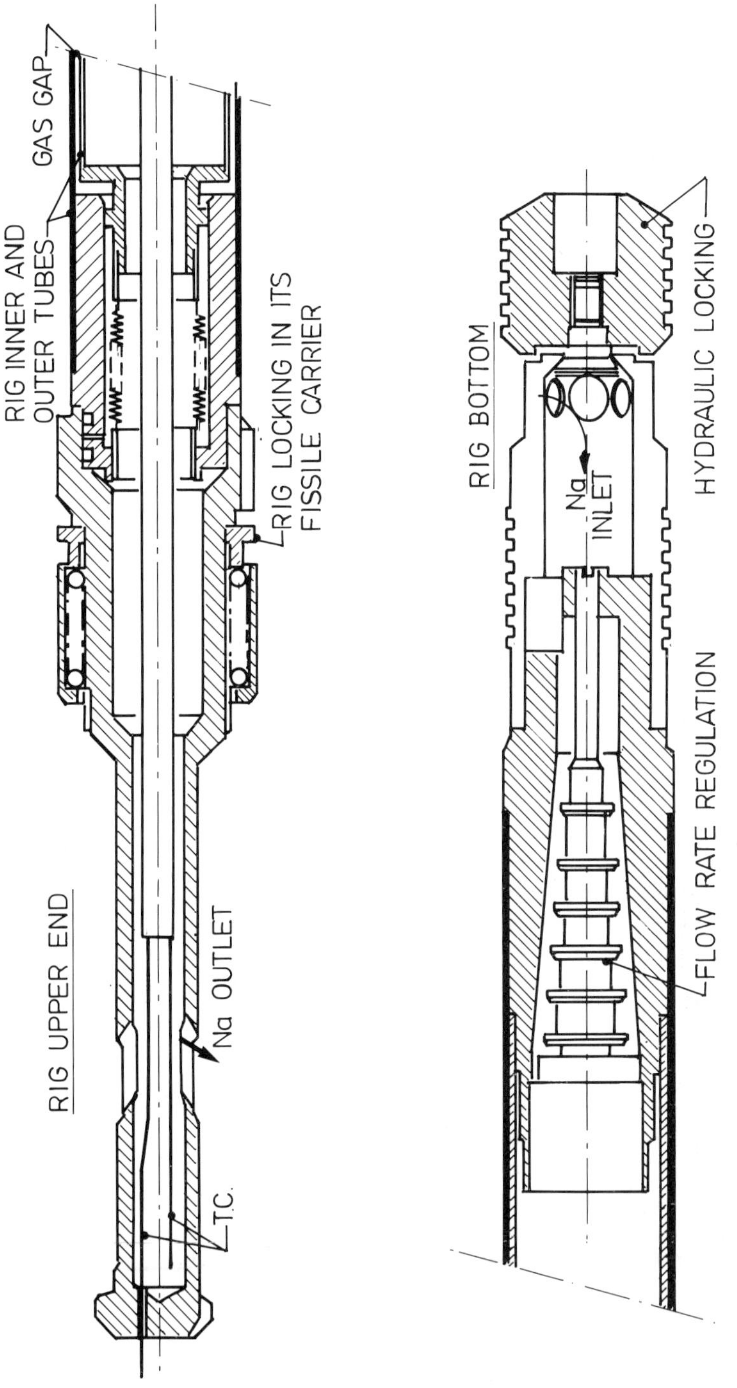

FIG.1. INNER RIG

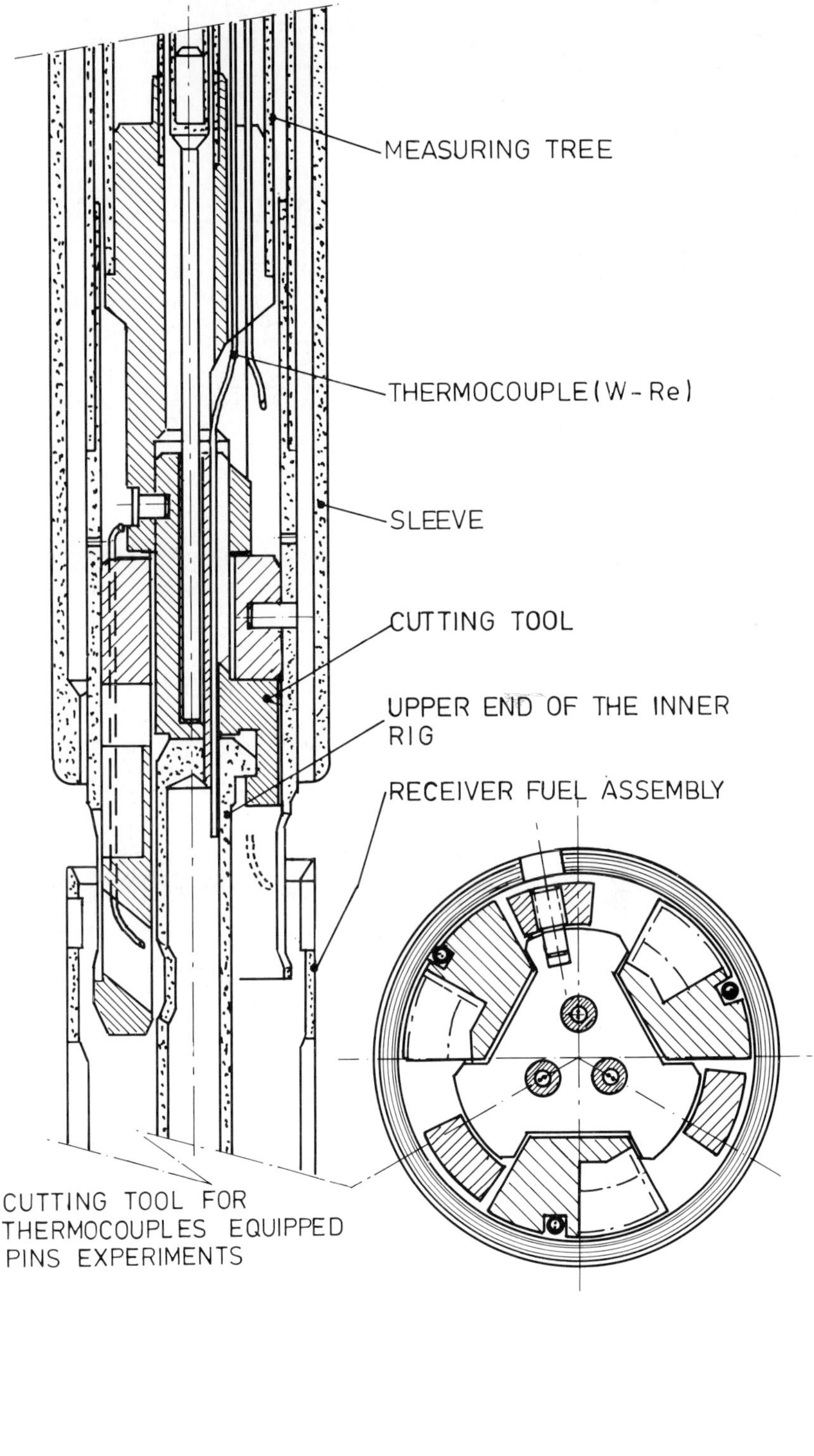

FIG.2. CUTTING TOOL FOR THERMOCOUPLES EQUIPPED PINS EXPERIMENTS

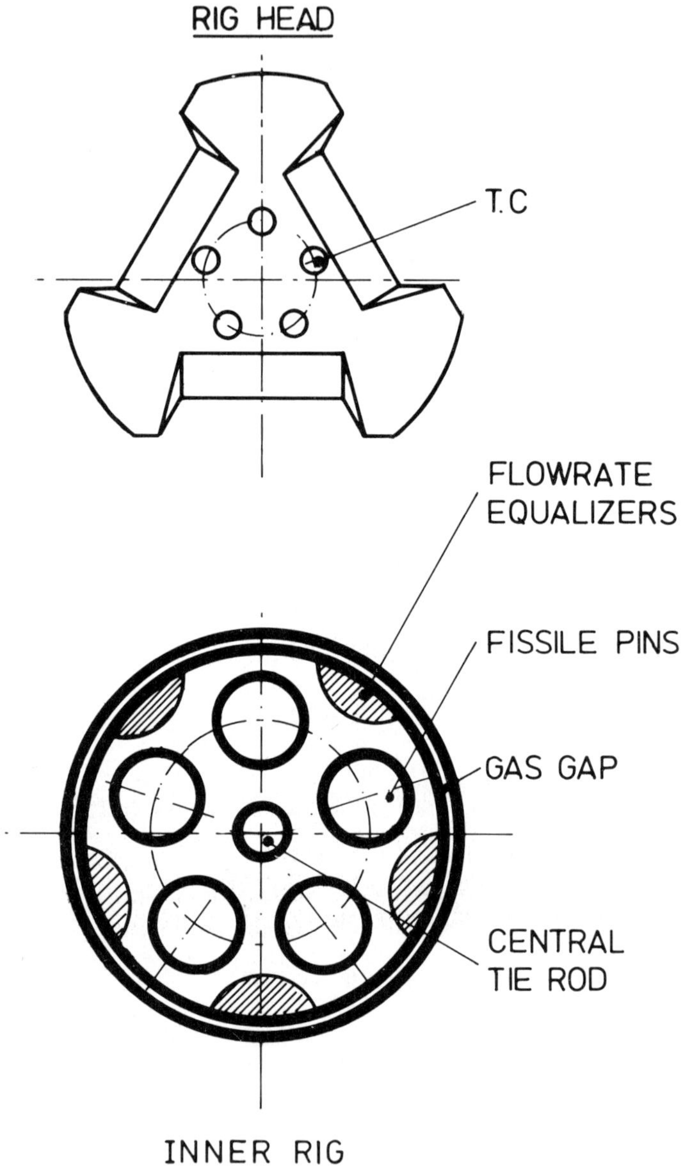

FIG.3. HORIZONTAL RIG CROSS SECTIONS

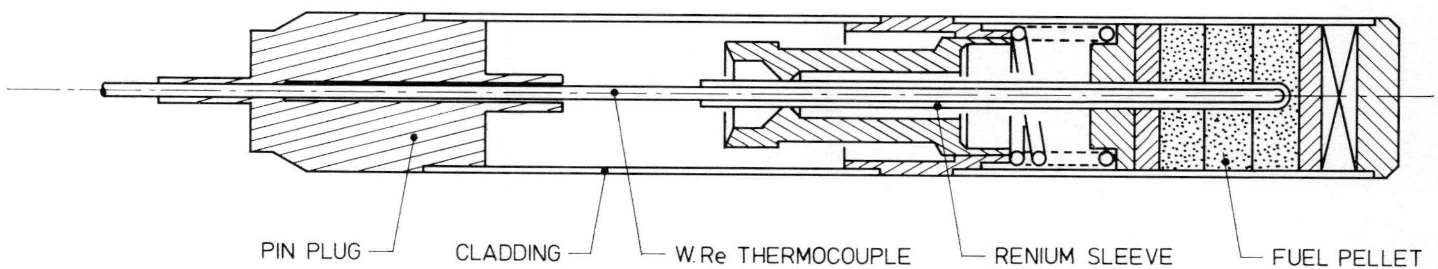

FIG.4. INSTRUMENTED FUEL PIN

REACTEUR RAPIDE EXPERIMENTAL RAPSODIE

MOYENS DE DETECTION ET LOCALISATION DES RUPTURES DE GAINE (DRG - LRG)

JC. Cauvin - J. Kermarrec - JP. Pagès

Centre d'Etudes Nucléaires de Cadarache (DER/SEER)

B.P.1 - 13115 St Paul Lez Durance - France

RESUME

A l'origine, en 1967, l'instrumentation du réacteur RAPSODIE était prévue pour détecter l'apparition des ruptures de gaine mais non pour les localiser. L'ensemble de ces moyens a été amélioré au cours du temps,en fonction de l'expérience acquise au travers d'une cinquantaine de ruptures de gaines spontanées ou artificiellement provoquées.

On décrit sommairement les principales méthodes utilisées que l'on peut ainsi regrouper :

- mesures sur le gaz de couverture (chambre d'ionisation - gammachromatographe - spectrométrie γ - spectrométrie de masse),
- mesures de sodium primaire (détection de neutrons différés - prélèvements pour analyses radiochimiques ..),
- méthodes spécifiques de LRG (traceurs gazeux - dispositif de sélection mécanique SRG - reniflage).

On dispose ainsi d'une panoplie de moyens complémentaires qui permet, d'une part, de minimiser l'incidence des ruptures de gaine sur le facteur de charge du réacteur, d'autre part, d'étudier, dans de bonnes conditions, le comportement de ces ruptures en pile.

1-INTRODUCTION

Au moment de sa mise en service, en 1967, le réacteur RAPSODIE était équipé d'une instrumentation prévue pour détecter l'apparition d'éventuelles ruptures de gaine des éléments combustibles. Très rapidement, il est apparu nécessaire d'améliorer et compléter cette instrumentation initiale,

P. von der Hardt and H. Röttger (eds.), Irradiation Technology, 299–308.

de manière à pouvoir non seulement détecter mais également caractériser et localiser les ruptures de gaine. C'est ainsi qu'on est arrivé progressivement à la mise en place des moyens de surveillance actuels, en fonction de l'expérience acquise au travers d'une cinquantaine de ruptures de gaine apparues naturellement ou artificiellement provoquées.

On se propose de décrire les principales méthodes utilisées que l'on peut ainsi regrouper :

- mesures sur le gaz de couverture,
- mesures sur le sodium primaire,
- méthodes spécifiques de localisation (LRG).

2 - MESURES SUR LE GAZ DE COUVERTURE

Sur RAPSODIE, le gaz de couverture est de l'hélium, remplissant un volume d'environ 2m3 (TPN) qui est renouvelé avec du gaz propre à un taux moyen de 0,5 m3/h.

Le circuit de prélèvement et d'analyse DRG-LRG est schématisé ci-après :

This illustration is printed from the best available original

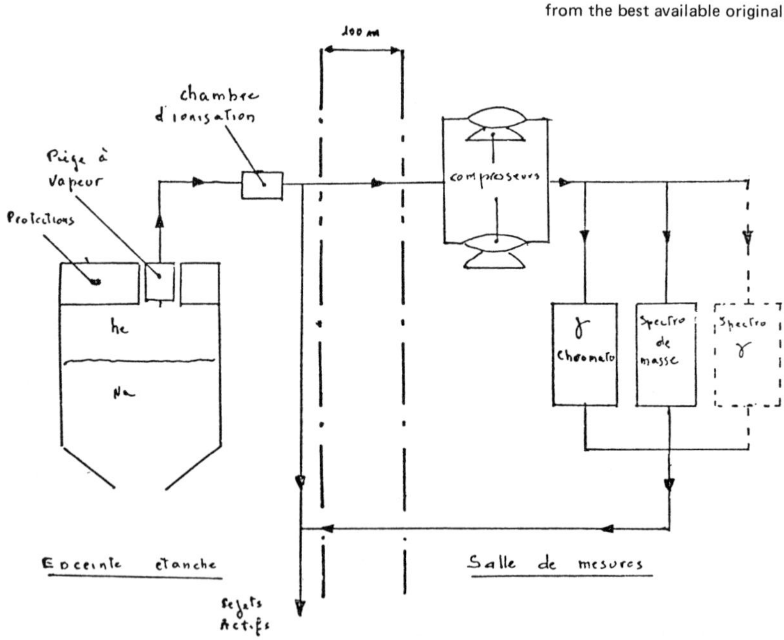

Ses principaux constituants sont :

- un piège à vapeur de sodium, régénérable par chauffage (situé dans les protections supérieures),
- une chambre d'ionisation à circulation (MX 32), mesurant l'activité globale ($\beta\gamma$) du gaz de couverture et précédée d'une ligne à retard destinée à ajuster le temps de transit (afin d'éliminer le Ne23 : $t \simeq 23s$),
- deux compresseurs en parallèle (Un en service, l'autre en secours), assurant la circulation d'hélium à un débit moyen de 0,3m^3h^{-1},
- un gammachromatographe, constitué d'une colonne chromatographique, séparant les corps entre eux (Hélium Krypton, Xénon), couplée avec une spectrométrie γ(Na I) analysant les raies caractéristiques des isotopes recherchés. Le fonctionnement, en ligne, de l'ensemble est automatique et cyclique (durée d'un cycle: 2x3mn). Les isotopes habituellement suivis sont : Xe 133 - Xe 135 - Xe 135m - Kr 88 - Kr 87 - Ne 23,
- un ensemble Enrichisseur-Spectromètre de masse (ENSPECT), destiné à la détection soit des isotopes stables de fission (Krypton et Xénon), soit d'isotopes stables de Xénon utilisés comme traceurs gazeux. L'ensemble est constitué d'un système de pièges à charbons actifs refroidis par de l'azote liquide et d'un spectromètre de masse du type quadripolaire. Il fonctionne, à la demande, de manière automatique et cyclique (durée du cycle réglable - en surveillance de routine, 1 mesure toutes les 24 heures).

  Il est prévu d'utiliser, à titre expérimental, une mesure en ligne par spectrométrie γ, au moyen d'une sonde au Germanium pur.

3 - MESURES SUR LE SODIUM PRIMAIRE

3.1 - Détection des neutrons différés (DND)

Sur chacune des boucles Nord et Sud du circuit de sodium primaire est installée une dérivation pour la détection par neutrons différés (DND). Le débit dérivé est de l'ordre de 1% du débit de la boucle.
Chacun des deux blocs de détection comporte un modérateur et un réflecteur en polyéthylène, une protection de plomb et six compteurs proportionnels à Hélium 3.
Le signal des chaînes de mesure (3 chaînes par bloc - 2 détecteur par chaîne) est utilisé directement pour la conduite du réacteur avec, en particulier, une action d'arrêt automatique sur dépassement d'un seuil.

Les informations DND sont également utilisées en vue d'une pré-localisation de l'assemblage non étanche; en effet, le rapport des signaux $\frac{DND/Nord}{DND/Sud}$ est fonction de la position de la source DND dans le coeur , au travers de la sectorisation hydraulique qui influe directement sur les temps de transit des produits de fission émetteurs de neutrons retardés. Cette méthode a été étalonnée au moyen d'une source d'uranium métallique déplacée dans le coeur et, empiriquement, au fur et à mesure de l'apparition des ruptures de gaine: elle permet généralement une pré-localisation par 1/6e du coeur.

3.2 - Prélèvement et analyse d'échantillons de sodium.

Des analyses radiochimiques du sodium primaire sont faites en laboratoire, à la demande, à partir de prélèvements d'environ 10g, effectués dans la cuve du réacteur au travers d'un canal expérimental (dispositif "TASTENA).

Le but de ces analyses en différé est de contrôler le relâchement des produits de fission et de la matière fissile au travers des ruptures de gaine. Les principaux isotopes régulièrement dosés sont : Cs 137 - Cs 134 - Cs 136 - I 131...

4 - METHODES SPECIFIQUES DE LOCALISATION (LRG)

4.1 - Traceurs gazeux

La méthode consiste à introduire dans les aiguilles combustibles, au cours de leur fabrication, un faible volume de Xénon de composition isotopique particulière obtenue par enrichissement de Xénon naturel en isotopes légers (Xe 124-126). En cas de rupture de gaine, ce gaz traceur stable est relâché dans le gaz de couverture où il est détecté au moyen de l'ensemble ENSPECT décrit au paragraphe 2.

Un traceur spécifique est caractérisé par le rapport de deux isotopes $\left(\frac{\text{Xe } 129}{\text{Xe } 124}\right)$ et on dispose d'une vingtaine de traceurs.

La méthode est utilisée couplée avec une méthode de partition complémentaire dite "méthode de l'âge". Celle-ci consiste à mesurer, au moyen de l'ENSPECT, le rapport entre un isotope stable de fission (Xe 134 - Xe 132 - Xe 136 ..) et un isotope de référence (Xe 133 ou isotope de Xenon traceur). Ce rapport est proportionnel au taux de combustion, donc à "l'âge " de l'élément combustible. On peut ainsi discriminer entre deux assemblages possédant le même traceur spécifique mais une histoire d'irradiation différente.

Le marquage systématique a été appliqué à RAPSODIE avec succès, depuis 1974 (un vingtaine de ruptures de gaine ainsi localisées); toutefois, certaines limites intrinsèques à la méthode ont été mises en évidence (cas de plusieurs ruptures de gaine simultanées - cas de ruptures de début de vie..)

4.2 - Dispositif de sélection mécanique (SRG)

La méthode consiste à renforcer le contraste entre les signaux DND Nord et Sud,en déviant vers la sortie Sud le jet de sodium sortant de l'assemblage testé. Pour celà, on utilise un "fourreau SRG" (Surveillance des Ruptures de Gaine) qui permet d'ausculter les 7 assemblages centraux et qui est schématisé sur la figure 1. En position SRG, le soulèvement de la douille de prélèvement permet au sodium de s'écouler vers la sortie Sud au travers d'un collecteur. Si ce sodium est contaminé en produits de fission émetteurs de neutrons retardés, on observe, sur les détecteurs DND décrits au paragraphe 3, un accroissement du contraste $\frac{\text{signal DND/Sud}}{\text{signal DND/Nord}}$. La durée moyenne d'une auscultation SRG est de 1 heure.

Après avoir été étalonné au moyen d'une source d'Uranium métal, ce dispositif a été utilisé en réacteur avec une efficacité totale; il a permis de localiser 5 ruptures de gaine. La figure 2 représente, à titre d'exemple, les signaux obtenus avec l'un de ces assemblages non étanches.

4.3 - Dispositif de reniflage

La méthode de "reniflage" , appliquée après mise à l'arrêt du réacteur, consiste à :

- solliciter l'assemblage combustible par soulèvement et dénoyage,
- provoquer ainsi, si l'assemblage est non étanche, une émission de gaz de fission qui est détecté au moyen d'une chambre d'ionisation.

Le dispositif utilisé est représenté dans son principe sur la figure 3. Il est composé de deux sous-ensembles :

- la hotte et le tube de reniflage, organes de manutention destinés à prendre l'assemblage et le soulever,
- le circuit de reniflage permettant, d'une part, de souffler du gaz (Hélium) dans le tube et, d'autre part, de prélever du gaz pour analyse.

Les principales séquences d'une opération de reniflage sont les suivantes :

- soulèvement de l'assemblage avec contrôle de sa puissance résiduelle,
- dénoyage de l'assemblage au moyen d'une chasse du sodium par de l'hélium et ce pendant une durée prédéterminée fonction de la puissance résiduelle de l'assemblage à contrôler,
- renoyage et dépose de l'assemblage,

- analyse de gaz au moyen d'une chambre d'ionisation.

La durée moyenne d'une opération de reniflage est d'environ 1 heure par assemblage.

La méthode a été utilisée avec une efficacité totale permettant de localiser 11 ruptures de gaine. La figure 4 représente, à titre d'exemple, un signal de reniflage positif.

5 - CONCLUSION

En conclusion, les méthodes qui viennent d'être décrites constituent une panoplie de moyens complémentaires répondant de façon satisfaisante au triple objectif recherché :

- détection des ruptures de gaine (chambre d'ionisation - gammachromatographe - ENSPECT),
- caractérisation (DND- gammachromatographe - spectrométrie γ),
- localisation (sectorisation DND - "âge" - traceurs - SRG - reniflage).

Ces moyens de surveillance permettent, d'une part, de minimiser l'incidence des ruptures de gaine sur le facteur de charge du réacteur, d'autre part, d'étudier dans de bonnes conditions le comportement en pile d'éléments combustibles non étanches.

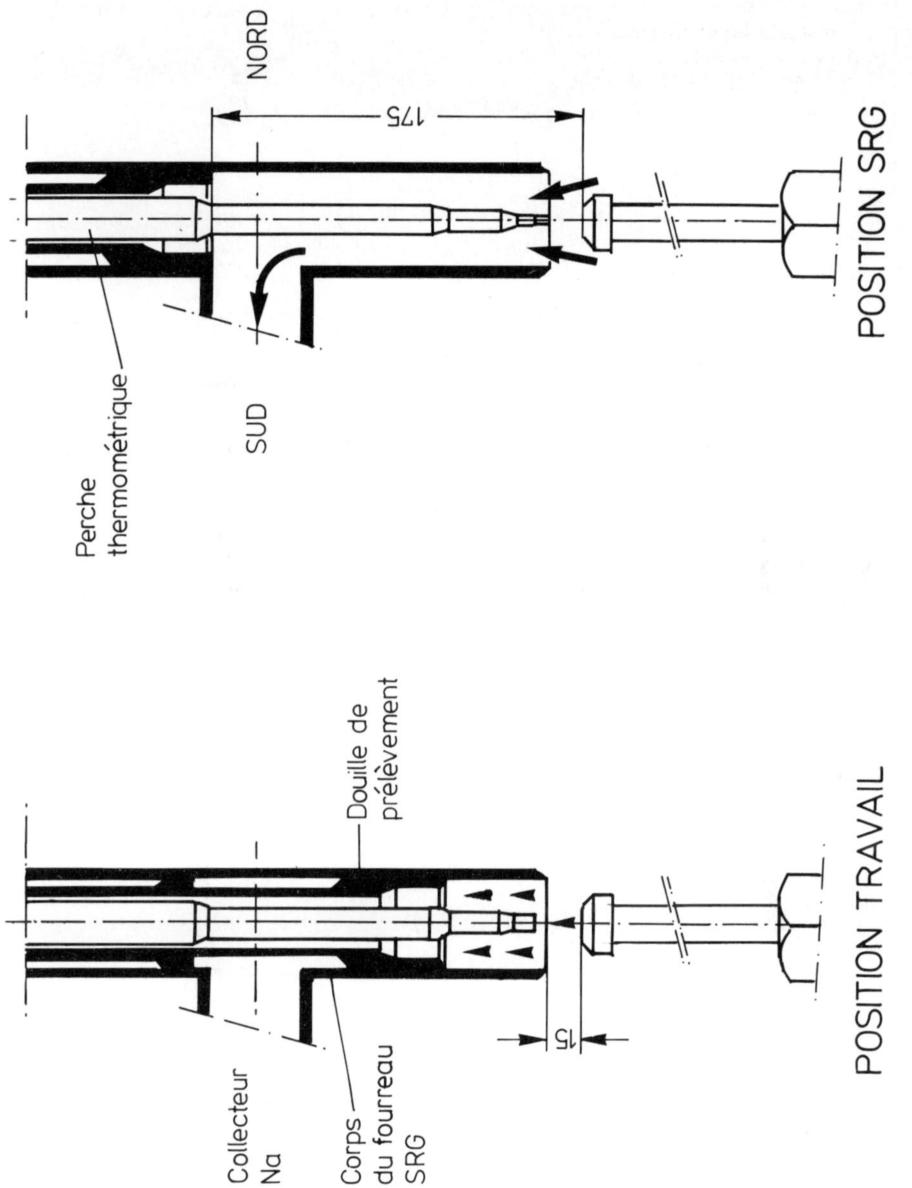

Fig. 1 Dispositif SRG - Schéma de Principe

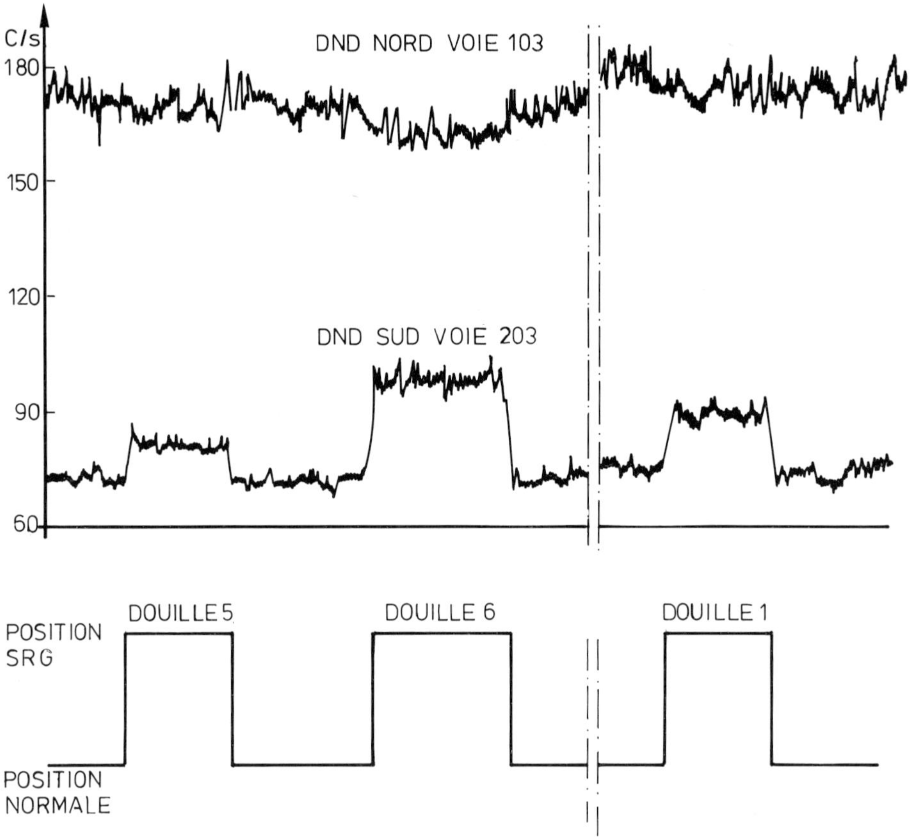

FIG. 2 EXEMPLE DE SIGNAL SRG POSITIF

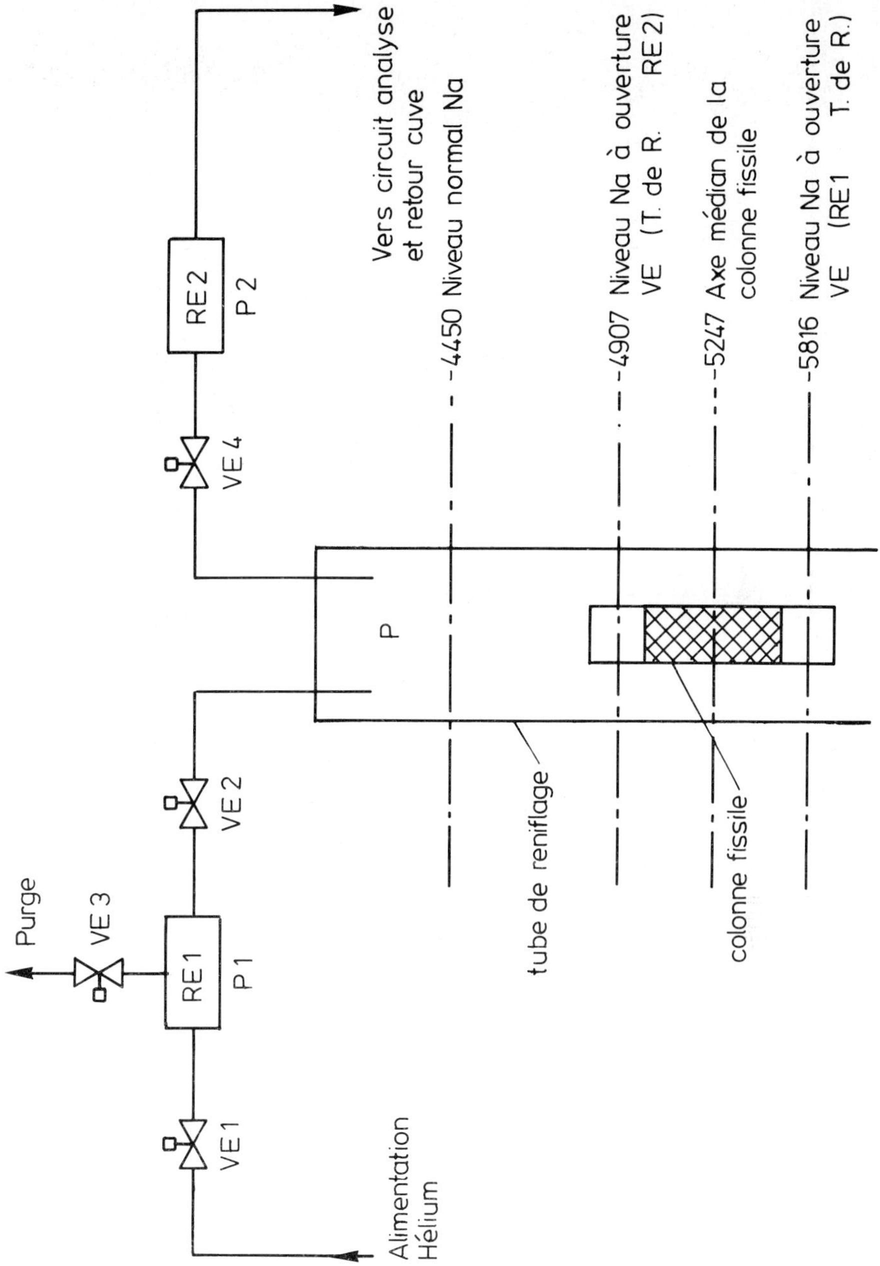

Fig. 3 Principe du Dispositif de Reniflage

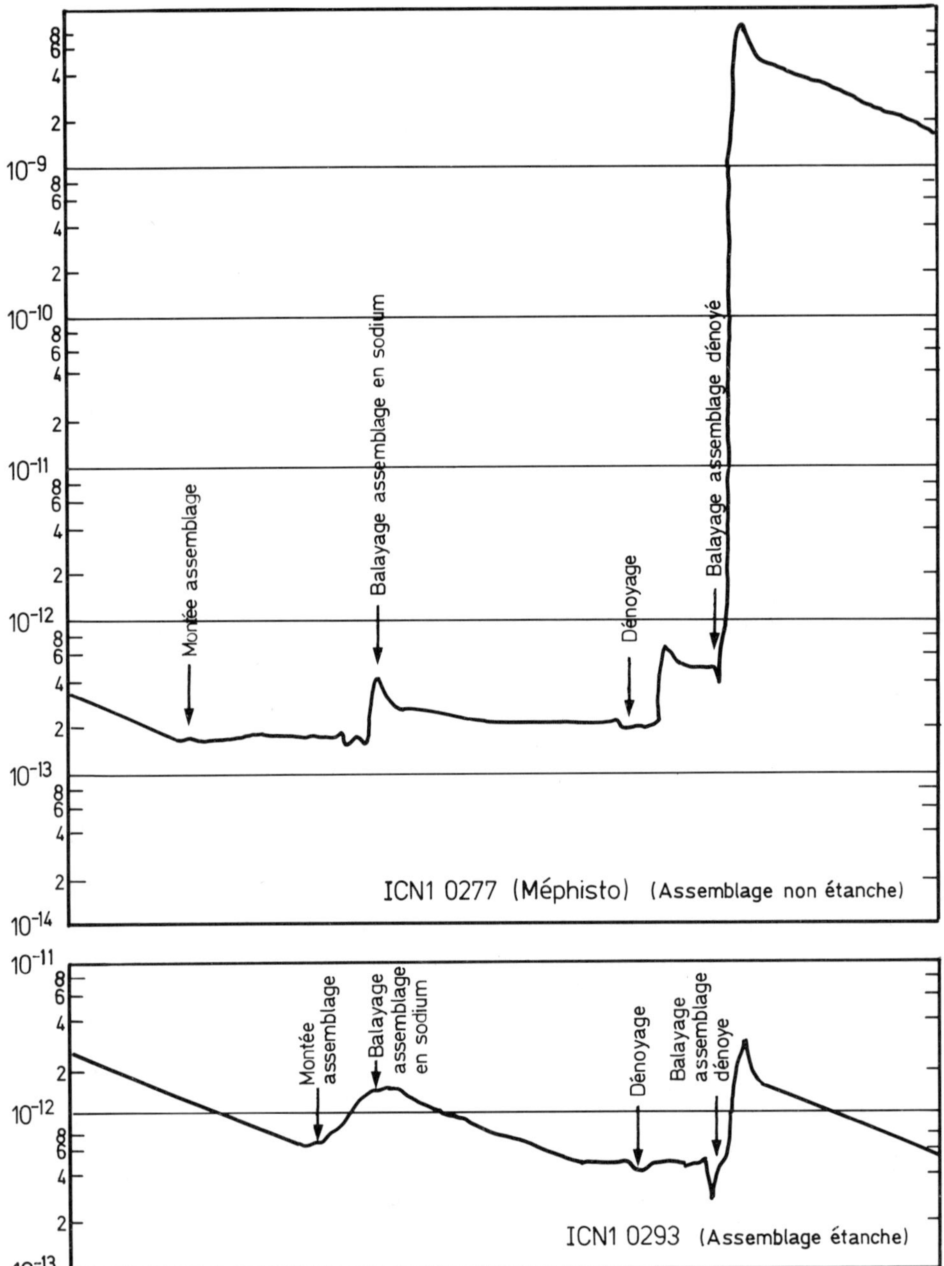

Fig.4 - Exemple de signal de reniflage positif

DEVELOPMENT OF THE UN-INSTRUMENTED IRRADIATION SUBASSEMBLY UNIS-B FOR THE EXPERIMENTAL FAST REACTOR JOYO

K. Ogata, H. Hyodo, K. Okada, Y. Ishida

Nippon Nuclear Fuel Development Co., Ltd. Oarai-machi, Ibaraki

Toshiba Corp. Toranomon Tokyo

PNC Akasaka Tokyo, Japan

ABSTRACT

The experimental fast reactor JOYO has been operated successfully at the rated powers of 50 and 75 MWt since April 1977 (JOYO Mark-I). Its core is now being changed so as to be suitable for irradiation tests as JOYO Mark-II. The un-instrumented irradiation subassembly, UNIS-B, is one irradiation device which has been developed for the Mark-II program. This device can be disassembled and reassembled for the convenience of interim examination. It has 6 compartments, each of which carries 5 test fuel pins, within a hexagonal wrapper tube forming the same outer profile as a driver fuel subassembly. Each compartment is composed of a double walled tube about 26 mm in diameter which provides a thermal insulating gas gap. Temperature and neutron fluence monitors are installed for the irradiation condition measurements.

1. INTRODUCTION

The experimental fast reactor JOYO is the first liquid metal cooled fast breeder reactor (LMFBR) in Japan. It has been operated successfully by the Power Reactor and Nuclear Fuel Development Corp. (PNC) at the rated powers of 50 MWt (first core) and 75 MWt (second core) since its initial criticality in April 1977 (JOYO Mark-I). Technical and engineering experience have been obtained for the development of a prototype fast breeder reactor. The JOYO core is now being changed so as to be suited to irradiation tests as JOYO Mark-II. The JOYO Mark-II core will be utilized as a fast neutron irradiation facility of fuels, materials, and other components for the LMFBR development program.

P. von der Hardt and H. Röttger (eds.), Irradiation Technology, 309–317.

Development activities of various irradiation devices required for irradiation tests have been carried out in conjunction with the reactor development. Design of their structures and thermal hydraulic characteristics have to ensure a fit into the JOYO Mark-II systems as well as into the post irradiation facilities. Supporting the design studies, fabrication tests and remote handling tests in a hot cell have been carried out with respect to the structures. Coolant flow characteristics, which have an affect on irradiation temperatures and fuel pin vibration, were evaluated by water and sodium flow tests.

This paper reviews the results of development of an un-instrumented irradiation subassembly B-type, UNIS-B, which is one irradiation device developed for the JOYO Mark-II program.

2. OUTLINE OF THE JOYO MARK-II PROGRAM

As a fast neutron flux irradiation facility, the JOYO core configuration and core elements are now being changed with the intention of obtaining both increased neutron flux and power density, and an enlarged neutron zone. The results of the above changes are shown in Table 1. Modifications and increase in the existing reactor system capability are also being carried out consistent with the irradiation test devices to be introduced.

Table 1 Main Core Parameters of JOYO

Parameter	Unit	Mark-I		Mark-II
		First	Second	
Reactor Output	MWt	50	75	100
Coolant Flow Rate	t/h	2200	2200	2200
Reactor Inlet Temp.	C	370	370	370
Reactor Outlet Temp.	C	435	470	500
Core Stack Length	cm	60	60	55
Core Volume (Max.)	l	294	304	250
Linear Heat Rate (Max.)	W/cm	210	320	400
Fuel Pin Diameter	mm	6.3	6.3	5.5
Pu Content	wt.%	18	18	30
^{235}U Enrichment	wt.%	23	23	12
Neutron Flux (Max.)	$n/cm^2.s$	$2.1x10^{15}$	$3.0x10^{15}$	$5.1x10^{15}$
Neutron Flux (Core av.)	$n/cm^2.s$	$1.2x10^{15}$	$1.9x10^{15}$	$2.6x10^{15}$

Table 2 Main Parameters of Test Fuel Pins

Parameter	Unit	Type I	Type II
Pu Content	wt.%	≤30	
^{235}U Enrichment	wt.%	≤41	≤35
Pellet Diameter	mm	~4.6	~5.4
Pellet Density	%TD	≤94	≤87
Fuel Pin Diameter	mm	5.4 - 5.8	6.4 - 6.8
Cladding Thickness	mm	0.3 - 0.5	0.4 - 0.6
Fuel Stack Length	mm	≤550	
Linear Heat Rate (Max.)	W/cm	≤450	≤530
Burnup (Max.)	MWd/kg	130	

Irradiation tests, while being conducted using several types of irradiation rigs, are mainly done using the un-instrumented irradiation subassembly, UNIS, B-type which is described below. A core materials irradiation rig, CMIR, is provided as a modification of the UNIS-B. Moreover, an instrumented test assembly, INTA, which can cotinuously monitor fuel behavior during irradiation is now under development.

Irradiation test fuel pins are designed and fabricated in accordance with test objectives. The main parameters of the authorized test fuel pins are listed in Table 2.

3. DESCRIPTION OF THE UNIS-B

3.1 Features of the Subassembly

The UNIS-B has been developed with the aim of creating an effective irradiation device compatible to many kinds of irradiation conditions and capable of interim examination. It has no instrument wiring for continuous monitoring, but temperature and neutron fluence monitors are equipped for measurements at interim and post irradiation examinations. Taking the UNIS-B out of the reactor core in the mid-course of irradiation, it can be disassembled in a hot cell for interim examination. After the examination of test fuel pins and monitors, it can be easily reassembled using manipulators in a hot cell and returned to the reactor core for further irradiation.

A typical UNIS-B subassembly is shown in Fig. 1. As it has the same outer profile as a driver fuel subassembly, it can be irradiated at any core position in place of such a subassembly. It can be transferred through the JOYO fuel handling system into and out of the reactor core.

The UNIS-B features 6 compartments arranged around a center tie rod within a hexagonal wrapper tube. An entrance nozzle, which provides a coolant sodium inlet and flow rate adjusting orifices, is welded to the lower end of the wrapper tube. A lower strainer is positioned in the entrance nozzle to rectify coolant flow. At the upper end, a handling head which has a support grid located at the top of the compartments is fixed by a locking nut screwed into the center tie rod. The handling head can be removed by unlocking the nut for interim examination.

Each compartment carries 5 test fuel pins supported by a fuel pin tie rod. Pin spacing is provided by a spacer wire wrapped spirally around each fuel pin. A compartment is composed of a double walled tube which provides a thermal insulating gas gap filled with an Ar and He gas mixture. Therefore, each compartment is thermally and physically independent. Coolant flow rate for each compartment is adjusted individually by a 2-stage orifice attached at the bottom of the compartment. The compartment inner tube has an axially fluted shaped tube as shown in the B-B cross section of Fig. 1. This is to reduce the radial temperature distribution induced by a disproportionate coolant flow rate around each fuel pin. The inner tube is made of 20 % cold worked 316 stainless steel. Its size is specified in accordance with the diameter of test fuel pins. The gas gap is spaced by a wire wrapping around the inner tube. A cap is screwed into the top of the compartment enabling hot cell manipulators to allow non-destructive removal of test fuel pins from the compartment for interim examination. Compartments are inserted in the wrapper tube and held down through springs by a support grid welded to the handling head.

Temperature monitors such as SiC and thermal expansion difference detector are equipped above the fuel pin bundle to measure outlet coolant temperature. Dosimetry wires are also equipped inside the center tie rod to monitor neutron fluence. These monitors are exchanged for new ones after interim examination.

3.2 Fabrication and Handling Tests

To verify the efficiency of structural design and fabrication technology, the UNIS-B subassemblies were trial manufactured for remote handling tests in hot cells. Some modifications on structures were made after the tests for the convenience of fabrication and remote handling. Fabrication procedures, such as welding

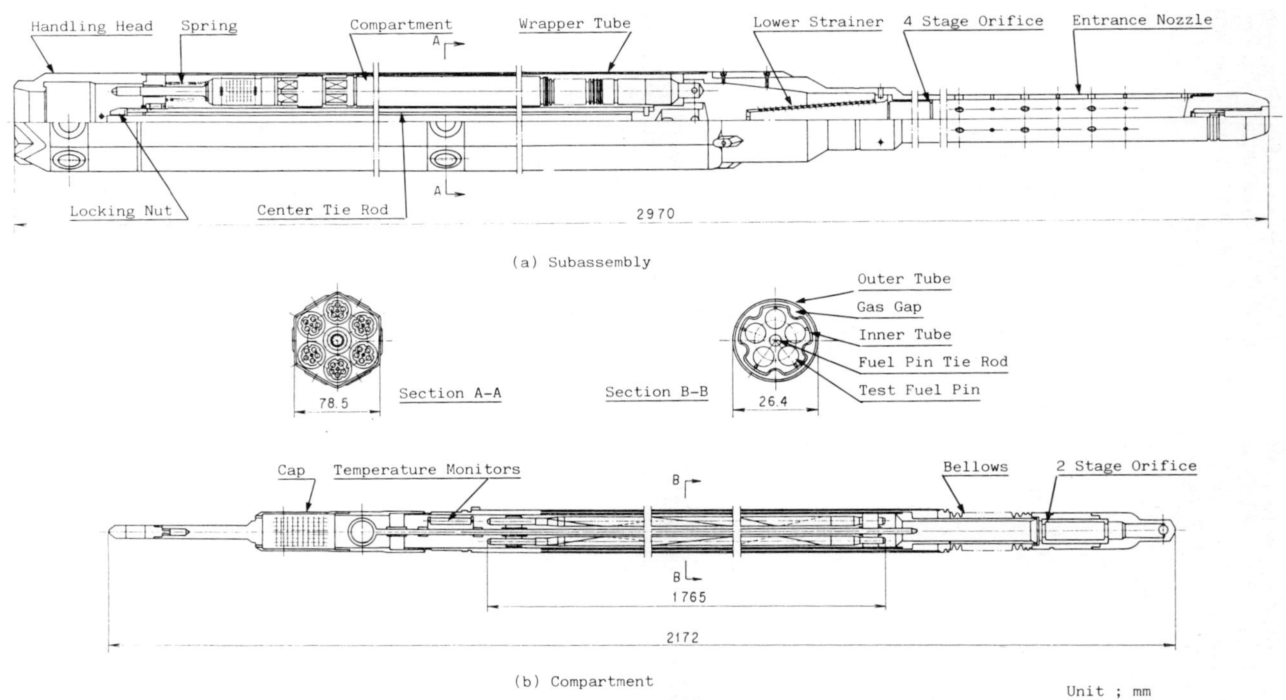

Fig. 1 Illustration of a Typical UNIS-B Subassembly and Its Compartment

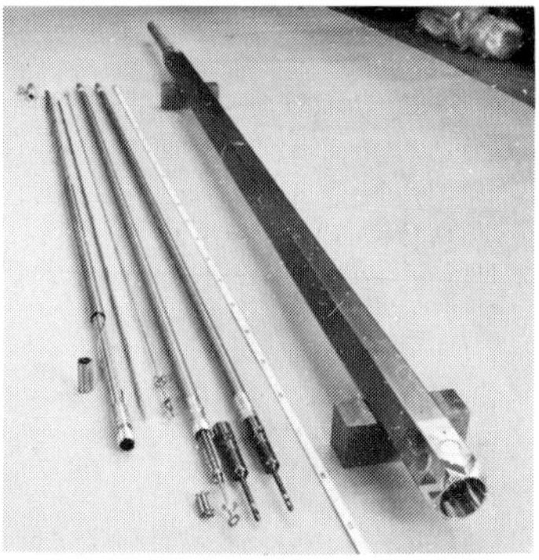

(a) Subassembly assemnled

(b) Handling head removed

Fig. 2 Appearances of the UNIS-B Trial Manufactured

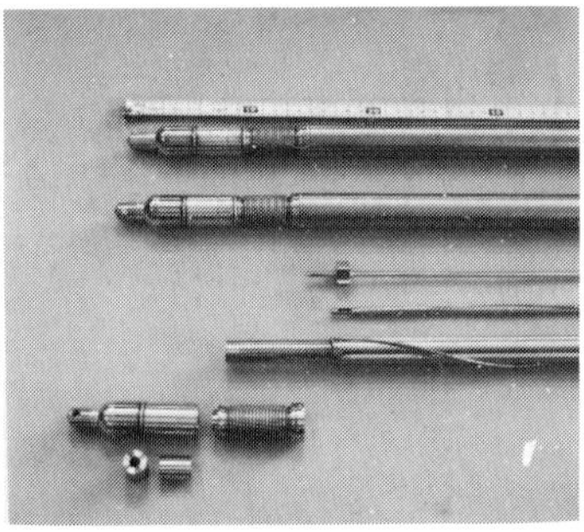

(a) Bottom

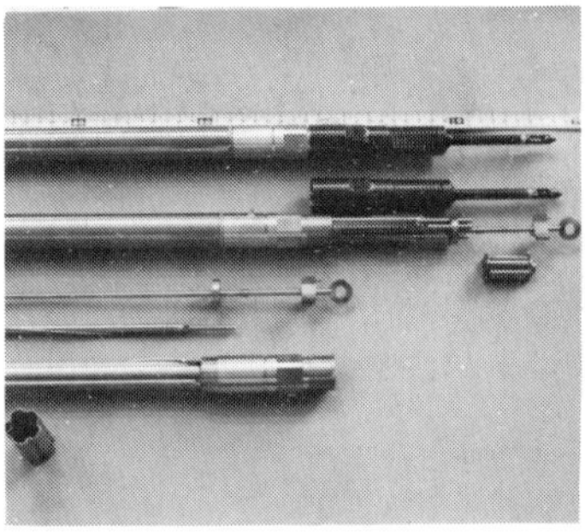

(b) Top

Fig. 3 Appearances of a Compartment and Its Elements

and inspection, were formulated through the tests. The axially fluted shaped tube (see cross section B-B in Fig. 1) was made by cold drawing from a circular tube. A simulated subassembly, improved through the tests, could be disassembled and reassembled satisfactorily for interim examination using manipulators in hot cells. Appearances of the subassembly and compartment are shown in Figs. 2 and 3.

3.3 Thermal Hydraulics

Coolant sodium for the UNIS-B is taken from the reactor primary coolant system through the entrance nozzle. The coolant flow is divided among the compartments for cooling of the test fuel pins, the outside area of the compartments, and the center tie rod. These flows meet at the top of the compartments and pass out through the handling head. Water flow tests to evaluate the coolant flow distribution, pressure drop, and flow rate control orifices were carried out using simulated subassemblies. The hydraulic design was made utilizing test results by the method of pressure balance calculations for the coolant flow circuit shown in Fig. 4.

A coolant flow rate adjustment is made using 2 types of orifices. One is a 4-stage orifice assembled in the entrance nozzle which adjusts the subassembly total flow making the pressure drop consistent with the reactor primary coolant pressure. A hydraulic interaction between orifice stages was observed during water flow tests because of the multi-stage orifice being assembled in a restricted space in the entrance nozzle. Therefore, its flow resistance was represented on the experimental data as a function of distance between orifice stages, thickness of orifice plate, and orifice diameter. The other type is a 2-stage orifice for the compartment coolant flow control, attached at the bottom of each compartment. The flow resistance coefficient of the 2-stage orifice was calculated as a function of orifice diameter on the basis of

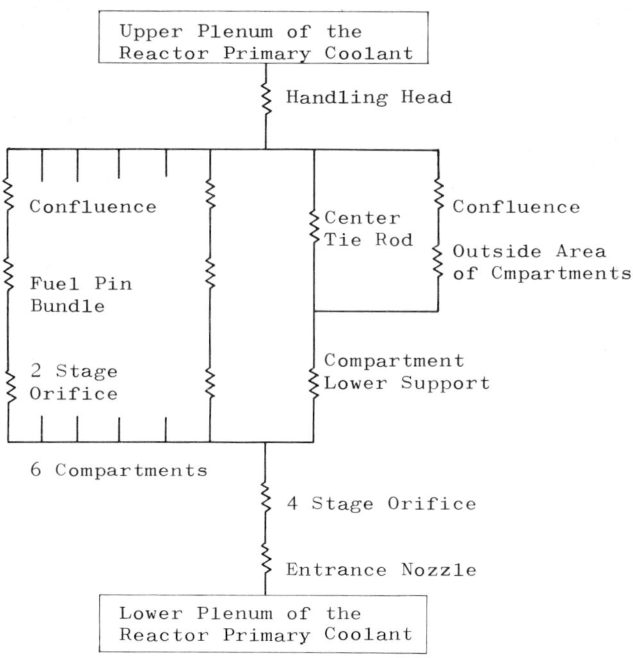

Fig. 4 Schematic Illustration of Coolant Flow Circuit

the water flow test results. Typical hydraulic design parameters are shown in Table 3.

Sodium flow tests were carried out to examine fuel pin vibration and sodium drain from the subassembly for the convenience of post irradiation handling. Flow turbulence induced vibration of the strainer, which was assembled in the entrance nozzle to rectify coolant flow, was observed. This problem was rectified by changing the strainer shape. No problems due to fuel pin vibration were observed during the tests. A small amount of sodium remained at the screw and interstices after taking the UNIS-B out of the sodium flow loop, however, this did not cause problems in removing sodium by steam and water using ordinary cleaning procedures in the hot cell.

Coolant temperatures were calculated using a subchannel analysis computer code developed as a part of the development program of an evaluation code system for irradiation plans and test rigs in the JOYO Mark-II core. Fig. 5 shows typical subchannel temperatures of coolant at the top of the core. The small temperature rise of the outside area of the compartments (Fig. 5 (a)) shows the thermal insulating effect of the compartment gas gap. Structural materials and fuel temperatures were also analyzed by the code.

Table 3 Hydraulic Design Parameters of a Typical UNIS-B

	Orifice Diameter (mm)	Coolant Flow Rate (kg/s)	Pressure Drop (kg/cm^2)
Compartment No. 1	7.36 x 2 Stages	0.312	1.80
No. 2	7.36 x 2 Stages	0.312	1.80
No. 3	7.36 x 2 Stages	0.312	1.80
No. 4	7.36 x 2 Stages	0.312	1.80
No. 5	7.36 x 2 Stages	0.312	1.80
No. 6	7.36 x 2 Stages	0.312	1.80
Outside Area	-	2.208	1.80
Center Tie Rod	-	0.061	1.80
Subassembly Total	23.42 x 4 Stages (Entrance nozzle)	4.141	3.31

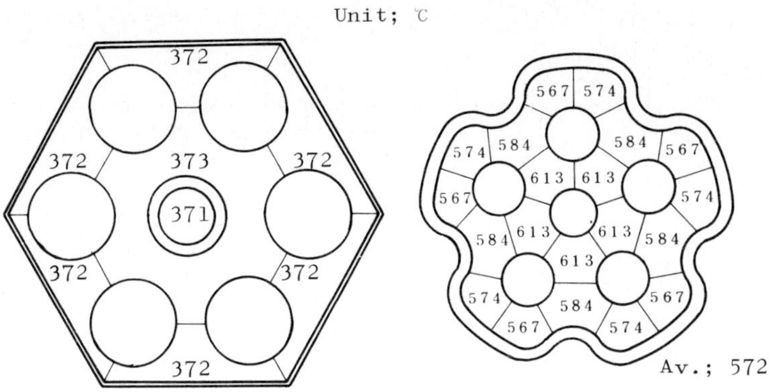

(a) Outside Area of Compartments (b) Compartment

Fig. 5 Coolant Temperatures Calculated at the Top of the Core (Coolant Inlet Temperature ; 370°C)

4. CONCLUSIONS

The UNIS-B has been developed for the JOYO Mark-II irradiation program with the aim of providing an effective irradiation device capable of irradiation under various kinds of conditions and of interim examination. Design studies, trial fabrication, remote handling tests, and water and sodium flow tests were performed.

The UNIS-B has 6 compartments each of which carries 5 test fuel pins. Temperatures and fuel pin design parameters for each compartment can be set individually. The UNIS-B can be disassembled and reassembled in hot cells after removal from the reactor core for interim examination. Temperature and neutron fluence monitors are equipped for measurements at interim and post irradiation examination.

The subassemblies for the practical use have been manufactured and prepared for the irradiation tests scheduled to be carried out during 1983. They are expected to make substantial contributions to the LMFBR development program in Japan.

ACKNOWLEDGEMENTS

The authors express their gratitude to Mr. K. Satoh, PNC, and Mr. A. Ito, Toshiba Corp. for investigating the hydraulic characteristics and to Messers Y. Kumaoka and S. Ishiyama, Toshiba Corp. for valuable advice. The authors also wish to thank the members of PNC who sponsored and contributed to the development studies described.

TRANSIENT OVERPOWER EXPERIMENT ON LMFBR FUEL PINS IN THE PETTEN HIGH FLUX REACTOR (HFR-TOP)

G. van der Koogh et. al.

Reactor Department

Netherlands Energy Research Foundation E.C.N.

Abstract

Within the framework of the German, Belgium and the Netherlands (DeBeNe) Research and Development Program on LMFBR's overpower experiments on single fuel pins will be carried out by the Netherlands Energy Research Foundation (ECN) in the Petten High Flux Reactor.
The pre-failure behaviour of pre-conditioned and pre-irradiated fuel pins is investigated in a small sodium loop. The experiments have to provide quantitative information to verify and improve advanced fuel pin models in computer codes.
The overpower transient in the loop is realised by a controlled decrease of the pressure in neutron absorbing BF_3 -gas in a gap around the fuel pin containing part of the loop. The sodium-water heat exchanger in the loop provides additional cooling during the transient by closure of an insulating gasgap due to thermal expansion of its hottest wall. The sodium pump is a small annular linear induction pump.
Operating experience with the prototype capsule TOP-01 is given.

P. von der Hardt and H. Röttger (eds.), Irradiation Technology, 319–333.

1. Introduction

Within the framework of the German, Belgium and the Netherlands (DeBeNe) Research and Development Program on LMFBR's overpower experiments on single fuel pins will be carried out by the Netherlands Energy Research Foundation (ECN) in the Petten High Flux Reactor of the Joint Research Committee of Euratom. The pre-failure behaviour of the pre-conditioned and pre-irradiated fuel pins is investigated in a small sodium loop. The experiments have to provide quantitative information to verify and improve advanced fuel pin models in computer codes.
In the TOP experiments, a typical LMFBR fuel pin will be subjected to overpowers, varying from 1 to 4 times nominal maximum power production, over risetimes from 0.5 to $\sim$ 10 seconds.
During the power transient the development of fuel elongation, fuel pellet temperature and fission gas pressure build up will be measured. An extensive post-irradiation investigation will be carried out. The irradiation target itself consists of a short section of a LMFBR pin, with blanket fuel at the ends. The irradiation facility consists of a small sodium loop, which will be irradiated in the poolside facility of the HFR.
The overpower transient in the loop is realised by a controlled decrease of the pressure in neutron absorbing BF_3 -gas in a gap around the fuel pin containing part of the loop, and is ended by a reactor scram. The sodium-water heat exchanger in the loop provides additional cooling during the transient by closure of an insulating gasgap due to thermal expansion of its hottest wall. The sodium pump is a small annular linear induction pump.
During the pre-irradiation of the pin, as well as during the transient a large number of measurements will be taken. An elaborate data collection and handling system will be used, the heart of which is a minicomputer, for control, data processing and safe guarding.

2. The irradiation facility

2.1 The loop and its main parts

The fuel pin is irradiated in axially flowing sodium, to obtain the correct axial temperature profile along the clad. So the irradiation facility is constructed as a small sodium loop, equipped with a small electromagnetic pump and a sodium-water heat exchanger.
Fig. 1 gives a cross section. The fuel pin containing part of the loop is called the test section.
The length of the test section is limited by the dimensions of the poolside facility. Therefore the maximum active fuel length is 250 mm.
The main specifications of the facility are :

fuel pin total length	: 355 mm
fuel pin diameter	: 7.6 mm
fissile zone length	: 250 mm
fissile material	: 20% enriched UO_2

linear fission power (stationary)	: 450 W/cm or 270 W/cm
max. linear power (transient)	: 1800 W/cm
transient time	: 0.5 - 10 sec.
inlet temperature	: 380°C
outlet temperature (stationary)	: 560°C or 480°C
outlet temperature (transient)	: < 1000°C
sodium flow rate	: 55 gr/sec.

2.2. The fuel pin

The irradiation target consists of a short section of an LMFBR type fuel pin, with blanket fuel at both ends. The active fuel length, 25 cm (20% enriched UO_2) is about ¼ of that of a typical LMFBR-pin. The axial power profile over the short pin resembles that of the LMFBR-pin, axially compressed by a factor 4.
Fuel stack elongation is measured by a differential transformer, mounted inside the enlarged clad at the upper end of the fuel pin. Clad elongation is measured by a similar transducer at the lower end of the pin, which end is free to expand axially.
The axial inelastic deformation of the clad is measured by means of a number of 15 μm deep V groves spaced 20 mm apart on the clad. After irradiation distances are measured again, giving indiction of the distribution of the local clad deformation. A high temperature thermocouple measures the temperature development in the third fuel pellet from the top during the transient. This thermocouple runs through the hollow core of the differential transformer and central holes in the blanket and fuel pellets into the third fuel pellet. A pressure transducer connected to the upper end of the clad by a short and wide minitube, measures the fission gas pressure build-up in the plenum.

2.3. Test section

In Fig. 2 the horizontal cross section of the test section is shown. The test section consists, from in- to outside, of : fuel pin, shroud tube, sodium containment wall, N_2 filled gasgap, second containment wall, BF_3 gasgap with Al heat conductor, Al heatsink and outer wall. Sodium enters the test section via the external sodium inletline at the bottom, flows upwards around the pin and leaves the test section at the upper end, just beneath the enlarged part of the clad.
The shroud tube forms the mounting for the thermocouples. These run down the outside of the shroud tube through nearly stagnant sodium. The sensitive ends of the thermocouples point through fitting holes in the shroud tube horizontally into the sodium part around the pin. The SS sodium containment wall is surrounded by a wide N_2 filled gasgap for thermal insulation.
The SS outer wall of this gap forms the second sodium containment as well as the inner wall of the BF_3 gasgap.
In the ∿ 5 mm wide BF_3 gasgap a finned Al heat conductor is installed over the full height. This conductor leads the heat, produced by γ-

heating of the inner wall and by the fission of B^{10}-atoms, through the fins to the outer wall, which is in contact with the reactor pool water.
The fins fit closely against an Al heatsink which is part of the outer BF_3 containment wall. The purpose of this heatsink is to solidify molten fuel by rapid cooling through its high heat conductance and heat capacity, eliminating the risk of fuel melting through all containment walls. In the sodium in- and outlet lines adjacent to the test section electromagnetic flowmeters are mounted. In the inletline a pressure transducer measures sodium pressure and pressure variations.

2.4 Sodium-water heat exchanger

The heat exchanger is described in detail in Ref. 1. During the transient additional cooling capacity is obtained automatically by closure of the gasgap between sodium and water due to thermal expansion of the inner wall.

2.5 Sodium pump

The sodium pump is described in detail in Ref. 1. The stator is reusable.

2.6 BF_3 system

As neutron absorbing gas BF_3, enriched in B^{10} content to > 90%, was chosen above He^3. Although BF_3 is corrosive when moist, poisonous, viscous and a bad heat conductor, it forms no longlived isotopes and is far cheaper than He^3. Furthermore, BF_3 is extremely soluble in water, so the underwater part can be single contained.
BF_3 is rapidly absorbed in absorption charcoal. A small container filled with charcoal is sufficient to absorb a large amount of BF_3, leading to a very low residual pressure in the shield.
The longest halflife of all isotopes formed during irradiation of BF_3 is F^{20}, halflife 11 seconds, decaying to stable Ne^{20}, with an easily measured γ of 1.6 MeV and a β particle.
($B^{10} + n \rightarrow \alpha + Li^7$ (stable) + 2.8 MeV and $Li^7 + n \rightarrow Li^8 \rightarrow$ prompt 2α). Due to the small cross sections of all reactions except B^{10} the specific activity of BF_3 stays very low.
Fig. 3 shows a diagram of the BF_3 system.
Near the BF_3 shield a very fast acting, large bore electromagnetic valve is situated for controlled release of the BF_3 pressure in the shield to the charcoal trap.
The valve, bore 10 mm, stroke 0.6 mm, opens and closes again in less than 0.02 seconds, giving $\sim$ 50 pressure decrease steps per second. This valve is not completely leaktight when closed. Hence a normal valve, situated outside the n- and γ flux fields is used as back-up during stationary irradiations.

3. Instrumentation

The "scientific" instrumentation consists in brief of :

a) A WRe 5 - WRe 26 thermocouple, embedded in the third fuel pellet from the top for check of computer calculations of fuel central temperatures and the development thereof during the transient.

b) A differential transformer, measuring the axial fuel stack elongation during both pre-irradiation and transient, with respect to the clad.

c) A differential transformer, measuring the clad elongation during both pre-irradiation and transient. This will be used to correct the stack elongation for clad elongation effects.

d) A differential pressure transducer, measuring the fission gas pressure build-up during both pre-irradiation and transient. The back of the transducer will be pressurized with He to keep the transducer in range. The He pressure is measured separately by another pressure transducer.

e) Ten chromel-alumel thermocouples ϕ 0.5 mm measure sodium temperatures adjacent to the fuel pin clad.

f) Eight chromel-alumel thermocouples measure the temperatures of the elongation- and pressure transducers, for correction of their temperature sensitivity.

g) Three cobalt self powered neutron detectors indicate changes in the neutron flux. One of these is used as source of neutron flux information for the controlled release of BF_3.

4. Data handling

The system consists of a minicomputer, PDP-11/34, an interface system and a batch of instrumentation amplifiers with build-in multiplexers. An extensive description is given in Ref. 2.

5. Irradiation procedure

Operation of the capsule is separated in two modes, pre-irradiation and transient, in the same facility. During pre-irradiation the capsule will be positioned in the psf, such that the required power production is reached without BF_3. After the required burn-up is reached, the capsule is withdrawn, neutrographed, reinstalled and brought to its original power production. Shield and charcoal filter are evacuated. Then the shield is filled with BF_3 until the desired power factor is obtained in the reverse direction. The capsule than is moved closer to the reactor until its original power production is restored. Thereafter the required BF_3 blow-down time is adjusted and last checks are done. The transient is initiated via a terminal in the reactor control room. It is ended by a reactor scram. After cooling down, and neutrography, the capsule is taken apart in the HFR dismantling cell. The testsection itself is shipped to the Hotcell Laboratory for PIE.

6. Irradiation history of the prototype capsule TOP-01

Before irradiation started, the sodium flowmeters (Ref. 1) had to be calibrated by means of an accurate thermal balance over the heat exchanger. The thermocouples measuring the cooling water temperature difference, however, were damaged during construction of the capsule and showed erratic readings. Then it was tried to obtain the sodium flowrate by measuring the travel time of a temperature disturbance through the circuit. (The disturbance was brought about by switching off the traceheaters in the heat exchanger). This resulted in a flowrate of about half the expected value. The same was true for the dummy experiment (Ref. 1) where the sodium flowrate obtained from a thermal balance was about twice that of travel time measurements. The reason was the very high heat capacity of the walls compared to that of the sodium. Computer calculations proved this point.

From all this, sufficient proof was available of adequate sodium flow for safe operation at half the nominal fuel pin power,

Two short irradiations were done in HFR cycle 82-05, at half nominal power, a.o. for testing of the self powered neutron detector (spnd) signals. One of these signals is used for control of the transient. The last irradiation was ended by a reactor scram, initiated by running a faked overpower transient on the Data Handling System.

During the subsequent reactor shutdown and holiday period, the calibration problem was solved by installation of two new thermocouple pairs in the heat exchanger in- and outlet pipes. To facilitate insertion through the bends in the pipes, the couples were mounted inside a stainless steel Bowden cable mantle. The hot welds were positioned in the centre of the pipes by a centering piece, at a distance of $\sim$ 3 mm from the in- and outlet headers of the heat exchanger (Fig. 4). The ΔT could now be measured with an accuracy better than 0.5%, providing the required sodium flow rate calibration.

The negative transient from $\frac{1}{2}$ $P_{nom} \rightarrow 0$ at the scram provided a lot of information about the behaviour of the diverse transducers.

In HFR cycle 82-07, a number of irradiations were carried out to check the flow calibration at diverse temperatures, and the capsule was irradiated at nominal power (480 W/cm).

Two transients were done from $\frac{1}{2}$ $P_{nom} \rightarrow P_{nom}$, without reactor scram, to study the BF_3 control system. The first was a fast blowdown, during which both BF_3 valves were opened and kept open. The blowdown was slower than expected from the trials in a 1:1 scale model of the system (Ref. 2). The second transient, from $\frac{1}{2}$ $P_{nom} \rightarrow P_{nom}$ in 8 seconds showed the same slow response, see § 7.1. During the first transient the HT thermocouple broke down, during the second one the fuel elongation transducer. The transducer could be used afterwards due to a trick (see § 7.5).

It was intended to end cycle 82-07 with a TOP-transient. During BF_3 filling the back-up valve leaked so heavely, that the proper transient was impossible. Again the reactor was scrammed by a more or less faked transient. During the reactor reloading period the failing valve was flushed by Helium. Eventually it operated properly again. On September

9 directly after the start of HFR cycle 82-08, the shield was filled with BF_3 and the capsule brought to power again. It was planned to do a fast transient from $P_{nom} \rightarrow 3.75\ P_{nom}$ in 1 second, using the almost linear part of the blowdown curves, closing the valves at a residual pressure of 8 bar (Fig. 5). The transient was carried out successfully. The back-up valve, however, failed to open properly, reducing the gas release rate too much. An overpower of only 1.4 x P_{nom} was reached at the moment the Data Handling System initiated a scram.
A histogram of the irradiations is given in Fig. 6.

7. Operation experience and preliminary test results

7.1 BF_3 system (Fig. 3)

The operation of the system is described in Ref. 2.
The 94% enriched BF_3 showed the expected neutron absorption characteristics. During the first small transient, a fast blowdown, the risetime of the spnd signals (÷ BF_3 density or pressure) was much slower than in the 1:1 scale model. The 8 second transient showed the cause to be excessive flow resistance in the part between shield and fast control valve. In Fig. 7 the spnd response is given versus time. The difference between the scale model and the real capsule is twofold : firstly, the three minitubes leaving the shield have a 90° welded bend, in the model the tubes were bend with a radius of ∿ 4 times the tube outer diameter. Secondly, the transition piece connecting the three minitubes to the much wider pipe leading to the control valve, is narrow with a blunt top angle. In the model this piece was longer, with a sharp top angle (see Fig. 8).
From Fig. 7 is evident that after opening of the control valve the pressure in the shield drops, and the valve closes. After closure, however, the shield pressure still drops sharply. What happened is that at opening the control valve the BF_3 pressure in the wide connection line drops immediately to almost zero. The shield pressure drops a little and the valve closes. The pressure in the shield and connection line equalizes, hence the pressure in the shield drops much farther than it should have. The ratio of shield volume and connection line volume corresponds very well with the measured pressure equalization, inferred from Fig. 7.
After the every transient the BF_3 system was thoroughly flushed with pure helium, to prevent corrosion. Nevertheless, the back-up valve failed to close properly during preparation for the final overpower transient. As the valve is located in the instrumentation connection box, which is closed by welding and situated under water just below the reactor top lid, it could not be dismounted. But by alternatively blowing high pressure helium through it from exhaust to inlet, and the reverse, suddenly the valve closed properly. The valve was checked repeatedly by opening and closing, and worked normally, as far as could be inferred from the pressure build-up and decay observed on the manometers in the gas panel (long minitubes). This valve, situated outside the γ and n flux field, is a type in which the main valve, whose teflon seat forms part of a plunjer, is operated

by a pilot valve.
During the final transient the main valve, although not leaking, failed to open, and the transient was determined by the resistance of the orifice of the pilot valve alone. This was evident from the slope of the spnd signal (Fig. 16) which was ∿ 1/12 of what it should have been. The flow resistance coefficient C_V for the pilot valve is 0.06, that of the total valve 0.76, about the same ratio.

7.2 Fission power of TOP-01 in the poolside facility

Fig. 9 shows the fission power production versus distance from the core box wall. Also the γ-power production is shown.
The pressure of BF_3 only decreases the thermal neutron flux, not the γ-flux. As the topics of interest are temperature dependent, the γ power production has to be taken into account when adjusting the stroke of the overpower transient.
The influence on the neutron flux of the BF_3 is evident from the figure. 37 Bar of 94% enriched BF_3 caused a neutron flux decrease by a factor of 4.5, and a total power decrease of 3.75 due to the γ-heat.

7.3 Flux distribution in the fuel pin

In Ref. 2 a method is described to measure the neutron flux distribution in the fuel pin. The same distribution has been calculated. The distribution shows an asymmetry between front and back of the fuel pin, especially during the pre-irradiation in a rather withdrawn position. This is also clear from Fig. 9, where the steep gradient is evident.
As this asymmetry was considered unacceptable, calculations were done at a TOP capsule with a grey neutron absorber between reactor core and the TOP-pin. The result is shown in Fig. 10. Mounting a 16 mm. thick strip of SS 316 directly in front of the capsule decreases the asymmetry from 15% to < 2%, in the pre-irradiation condition.
After BF_3-filling and power adjustment the st.st.strip is removed and power production readjusted, prior to the transient.

7.4 Fuel temperature measurements

Fig. 11 gives the measured fuel temperaturedrop over the third fuel pellet from the top of the pin, versus thermal power production in the pin. During the second approach to $\frac{1}{2}$ P_{nom} the thermocouple failed at a temperature of ∿ 2200°C, as was to be expected. Consequently the start-up of the next capsules will be carried out in many small steps, adjusted to the fuel temperature rise, avoiding too high thermal stress in the hot-weld and allowing time for settling of the fuel thermal conductance and gapclosure.

7.5 Fuel and clad elongation measurements

As mentioned before, the fuel elongation transducer failed, due to a broken connection in one of the two coils. The two coils form part of a bridge circuit, see Fig. 12. The bridge was now balanced with a coil of an identical transducer, temporarely wired out-of-pile into the measuring circuit. An adjustable magnetic core was used to bring the bridge to balance. The sensitivity of the fuel elongation transducer decreased by a factor 1.83, the linearity over the desired range decreased from < 1% to ∿ 2%.

In Fig. 13 the response is given of the clad elongation transducer versus clad temperature. The slope of the curve corresponds accurately with the thermal expansion coefficient of the st.st. employed. Plastic deformation did not occur, due to the low temperatures and the small burn-up during these quasi steady state conditions. The behaviour of the fuel elongation transducer is given in Fig. 14. It should be borne in mind that the clad expands downwards and the fuel expands upwards : the net fuel expansion is the algebraic sum of the two transducers responses. As shown, this measured expansion corresponds quite well to the values expected from the fuel temperatures.

7.6 Transient behaviour of the facility

Except for the problems already described, the response of the loop to all transients, positive and negative, were very close to that of the out-of-pile dummy experiment (Ref. 1). The selfregulating properties of the heat exchanger were not yet employed to the full, but from the measurements can be concluded, that it works as well as in the dummy. The high heat capacities of the st.st. walls smoothe the temperature rises appreciably, all temperatures during overpower transients stay lower than expected from earlier calculations.

7.7 Transient behaviour of the transducers

- Fission gas release could not be detected, due to the low burn-up and the low overpower reached during the final transient.
  The elongation transducers worked as well as could be expected during the transients, both negative and positive.
- The high temperature thermocouple failed during start-up as it reached more than 2200°C.
- Cobalt spnd's
  In Fig. 15 the response of a Co. spnd is shown during the approach to power prior to the transients. The linearity of the spnd signal versus neutron flux is excellent, and the stability is such that under stationary irradiation conditions the signal change is less than 0.5% per 24 hours. There is some temperature and γ sensitivity, but at neutron flux levels > 10^3 n/cm^2 sec these effects are neglectable. Reproducability is also very good. At every irradiation the curve spnd signal- versus fission power matched to within 5%, which is the accuracy of the thermal power measurement. Their immediate response to neutron flux changes makes the Co-spnd's a simple, cheap and useful signal source for the overpower control mechanism.

8. Status of the TOP project

Before the HFR vessel exchange in 1983, TOP-2 and -3 will be irradiated and "transiented". Furthermore, a "steady state" experiment will be carried out. This is a simple capsule, in which a TOP fuel pin will be irradiated under the same circumstances as the TOP-s, but without transient. This fuel pin will be used as reference for the post-irradiation results.

List of references :

Ref. 1 Out-of-pile experiments in support of the HFR-TOP-Irradiation.
J.M. Plaum, C. Croon, I. Monster, H. Manuel, J.E. de Vries
Paper presented at the International Topical Meeting on Irradiation Technology
Grenoble, France, September 28th - October 1st, 1982

Ref. 2 Transient Overpower Experiment on LMFBR Fuel Pins in the Petten High Flux Reactor (HFR-TOP)
G. v.d. Koogh et al.
Paper presented at the Conference on Fast, Thermal and Fusion Reactor Experiments,
Salt Lake City, Utah, U.S.A., April 12th - 15th.

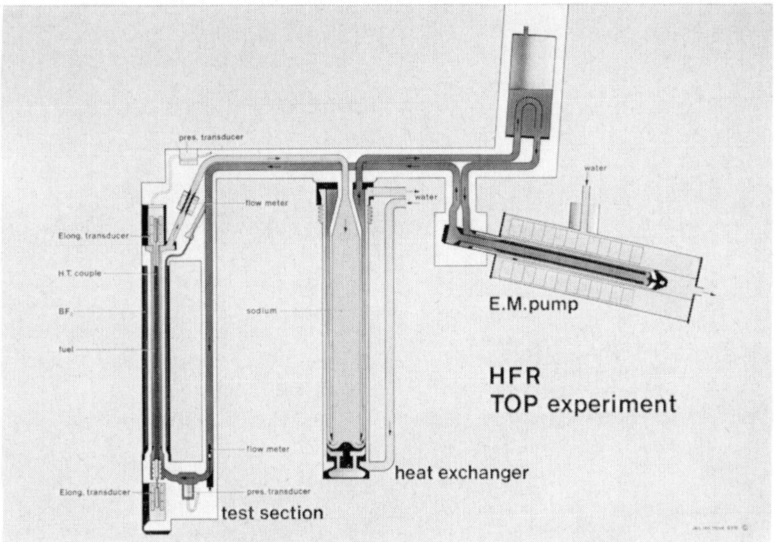

FIG. 1

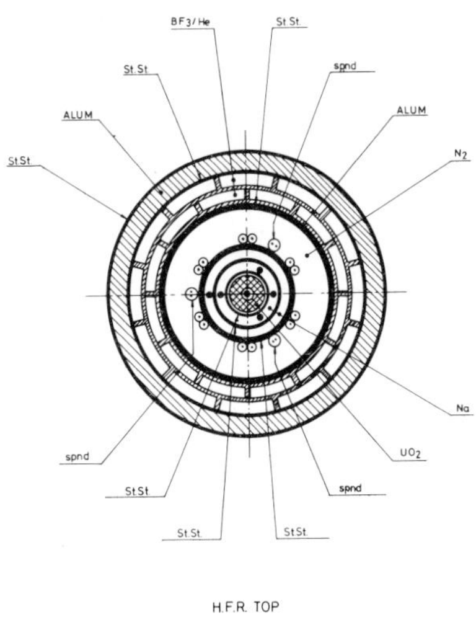

FIG. 2: CROSS SECTION OF THE TESTSECTION.

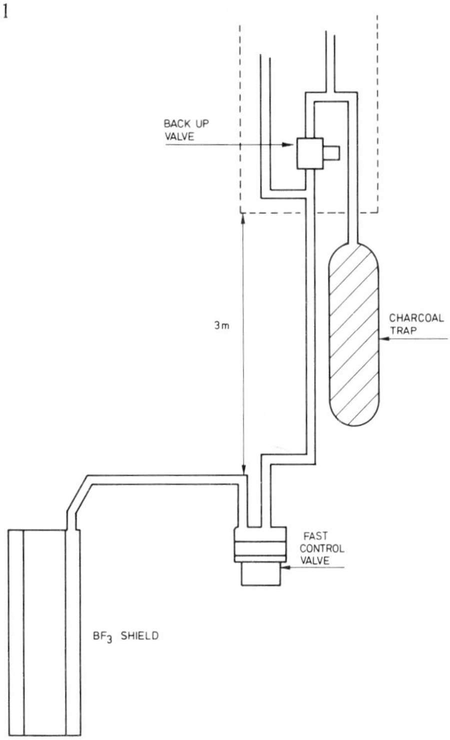

FIG. 3. TOP BF_3 SHIELD.

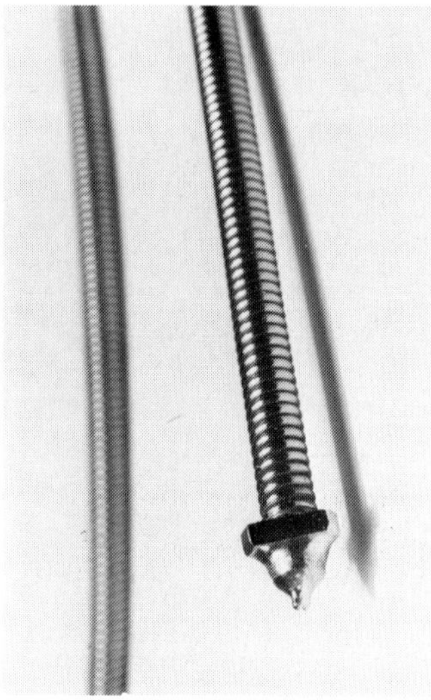

Fig. 4 THERMOCOUPLE INSERTION PIECE

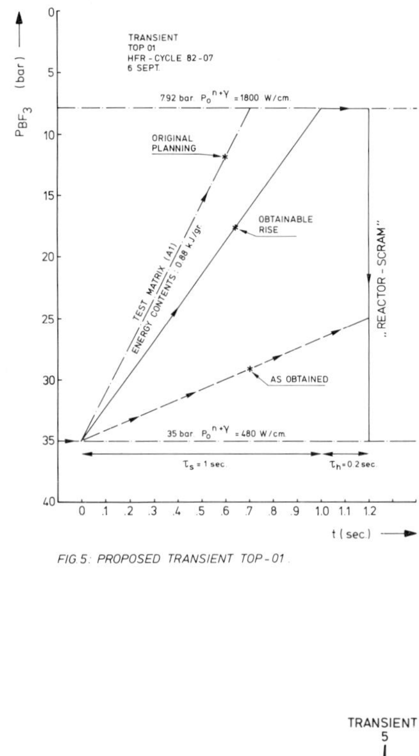

FIG. 5: PROPOSED TRANSIENT TOP-01.

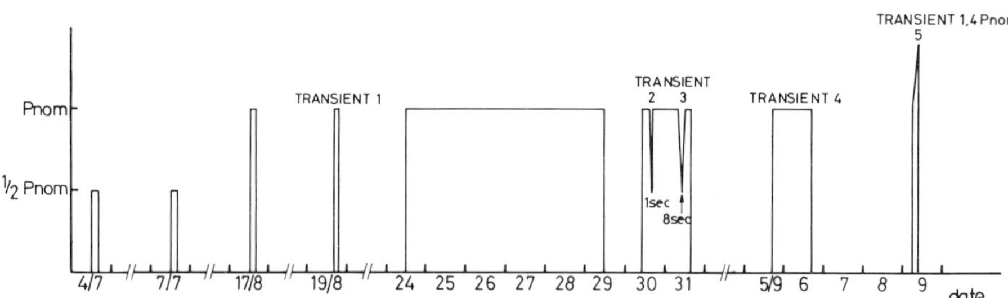

FIG. 6: HISTOGRAM OF TOP_01 IRRADIATIONS.

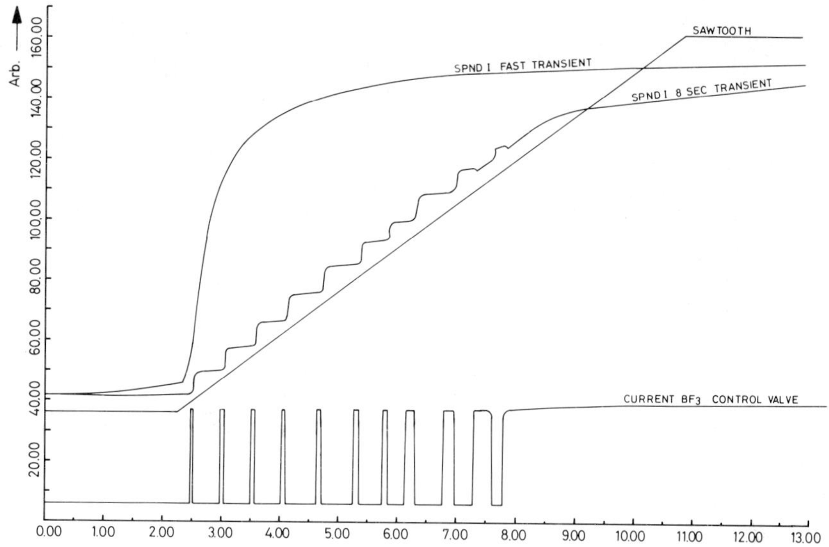

FIG. 7 TOP 01 TRANSIENTS FROM ½ Pnom. → Pnom.

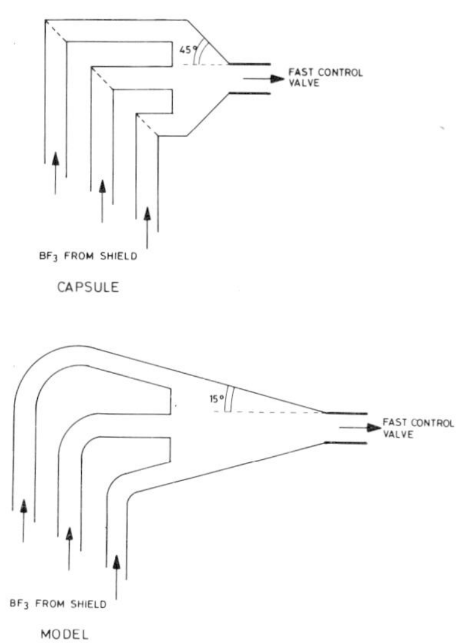

FIG. 8: BENDS AND TRANSITION PIECE BF_3 SYSTEM

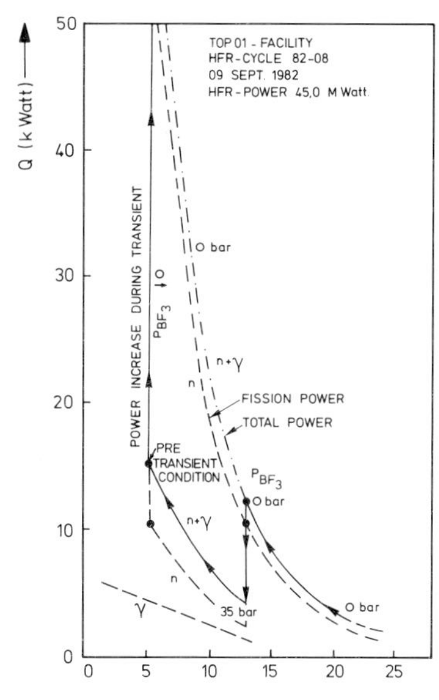

FIG. 9. POWER GENERATION DURING A TOP-TEST.

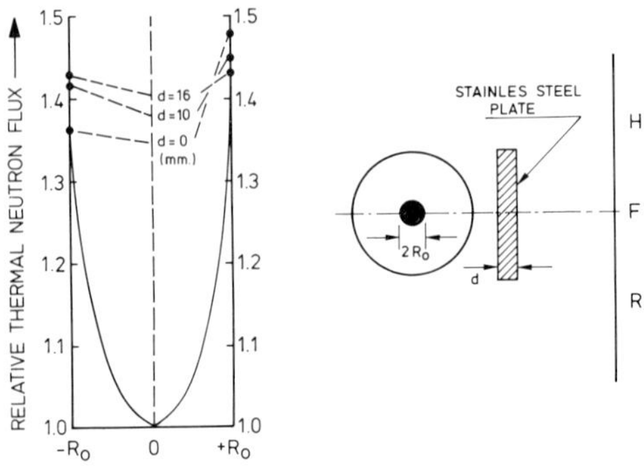

FIG.10: ASYMETRY - CORRECTION RADIAL NEUTRON FLUX DISTRIBUTION IN FUEL PIN.

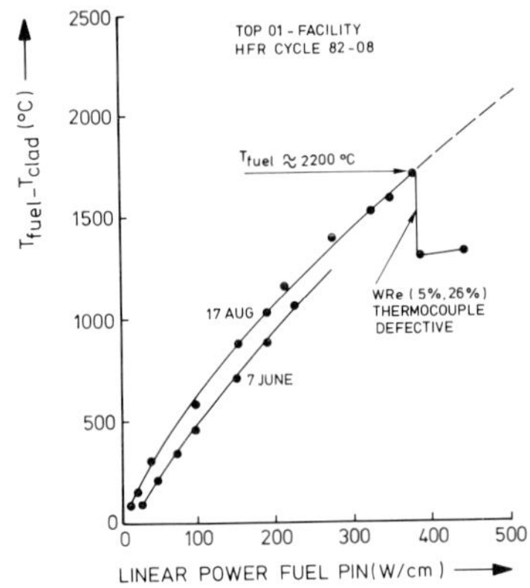

FIG 11 TEMPERATURE DROP FUEL - PIN VERSUS POWER (STEADY).

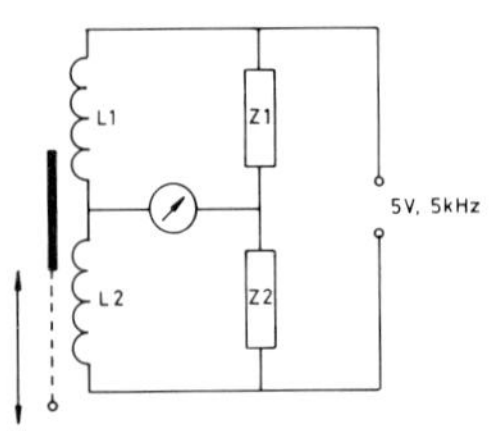

NORMAL SITUATION

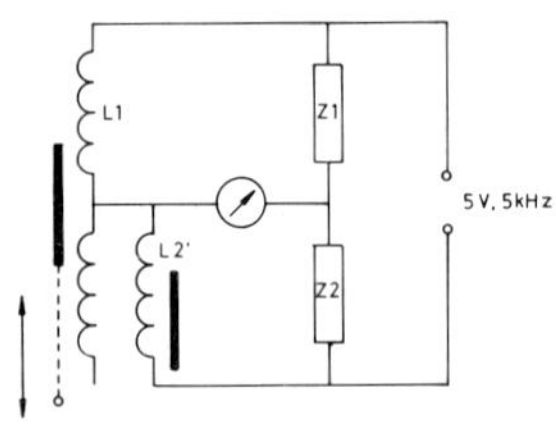

NEW SITUATION

FIG. 12: BRIDGECIRCUIT FUEL ELONGATION TRANSDUCER

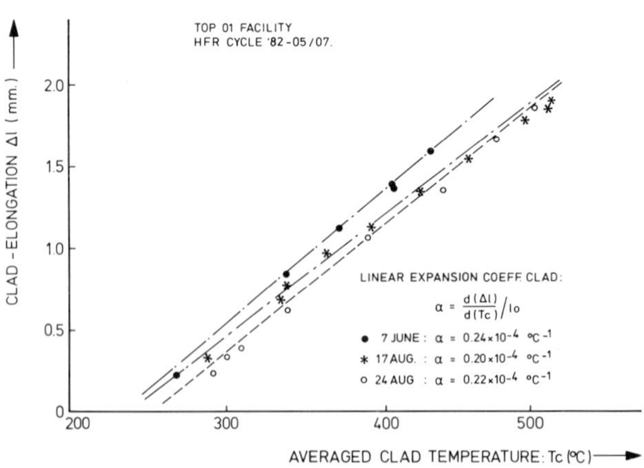

FIG.13: CLAD ELONGATION MEASUREMENTS (STEADY).

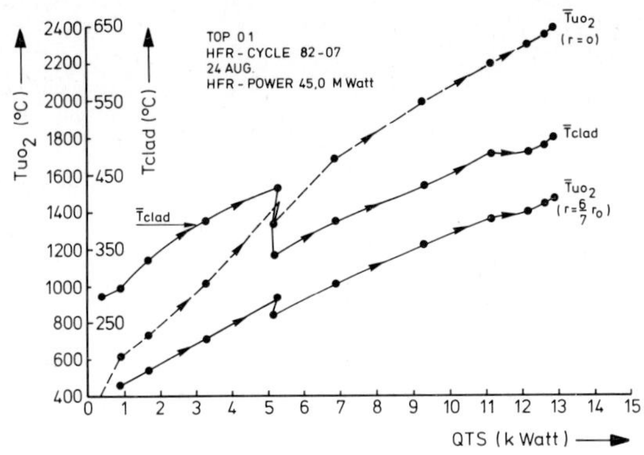

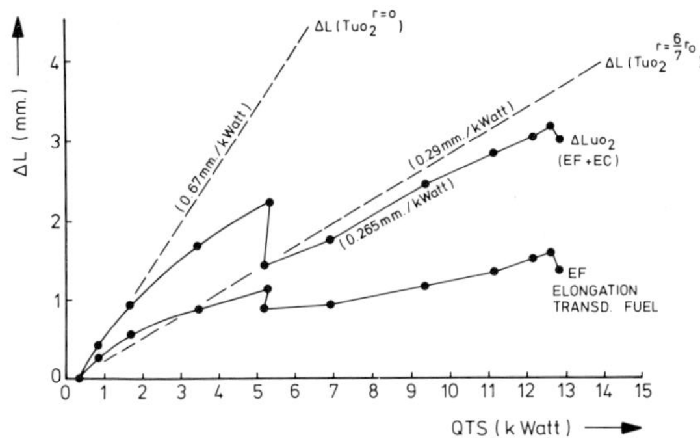

FIG.14: FUEL ELONGATION MEASUREMENTS (STEADY) COMPARED WITH EXPECT. FROM FUEL-TEMPERATURES.

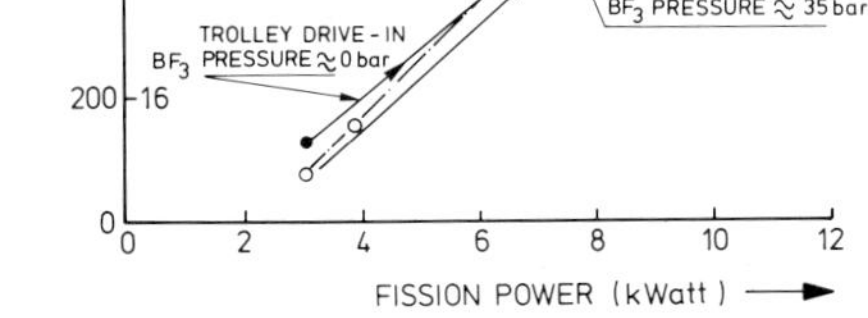

FIG.15: SPND (COLLECTRON) - RESPONSE VERSUS FISSION-POWER DURING TOP - EXP. PROCEDURE.

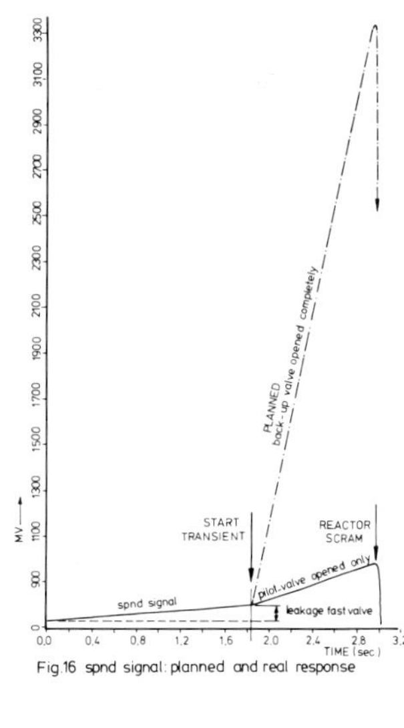

Fig.16 spnd signal: planned and real response

PAPRICA - Code de gestion et de calcul du coeur du Réacteur RAPSODIE

L. Goutagny
G. Gache } CEA/DER/SEER

G. Lecardinal CISI

Centre d'Etudes Nucléaires de Cadarache
BP 1 - 13115 St Paul Lez Durance - France

RESUME

Comme pour tout réacteur à caractère expérimental très marqué tel que RAPSODIE, il est nécessaire d'effectuer un suivi des caractéristiques de l'installation (puissance, débit de sodium, réactivité, température) et des assemblages ou expériences en irradiation (puissance, température, TCF, Dpa, fluence).

La multiplication des expériences et leur diversité (nature du combustible, taux de combustion) ont conduit à développer un outil de calcul performant, tout en restant souple d'emploi et relativement peu coûteux.

Le code PAPRICA doit répondre à cet objectif. On décrit les différents modèles de calcul utilisés (diffusion neutronique à deux dimensions, chaîne d'évolution à 11 isotopes, bilan thermique, etc..).

On compare, ensuite, les résultats obtenus avec les mesures effectuées sur plusieurs configurations du coeur du réacteur.

1-INTRODUCTION

Le coeur du réacteur RAPSODIE, du fait de sa vocation expérimentale, est couramment constitué pour moitié d'assemblages nourriciers standards, pour moitié d'expériences de nature et d'origine très diverses, contenant du combustible ou non. Ces expériences, en irradiation, peuvent être montées dans des assemblages géométriquement identiques aux assemblages nourriciers ou dans des capsules constituées de tubes venant s'insérer dans un assemblage standard porteur.

P. von der Hardt and H. Röttger (eds.), Irradiation Technology, 335–344.

Le code PAPRICA de gestion du coeur de RAPSODIE répond aux deux objectifs suivants :

- Prévisions pour la conduite du réacteur : bilan de réactivité pour les changements de configuration, perte de réactivité par usure du combustible, températures de sortie des assemblages du coeur...
- Fourniture, aux expérimentateurs, des caractéristiques d'irradiation de leurs expériences : taux de combustion, flux intégré, Dpa, évolution des isotopes lourds, etc..

On s'est donc orienté vers un calcul neutronique direct couplé avec des modules d'évolution et de bilan thermohydraulique. Pour éviter à l'exploitant des manipulations fastidieuses inhérentes à ce type de calcul, l'enchaînement des modules a été entièrement automatisé.

2 - DESCRIPTION DU CODE

L'enchaînement général des calculs est décrit sur la figure 1. La structure du code est entièrement modulaire et, de ce fait, l'utilisateur a toute liberté pour définir des enchaînements différents de celui de la figure 1, tant en ce qui concerne la nature des fonctions réalisées que leur ordre chronologique. Des sauvegardes existent à toutes les étapes pour permettre d'arrêter le calcul à l'endroit souhaité et, éventuellement, de le poursuivre ultérieurement.

2.1. Tri des assemblages

Une bibliothèque d'assemblages utilisables a été constituée; elle contient les caractéristiques de fabrication nécessaires aux calculs (masses d'isotopes lourds, débits, etc...). Par un module de tri, on extrait de la bibliothèque les assemblages présents dans le coeur.

2.2. Description de la géométrie du coeur

Le coeur est décrit en géométrie hexagonale (voir figure 2) à deux dimensions planes pour 12 couronnes d'assemblages. Le maillage de calcul comprend 7 points par hexagone (6 sommets et le centre). Pour les 8 premières couronnes du coeur, à chaque cellule hexagonale, on fait correspondre l'assemblage obtenu par le tri précédent. La couronne 9 est remplie par un milieu correspondant aux assemblages réflecteurs en nickel, les couronnes 10 à 12 par un milieu représentatif des assemblages fertiles de couverture.

2.3. Calcul des compositions et sections efficaces

Les compositions des assemblages sont calculées pour une température moyenne de 450°C; la liste des corps pris en compte est la suivante :
Fe , Cr, Ni, Mo, Si, Mn, C, B, Ta, O, N, Na, U234, U235, U236, U238, Pu238, Pu239, Pu240, Pu241, Pu242, Am241, PF (Produit de fission).

Les sections efficaces de ces divers corps sont calculées dans un découpage énergétique à 12 groupes dont les bornes sont les suivantes :
14,5MeV - 3,68MeV - 1,35 MeV - 498 KeV - 183 KeV - 67,4 KeV - 40,9KeV - 24,8 KeV - 9,12KeV - 3,36KeV - 1,23KeV - 454 eV - 101eV - 0.

2.4. Bilan neutronique

Le bilan neutronique est calculé avec l'approximation de la diffusion dans le découpage énergétique décrit en 2.3.
L'équation à résoudre est la suivante:

$$\left\{- \mathrm{div}\left[Di \ \mathrm{grad}\ \phi_i\right] + \sigma_i \phi_i = \frac{X_i}{\lambda} \sum_{j=i}^{12} Fj\,\phi j + \sum_{j<i} \sigma ji\,\phi j\right\} i = 1,12$$

Di: coefficient de diffusion pour le groupe i
σi: section efficace macroscopique totale pour le groupe i
Fj: production de neutrons par fission pour le groupe j
σij: section efficace de transfert du groupe i vers le groupe j
Xi: spectre de fission .

Les valeurs recherchées sont : ϕi flux du groupe i
λ coefficient de multiplication.

La discrétisation des équations est effectuée par différences finies.

Les puissances sont calculées pour chacun des assemblages, sans tenir compte du transport du rayonnement

$$Pi = FiVi \sum_{g=i}^{12} \sum_{j=i}^{24} (Ef_j\, \sigma f_j + Ec_j\, \sigma c_j)_g \ Nji\ \overline{\phi}ig$$

Pi : puissance de l'assemblage i
Vi : volume de l'assemblage i
Fi : facteur d'aplatissement axial de puissance pour l'assemblage i
$(Efj, \sigma fj)_g$: énergie et section de fission pour l'isotope j dans le groupe g
$(Ecj, \sigma cj)_g$: énergie et section de capture pour l'isotope j dans le groupe g
Nji: concentration de l'isotope j dans l'assemblage i
$\overline{\phi}ig$: flux moyen dans l'assemblage i pour le groupe g.

2.5. Bilan thermohydraulique

Comme on peut le voir sur la figure 3, la cuve du réacteur est séparée en deux zones distinctes:

- la zone haute pression (8 premières couronnes du coeur) où la perte de charge appliquée aux bornes de la cuve est identique à celle appliquée entre pieds et têtes d'assemblage,

- la zone basse pression (couronnes 9 et suivantes) où la perte de charge appliquée aux bornes de l'assemblage est très fortement diminuée par l'interposition d'un diaphragme.

Le modèle, très simplifié, servant au calcul de ce bilan thermohydraulique part de l'hypothèse qu'un assemblage a une loi de perte de charge du type suivant :

$$\Delta P = AQ^a$$

ΔP : perte de charge
Q : débit
A,a: constantes mesurées.

On admet que pour toute zone de la cuve comportant des combinaisons d'éléments en parallèle (zone HP), en parallèle et série (zone BP), cette forme de loi reste valable pour la zone concernée. On détermine, ainsi, des lois de perte de charge pour les zones HP, BP et le diaphragme:

$$\Delta P_{HT} = A_{HT}\, Q_{HP}^{\ a_{HP}}$$

$$\Delta P_{BP} = A_{BP}\, Q_{BP}^{\ a_{BP}}$$

$$\Delta P_{D} = A_{D}\, Q_{BP}^{\ a_{D}}$$

ΔP_{HP}, ΔP_{BP}, ΔP_D pertes de charge des zones HP, BP et diaphragme .

Q_{HP}, Q_{BP} débits des zones HP et BP.

La perte de charge haute pression est égale à la somme de la perte de charge aux bornes du diaphragme et aux bornes des assemblages en zone basse pression. Par ailleurs, le débit passant dans la cuve Q_R,mesuré par débitmètre à la pompe, est la somme des débits passant en zone HP et BP; il en résulte l'équation suivante à résoudre :

$$O = A_D\, Q_{BP}^{\ a_D} + A_{BP}\, Q_{BP}^{\ a_{BP}} - A_{HP}\, (Q_R - Q_{BP})^{a_{HP}}$$

On obtient, ainsi, les débits Q_{BP} et Q_{HP}, les pertes de charges ΔP_{HP} et ΔP_{BP} et, en conséquence, les débits des assemblages.

On obtiendra la température brute de sodium, à la sortie de l'assemblage par la relation :

$$\frac{Pi}{Qi\,\rho\, C} = \Delta Ti \qquad Tsi = Te + \Delta Ti$$

ΔTi : accroissement de température du sodium dans l'assemblage i

Pi,Qi : puissance et débit de sodium de l'assemblage i
ρ, C : masse volumique et chaleur spécifique du sodium
Tsi : température de sortie du sodium de l'assemblage i
Te : Température d'entrée du sodium dans la cuve

2.4. Calcul d'évolution

Si Ni(t) est la concentration au temps t de l'isotope i, l'évolution des concentrations des isotopes considérés est régie par le système d'équationssuivant :

$$\begin{cases} \dfrac{d\,\vec{N}(t)}{dt} = A\vec{N}\ (t) \\ \vec{N}\ (o) = \vec{No} \quad \text{valeurs initiales} \end{cases}$$

Le module d'évolution utilisé laisse toute liberté à l'utilisateur pour définir la matrice d'évolution A. Dans notre cas, la matrice d'évolution est décrite dans la figure 4 et le vecteur $\vec{N}$(t) a 11 composantes qui sont les concentrations des isotopes suivants :

U234, U235, U236, U238, Pu238, Pu239, Pu240, Pu241, Pu242, Am241, PF.

Dans un but de simplification des calculs et d'économie, des isotopes tels que Np238 et Np239 ont été négligés, ce qui n'a pas d'effet sur la qualité du calcul, car le combustible utilisé contient des concentrations très faibles en U236 et U238. De même, toutes les filiations,par réactions (n,2n) en décroissance α, ont été négligées.

3 - RESULTATS OBTENUS

En vue de la qualification du code, on a entrepris le calcul de toutes les configurations du coeur réalisées dans RAPSODIE depuis la 50ème campagne d'irradiation (janvier 79); les résultats obtenus ont été systématiquement comparés aux mesures correspondantes.

A titre d'exemple, on indique, en figure 2, les rapports de ΔT calculés et mesurés, obtenus pour le plan de chargement 222 de la 50ème campagne d'irradiation de RAPSODIE. On peut constater un accord satisfaisant entre mesures et calculs.

Toutefois, deux difficultés principales ont été mises en évidence , concernant , d'une part, l'ensemble des calculs pour les positions voisines des barres de contrôle et, d'autre part, la prévision des températures de sortie des capsules compte tenu de la thermohydraulique particulière de ces dispositifs.

L'écart calcul-mesure sur l'évaluation de la perte de réactivité par usure du combustible est de 10 à 12%. Sur cet aspect, on se heurte à la mauvaise connaissance expérimentale des conditions thermiques des bouchons qui influent très directement sur la cote apparente des barres de contrôle et conduisent à des incertitudes importantes sur l'évaluation de l'usure du combustible.

4 - MODALITES DE MISE EN OEUVRE DU CODE

Le code PAPRICA de suivi du coeur de RAPSODIE est implanté actuellement sur ordinateur IBM 370/68 et sur cette machine, sans les séquences de tri et de mise à jour des bibliothèques les performances obtenues pour un cycle complet de calcul sont les suivantes :

. encombrement : 1000K
. temps : ∿ 180 sec.
. coût : ∿ 800 F.

5 - CONCLUSIONS

En conclusion, le code PAPRICA apparaît comme un outil performant, adapté à la gestion du chargement expérimental très diversifié du réacteur RAPSODIE. A ce jour, la validation globale du code a été acquise à partir de comparaisons calcul-expérience, mais il est prévu de poursuivre les qualifications systématiques à partir des configurations de coeur réalisées. Par ailleurs, des développements sont prévus concernant les points suivants :

- calculs d'échange de chaleur par conduction entre assemblages,
- évaluation des échanges de puissance par rayonnement entre coeur et réflecteur nickel,
- amélioration des connaissances du régime hydraulique des capsules, car actuellement les mesures de température ne sont pas représentatives des puissances.

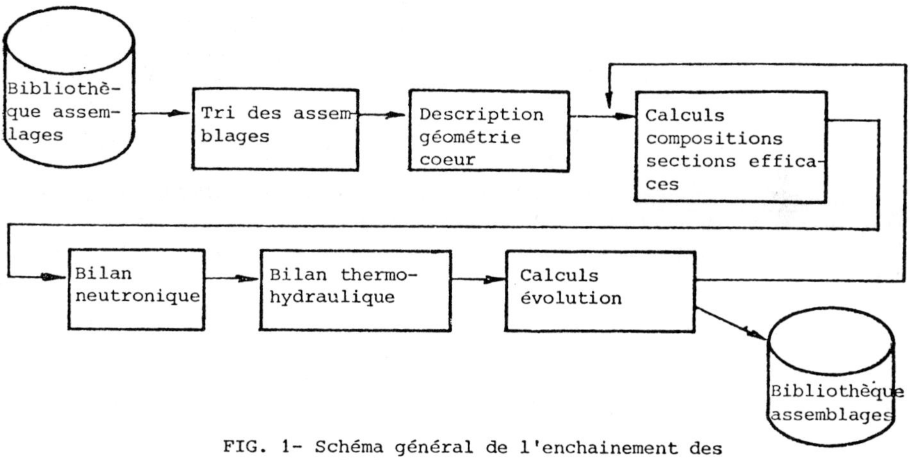

FIG. 1- Schéma général de l'enchainement des calculs.

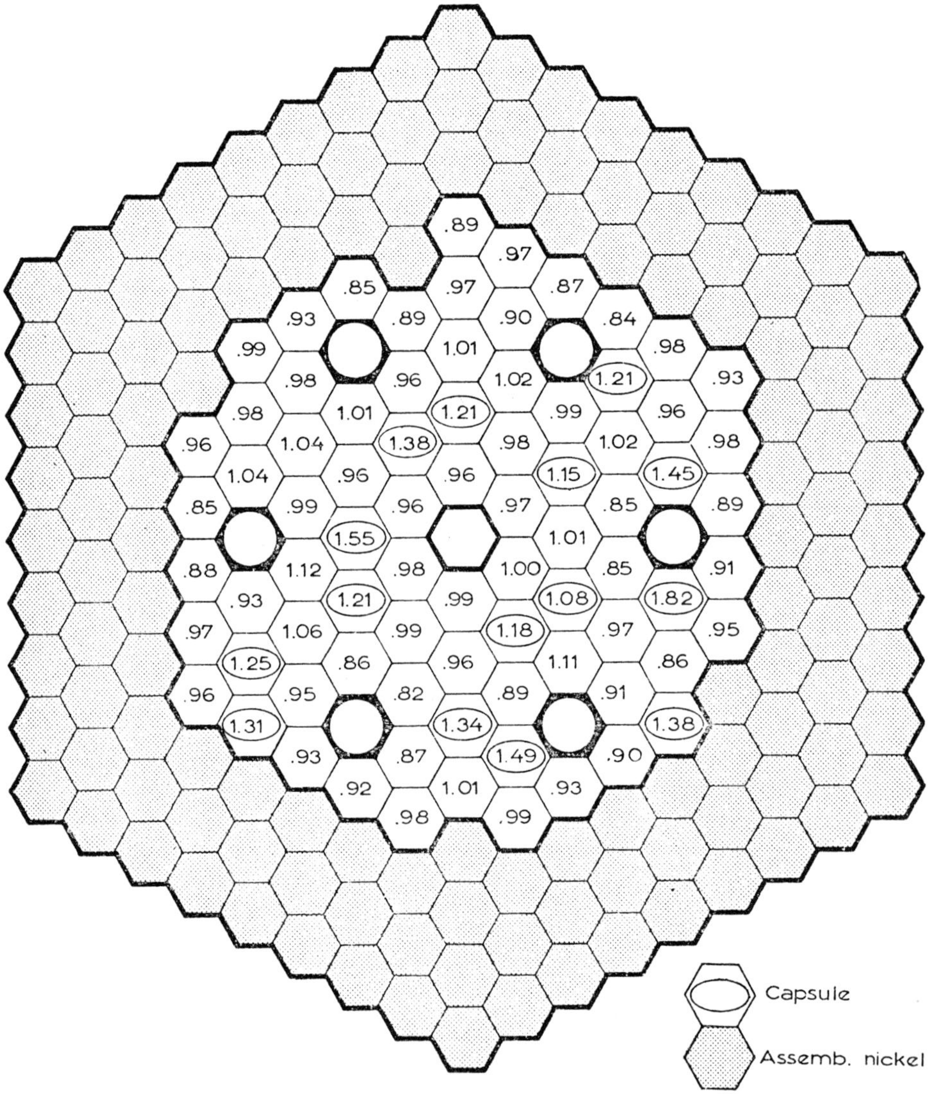

FIG. 2 - Températures

Δ T CALCULE/Δ T MESURE

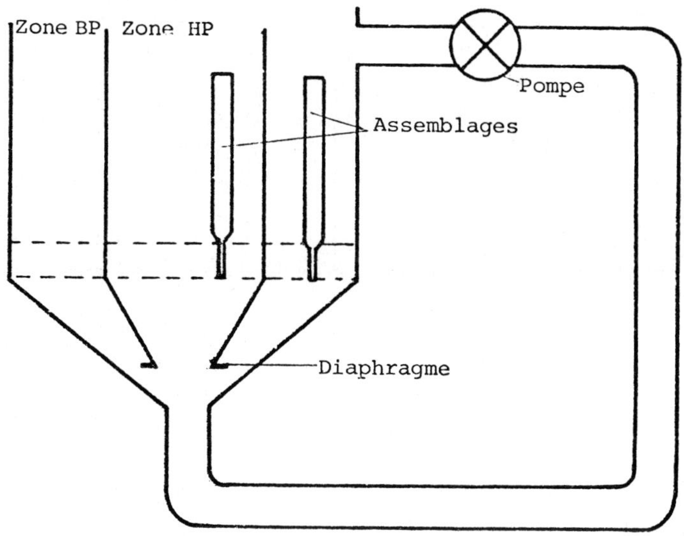

FIG. 3 - Hydraulique de la cuve

	U234	U235	U236	U238	Pu238	Pu239	Pu240	Pu241	Pu242	Am241	PF
U234	$-\bar{\sigma}_c\phi$ $-\bar{\sigma}_f\phi$										
U235	$\bar{\sigma}_c\phi$	$-\bar{\sigma}_c\phi$ $-\bar{\sigma}_f\phi$									
U236		$\bar{\sigma}_c\phi$	$-\bar{\sigma}_c\phi$ $-\bar{\sigma}_f\phi$								
U238				$-\bar{\sigma}_c\phi$ $-\bar{\sigma}_f\phi$							
Pu238			$\bar{\sigma}_c\phi$		$-\bar{\sigma}_c\phi$ $-\bar{\sigma}_f\phi$						
Pu239				$\bar{\sigma}_c\phi$	$\bar{\sigma}_c\phi$	$-\bar{\sigma}_c\phi$ $-\bar{\sigma}_f\phi$					
Pu240						$\bar{\sigma}_c\phi$	$-\bar{\sigma}_c\phi$ $-\bar{\sigma}_f\phi$				
Pu241							$\bar{\sigma}_c\phi$	$-\bar{\sigma}_c\phi$ $-\bar{\sigma}_f\phi$			
Pu242								$\bar{\sigma}_c\phi$	$-\bar{\sigma}_c\phi$ $-\bar{\sigma}_f\phi$		
Am241								λ_{β^-}		$-\bar{\sigma}_c\phi$ $-\bar{\sigma}_f\phi$	
PF	$2\bar{\sigma}_f\phi$	$2\bar{\sigma}_f\phi$	$2\bar{\sigma}_f\phi$	$2\bar{\sigma}_f\phi$	$2\bar{\sigma}_f\phi$	$2\bar{\sigma}_f\phi$	$2\bar{\sigma}_f\phi$	$2\bar{\sigma}_f\phi$	$2\bar{\sigma}_f\phi$	$2\bar{\sigma}_f\phi$	0

FIG. 4 - Matrice d'évolution

DIE KNK II ALS BESTRAHLUNGSREAKTOR

W. Kathol

Kernforschungszentrum Karlsruhe GmbH
Projekt Schneller Brüter
Postfach 3640, D-7500 Karlsruhe

Das Versuchskernkraftwerk KNK II (Kompakte Natriumgekühlte Kernreaktoranlage mit Schnellem Kern) im Kernforschungszentrum Karlsruhe wurde als technologische Vorstufe zu der Prototypanlage SNR 300 errichtet. Bei der Planung der Anlage wurden daher keine Vorkehrungen für ihre eventuelle Verwendung als Experimentier- und Bestrahlungsreaktor getroffen. Trotzdem ist es inzwischen gelungen, das Kraftwerk in großem Umfang für Versuche zu nutzen und Einrichtungen für ihren Einsatz als Bestrahlungsreaktor zu entwickeln. In diesem Beitrag wird über die durch die Anlage vorgegebenen Randbedingungen für Bestrahlungsexperimente und über das laufende und geplante Bestrahlungsprogramm berichtet.

ANLAGENBESCHREIBUNG UND BETRIEBSERFAHRUNGEN

Die KNK ist ein Versuchskernkraftwerk mit einer elektrischen Leistung von 20 MW. Erstes Ziel der 1961 begonnenen Vorarbeiten für dieses Projekt war es, eine besonders kompakte thermische Reaktoranlage zu bauen. Daher wurde Natrium als Wärmeübertragungsmittel gewählt, für die Moderation sorgte Zirkonhydrid. Mit der Errichtung der Anlage konnte 1966 begonnen werden; sie wurde von 1971 bis 1974 erfolgreich mit einem thermischen Core betrieben.

Noch vor Baubeginn hatte sich die Zielsetzung der KNK geändert, die Anlage war zu einem Teil des deutschen Programms zur Entwicklung schneller Brutreaktoren geworden. Aus terminlichen Gründen wurden jedoch die bereits erstellten Planungsunterlagen beibehalten. Zunächst sollte die Natriumtechnologie mit einem

P. von der Hardt and H. Röttger (eds.), Irradiation Technology, 345–354.

thermischen Core unter Reaktorbetriebsbedingungen getestet werden, anschließend war der Übergang auf einen schnellen Kern vorgesehen. Das erste schnelle Core ging dann 1977 in Betrieb.

Die Bestrahlungsergebnisse des inzwischen abgebrannten Erstkerns sind sehr zufriedenstellend. Der Spitzenabbrand in der Testzone liegt mit ca. 100.000 MWd/t_{SM} wesentlich höher als der ursprüngliche Auslegungswert und um 20 % über dem Zielabbrand des SNR 300. Ein gleichermaßen positives Ergebnis kann damit auch für dessen Brennelemente erwartet werden, da in der Testzone der KNK II weitgehend SNR 300 ähnliche Bestrahlungsbedingungen eingestellt wurden. Zwei bei niedrigen Abbränden defekt gewordene Brennelemente brachten Erfahrungen bei der Detektion, Handhabung und Zerlegung in Hinblick auf die Nachbestrahlungsuntersuchung und Wiederaufarbeitung.

Durch die Fortführung der ursprünglichen Anlagenkonzeption (Kraftwerksbetrieb, kompakter Aufbau, thermisches Core) wurden wichtige geometrische und thermohydraulische Parameter nicht immer optimal für die Gestaltung der schnellen Cores sowie des später begonnenen Versuchs- und Bestrahlungsprogramms festgeschrieben. Genannt seien beispielsweise die begrenzte Zugänglichkeit zum Kern nur über wenige Bohrungen im Reaktortankdeckel, die Reaktorleistung oder der maximale Kühlmitteldurchsatz.

Um bei der vorgegebenen thermischen Reaktorleistung von 58 MW Brennelemente mit für einen schnellen Brutreaktor repräsentativen Daten wie z.B. einer Stablängenleistung von 435 W/cm erproben zu können, wurde der schnelle Kern der KNK II als Zweizo-

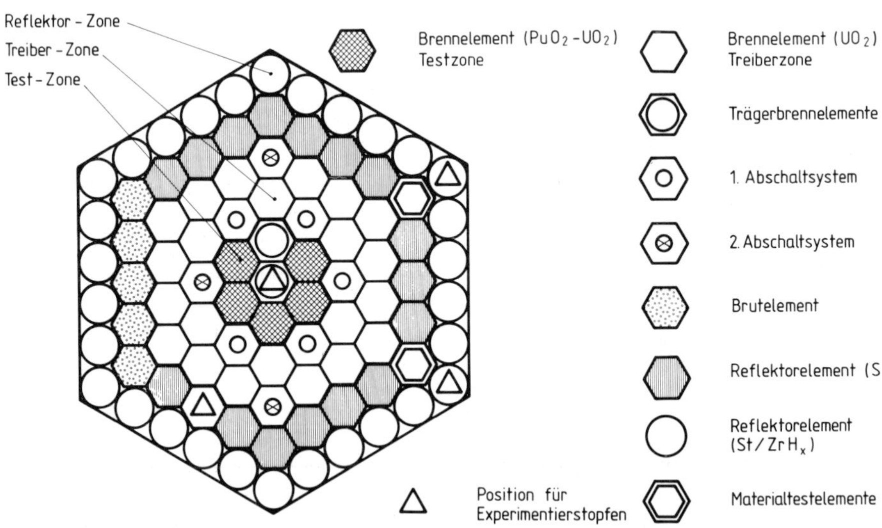

Abb. 1: Corequerschnitt der KNK II

nencore ausgelegt. Eine Testzone bestehend aus 7 Testbrennelementen ist von einer Treiberzone mit 22 Elementen umgeben. Die Reaktivitätsregelung und Abschaltung erfolgen über 5 Stellelemente. Zusätzlich ist eine vom ersten Abschaltsystem unabhängige, aus 3 Absorberelementen bestehende zweite Abschalteinheit vorhanden. Der Kern wird radial von Stahl-Zirkonhydrid-Reflektorelementen sowie von 5 Brutelementen abgeschlossen.

Aus dem Corequerschnitt (Abb. 1) geht auch hervor, über welchen Positionen des Cores sich eine Durchführung im Drehdeckel des Reaktortanks befindet.

Die Wärmeabfuhr aus dem Kern erfolgt über je zwei Primär- und Sekundärsysteme mit Natrium als Wärmeübertragungsmittel an das tertiäre Wasser-Dampf-System. Der nicht aktivierte Sekundärkreislauf verhindert bei einem Dampferzeugerschaden die Reaktion von aktiviertem Natrium mit Wasser (Abb. 2).

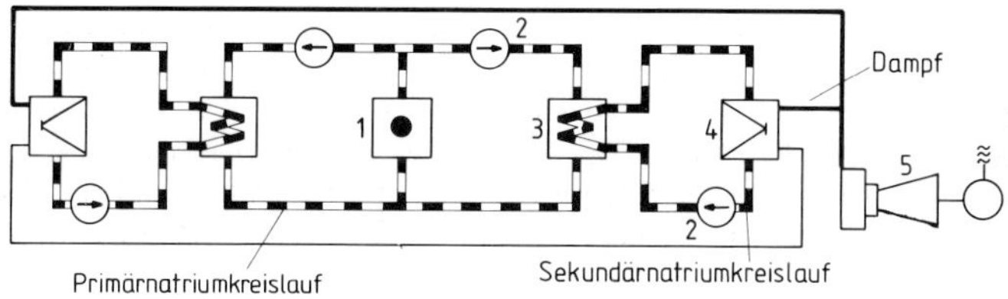

1 Reaktor
2 Kühlmittelpumpen
3 Zwischenwärmetauscher
4 Dampferzeuger
5 Turbine mit Generator

Abb. 2: Hauptfließschema der KNK II

Einige charakteristische Daten der Anlage sind:

Thermische Leistung	58 MW	Mittlerer Kerndurchm.	824 mm
Elektrische Leistung	20 MW	Kernhöhe	600 mm
Kühlmitteltemperaturen		Brennstoffeinsatz	
Primärsystem	525/360°C	Testzone UO_2/PuO_2 ca.	160 kg
Sekundärsystem	504/322°C	Anreicherung Uran	63 %
Natriumdurchsatz		Anteil Plutonium	30 %
Primärsystem	2x498 t/h	Treiberzone UO_2 ca.	525 kg
Sekundärsystem	2x451 t/h	Anreicherung Uran	59/43,5 %
Frischdampfzustand	80at,485°C	Mittlere Leistungsdichte	
Reaktortankdurchmesser	1900 mm	Testzone	183 MW/t
Hauptkühlmittelleitung	200 mm	Treiberzone	60 MW/t

Das Neutronenenergiespektrum der KNK II (Abb. 3 und Abb. 4) hat in der Testzone ein Maximum bei $\bar{E}_n$=0,85 MeV. Der schnelle Neutronenfluß (E_n>0,1 MeV) erreicht $\Phi_{max}=1,6\cdot10^{15}$ cm^{-2} s^{-1}. Die KNK II bietet somit die Möglichkeit, die an thermischen Materialtestreaktoren begonnene Entwicklung von Strukturmaterialien für schnelle Reaktoren zu ergänzen. Mit dem ersten Core wurden strahleninduzierte Schädigungen von ca. 30 dpa-NRT erreicht. Dies entspricht der zu erwartenden Schädigung corenaher Strukturteile in einem schnellen Brutreaktor.

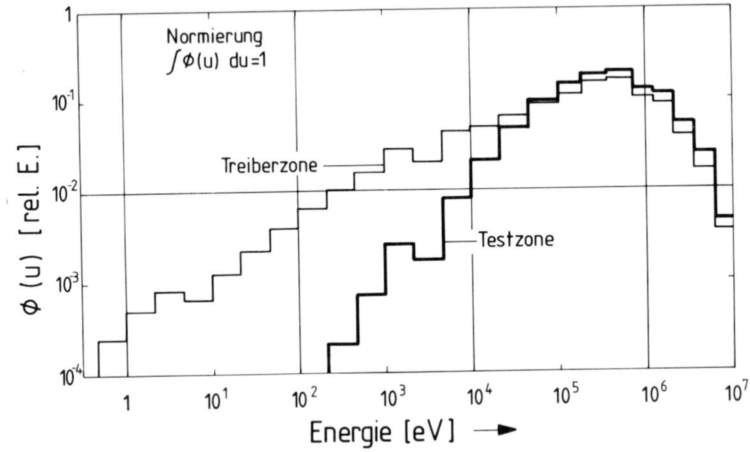

Abb. 3: Spektrum der Neutronenenergie in Test- und Treiberzone

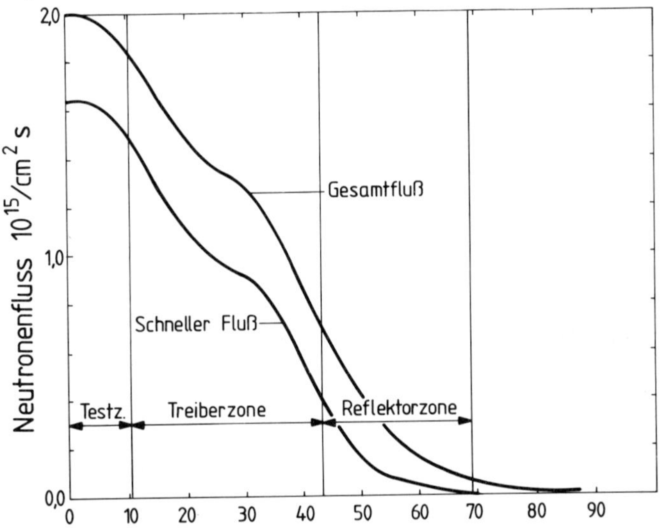

Abb. 4: Radialer Flußverlauf im zweiten Kern KNK II/2

Die KNK II wird normalerweise bei Vollast betrieben. Auf Wunsch der an der Anlage tätigen Experimentatoren werden jedoch auch Teillastbetrieb und Laständerungen gefahren. Regelmäßige Abschaltphasen, abgesehen vom Corewechsel, sind nicht vorgesehen. Den Auslegungen der Bestrahlungsversuche wurde eine Zyklusdauer von 120 (Vollast-)Tagen zugrundegelegt.

BESTRAHLUNGSEINRICHTUNGEN UND BESTRAHLUNGSPROGRAMM

Die für Bestrahlungsexperimente in der KNK II entwickelten Einrichtungen sind weitgehend durch die Konzeption der Anlage bestimmt (Abb. 5):

- Bis zu vier Bestrahlungsrigs (Materialbestrahlungsstopfen) können in den Reaktordrehdeckel eingesetzt werden.

- Oberhalb des eigentlichen Corebereichs können mit speziellen Bestrahlungsstopfen Kurzzeitbestrahlungen (Isotopenbestrahlungen) durchgeführt werden.

- Zwei Materialtestelemente besetzen die Positionen von Reflektorelementen.

- Zwei ringförmige Trägerbrennelemente in der Testzone können kleine Brennelementbündel oder Materialbestrahlungseinsätze (auch verlängerte Materialbestrahlungsstopfen) aufnehmen.

- Anstelle von Brennstäben werden einige Probenrohre mit Materialproben in die Testbrennelemente assembliert. Brennelement-Strukturteile werden ebenfalls für Materialbestrahlungen eingesetzt.

Die Bestrahlung von Kernbrennstoffen und Absorbermaterialien erfolgt im Brennelement- bzw. Absorberelementverband. Hervorzuheben ist hier die Entwicklung druckloser Absorberstäbe.

Materialbestrahlungsstopfen (Experimentierstopfen)

Die für die Coreüberwachung benötigten Thermoelemente sind in der KNK II in einer Verteilerplatte oberhalb der Brennelemente fixiert; lediglich über den Corepositionen 100 (Zentralbrennelement) und 511, 621, 626 (Reflektorelemente) werden Einzelthermoelementzuführungen verwendet. Diese Zuführungen sind ca. 6 m lange Stopfen, die durch den Reaktordeckel geführt werden und unmittelbar über den Elementaustrittsöffnungen enden. Für erweiterte Meß- oder für Bestrahlungsaufgaben können die Einzelthermoelementzuführungen durch modifizierte Stopfen (Experimentier- bzw. Materialbestrahlungsstopfen) ersetzt werden. Solche Stopfen gehören zu den wichtigsten Experimentiereinrichtungen der KNK II, da nur sie den Zugang in den Reaktortank ermöglichen. Zur Erweiterung ihres Verwendungsbereichs wurde in die Coreposition 511 ein besonderes Reflektorelement eingesetzt, das es erlaubt, mit einem verlängerten Experimentierstopfen bis zur Coremitte zu gelangen. Für die Position 100 ist aus dem gleichen Grund ein ringförmiges Trägerbrennelement vorgesehen.

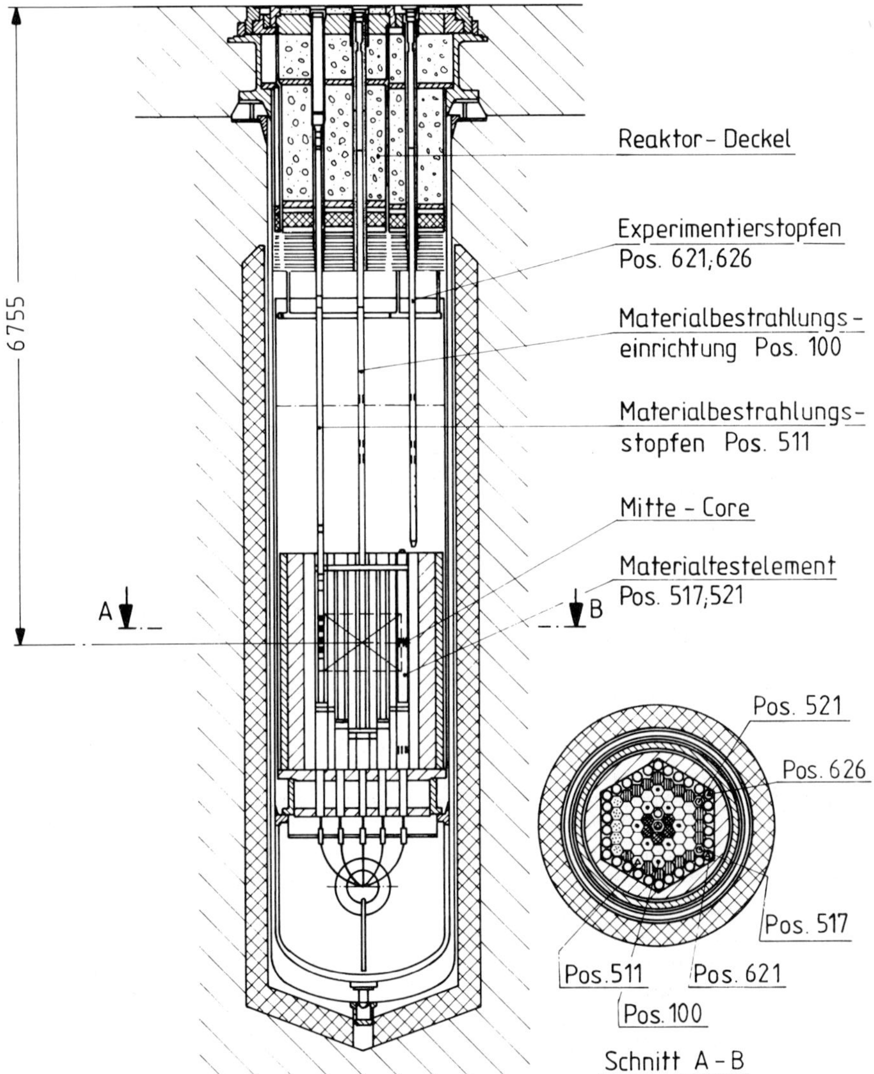

Abb. 5: Reaktorlängs- und Corequerschnitt der KNK II mit Anordnung der Experimentierstopfen

Bisher wurde eine größere Zahl von Experimentierstopfen für verschiedene Meßaufgaben eingesetzt. Ein erster Materialbestrahlungsstopfen mit elektrisch beheizten Natriumkapseln zur Bestrahlung von Strukturmaterialproben bei Temperaturen zwischen 450°C und 650°C und bis zu Neutronendosen von $5 \cdot 10^{21}$ cm^{-2} (E>0,1 MeV) wird zur Zeit gefertigt.

Während die meisten Experimentierstopfen für spezielle Aufgaben konstruiert und gebaut werden, stehen inzwischen mehrere Stopfen für Bestrahlungen oberhalb des eigentlichen Corebereichs bereit. In diesen Positionen können nur geringere Fluenzen erreicht werden (Kurzzeit- bzw. Isotopenbestrahlungen), dafür ist der Bestrahlungsraum jedoch auch während des Reaktorbetriebs zugänglich. Voraussetzung für die Bestrahlungen ist jedoch, daß die Proben der hohen Temperaturen des Primärnatriums (525°C) standhalten.

Materialtestelemente

Schon mit dem thermischen KNK I-Core wurde ein Materialtestelement (MTE 1), das zunächst vornehmlich Proben des Reaktortankmaterials enthielt, bestrahlt. Leicht modifiziert wurde dieses Element für KNK II übernommen. Zusammen mit dem Zweitcore wird jetzt ein zweites Materialtestelement (MTE 2), das konstruktiv weitgehend dem ersten entspricht, bereitgestellt.

Die Materialtestelemente sind in ihrer äußeren Kontur den Brenn- bzw. Reflektorelementen gleich. Ihr herausnehmbarer Innenteil enthält, aufgeteilt in mehrere Etagen, Raum für die zu bestrahlenden Proben. Die erste Beladung des MTE 2 besteht aus Zug- und Kerbschlagproben in offenen sowie in wärmeisolierten Kapseln. Für Untersuchungen zum Brennstoffkreislauf werden ferner Probenrohre mit aktinidenhaltigen Pellets eingebaut. Das MTE 1 ist weiterhin mit nicht gekapselten Materialproben belegt.

Da die beiden Materialtestelemente während der Bestrahlung nicht mit Meß- oder Steuerleitungen erreicht werden können, ist die Bestimmung der Bestrahlungstemperatur nur auf indirektem Wege, z.B. über die Auslegungsrechnung möglich. Lediglich zur Ermittlung der maximalen Bestrahlungstemperatur werden plastisch verformte Temperaturaufnehmer, die im Kernforschungszentrum Karlsruhe entwickelt wurden, eingesetzt. Zur Bestimmung der Neutronenflüsse, differenziert nach der Neutronenenergie, sind spezielle Dosimeterkapseln vorhanden.

Soweit die Bestrahlungstemperatur oberhalb der des Natriums liegen soll, wird dazu die Aufheizung der Proben durch die Gammastrahlung benötigt. Diese ist jedoch direkt abhängig von der Reaktorleistung. Durch den Einsatz von Wärmerohren (heat pipes) kann eine Regelung der Bestrahlungstemperatur, in weitem Bereich unabhängig von der Reaktorleistung, erreicht werden. Der Einsatz von Bestrahlungskapseln mit gasgepufferten Wärmerohren wird vorbereitet und ist für eine der ersten Nachladungen des MTE 2 vorgesehen.

Trägerbrennelemente

Da die KNK II als Testbett für Schnellbrüter-Brennelemente ausgelegt ist, wurde der Brennstoff zur Erreichung der gewünschten hohen Leistungsdichte in der Testzone konzentriert. Der Reaktivitätsbeitrag eines jeden Testelements ist dadurch jedoch so hoch, daß allenfalls auf einige Brennstäbe verzichtet werden kann. Für Bestrahlungsexperimente wurden daher ringförmige Trägerbrennelemente entwickelt. Es handelt sich dabei um Brennelemente, denen nur die inneren Brennstabreihen fehlen. Aus thermohydraulischen und sicherheitstechnischen Gründen ist der so im Zentrum des Elementes entstandene Bestrahlungsraum durch einen kleinen Hüllkasten vom Brennstabbereich abgegrenzt. Für das erste Folgecore der KNK II werden zur Zeit zwei Trägerbrennelemente gefertigt, von denen eines in die Mitte, das andere auf eine Kranzposition des Cores gestellt wird (Abb. 6).

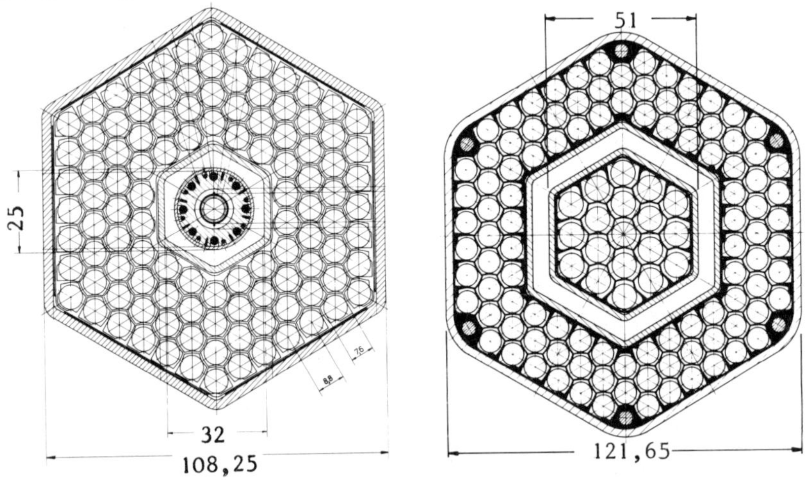

Abb. 6: Querschnitt durch die Trägerbrennelemente mit Materialtesteinsatz bzw. Brennstabbündel

Das zentrale Element, bei dem auch noch der Brennelementkopf geändert wurde, ist für einen Materialtesteinsatz, der wie ein Experimentierstopfen durch den Reaktordeckel geführt wird, vorgesehen. In dem Testeinsatz werden Hüllrohrproben bei hohen Temperaturen und unter Innendruckbelastung bestrahlt. Die Proben befinden sich in natriumgefüllten Kapseln und werden über innenliegende Miniatur-Mantelheizleiter erwärmt.

Das Trägerbrennelement auf der Kranzposition enthält ein integriertes kleines Brennstabbündel mit Uran-Plutonium-Karbid als Brennstoff. Dieses 19 Stab-Bündel bildet den vorläufigen Abschluß der seit langem vom Kernforschungszentrum Karlsruhe durchgeführten Karbidbrennstoff-Entwicklung. Der Einsatz eines demontierbaren Brennelements, bestehend aus einem Träger und einem daraus im Reaktor lösbaren Kleinbündel, erfordert Änderungen der Handhabungseinrichtungen der KNK II. Demontierbare Brennelemente sind für den Test von nach verschiedenen Verfahren hergestelltem Plutonium-Mischoxid-Brennstoff vorgesehen.

Bestrahlung mit den Brennelementen

In der KNK II werden in Verbindung mit den Brennelementen verschiedene Strukturmaterialien für Hüllrohre, Brennelementkästen usw. verwendet. Die Hüllrohre des Erstcores bestehen aus den Werkstoffen 1.4981, 1.4981 kv, 1.4970, 1.4970 kv, 1.4988. Die Strukturmaterialien der Folgecores werden ähnlich variiert. So sind z.B. unbelastete Zwischenhülsen aus den Werkstoffen 12XRN72, PE16, Incoloy 800, 1.4980, 1.4864 usw. gefertigt. Damit bietet die KNK die Möglichkeit, in großem Umfang Hüllrohr-Strukturmaterialien zu testen und damit die statistische Auswertbarkeit der bisher vorliegenden Ergebnisse zu verbessern.

Für weitere Bestrahlungen werden Probenrohre herangezogen, die im Brennelementbündel auf der Position von Brennstäben assembliert werden. Die Probenrohre bestehen aus einem Brennstabhüllrohr mit Endstopfen, beide ebenfalls aus neu zu untersuchenden Werkstoffen. In den Proben befinden sich, gehalten durch einen Probenträger, Rundzug-, Zylinder- und Kerbschlagproben. Für das Zentralbrennelement des Zweitkerns wurden sechs Probenrohre gefertigt.

GENEHMIGUNGSVERFAHREN

Das Versuchskernkraftwerk KNK II unterliegt wie jedes Kernkraftwerk in Deutschland voll den sich aus dem Atomgesetz ergebenden Vorschriften. Diese sind auch für den Einsatz von Bestrahlungseinrichtungen verbindlich.

Für jede Komponente, die in den Reaktortank eingebracht und dort betrieben werden soll, müssen zuvor erstellt werden:

- eine Sicherheitsbetrachtung, in der die Sicherheit der Anlage in allen normalen und speziellen Zuständen von Reaktor und Versuchseinrichtung gezeigt wird;

- eine Spezifikation, die die Bestrahlungseinrichtung mit ihren Konstruktions- und Abnahmevorschriften beschreibt.

Beide Papiere werden im Auftrag der Genehmigungsbehörde vom Technischen Überwachungsverein begutachtet. Der TÜV überwacht dann weiter die Konstruktion, die Fertigung und den Einbau der Komponenten in die Anlage.

Für die Sicherheit der Anlage ist eine umfangreiche Qualitätskontrolle auch bei Versuchseinrichtungen erforderlich. Verzögerungen bei der Versuchsabwicklung, die sich gelegentlich noch aus dem formalen Ablauf des Genehmigungsverfahrens ergeben, müssen nach und nach überwunden werden. Insgesamt stellen die Genehmigungsverfahren keine wesentlichen Hindernisse bei der Durchführung des Versuchsprogramms an der KNK II dar.

AUSBLICK

Im Rahmen der durch die Reaktoranlage gegebenen Möglichkeiten wird die KNK II jetzt mit den beschriebenen Einrichtungen optimal für Materialbestrahlungen eingesetzt. Darüber hinaus werden jedoch detaillierte Untersuchungen darüber durchgeführt, welche Änderungen an einem Folgecore erforderlich wären, um die Anlage flexibler für Experimente nutzen zu können.

Die KNK II ist mit fünf Einheiten des ersten und drei des zweiten Abschaltsystems ausgerüstet. Die Erfahrungen bei der Fertigung dieser Abschalteinheiten und die Fortschritte, die inzwischen bei der Analyse der Sicherheit Schneller Brutreaktoren erreicht wurden, legen es nahe, auf eine oder zwei Einheiten zu verzichten; dies natürlich ohne die Sicherheit der Anlage zu beeinträchtigen. Durch eine solche Änderung würde zusätzlicher, relativ großer Bestrahlungsraum im Corebereich verfügbar. Weiter könnten anstelle der damit ebenfalls entfallenden Stellstabantriebe Experimentierstopfen durch den Reaktordeckel geführt werden.

Ein wesentlich größerer Eingriff in die Konzeption der KNK II wäre der Übergang zu einem Einzonencore, bei dem die Gesamtleistung von 58 MW annähernd gleichmäßig über das Core verteilt wird. In einem Einzonencore können Schnellbrüter-ähnliche Bestrahlungsbedingungen nur noch in kleinen Brennstabbündeln erreicht werden. Da der Reaktivitätsbeitrag der einzelnen Brennelemente jedoch geringer als in der Testzone eines Zweizonenkerns ist, verhält sich das Einzonencore flexibler gegen zusätzliche, für Experimente durchzuführende Änderungen.

TRANSIENT EXPERIMENTS ON LMFBR FUEL PINS IN THE REACTOR BR2 WITH THE VERSATILE SODIUM LOOP "VIC"

Jean DEKEYSER

S.C.K./C.E.N.

B - 2400 MOL, Belgium

ABSTRACT

The "VIC" sodium loop is a compact and versatile irradiation rig for the reactor BR2, devised for testing LMFBR fuel pins - single rods and small clusters - under transient operating conditions. Basically two types of transient experiments can be carried out, either separately or combined, in order to explore the fast breeder fuel operating range before damage occurs : fission power transients with rates up to 15 %/s and the mismatch of sodium cooling flow with transient outlet temperatures. Alteration of the fission power is achieved by a variable neutron absorbing screen, containing the gaseous ^{3}He isotope.

Interesting features of the VIC loop design are also its ease for neutron spectrum tayloring, using an appropriate Cd screen, and its ability for testing active rods, pre-irradiated in fast reactors.

In-pile functional testing of the prototype loop, loaded with a fresh KNK II/Mk 1 fuel pin, and including three programmed fast-power transients, had been successfully accomplished in early 1982.

The paper describes the principal design characteristics of the "VIC 1" loop (single pin testing rig) and depicts the irradiation experience with the prototype rig.

P. von der Hardt and H. Röttger (eds.), Irradiation Technology, 355–367.

PRINCIPAL USES OF THE VIC LOOP

The in-pile sodium loop VIC (Variable Irradiation Conditions) is conceived for performing transient operating experiments on LMFBR fuel pins in the reactor BR2[1], whether on a single pin (VIC 1 type) or even on a small cluster, comprising a maximum of 7 pins (VIC 7 type). Construction of the in-pile section follows the line of an integrated compact loop with forced convection sodium cooling. Its proper design offers a versatility of experimental transient scenarios which can be set-up to explore the fast breeder fuel operating range before damage occurs and which cover a wide range of off-normal operation modes to be expected during the fast breeder reactor life. Basically two kinds of transient experiments can be carried out in the same VIC test section, either separately or combined :

- fission power transients (ramping, transient overpower or cyclic runs) with rates up to 15 %/s of nominal power level ;
- mismatch of the sodium cooling flow, yielding transient Na outlet temperatures with ramping rates up to 10 K/s.

Up to now the prototype loop "VIC 1-A1", loaded with a single fresh fuel pin (KNK II/Mk 1 reactor design, dia. 6 mm, length 1,6 m), has been successfully tested in the BR2 reactor. Irradiation was done under thermalized neutron flux. Besides the testing of fresh fuel, also active rods, partially burnt in a fast reactor (over 10 at %), can be loaded into the VIC rig.

TABLE I - Fuel specification range for VIC 1 loop

Characteristic	Possible range	Prototype loop VIC 1-A1 (KNK-II/Mk 1)
- Rod diameter	6 mm (rig type VIC 1-A) 7.6 mm (rig type VIC 1-B)	6 mm
- Active length	up to 65 cm	60 cm
- Total length	up to 200 cm	160 cm
- Chemical composition (other ranges could be accepted)	Pu : 15 - 30 % U : 85 - 70 %	30 % 70 %
- Isotopic composition	Pu fissile fraction : 75-91 % $^{235}U/U$: 37-93 %	84 % 83 %
- Nominal heating rate	400 - 600 W/cm*	500 W/cm
- Initial burn-up	up to 10 at %	fresh pin

(*) Under fast or thermal neutron flux.

Irradiation can be done under thermal neutron flux as well as in a representative fast neutron spectrum, using an appropriate Cd screen surrounding the test section. Table I summarizes the applicable fuel specification range for the VIC 1 loop in general and the actual references for the prototype VIC 1-A1 in particular.

LOOP DESIGN CHARACTERISTICS

General Design

A schematic outline of the VIC 1 loop and its peripheral control equipment is shown in Fig. 1. The irradiation device is an integrated test rig, compactly taylored to fit into any standard channel (I.D. 84 mm) of the BR2 reactor core matrix[2]. It contains in its lower active part one LMFBR fuel pin and houses in the upper part a small electromagnetic sodium pump that circulates the liquid metal from top to bottom. The sodium passes upward in the central tube, along the fuel rod, through a recuperative heat exchanger section to the cooler and further to the pump. At the top end of the pump the sodium flow is reversed and flows down through two devided tubes, by-passing the cooler, entering into the sodium expansion vessel and descending further in the outer annulus around the fuel wrapper tube.

The integrated cooler is a heat exchanger between the up-flowing Na and the reactor cooling water. Its construction comprises a cylindrical copper block press-fitted onto the central sodium tube and into an outer enveloping tube. Heat flow passes radially by conduction through the copper from the sodium inside the central tube to the outer cooling wall. This wall is surrounded by a stagnant water annulus where the heat flow crosses to the opposite wall which in turn is cooled at the outside by the BR2 cooling water. The BR2 water enters the cooling sleeve of the rig above the reactor core matrix and leaves at the foot end below the core, using thus the available pressure difference of about 2,8 bar. The copper block heat exchanger system assures a double containment barrier between sodium and water. The height of the water column in the annular gap can be changed, so that the cooling capacity of the loop can be automatically controlled in order to operate, for instance, at a constant sodium inlet temperature despite of fuel rod power variations (units TRA/1 and LRC/1 in Fig. 1).

Alteration of the fuel rod heating rate is realized by varying the pressure of the neutron absorbing ^{3}He gas contained in an annular gap around the test section. By means of the out-of-pile control station "VNS 2" (Variable Neutron Screen)[3] which regulates the ^{3}He pressure, the fuel rod power can be automatically controlled and power transients can be run with respect to amplitude and rate (unit PRC/1).

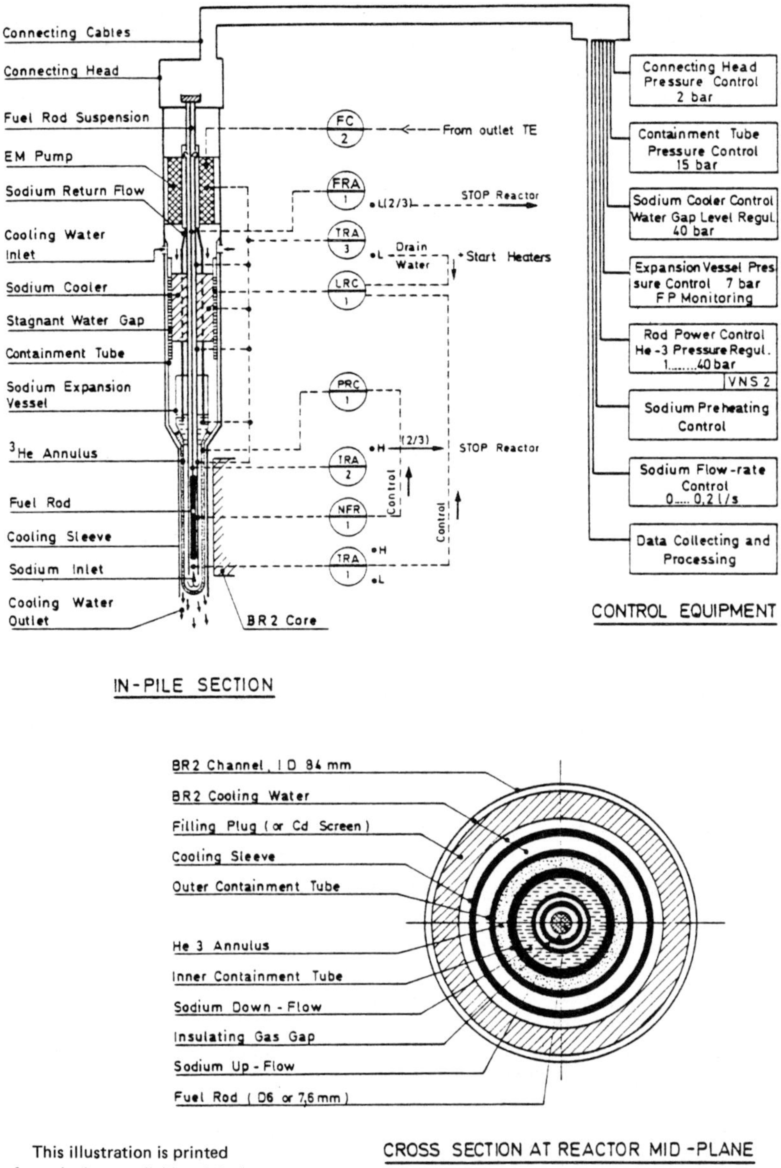

This illustration is printed from the best available original

Fig. 1 - VIC 1 LOOP (Variable Irradiation Conditions) Schematic Outline SCK-CEN F601

The flow rate of the sodium coolant is regulated by the voltage supplied to the electromagnetic pump. Automatic Na flow-rate control can be performed for programmed time dependent variations of sodium outlet temperature, according to specific cooling deficiency conditions (unit FC/2).
Gas pressure and leak tightness are supervised in the different separate compartments of the in-pile section. Also gas samples can be drawn from the plenum of the Na expansion vessel for checking on fission gas activity. For avoiding overheating and possible destruction of the fuel rod, the reactor will be automatically scrammed in case of too high Na outlet temperature and in case of flowrate deficiency (units TRA/2 and FRA/1).
The sodium of the in-pile section is warmed-up and liquefied by several separate heating wire circuits. With increasing temperatures of the sodium due to nuclear heating, the annular gap in the cooler is gradually filled with water and the electric heating is run down. In case of reactor stop, this procedure is reversed. The temperature difference of the sodium over the fuel rod section serves, together with the sodium flow-rate, to determine the nuclear heating rate under steady state operation. An integrated sodium flow meter is mounted downstream of the fuel pin (FRA/1). Fast variation of rod power is measured by prompt reacting self-powered neutron detectors, being previously calibrated (NFR/1).
The fuel rod is introduced into the central tube of the VIC 1 in-pile section via the top end of the rig. It is supported by a suspension train which carries instrumentation leads of the loop and, in case, of the fuel pin itself. This loading procedure can also be applied for an active, preirradiated fuel rod. It reduces the necessary remote handling manipulations to a minimum and allows direct access to the rig head for closing and sodium filling while only the foot end is protected by biological shielding.
Relevant operating characteristics and testing features of the prototype loop are listed in Table II.

Test and Process Instrumentation

As an example of the in-pile instrumentation, the VIC 1-A1 test section is equipped with 32 thermocouples (K-type) of which nine are placed into a special spacer cage centred around the fuel pin, thus measuring the Na temperature near the cladding. Three Na flow rate probes (permanent magnet type) are mounted inside the central rig tube downstream of the fuel section. Also two self-powered neutron detectors (Co-type) are fitted onto the wrapper tube at the fuel mid-plane level. The pressure of the Na cover gas and the ^{3}He screen is measured by connected pressure transducers, mounted in the rig head.

TABLE II - VIC 1-A1 loop operating and testing features

	Operating range	Nominal
- Sodium flow rate	50-200 g/s	130 g/s
- Sodium velocity along fuel pin	1 - 5 m/s	3 m/s
- Na temperature, inlet section	350-500 °C	450 °C
outlet section	max. 830 °C	600 °C
- Na absolute pressure	≤ 10 bar	8 bar
- Heat dissipation capacity of the sodium cooler	max. 35 kW	5 kW
- Fuel rod fission power	max. 43 kW	24 kW
- ^{3}He absolute pressure	1,2 - 38 bar	34 bar
- Overpower capacity by ^{3}He pressure control	up to 220 % (★) versus nominal	
- Rate of power transients	up to 15 %/s positive and 10 %/s negative ramping (★)	
- Rate of Na outlet temperature ramping	up to 10 K/s (+ and -)	
- Unperturbed thermal neutron flux (BR2 reflector channel)	$1 - 2 \cdot 10^{14}$ n/cm^{2} · s	

(★) Results obtained under thermal neutron flux only.

The out-of-pile equipment comprises conventional command, monitoring and protection systems as well as automatic control devices being operated during the transient tests. In particular, the PID controllers for the ^{3}He pressure and for the Na flow rate are combined with programmable setpoint generators versus time, where a choice of input signals is available : the ^{3}He pressure itself, a self-powered neutron detector signal or a relevant loop temperature, e.g. measured at the Na outlet section.
Automatic, setpoint following, variation of the fuel power is realized by means of a ^{3}He pressurizer bellows (contents 45 l) and is used for the slower power ramps or cycling : 1 to 30 %/min. Fast transient power variations (1 to 15 %/s) are achieved by controlled blow-off of the ^{3}He screen towards a low pressure dump tank for the positive power ramps and by re-pressurization from high pressure ^{3}He flasks for the negative power ramps. These devices are part of the double contained equipment VNS 2, which also assures a small ^{3}He gas flow with purification from the tritium being produced by neutron capture[3].
Two digital data acquisition systems are connected for recording and storing the steady state and the fast transient data. In stationary regimes a batch of maximum 100 signals from the test and process instrumentation is continuously collected (each 20s) by a desktop computer, which also calculates the fuel power and

dumps statistically treated information on magnetic tape at regular intervals (e.g. each 2 hours). At any time data sheets and selected graphs can be printed. In parallel, and in particular during the transient tests, a medium speed mini computer scans and stores the proper test instrumentation, capable of measuring a batch of maximum 32 channels up to 100 times per second. All fast data are stored on magnetic disc, having 1 hour autonomy when measuring at the highest speed.

Special Features

Between the in-core part of the VIC 1 enveloping tube and the inner wall of the BR2 reactor channel inside the beryllium matrix, sufficient space is available to fit a filling plug of any selected material (Fig. 1) : plain aluminum (as for the prototype loop), stainless steel or a stronger neutron absorbing screen like boron carbide or cadmium. This feature offers the flexibility required for tayloring the incident neutron spectrum. In particular, when representative transient irradiation tests on fast breeder fuel are specified, a Cd screen is used which geometry is optimized to compromise the following contradictory conditions :

- sufficient amplitude of rod power variation, using the ^3He pressure regulation technique which is most sensitive for thermal neutrons;
- small radial depression of the fission power density in the fuel pellets, requiring a minimal thermal component in the neutron spectrum.

Neutronic calculations and mock-up measurements with an optimized Cd screen in the low flux reactor BR 02 (zero power copy of BR2) yield the following results[4] :

- the obtainable amplitude of rod power variation, when varying the ^3He pressure between 40 and 1 bar abs., is 150 % (reference of 100% being at 40 bar) ;
- the radial fission power depression (pellet surface versus center ratio) ranges from 1,6 to 2,4 with corresponding ^3He pressures of 40 to 1 bar abs.

Another feature of the VIC loop design is the ability for loading active pins which have been pre-irradiated in a fast reactor. The fuel test train with connected pin at its bottom is lowered into the central tube of the rig after the test section has been completely assembled and commissioned. Loading operations are done in the BR2 hot cells. Afterwoods, the Na filling takes place in the storage canal of the BR2 complex where the active part of the rig is biologically shielded by water, while the rig head with the instrumentation connections remains accessible. No other essential requirements are imposed on the candidate test pins than those listed in Table I, except the need for receiving cleaned and undamaged pins. Eventually the upper end plug must be somewhat adapted for being suspended to the test train.

IRRADIATION EXPERIENCE

Prototype Loop Experiment

The prototype loop "VIC 1-Al" comprises all the general features described above. It contains a single fresh fuel pin from the original KNK II reactor type (Mk 1, dia. 6 mm), with characteristics listed in Table I. Construction of the in-pile section and all the associated out-of-pile equipment was done in frame of the so-called "Transients Programme" on the fuel element of the DEBENE fast breeder reactor SNR 300[5].

Irradiation in BR2 occupied two reactor cycles, of 3 weeks each, the first in mid 1981 for carrying out the in-pile functional tests, and the second in early 1982 for running a specified programme of transient power experiments. The rig was equipped with an Al channel plug and loaded in a standard reflector channel of the BR2 core, thus being irradiated under essentially thermal neutron flux. The experience obtained during these periods is summarized hereafter.

In-Pile Functional Testing

During the first irradiation period, the prototype has been submitted to a series of functional tests in order to obtain confirmation of the important design parameters : thermodynamic, neutronic and mechanical design ; layout and behaviour of the in-pile instrumentation ; performance of the out-of-pile devices with emphasis on the loop command, control and protecting systems. These tests have been correctly concluded.

Experience has been obtained on the performance of the integrated sodium cooler as a device for stabilizing the Na inlet temperature at any desired value between 250 °C and 500 °C at nominal rod power. The relation between the fuel power, as calculated from the loop thermal balance, and the ^{3}He screen pressure was measured at 20 % reactor power. As seen in Fig. 2 (curve a), a power amplitude ratio of 2,2 versus nominal power level (obtained at 32,5 bar) has been achieved for the lowest ^{3}He pressure. For convenience, the screen pressure is converted to the reference temperature of 270 °C, being approximately the nominal average screen temperature. An identical calibration curve has been measured in the second irradiation period. Just for comparison, the curve b in Fig. 2 represents the calculated variation of rod power versus ^{3}He pressure to be expected when irradiating under Cd screen[4] (future VIC experiments).

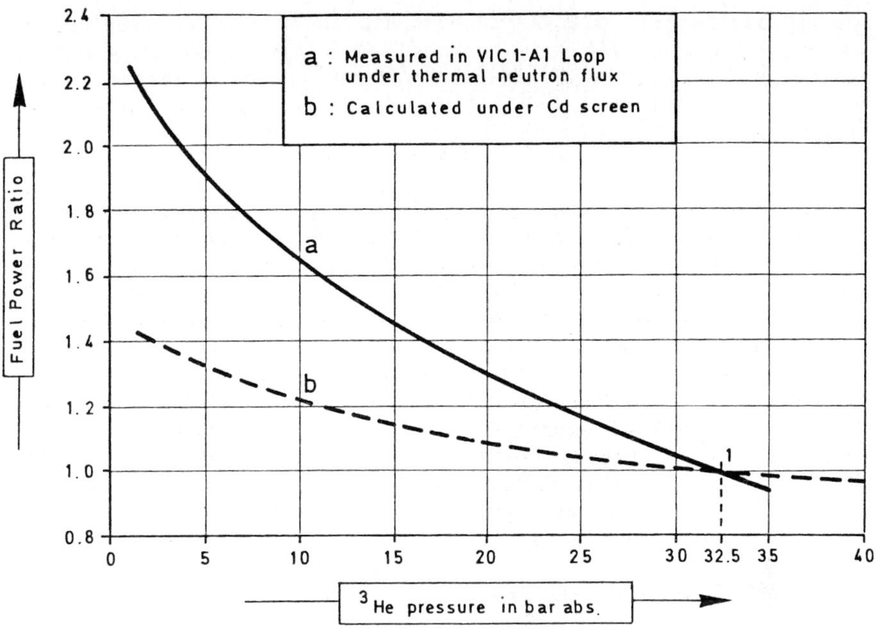

Fig. 2 - CALIBRATION OF FUEL POWER VERSUS ^{3}He PRESSURE

F 661

Performance of Power Transients

The goals of the second irradiation period were dual :

- to demonstrate the dynamic performance of the ^{3}He power cycling system and to experience the transient thermodynamic response of the in-pile section ;
- to submit the test fuel pin to a meaningful series of power transients as specified by the experimenter.

Three peakwise power transients were run, resp. up to 120, 140 and 160 % of nominal power level. Each transient was characterized by a positive ramping rate of 2 %/s, almost zero delay time at high power level and fast return to nominal, except for the 160 % transient which was reversed by programmed reactor scram. In between two consecutive transients the fuel pin was given 4 days to stabilize at nominal power rate. The calibration curve a in Fig. 2 was used to determine at forehand the residual ^{3}He pressure corresponding to the different peak power levels.
Practical experience from the 120 % transient allowed to correct the aperture of the control valve in the ^{3}He depressurisation line. This aperture determines the speed of depressurization, which is run during a pre-set time, and thus determines also the final ^{3}He pressure in the screen and therefore the peak power level. In turn, the blow-off time is defined by the required

ramping parameters, e.g. 20 seconds are to be set for ramping the power from 100 to 140 % at a speed of 2 %/s. Although a test case, the 120 % transient parameters approached relatively well the specified values. The 140 % and 160 % transient parameters were very close to those specified.
A comparison between the realized and the specified parameters for the 160 % power transient is listed in Table III. By means of the "Fast Cycling Device" of the ^3He power control system (VNS2) the time dependent variation of the ^3He pressure could be precisely established with respect to amplitude and rate of the power ramp. The peak level of 816 W/cm was a little higher than required, although the overshoot represents only a minor fraction of the imposed power increase. Also the power maxima as obtained from the final ^3He pressure after blow-off using Fig. 2, as well as from the collectron signal increments, are quite consistent within the limits of experimental accuracy. (Cf. Table III at the lines "From ^3He pressure" and "From collectrons NFE 1, NFE 2"). Furthermore the curvature of the collectron signals during the ascending power ramp was quite straight, meaning that the rod power also must have increased linearly with time.

TABLE III - VIC 1-A1 Experiment - POWER TRANSIENT 160 %

TRANSIENT PARAMETERS	SPECIFIED	REALIZED
- Initial linear power - Maximum linear power	X_0 = 500 W/cm X_{max} = 800 W/cm	500 W/cm From ^3He pressure : 816 W/cm From collectrons NFE 1 : 810 W/cm NFE 2 : 818 W/cm
- Rise time to X_{max} - Power increase % X_0/s curvature - Power reduce time	30 s 2 %/s linear minimal : by reactor scram	30,0 s 2,11 %/s linear (from collectrons) reactor scram (< 10 %X_0 in 0,6 s)
- Na inlet temp. initial max. - Na outlet temp. initial max.	450 °C < 500 °C 600 °C --	450 °C 455 °C (after 45 s) 603 °C 695 °C
- Na flow rate	nominal	133 g/s
- ^3He pressure initial minimal	-- --	32,8 bar abs. 11,1 bar abs.

For each fast transient the negative power ramp phase, achieved by repressurizing the ^3He screen, was started 2s after the foregoing blow-off. Such was to allow intermediate pressure stabilization in the screen. This phase was performed at high speed by having the in-line control valve completely opened. In case of the 140 % transient the initial rate of negative power ramping was 8 %/s. The 160 % transient however was reversed by programmed reactor scram to reduce rod power as fast as possible. Here also the ^3He screen was repressurized to back-up eventual failure of the timed reactor scram.

An illustration of the 160 % power transient is shown in Fig. 3. This figure is a format reduced copy of the original "Medium Speed System" computer plot, representing the time dependent evolution of several relevant measuring signals. The scanning frequency was 100 measurements per second and per channel ; recording of two minutes real time is shown. As a means for exactly defining the different transient phases, the opening and closing of two magnetic valves, MV 01 and MV 04, were recorded : 5 V-level for closed and 0 V-level for open valve. These valves are mounted in the ^3He connecting lines of VNS 2 in such way that the ^3He gas is blown-off through valve MV 01 and then is repressurized through valve MV 04 after closing MV 01. The corresponding switching of these valves and variation of the ^3He pressure, labeled PE 30, is seen in Fig. 3. Instantaneous rod power is identified by the collectron signal NFE 1, which goes up linearly with time during 30s and then drops sharply at the instant of reactor scram. The power rise from 500 to 810 W/cm clearly synchronizes with the decrease phase of ^3He pressure.

Three thermocouple signals in Fig. 3 show the variation of sodium temperatures along the fuel pin : TE 3 at inlet section, TE 7 at mid-plane of the fissile stack and TE 10 at outlet section. The inlet temperature remains unaltered, while temperatures at the higher pin levels increase together with the fuel power (or NFE1). Shortly after scram the Na temperature profile along the pin vanishes and practically coincides with the inlet temperature which further drops slowly as the test section cools down. After about 10 minutes (not shown in Fig. 3) the Na temperatures throughout the loop go down under a pre-set level at which time all electrical heaters in the in-pile section are automatically switched on in order to stabilize the loop at 150 °C. Meanwhile the stagnant water gap in the sodium cooler has been emptied. The Na flow rate, labeled FE 1, remained at constant nominal level during and after the transient.

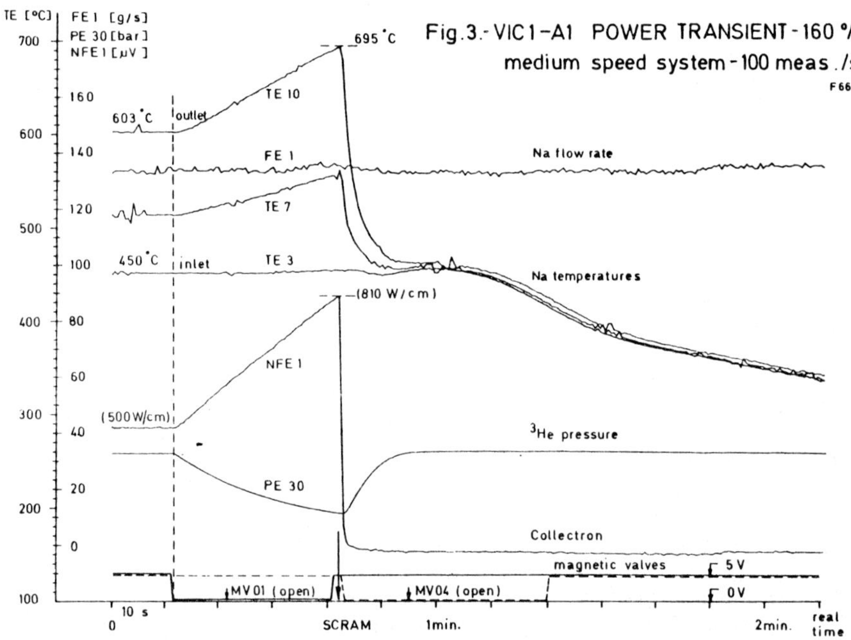

Fig.3.- VIC 1-A1 POWER TRANSIENT - 160 %
medium speed system - 100 meas./s

This illustration is printed
from the best available original

FUTURE VIC LOOP EXPERIMENTS

Transient testing of LMFBR pins in VIC loops in the reactor BR2 is planned to continue at least till the end of 1983 and probably beyond this date for different experimenters. At present two VIC 1 loops are under construction in the frame of the "Transients Programme" for the DEBENE fast breeder reactor SNR 300. Two types of fuel pins for this reactor will be tested : Mk 1 (dia. 6 mm) and Mk 2 (dia. 7,6 mm), covering both fresh rods and active rods, pre-irradiated in the test reactor KNK II (Federal Republic of Germany). Irradiation in BR2 will be done under fast neutron flux inside Cd screen and the test programmes will emphasize fuel power transients as well as sodium cooling mismatch transients, according to the experimental programme defined in ref. 5.
Beside the experiments with single fuel pins (VIC 1 loop type), also the perspective on transient testing of small clusters with up to 7 pins (VIC 7 loop type) is stressed as being feasible in the reactor BR2.

REFERENCES

1. J-M. BAUGNET, Ch. DE RAEDT, F. LEONARD, F. MOONS, G. VANMASSENHOVE, "The BR2 Materials Testing Reactor - Its Capability for Fast, Thermal, and Fusion Reactor Experiments", Fast, Thermal and Fusion Reactor Experiments Conference (Salt Lake City, U.S.A., April 12-15, 1982).
2. J-M. BAUGNET, W. BOEYKENS, F. LEONARD, "The BR2 Materials Testing Reactor - Past and Future", Seminar on Research Reactor Operation and Use (Jülich, Germany - September 14-18, 1981), IAEA-SR-77/36.
3. A. FALLA, "Expérience Acquise dans l'Utilisation de l'Absorbant Neutronique ^{3}He pour l'Exécution de Transitoires de Puissance de Fission", Paper of this conference in Grenoble.
4. Ch. DE RAEDT, N. MAENE, R. MENIL, J. DEBRUE, "Measurements and Calculation Techniques Utilized for the Neutronic Characterization of BR2 Irradiation Devices of the Variable Neutron Screen Type - Application to the "VIC" SODIUM LOOP", Paper of this conference in Grenoble.
5. H. PLITZ, "Experimental Program to Investigate LMFBR-Fuel Pin Behaviour During and Post Operation Transients", in Proc. ANS Topical Meeting - Reactor Safety Aspects of Fuel Behaviour, American Nuclear Society (1981), p. 1-211.

PROGRAMME DE PREQUALIFICATION EN PILE D'UN COMBUSTIBLE A OXYDE MIXTE FABRIQUE PAR UN NOUVEAU PROCEDE (GSP)

G. MARINUCCI, A. NOBILI, M. LANCHI, R. CARACCHINI (ENEA)

G. DUPONT, J. GALTIER (CEA)

1) INTRODUCTION

Le dévelopment et la mise au point d'une nouvelle technique de fabrication de combustible produit suivant la méthode GSP (Gel Supported Précipitation) a fait l'objet d'un accord de coopération entre ENEA, AGN et CEA.

La première phase du programme de qualification prévoit, dans le cadre de cet accord, trois expériences d'irradiation dans le reacteur SILOE pour étudier certains caractéristiques du combustible GSP par comparaison au combustible déjà essayé, du TIPE FBR, que on peut appeller combustible standard.

Les trois expériences prévues (appelées P3, LUGEL et DIGEL) permettront d'analyser le comportment thermique et la déformation d'une aiguille combustible jusqu'à un burn up élevé (60000/80000 $MWJT^{-1}$), et de fournir des indications utiles pour les futures expériences prévues dans le programme de qualification.

Les dispositifs d'irradiation utilisés, d'un type standard, ont été mis au point par le Service des Piles du CEN Grenoble.

La fabrication du combustible et le montage des aiguilles combustibles ont été réalisés à ITU Karlsruhe par des techniciens de Euratom et AGN avec la collaboration de techniciens de ENEA.

P. von der Hardt and H. Röttger (eds.), Irradiation Technology, 369–378.

2) DESCRIPTION DES DISPOSITIFS EXPERIMENTAUX

21) Caractéristiques générales

Les dispositifs utilisés pour l'irradiation des trois expériences comprennent deux parties (figure 1).

1) Partie en pile constitutée d'une partie basse qui contient l'aiguille combustible sous flux et d'une partie haute qui assure le transfert des différentes functions. La partie basse est positionnée dans un fourreau sur une boite à eau mobile qui permet de régler le flux donc la puissance engendrée dans le combustible de l'aiguille et d'évacuer cette puissance. La partie basse est constituée d'un four CYRANO standard qui comporte le dispositif calorimétrique d'étalonnage et de mesure de la puissance engendrée dans l'échantillon.

2) Partie hors pile constituée des parties suivantes:
. des baies électriques qui assurent l'alimentation du four, le contrôle et la mesure des différents paramètres, la sécurité du dispositif et du réacteur,
. une liaison avec le calculateur GENEPI, pour le transfert des données expérimentales,
. une liaison avec la baie qui assure l'enregistrement des données de dosimétrie.

22) Expérience P3

On utilise un dispositif PENELOPE qui permet de mesure directement pendant l'irradiation le profil de la température centrale de l'aiguille combustible au moyen d'un système à thermocouple rétractable. Ce système permet de remplacer le thermocouple en cas de rupture jusqu'à ce que le burn up préfixé soit atteint. La mesure, du type différentiel, est réalisés simultanément sur un lot de pastilles GSP et un autre de pastilles standard.

La figure 2 montre un schéma de la partie basse du dispositif PENELOPE. Cette partie basse est constituée mécaniquement de deux parties:

1) Un four CYRANO standard.

2) Un porte échantillon PENELOPE qui comprend:
. le diffuseur thermique contenant l'aiguille combustible équipée d'un doigt de gant,
. le mécanisme de manouvre du thermocouple,
. le capteur de force,

le capteur de position.

23) Expérience LUGEL

On utilise un dispositif CYRSUM CONTAINTE qui permet de mesurer directement pendant l'irradiation la dilatation axiale d'un lot de pastilles combustibles GSP au moyen d'un système à cavité résonnante.

La figure 3 montre un schéma de la partie basse du dispositif. Cette partie basse est constituée mécaniquement de deux parties:

1) Un four CYRANO standard,

2) Un porte échantillon LUGEL qui comprend:
. le diffuseur thermique contenant l'aiguille combustible,
. le palpeur de mesure,
. le soufflet de mise en contrainte,
. la cavité résonnante,
. la guide d'ondes.

24) Expérience DIGEL

On utilise un dispositif, appellé CYPRO, qui permet de determiner l'interaction mécanique entre la gaine et le combustible sur toute la longuer de l'aiguille remplie pour moitié de pastilles standard et pour moitié de pastilles GSP. Il est ainsi possible d'avoir dans des conditions d'expériences définies, une comparaison du comportement des deux types de combustible.

On remarque la possibilité de faire des mesures de profil de trois façons differentes et indépendantes, en déplaçant l'aiguille en rotation, en translation ou suivant une hélice.

La figure 4 montre un schéma de la partie basse du dispositif DIGEL. Cette partie basse est constituée mécaniquement de deux parties:

1) Un four CYRANO standard,

2) Un porte échantillon DIGEL qui comprend:
. le mécanisme de manouvre,
. le diffuseur thermique contenant l'aiguille combustible,
. le capteur à jauges,

. le système de palpage,
. le calibre de mesure.

3) CARACTERISTIQUES DES ECHANTILLONS

La figure 5 représente un schéma simplifié des aiguilles combustibles utilisées et indique les caractéristiques principales du combustible et de la gaine pour chaque aiguille.

4) CONDITIONS D'IRRADIATION

Comme il a déjà été dit, les trois dispositifs sont placés dans une position d'irradiation à flux variable avec la possibilité d'ajuster la puissance engendrée dans les aiguilles combustibles en déplaçant les dispositifs par rapport au coeur du réacteur SILOE.

Les conditions d'irradiation des trois dispositifs sont définies en tenant compte des valeurs de référence de la filière FBR
. puissance linéaire 500 - 550 $W.cm^{-1}$
. température intérieure gaine 550 - 650°C -

Les résultats expérimentaux attendus pour des trois expériences donneront une première indication soit du comportement global bu combustible GSP par rapport au combustible standard, soit des phénomènes physiques particuliers qui auront lieu pendant l'irradiation des aiguilles.

En particulier l'expérience P3 permettra la mesure de la distribution de température dans l'aiguille jusqu'à un burn up élevé (60000/80000 $MWJT^{-1}$). L'expérience LUGEL permettra une première évaluation des mouvements du combustible dans sa gaine qui ont lieu au bébut de vie (densification et fracturation du combustible) ou à moyen terme (réarrangement). Enfin, l'expérience DIGEL fournira une mesure de l'interaction mécanique entre la gaine et le combustible et de son évolution dans le temps. Les mesures effectuées aux intercycles permettront d'évaluer la déformation de la gaine par rapport au combustible.
Pour definir le programme d'irradiation il sera interessant de faire suivre le meme historique de puissance aux dispositifs LUGEL et DIGEL pour le mesure mecaniques.
Pour cela ils seront places cotè a cotè dans la meme boite mobile. Le dispositif P3, sèra irradiè quelque cycles avant les autres de façon à-mieux connaitre l'etat termique du combustible.

REFERENCES

(1) ESPERIENZA D'IRRADDIAMENTO DI COMBUSTIBILE OTTENUTO DA POLVERI SOL-GEL
A. NOBILI
CNEN RTI/COMB/TECN (79) 95

(2) PROGRAMMA D'IRRAGGIAMENTO DI COMBUSTIBILE OTTENUTO DA POLVERI SOL-GEL
SPECIFICHE DI FABBRICAZIONE DEI CAMPIONI E RELATIVI CONTROLLI
R. CARACCHINI - M. LANCHI - G. MARINUCCI
CNEN RTI/COMB/TECN (80)95

(3) ANABASI - CODICE DI CALCOLO PER LE PREVISIONE DEL COMPORTAMENTO DI UNA BARRETTA COMBUSTIBILE SOTTO IRRAGGIAMENTO
A. CALZA BINI et AL.
CNEN RTI/LTC (74) 1

(4) DOSSIER 2 ESPERIENZE D'IRRAGGIAMENTO P3
R. CARACCHINI - G. MARINUCCI
IN PUBBLICAZIONE

(5) DOSSIER 2 ESPERIENZE D'IRRAGGIAMENTO LUGEL
R. CARACCHINI - G. MARINUCCI
IN PUBBLICAZIONE

(6) DOSSIER 2 ESPERIENZE D'IRRAGGIAMENTO DIGEL
R. CARACCHINI - G. MARINUCCI
IN PUBBLICAZIONE

(7) "CAPSULE CYRANO" - METHODE DE MESURE CALORIMETRIQUE DE LA PUISSANCE EXTRAITE D'UN CRAYON COMBUSTIBLE
CENG-Pi/SEDTI 170-135/72

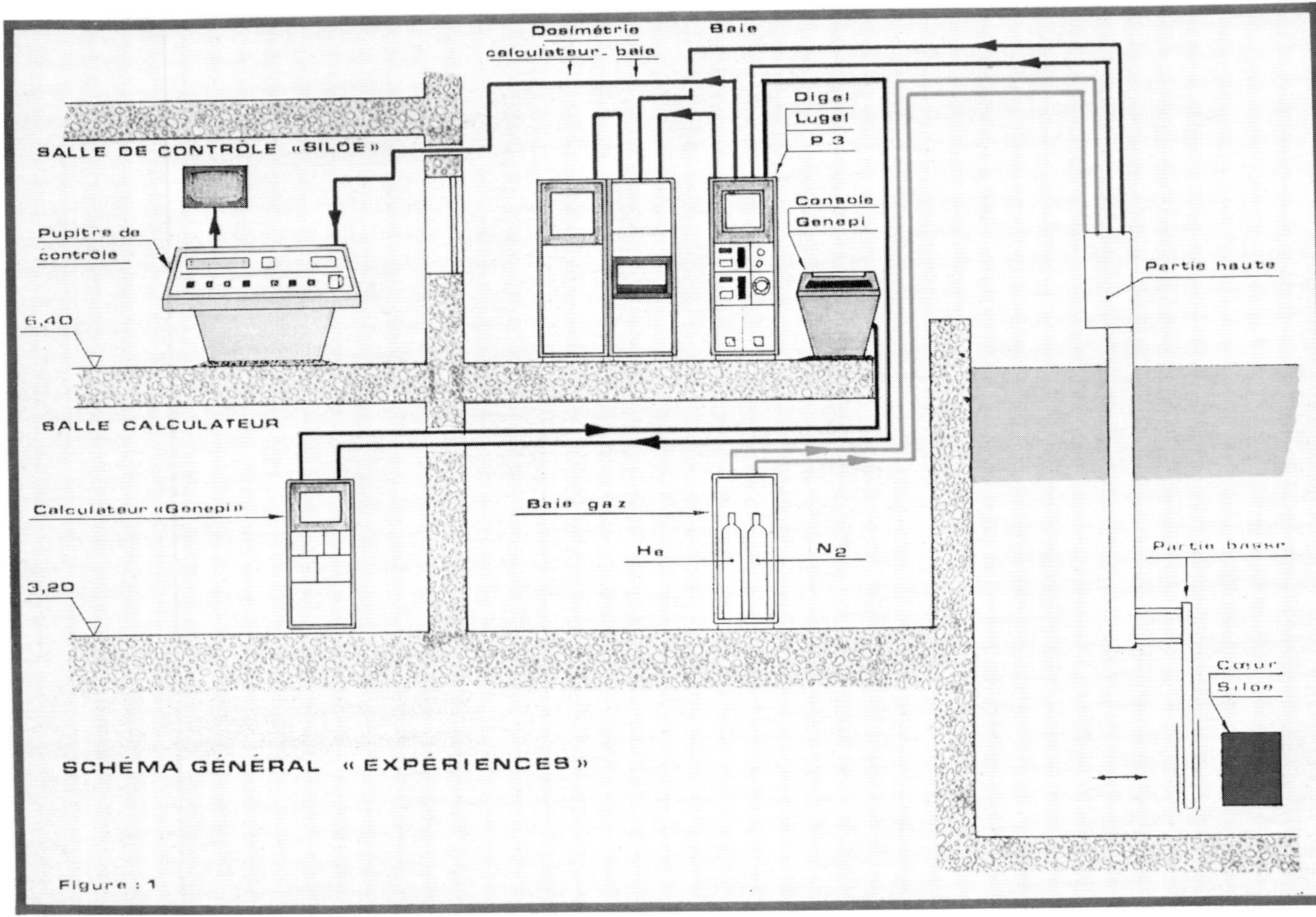

SCHÉMA GÉNÉRAL «EXPÉRIENCES»

Figure : 1

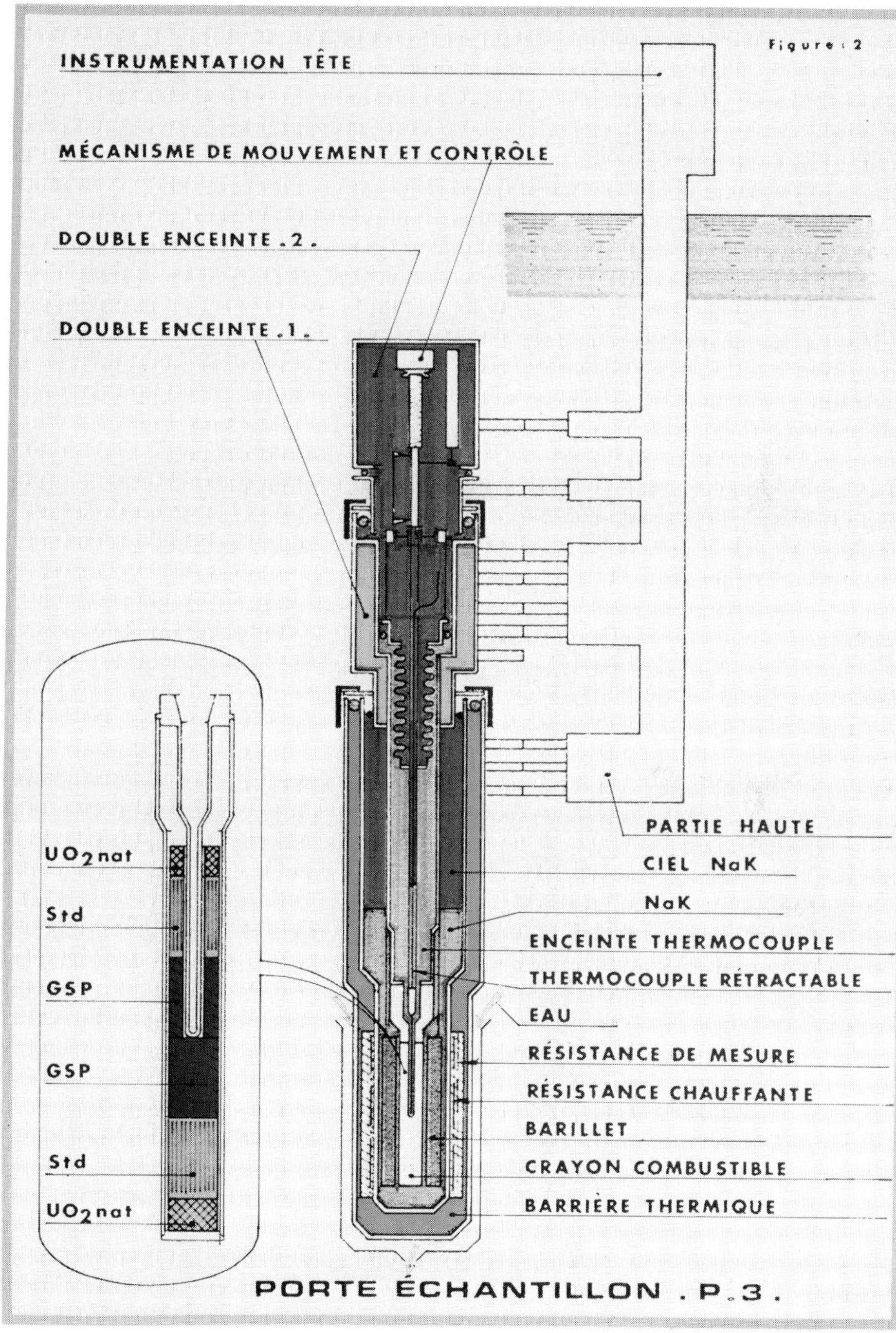

PORTE ÉCHANTILLON . P.3 .

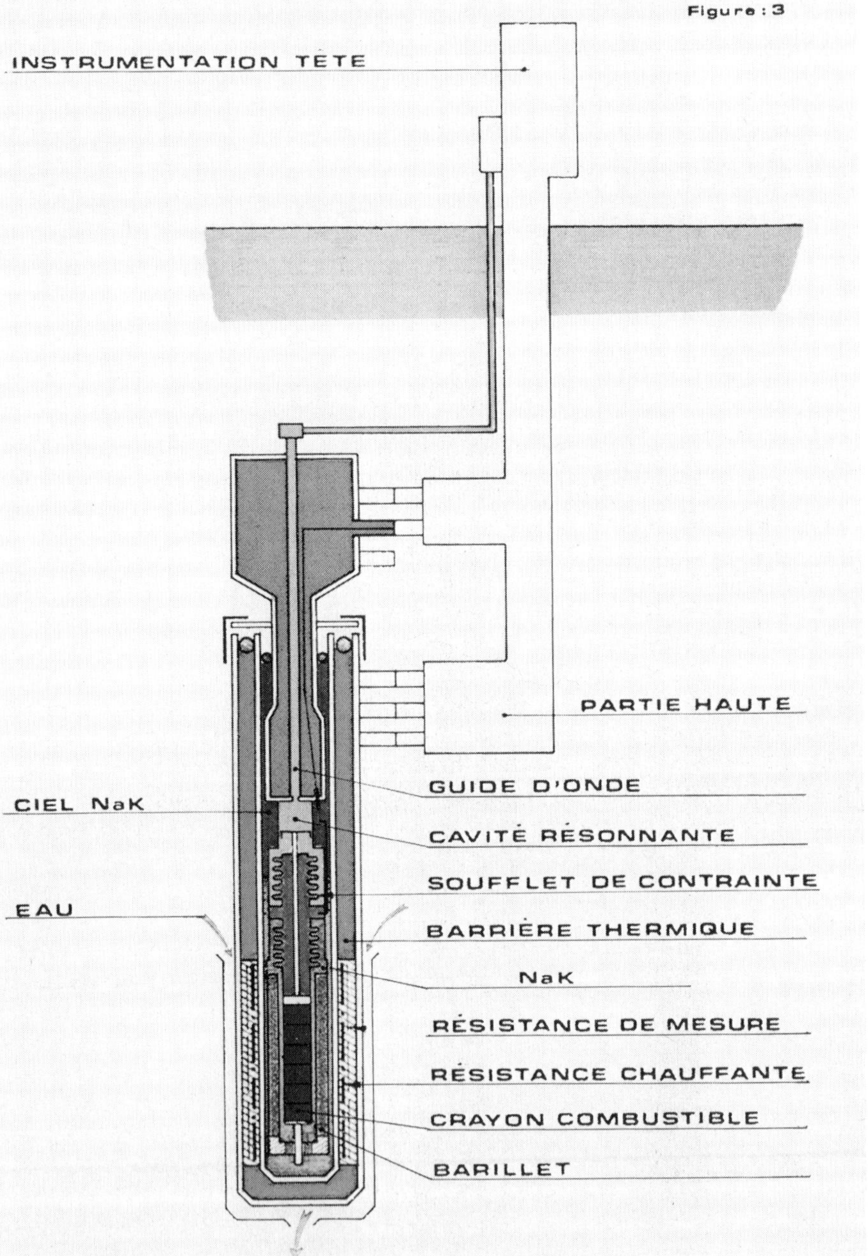

PORTE ÉCHANTILLON CYRSUM CONTRAINTE

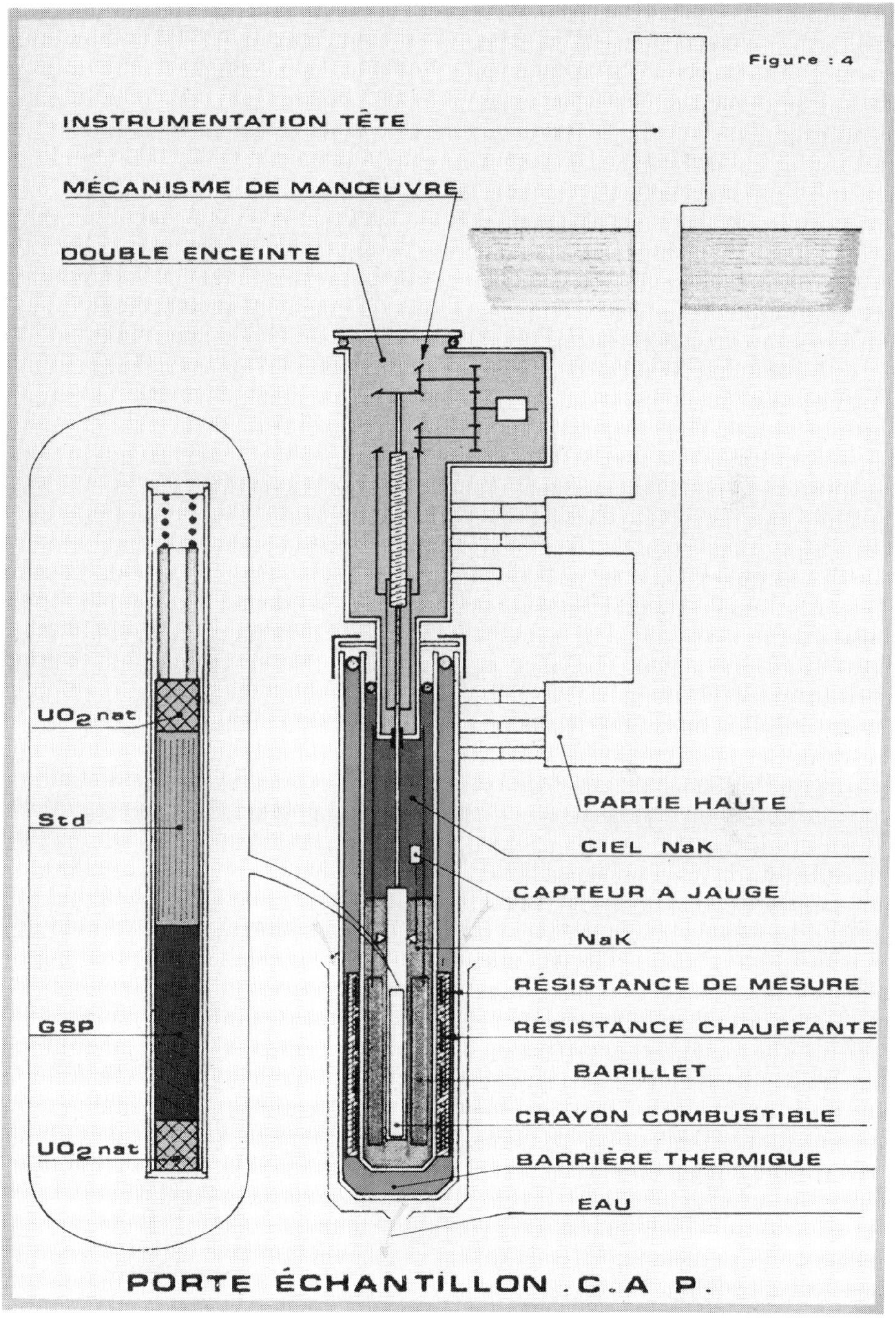
Figure : 4
INSTRUMENTATION TÊTE
MÉCANISME DE MANŒUVRE
DOUBLE ENCEINTE
UO2 nat
Std
GSP
UO2 nat
PARTIE HAUTE
CIEL NaK
CAPTEUR A JAUGE
NaK
RÉSISTANCE DE MESURE
RÉSISTANCE CHAUFFANTE
BARILLET
CRAYON COMBUSTIBLE
BARRIÈRE THERMIQUE
EAU
PORTE ÉCHANTILLON . C . A P.

Figure : 5

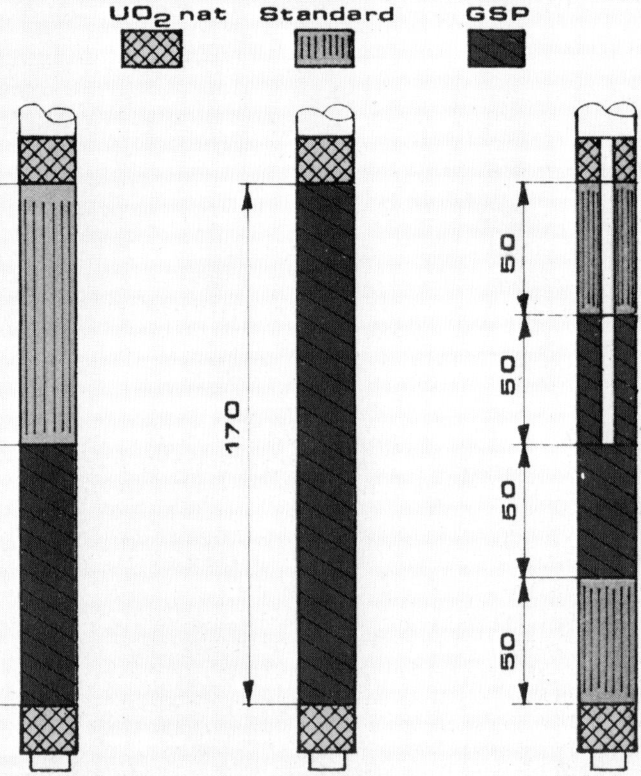

	DIGEL	LUGEL	P.3
Gaine			
matière	AISI.316M	AISI.316M	AISI.316M
Ø extèrieur	6,7	6,7	6,7
Ø intèrieur	5,8	5,8	5,8
Combustible			
pastille	pleine	pleine	percée
Ø pastille	5,7	5,7	5,6
Ø trou			2
densité GSP. %	94,8	94,8	94,8
densité Std. %	91		91
enrich^t en Pu.%	20	20	20
rapport O/M	1,97	1,97	1,97

LE PROGRAMME D'IRRADIATION GOCAR

M. Campana

Commission des Communautés Européennes
Centre Commun de Recherche
Etablissement de Karlsruhe
Institut Européen des Transuraniens
Postfach 2266
D-7500 Karlsruhe
République fédérale d'Allemagne

1. INTRODUCTION

L'Institut Européen des Transuraniens, dans le cadre de ses programmes pluriannuels, a entrepris d'étudier les combustibles pour réacteurs rapides et en particulier leurs limites d'utilisation. Ces études se sont concentrées principalement sur les combustibles dits avancés: carbures, carbonitrures et nitrures mixtes uranium-plutonium. Le but de ces recherches était d'arriver à une description à la fois qualitative et quantitative du comportement en réacteur d'une aiguille fissile d'un type de combustible connu, sous des conditions d'irradiation données. Une des plus impératives limites d'utilisation est causée par l'effort mécanique exerçé sur la gaine par le gonflement du combustible, ce qui peut influencer l'intégrité et donc la sécurité d'un assemblage fissile.

Le gonflement du combustible sous irradiation constitue l'un des problèmes les plus ardus auxquels aussi bien le physicien que le technologue aient à faire face. Le processus de gonflement qui dépend d'un grand nombre de paramètres - composition, structure, conditions d'irradiations - peut se caractériser en première approximation par une modification de la géométrie du combustible. Le but de la série d'expériences GOCAR - GOnflement CARbures - est d'étudier de manière approfondie les divers facteurs qui occasionnent ce gonflement.

Cet exposé sera consacré principalement à l'aspect technologique du programme GOCAR, conformément à l'objectif de cette conférence. Il était dans la tradition de l'Institut d'entreprendre une

P. von der Hardt and H. Röttger (eds.), Irradiation Technology, 379–389.

telle étude, vu le grand nombre d'irradiations sous conditions sévères réalisées auparavant aussi bien en flux rapide que thermique. D'autre part le combustible étant préparé dans nos laboratoires, il était possible de le spécifier avec précision et d'en connaître sa structure et ses propriétés avant irradiation de façon détaillée. De même les laboratoires chauds de l'Institut permettent les examens approfondis du combustible après irradiation, cette tâche constituant la part la plus importante du programme GOCAR.

2. GONFLEMENT. TEMPERATURE CRITIQUE /1/

Si l'on veut examiner le phénomène du gonflement d'une manière approfondie, on est amené à en décomposer l'étude en plusieurs étapes partant d'une observation macroscopique, au niveau de l'aiguille, pour aboutir à une observation microscopique située au niveau du grain de combustible. Nous avons été conduits à définir 3 types de gonflement:

- le gonflement géométrique, ou technologique, qui s'applique à la pastille combustible,
- le gonflement local, qui s'applique à un volume représentatif de la structure initiale du combustible, quelques grains et les joints associés,
- le gonflement microscopique qui s'applique à un volume inférieur au grain.

On voit que chacun de ces gonflements est contenu dans le précédent, c'est donc le gonflement microscopique qui doit être considéré comme le phénomène fondamental. Une de ses causes est la précipitation des gaz contenus lors de la fabrication en bulles dont le diamètre varie de 4 à 400 nm.

Des examens de combustibles irradiés, ainsi que des essais de répliques par microscopie électronique, effectués à l'Institut, ont mis en évidence qu'il existe une température au delà de laquelle le gonflement d'abord de nature microscopique change de type et se transforme en un gonflement au niveau des pores.

Le problème de base se formule ainsi: dans quelles conditions les bulles microscopiques se transforment-elles en pores de plus grandes dimensions, autrement dit quels sont les facteurs qui gouvernent la croissance de ces bulles et à quelle température, dite critique, apparaît ce phénomène. Au premier abord interviennen la composition du combustible, le taux de fission, le taux de combustion, la température du combustible, etc...

On voit, à l'énonçé de ces paramètres, se dégager les spécifications d'irradiation qui devaient permettre de déterminer la température critique dans divers types de combustible.

	Programme GOCAR . Spécifications
Combustible	Composition chimique: $(U,Pu)C$ $(U,Pu)N$ $(U,Pu)C_{0,8}N_{0,2}$ $(U,Pu)C_{0,85}O_{0,15}$ Composition isotopique: $\frac{Pu}{U+Pu} = 0,15$ $\frac{Pu_{239}}{Pu} = 0,7 \quad \frac{Pu_{240}}{Pu} = 0,3$ Densité combustible: 95 % d.th. Dimensions pastilles: diamètre extérieur: 6 mm diamètre intérieur: 1,7 mm hauteur : 8 mm Longueur colonne combustible :40 mm Jeu diamétral combustible-gaine :30-40μm Pastille isolement UC, UN 1 pastille à chaque extrémité
Gaine	Matériau : Nb-1% Zr Diamètre intérieur: 6 mm Diamètre extérieur: 7-13 mm Gaz de remplissage: hélium
Conditions d'irradiation	Puissance linéaire maximum : 650 w/cm Puissance spécifique 200 w/g Température surface combustible: 700-900-1100-1300°C Δt centre-surface : 180-200°C température centrale : 900-1100-1300-1500°C Taux de combustion : 3-4 a/o FIMA

3. SPECIFICATIONS GENERALES /2/

Le domaine de température s'étendra de 700 à 1500°C, découpé en 4 essais qui seront conduits à des températures fixes de surface du combustible de 700, 900, 1100 et 1300°C; chacun de ces paliers sera stabilisé à ± 15°C tout au long de l'irradiation. La différence de température entre surface et centre sera comprise entre 150 et 200°C . Ainsi nous obtiendrons aux fins d'examens, et pour chaque composition choisie, un combustible irradié à une température connue et variant de façon continue de 700 à 1500°C.

A cause de l'instrumentation importante, les expériences GOCAR n'étaient pas réalisables en réacteur rapide. Néanmoins afin de ne pas trop s'éloigner des conditions en flux rapide, nous nous sommes imposés une dépression de flux dans le combustible relativement faible.

Pour cela, le diamètre des pastilles a été limité à 6 mm, l'enrichissement en Pu à 15 %, le plutonium ne contenant que 70% de l'isotope fissile 239. Tenant compte de la bonne conductibilité thermique des combustibles avancés, une puissance de 650 w/cm est nécessaire afin d'obtenir une différence de température d'environ 200°C entre centre et surface. Quatre compositions ont été choisies: MC, $MC_{0,8}N_{0,2}$, MN, $MC_{0,85}O_{0,15}$. Les spécifications prévoyaient la mesure de la température de surface du combustible, cela était difficile, vu les dimensions de la pastille. Après avoir envisagé différentes possibilités de placement du thermocouple, nous avons décidé de le loger au centre de la pastille. Bien sûr, ce n'est plus la température de surface qui est mesurée, mais avec une bonne connaissance de l'intégrale de conductibilité, nous avons pensé pouvoir effectuer une corrélation précise entre ces 2 températures.

4. DISPOSITIFS D'IRRADIATION

La réalisation des expériences en pile a été confiée au CEN Grenoble qui nous a proposé 2 types de dispositifs:

- une version basse température, paliers 700-900°C
- une version haute température, paliers 1100-1300°C

car il a semblé difficile qu'un même dispositif puisse réguler la température de surface soit à 700, soit à 1300°C.

La version basse température, de type classique, est constituée par une capsule à double paroi. Le tube intérieur, contenant les 2 crayons, est rempli de NaK; entre les 2 parois un mélange gazeux He/N_2 permet la stabilisation de la température.
Le crayon ne présente aucune particularité, sinon le matériau de gainage. En effet, il était nécessaire que la gaine exerce une pression de l'ordre de 100 bar sur le combustible, et ce à des températures $> 1100°C$. Nous avons choisi un alliage Nb-1%Zr comme matériau de gainage pour d'une part ses propriétés mécaniques à haute température, et d'autre part ses facilités d'usinage et de soudage.

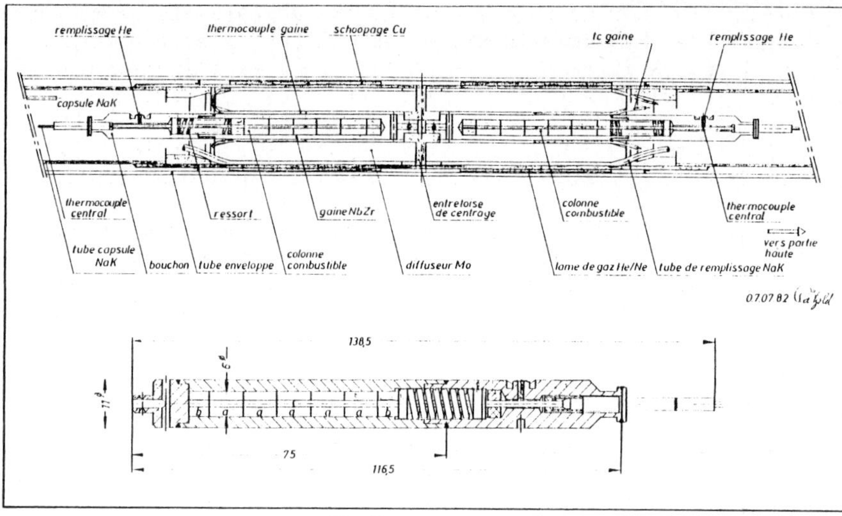

This illustration is printed
from the best available original

Fig.1. : GOCAR, version basse température

La version haute température est un dispositif original dit BTV, barrière thermique variable, constitué par un cône mâle mobile que l'on déplace verticalement par rapport à un cône femelle fixe; l'espace variable entre les 2 cônes est rempli d'un mélange gazeux He/N_2. L'assemblage fissile est formé de 2 1/2 cônes articulés sur un joint de cardan. L'intérieur des gaines est revêtu d'une couche de W - épaisseur de $\sim$20 µm - déposée par voie gazeuse et destinée à éviter une réaction combustible-niobium.

La mesure de la puissance est obtenue grâce aux signaux de 4 collectrons qui ont été étalonnés à l'aide d'une maquette dosimétrie. Chaque expérience a permis l'étude de 2 compositions. Les versions BT ont atteint pour la plupart le taux de combustion prévu, de l'ordre de 3 à 4 a/o, par contre les versions HT ont dû être abrégées.

This illustration is printed from the best available original

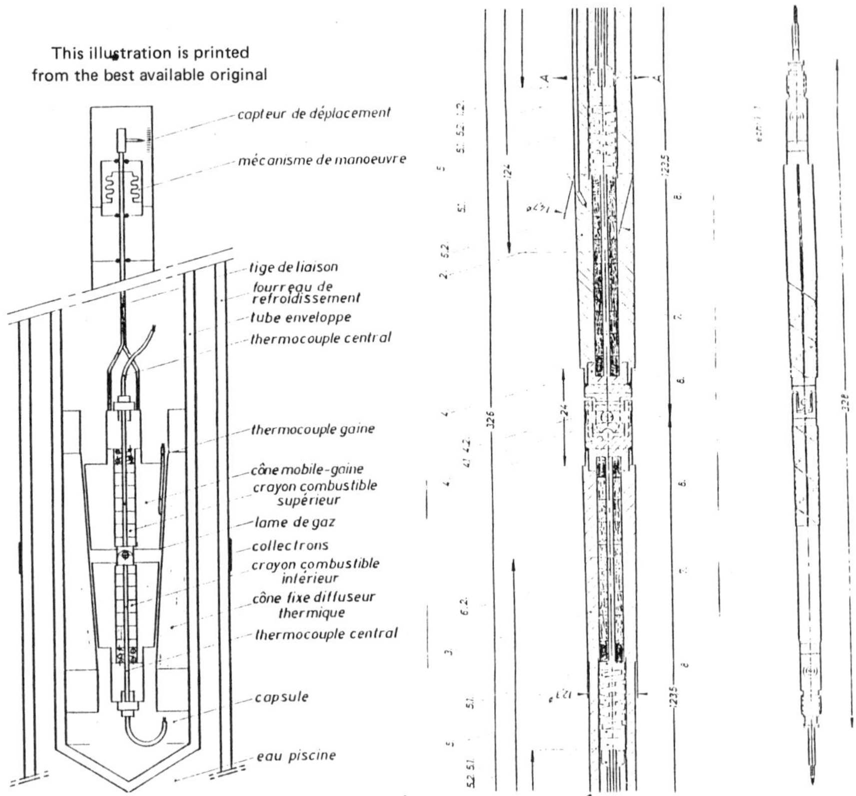

Fig.2.: GOCAR, version haute température

5. DEROULEMENT DES IRRADIATIONS

Depuis 1976, une douzaine d'expériences ont été effectuées dans SILOE.

L'historique de GOCAR I, crayon MC - fig.3 - permet quelques remarques générales qui s'appliquent à l'ensemble du programme. On voit que la puissance linéaire, la température centrale restent stables suivant les spécifications, ce qui entraîne une augmentation de la température de gaine. Le comportement d'un crayon GOCAR est ainsi à l'opposé de celui d'une aiguille irradiée normalement, c'est-à-dire, où la température centrale, élevée en début de vie décroit rapidement, la température de gaine restant sensiblement constante. L'instrumentation du dispositif nous a permis d'étudier le transfert thermique combustible-gaine, que l'on caractérise en général par la conductance thermique h ($W/cm^2.°C$). Les valeurs élevées de h, obtenues au cours des irradiations GOCAR, montrent, par rapport aux oxydes, un gonflement plus important dû à un relâchement moindre des gaz de fission et un manque de plasticité, ce qui contribue à un bon contact entre le combustible et la gaine.

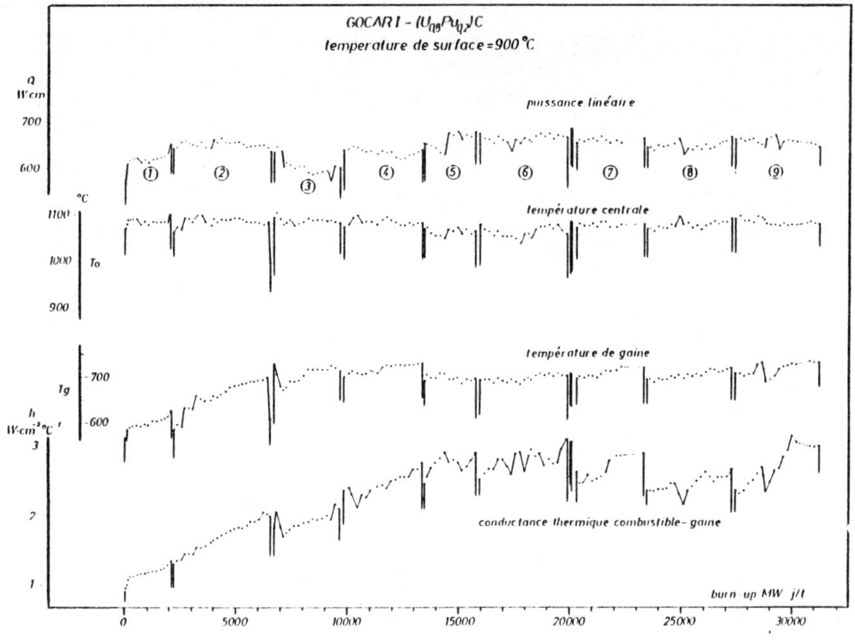

Fig.3.: GOCAR I, historique crayon MC

La fig.4 montre la distribution du taux de combustion à travers une pastille MCN. L'accord entre valeurs calculées et mesurées est satisfaisant.

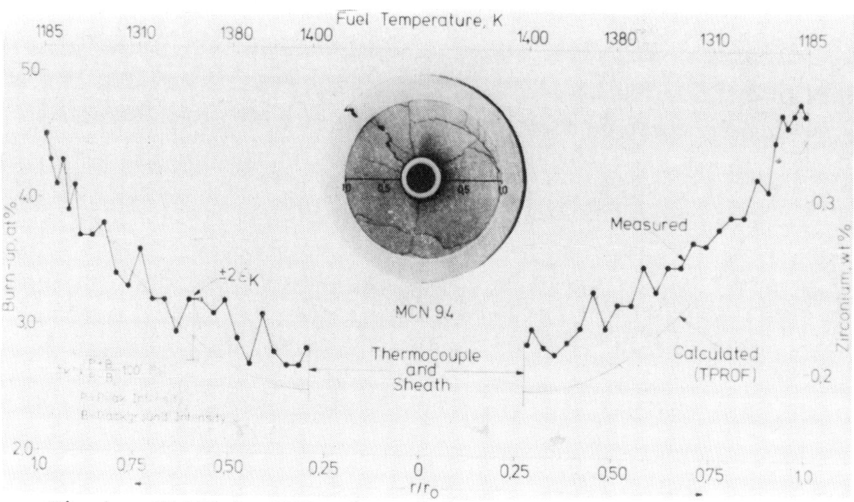

Fig.4.: GOCAR I, MCN, répartition du taux de combustion /3/

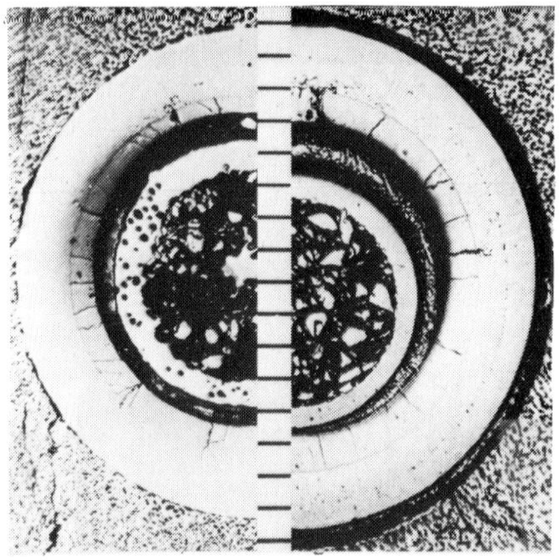

Sur les coupes radiales des crayons MC et MCN au niveau du trou central -fig.5- on observe des gonflements nettement différents. Le jeu de montage a été rempli par le carbure alors qu'il subsiste en partie dans le carbonitrure.

Fig.5.: GOCAR I, MC, MCN, section trou central /4/

Sur la figure 6 qui représente une coupe en profondeur d'une pastille MC, on voit les bulles se former, grossir, précipiter dans les pores, entraînant ainsi un changement de type de gonflement.

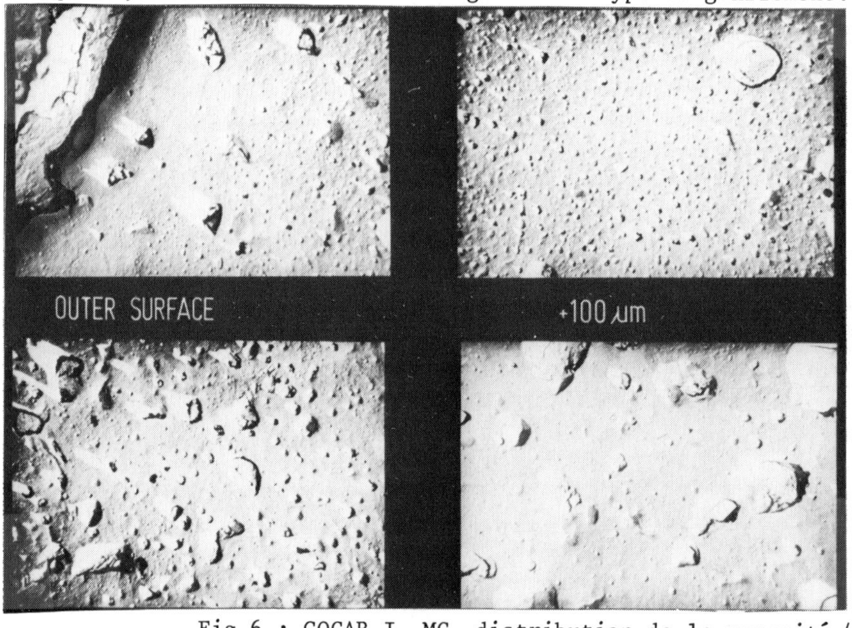

Fig.6.: GOCAR I, MC, distribution de la porosité /5/

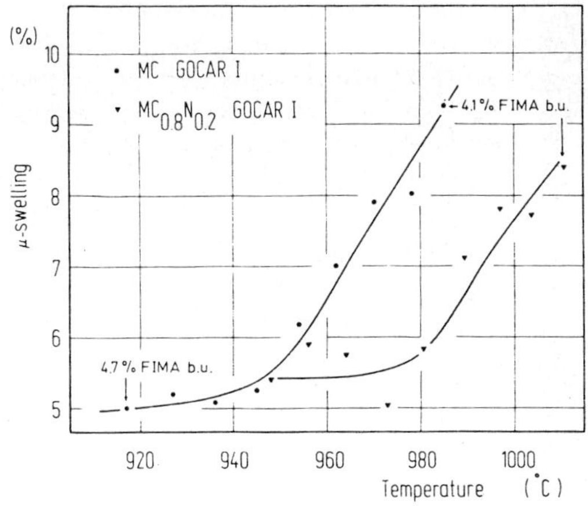

Ces observations, traduites en courbes -fig.7- donnent l'évolution du gonflement en fonction de la température, son changement est net à des températures de 950°C pour MC et de 980°C pour MCN, valeurs relativement basses.

Fig.7.: GOCAR I, g = f(T) /6/

On remarque sur l'historique de GOCAR IV -fig.8- une augmentation rapide de la conductance h, de MN, jusqu'à une valeur de 6 W/cm².°C, ce qui implique un excellent transfert thermique combustible-gaine.

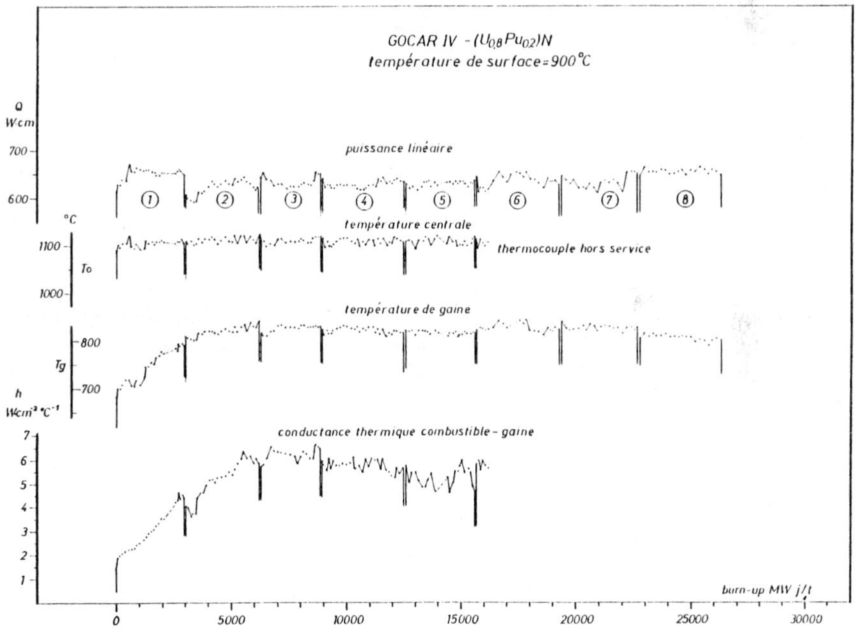

Fig.8.: GOCAR IV, Historique crayon MN

Sur la coupe du crayon MN -fig.9- on observe que les jeux de montage ont été remplis par le combustible. La zone de transition gaine-combustible -fig.10- montre une interpénétration des 2 matériaux, d'où un contact parfait entre niobium et nitrure.

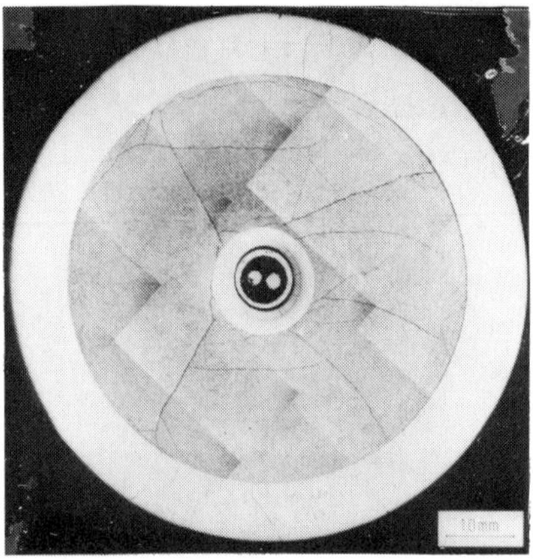

Fig.9.: GOCAR IV, coupe MN /7/

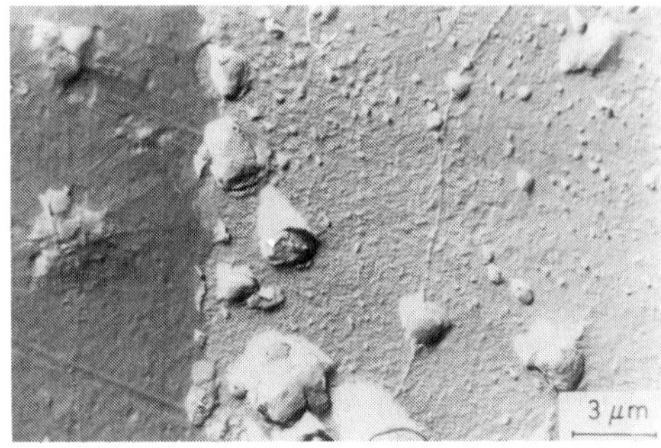

Fig.10.: MN, zone combustible-gaine /8/

Sur l'historique de GOCAR IIbis -fig.11- version haute température avec dispositif à barrière thermique variable - on voit que cette expérience a été stoppée à 1 a/o, les 2 thermocouples à coeur étant hors d'usage.

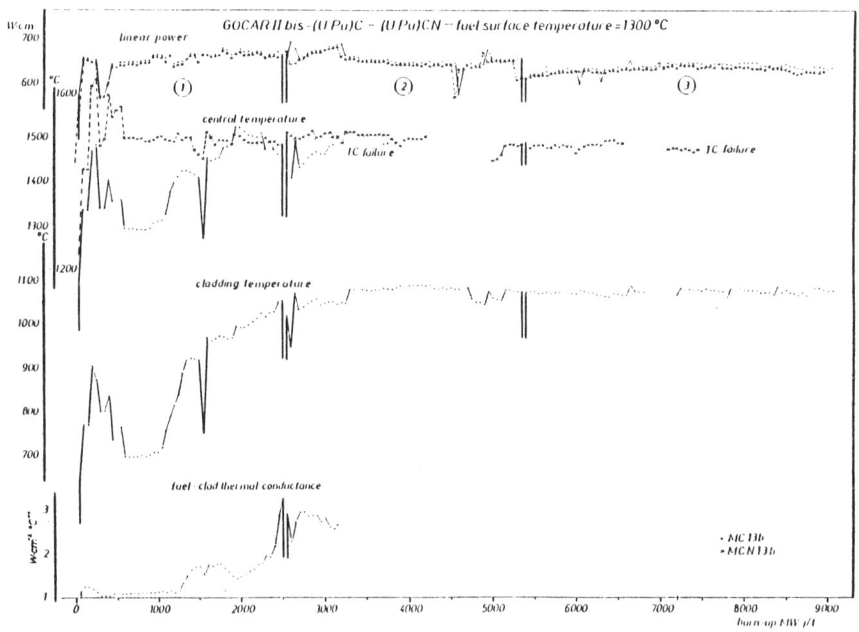

Fig.11.: GOCAR IIbis, historique MC, MCN

On peut se faire une idée du gonflement par l'augmentation de la température de gaine, de l'ordre de 400°C au cours du 1er cycle, ce qui montre aussi la flexibilité du dispositif BTV.

Les macrographies des crayons MC13 et MCN13 - fig.12 - expliquent ce que laissait supposer l'historique : rattrapage du jeu gaine-combustible, trou central disparu, gonflement géométrique > 10%, restructuration complète, déformation de la couche de W destinée à éviter une interaction avec la gaine.

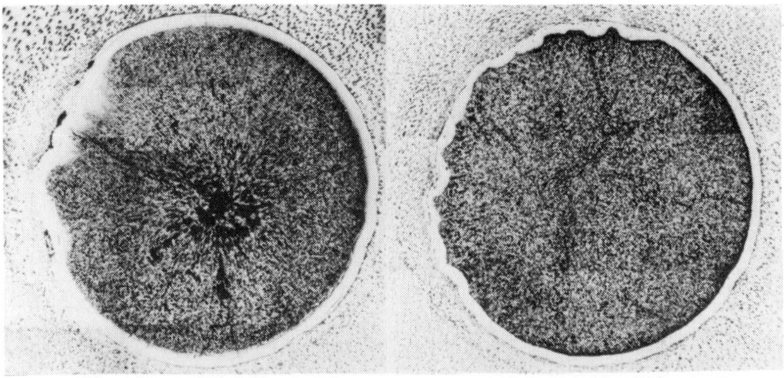

Fig.12.: GOCAR IIbis, macrographies MC, MCN /9/

Seul un bref aperçu des examens exécutés sur les combustibles GOCAR a pu être donné. Il aura néanmoins montré l'ampleur de la recherche approfondie, les moyens importants mis en oeuvre, la place prépondérante du programme GOCAR dans l'étude des combustibles avancés.

Ce programme est en voie d'achèvement, les 2 dernières irradiations sont en cours, nul doute qu'elles se poursuivront de façon satisfaisante comme les précédentes, grâce à l'excellente coopération maintenue entre le CEN Grenoble et l'I.E.T.

Références

/1/ H. Blank, some background information to the irradiation experiments GOCAR, internal note, Sept. 73
/2/ M. Campana, Exp.GOCAR, note préliminaire, NT 46/ESSIR, Nov.74
/3/ F. Caligara, C.T. Walker, Joint Research Centre, European Institute for Transuranium Elements, Karlsruhe Establishment, Programme Progress Report, January-June 79, p.13
/4/ M. Coquerelle, Ibid, July-December 78, p.15
/5//8/ I. Ray, communication personnelle
/6/ C. Ronchi, Ibid, July-December 78, p.16
/7/ M. Coquerelle, communication personnelle
/9/ M. Coquerelle, Ibid, July-December 80, p.15

EBR-II TRANSIENT OPERATION AND TEST CAPABILITIES

B. R. Seidel, D. C. Cutforth, and G. L. Lentz

Argonne National Laboratory, P.O. Box 2528,
Idaho Falls, Idaho 83401

and

J. D. B. Lambert

Argonne National Laboratory, 9700 South Cass Avenue,
Argonne, Illinois 60439

I. INTRODUCTION

Experimental fuel pins intended for eventual use in LMFBR's have been irradiated for many years in fast test reactors around the world. A wealth of data has been obtained on their performance under steady-state conditions, and fuel-pin performance codes have been developed to predict their behavior.[1,2] In parallel, safety tests of fuel pins to explore behavior under accident conditions have been performed in transient reactors like TREAT in the US, and CABRI in France. These two types of testing generally have had different aims and have tended to produce results which are not reconcilable with a common modeling code, such as a LIFE[3] or COMETHE,[4] in the middle ground between normal and off-normal conditions.

But as the licensing and commercialization of LMFBR's approaches, the attention and needs of the fuel-pin designer and licenser have focused on this middle ground between "steady-state" and "accident" testing of fuel pins and subassemblies. In the US this middle ground has come to be known as Operational Reliability Testing (ORT). It relates to the off-normal conditions that may be anticipated to occur sometime during the operating life of a commercial reactor, and for which data are now sorely needed. At EBR-II such testing involves, but is not limited to, pins that have already failed (run-beyond-cladding-breach testing);

P. von der Hardt and H. Röttger (eds.), Irradiation Technology, 391–403.

pins that experience mild-overpower or duty-cycle transients during irradiation (operational-transient testing); and the more general areas of the thermal-hydraulic testing of naturally distorted clusters of pins under both normal and natural-convection modes of cooling (shutdown-heat-removal testing). Combinations of these conditions may lead to "local faults" in subassemblies, which are a lingering concern in licensing.

Preparations and now capability for operational reliability testing at EBR-II were the subject of papers at recent conferences.[5–13] This paper updates the status of those preparations to the present time when the ORT program is about to begin.

II. OPERATIONAL RELIABILITY TEST STRATEGY

The ORT issues can be addressed in the context of scale - dealing appropriately with the performance of pins, the assembly, or the whole plant. The concerns related to pins have been associated with steady-state (SS) performance, operational-transient (OT) reliability, and run-beyond-cladding-breach (RBCB) safety. The concerns related to assemblies deal with distorted bundles and heat removal. The shutdown-heat-removal testing (SHRT) and local-faults testing (LFT) address response of the assemblies under normal and off-normal events. SHRT must also address upset conditions which affect the whole plant, that is, the dynamics between the heat generating and removal systems. Components outside the primary system, particularly pumps and steam generators, have also been recognized as potential sources of upset conditions; therefore, component testing (CT) of near-full-scale working models is necessary to assure successful operation of a commercial plant. In addition, computer-controlled operations will provide sensor validation and prompting of prioritized procedures for operator response to off-normal events. This interface between the operator and the system, termed man-machine interface (MMI), must also be developed and tested.

Each of these activities focuses on the objectives of a commercial LMFBR. Testing in each of the above areas requires a facility that can demonstrate acceptable performance, reliability, and safety of its driver fuel under real normal and off-normal operating conditions. EBR-II has been prepared to encompass all of the above missions with (a) new special facilities for testing; (b) increased instrumentation; (c) upgraded control systems; (d) tailored core environments; and (e) qualification of the fuel, reactor, and plant systems for the expanded requirements.

It has been a challenge to define a strategy for reactor operations that provides conditions for all the programs. The current strategy provides for dedicated periods of steady-state operation for the SS, RBCB, CT, and MMI tests and OT preconditioning; dedicated periods of transient testing for the RBCB, OT,

SHRT, CT, and MMI tests; and also dedicated periods of off-normal plant tests for SHRT. The core configuration will be altered to accommodate each mode of testing. The nominal steady-state reactor power will be 57.25 MWt with the nuclear center of the core offset two rows from the geometric center to accommodate load-following tests of fueled assemblies and preconditioning of OT assemblies. Mild overpower transients (15%) can be conducted with only minor changes to the steady-state core. For 60%-overpower-transient tests, the core will be reduced and centered to peak the power for the test, which will be near the core center. Most transient test periods will be of short duration, and steady-state experiments would be removed during them. Such is the strategy for SHRT testing also.

Although low-ramp-rate transient tests (about 1.6% power increase per second from 24 to 62.5 MWt) were conducted in February 1981, and high-ramp-rate transient tests at rates ten times greater will be conducted in November 1982, the new operating strategy to fulfill all missions will not begin until early 1983. During these first two transient test periods, the core size was reduced to increase specific power in the center of the core. During both test periods, however, experimental oxide fuel and blanket pins were tested in addition to several test subassemblies designed to qualify the metallic driver fuel. With the EBR-II driver-fuel supply extending into the next decade, a small SS mission is expected to continue, allowing EBR-II to continue supplying electricity to the grid. The RBCB testing will continue to 1985; the OT testing period will extend from 1983 to 1986; the SHRT will continue through 1986; the CT testing of large models will begin in 1986; and the MMI mission will continue as long as needed.

Two major areas of activity in preparing EBR-II, in addition to plant modifications, have been (a) development of the core configurations required to obtain the necessary fission rates in the special open-loop test facilities, and (b) the overall effort to qualify the plant, including its driver fuel. These are now described.

Core Configuration

Core configurations must meet the diverse conditions required for the various testing porgrams. Nominal reactor power must be maintained at 57.25 MWt to accommodate SS experiments and to produce electricity. However, power must not exceed 62.5 MWt under other conditions to stay within present safety limits. The primary flow must also be limited to the historical nominal 100% value. To maintain continuity in RBCB tests, the reaction rates beneath the breached-fuel test facility (BFTF), described below, must not exceed their pre-ORT values. Pretransient conditions for the OT tests must match previous steady-state power and temperature conditions. And finally, load-following experiments must also not have to be reconstituted to meet test conditions.

Three modes of operation have therefore been defined: (a) steady state with an offset core, (b) a core for 15% overpower transients, and (c) special core loadings to achieve 60% overpower transients and SHRT tests. The offset core is necessary to provide a 30% change in reaction rates for duty-cycle tests simulating loss of one loop in a large plant. For these tests, pairs of experiments will be interchanged between high- and low-power positions in successive runs. In EBR-II a test assembly can be positioned anywhere in grid rows one to five without modification. The nuclear center of the core, therefore, has been shifted from the center of the grid to a point two rows over, as shown in Fig. 1.

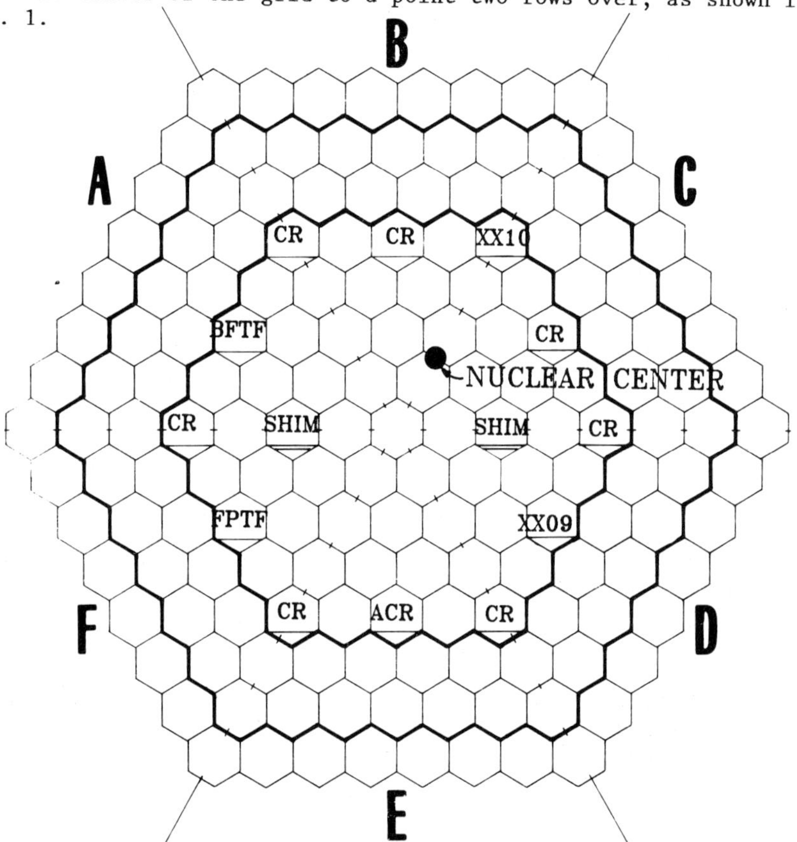

Fig. 1. EBR-II Core Positions for Control Rods (CR), Shim Rods (Shim), Automatic Control Rod (ACR), Special Facilities (BFTF, FPTF, XX09, and XX10) and the Offset Nuclear Center.

The 15% overpower transients will be accommodated by relocating the experiment from a low-reaction-rate position to a high-reaction-rate position and by beginning the transient at a reduced power level. The experiment will be orificed internally, and flow will be controlled by maintaining the prescribed coolant pressure drop. Since side-by-side irradiations are involved, with some pins subjected to transients while others are not, the pretransient temperature and power conditions must be the same as during steady-state operation. These OT tests will be repeated several times throughout the lifetime of the pins (without reconstitution) to evaluate the effect of repeated protected transients.

For extended transients (60% overpower), the core size will first be reduced and then reactor power will be increased to accomplish each transient. The SHRT tests will be conducted in relatively short test periods at about six-month intervals. The timing for these tests is being arranged to minimize fuel handling.

Plant Preparation

Qualification of the driver fuel for transient operation began in 1979. Reference 5 describes the early phases of that effort, which included out-of-reactor transient testing of irradiated cladding and fuel to determine transient mechanical properties of the metallic driver fuel and the kinetics of phase transformation.[12] In-reactor testing of driver fuel included: localized power-change tests during steady-state operation;[13] load-following tests by interchange of subassemblies between high- and low-power positions at every reactor cycle; and dedicated transient operation of standard and elevated-temperature subassemblies.[5]

The driver fuel was subjected to repeated transients at a power increase of ∿1.6% initial power per second from 24 to 62.5 MWt. After a hold at peak power for 720 seconds, the power was decreased at the same rate which was constrained by the maximum standard control-rod drive speed. The transients were repeated three times a day every other day until 56 transients were obtained. Of the four test driver-fuel subassemblies, two were in row 2, and two were in row 5. In each row, one subassembly had a peak cladding temperature of 600°C, and the other a peak cladding temperaure of 630°C. These temperatures corresponded to the nominal and allowed peak temperatures for driver fuel under transient conditions. The driver-fuel pins in the test subassembly under the BFTF were operated at a peak cladding temperature of 660°C, which corresponds to the value allowed with uncertainties included. The peak cladding temperature of the driver-fuel elements in a test control rod was 600°C. The fueled control rod was inserted into the core before each transient and withdrawn after each transient to obtain even greater power increases than could be obtained on fuel within the core undergoing a power transient only.

The transient testing was very successful; no cladding breach was generated, and those pins which were intentionally operated RBCB through 56 transients exhibited no degradation. Only one of these three breached pins continued to release fission gas. Examinations showed that the driver fuel is fully capable of sustained steady-state and transient operation at low ramp rates.

To perform tests at high-ramp-rates, the control system has been modified and made automatic. The mechanical portion of the modification consisted of replacing the existing constant-speed control-rod drive motor and reduction gear with a pulse-width modulated DC servo-drive motor and new reduction gearing. The control portion of the modification has two modes of operation. In the "manual" mode, the speed of the new control-rod drive is limited by design to the same speed as that of the other control rods and the reactor operator may use it interchangeably with any other control rod. In the "automatic" mode, the speed of the control-rod-drive is controlled by a computer. Control is by computer comparison of a preprogrammed transient power profile and an actual power feedback from one of the nuclear detectors.

There are several noteworthy features of the new system. First it was designed so that any faults that could arise from operation of the drive system in either manual or automatic modes would fall within the capabilities and safety envelope of the existing reactor shutdown system. Second, the control software is structured so that the transient profiles can be entered and verified by the operating crews under the constraints of a test procedure. To verify controller stability, the system has been subjected to extensive ex-core testing with a control-rod-drive mockup, supplemented by in-core testing with the reactor shut down to further refine the ex-core test results. Automatic systems are also planned for controlling flow in the secondary sodium loop and for controlling feedwater supply.

The normal EBR-II protection system will prevent any significant degradation for all anticipated and unlikely events associated with transient testing. Also, analyses of extremely unlikely events show that, for example, a combination of simultaneous failure in the automatic-reactor-control, reactivity-control, and reactor shutdown systems would not promote conditions outside bounds established for EBR-II in its original design.

The reliability of the major EBR-II components was also assessed to qualify the plant for transient operation.[14] The maximum stresses and strains in the components were found to satisfy the limits allowed by the appropriate design code, and the structural damage due to creep was always found to be small. The most critically stressed component will be the intermediate heat exchanger (IHX), as expected, and the worst event will be reactor scram from high power. But based on conservative estimates, the cumulative damage will be about one-half lifetime for a projected 1000 duty cycles on the IHX.

In summary, the operating strategy will accommodate all testing missions, diverse as they are; and plant qualifications are near completion. A demonstration of repeated high-ramp-rate transient tests in November 1982 will climax preparations. Transient testing of oxide test pins will immediately follow this final period of driver-fuel qualification.

III. TEST CAPABILITIES, CURRENT STATUS, AND RESULTS

System capabilities developed for EBR-II's past steady-state mission will be either applied directly or enhanced for the operational reliability missions. These were summarized at an earlier conference,[8] and only recent highlights will be reported here.

Reactor Systems

A wide variety of tests can be conducted in the reactor with regular open-core irradiation subassemblies. These subassemblies are very flexible in design, and by appropriate choice of power and flow, an experiment can be conducted for minimal cost.

An instrumented fuel subassembly that was irradiated to breach and beyond has been recently potted to preserve its pin-bundle geometry and sectioned to correlate it with the observed operational temperatures. The temperature distribution observed during operation was consistent only with suspected distortion of the bundle. Two new instrumented subassemblies, fueled (XX09) and nonfueled (XX10), have been built and will be installed shortly to monitor temperature during transient operation and to characterize the dynamics of upset conditions, particularly loss-of-flow events -- the main thrust of the SHRT program.

Two in-core test faclities, the BFTF and the fuel-performance-test facility (FPTF), have been installed to accommodate special fuel irradiations and have been already used. The facilities penetrate the reactor cover in control-rod-drive positions and mate with an experimental subassembly in the grid postiion directly below. Since each facility is an open loop, easy fuel handling and rapid exchange of experiments is facilitated.

Both facilities contain instruments to measure coolant flow and temperature and delayed-neutron (DN) activity. The BFTF is designed to monitor breached pins and to sample sodium from a subassembly for fission products and fuel particles. The FPTF has no fission-product-deposition sampler, but contains a valve that controls flow in the test subassembly during reactor operation to simulate flow and temperature transients and provide very-high-temperature capability. The total coolant flow from the test subassembly enters the FPTF, whereas there is a small leakage in the BFTF. Thus, delayed neutrons from breached pins beneath the BFTF are also detected by the primary-system detector. Since the outlet from the FPTF is outside the reactor vessel, but into the primary tank, DN's are not measured by the primary-system detector.

DN Release from Breached Pins. Measuring the volatile DN-emitting fission products that are released from a breached oxide pin may, in principle, be used as an indicator of how severe the cladding breach really is in-reactor. A technique has been developed at Argonne in recent years to obtain, on-line, more information about a breached pin than just the magnitude of the DN signal from it.

DN precursors released to the primary sodium of EBR-II, whether from the BFTF or an open-core test, are finally measured by the fuel-element rupture detector (FERD), which is located on a bypass loop of primary sodium obtained from the outlet of the IHX. The normal transit time for DN's to reach FERD from an open core test is 19 seconds, 6 seconds of which is spent in the FERD loop (29 seconds and 6 seconds, respectively, for a BFTF test). The time in the FERD loop can be increased by lowering voltage to the FERD pump and decreasing its flow. In this way the total transit time may be increased at will to "age" the DN signals, i.e., to alter the relative activities of the various half-life families of DN emitters which are present in the sodium. General experience in this area was described at a recent conference.[15]

This technique, which was developed using a recoil source of fission products (FPS) and which is fully described elsewhere,[16] is being used to unfold information about the condition of a breached pin; specifically its "holdup time" and its "equivalent recoil surface area" for release of DN's. This technique involves inserting the results of FERD-loop flow-reduction (FFR) tests into a computer program called GIRAFFE, which treats the release, transport, and decay of the parent and daughter emitters of DN's in the primary sodium. The code then assumes that release of DN precursors from exposed mixed-oxide fuel is proportional to the amount remaining in the fuel, with appropriate correction for decay during the process.

Output from the code includes estimates of "the equivalent recoil area" of the source; its "hold-up" or aging time for DN release; and the "enhancement factor" for DN release. These terms are explained in Ref. 15. This enhancement factor corresponds with the "k" factor as used by Europeans, specifically German investigators.[17]

Values of the enhancement factor found for breached oxide pins in EBR-II to date have typically been in the range of 100-300, and equivalent recoil area in the range of 50-100 cm^2. Although mechanisms to explain such high enhancement factors are sought on both sides of the Atlantic, one conclusion reached so far is that breached oxide pins are readily detected and monitored by DN-monitoring systems in LMFBR's. The exact causes of such high DN signals, and the corresondence between the character or magnitude of a signal and the breach condition, are the subject of continuing research. So also is the correlation between a DN signal from a breached pin and the contamination it may produce.

Contamination Release from Breached Pins Under BFTF. By far the largest contaminants of the EBR-II primary circuit from breached oxide pins have been found to be the fission gases themselves and the volatile fission products iodine and cesium. The radioactive gases collect eventually in the cover-gas space, from which they must be purged, and iodine and cesium remain in solution in the primary sodium. The major iodine isotope, ^{131}I, has only an 8-day half-life and has not been greatly troublesome. The longer-lived ^{134}Cs and ^{137}Cs isotopes, however, had built up continuously in the primary sodium and had begun to affect normal operations, including fuel handling.

Both major sources of circuit contamination at EBR-II -- fission gases and cesium resulting from RBCB testing -- have been very effectively brought under control by a cryogenic cover-gas cleanup system, which was installed in 1977, and a cesium trap, which was placed in the primary sodium in 1978. The latter device drastically reduced the cesium activity which had accumulated to that time from earlier endurance tests of pins and some preparatory RBCB tests.

Surveillance of primary sodium during more recent RBCB tests has not shown large release of less-volatile fission products from breached pins. Trace quantities of ^{95}Zr-^{95}Nb and the cesium isotopes have only occassionally been observed. Contamination of the sodium by uranium and plutonium has seldom been observed above the respective detection limits of 10^{-9} g/g of sodium and 4×10^{-12} g/g of sodium.

Gamma scanning of the first sample train from the BFTF did, however, show that some small quantities of these contaminants were released into the hot leg of EBR-II, even if they could not be detected at the cold-leg sampling point (in a great dilution). These results were summarized in a recent paper[18] and are given here:

A gamma scan of the BFTF containing the sampler was performed shortly after its removal from the reactor. The scan showed that

^{140}Ba-^{140}La was the dominant contaminant, with large deposits above (downstream from) the sampler.

After decay of the ^{140}Ba-^{140}La activity, ^{95}Zr-^{95}Nb was the principal gamma activity remaining in the sampler. Other radionuclides found in small quantity included ^{141}Ce, ^{144}Ce, ^{154}Eu, ^{106}Ru, and ^{54}Mn. Some ^{137}Cs was found at two isolated locations, but the flowing sodium swept most of it through the sampler. Much of the radioactivity was loosely adherent and smearable. Over 80% of the gamma-emitting contamination in the sampler was deposited on the lower 50 cm of the filter section of the sampler, and alpha- and gamma-activity measurements showed that this activity was associated with a minute release of fuel particulates. Plutonium concentration on the surfaces downstream from the filter was small compared to deposition on surfaces in the outer annulus in which the sodium was not filtered. This suggests that the filter was effective in retaining the bulk of the plutonium contamination incident on it.

The release and transport behavior of these contaminants could be interpreted in terms of two basic mechanisms: particulate entrainment and solution of volatile, short-lived precursors. Examination, of the sampler indicates that for uranium, plutonium, ^{95}Zr-Nb, ^{106}Ru, ^{103}Ru, ^{144}Ce, and ^{154}Eu, particulate entrainment was the dominant mechanism. The distributions of ^{141}Ce and ^{140}Ba-La deposits indicated that both mechanisms might be significant in the transport of those species.

In summary, therefore, the BFTF sampler was found to provide not only a sensitive measure of the magnitude of contamination release from a breached fuel pin, but also detailed information concerning its nature and transport behavior. A series of such tests is now under way in the RBCB program; further results will be reported at a later date.

Driver-fuel Testing Under the FPTF

The first test performed under the FPTF was irradiation of the fission-product source for calibration of its DN detector. Flow reductions in the facility were generated with the programmable valve controller to characterize signal strength with flow. Since both DN concentration in the coolant and transit time are both proportional to flow rate, the activity at the detector was found to increase and then decrease with flow rate. Successive irradiations were then performed to calibrate the remaining FPTF sensors. In particular, tests with both fueled and nonfueled orificed subassemblies were conducted to certify the flow and temperature indications. With the use of thermal-expansion-difference (TED) passive temperature monitors in the fueled test subassemlbies, the thermal-hydraulic codes relating peak cladding temperatures (PCT) in the tests to the temperatures measured at the inlet of the FPTF were calibrated to account for intersubassembly heat transer between the core region and the top of the subassembly. It has been found that the PCT can be controlled more precisely by monitoring the FPTF thermocouple indications instead of the FPTF flowmeters, which are temperature sensitive and temperature compensated.

The first driver-fuel test underwent a loss-of-flow transient with delayed reactor scram. The PCT at the top of the fuel column during the event was not greater than the nominal steady-state temperature, but the temperature at the midplane was elevated. Although the neighboring subassemblies were cooled by the action of the auxiliary pump (about 5% flow), there was no forced cooling of the test subassembly beneath the FPTF. From analysis of the TED results, a natural convective flow of 1.5% was deduced to occur.

Repeated overtemperature transients simulating high-ramp-rate overpower transients (∿16% power increase per second) are being conducted on driver-fuel test pins under the FPTF during September and October 1982 in preparation for the whole-core power transients scheduled to begin in November. Reliability of the driver fuel after 80 overtemperature transients is expected to assure satisfactory performance during the 15 whole-core power transients scheduled for November.

Following the transient testing, metallic driver fuel will be irradiated under the FPTF at elevated temperatures to generate eutectic alloy with the stainless steel cladding. If the in-reactor eutectic penetration rates are observed to be as low as demonstrated by hot-cell testing of irradiated pins, then either the safety margin for protection during extended overpower transients can be expanded or the overpower capability can be extended and still assure driver-fuel reliability.

<u>Plant System</u>

The monitoring and diagnostic systems supporting the ORT program include DN detection, gamma spectroscopy of the argon cover gas for gaseous fission products, and chemical analysis of the primary sodium. Online mass spectormetry also provides for continuous analysis of tag gas. Radioactivity is controlled by a cryogenic cover-gas cleanup system and cold traps to remove activity from the primary coolant.

Additional instrumentation has been added to the secondary and steam systems to enable better control during transient operation. Plans include automatic controllers for those systems. Their development is one focus of the effort of the MMI activity to provide sensor validation and control.

IV. CONCLUSIONS

EBR-II is now almost completely qualified to provide transient capability. This capability ranges from steady-state to high-ramp-rate transient conditions. The reactor and plant systems have been enhanced with new facilities, and with new control and monitoring systems. Capability has been demonstrated for low-ramp-rate transients and soon will be for transients at high ramp rates. Significant reassurance of the inherent safety and reliability of EBR-II reactor systems has already been obtained. The wide range of missions implemented at EBR-II demonstrates not only the flexibility and adequacy of the EBR-II system itself, but also the inherent benign performance and reliability of large breeder reactors.

REFERENCES

1. Proceedings of the ANS Conference on Fast Reactor Fuel Element Technology, New Orleans, LA, April 1971.

2. Proceedings of the ANS Conference on Fast Breeder Reactor Fuel Performance, Monterey, CA, March 1979.

3. J. S. Roth, et al, "Modeling of Fast Reactor Fuel Transient Behavior Using the LIFE Code," ibid, 642.

4. J. VonVliet, et al, "LMFBR Fuel Pin Modeling with the Computer Codes COMETHE III-J and IAMBUS," ibid, 655.

5. J. A. Buzzell, et al, "Transient Performance of EBR-II Driver Fuel," Proceedings of the ANS/ENS Topical Meeting on Reactor Safety Aspects of Fuel Behavior, 1-415, August 2-6, 1981, Sun Valley, Idaho.

6. J. I. Sackett and G. H. Golden, "Safety and Operating Experience at EBR-II: Lessons for the Future," 16th Intersociety Energy Conversion Engineering Conference, August 9-14, 1981, Atlanta, Georgia.

7. J. D. B. Lambert, et al, "Operational Safety and Reliability Research at EBR-II," Paper 81-JPGC-NE-11 of the Joint ASME/IEEE Power Generation Conference, October 4-8, 1981, St. Louis, Missouri.

8. J. I. Sackett and D. C. Cutforth, "Experimental Capability at EBR-II for Off-normal Testing of Fast Reactor Fuels and Systems," ANS Topical Meeting on Fast, Thermal, and Fusion Reactor Experiments, April 12-15, 1982, Salt Lake City, Utah.

9. J. A. Pardini, et al, "The Experimental Breeder Reactor-II (EBR-II) Fuel Performance Test Facility (FPTF)," ibid.

10. R. H. Olp, et al, "The Experimental Breeder Reactor-II (EBR-II) Breached Fuel Test Facility (BFTF)," ibid.

11. J. Poloncsik, et al, "The Experimental Breeder Reactor-II (EBR-II) Instrumented Subassemblies, INSAT XX09 and INSAT XX10," ibid.

12. B. R. Seidel, et al, "Phase Transformation of EBR-II Metallic Driver Fuel During Transient Operation," Trans. ANS 38, 282 (1981).

13. J. L. Welker and B. R. Seidel, "Power Change Testing of Metallic Driver Fuel," Trans. ANS 38, 281 (1981).

14. L. K. Chang and M. J. Lee, "Thermal-Structural Response of EBR-II Major Components Under Reactor Operational Transients," to be presented at the 7th International Conference on Structural Mechanics in Reactor Technology, August 22-26, 1983, Chicago, Illinois.

15. J. D. B. Lambert et al, "US Delayed-neutron Monitoring Experience," IAEA Conference on Fuel Failure Detection Monitoring System in LMFBR's, Karlsruhe, May 1981.

16. K. C. Gross and R. V. Strain, Delayed Neutron Signal Considerations in a Fast Reactor," Nucl. Sci. & Eng. 76, 163 (1980).

17. S. Jacobi and G. Schmitz, "Release of Fission Products from Artifically Defected LMFBR Oxide Fuel Pins," Proceedings of the ANS Conference on Fast Breeder Reactor Fuel Performance, March 1979, Monterey, California, 607.

18. R. P. Colburn and H. P. Maffei, "Contamination Transport in the Primary Coolant System During a Breached Fuel Pin Test in EBR-II," Trans. ANS 35, 225 (1980).

EXPERIENCES PAHR A MELUSINE

D. ROUSSEAU, P. DEREYMEZ, H. GUYON, E. JUNOD,
M. PLOUJOUX, F. TOURNEBIZE
CEA-CEN/Grenoble - Service des Piles

H.BACKS
JRC,Ispra Establishment

INTRODUCTION

Parmi toutes les études relatives à la Sureté des Réacteurs à Neutrons Rapides, un Programme Européen d'Analyse du Refroidissement des Lits de Débris a été défini et est, à l'heure actuelle, conduit par le CCR/EURATOM (ISPRA).

Ce programme dénommé PAHR (Post Accident Heat Removal) est un programme expérimental en pile qui nécessite donc la définition et la mise au point d'un dispositif complet d'irradiation répondant à des spécifications précises ainsi que la conduite d'un programme de soutien hors pile pour résoudre des problèmes phénoménologiques et technologiques.

Après une étude de faisabilité détaillée comprenant des calculs neutroniques, thermiques et thermodynamiques effectuée durant le premier semestre 1980, le CCR/EURATOM a confié la réalisation de la moitié de ce programme au Service des Piles de Grenoble.

OBJECTIFS

Après un accident hypothétique du coeur d'un réacteur à neutrons rapides, le combustible et les matériaux de structure constituant ce coeur peuvent fondre partiellement. Un certain nombre d'essais hors pile (1) montrent alors qu'un phénomène d'interaction thermodynamique entre ces matériaux fondus et le sodium caloporteur peut se produire.

Un des principaux résultats de cette interaction est la for-

P. von der Hardt and H. Röttger (eds.), Irradiation Technology, 405–413.

mation de débris de combustible et d'acier inoxydable, de forme et de diamètre variés.

Selon les experts, ces débris sédimentent alors sous le champ de gravité, sur les différentes structures internes de la cuve primaire du réacteur, notamment le récupérateur, installé au fond de la cuve, dont la face inférieure est refroidie par convection naturelle du sodium.

Ainsi, quelques secondes après l'accident, un lit de débris noyé dans du sodium s'est formé sur un plateau d'acier. Ce lit de débris est le siège du dégagement de la puissance résiduelle due à la décroissance radioactive des produits de fission et des actinides encore présents dans le combustible après l'accident.

Le programme PAHR a alors pour principal objectif le comportement thermomécanique du lit en prenant en compte les paramètres les plus importants pouvant régir les phénomènes. Pour être plus précis, les questions auxquelles on cherchera à donner une réponse peuvent être résumées comme suit :

1) Quel est le seuil de puissance auquel apparaît l'assèchement du lit ? Quel est l'influence sur ce seuil de la granulométrie du combustible et de l'acier, de la quantité relative d'acier présent dans le lit, de la température du plateau d'acier et de la mer de sodium, du degré de stratification du lit résultant des conditions de sédimentation des débris ?

2) Pendant l'étude de l'extension de la zone asséchée, on s'attachera à définir précisément à quel moment on atteint dans le lit une température correspondant au point de fusion de l'acier (1400°C). L'acier étant fondu, dans quelles conditions migre-t-il éventuellement dans les porosités du lit et quelles sont les conséquences de cette migration sur le partage du flux de chaleur vers le haut et vers le bas et sur l'évolution de l'état du lit ?

3) A quel niveau de puissance atteint-on éventuellement le point de fusion du combustible ? Quel est alors le comportement de ce combustible et quel est la nature du comportement thermomécanique du lit dans ce stade final de l'évolution ?

De façon générale, sachant que les connaissances sur l'assèchement sont déjà très élaborées (2), notre but sera de porter une attention particulière sur le comportement du lit au-delà du point de fusion de l'acier jusqu'à la fusion partielle du combustible.

PROGRAMME EXPERIMENTAL

Le programme expérimental qui sera conduit dans le Réacteur Mélusine comprend 3 essais choisis parmi la matrice d'origine de 6 essais, et 1 essai de qualification appelé Eo.

Ce premier essai a pour objectif principal et prioritaire de qualifier le dispositif sur les plans neutronique et thermique, le matériel hors pile et son fonctionnement, les procédures d'irradiation et de démantèlement qui sont originales et qui doivent tenir compte de la complexité des expériences.

La caractéristique principale de cet essai réside dans l'utilisation d'un lit constitué de billes de Carbure de Tungstène, matériau qui offre les avantages suivants :

. densité suffisamment élevée pour obtenir un auto-échauffement gamma important en pile même si la dépression de puissance dans le lit est alors importante.
. conductivité suffisamment faible pour pouvoir atteindre 1400°C après assèchement au centre d'un lit haut (250mm) et peu poreux ($\varepsilon \lesssim 30\%$)

Le dispositif sera alors identique pour tous les essais de façon à ne pas introduire de perturbations inconnues tout au long du programme.

Durant le deuxième essai, on cherchera à atteindre une température de 1600°C, supérieure donc de 200°C au point de fusion de l'acier. On caractérisera alors au cours de cet essai :

. le seuil d'apparition de l'assèchement
. la progression de la zone asséchée
. la distribution des températures dans le lit et l'apparition de phénomènes tels que pénétration d'un régime de convection naturelle dans le lit et création éventuelle de chenaux de vapeur de sodium.
. le comportement de l'acier juste au dessus de son point de fusion et les conséquences de son mouvement éventuel sur la distribution des températures et des flux de chaleur.

Le troisième essai sera conduit jusqu'à 1800°C. Après avoir confirmé les résultats de l'essai précédent, on caractérisera les phénomènes suivants :

. le comportement d'une importante quantité d'acier fondu. Dans le cas où il migre, suit-il un champ de gravité ou des gradients de température ? Comment se répartissent les flux de chaleur après resolidification de l'acier ?
. le comportement des particules dans les zones que l'acier a quit-

tées. Le lit s'effondre-t-il alors sur lui-même localement, ce qui conduit à l'apparition de cavités vides empêchant tout échange par conduction ?

Enfin, le quatrième essai sera conduit jusqu'à la fusion éventuelle du combustible. Dans ce cas, nous chercherons à mettre en évidence :

. les mouvements du combustible.
. la resolidification du combustible et son nouvel emplacement.
. l'influence de l'acier sur les phénomènes précédemment cités.

Les paramètres importants ont été retenus suivant les deux critères prioritaires suivants :

. représentativité pour transposition des résultats au cas réacteur.
. bonne maitrise pour pouvoir assurer la constance de certains de ces paramètres tout au long du programme.

Les paramètres qui sont déterminés à l'heure actuelle et qui servent de base aux différents calculs menés pour le projet sont :

. la composition du lit

Combustible

- nature = UO2 (naturel ou légèrement enrichi)
- quantité = 1000 kg/m2, soit environ 20 kg pour un lit de 160 mm de diamètre
- granulométrie = étendue, c'est-à-dire que le diamètre des particules va de 40µm à 4000µm avec une certaine loi de distribution
- porosité = 43 % (estimée) avec une distribution de la granulométrie en fonction de la hauteur homogène.

Acier

- nature = 316 L "lavés" préalablement avec du sodium
- quantité = 300 kg/m2, soit environ 6 kg pour un lit de 160mm de diamètre
- granulométrie = 2 fois la granulométrie de l'UO2 limitée à 3000µm

Sodium

- quantité = 7 kg environ
- pureté = 2 ppm d'O2

. la géométrie du lit

- diamètre = 150 mm $\leqslant \phi \leqslant$ 160 mm. Le diamètre exact sera fixé en fonction de l'épaisseur de la paroi du creuset.
- hauteur = 233mm pour une porosité de 43% ou 190 mm pour une porosité de 35%.

La porosité sera fixée par le chargement du lit lui-même.

. la puissance
La puissance maximum que l'on atteindra est de 3,2 W/g UO_2

REALISATION DU PROGRAMME

Pour conduire à bon terme un tel programme, il est nécessaire de définir
. la capsule (CELIA) et son instrumentation
. les conditions de fonctionnement du réacteur et le facteur de couplage
. les matériels hors pile et leurs conditions de fonctionnement.

La capsule CELIA

On trouvera en figure 1 la vue schématique de la capsule CELIA (Capsule d'Etudes de LIts Asséchés).

Les principaux composants de ce dispositif sont :
. le creuset fabriqué avec un matériau compatible avec le sodium liquide et vapeur, l'acier inoxydable liquide et vapeur et l'UO_2 liquide (voir conclusion). Le diamètre extérieur de ce creuset est de 172 mm.
. une barrière thermique étanche, constituée de 12 viroles concentriques de molybdène ou de tungstène destinée à couper les pertes radiales par rayonnement et conduction.
. une enceinte de confinement, barrière physiologique qui vient s'appuyer sur le tube de pression en cas d'interaction.
. le tube de pression, dimensionné pour contenir la source de pression prenant naissance dans le lit en cas d'interaction combustible-sodium.
. le tube de sécurité qui a les mêmes caractéristiques mécaniques que le tube de pression.
.le circuit sodium supérieur qui refroidit la mer de sodium et permet de réguler sa température
. les deux branches d'un circuit d'eau pressurisée, une branche refroidissant directement le plateau par l'extérieur du tube de pression, l'autre branche refroidissant le circuit sodium supérieur.

L'instrumentation qu'il est prévu d'installer pour la gestion du contrôle commande et l'acquisition des données est constituée des capteurs suivants :
. 130 thermocouples dont une grande majorité sont dans les structures latérales. Ce poste ne comprend pas environ 20 thermocouples placés directement dans le lit
. 3 débitmètres électromagnétiques sur le circuit sodium
. 4 débitmètres sur la boucle à eau
. 28 capteurs de pression surveillant 12 enceintes gazeuses différentes

CELIA DEVICE : LOWER PART

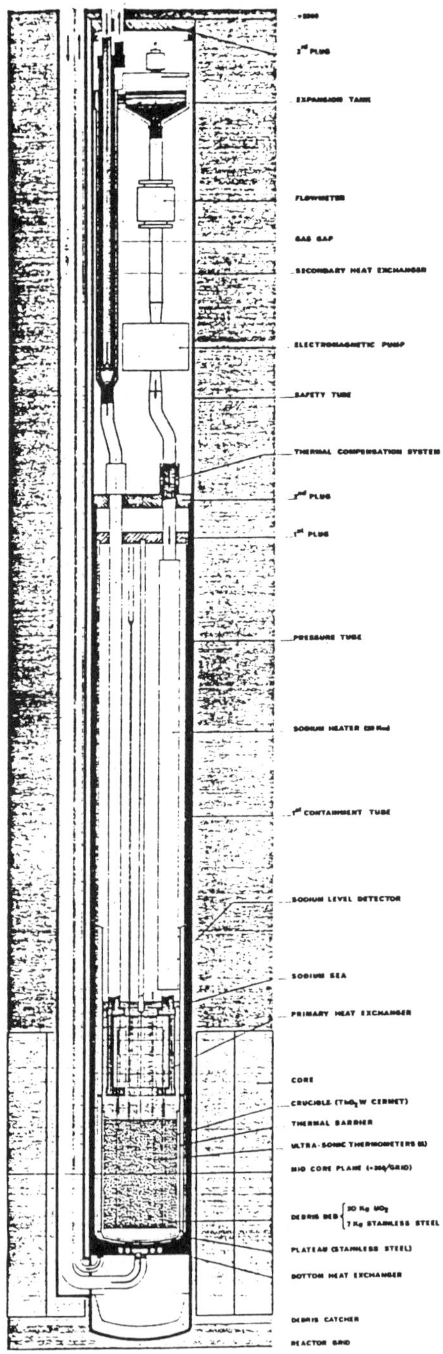

This illustration is printed
from the best available original

. une sonde de mesure de niveau libre du Sodium
. 8 sondes autogénératrices d'électrons
. 6 thermomètres à ultra-sons

On prévoit en outre d'effectuer des mesures de bruits thermique et acoustique, ainsi que des mesures par neutronographie et spectrométrie gamma après irradiation.

Les calculs qui sont menés dans le cadre du projet sont de différente nature :
. les calculs neutroniques fournissent le niveau et la répartition de la puissance neutronique et gamma dans le lit et dans les structures ainsi que les différentes données nécessaires au pilotage du réacteur. Les méthodes de calcul ont été validées par les résultats de l'irradiation d'une maquette neutronique irradiée à SILOETTE en janvier 1982.
. les calculs thermiques sont faits avec un code qualifié de conduction (2d) et permettent de définir le comportement thermique du dispositif. Les calculs font ressortir une influence mutuelle dispositif-lit très importante, ce qui permet de dire qu'il sera difficile d'interpréter les résultats sans un outil de calcul 2d. Les calculs de l'échangeur inférieur, point crucial du dispositif sont validés par une maquette hors pile.
. les calculs de contraintes thermiques et de déplacements montrent que le composant le plus sollicité est le creuset. Sa mise au point demande et demandera donc la plus haute attention tant du point de vue compatibilité que du point de vue thermomécanique.
. Parallèlement des calculs d'évaluation de la source de pression due à une interaction combustible-sodium ont été menés dont les résultats ont permis de vérifier la bonne tenue de la capsule en dynamique avec une marge de sécurité suffisante.
. enfin des calculs thermiques en transitoire sont en cours dans divers cas de fonctionnement incidentel qui font ressortir une lenteur inhabituelle des phénomènes due à la grande masse thermique mise en oeuvre au regard des faibles puissances injectées.

Pour terminer cette rapide présentation du dispositif, on insistera sur l'originalité de la méthode de remplissage du lit en sodium qui permet d'éviter au maximum que ne restent des gas piégés dans les porosités qui perturberaient de façon importante les essais. Cette méthode consiste à descendre le lit de particules dans un bain de sodium. Le sodium pénétre donc dans le lit de bas en haut avec un contrôle visuel et rigoureux de la non perturbation du lit occasionnée par ce remplissage.

Le réacteur

Le réacteur MELUSINE où seront faites les expériences est installé au Centre d'Etudes Nucléaires de Grenoble et le Service des Piles a la responsabilité de son fonctionnement. C'est un réacteur de type piscine dont la puissance nominale est de 8MW. Le combustible est un alliage U-Al (26%U) où l'uranium est hautement enrichi (93%). L'élément combustible standard est composé de 23 plaques combustibles gainées d'Aluminium.

La hauteur du coeur est de 735 mm, sa hauteur active étant de 600 mm centré sur un plan situé à 368 mm de la grille du réacteur. Le débit primaire est de 600 m3h-1, l'échauffement de l'eau étant de 11,5°C pour une température moyenne de celle-ci de 40°C.

Pour les expériences PAHR, nous utiliserons une configuration de coeur spéciale aménageant une cavité centrale d'irradiation de 3 trous x 3 trous permettant un diamètre extérieur du dispositif de 220 mm. Ce coeur est constitué de 27 éléments standard et 5 éléments de contrôle.

Le matériel hors pile

Il est constitué de trois postes principaux :

. une boucle à eau pressurisée (40b-250°C). Le débit principal de cette boucle à eau est divisé en deux par une vanne trois voies. Comme on l'a déjà indiqué, la première branche refroidit directement le plateau par un échangeur situé à l'extérieur du tube de pression, la seconde branche refroidissant le circuit sodium supérieur par un échangeur comportant une enceinte gazeuse contrôlée séparant le sodium et l'eau. Cette enceinte permet en outre de faire varier le coefficient d'échange par la nature du gaz présent dans cette enceinte.

. le contrôle commande. Il met en oeuvre dans l'état actuel du projet 150 capteurs. En outre, à cause de la très faible fiabilité du fonctionnement des capteurs immergés dans le lit, ces derniers ne seront pas utilisés pour assurer l'intégrité du dispositif.

Les principales fonctions du contrôle commande sont :

- d'assurer le confinement du lit en cas d'interaction combustible sodium
- d'assurer le confinement du lit malgré l'extrême difficulté de confiner de l'acier liquide à haute température
- d'assurer le refroidissement du lit en toutes circonstances, notamment en l'absence d'alimentation électrique
- de détecter toute fuite d'une quelconque enceinte vers une autre et déclencher toute séquence automatique, qu'une analyse préalable approfondie aura définie, nécessaire au maintien de l'intégrité globale du dispositif.

. la centrale de mesures doit permettre de fournir les données nécessaires à l'interprétation correcte des essais. Ses principales caractéristiques sont ;

- nombre d'entrées 150 voies
- vitesse maximum d'acquisition : 3000 mesures/seconde, soit une acquisition de 150 voies toutes les 50 msec.
- capacité maximum de stockage des données : 3 heures d'enregistrement à cadence maximum.

En outre, la centrale élabore en ligne toutes les données nécessaires à une conduite correcte de l'expérience.

CONCLUSION

Les études de projet ont commencé le 1.9.82. Les dates prévues à cette époque et que rien ne permet de remettre en cause de façon importante aujourd'hui sont les suivantes :

. essai Eo : Mai 1983
. 2ème essai : Novembre 1983
. 3ème essai : Août 1984
. 4ème essai : Mai 1985

Il est bien évident que le succés de ce programme dépend du résultat d'un certain nombre d'actions hors pile conduites par le CCR/EURATOM/ISPRA visant à résoudre deus types de problèmes :

. des problèmes phénoménologiques. On citera à ce titre les exemples de l'étude de la migration de l'acier sous haut gradient thermique au CEA et de l'étude du comportement physico-chimique du combustible noyé dans du sodium à l'UKAEA.
. des problèmes technologiques. Les études visent à mettre au point des solutions fiables et éprouvées pour répondre à certaines spécifications du dispositif. On citera notamment :

- un programme de développement d'un cermet ThO2-W et de mise au point d'une fabrication industrielle d'un creuset réalisé avec ce cermet, mené au CEA
- un programme de mise au point d'une fabrication de mini-tubes longs (250 mm) en ThO2 destinés à la protection des sondes en tungstène des thermomètres à ultra-sons contre l'attaque par l'acier liquide.

REFERENCES

(1) M. AMBLARD, G. BERTHOUD, UO2-Na Interaction : the Correct 2 experiments, FCI 4 SLASH P22, 4th CSNI Specialist Meeting on Fuel Coolant Interaction In Nuclear Reactor Safety, Bournemouth, April 2-5, 1979.

(2) C. LE RIGOLEUR, Particle Bed dry out in Sodium, Proceeding of the Fourth Post Accident Heat Removal Information Exchange Meeting, VARESE, Italy, October 10-12, 1978, European appl. Res. Rept. - Nucl. Sci. Technol., Vol 1 n°6, 1979, pp 67-84, Harwood Academic Publishers.

SESSION VII
INSTRUMENTATION

LES MESURES DND FAITES DANS LE CADRE DES IRRADIATIONS DE LA FILIERE A NEUTRONS RAPIDES

P. VULLIEZ CEA/CEN Grenoble - Service des Piles
M. PLUCHERY CEA/CEN Grenoble - DMG/SER

CEA-CEN/G 85X 38041 GRENOBLE CEDEX

Des aiguilles fissiles ou fertiles préirradiées dans les réacteurs Rapsodie ou Phénix sont réirradiées dans les boucles à thermopompes à Siloé.

Le dispositif à thermopompes est une boucle à sodium qui, au cours du temps et en fonction de l'évolution de l'instrumentation a été progressivement modifiée. La version actuelle est du type à circuit dérivé.

1) DISPOSITIF D'IRRADIATION (fig. 1)

La hauteur totale de la boucle est d'environ 10 m. On distingue 3 zones :

. partie basse
. décrochement biologique
. partie haute

Partie basse :

. section sous flux
. circuit primaire - pot de vidange
. circuit secondaire
. connexions
. support boite à eau

Décrochement biologique :

. connexions électriques, pneumatiques et hydrauliques
. boitier produits de fission partie basse

P. von der Hardt and H. Röttger (eds.), Irradiation Technology, 417–427.

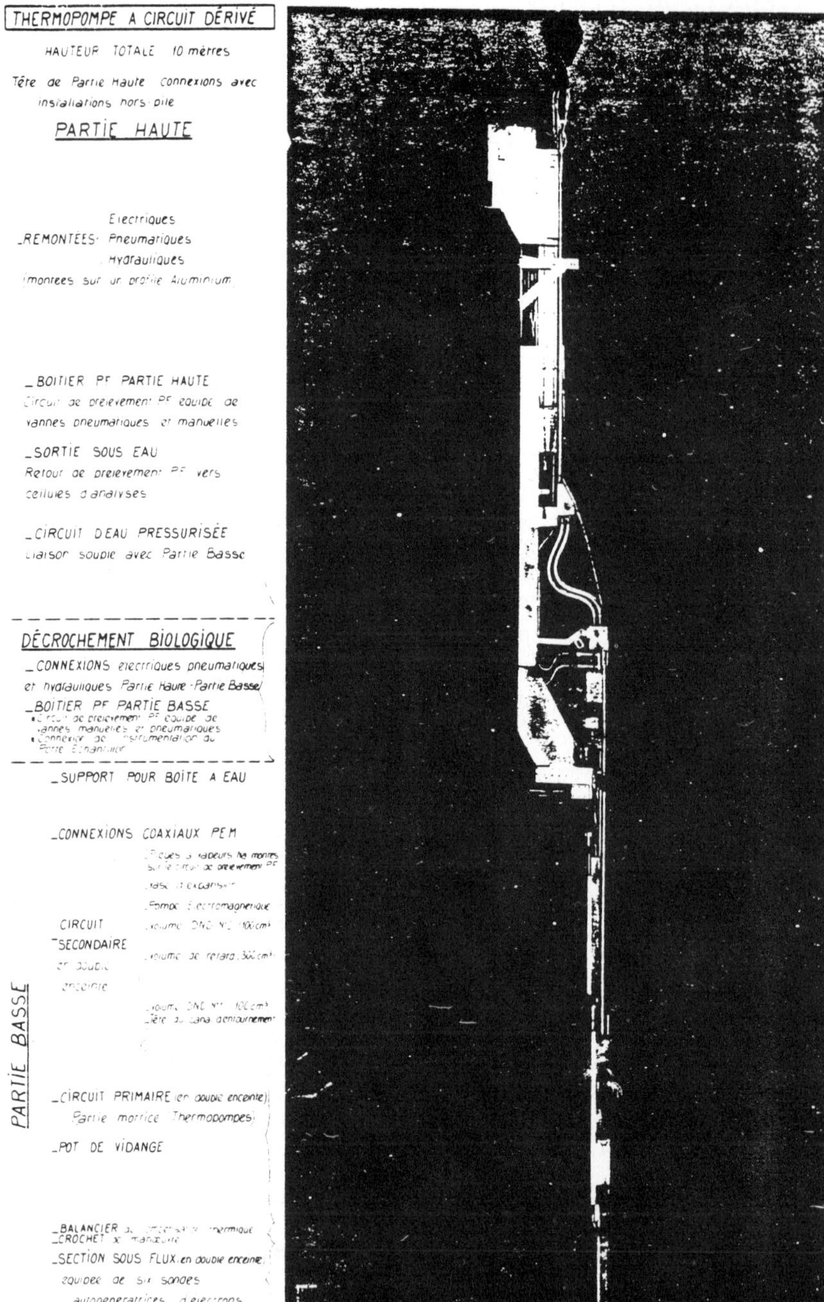

This illustration is printed from the best available original

Figure 1

Partie haute :
. circuit d'eau pressurisée
. sorties sous eau
. boitier produits de fission partie haute
. remontées électriques, pneumatiques, hydrauliques

2) INSTRUMENTATION DND - CIRCUIT DE MESURE DND SCHEMA DE PRINCIPE PARTIE BASSE (fig. 2)

Circuit primaire. Ce circuit comprend :
. la partie sous flux au niveau coeur du réacteur dans laquelle est placée l'aiguille combustible
. les éléments thermopompes
. le pot de transfert

Circuit dérivé :
Une partie du sodium du circuit primaire est dérivé dans une tuyauterie annexe. Dans cette branche, la circulation du sodium est réalisée à l'aide d'une pompe électromagnétique (P.E.M.). Dans le sens de circulation du sodium, on distingue successivement :
. 3 volumes hélicoïdaux
. la pompe électromagnétique
. un vase d'expansion
. un piège à oxydes

Au droit des volumes DND n° 1 et 2 sont placées les chambres de mesures des neutrons différés. Le volume hélicoïdal central est un volume de retard qui permet d'augmenter le temps de transit entre les 2 volumes DND.

3) SYSTEME DE COMPTAGE

Les capteurs utilisés pour les mesures DND sont des chambres à fission à dépôt Uranium 235. Le "bloc DND" est constitué par l'eau de la piscine du réacteur. L'intérêt des chambres à fission est leur faible sensibilité au rayonnement gamma du réacteur et du sodium de la boucle. Le bruit de fond est dû aux neutrons du réacteur et aux photoneutrons. Le compteur placé le plus bas a un niveau de bruit de fond plus important, mais aussi un signal plus élevé (temps de transit plus court). La sensibilité des chambres est de 0,1 ou 0,01 c/s/(n/cm2/s)

4) SYSTEME D'ACQUISITION

Les impulsions des compteurs sont traitées et mises en forme par une électronique comportant : amplificateur, discriminateur, sorties sur échelle de comptage et ictomètres.

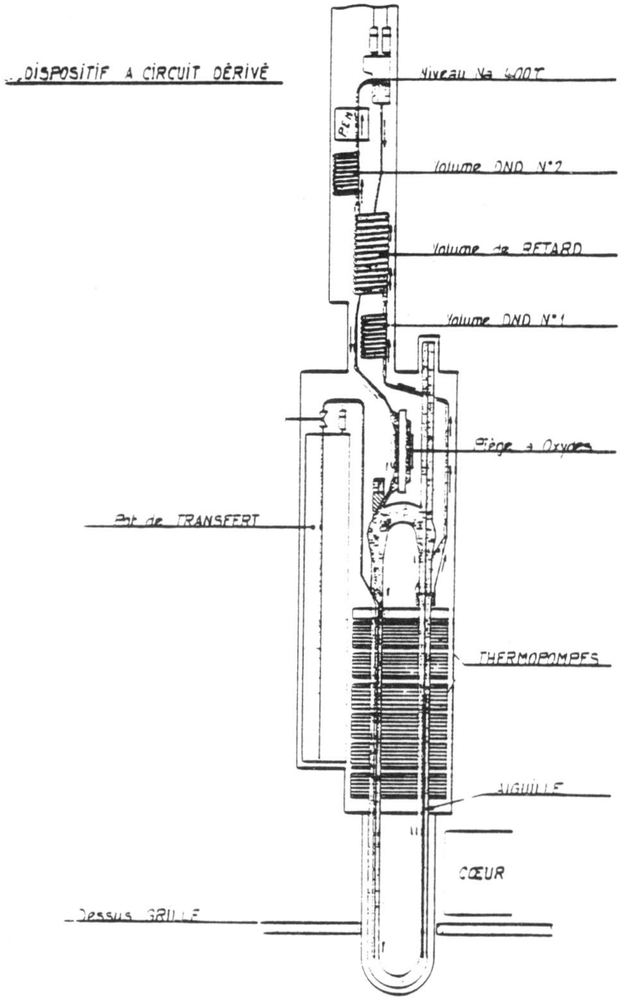

Figure 2

This illustration is printed
from the best available original

On attaque ensuite différents systèmes d'acquisition :
. enregistrement papier
. imprimante
. bande magnétiques
. module interface calculateur

Le système d'acquisition permet un suivi en continu des signaux DND qui peuvent être comparés à l'activité en produit de fission mesurée par ailleurs.

5) COMPARAISON ACTIVITE GAS - SIGNAL DND (fig. 3)

Nous montrons ici un exemple de rupture. Nous distinguons l'apparition de la rupture par évolution de l'activité totale et la montée du signal DND. Le suivi de cette rupture a duré 4 jours. Les taux de comptages DND compteurs 1 et 2 sont très importants.

6) VARIATION DU SIGNAL DND EN FONCTION DES CONDITIONS D'IRRADIATION

Lors d'un cycle d'irradiation, la puissance dégagée dans l'aiguille peut varier, soit par modification du niveau de puissance du réacteur (chutes de barres), soit par modification du flux neutronique incident (déplacement de la boucle par rapport au coeur).

Nous donnons ici (figure 4) un exemple de variations de signaux DND pour ces deux types de conditions de fonctionnement. A noter que sur cette figure le "signal" DND est exprimé en grandeur réduite (cm2 équivalent recul). La bonne cohérence entre les voies DND 1 et 2 montre que dans ce cas le modèle de l'émission en recul prise pour le calcul est correct.

7) VARIATION DES CONDITIONS DE FONCTIONNEMENT DE LA BOUCLE ANALYSE DES SIGNAUX DND (fig. 5)

Il est possible de faire varier le débit de sodium dans le circuit dérivé dans de larges proportions sans perturber sensiblement le circuit primaire.

On peut obtenir l'arrêt complet de la circulation du sodium dans la boucle en dérivation en arrêtant la pompe électromagnétique et en la faisant fonctionner légèrement en inverse pour compenser la poussée résiduelle due à la boucle primaire. Dans cette situation, le sodium présent dans chacun des volumes de comptage contient une certaine concentration de produits de fission émetteurs de neutrons différés qui est caractéristique de l'état

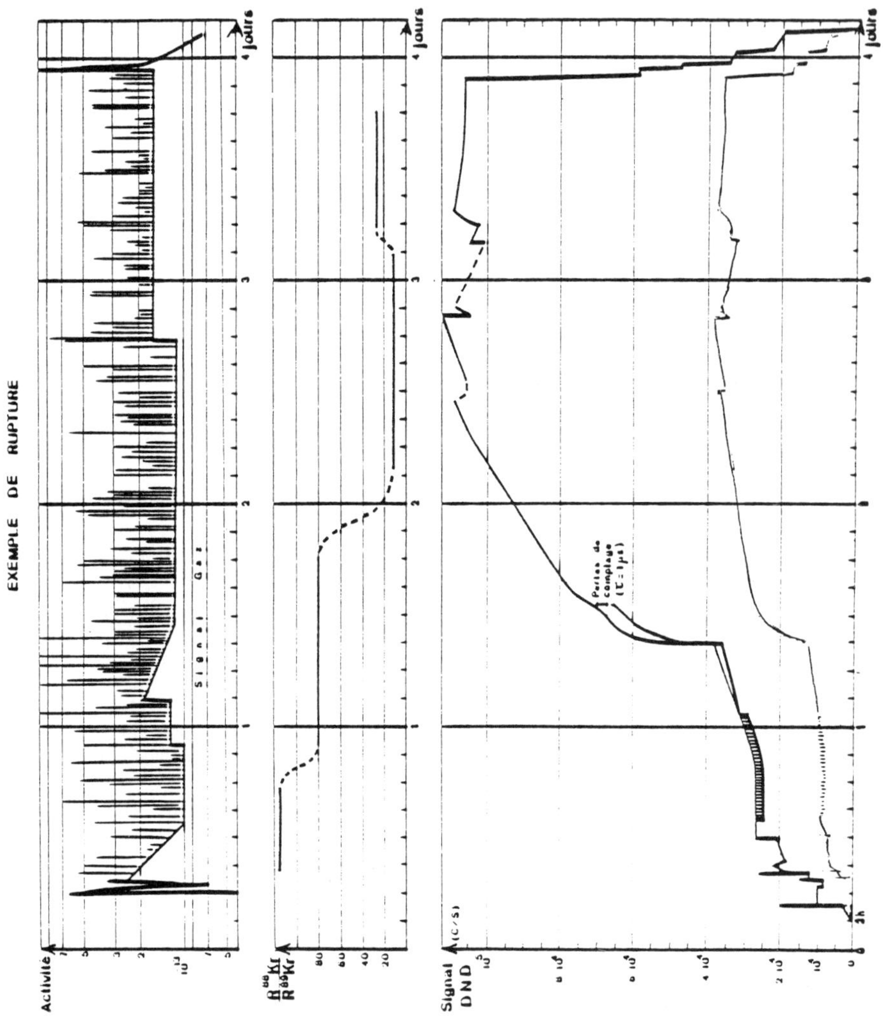

Figure 3

This illustration is printed
from the best available original

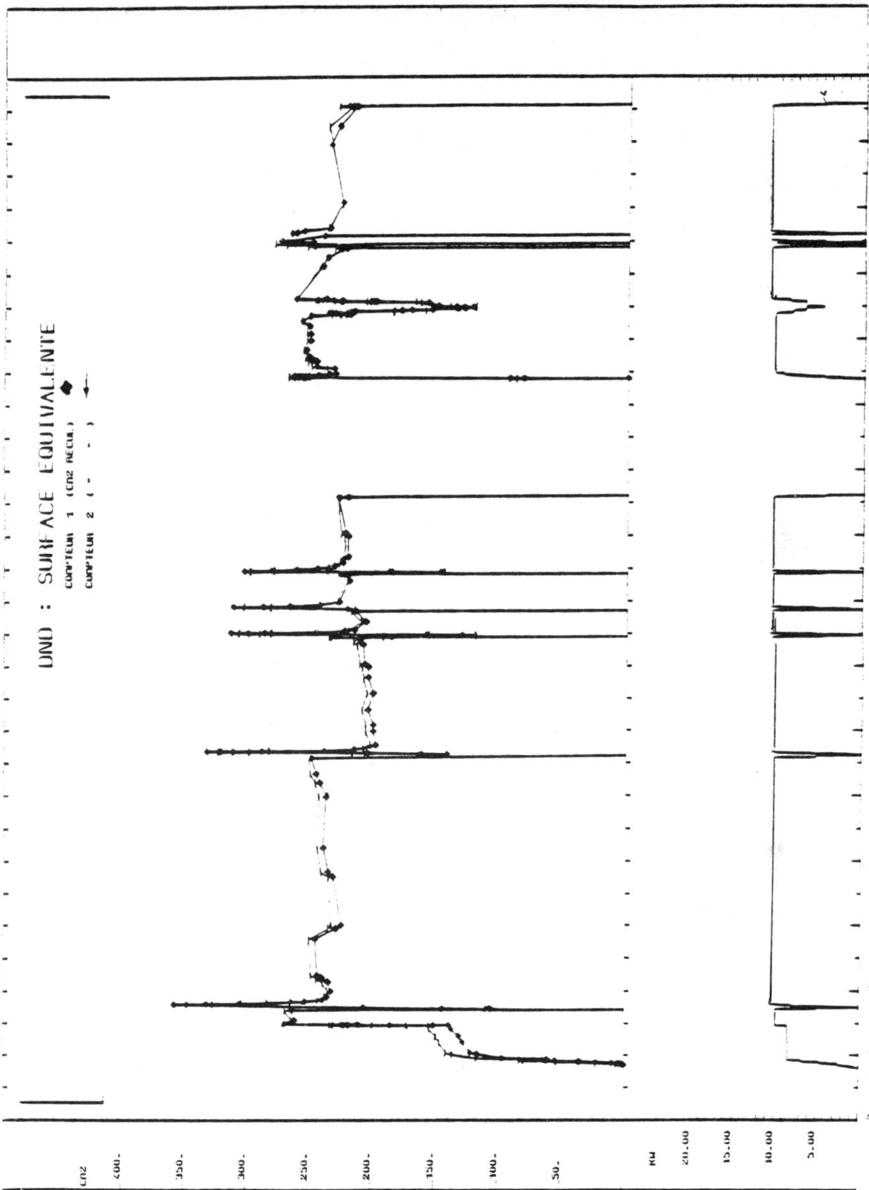

Figure 4

This illustration is printed
from the best available original

d'équilibre de la boucle en régime permanent. L'enregistrement de l'évolution du signal en fonction du temps (décroissance) permet après traitement numérique de préciser les proportions des différents émetteurs.

Ici, un exemple d'enregistrement et sa décomposition en 4 groupes d'émetteurs de périodes radioactives :

. 55,7 s
. 22,7 s
. 6,2 s
. 2,3 s

8) ORGANIGRAMME DE LA CHAINE DE TRAITEMENT (fig. 6)

Sur ce synoptique, nous représentons les différentes étapes du traitement de l'information DND.

Données de bases :

. mesure du signal des chambres Cp/s
. paramètres de l'irradiation
  - puissance échantillon ou aiguille P (W/cm^3)
  - débit primaire Q1 (cm^3/s)
  - débit secondaire Q2 (cm^3/s)

Calculs :

Les calculs prennent en compte les mesures et un étalonnage.

Signal DND :

Le signal DND est exprimé en cm^2 équivalent recul.

NB = la connaissance de la composition en famille d'émetteurs est intéressante pour permettre la transposition en réacteur (temps de transit différents).

9) ETALONNAGE
MESURE DE LA SENSIBILITE DES COMPTEURS (fig. 7)

La sensibilité des compteurs, réponse en coups par seconde pour un neutron émis dans le volume DND, est déterminée expérimentalement.

On réalise une maquette neutronique équivalente de la boucle. A l'intérieur de cette maquette, on déplace une source de neutrons. La réponse des compteurs peut ainsi être corrélée à la position de la source dans les volumes de comptage, de retard et les tuyauteries annexes (vases d'expansion, retour sodium) et être prise en compte dans les calculs. La source de neutrons est :

. soit une source constante connue dont le spectre d'émission de neutrons est voisin de celui des neutrons différés (source Pu.Li)

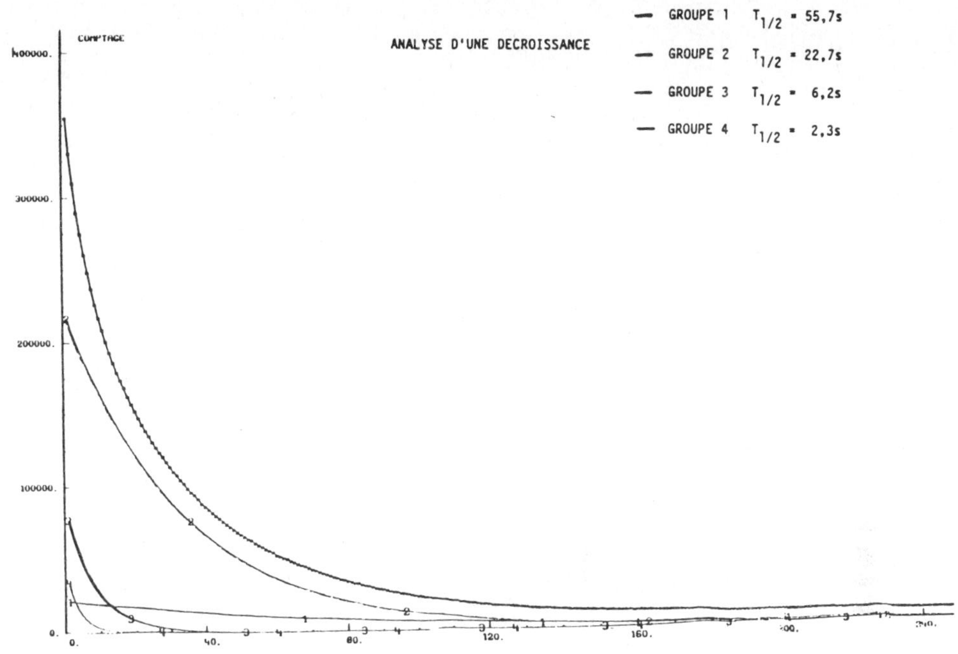

Figure 5

These illustrations are printed from the best available originals

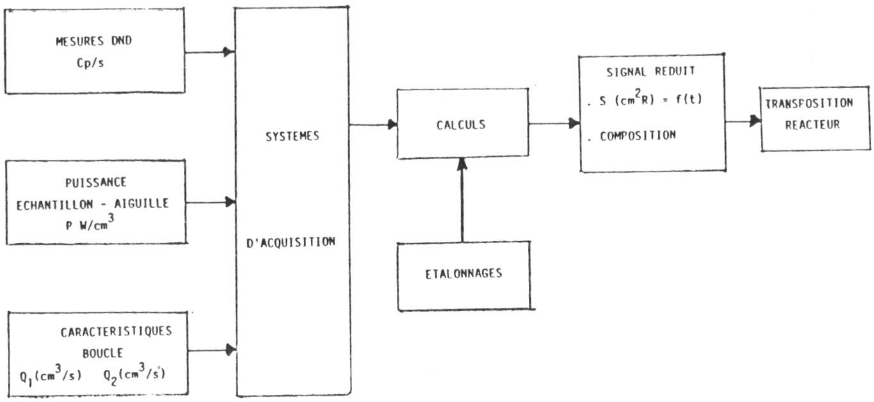

Figure 6

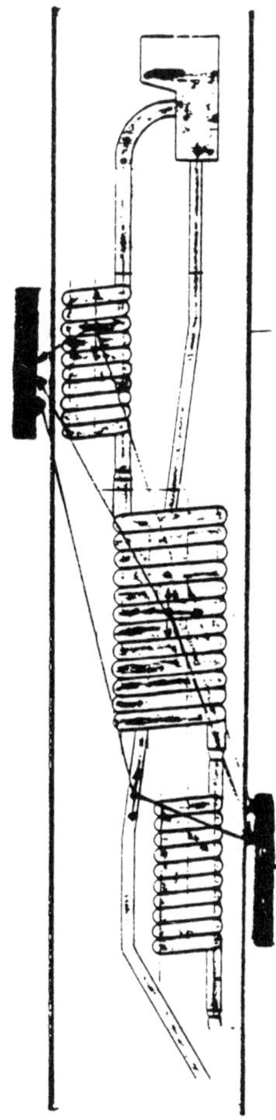

** ETALONNAGE OND **

Figure 7

This illustration is printed
from the best available original

. soit une source de neutrons différés constituée d'un fragment de combustible (U5 ou Pu9) irradié et transféré rapidement ($\simeq$ 500 ms) au niveau du bloc de comptage

Un ensemble de mesures comparatives avec différentes sources de neutrons (californium, Am.Be, Pu.Be, Pu.Li et neutrons différés) a permis de chiffrer l'effet du spectre d'émission des neutrons sur la sensibilité des chambres.

CONCLUSION

Les expériences réalisées à Siloé, à l'aide de cette instrumentation ont permis d'étudier une large gamme de signaux émis par diverses ruptures d'aiguilles (de quelques cm2 équivalent recul à 300 cm2 équivalent recul) réf. (1).

La souplesse de fonctionnement du circuit dérivé dont le débit peut varier de 0 à 70 cm3/s est bien adaptée pour l'analyse de la composition des émetteurs de neutrons différés présents dans la boucle.

Une prochaine version du dispositif permettra d'augmenter la plage de variation de la vitesse du sodium dans le circuit primaire au niveau de l'aiguille et d'étudier son effet sur l'émission des émetteurs de neutrons différés.

REFERENCE

(1) Programme Commun Franco-allemand pour étudier le fonctionnement d'un surgénérateur en présence d'aiguilles combustibles ruptées.
P. MICHAILLE, et Al.

Conférence internationale sur la Sureté des Réacteurs Rapides à Métal Liquide et ses aspects conception et fonctionnement.
Lyon 18/23 Juillet 1982

THE DELAYED NEUTRON DETECTION METHOD FOR CLADDING TUBE MONITORING AND EXPERIMENTAL RESULTS

Siegfried Jacobi

Kernforschungszentrum Karlsruhe GmbH
Institut für Reaktorentwicklung
Postfach 3640, 7500 Karlsruhe

ABSTRACT

The goal of in-pile experiments involving fuel rods or fuel rod bundles frequently consists in the observation of cladding failure. The following information is required: time of cladding failure, number of failed rods, extent of failure and its development with the time. Since optical methods of observation cannot be applied, the detection of fission products released into the coolant upon a cladding failure constitutes an eligible means. Fission product nuclides are preferred whose daughter products emit delayed neutrons. The method and the results obtained by it are described by the example of Mol 7C: A monitor placed beside the head section of the loop detects the neutrons and thus the free fuel surface. This report deals with the design of the monitor, its layout and calibration as well as with the interpretation of the measurement signals. Above all by the simultaneous evaluation of the temperature and flow signals the occurrence and development with time of the following events can be analyzed: breaking up of the claddings due to dryout, partial melting of the claddings and local loss of the bundle integrity. This is confirmed by micrographs; but some problems arise in signal interpretation.

INTRODUCTION

In in-pile tests involving single fuel rods or rod bundles a distinction is made between two types: (1) performance tests by which it is proved that a given target burnup can be attained without cladding failure; (2) a group of experiments in which cladding failures are actually expected. One objective of these tests consists e.g. in the observation in experiments of the bundle behavior in case of disturbances

P. von der Hardt and H. Röttger (eds.), Irradiation Technology, 429–438.

in cooling [1, 2]. Moreover, this group of tests include studies on the operating behavior of defective fuel elements [3, 4]. In all cases indicated it is necessary to obtain out-of-pile the most precise information possible, with the shortest possible time delay, about the size, type and development with time of a cladding failure.

As in monitoring an entire sodium-cooled reactor core, measurement of delayed neutrons in sodium, the so-called Delayed Neutron Detection (DND), has proved to be particularly suited to fulfil this requirement. Unlike in reactor kinetics, however, special attention must be paid here to the behavior of the precursors of the delayed neutron emitters. The reason is that not the emitters of delayed neutrons but the precursors of the emitters must be transported through the fuel to the free fuel surface facing the sodium and then, within the sodium, to the point of measurement (Fig. 1). For delayed neutron detection neutron counters or neutron ionization chambers have been provided at the coolant pipe and/or at a bypass.

This method and the results obtained by it are described by the example of Mol 7C - a series of in-pile blockage experiments involving UO_2 and $(U, Pu)O_2$, respectively, and sodium as the coolant - with the express instruction that this method is likewise suited for other fuel and other coolants [2].

THE TASK

In conformity with the objective of the Mol 7C experiments the task has been a very comprehensive one. Very small cladding failures and also modifications in the chromium coated UO_2 spheres serving as the blockage were to be detectable during the phase preceding the transient, the same as extremely great failures occurring during and after the transient. The term transient is used here to mean the transition

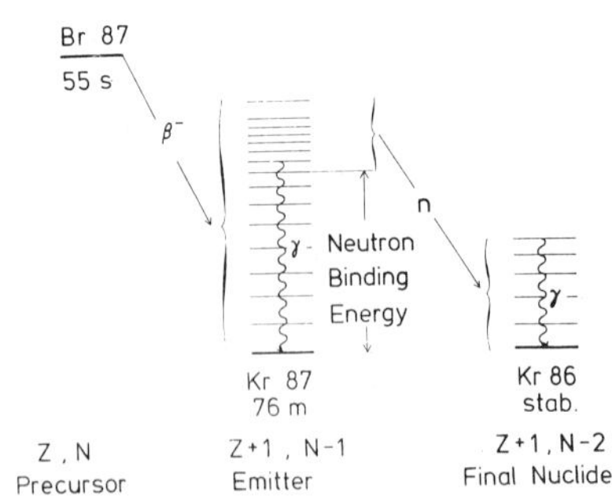

Fig. 1 Emission of delayed neutrons by the example of Br-87 - Kr-87

from the intact bundle into the failed condition as a result of the intentionally initiated local disturbance in cooling. Moreover, both extremely slow and very fast changes in cladding failures had to be anticipated which means that extremely different scopes of failure and rates of change were to be recorded.

DESIGN OF INSTRUMENTATION

The design of the DND monitor started from the neutron flux expected within the monitor. The limit points were the minimum flux - caused by the uranium surface contamination in the bundle - and the maximum flux - caused by an assumed most severe destruction of the bundle. This range served as the basis for a nomogram in which only horizontal connecting lines are admitted for the assignment to the individual scales (Fig. 2). After selection of a suitable neutron counter it was possible to enter the scale for the pulse rate. Since the upper flow region could be measured only with a neutron ionization

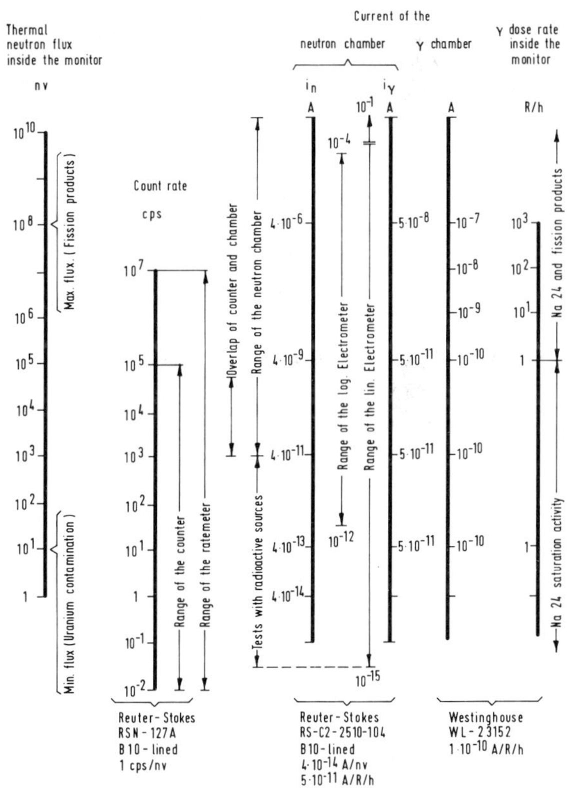

Fig. 2 Design of instrumentation

chamber, it had to be ensured that the ranges of measurement of the counter and chamber overlapped sufficiently. This requirement was fulfilled by the choice of a suitable ionization chamber. The question whether a compensated or non-compensated neutron ionization chamber should be used was solved, proceeding from the right margin in the nomogram. The basis used was the relation between the neutron activity concentration and the gamma radiation in the monitor: For an extremely severe destruction of the bundle a thermal neutron flux of approximately 10^8 nv was determined. The resulting gamma flux in the monitor generated by the fission products passing by amounts in this case to approximately 10^3 R/h. These two values are correlated to each other and, therefore, they lie on a horizontal connecting line in Fig. 2. When all circumstances to be anticipated had been taken into account, it appeared that use of a compensated neutron ionization chamber was not an absolute necessity. On the other hand, measurement of the gamma flux in the monitor seemed reasonable with a view to a possibly necessary correction of the measured values of the neutron flux in the chamber.

When designing the measuring configuration as a whole the expenditure had to be kept within fixed limits. Use of a computer with all the peripheral and programming problems associated was ruled out. The simplest commercially available standard electronic system was selected: ratemeter and electrometer amplifier with outputs for analog recorder and magnetic tape unit. Linear amplifiers were chosen for the slow and small signal changes, logarithmic amplifiers for very fast signal changes. In this way, six measuring channels were available: one linear and one logarithmic channel each for the neutron counters, neutron ionization chambers and gamma ionization chambers. In the evaluation the most favorable channel could be selected. This principle has proved to be useful in Mol 7C/1 to 3 and will therefore be retained for Mol 7C/4 and 5 as well.

PLACE OF MEASUREMENT AND MONITOR CONCEPT

In the experiments Mol 7C/1 to 3 the DND monitor was arranged at the greatest possible distance of the core beside the head of the loop /5/. At the same level the upper sodium pump has been installed in the loop. The sodium in the pump served as the sample volume for the emission of delayed neutrons.

In contrast to the three preceding experiments, it is planned to perform the experiments Mol 7C/4 and 5 with burnt up KNK II fuel rods. The resulting reconstruction of the loop implied a new monitor design and a different place of measurement (Fig. 3). The highest point exposed to fresh sodium flow is now the lower part of the expansion tank (Fig. 4 and 5). As in Mol 7C/1 to 3, the monitor housing has an internal cadmium sheet lining of 1 mm thickness. It absorbs photo-neutrons

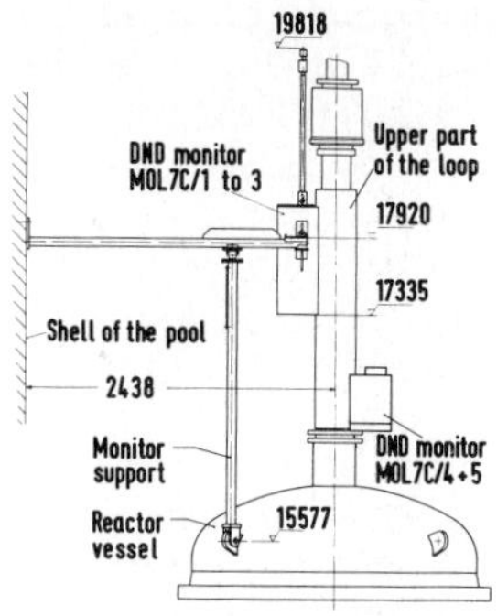

Fig. 3 DND monitor Mol 7C/1-3 and Mol 7C/4-5 in the pool of BR 2 beside the Mol 7C loop. Dim. in mm

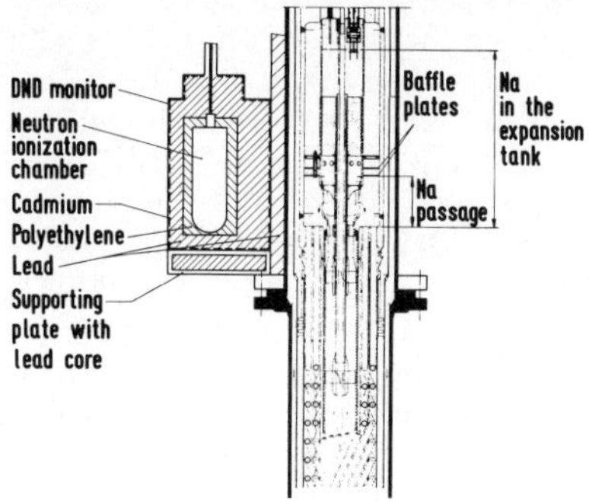

Fig. 4 DND monitor Mol 7C/4-5 on the supporting plate beside the expansion tank

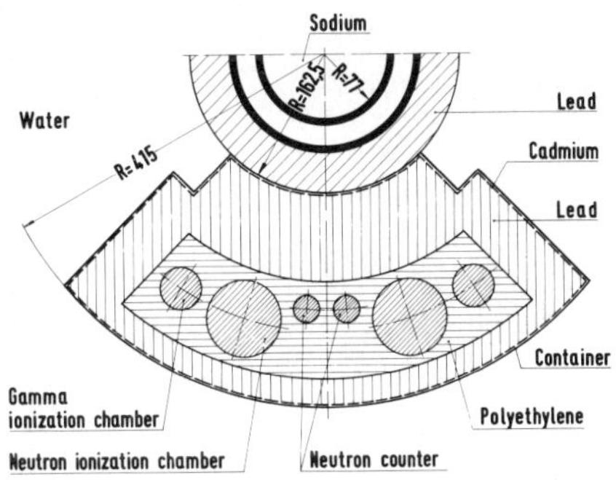

Fig. 5 DND monitor Mol 7C/4-5. Cross section in the level of the expansion tank. Dim. in mm

generated in the pool water by the gamma radiation of Na-24 or by core radiation and moderated in the pool water. The lead in the monitor serves as a shielding against external gamma radiation so as to

- reduce the generation of photo-neutrons in polyethylene,
- protect the counters against an excessively high gamma radiation level, and
- reduce the current of the neutron ionization chamber due to the gamma sensitivity inherent in the ionization chamber.

Polyethylene as a neutron moderator increases in the usual way the neutron detection sensitivity of the neutron detectors.

Contrary to the detectors indicated in Fig. 2, the modifications for Mol 7C/4 and 5 are the following:

- Extremely constricted space imposed the installation of a smaller sized neutron ionization chamber: Westinghouse WL 8105, neutron sensitivity = 2.5×10^{-14} A/nv, gamma sensitivity = 2.5×10^{-12} a/R/h.
- Regarding the counter tubes only the type name has been changed into RS-p7-0805-127 while the specification have remained the same.

NUCLEAR CALIBRATION

The objective of calibration consists in understanding the functional dependence of the measuring variables - in this case counter pulse rates and ionization chamber currents - on the free fuel surface in contact with sodium. A mockup of the loop head with pump dummy was a greatly simplified simulation of reality in neutron physical terms without use of sodium. A neutron source simulated the neutron emitting sodium. The measurements were performed with the original DND monitor Mol 7C/1 to 3 and its neutron counter channel. First the neutron sensitivity η_n was calculated:

$$\eta_n = \frac{\text{counting rate (cps)}}{\text{source strength (neutron s}^{-1})} = 5.8 \times 10^{-5} \text{ counts/neutron}$$

As the next step the relation between the pulse rate and the free fuel surface was determined with the code FICTION I calculated according to the recoil model (only release by recoil from the fuel surface facing the sodium) / 6 /. The Mol 7C-specific data entered in Table 1 were used for this purpose.

Table 1: Parameters for Calculating the Relationship Existing Between Pulse Rate and Free Fuel Surface

Specific rod power	400 W/cm
Specific fissile material volume	0.203 cm^3/cm
Fissuring factor)[1]	1
Coolant flow	3.77×10^3 gs^{-1}
Coolant transit time until arrival at point of measurement	1.9 s
Coolant circulation time	5.8 s
Sensitivity of detector	5.8×10^{-5} counts/neutron
Volume of sodium probe	1.39×10^3 cm^3

)[1] fissuring factor = $k = \frac{\text{effective fuel surface}}{\text{geometric defective area}}$

The result is the response sensitivity to free fuel surfaces in case of use of the counter tube channel:

$$\eta_{fc} = 5.9 \times 10^2 \frac{cps}{cm^2 R}$$ R = recoil; no operand here but only an indication of the recoil model

The transition from the counter tube channel to the ionization chamber channel was effected with the help of the neutron sensitivities of these two detector types. This gives the response sensitivity to free fuel surfaces in case the ionization chamber channel is used:

$$\eta_{fi} = 2.4 \times 10^{-11} \frac{A}{cm^2 R}$$

RESULTS

From the multitude of results derived from previous experiments three phenomena of particular interest will be discussed below. The blockage detection proper and associated problems of signal interpretation, part of them still unsettled, has been treated in detail in /5/.

Rise in Counting Rate During the Preirradiation Phase

In the experiment Mol 7C/3 an active blockage was simulated by UO_2 spheres of 0.5 mm diameter /2/. Since under all circumstances plugging of the small cross sections of the cooling channels was to be avoided in the piled up spheres, the spheres had been coated by a 10 µm thick chromium layer in order to protect them from chemical reactions with the sodium or its impurities. Nevertheless, a steady rise in the counting rate was recorded during the preirradiation phase (Fig. 6). The only interpretation was that in some spheres the chromium coating had failed so that fission products were detected in the sodium.

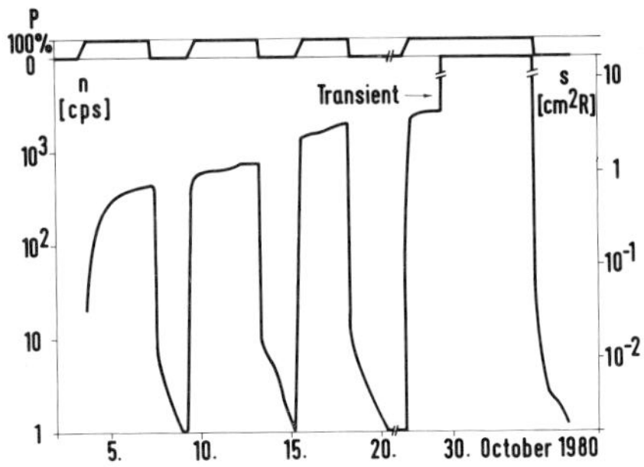

Fig. 6 Power P of BR 2, rise of the counting rate n and free fuel surface s during the preirradiation phase

Gas Bubbles in the Sodium

At the onset of the transient in experiment Mol 7C/1 an interesting phenomenon was observed: By interruption of sodium cooling in the blockage the transient was initiated. The beginning of temperature enhancement within the blockage was denoted t = 0. 12.6 s after that time a steep rise in the counting rate was recorded with the logarithmic counter channel (Fig. 7). Subsequently and also after attainment of the maximum counting rate violent temporary fluctuations were observed. This must lead to the conclusion that gas bubbles emanating from the filling gas simulating fission gas of the fuel rods gave rise to temporary displacement of parts of the sodium upstream of the DND monitor. The very steep transition to the saturation value is worth noticing. It shows the extremely short response time of the entire measurement equipment which in this form had not been measured by nuclear methods before the experiment.

DND-Signal as an Indicator of Open Fuel

Fig. 8 shows by way of example the interplay of the major signals during and after the transient of experiment Mol 7C/1. After interruption of sodium cooling within the blockage the onset of temperature rise in the blockage is recorded at time t = 0. After attainment of the sodium boiling temperature the first rods break up due to dryout. The associated gas release and material relocation cause fluctuations in flow and - delayed by the transit time - a rise by four orders of magnitude of the DND signal from the logarithmic neutron ionization chamber channel. The first major rise of the DND signal is assigned to the breaking up of the claddings. When the melting temperature of steel was attained a temporary signal plateau was recorded and a further rise to about 10^4 cm^2 R was the clear indication of partial melting of the cladding tubes followed by the loss of integrity of some rods within and downstream of the blockage (Fig. 9).

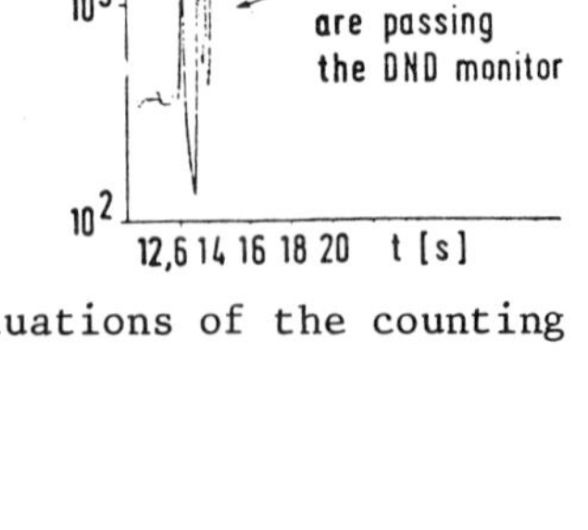

Fig. 7 Mol 7C/1. Fluctuations of the counting rate n after initiating the transient

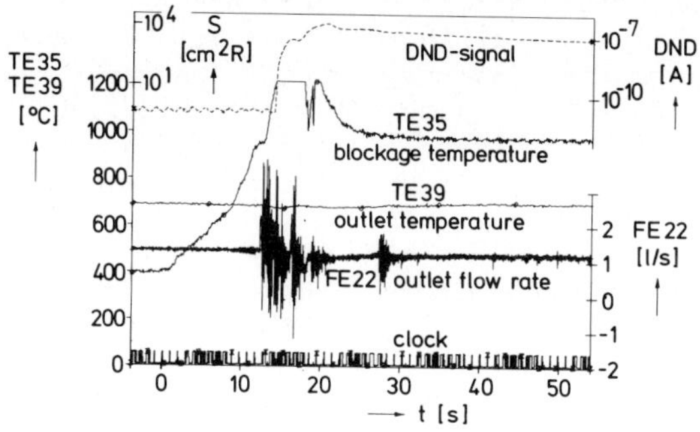

Fig. 8 Signals of Mol 7C/1 during the transient. Sodium flow rate FE 22, sodium temperature in the blockage TE 35 and at the outlet of the bundle TE 39, DND signal as current of the chamber [A] and as open fuel surface s [cm^2R]

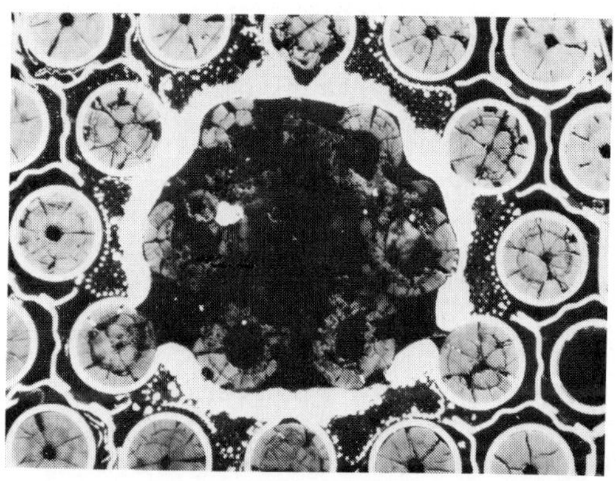

Fig. 9 Mol 7C/1. Cross section of the blockage after the transient. Molten and resolidified cladding material, blockage cage with steel spheres and spacer grid

SIGNAL INTERPRETATION AND PERSPECTIVES

In the experiments carried out a good qualitative observation of failure development and time resolution was demonstrated although the quantitative evaluation of the damage still poses problems: On account of the still insufficient description of fission product release from the fuel by modeling theory the recoil model is frequently used as the simplest approximation. But treating the signals in this way gives calculated surfaces which are greater by the factor 10 to 500 than the observed free surface. It has been shown that diffusion and transport of the DND precursors promoted by fission gas, above all through pores and at grain boundaries, contribute considerably to signal formation /¯5_7. The influence of this highly fuelspecific phenomena, however, still calls for systematic basic experiments specially adapted to these problems accompanied by a treatment based on modeling theory. Electron micrographs show that at least in case of fresh fuel the sintering temperature plays a major role as regards the remaining coherent voids between the grains /¯7_7.

Another phenomenon not yet understood is the signal development after the transient. In this phase the fuel temperature falls again and the fission product fraction release should follow this course /¯8_7. In this way, a distinct signal reduction after the transient could be expected to take place without noticeable delay. But contrary to expectations, only a very slow reduction to approximately half the maximum value has been observed in the three experiments performed so far; see e.g. Fig. 8. Regarding this range of transient release of precursors to the emitters of delayed neutrons a more thorough evaluation is still needed.

LITERATUR

/¯1_7 T.H. Braid et al.: DN Detection during SLSF Tests, Conf. on Fast, Thermal and Fusion Reactor Experiments, April 12-15, 1982, Salt Lake City, Utah

/¯2_7 W. Kramer et al.: Comparative Discussion of Three Mol 7C In-Pile Blockage Experiments. Intern. Topical Meeting on Liquid Metal Fast Breeder Reactor Safety. July 19-23, 1982, Lyon

/¯3_7 S. Jacobi, G. Schmitz: Open Fuel in KNK II for Investigating Cladding Failures and their Possible Consequences, Ibid

/¯4_7 P. Michaille et al.: Joint French-German Program to Investigate Continue LMFBR-Operation with Defective Fuel Pins, Ibid

/¯5_7 S. Jacobi: Behaviour of Fuel Fragment Blockages - Detection by Delayed Neutrons and Problems of Signal Interpretation. Topical Meeting on Reactor Safety Aspects of Fuel Behavior, August 2-6, 1981, Sun Valley, Idaho

/¯6_7 G. Hoffmann et al.: Fuel Failure Detection and Location in Fast Breeder Reactors. In: S. Jacobi, Editor: Fuel Failure Detection and Location in LMFBRs. Proc. IAEA Specialists Meeting, May 11-14, 1981, Karlsruhe, KfK 3203

/¯7_7 G. Karsten, KfK, persönliche Mitteilung, August 1982

/¯8_7 E.R. Weber et al.: Laboratory Studies on Melting and Gas Release Behavior of Irradiated Fuel, IAEA Meeting of Specialists on Fuel Failure Mechanisms, May 12-16, 1975, Seattle

EXPERIENCE ACQUISE DANS L'UTILISATION DE L'ABSORBANT NEUTRONIQUE ^{3}HE POUR L'EXECUTION DE TRANSITOIRES DE PUISSANCE DE FISSION

André FALLA

C.E.N./S.C.K.

B 2400 MOL BELGIQUE

ABSTRACT

To meet an increasing interest for fuel irradiations in BR2 reactor under transient conditions of fission power, the isotope ^{3}He of helium gas is utilized as a variable neutron absorbent in various irradiation rigs.

Fast and slow ^{3}He pressure variations are carried out between 38 bar and 1 bar giving rise to fission power transients in the ratio of 1 to 1.75 for a ^{3}He screen thickness of 2 mm.

The following aspects are dealt with in this paper :

- the slow ^{3}He pressure control by means of a metal bellow pressuriser ;
- the fast ^{3}He pressure control using high and low pressure tanks ;
- the circulation and purification system using titanium sponge as a tritium absorber ;
- the tritium release monitoring system.

Observations are also reported with respect to the ^{3}He neutron absorbing efficiency, the tritium leakrate, the tritium purification efficiency, and the tritium decontamination of ^{3}He circuits.

P. von der Hardt and H. Röttger (eds.), Irradiation Technology, 439–449.

1. INTRODUCTION

La réaction de capture des neutrons par l'isotope gazeux ^3He

$$^3He + ^1n \longrightarrow ^1H + ^3H$$

est utilisée pour modifier la puissance de fission dissipée dans des dispositifs d'irradiation. La section efficace de capture du neutron par ^3He est de 5,3 x 10^{-21} cm^2 pour une energie du neutron de 0.025 eV et est inversément proportionnelle à la vitesse du neutron pour les domaines d'énergies thermique et épithermique.

La première installation utilisant le gaz ^3He comme absorbant neutronique pour l'exécution de transitoires de puissance dans des dispositifs d'irradiation, a été mise en service au CEN/SCK-Mol en octobre 1977.

Une deuxième installation, plus performante, contrôlant 3 dispositifs d'irradiation et permettant des transitoires rapides (jusqu'à 15 % par seconde) a été mise en service en juillet 1981. Cette installation est utilisée pour deux programmes d'irradiation dont la description est exposée dans les références 1 et 2. Les éléments repris dans la présente note se rapportent essentiellement à cette installation n°2, désignée sous le sigle VNS 2 ("Variable Neutron Screen nr 2").

2. DESCRIPTION D'UNE INSTALLATION VNS

2.1. Circuits de gaz

2.1.1. La figure 1 représente un schéma simplifié de l'installation VNS 2. Pour la clarté un seul circuit d'écran ^3He y est représenté.
Dans les grandes lignes, l'installation comprend :

- trois circuits d'écran ^3He connectés chacun à un dispositif en pile ;
- un pressuriseur à soufflet métallique permettant d'ajuster la pression dans chaque circuit d'écran et d'effectuer des transferts de gaz entre les divers circuits et réservoirs. L'azote est utilisé comme gaz moteur pour la pressurisation ;
- un réservoir de 50 litres à basse pression ^3He (1 bar absolu) pour les dépressurisations rapides des circuits d'écran
- un réservoir de 6 litres à haute pression ^3He (38 bar absolu) pour les augmentations rapides de pression dans les circuits d'écran ;

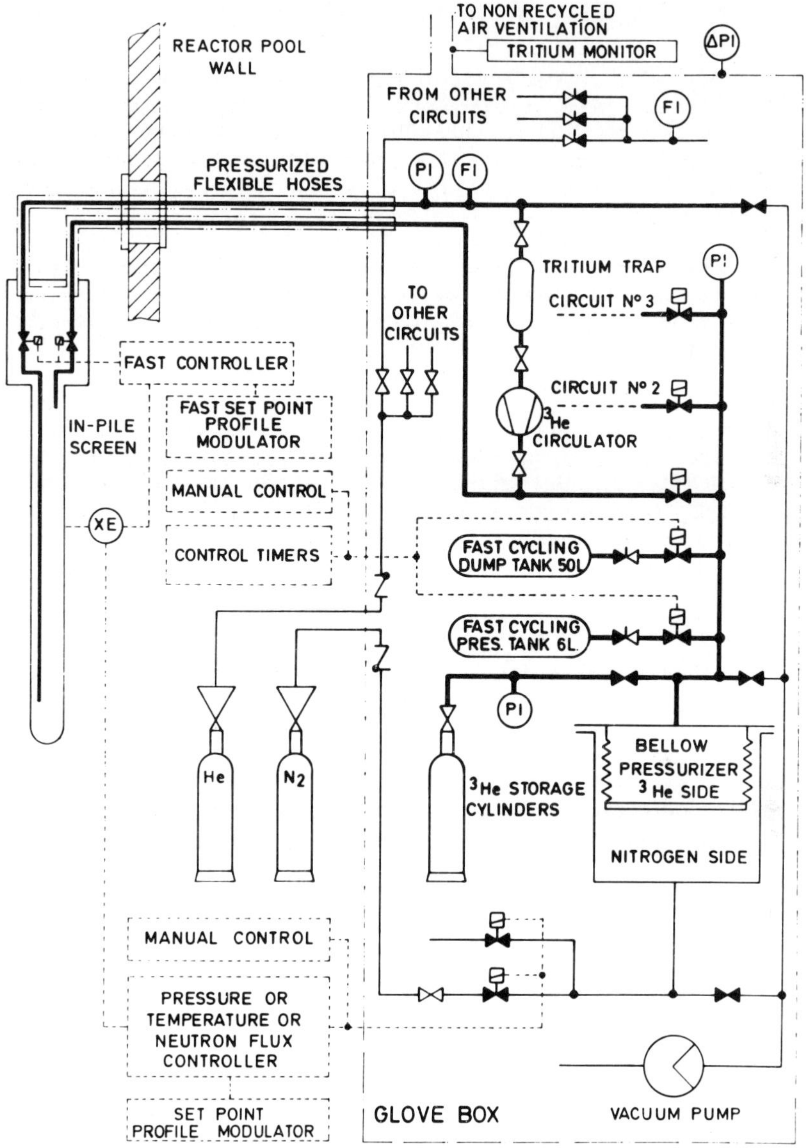

FIG. 1 SIMPLIFIED FLOW DIAGRAM OF THE FISSION POWER TRANSIENT DEVICE VNS (VARIABLE NEUTRON SCREEN) USING ^{3}He GAS AS NEUTRON ABSORBENT

- des réservoirs de stockage du gaz ^3He ;
- une installation d'évacuation par pompe à vide, connectable aux divers circuits ;
- tous les circuits ^3He sont entourés d'une deuxième enveloppe en communication avec la ventilation d'air non recyclé du bâtiment du réacteur BR2 ;
- la deuxième enveloppe des lignes de gaz ^3He entre l'installation hors pile ("Glove box") et le dispositif en pile est rincée en continu par un débit d'hélium sous une pression de 2 bar. La deuxième enveloppe des autres parties des circuits ^3He est en dépression (10 à 20 mm H_2O) par rapport à l'ambiance du bâtiment réacteur.
  Cette deuxième enveloppe est en fait une boîte à gants ;
- tous les rejets de gaz (des circuits ^3He, du système de pressurisation à l'azote, du rinçage des deuxièmes enveloppes, de la pompe à vide) sont collectés par la ventilation de la boîte à gants et un prélèvement continu (150 cm^3/min) d'air de ventilation est envoyé à un détecteur tritium.

2.1.2. Le tritium produit sous forme de molécules mixtes HT est retenu dans une trappe contenant 100 g de mousse de titane chauffée à 430 °C. Une trappe est remplacée lorsque, par calcul, la quantité absorbée de tritium est évaluée à 10 000 Ci. La circulation du gaz ^3He (20 à 300 cm^3 T.P.N./min suivant la pression de travail) est obtenue par un cylindre chauffé électriquement engendrant des pulsations de pression et par des clapets forçant le déplacement du gaz dans le sens adéquat.

2.1.3. Les circuits ^3He sont réalisés en tubes d'acier inoxydable 4 mm int. / 6 mm ext., en grande partie assemblés par soudure. Les éléments non soudés sont connectés par raccords métalliques biconiques ou, en cas de démontages fréquents (trappes), par raccords à joints toriques. La robinetterie des circuits ^3He est pourvue d'une étanchéité extérieure par soufflet métallique.

2.2. Contrôle de la pression ^3He et exécution des transitoires

Les divers modes de contrôle et d'ajustement de la pression du gaz ^3He dans les écrans absorbants sont les suivants (figure 1) :

2.2.1. Le pressuriseur à soufflet est mis en communication avec le circuit d'écran concerné et la pression est modifiée par action manuelle sur les électrovannes d'admission et d'émission d'azote.

2.2.2. Le pressuriseur à soufflet métallique est mis en communication avec l'écran concerné et la pression d'azote est automatiquement adaptée pour suivre un profil préprogrammé de transitoire de puissance de fission dans le dispositif en pile. La puissance de fission peut être injectée (XE) au niveau du régulateur sous forme soit d'une différence de température, soit d'une pression, soit d'un signal de détection de flux neutronique instantané. Le système peut réaliser des transitoires de puissance nécessitant des taux maxima de variations de pression ^{3}He de 0.1 bar/s.

2.2.3. Le circuit écran concerné est mis en communication avec soit le récipient de 50 litres (réduction de pression), soit le récipient de 6 litres (augmentation de pression), par ouverture des électrovannes correspondantes et via des vannes de réglage de débit. Ces dernières ont été préréglées pour donner la pente moyenne requise au transitoire de puissance. A titre indicatif, une augmentation de 5 bar à 35 bar peut être obtenue en 4 secondes et une réduction de 35 bar à 10 bar peut être obtenue en 5 secondes.

2.2.4. Le circuit d'écran est d'abord mis en communication avec l'un des récipients de 50 l ou de 6 l, mais l'écran reste toutefois isolé par des électrovannes situées dans le dispositif en pile, le plus près possible de la partie active de l'écran. L'ouverture de ces électrovannes est ensuite commandée par un régulateur rapide de façon à réaliser un profil préprogrammé de puissance de fission. En principe, à condition de disposer de l'élément de mesure de la puissance de fission (XE) à réponse suffisamment rapide, ce mode de contrôle permet des transitoires presqu'aussi rapides que le mode précédent décrit au § 2.2.3.

2.3. Performances et caractéristiques principales

Le tableau ci-après et le diagramme de la figure 2 résument les performances et caractéristiques de l'installation VNS 2.

- Gamme de pression ^{3}He bar abs. 1 à 38

- Facteur d'attenuation de la puissance de fission $= \dfrac{\text{puissance de fission à 1 bar abs}}{\text{puissance de fission à 38 bar abs}}$

Ecran de 2 mm d'épaisseur et dispositif de type "Pressurized Water Capsule" (réf 1)	1.75
Ecran de 3.5 mm d'épaisseur et dispositif de type "VIC" sans écran cadmium (réf 2)	2.50

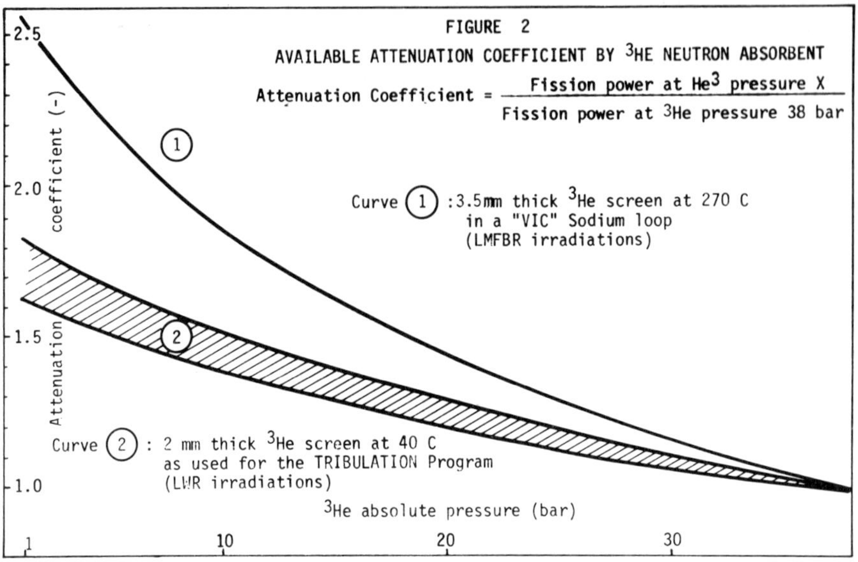

- Vitesse maximum des transitoires de puissance :

-pour un accroissement de puissance en %/s de la puissance initiale	15
-pour une réduction de puissance en %/s de la puissance initiale	12

- Volume de l'écran en pile litre 0.2 à 1

3. SURVEILLANCE DES REJETS DE TRITIUM DANS LA VENTILATION DU BATIMENT DU REACTEUR BR2

3.1. La production de quantités relativement grandes de tritium libérable(jusqu'à 80 Ci/j, par circuit d'écran He) a nécessité la mise en place de dispositifs de surveillance des rejets de tritium dans le système de ventilation du bâtiment du réacteur BR2.

3.2. Ces dispositifs de surveillance sont de trois types (références 3 et 4) :

3.2.1. Des détecteurs à chambre d'ionisation pour la mesure continue de la teneur totale en tritium (tritium oxydé HTO + tritium non oxydé HT). Ces détecteurs sont alimentés par un mélange composé, en volume, de 1/3 d'air à analyser et 2/3 de méthane.

Les détecteurs situés sur la ventilation de la boîte à gants (figure 1) des installations VNS sont de ce type. L'absence d'un système de compensation et d'un blindage contre le rayonnement ambiant limite la sensibilité de ces détecteurs à 10^{-4} Ci/m^3 d'air. Cette sensibilité est toutefois suffisante pour suivre les rejets tritium de chaque installation en boîte à gants.

Deux détecteurs de même principe, mais plus sensibles et équipés d'un système de compensation et d'un blindage, assurent la surveillance de l'air ambiant du bâtiment réacteur. Leur seuil de sensibilité se situe à 10^{-7} Ci/m^3 d'air.

Ces détecteurs d'ambiance sont chacun pourvus de deux niveaux d'alarme. Le premier niveau situé à 2.5×10^{-6} Ci/m^3, soit 50 % de la tolérance*, déclenche une action sur la ventilation du bâtiment réacteur qui consiste à interrompre le recyclage partiel de l'air de ventilation.

Le deuxième niveau est situé à 10^{-4} Ci/m^3, soit 20 fois la tolérance et, en cas de coïncidence d'alarme entre les deux détecteurs, il y a ordre d'évacuation du personnel occupé dans le bâtiment réacteur et isolation de ce bâtiment du point de vue de la ventilation.

3.2.2. On condense l'humidité de l'air ambiant sur des plaques réfrigérées. L'activité tritium de l'eau recueillie est mesurée. La mesure est discontinue et fournit une teneur moyenne en tritium oxydé (HTO) dans l'air sur la période de condensation. Le système est très mobile. Les taux de comptage minima observés dans le bâtiment réacteur sont de l'ordre de 150 Bq/cm^3 d'eau, correspondant à 4×10^{-8} Ci/m^3 (référence 3).

3.2.3. On prélève en continu un débit d'air de 1 m^3/h au pied de la cheminée de ventilation du bâtiment du réacteur. Ce débit d'air traverse un dispositif de barbottage à l'eau. Cette eau est prélevée périodiquement et son activité en tritium est mesurée. Ce système permet d'intégrer la quantité de tritium oxydé relâchée par l'ensemble des sources présentes dans le bâtiment. Il semble cependant que ce dispositif ne réponde pas efficacement à des rejets de courte durée (quelques minutes) et un programme d'essais est en cours pour étudier cet aspect.

---

*La tolérance pour le tritium oxydé est de 5×10^{-6} Ci/m^3 pour des personnes professionnellement exposées pendant 40 h par semaine.

4. OBSERVATIONS

4.1. Rejets normaux et accidentels de tritium par les circuits ^3He

4.1.1. Dans leur état neuf, l'étanchéité des circuits ^3He a été contrôlée par reniflage au spectromètre de masse à hélium et par un test de longue durée sous pression d'hélium. On peut estimer que le taux de fuite total d'un circuit est alors inférieur à 10^{-5} bar cm^3/s.
L'étanchéité des circuits ^3He est testée à nouveau avant remise en service au début de chaque cycle du réacteur BR2, soit une fréquence de 1 à 2 contrôles par mois. Les moyens de détection de fuite et la procédure suivie pour les tests de pression permettent de dire que le taux de fuite reste inférieur à 5×10^{-4} bar cm^3/s.
Pour de tels circuits ^3He, en cours d'irradiation à 30-35 bar les rejets observés de tritium vers la ventilation de la boîte à gants se situent entre 0.1 et 0.3 Ci/jour, dont la majeure partie (plus de 90 %) (réf. 4) est sous forme oxydée HTO.

4.1.2. Des rejets accidentels de tritium ont eu lieu. Le rejet le plus important à ce jour, 15 à 25 Ci, s'est produit en moins d'une minute. Le tritium s'est répandu dans la boîte à gants et deux heures après ce rejet, le taux de comptage tritium dans la ventilation de la boîte à gants était à nouveau normal. Il n'y a pas eu de conséquence décelable sur les dispositifs de détection tritium dans l'air ambiant du bâtiment réacteur.
Par contre l'incident, relaté au § 4.3., a eu certaines conséquences pour l'ambiance du bâtiment réacteur.

4.2. Teneur en tritium du gaz ^3He

4.2.1. La teneur en tritium du gaz ^3He dans les circuits d'écran est régulièrement contrôlée par prélèvement d'échantillons dont on détermine la concentration en tritium non oxydé (molécules HT) et oxydé (molécules HTO). La teneur en tritium semble avoir atteint une valeur constante de 0.8 Ci (HT + HTO)/litre T.P.N. de ^3He dont 95 % sont sous forme de HTO. Cette teneur correspond donc à une concentration minimum de 600 Vpm de vapeur d'eau tritiée.

4.2.2. Cette vapeur d'eau tritiée a probablement deux origines :
- oxydation de HT par les traces d'oxygène présentes dans les circuits (en particulier dans la mousse de titane) même après évacuation par la pompe à vide ;

-oxydation de HT par l'oxygène présent, à l'origine, dans le gaz ^{3}He (le fournisseur de ce gaz garantit $[O_2] < 100$ Vpm).
L'oxydation de la molécule HT se ferait dans la mousse de titane à 430 °C.

4.2.3. La présence de cette vapeur d'eau tritiée n'est pas tolérable pour les raisons suivantes :
-il y a accumulation de 300 Ci de tritium libérable dans l'ensemble des circuits ^{3}He ;
-cette vapeur d'eau engendre une contamination de surface importante des circuits ^{3}He (cfr. § 4.4.).
Pour réduire cette teneur en vapeur d'eau tritiée, un sécheur à tamis moléculaire a été mis temporairement en place sur la ligne de gaz ^{3}He vers le pressuriseur. Les effets de cette nouvelle disposition ne sont pas encore connus.

4.3. Teneur en tritium de l'air ambiant du bâtiment réacteur (références 3 et 4)

4.3.1. Depuis le remplacement de la matrice beryllium du réacteur BR2 (1979-1980) la concentration en tritium dans l'air ambiant du bâtiment réacteur se situe normalement à des valeurs de 1 à 2 % de la tolérance (cfr. note relative au § 3.2.1.).

4.3.2. Cependant l'incident relaté ci-après a provoqué une augmentation sensible de cette concentration jusqu'à 60 % de la tolérance. La source de la contamination a été localisée grâce aux prélèvements par condensation sur plaque réfrigérée (cfr. §3.2.2.) et aux frottis pris sur les surfaces suspectées.
Cette contamination trouve son origine dans la coïncidence des incidents suivants :
-une défection non détectée de la circulation du gaz ^{3}He a donné lieu à une accumulation anormale de tritium (plusieurs centaines de Curie) dans un circuit d'écran ;
-un coupleur auto-obturant, situé sur ce circuit d'écran, a été à l'origine de fuites intermittentes vers la deuxième enveloppe pressurisée, réalisée partiellement en chlorure de polyvinyle (C.P.V.). Quelques 10 à 15 Ci de tritium se seraient ainsi accumulés dans cette enveloppe ;
-cette deuxième enveloppe, par conception du dispositif d'irradiation, n'est pas rincée en continu par un débit de gaz et le tritium a finalement diffusé au travers de l'enveloppe en C.P.V.
Le relâchement de tritium vers le bâtiment du réacteur, dont le taux en début de période devait être de l'ordre de 1 Ci/jour, a nécessité l'arrêt du recyclage partiel de l'air de ventilation pendant 19 jours. Malgré le rinçage continu de la deuxième enveloppe incriminée, il s'est écoulé un délai de 12 semaines

avant de revenir à un niveau quasi normal (soit quelques % de la tolérance) de la contamination de l'air ambiant du bâtiment réacteur.

4.3.3. Les enseignements tirés de cet incident sont les suivants :

- des sollicitations (mouvements des lignes de gaz) sur les coupleurs auto-obturants provoquent des fuites intermittentes intolérables pour leur présente utilisation sur des circuits ^{3}He. Prévus à l'origine pour faciliter les déconnexions de circuits contaminés, ces coupleurs ont été remplacés par des raccords à visser ;
- la deuxième enveloppe doit être parcourue en continu par un débit de gaz. Ceci permet, grâce aux rejets de tritium, de détecter une perte d'étanchéité même intermittente sur le circuit ^{3}He et évite le séjour incontrôlé de tritium dans la deuxième enveloppe ;
- les matières plastiques (flexibles, isolants électriques, joints et bourrages d'étanchéité) montrent une grande capacité d'absorption de tritium (en particulier sous forme oxydée) dont la désorption ne se fait que très lentement.

4.4. Décontamination de circuits ^{3}He

Le démantèlement d'un dispositif en pile incluant un circuit ^{3}He doit être précédé d'une décontamination aussi complète que possible et d'une mesure du taux de relâchement tritium afin de mieux contrôler les rejets tritium pendant les opérations. Une technique efficace et sûre de décontamination allégerait les précautions à prendre lors de travaux sur les circuits ^{3}He.
Par rinçage des circuits contaminés à l'hélium naturel, à la température de la piscine (25 à 40 °C), le taux de désorption du tritium d'une surface métallique ne diminue que très lentement. A titre indicatif, on a constaté une réduction du taux de désorption de 0.5 Ci à 0.2 Ci/jour$\cdot$m^2 pour une durée de rinçage de 25 jours.
Plus de 90 % du tritium désorbé par les parois métalliques est sous forme oxydée. L'utilisation d'un sécheur (cfr. § 4.3.3.) permet d'espérer une certaine réduction de la contamination des parois.
Une extraction par la pompe à vide ne permet pas une décontamination plus efficace que par rinçage, vu les pressions partielles très faibles (quelques μbar) de HT et HTO.
La mise au point d'une technique de décontamination doit donc être poursuivie.

REFERENCES

1. H. NACKAERTS — SCK/CEN - MOL
Operational transients on LWR fuel in BR2 reactor - TRIBULATION Program
Int.Top. Meeting on Irradiation Technology - Grenoble Sept. 1982.

2. J. DEKEYSER — SCK/CEN - MOL
Transient experiments on LMFBR fuel pins in the reactor BR2 with the versatile sodium loop VIC.
Int.Top. Meeting on Irradiation Technology - Grenoble Sept. 1982.

3. P. GOVAERTS — SCK/CEN - MOL
Tritiumbesmetting BR2.
Internal note 14.02.1979.

4. P. GOVAERTS — SCK/CEN - MOL
Tritiumbesmetting in het reaktorgebouw BR2 in 1981 - Overzicht.
Internal note 03.12.1981.

MESURES DES TEMPERATURES D'IRRADIATION ET ETUDE DES SURCHAUFFES LOCALES DE GAINE EN ACIER AUSTENITIQUE, PAR DIFFRACTION X

P. ROUSSET* et R. CADALBERT**

CENTRE D'ETUDES NUCLEAIRES DE SACLAY, *DTech-SELECI, **DTech-SRMA

91191 GIF-sur-YVETTE CEDEX

INTRODUCTION

Le développement des éléments combustibles de réacteurs à Neutrons Rapides, conduit le C.E.A. à effectuer de nombreuses irradiations d'aciers du type 316 en particulier, dans Rapsodie et Phénix; la connaissance précise et le contrôle des températures d'irradiation posent un problème important, tant pour les gaines expérimentales que pour les gaines d'aiguilles combustibles utilisées dans ces réacteurs. La maitrise de ce paramètre est indispensable pour l'amélioration de la connaissance des propriétés sous irradiation de ces matériaux.

Les capsules et les assemblages n'étant pas ou peu instrumentés, différentes techniques ont été développées ces dernières années pour déterminer les températures locales d'irradiation dans les matériaux de structure, telles que désorbtion du Kr85, détecteurs SiC ou à point de fusion etc... mais elles présentent des difficultés d'exploitation, en particulier l'impossibilité de placer certains de ces détecteurs dans des aiguilles combustibles. En conséquence, nous avons été amenés dans ce but, à utiliser la relation qui existe entre la température d'irradiation et la densité de défauts présents dans les aciers austénitiques.

I - PRINCIPE DE LA METHODE

I.1 Généralités

La largeur des raies de diffraction X permet facilement de caractériser le degré de distorsion du réseau cristallin et ainsi d'analyser quantitativement la densité de défauts créés soit par

P. von der Hardt and H. Röttger (eds.), Irradiation Technology, 451–459.

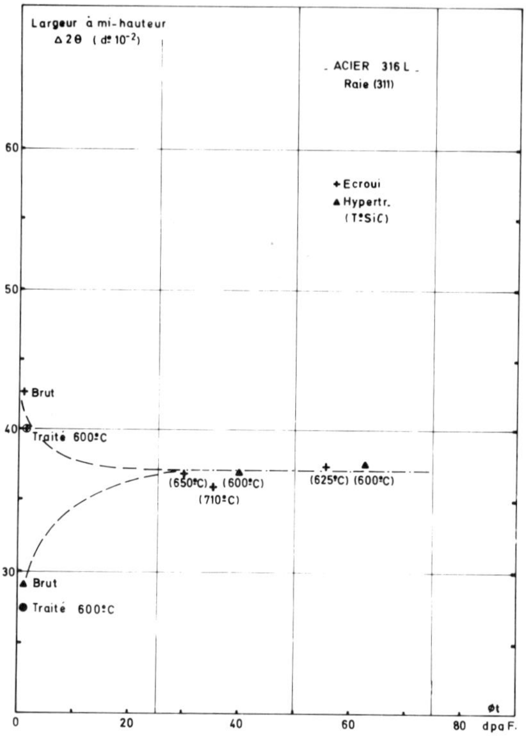

EVOLUTION DE LA DENSITE DE DEFAUTS AVEC LA DOSE D'IRRADIATION A TEMPERATURE ELEVEE. Fig. 1

déformation plastique [1] soit, dans le cas présent, par irradiation aux neutrons. Après une faible dose d'irradiation, (quelques dpaF), on observe qu'il s'établit, dans les aciers austénitiques une densité de défauts d'équilibre stable [2-3], qui ne dépend plus essentiellement, pour un alliage donné et dans un certain domaine de flux rapide, que de la température, ceci quelle que soit la densité de défauts initiale, de l'état hypertrempé ou écroui (fig. 1). Des mesures de diffraction X ont montré que la largeur de raie -, qui caractérise cette densité de défauts -, variait fortement avec la température entre 400 et 650°C environ, ce qui correspond pratiquement aux températures normales d'utilisation des éléments combustibles rapides. Pour appliquer cette méthode, il était alors nécessaire d'effectuer un étalonnage afin de disposer de la relation entre la largeur des raies de diffraction X, soit $\Delta 2\theta$ et la température d'irradiation.

I.2 Etalonnage de la méthode

Les résultats obtenus nous ont donc conduits à appliquer cette technique à la détermination de la température de fin d'irradiation d'éléments de structure en acier et en particulier des gaines en acier 316 : la température de référence choisie pour l'étalonnage, la plus "commode", a été la température calculée par les codes thermohydrauliques sur les assemblages combustibles. Ceci permet de mieux comparer les résultats sur échantillons de gaine et sur aiguilles combustibles : les aiguilles choisies pour celà, sont des aiguilles centrales dont la thermohydraulique est la mieux connue. La première courbe d'étalonnage a été réalisée sur une aiguille d'un assemblage de surveillance de Phénix en acier 316 hypertrempé, ayant atteint 60 dpa au flux maximal (§ I.4).

I.3 Technique utilisée

L'appareil est un diffractomètre à double diffraction, avec monochromateur AR en Al (111), associé à un compteur proportionnel blindé : ces caractéristiques permettent de réduire fortement le bruit de fond γ dû à l'échantillon et la largeur spectrale du rayonnement X ($L\alpha$ du W). La raie utilisée est la raie aux grands angles d'indices(420), la sensibilité augmentant avec $Tg\theta$ (θ = angle de Bragg).

Les échantillons de gaine sont prélevés par tronçonnage d'aiguilles et vidés du combustible par ultra-sons, en prenant soin de ne pas introduire d'écrouissage parasite; puis ils sont nettoyés par un léger polissage électrolytique en bain perchlorique, l'épaisseur enlevée étant <1/100 mm. Pour les échantillons de capsules d'irradiation, ils sont simplement polis par électrolyse.

I.4 Résultats de l'étalonnage

On a étudié l'évolution axiale de la largeur de raie d'une gaine en 316 Hyp, qui après conversion en température, a conduit à la courbe figure 2, représentant l'évolution de $\Delta 2\theta$ avec la température extérieure gaine calculée, Teg (°C).

a) On y distingue deux domaines très différents : une partie quasi linéaire vers les moyennes et hautes températures et un maximum vers les basses températures (en bas de colonne fissile). Ce maximum du $\Delta 2\theta$ mesuré, pouvant être dû à plusieurs causes, n'est pas totalement expliqué pour le moment. Or on sait par expérience, que $\Delta 2\theta$ décroît régulièrement avec la température, au moins à partir de 400°C. On a donc été amené à extrapoler linéairement cette courbe vers les basses températures, afin de pouvoir l'utiliser dans toute la gamme de mesure prévue, ceci en attendant de pouvoir obtenir une courbe expérimentale complète.

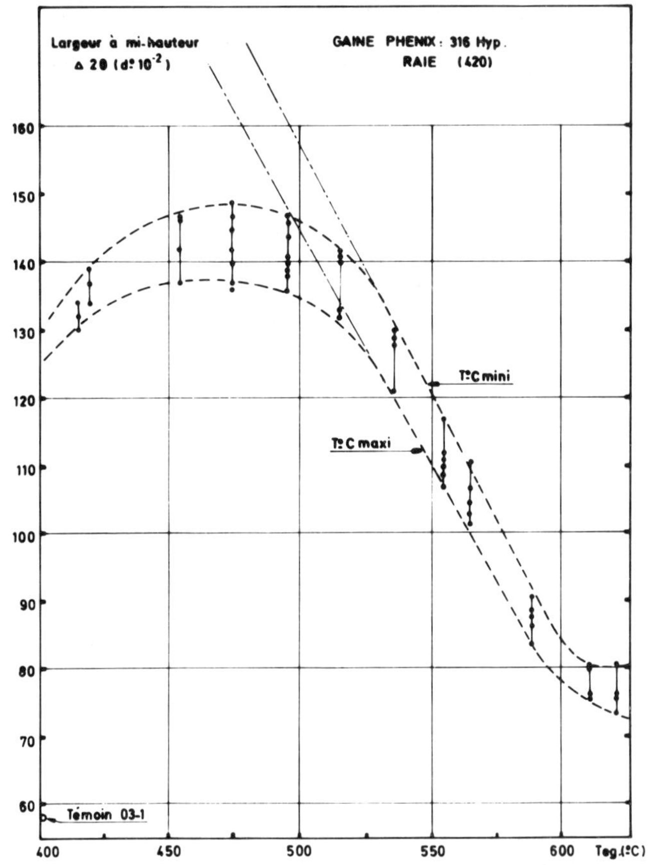

EVOLUTION DE LA LARGEUR DE RAIE AVEC LA TEMPERATURE EXTERIEURE - GAINE PHENIX 316 Hyp. Fig. 2

b) Ces mesures ont mis en évidence un gradient circonférentiel de largeur de raie, variant légèrement le long de l'aiguille (12 à 17°C dans la partie linéaire); celui-ci peut être attribué, en l'absence de tout écrouissage parasite, à un gradient de température dans une même section. Ces écarts de température se retrouvent dans tous les cas examinés, comme on le verra plus loin, dans les aiguilles centrales et les échantillons de gaine, en principe exempts de gradient thermique radial.

c) La valeur initiale de $\Delta 2\theta = 58/100°$, pour le témoin avant irradiation (316 Hyp à 1150°C), montre l'augmentation très importante de $\Delta 2\theta$ après irradiation à basse température. La pente de la courbe $\Delta 2\theta = f$ (T°) est telle qu'on peut mettre en évidence des variations relatives de températures très faibles de 2°C environ; mais la précision absolue de la méthode est en réalité moins grande et difficile à évaluer actuellement : validité des codes, mauvaise précision dans les fortes largeurs de raie, contrôle de la profondeur de polissage etc ...

II APPLICATIONS

II.1 Etude des profils thermiques des gaines dans un assemblage

Indépendamment des mesures indirectes des températures d'irradiation développées ci-dessous, cette méthode permet d'effectuer une étude fine du profil thermique des différentes aiguilles d'un assemblage, en particulier pour vérifier les anomalies de refroidissement ou préciser le fonctionnement d'aiguilles particulières, par exemple les aiguilles d'angle. Cette technique a été appliquée au cas de 3 aiguilles d'un assemblage Phénix en 316 Hyp. irradié à 70 dpa maximum, qui a été tourné de 180° à mi-vie : une en position centrale et deux en position semi périphérique et diamètralement opposées.

La comparaison des largeurs de raie obtenues à une côte donnée, en haut de colonne fissile, a permis de déterminer l'orientation finale de l'assemblage par rapport au coeur du réacteur.

La figure 3 représente les 3 profils axiaux de largeur de raie $\Delta 2\theta$, en fonction de la température calculée Teg; on peut en déduire les observations suivantes :

a) les 2 aiguilles extérieures ont des profils thermiques identiques entre eux, mais assez différents de celui de l'aiguille centrale, surtout dans la moitié inférieure de la colonne fissile : on peut penser que les conditions de refroidissement au bas de la colonne fissile, sont assez différentes entre les 2 types d'aiguilles, même si un effet de recuit ultérieur a pu modifier le profil réel de densité de défauts après irradiation.

b) on observe également ici, un certain gradient circonférentiel de température le long des 3 aiguilles, ce qui permet en outre de mettre en évidence des surchauffes éventuelles lors de contacts gaine-gaine. Ces résultats montrent l'intérêt de la méthode pour étudier les conditions de refroidissement dans un assemblage.

II.2 Etude des surchauffes locales : mesure des points chauds

La sensibilité de la méthode et sa finesse d'exploration sur des gaines de géométrie Phénix (Ø 6,55 mm), en font une technique originale et parfaitement adaptée à l'étude des surchauffes locales consécutives à des contacts linéaires gaine/gaine ou gaine TH.

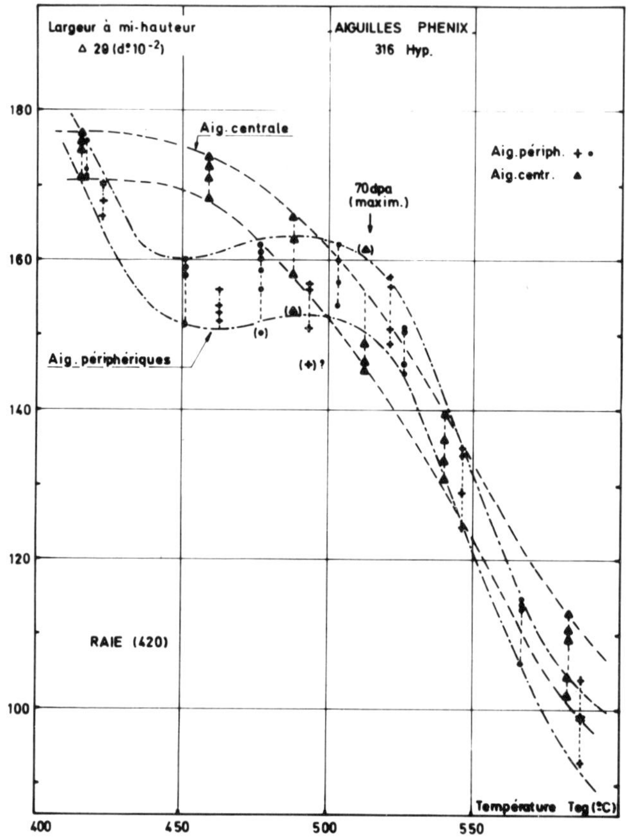

EVOLUTION DE LA LARGEUR DE RAIE AVEC LA TEMPERATURE CALCULEE (CODE PHENIX) Fig. 3

Plusieurs cas ont été examinés dans ce sens, dont nous présentons un exemple caractéristique : il s'agit de 2 tronçons de gaine de hauteur 10 mm prélevés dans un assemblage enrobé puis tronçonné, ce qui a permis la mise en évidence des contacts mécaniques gaine/gaine : la génératrice du contact ayant pu être repérée avant désenrobage, elle a servi de référence pour les mesures en diffraction X.

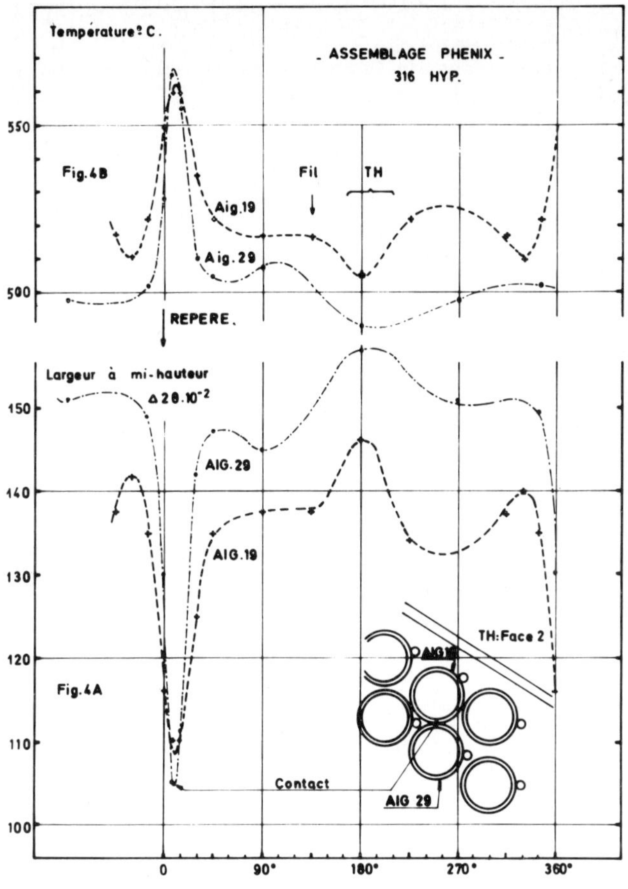

MISE EN EVIDENCE D'UN POINT CHAUD AU DROIT D'UN CONTACT GAINE-GAINE DANS UN ASSEMBLAGE PHENIX - Fig. 4

La figure 4a en bas, illustre les variations très importantes de la largeur de raie exactement au droit du contact pour les 2 tronçons; à partir de ces valeurs de $\Delta 2\theta$, converties en terme de températures calculées, on a pu obtenir (fig. 4b en haut) l'évolution circonférentielle de température avec des points chauds très nets sur environ 30° d'angle, ce qui correspond à des surchauffes locales de 75 - 80° et 55 - 60° respectivement, au-dessus des températures minimales. Ces résultats sont en très bon accord avec les estimations effectuées par les codes thermo-hydrauliques; les fluctuations radiales sont dans ce cas d'environ 20°C, hors point chaud.

II.3 Mesures des températures d'irradiation en capsules

De nombreuses mesures de température sur des gaines expérimentales irradiées en capsules sont effectuées, afin de contrôler les températures calculées et de vérifier le fonctionnement thermique de ces capsules : la plupart des irradiations récentes ou en cours concernant surtout la nuance 316 Ti - ε 20 %, l'exploitation de ces mesures est faite provisoirement à l'aide de la courbe d'étalonnage propre au 316 Hyp. (voir § 1.4). La figure 5 montre les résultats obtenus sur des tubes de gaines en 316 Ti-ε , irradiés dans une capsule à lame de gaz et permet de comparer les profils de température calculée (de 550 à 650°C), avec les températures mesurées par RX : on voit que le profil mesuré est beaucoup moins accusé que celui visé et qu'il existe un certain écart entre les deux courbes. Cependant, l'utilisation ultérieure d'une courbe étalon propre au 316 Ti-ε , permettra certainement d'après nos résultats récents, de réduire l'écart observé, en particulier pour les hautes températures, où la densité de défauts est plus élevée pour le 316 Ti-ε que pour le 316 Hyp.

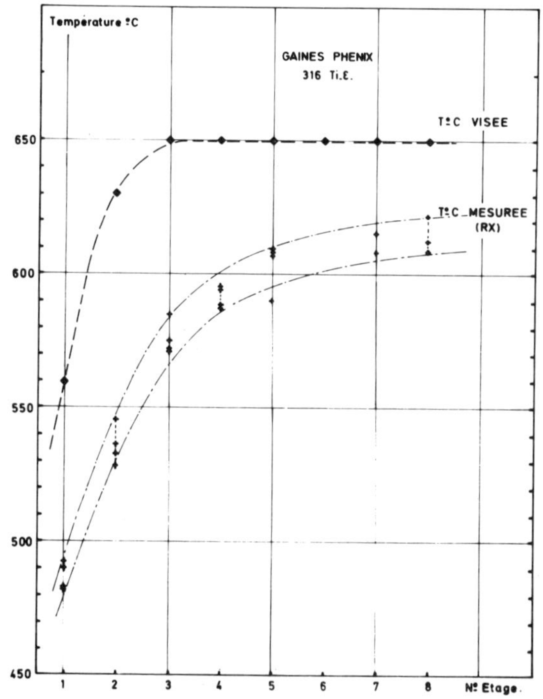

EVOLUTION DES TEMPERATURES AVEC LA COTE AXIALE (CAPSULE)
GAINES PHENIX 316 Ti-ε - Fig. 5

CONCLUSIONS

Ces quelques exemples montrent le grand intérêt des mesures de largeurs de raie de diffraction X, sur des gaines en acier austénitique irradiées en neutrons rapides, d'une part pour l'étude des profils thermiques des aiguilles combustibles et des surchauffes éventuelles, d'autre part pour la mesure des températures d'irradiation en capsules. Cette méthode très sensible est simple à mettre en oeuvre sur un diffractomètre à compteur, en cellule blindée, associé à un calculateur et un porte-échantillon automatique, permettant de tracer rapidement des cartes de répartition de température. Elle est aussi largement applicable aux autres composants du coeur tels que tubes hexagonaux (TH), barres de commande etc ...

BIBLIOGRAPHIE

1. R. CALDALBERT, J.L. BARON
   Communication Journées d'Automne de Métallurgie - PARIS (1977)

2. N. AZAM, J. DELAPLACE, L. LE NAOUR
   Conférence d'Ajaccio - JUIN (1979)

3. Communication personnelle (1979)

DEVELOPPEMENT D'EQUIPEMENTS ET D'EXPERIENCES HORS PILE COMME SUPPORT D'IRRADIATIONS EN REGIME TRANSITOIRE

M. Coquerelle, P. Knappik, C. Perrier, R. Schreiber

Commission des Communautés Européennes
Centre Commun de Recherche
Etablissement de Karlsruhe
Institut Européen des Transuraniens
Postfach 2266
D-7500 Karlsruhe
République fédérale d'Allemagne

1. INTRODUCTION

La compréhension des déformations mécaniques subies par une aiguille combustible irradiée dans des conditions de régime transitoire nécessite la caractérisation physico-chimique du combustible expérimenté. Dans ce contexte, l'accent sera mis sur la détermination quantitative des produits de fission volatiles et leur distribution dans le combustible avant et après l'excursion de température. En particulier, il sera intéressant de voir si, dans le domaine de température considéré ces produits de fission sont maintenus en solution dynamique ou non. Les informations acquises dans le cadre d'un tel travail s'adressent aux modélistes qui calculent la cinétique des processus régissant les variations de volume du combustible et la migration des produits de fission. La quantité de matériau à analyser, liée au nombre d'irradiations en régime transitoire sera toujours limitée et la possibilité de simuler sur combustible irradié des régimes de température se rapprochant des conditions en réacteur augmenterait de façon sensible le potentiel d'analyses et justifie, de ce fait, des développements expérimentaux dans ce sens.
Ces expériences de simulation ne mettent en jeu que de faibles quantités de matériau - 10 à 25 mm de colonne de combustible - sur lesquelles sera créé un gradient de température se rapprochant de celui obtenu en réacteur dans des conditions réelles. Une autre approche expérimentale est d'exécuter des recuits thermiques isochrones. Les informations expérimentales découlent de la comparaison de la caractérisation avant et après la simulation. La validité de ce travail suppose d'une part, une connaissance relativement précise du régime de température durant la simulation et, d'autre part,

P. von der Hardt and H. Röttger (eds.), Irradiation Technology, 461–473.

la disponibilité de laboratoires de microscopie optique, électronique, de spectrométrie de masse et gamma de haute qualité.

Il était donc nécessaire de disposer de techniques de traitements thermiques permettant de simuler les conditions rencontrées en réacteur et aussi de moyens modernes d'analyse assurant la détermination de la mobilité des produits de fission volatiles. En conséquence, des améliorations ont dû être apportées à l'équipement standard de notre laboratoire α, β, γ dans deux domaines:

- L'équipement de notre cellule de traitement thermique a dû être repensé et une nouvelle installation comprenant deux fours a été décidée:

  . un four à chauffage électrique direct (DEH) permettant l'obtention de gradients thermiques semblables à ceux existant dans le combustible lors d'excursion de température dans le réacteur. Dans ce cas, le combustible irradié sert de résistance chauffante et est placé entre deux électrodes en tungstène connectées au générateur; l'installation permet d'atteindre la fusion centrale du combustible. Au cours du recuit, opéré en atmosphère hélium, la température extérieure du combustible est évaluée en deux positions par pyrométrie; la puissance injectée est aussi mesurée. Ces recuits thermiques ont nécessité la mise au point de la préparation d'échantillons de combustibles irradiés à partir d'aiguilles soit par extrusion mécanique soit par dissolution chimique ou électrolytique.
  Simultanément l'analyse des gaz de fission relâchés durant l'expérience peut se faire par spectrométrie γ ou par spectrométrie de masse.

  . Parallèlement à ces expériences sous gradient de température, un four à induction permet le recuit de combustible, à température constante sous atmosphère contrôlée jusqu'à la température de fusion du combustible (500 Khz, 25 KVa sur la bobine). Durant ces traitements le relâchement des gaz de fission peut également être mesuré par spectrométrie γ ou par spectrométrie de masse.

- Notre laboratoire disposait déjà des techniques permettant les mesures de distribution de porosité par microscopie optique quantitative et par microscopie électronique à partir de répliques, c'est-à-dire avec une limite de résolution située vers 400 A° /1,2,3/. Un élargissement de la gamma d'analyse apparut nécessaire et l'achat d'un microscope électronique à balayage fut décidé. Le rôle de cet instrument est de compléter les informations acquises en microscopie optique sur les variations morphologiques subies par le combustible, en particulier, les relations entre le grain, le joint de grain et la porosité et son mode d'interconnexion.

Ces installations sont opérationnelles depuis quelques mois et ce rapport sera principalement consacré, pour des raisons évidentes, aux dispositifs expérimentaux en cellule chaude et commentera, par la suite, les essais qui ont été réalisés durant la phase initiale d'exploitation.

2. EQUIPEMENT EXPERIMENTAL

2.1. Traitements thermiques

2.1.1. Principe de l'installation

L'installation comporte une cellule α, β, γ et une boîte à gants (Fig.1). La cellule α, β, γ contient les fours à chauffage électrique direct et à induction placés sur une console commune en position arrière du caisson α. Ce positionnement était imposé, d'une part par la nécessité de réduire les pertes en énergie pour l'alimentation de la bobine du four à induction, en limitant au maximum la distance oscillateur-bobine et, d'autre part, par le désir de pouvoir intervenir manuellement, le cas échéant, via la fenêtre arrière dans la mesure où le niveau d'activité gamma l'autorise.

L'ensemble fours est desservi par un circuit fermé d'hélium utilisé pour le conditionnement d'atmosphère, comme gaz de refroidissement ou encore comme gaz vecteur pour produits de fission gazeux. En aval des fours, sur le circuit He se trouvent:

a) des pièges permettant la rétention de produits de fission solides tels que Cs, Ru, Rb ou I_2

b) hors de la cellule α, β, γ une boîte à gants munie d'un flow counter permettant la mesure en continu du Kr et du Xe ainsi qu'un piège à azote liquide en vue de collecter l'ensemble des produits de fission gazeux relâchés durant les traitements thermiques. L'analyse qualitative et quantitative des produits de fission gazeux peut être effectuée au moyen d'un spectromètre de masse dont le quadrupole est installé dans la boîte à gants (Fig.1).

2.1.2. Four à chauffage électrique direct (D.E.H.)

Des installations similaires dont le principe consiste à utiliser le combustible comme résistance électrique ont déjà été développées en laboratoire froid à l'Institut Européen des Transuraniens par Sari /4/, à la KfK par Alexas /5/ et en laboratoire chaud aux USA par Wrona /6/ et au CEA par Combette et Krieger /7/. Chacun de ces fours est l'application de considérations théoriques développées par Freund et Schikarski /8/ qui les premiers ont décrit les possibilités de simulation hors pile des régimes thermiques rencontrés par les combustibles en réacteur. L'appareillage conçu pour nos cellules chaudes découle de l'installation de Sari. Les changements que nous y avons apportés sont dictés par les problèmes associés à l'état mécanique de l'échantillon irradié, à son positionnement dans le four avec le manipulateur et aussi à la récupération des produits de fission gazeux; aucune modification n'a été apportée au générateur.

- Alimentation

Le générateur doit tenir compte du niveau de température maximum à atteindre et aussi des propriétés de résistance spécifique des combustibles à analyser. Celles-ci diffèrent totalement suivant qu'il s'agit d'oxyde ou de carbure. Pour couvrir la gamme des besoins, nous disposons de 4 régimes de puissance.

Dans le cas de l'oxyde, la résistance spécifique diminue exponentiellement avec la température, le coefficient de température $\alpha = 1/\rho \frac{d\rho}{dT}$, étant négatif, le traitement thermique s'effectue en plusieurs étapes. L'échantillon subit un préchauffage obtenu par une tension d'amorçage d'environ 300-400 V, qui chute rapidement à 50 V avec un courant d'environ 2,5 A. Au cours de cette étape la résistance spécifique de UO_2 chute de 10^9 à 10^{-1} $\Omega.cm^{-1}$. Ce stade étant atteint, on passe à un deuxième régime de puissance avec une tension d'entrée de 40-50 V et un courant de 4-5 A, ce qui permet d'obtenir des températures de 1200°C sur l'échantillon. Si des températures plus élevées sont souhaitées, 2 régimes de puissance restent à la disposition de l'opérateur, l'un caractérisé par une tension de 60 V et un courant de 250 A, l'autre par une tension de 15 V et un courant de 2000 A.

- Contrôles

En cours d'opération, la tension entre les électrodes et le courant passant à travers l'échantillon sont contrôlables et enregistrables en permanence. Les seuls contrôles de température consistent en la mesure de la température en surface de l'échantillon en deux positions diamétralement opposées, par pyrométrie optique. Les pyromètres utilisés sont une version spéciale du type IS2 de la firme Gulton où l'électronique est séparée de l'optique et est installée hors cellule. Les mesures s'effectuent dans le domaine spectral s'étendant de 0.4 - 1.1 µm au moyen d'un détecteur en silicium. Les domaines de température mesurables s'étendent de 750 à 1500°C ou de 1100 à 3000°C. La surface mesurée sur l'échantillon est d'environ 5 mm². En cellule, l'optique du pyromètre est protégée par un blindage en Pb de 80 mm d'épaisseur; l'ensemble est fixé sur une table à 2 mouvements de liberté (selon l'axe et le rayon de l'échantillon) télécommandables.

- Programmeurs

2 programmeurs, l'un optique (firme Hartmann und Braun) et l'autre digital (firme IEA) permettent la programmation de la puissance en fonction du temps.

- Préparation de l'échantillon

La préparation de l'échantillon constitue certainement la partie la plus aléatoire des différentes étapes expérimentales nécessaires pour ce genre d'essai. Elle consiste à préparer une section de colonne de combustible sans matériau de gainage et à la placer dans l'enceinte du four. Nous avons développé trois méthodes pour récupérer une section de combustible irradié:

a) Extrusion mécanique

Une section d'aiguille combustible est tronçonnée à une longueur ad hoc. Cette section est fixée dans un guidage aligné sur l'axe du piston d'une presse hydraulique dont l'avance est contrôlable. Lors de l'avance progressive du piston, l'échantillon est extrudé de la gaine et est transféré directement dans le tube en quartz du porte-échantillon. Avant d'en effectuer le transfert dans l'enceinte du four, il convient d'éliminer les modifications de résistance de l'échantillon créées pour la présence de fissurations. On peut pallier partiellement cet inconvénient en remplissant de poudre UO_2 les fissures provenant de l'irradiation. La vibration par ultra-son nous est apparue comme une méthode efficace pour effectuer cette opération.

b) Dissolution de la gaine

Dans le cas où l'extrusion mécanique du combustible se révèle impossible, deux méthodes de remplacement peuvent être utilisées; elles consistent à dissoudre le matériau de gainage soit chimiquement soit électrolytiquement.

1. Dissolution chimique en milieu méthanol - Br_2, en utilisant un dispositif à circulation continue du solvant en circuit fermé
2. Dissolution électrolytique en milieu nitrate.

2.1.3. Four à induction

Cette installation répond à deux besoins:

a) réaliser des recuits thermiques à température constante et suivre simultanément le relâchement du gaz de fission
b) déterminer la quantité de gaz de fission retenue dans des combustibles irradiés.

- Alimentation

Il s'agit d'un générateur 500 Khz, 51 KVA de la firme Hüttinger assurant une puissance maximum de 25 KVA sur la bobine. Dans ce type d'alimentation, l'oscillateur est séparé du générateur et en vue d'éviter des pertes de puissance intolérables, il convient de réduire au maximum la distance oscillateur-bobine (1-1,5m), ce qui dans notre cas particulier a imposé le positionnement de la bobine en position arrière du caisson α et celle de l'oscillateur entre la fenêtre arrière du caisson α et la protection arrière γ de la cellule α, β, γ(Fig.1). Ce conditionnement a nécessité le développement d'un passage coaxial étanche α fixé sur la fenêtre arrière du caisson α. L'ensemble générateur, oscillateur et bobine est refroidi par un circuit fermé à eau de la firme Wischman, débitant au maximum 13 m^3/h.

Enceinte du four. Porte-échantillon

L'ensemble creuset-bobine se trouve dans une enceinte cylindrique Ø = 280 mm, h = 200 mm comportant quatre fenêtres en quartz, trois suivant des positions radiales situées à 90° et une en position axiale supérieure. Cette enceinte peut travailler sous vide primaire ou sous atmosphère contrôlée. Le porte-échantillon est constitué d'un creuset en tungstène de Ø = 15 mm, h = 40 mm posé et centré sur un bloc d'isolation en nitrure de bore; il est associé à une tige mobile, déplaçable permettant le transfert du creuset hors de l'enceinte au niveau de la fenêtre axiale en position supérieure.

Mesure de la température

La température de l'échantillon est assurée par un pyromètre à travers la fenêtre axiale supérieure. Il s'agit d'un pyromètre à détecteur PbS de type IP1, de la firme Gulton, à double domaine de température - 600-1500°C et 1000-3000°C opérant dans le domaine spectral compris de 1 à 3 µm.

2.2. Microscopie électronique à balayage JSM-35C pour examens de matériaux irradiés

2.2.1. Conditionnement alpha, gamma

L'installation sera conçue pour permettre l'examen d'échantillons obtenus généralement par fractographie, non décontaminés et susceptibles d'entraîner une contamination alpha, gamma lors des transferts et des analyses. Une hypothèse de base était, donc, d'admettre l'éventualité d'une contamination α, β, γ de la colonne du microscope.
Nous avons admis que la flexibilité maximum d'utilisation ne pouvait être assurée qu'en plaçant le microscope entièrement dans une boîte à gants. Ce choix s'accompagnait d'un avantage marquant sur le plan mécanique, car il évitait l'adaptation complexe d'un transfert d'échantillons vers la chambre du microscope.
Les seules modifications à signaler de la version standard du microscope sont donc a) la commande et le contrôle des mouvements de la table goniométrique et b) le blindage du détecteur.

2.2.2. Boîte à gants. Protection alpha

L'ensemble du microscope est installé dans une boîte à gants étanche (Fig.3). Celle-ci comporte dans sa partie inférieure quatre consoles en équerre qui supportent le microscope. En position médiane, une table sépare la colonne et la chambre du microscope, des accessoires vides, alimentation électrique, filtre et vannes qui sont placés dans la partie inférieure de la boîte (Fig.3). Cette table sert aussi de plan de travail pour les manipulations d'échantillon.

Dans la partie supérieure, un pont roulant facilite les manipulations de démontage de la colonne, en cas de nécessité de décontamination. Cette opération se déroule dans une deuxième boîte à gants située dans le voisinage du microscope et les transferts nécessaires se font par la paroi arrière de la boîte à gants du microscope, par un système double couvercle La Calhène.

2.2.3. Protection biologique gamma

La protection biologique gamma était, en conséquence, déterminée par le choix du conditionnement alpha. La boîte à gants sera entourée de 2 murs latéraux mobiles de 15 cm d'épaisseur. La face avant de la protection gamma est constituée d'un mur de briques en Pb de 10 cm d'épaisseur, muni de deux hublots et de deux pinces. Le microscope étant adossé à la face arrière d'une cellule α, β, γ chaude du laboratoire α, β, γ le problème de la paroi arrière ne se posait pas.

2.2.4. Modifications du microscope

Blindage des détecteurs

En raison de l'émission β, γ des échantillons analysés, une interaction potentielle de ces radiations avec l'électronique du détecteur se situe à deux niveaux: d'une part, les β émis par l'échantillon peuvent interférer avec les électrons secondaires responsables de la formation de l'image et, d'autre part, les gamma peuvent entraîner des dégâts de l'électronique, en particulier, perturber le fonctionnement du photomultiplicateur. Ces différentes interactions conduisent à une augmentation de bruit de fond et à une altération de la qualité et de la résolution de l'image. Il s'agit donc de les limiter dans la mesure du possible. Etant donné que la trajectoire des électrons secondaires est concave, un blindage peut donc être placé entre l'échantillon et le collecteur; son rôle est d'absorber le plus possible les β et γ de haute énergie. Cette solution technique a été développée par la firme JEOL et nous la décrivons ci-après.
Un blindage en Pb est placé d'une part, à l'intérieur de la chambre du microscope entre l'échantillon et le collecteur et, d'autre part, à l'extérieur de la colonne, enveloppant le scintillateur. Le design du blindage est conçu de façon à réduire l'activité γ d'un facteur 10^3, tout en maintenant une efficacité de collection des électrons secondaires d'un niveau comparable à celui d'un détecteur conven-

tionnel. Ce but est atteint par l'insertion d'électrodes au niveau du collecteur. Une possibilité technique supplémentaire permettant d'améliorer le signal donné par électrons secondaires sera obtenue en variant la tension positive sur le collecteur. L'efficacité de ce système a été testée avec succès sur des échantillons MC et MO_2 irradiés. Dans son design actuel, le détecteur est à même d'effectuer des analyses sur des échantillons présentant une activité maximum de 4 curies 1 MeV, avec une résolution de 400 Å.

Commande et contrôle des mouvements de la table goniométrique

Tous les mouvements de la table goniométrique x, y, z, la rotation et l'inclinaison seront télécommandés par des micromoteurs dont les déplacements sont contrôlés par des transducteurs potentiométriques.

Echantillon

Préparation et métallisation de l'échantillon

Les échantillons seront obtenus par fractographie et fixés sur un porte-échantillon disposant d'un mors mobile; une décontamination de l'échantillon n'est pas prévue et seule une attaque ionique sera, selon la nature de l'information désirée, effectuée avant la métallisation. Cette opération pourra être réalisée au moyen de deux dispositifs soit par sputtering soit par évaporation d'une feuille d'or.

Le premier n'est rien d'autre que le canon à ions que nous utilisons pour l'attaque cathodique des échantillons métallographiques. Dans le cas présent, l'échantillon métallographique est substitué par une feuille d'or qui est bombardée par des ions argon. Un porte-échantillon supplémentaire a été installé, selon une orientation adéquate vis-à-vis de la feuille d'or, pour permettre la métallisation de l'échantillon. La même opération peut être effectuée dans une cloche à évaporation de la firme Polaron conçue pour le travail en cellule chaude.

3. Résultats expérimentaux obtenus à partir de simulations thermiques

Le principe de base de ces expériences hors pile est d'apporter des informations complémentaires aux irradiations dans le but de déterminer les variations de structure des combustibles soumis à des conditions de régime transitoire. La notion de complémentarité peut s'exprimer en respectant trois critères:

- meilleure connaissance des conditions thermiques expérimentales
- élargissement du domaine de température investigué
- exécution de recuits sur des combustibles irradiés bien caractérisés correspondant à une gamme paramétrique, puissance linéaire, burn-up, régime de température aussi large que possible. Ce tra-

vail a démarré depuis quelques mois et nous nous limiterons dans ce rapport à la démonstration des possibilités expérimentales de ces développements en citant quelques résultats caractéristiques.

3.1. Traitements thermiques isochrones

A titre d'exemple, nous ferons appel à une étude effectuée dans le cadre du programme gonflement des combustibles avancés qui comprend deux volets. D'une part, des irradiations de type GOCAR et GOCART /9, 10/, effectuées dans le réacteur SILOE; d'autre part, des études de variation de morphologie du combustible, en particulier, de la dimension du grain, de la distribution de la porosité, du Xe, Cs après avoir été soumis à des traitements thermiques isochrones et à des montées en température variant de 10 à 100°C/min. Après chaque recuit, le combustible fut examiné en microscopie optique éléléctronique. Les profils radiaux de Xe et Cs ont été déterminés par microsonde.
Ce type d'étude a permis de préciser l'évolution de la structure du carbure mixte après des étapes transitoires s'étendant de 700 à 1300°C. Trois processus principaux ont pu être mis en évidence:

- la précipitation du gaz sous forme de bulles intragranulaires (fig. 4)
- l'augmentation de la dimension moyenne des bulles intergranulaires et la diminution de leur concentration
- pas de variation significative de la porosité de départ.

Ces résultats s'expliquent difficilement à partir des connaissances actuelles de la cinétique des combustibles. Ronchi /11/ a appliqué les codes TPROF/EUGES/ARIES à l'étude de la précipitation et de la libération des produits de fission volatiles et les prédictions qu'il a pu formuler sur le comportement des produits de fission sont développées dans la référence 11.

3.2. Simulation de régime transitoire par chauffage électrique direct

Comme dans le cas précédent, il s'agit d'associer à des expériences d'irradiation fonctionnant en régime transitoire, des simulations thermiques en utilisant l'installation de chauffage électrique direct qui a été décrite précédemment.
Nous sommes actuellement au stade de la mise au point de l'installation et de la détection de ses limites d'exploitation. L'obstacle majeur nous paraît être la difficulté d'obtenir des conditions de chauffage homogène sur tout le volume de l'échantillon; il en résulte, évidemment, un certain degré d'incertitude lors de l'interprétation des analyses. Malgré ces réserves, les premiers résultats autorisent certains espoirs à condition de limiter les objectifs.
Les figs. 5 et 6 illustrent un aspect des résultats expérimentaux qui peuvent être acquis à partir de ces simulations; dans le cas

présent, il s'agit d'examens de morphologie par microscopie électronique par balayage. La figure 5 démontre les phénomènes de coalescence de porosité intergranulaire dans les zones à grains equiaxes et à grains colonnaires pour des températures de combustible en fin de régime transitoire supérieures à 1970°C. Ces mécanismes responsables de l'accélération du relâchement des produits de fission volatiles et du gonflement du combustible ont été mis en évidence dans des combustibles oxyde mixte de l'expérience d'irradiation TRANSON /9/, soumis à 6 transitoires successifs, la puissance linéaire augmentant de ce fait graduellement de 390 à 632 W/cm. En parallèle, la figure 6 montre des modes de restructuration identiques obtenus sur un oxyde mixte traité en chauffage électrique direct soumis à 2 transitoires successifs à 1200 et 1300°C, comme température en surface des combustibles, pour une durée de 2 minutes.

Références

/1/ COQUERELLE M., Proc. of 15th Conf. on Remote Systems Technology, A.N.S. Meeting, (1967) 246

/2/ COQUERELLE M., PICKERING S., RAY I.L.F., WALKER C.T., WHITLOW W.H., Int. Conf. on Fast Breeder Reactor, Fuel Performance, Monterey, California (1979)

/3/ COQUERELLE M., WHITLOW W.H., Proc. of Conf. Post-Irradiation Examination, Grange over Sands, B.N.E.S. London (1981) 161

/4/ QUICK F. and SARI C., J. Nucl. Mat. 47 (1973) 219

/5/ ALEXAS A., KfK 2658 (1981)

/6/ WRONA B.J. and ROBERTS J.T.A., Nucl. Technology, 20 (1973) 114

/7/ KRIEGER B., DUCAMP S., COMBETTE P., Proc. of ANS Meeting Reactor Safety Aspects of Fuel Behaviour, Sun Valley Idaho, (1981)2, 237

/8/ FREUND D. and SCHIKARSKY W., KfK 1031 (1970)
for Transuranium Elements, Karlsruhe, Progress Report n° 33, January-July 1982), in preparation

/9/ CAMPANA, M., this meeting

/10/ European Institute for Transuranium Elements, Karlsruhe, Progress Report n°33, January-July 1982, in preparation

/11/ BLANK H., COQUERELLE M., RAY I.L.F., and RONCHI C., Proc. of ANS Meeting Reactor Safety Aspects of Fuel Behaviour, Sun Valley, Idaho (1981) 1, 403.

Fig.2. Console avec four à induction (à droite) et four à chauffage électrique direct (à gauche)

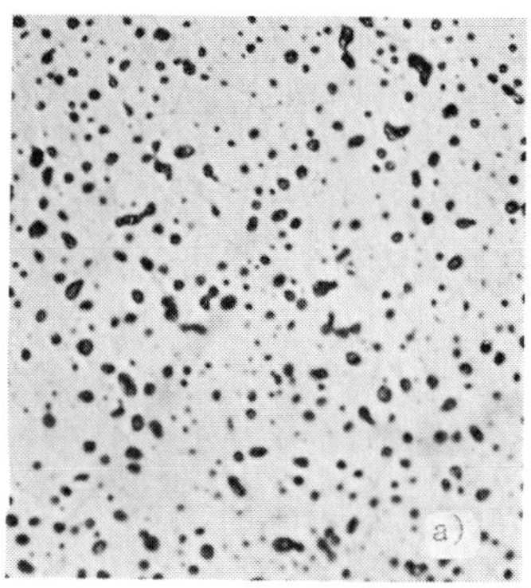

Fig.4. Combustible (U,Pu)C

a) irradié à 700-900°C

b) recuit en transitoire à 1200°C

c) recuit en transitoire à 1300°C

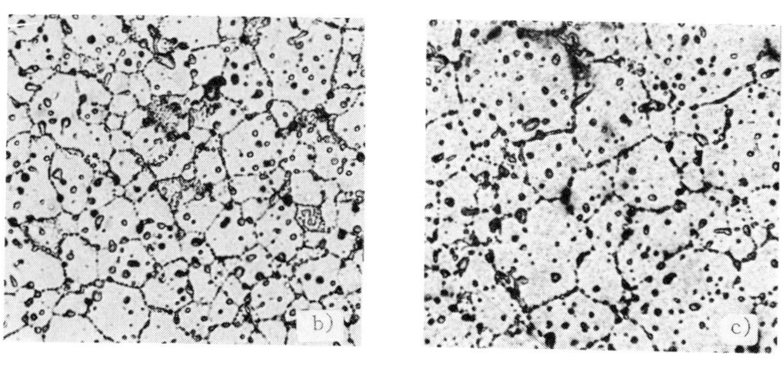

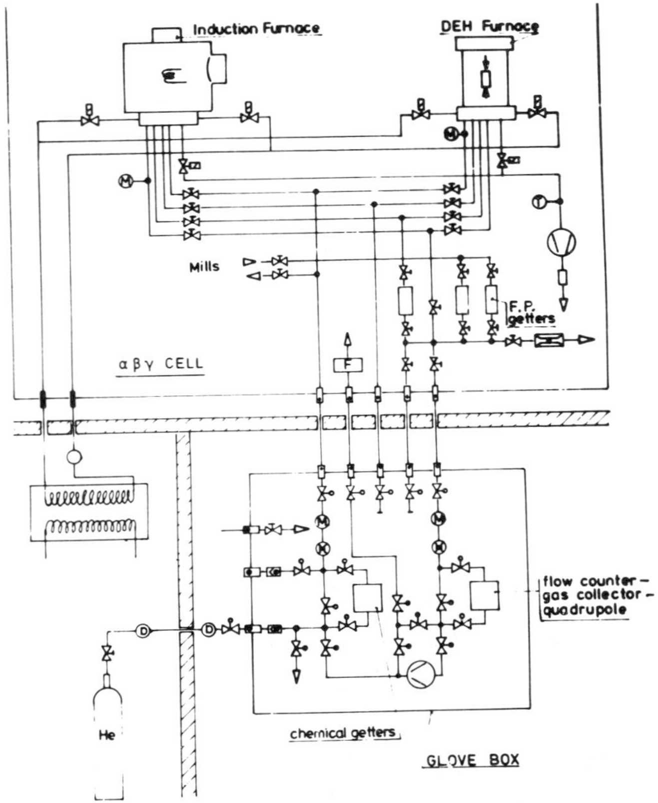

Fig.1. Implantation de la cellule α, β, γ de traitement thermique

Fig.3. Microscope électronique à balayage installé en cellule α, β, γ.

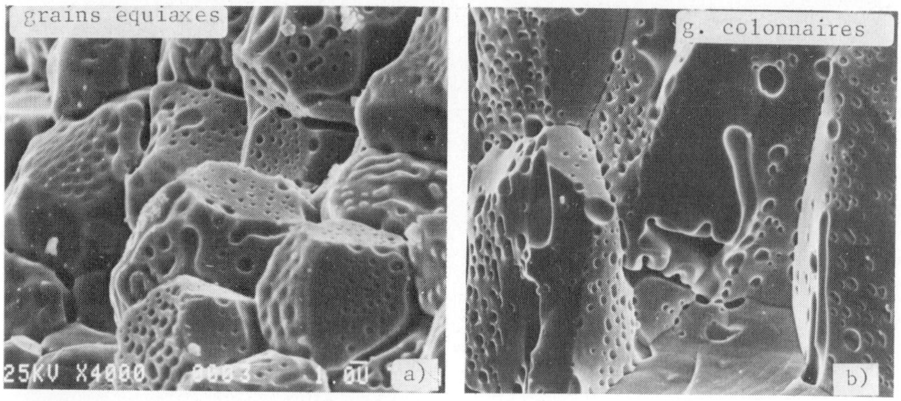

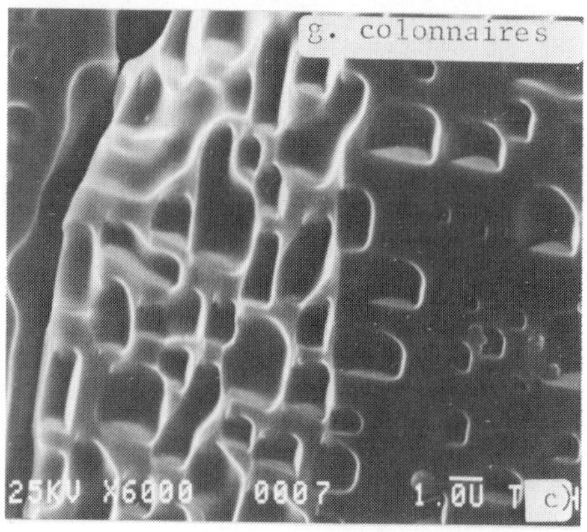

Fig.5. Irradiation TRANSON $(U,Pu)O_2$ irradié en transitoire

a,b) coalescence de porosité
c) micropores lenticulaires

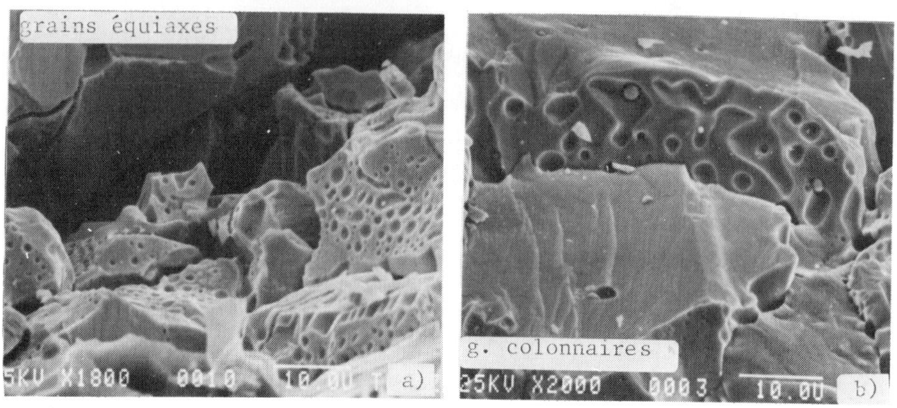

Fig.6. $(U,Pu)O_2$ irradié soumis à des transitoires dans un four D.E.H. Coalescence de porosité.
a) grains équiaxes b) grains colonnaires

MEASUREMENT AND CALCULATION TECHNIQUES UTILIZED FOR THE NEUTRONIC CHARACTERIZATION OF BR2 IRRADIATION DEVICES OF THE VARIABLE NEUTRON SCREEN TYPE. APPLICATION TO THE "VIC" SODIUM LOOP.

Ch. DE RAEDT, N. MAENE, R. MENIL, J. DEBRUE

S.C.K./C.E.N.
2400 MOL, Belgium

ABSTRACT

A description is given of the techniques utilized for the neutronic characterization of BR2 irradiation devices in the design stage. These are applied to the "VIC" sodium loop with cadmium screen, presented elsewhere in this symposium.
By calculation, the in-pile geometry of the loop was optimized in order to keep an adequate power level in the LMFBR test fuel pin, to reduce the radial fission density depression and to obtain a sufficient power cycling capacity when varying the ^{3}He pressure. Hence were defined the type and position of the fixed neutron screen, the thickness of the ^{3}He variable pressure screen and of the water gap inside the fixed screen.
The calculations concerning the power cycling capacity of the VIC loop were checked on a mock-up irradiated in BR02, the zero power model of BR2. The location of the test channel in an adequate core configuration ensured the required fuel pin power while maintaining the specifications of the other BR2 experiments.

INTRODUCTION

The neutronic characterization of BR2 irradiation devices of the variable neutron screen type, such as the VIC (Variable Irradiation Conditions) loop described in ref. 1, which aim at the cycling of the fission power in test fuel pins, concerns mainly the following quantities :

P. von der Hardt and H. Röttger (eds.), Irradiation Technology, 475–485.

i. the absolute level of the nominal fission power in the irradiated fuel pins, taking into account the nominal operation power of the reactor, imposed by the whole of the experimental loading

ii. the cycling power, i.e. the relative variation of this power by decrease of the pressure (i.e. of the absorption) of the ^{3}He gas constituting the variable neutron screen

iii. the radial power depression in the fuel pins. This depression influences the radial temperature profile which should be as near as possible to that prevailing in reactors (either of the LWR or - in the case of VIC - of the LMFBR type) of which fuel is examined. In general fixed neutron absorbing screens (Cd, B_4C, thick stainless steel ...) are needed to maintain the power depression within acceptable limits.

To determine design estimates and, if necessary, to adapt the irradiation device in order to meet the requirements concerning these quantities, which are fixed by the experimenter, a good knowledge of the neutron flux spectrum is further needed. Indeed,

- for technological reasons the power cycling is performed by varying the pressure of a low energy neutron absorbing gas viz ^{3}He, as mentioned above; the neutron flux should not be too strongly suppressed by the Cd, B_4C or thick stainless steel screen in the relevant flux energy region
- the absolute fission power level and the radial fission power depression in the fuel pins both increase with increasing thermalization of the neutrons incident on the fuel pin.

These effects generally ask for a combination of succeeding layers of thermal neutron absorbing and neutron rethermalizing materials in the lay-out of the irradiation device, which ensure the optimal shaping of the flux spectra.

A general survey of calculations and related measurements will be discussed in the present note. It will be shown how these techniques were applied to the VIC irradiation device.

CALCULATION METHODS AND EXPERIMENTAL TECHNIQUES EMPLOYED IN GENERAL

The basic code used to determine the neutron flux chart in BR2 irradiation devices and their surroundings is the one-dimensional neutron transport code DTF-IV (ref. 2), containing several improvements made at CEN/SCK and flanked by preprocessing and postprocessing routines, the whole called the MULCOS code system. This system is used together with a 40 energy group cross-section library, including self-shielding factors (ABBN formalism), elaborated at CEN/SCK for the study of coupled fast-thermal systems.
Code and library are used since many years for the neutronic characterization of BR2 irradiation devices both with one or with many (up to thirty) fuel pins. The results obtained are in general satisfactory : see e.g. refs 3, 4 and 5.
The calculations yield the neutron flux chart in BR2 irradiation devices in arbitrary ("floating") units. This flux chart is normalized by substituting in the calculations for the studied device, devices with known absolute irradiation characteristics (flux or fission density levels), as explained in ref. 6. When this substitution method cannot be applied, because e.g. the reactor configurations adopted for the newly studied devices do not at all correspond to those surrounding the known devices, the neutron flux charts in the devices are normalized to the calculated - and, on the other hand, known - power levels in the fuel elements surrounding the devices. This normalization method, which has to be given a larger uncertainty margin than the substitution method, had to be applied e.g. in the VIC calculations.

Another branch of calculations concerns the determination of the gamma heating distribution inside devices irradiated in BR2. The calculation methods utilized are explained briefly in ref. 4 but were not applied to the present study of VIC as the gamma heating is relatively small in this loop.

In parallel with the calculations, neutronic measurements are carried out on nuclear mock-ups of the devices in the BR02 reactor, the zero power mock-up of BR2. As an example, fig. 1 gives a cross-section of the VIC irradiation device and of its nuclear mock-up in BR02.
The most important objective is generally the prediction of the absolute fission power level in the fuel pin irradiated. To this end fissile alloyed disks are irradiated in BR02 in the empty mock-up between the fuel pellets. They are previously calibrated with the aid of absolute NBS fission chambers. Before the start of the irradiation of the device in

BR2, irradiations of activation foils in the empty mock-up in BR02 and in the same mock-up in BR2 during an absolute low power run (in the channel where the device will be loaded subsequently) will provide the necessary link between the two reactors. Another method consists in simultaneously irradiating in BR02 two pins, one loaded in the mock-up of the new device and the other loaded in the mock-up of a device for which a lot of data about the irradiation conditions in BR2 are available : see fig. 4, discussed further on. A relative gamma-scanning measurement of the two pins will then also allow to link the BR2 reactor power to BR02.
Following measurements may also be carried out in BR02 in order to calibrate or check the calculations :

- axial fission distribution in one pin or/and radial fission distribution from pin to pin (for loops containing a bundle of pins), determined by gamma-scanning,
- radial fission density distribution inside the fuel pin, by Solid State Track Recorders,
- gamma heating contribution, using Thermo Luminescent Dosimeters,
- radial gradient through the mock-up, with activation foils,
- spectral flux characterization, by means of NBS fission chambers, activation foils, P.I.N. diodes (Si damage) and ^{6}Li neutron spectrometers.

In several of these areas, calculations and measurements are closely linked, as is explained e.g. in ref. 5.
The reactivity variation due to the loading or unloading of the device or parts of it may be determined in BR02 by means of :

- the source jerk technique,
- a reactor period determination,
- calibrated control rods,
- the pulsed neutron technique.

DETERMINATION, BY NEUTRONIC CALCULATIONS, OF THE OPTIMUM GEOMETRICAL LAY-OUT OF THE DEVICE

The calculations made for the optimization of the VIC device led to following conclusions :

- a thermal neutron absorbing screen (Cd or thick stainless steel) is needed to reduce the radial fission power depression in the fuel pin and to keep the absolute level of the nominal fission power within acceptable limits, taking into account, as already mentioned above, the nominal operation power of the reactor, imposed by the whole of the experimental loading. The exact required fission power level of the fuel pin can be obtained eventually by the choice of the position and environment of the irradiation channel in the reactor

- with ^3He as variable neutron absorbing screen, Cd is to be preferred to thick stainless steel for the fixed thermal neutron absorbing screen
- inside the cadmium screen, one or more water layers with a total thickness of at least 6.5 mm are needed to obtain a cycling power of 1.4 - the minimum value required by the experimenters - for ^3He pressures varying from 36 to 1.2 bar at 270 °C
- under these conditions the thickness of the ^3He variable neutron screen should be of the order of 3.5 mm, at least, to obtain the cycling power desired.

Finally, a loop lay-out, with cadmium screen, was adopted, as indicated on fig. 1, left hand side. The total thickness of the water layers inside the cadmium screen amounts to 7.75 mm and the ^3He gas layer thickness to 3.55 mm.

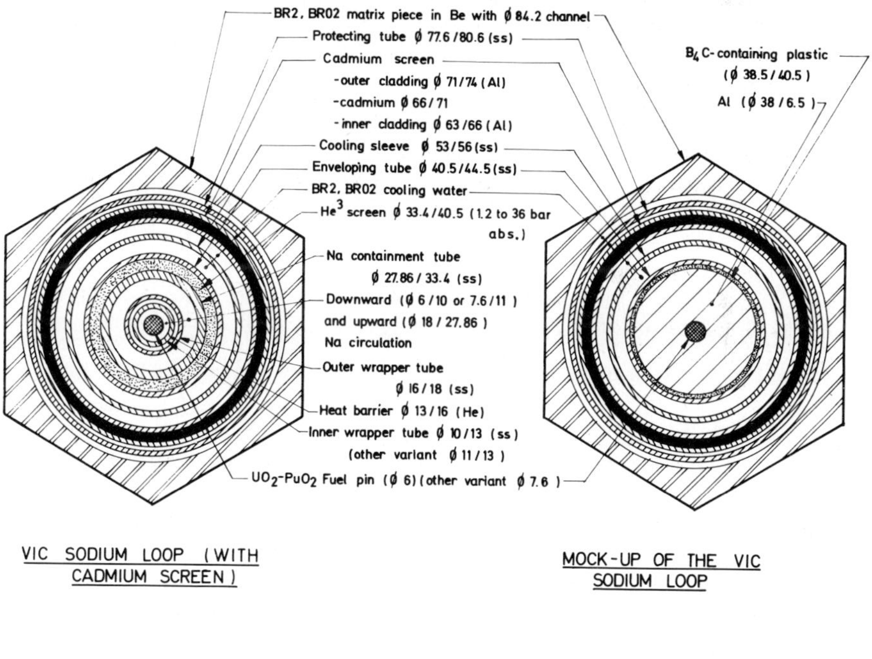

Fig. 1. In-pile cross-sections of the VIC loop with cadmium screen designed for irradiation in BR2 and of its mock-up for BR02.

It was assumed that this solution yielded the best compromise between two requirements, on the one hand, the calculated cycling power, viz 1.45 (when the ^3He pressure varies from 36 to 1.2 bar), which was considered as having a sufficient safety margin with respect to the 1.4 value imposed by the experimenters, and, on the other hand, the radial power depression in the fuel pin ($p_{border\ of\ pin}/p_{axis\ of\ pin}$ is about 2.4 at low pressure, about 1.7 at high pressure, which values appeared to be acceptable). The characteristics of the VIC device are visualized on fig. 2, as a function of the ^3He pressure.

DETERMINATION OF THE MAIN NEUTRONIC QUANTITIES OF THE DEVICE FINALLY RETAINED

In the preceding paragraph it was shown how the design calculations led to the determination of a device with cycling power and radial depression in the fuel pin that were acceptable to the experimenter. These quantities still had to be checked experimentally and an appropriate reactor position for the device had to be determined in order to meet the requirements concerning the absolute power level (450 W/cm at nominal power, i.e. for 36 bar ^3He pressure).
The experimental work was carried out in BR02, as mentioned above, with the loop mock-up shown on the right hand side of fig. 1.
First the ability of the calculations to determine the cycling power was checked : fission density measurements (with fission chambers and with U-Al detectors) were carried out in the mock-up without fuel pin, for various compositions of the zone corresponding to the variable neutron screen viz void (or air) and one and two 1 mm thick layers of B_4C dispersed in a polyurethane ($NH_2CO_2C_2H_5$) carrier : cases a, b and c. The concentration of the B_4C was chosen in such a way that one layer simulated approximately the ^3He screen at 36 bar (in fact : at 42.7 bar) and 270 °C. Measurements were also performed with one and two layers of pure polyurethane in order to determine its moderating (and hence fission power increasing) properties : cases d and e. Indeed, the influence of this moderation has to be taken into account when interpreting the influence of B_4C in the measurements b and c. Therefore, in addition to the calculations performed in parallel to the BR02 measurements, for the same device geometry and compositions, calculations (cases f and g) were made for one and two layers of boron without polyurethane carrier, i.e. for the ideal case, which could not be realized in BR02. The results of all measurements and calculations are indicated in table I. The influence of the polyurethane is small when B_4C is also present : cf. cases b, c with cases f, g.

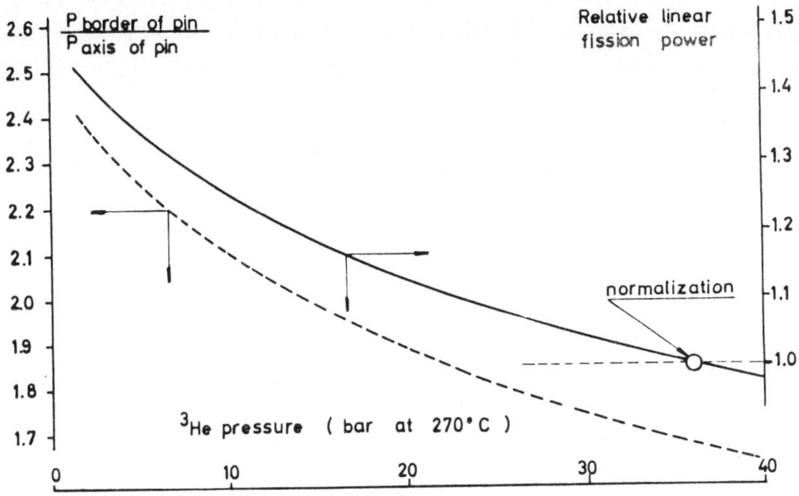

Fig. 2. Calculated relative linear fission powers and radial fission power depressions in a 6 mm UO_2-PuO_2 fuel pin irradiated in the VIC device, as a function of the ^{3}He pressure.

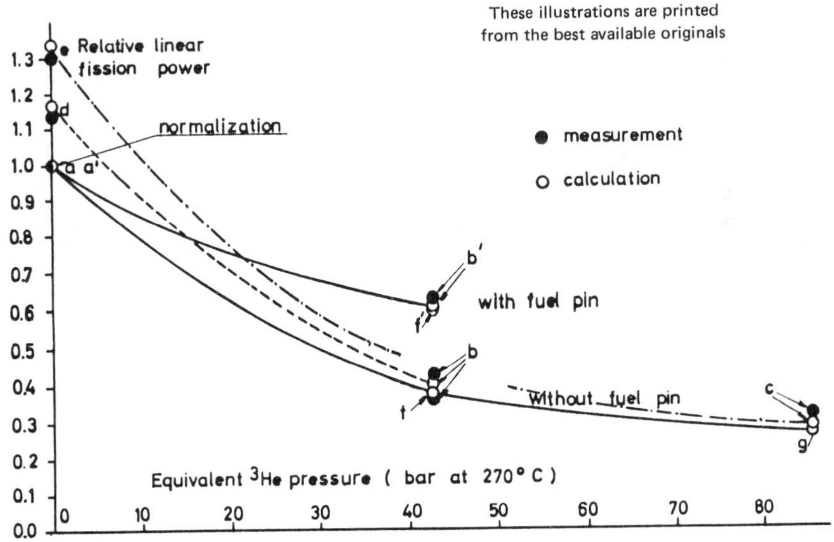

Fig. 3. Validation of the calculation methods : measured and calculated linear fission powers in the mock-up of the VIC device in BRO2, as a function of the equivalent ^{3}He pressure (the curves are only eye-guides).

Table I. Comparison of measured and calculated fission power densities in the VIC mock-up without fuel pin

Case N°	Composition of the zone simulating the variable neutron screen	Fission chamber results	U-Al detector results	Calculated results
a	void or air (reference case)	1.00	1.00	1.00
b	one layer of borated polyurethane	0.42	0.36	0.38
c	two layers of borated polyurethane	0.31		0.28
d	one layer of pure polyurethane	1.13		1.16
e	two layers of pure polyurethane	1.30		1.34
f	one layer of boron without carrier			0.37
g	two layers of boron without carrier			0.26

Next, measurements were performed in the mock-up with fuel pin, viz measurements of the pin activity for two - three in the calculations carried out in parallel - compositions of the zone simulating the variable neutron screen : table II. The fuel pin utilized for these measurements was a UO_2-PuO_2 fuel pin having a composition rather similar to that foreseen for the actual VIC irradiation.

Table II. Comparison of measured and calculated fission power densities in the VIC mock-up with fuel pin

Case N°	Composition of the zone simulating the variable neutron screen	Fuel pin activity	Calculated results
a'	void or air (reference case)	1.00	1.00
b'	one layer of borated polyurethane	0.64	0.62
f'	one layer of boron without carrier		0.61

The values indicated in tables I and II are visualized in fig. 3 with relative fission power densities drawn as a function of the equivalent ^{3}He pressure at 270 °C. The results may be considered as a confirmation of the calculation method. Hence the values, mentioned in the preceding paragraph (fig.2), of the cycling power calculated for the actual VIC device (fig. 1, left hand side) may be accepted with good confidence.

When performing the calculations leading to the choice of the optimum VIC geometrical lay-out, it was found that, for the given composition of the fuel pins (highly enriched UO_2-PuO_2), in the high flux channels of the configuration existing at that time and for BR2 operating at nominal power, the absolute power level in the fuel pins would be quite a lot higher, even with cadmium screen in the device, than the design value of 450 W/cm. On the other hand, lower flux irradiation channel positions, away from the reactor centre, were unsatisfactory because of the strong flux gradient across them. Hence it was already known before the BR02 irradiation campaigns started that a re-design of the reactor core configuration was necessary, yielding a lower flux channel for VIC with acceptable flux gradient across it. Another very important factor intervening in the re-design of the core configuration was the strong antireactivity of the VIC device due to the large cadmium screen surrounding the device.
Many configuration studies were carried out in BR02, leading eventually to the configuration indicated in fig. 4. Thanks to the presence of the beryllium plug between the channel containing the VIC device and the H1 channel, the absolute flux level in VIC is depressed while the gradient across the device remains rather small.
In this configuration the absolute power level in the VIC fuel pin was determined by combined calculations and measurements, the whole being normalized to the fission power of a fuel pin in a reference capsule, viz the FAFNIR cadmium screened capsule which was irradiated simultaneously with the VIC mock-up. Several such FAFNIR devices were irradiated in the past in BR2 and their characteristics are well known. It was finally found that a linear power of about 400 W/cm would be obtained in a 6 mm UO_2-PuO_2 fuel pin irradiated in the VIC device in BR2 and a linear power of about 500 W/cm in a 7.6 mm UO_2-PuO_2 fuel pin. The relatively small deviation from the 450 W/cm will be further reduced by a judicious choice of the burn-up level in the fuel elements surrounding the VIC device during the actual BR2 irradiation and, finally, by adapting the nominal ^{3}He pressure in the device, which may be varied slightly around 36 bar.

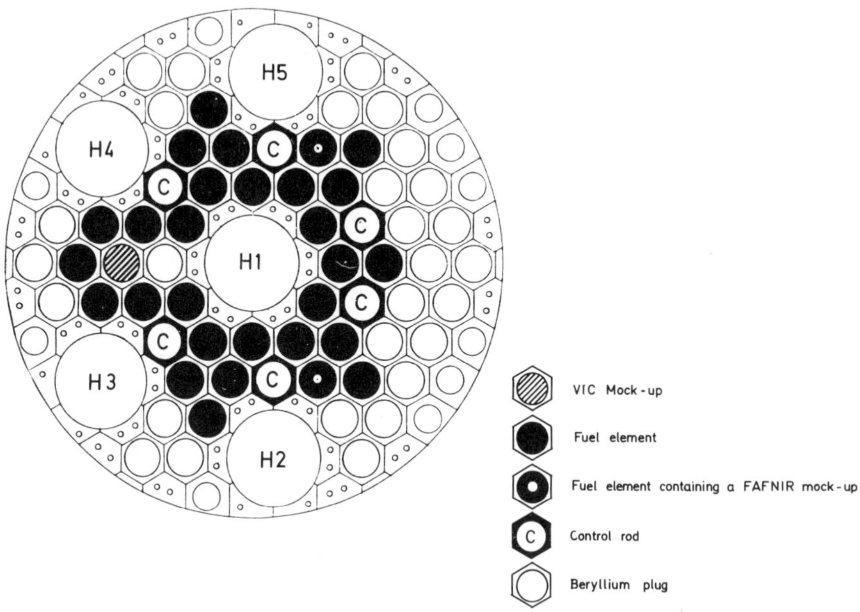

Fig. 4. BRO2 core configuration, used for the experimental determination of the VIC characteristics.

As to the radial power depression in the fuel pin, no specific measurements were performed for VIC. In the past, for other cadmium screened devices, calculations have been compared with solid state track recorder measurements on disks placed between polished pellets and gave confidence to the calculations.

CONCLUSIONS

By combining calculations, measurements in BRO2 on a mock-up and data from preceding BR2 irradiations, the neutronic characteristics of newly designed devices can be determined at an early stage in the project with a good degree of confidence and, when necessary, the device geometrical lay-out and the reactor configuration may be adapted to meet the requirements.

REFERENCES

1. J. DEKEYSER. Transient experiment on LMFBR fuel pins in the reactor BR2 with the versatile sodium loop "VIC". To be presented at this conference.

2. K.D. LATHROP. DTF-IV, a FORTRAN-IV program for solving the multigroup transport equation with anisotropic scattering. LA-3373 (Nov. 1965).

3. J. DEBRUE, G. DE LEEUW-GIERTS, S. DE LEEUW, Ch. DE RAEDT, A. FABRY, L. LEENDERS, N. MAENE, R. MENIL. Dosimetry work in connection with irradiations in the high flux materials testing reactor BR2. First ASTM-EURATOM symposium on reactor dosimetry. Petten, Sept. 22-26, 1975.

4. Ch. DE RAEDT, L. FOUARGE. Calculation methods applied to BR2 irradiations. Third ASTM-EURATOM symposium on reactor dosimetry. Ispra, Oct. 1-5, 1979.

5. Ch. DE RAEDT, L. LEENDERS, H. TOURWE, H. FARRAR IV. Dosimetry work and calculations in connection with the irradiation of large devices in the high flux materials testing reactor BR2. Fourth ASTM-EURATOM symposium on reactor dosimetry. Washington, March 22-26, 1982.

6. J. DEBRUE, Ch. DE RAEDT, N. MAENE. Prediction of the irradiation characteristics of fast reactor fuel pins tested in cadmium screened loops in BR2. Methods and application to the MOL 7B sodium loop. BLG 488 (Feb.1974).

SESSION VIII
GAS COOLED REACTORS
AND FAST BREEDER REACTORS

A CALORIMETRIC DEVICE FOR IRRADIATION OF FBR FUEL PINS

F. MOONS

C.E.N./S.C.K. Nuclear Research Center

B - 2400 MOL - BELGIUM

ABSTRACT

The FFC (Fast Flux Capsule) type rigs exhibit the following characteristics :

- temperature regulation
- calorimetric heating rate measurement
- cladding rupture monitoring
- irradiation in fast reactor representative circumstances
- and a considerable potential for sophisticated instrumentation.

They mainly consist of a copper capsule of cruciform cross section surrounded by a Cd-screen. The copper capsule contains in its central hole the fuel pins which are immersed in liquid metal (NaK). The gas plenum of the copper capsule can be connected to a sweeping circuit for fission gas detection in case of fuel cladding rupture.

By means of thermocouple pairs in the four wings of the copper cross, the temperature gradients are measured and hence the fuel rod power. The cladding temperature can be extrapolated from the measurements or directly registered by thermocouples attached to the cladding.

The Cd-screen around the cross creates a neutron environment representative for fast reactor application.

P. von der Hardt and H. Röttger (eds.), Irradiation Technology, 489–498.

CONTENTS

1. Introduction

2. Capsule description

3. Out-pile control equipment

4. Capsule characteristics or possibilities

4.1. Temperature regulation
4.2. Calorimetric heating rate measurement
4.3. Cladding temperature determination
4.4. Cladding rupture monitoring
4.5. Irradiation in fast reactor representative circumstances
4.6. Potential for additional instrumentation

5. Capsule limitations or drawbacks

5.1. Temperature regulation
5.2. Calorimetric heating rate measurement
5.3. Cladding temperature determination
5.4. Cd-screen consumption

6. Conclusion

1. INTRODUCTION

The Fast Flux Capsule (FFC) - type rigs are a logical evolution in the design of rigs with an internal copper capsule of cruciform cross section, mainly developped for fuel rod irradiation (see fig. 1).

It started with the CIRCE - Calorimetric IRradiation Capsule Experiment - type rigs.

A copper capsule of cruciform cross section contains in its central hole the fuel pins which are immersed in liquid metal (NaK). The copper capsule is surrounded by a stainless steel enveloping tube as barrier against the reactor cooling water, flowing outside. The thermal link between copper capsule and enveloping tube is realised by an extrusion process. At the inside wall, the copper capsule contains a coextruded stainless steel tube as protection against the hot NaK.

By means of thermocouple pairs in the four legs of the copper cross, the temperature gradients are measured and hence the fuel rod power. The cladding temperature can be extrapolated from these measurements or directly registered by thermocouples attached to the cladding.

The gas plenum of the copper capsule is connected to a sweeping circuit for fission gas detection in case of fuel cladding rupture. Several fuel pins may be loaded into one capsule and by means of gas tagging, a ruptured pin can be identified.

The basic design of the CIRCE device offers numerous possibilities for supplementary and particular instrumentation or operating amendments. The "S" type version for instance was developed for measuring the swelling of a fuel pin during irradiation.

The CFC - Compatibility Fuel Cladding - irradiation devices are a further step (see fig. 1).

The main difference with the CIRCE device consists in a small gas gap which is realised between the four wings of the copper cross and the stainless steel enveloping tube. By means of well defined spacer pins attached to the cruciform copper capsule, it takes a concentric position in the surrounding tube. The gas gaps at the wings of the copper capsule (of the order of a tenth of a mm thickness) create a heat transfer barrier for the fission power developed by the fuel pins inside the copper capsule.

A FFC - Fast Flux Capsule - capsule is a modified CFC, specially designed for testing FBR (Fast Breeder Reactor) fuel pins. The outer stainless steel enveloping tube is replaced by a multilayer (coextruded stainless steel-cadmium-stainless steel tubes) tube. The cadmium screen creates for the irradiated fuel pin a FBR-representative neutron environment.

2. FFC - CAPSULE DESCRIPTION

A FFC-capsule is a calorimetric temperature controlled fuel pin capsule, developed with the objective to make available for different experimenters a rig, allowing the irradiation of prototype fast reactor fuel pins at a constant specified temperature of the cladding in liquid metal environment.

It serves for the irradiation of fast reactor fuel pins in epithermal and fast neutron flux while the thermal neutrons are screened of by the cadmium. The outside diameter of the rig is 34 mm and it is loaded into the central hole of a five-shell BR2 driver fuel element (see fig. 2).

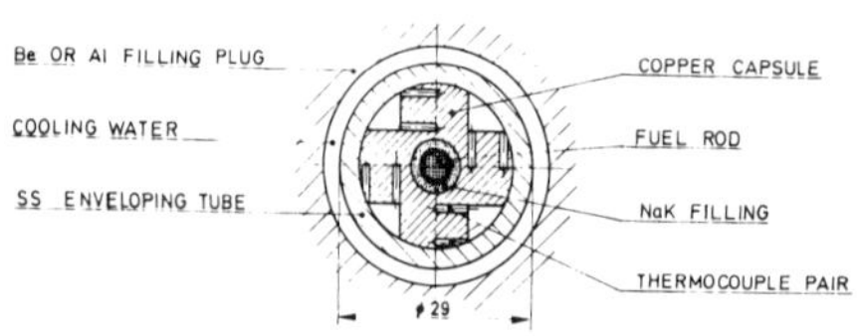

"CIRCE„ (Calorimetric Irradiation Capsule Experiment)

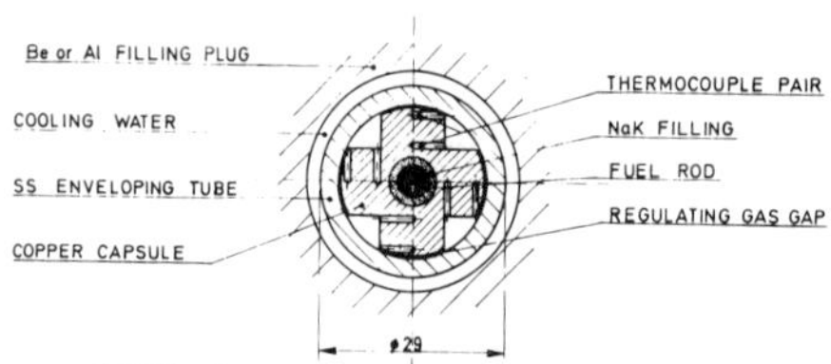

"CFC „ (Compatibility Fuel Cladding)

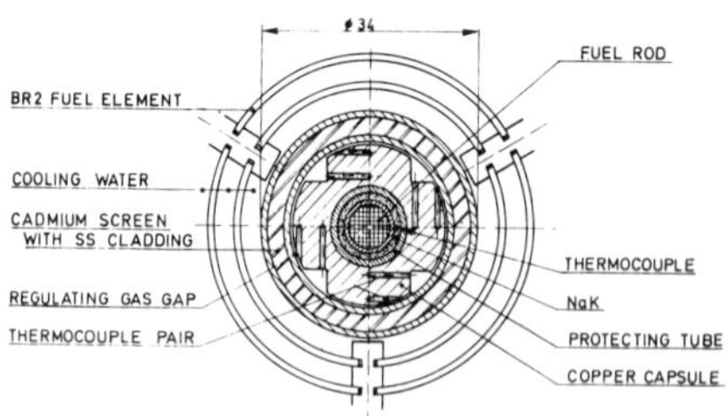

"FFC„ (Fast Flux Capsule)

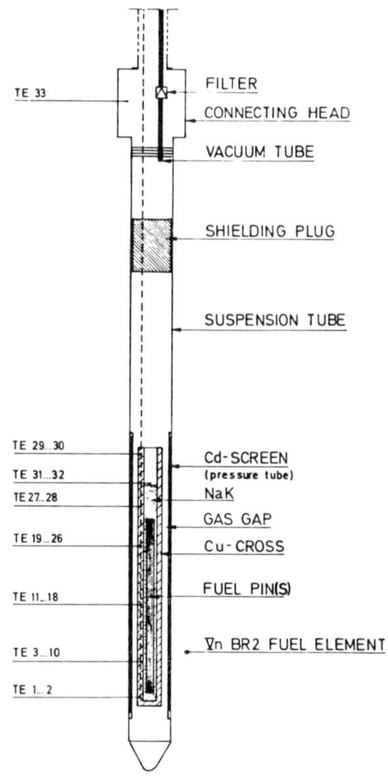

FFC in-pile section

The rig consists of a copper capsule with cruciform cross section containing in its central hole, the fuel pin which is immersed in liquid metal (NaK). At the inside wall, the copper capsule is covered by a layer of Nickel or a coextruded stainless steel tube as protection against the hot NaK.

The copper capsule is surrounded by a cadmium tube, sandwiched between two stainless steel tubes. This enveloping multilayer tube is cooled on the outside by the BR2 water. A small radial gas gap of about 0,1 to 0,25 mm between the four ribs of the copper cross and the enveloping tube allows to control the fuel pin cladding temperature by varying the gas pressure in the gap between about 1 and 0,01 bar. The usual range of the specified cladding temperature is 400 to 750 °C for fuel pin powers between 300 and 500 W/cm. The fuel pin power and the cladding temperature are deduced from the temperatures measured by the thermocouple pairs in the four legs of the copper cross. This can be repeated at several axial levels.

3. OUT PILE CONTROL EQUIPMENT (see figure 3)

The "FFC" devices are operated by so-called TCV - Temperature Control by Vacuum - control panels.

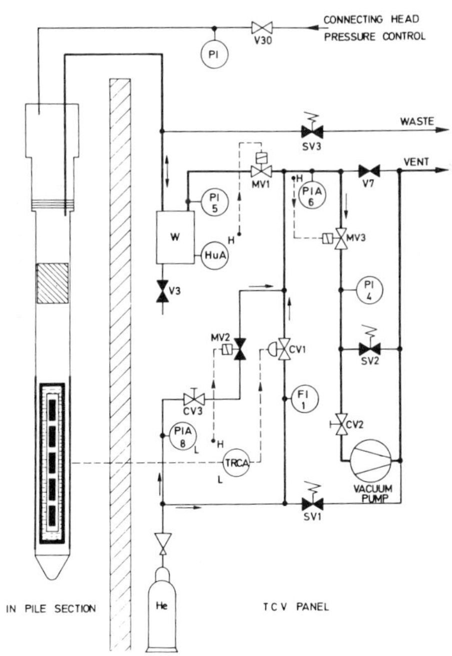

A vacuum pump creates a permanent depression at the gas outlet of the rig. In function of a given set point temperature, an automatic feed valve will open or close to adjust the quantity of He-atoms in the gas gap and thus decrease or increase the thermal resistance of the gas gap.

In case of an incident (experiment to hot, radiation level in vacuum loop to high), the vacuum pump will be automatically separated from the rig, and He at a pressure of 2 bar, will be injected in the temperature regulating gas gap, to improve the heat conduction and bring down the internal temperature.

The usual operating gas pressures are ranging from 1 to 0,01 bar. Several hundred degrees Kelvin of temperature control range can be covered by the regulating system.

4. CAPSULE CHARACTERISTICS OR POSSIBILITIES

The FFC-type rigs exhibit the following characteristics :
- temperature regulation
- calorimetric heating rate measurement
- fuel pin cladding temperature determination
- cladding rupture monitoring
- irradiation in fast reactor representative circumstances
- considerable potential for sophisticated instrumentation.

4.1. Temperature regulation (see ref. /2/)

The set point temperature of the fuel pin cladding is reached by reducing the thermal conductivity in the Helium gas gap between the copper cross ribs and the enveloping tube. The decrease of the conductivity is obtained by lowering the helium pressure around the copper capsule. This control pressure has to decrease continuously during the irradiation time due to the swelling of the copper that makes the gap smaller. The capsules are designed so that the control pressure would decrease between 250 and 50 mbar over the irradiation period in order to keep the cladding temperature at its constant set point value.

The pressure of 250 mbar gives a thermal conductivity of about 80% of the macroscopic value. The pressure of 50 mbar however reduces the helium conductivity by a factor of 10. This compensates the higher heat transfer rate through the smaller gas gap (narrowing of the hot gas gap from 0,1 mm to the order of 0,01 mm is admissible).

The heat transfer through the small gas gap between the copper legs and the enveloping tube, at reduced pressure, can no longer be calculated by using simply the macroscopic thermal conductivity of the gas.

A calculation model that is commonly used for determining the heat transfer in small gaseous gaps assumes that the apparent gap width S is increased by two "Extrapolation widths" g, in order to take account of the reduced thermal conductivity of the gas :

$$\text{Heat flux } \left[\frac{W}{m^2}\right] = \frac{\lambda m}{S + 2\cdot g} \cdot \Delta\,\theta = \frac{\lambda^\star}{S} \cdot \Delta\,\theta$$

with λ_m = macroscopic thermal conductivity at mean gas temperature in $\left[\frac{W}{m\ °C}\right]$

S = hot gas gap in [m]

$\Delta\theta$ = temperature gradient in the hot gas gap [°C]

$\lambda^\star$ = reduced thermal conductivity in $\left[\frac{W}{m\ °C}\right]$

The value of g depends on the pressure in the gas gap. Formulas for determining this value are given in ref. /4/.

The copper capsule remains controllable as long as the hot regulating gap is bigger than about 0,01 mm.

4.2. Calorimetric heating rate measurement

The linear fuel pin power in the capsule can be determined by :

$$p = k^\star \cdot \Delta\, t^\star$$

with $k^\star$ = coefficient depending on the ratio between the BR2 gamma heating and the fuel pin power

$\Delta t^\star$ = mean temperature gradient, measured by the four thermocouple pairs in the copper capsule.

The design of the rig is made in a way that there is no axial heat flow in the copper capsule at nominal fuel pin power and nominal BR2 gamma heating. This is achieved by chosing different gas gaps along the capsule resulting in a constant axial copper temperature. The axial heat flow in the fuel pin itself, in the NaK filled annulus and in the protecting tube may be considered as negligible.

4.3. Cladding temperature determination

The fuel pin cladding temperature can be determined or by calculation, starting from the Cu-cross thermocouple measurements or by locating a thermocouple in the NaK annulus and attaching it to the fuel pin.

4.4. Cladding rupture monitoring

The gas plenum of the copper capsule can be connected to a sweeping circuit for fission gas detection in case of fuel cladding rupture. Several fuel pins may be loaded in one capsule and by means of gas tagging, a ruptured pin can be identified.

4.5. Irradiation in fast reactor representative circumstances

The fuel rod(s) being completely shielded against thermal neutrons, by the surrounding Cd-screen, show(s) a flat fission rate distribution over the fuel cross section, typical for fast reactor conditions.

4.6. Potential for additional instrumentation

- fuel pin centre temperature measurement by means of high temperature thermocouples ;
- pressure transducers connected by capillary tubes to the fuel pin gas plenum for fission gas pressure build-up determination ;
- self-powered neutron detectors or collectrons as well as activation monitors can be installed for neutron flux and fluence measurements ;
- the temperature gradient across the cadmium screen may be measured by incorporated thermocouple pairs for ab additional evaluation of the fuel rod power.

5. CAPSULE LIMITATIONS OR DRAWBACKS

5.1 Temperature regulation

- It must be avoided to regulate the capsule at set point temperature for linear fuel powers lower than 60% of nominal power because in this case the radial thermal dilatation of the copper cross becomes too high. This can cause plastic deformations at the location of the centering pins.
- For the choice of the regulating helium gap, one must take account of a certain swelling of the copper cross by the irradiation.

5.2. Calorimetric heating rate measurement

- The fuel pin power must be determined by the mean value of the temperature differences measured by four thermocouples pairs, one in each leg of the copper cross. This is necessary because of the fact that the individual temperature differences can vary by $\pm$ 60% around the mean value if the copper cross is completely eccentric in the enveloping tube.
- It is necessary to take account of the ratio between the fuel pin fission power and the BR2 gamma heating rate for determining the fuel pin power. This ratio is responsible for the possible axial heat flow in the copper cross which influences the temperature difference ΔT measured by the thermocouple pairs. The temperature difference becomes bigger if the gamma heating rate is higher than supposed and becomes smaller if the gamma heating rate is lower. The gas gap along the capsule is chosen so that there is no axial heat flow for nominal conditions.
- The temperature difference of the thermocouple pairs is influenced by the helium pressure regulation ranging from 1 to 0,01 bar for a constant fuel pin power. This happens because of the variation of the relative distribution of the heat transfer through the gas gap over the length of the capsule. Only the heat transfer opposite to the fissile material is influenced by the pressure regulation (small gap).
  The heat transfer opposite to the breeder material or rod plenum is not influenced because the helium gap is here of the order of millimeters.

5.3. Cladding temperature determination

- An eccentric position of the copper cross inside the enveloping tube does not influence in an important way the temperature distribution over the circumference of the fuel pin because of the very good thermal conductivity of copper.
- An eccentric position of the fuel pin in the copper capsule of the order of 0,3 mm can give a deviation of $\pm$15°C over the circumference of the fuel pin cladding.
- A small radial displacement of the thermocouple in the NaK annulus attached to the fuel pin can lead to an important underestimation of the cladding temperature.

5.4. Cd-screen consumption

The Cd-screen, being an integral part of the capsule can create a problem for long lasting (more than 1 year) irradiations.

Experience with Cd-screen irradiations at BR2 reveals that a sheet of 1 mm of Cd becomes transparent after a thermal neutron fluence of about 3,5 10^{21} n/cm^2 (see ref. /6/).

An irradiation in an FFC-capsule lasting two effective years, requires a screen of at least 6 mm thickness. For such dimensions, capsule design becomes difficult due to

- high (gamma-)heat production in the Cd-screen ;
- fabrication difficulties in Cd-tube fabrication ;
- restricted dimensions for Cu-cruciform capsule and thus small ΔT values for thermocouple pairs in the Cu-cross wings.

6. CONCLUSION

A calorimetric temperature controlled fuel pin capsule has been designed for the irradiation of fast reactor fuel pins in epithermal and fast neutron flux.

The major pecularities of the capsule are :

- the NaK-environment ;
- calorimetric temperature control by means of thermocouple pairs in a copper cross capsule ;
- integrated cadmium-screen for spectrum tailoring.

REFERENCES

/1/ J.M. BAUGNET, W. BOEYKENS, F. LEONARD
The BR2 materials testing reactor. Past and future.
Seminar on research reactor operation and use.
Jülich (Germany) - September 14-18, 1981 IAEA-SR-77/36.

/2/ W. DETAVERNIER, W. HEBEL
Thermodynamic design of a calorimetric temperature controlled fuel pin capsule.
26th Plenary Meeting of the "Irradiation Devices Working Group". GKSS Research Centre Geesthacht (Germany) - October 8-10, 1980.

/3/ W. HEBEL, J. PLANQUART, G. VANMASSENHOVE
Irradiation devices for BR2 reactor. Design and operating characteristics. CEN/SCK - Technology and Energy Department - March 1976.

/4/ W. HEBEL
Temperaturregelung von Bestrahlungsproben durch Gasdruckveränderung. EUR. 4478 d - 1970.

/5/ Max JAKOB
Heat Transfer. Vol. I Chapter II (Convection)
New York, Wiley ; London, Chapman.

/6/ H. LENDERS
SCK/CEN Internal note Dep.BR2 HL/mc nr 72/0624 - 23.6.78.

IRRADIATION FACILITY FOR TEMPERATURE TRANSIENT AND CYCLING EXPERIMENTS OF HTR FUEL ELEMENTS

R.Conrad[*)],R.Keller[**)]

*) JRCPetten Establishment, HFR Division
**) JRCPetten Establishment,Infrastructure
Petten,The Netherlands

ABSTRACT

Accident analysis of High Temperature Reactor(HTR) spherical fuel elements is focussed on investigation of temperature transient conditions after failure of the main heat sinks,either at system pressure or after depressurization,under different fuelling schemes,e.g. single pass(OTTO) or multiple pass(Modul-HTR).

JRC Petten has developed an irradiation facility to simulate temperature transients for HTR fuel elements under realistic operating conditions with simultaneous analysis of gaseous fission product release rates. Typical transient parameters range to about 1600°C within 100 hours.The tests can be performed under steady nuclear conditions.The irradiation facility covers requirements for power ramp tests.

This paper describes the irradiation facility and presents a thermal analysis for early and late in life case studies.

P. von der Hardt and H. Röttger (eds.), Irradiation Technology, 499–508.

INTRODUCTION

The new concepts of future HTR's are based on smaller units which underline the inherent safety aspects of this reactor type. The Modul HTR /1/, based on the AVR[1] is one concept. The 450 MWe and the 900 MWe are other concepts based on the THTR[2]. Common to all new concepts is the variety of applications, e.g. electricity, process heat, district heating. Any new system must be subjected to a demonstration test as part of the licensing procedure. The tests should show that a number of safety criteria are met with respect to fuel element behaviour under postulated temperature transients and/or cycling for early and late in life conditions and specific fuelling principles.
JRC Petten has therefore developed an irradiation facility for temperature transients simulations, TETRASI[3], based on existing facilities which were originally designed for full size fuel elements testing at steady state conditions with continuous fission gas release measurements.
Although, defined scenario's for temperature transients of new HTR's are not yet available, one can assess from safety analyses, performed for existing plants, a parameter range which permits a feasibility study of a transient irradiation facility, e.g. fuel temperature increase to 1600°C between 10 and 100 hours.
The JRC Petten irradiation facility, TETRASI, basically consists of a twin capsule rig, connected to existing sweep loops for process control and fission gas analysis. Each capsule contains one or two full size fuel elements and is independently controlled. Temperature and power control are achieved by vertical displacement of the complete in-core section (stroke 630 mm) and an independent displacement of the upper capsule (stroke 250 mm). Temperature transients and/or cycling are achieved with adaptable gas mixtures (helium-neon) and by variable control of gas gaps in the capsules. One capsule is designed for continuously variable control of gas gaps. The other capsule provides a variable gas gap with a step characteristic. Temperature control and transients are automatically performed to specific programmes.
This report outlines the design of the irradiation rig TETRASI, developed by JRC Petten and gives a thermal analysis for different case studies.

---

1) AVR = Arbeitsgemeinschaft Versuchs-Reaktor
2) THTR = Thorium-Hoch-Temperatur-Reaktor
3) TETRASI= TEmperature TRAnsient SImulation Rig

EXPERIMENTAL REQUIREMENTS

The design of the TETRASI-Rig is based on experimental requirements as described in table 1. The assumed experiments are accelerated tests with respect to power reactors, to demonstrate performance of reference fuel elements under normal and transient operating conditions. The fluence/burn-up correlation is chosen according to proof test conditions. The considered fuel element for this study is the reference spherical HTR fuel element with low enriched uranium 500 µm coated fuel particles with TRISO coating and 1 g U-235 loading.

The new HTR concepts, i.e. Modul-HTR and 450/900 MWe, with their specific fuelling schemes are considered for the different case studies. Fuel temperatures in a Modul-HTR with multiple pass of fuel elements vary between 750-950 and 1050°C. Demonstrable temperature transients are expected up to about 1250°C. Fuel temperatures for the 450/900 MWe plants with single pass fuelling (OTTO= One through then out) will steadily increase from approx. 700°C until approx. 1050°C, and remain at about 1000°C until full dose is reached. Temperature increase during transients up to about 1600°C is expected.

Temperature transients should be performed at BOL[1)] and EOL[2)] conditions, preferable in core positions with similar nuclear characteristics.

---

1) BOL = Beginning of life
2) EOL = End of life

TABLE 1: EXPERIMENTAL REQUIREMENTS

no.	Parameter	Requirements
1	Neutron fluence (E>16 fJ)	between 3.10^{25} and 6.10^{25} m^{-2}
2	Burn-up	between 8 and 12%FIMA(Fissions per initial atoms)
3	Fission power	up to about 3500 W
4	Fuel temperarure	MODUL-HTR: Simulation of multiple pass fuelling with cycled temperarures per operating cycle,i.e. 750°C,950°C,1050°C and superposed transient at BOL and EOL conditions between 1250°Cand 1500°C during approx.10 hours(FIG.2). 450/900 MW_e-HTR : Simulation of OTTO fuelling with temperatures from 700°C to 1050°C and transient at BOL and EOL conditions up to about 1600°C during approx.100 hours (FIG.1).
5	Fission gas release	Continous analysis of Krypton and Xenon isotopes
6	Specimen enviroment	High purity inert gas;Helium and Neon.

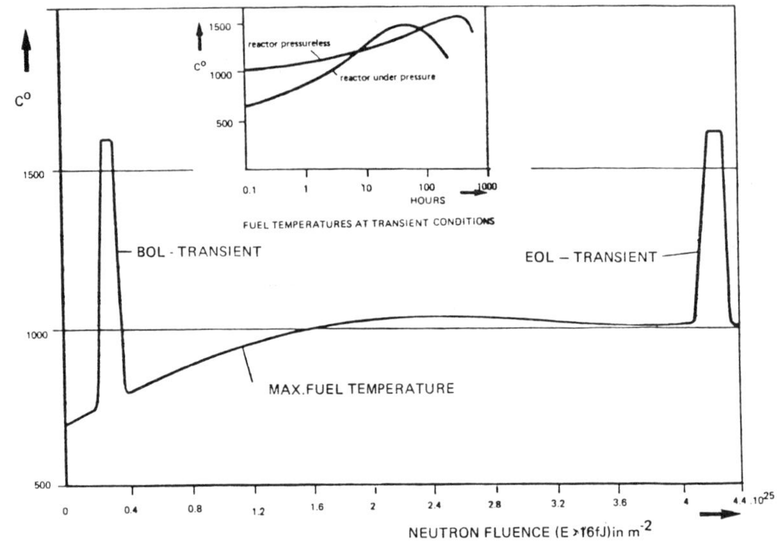

FIG.1: FUEL TEMPERATURES FOR 450/900MW$_e$HTR with BOL-andEOLTEMPERATURE TRANSIENTS vs. NEUTRON FLUENCE

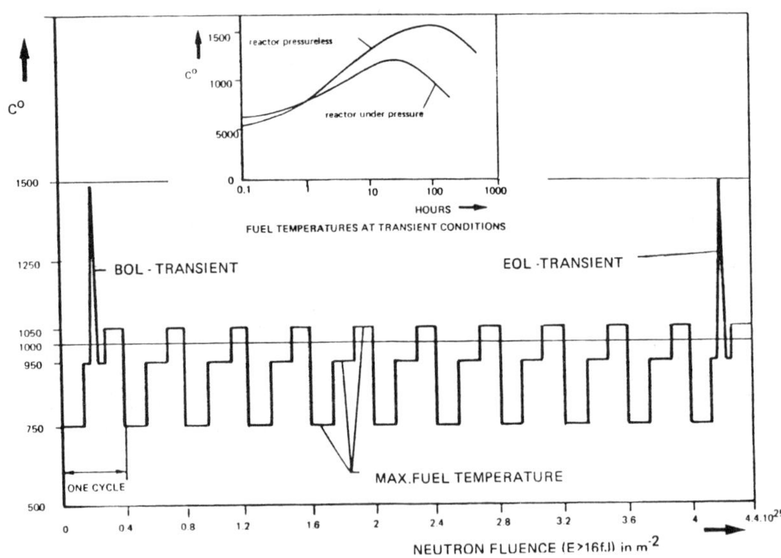

FIG.2: FUEL TEMPERATURES FOR MODUL-HTR with BOL-and EOL TEMPERATURE TRANSIENTS vs.NEUTRON FLUENCE

DESIGN OF IRRADIATION FACILITY

Temperature transients and cycling for full size HTR fuel elements under realistic conditions as mentioned in the foregoing section require a specially designed rig to realize temperature variations between 500°C and 900°C without changing reactor core position. The rig described is based on a proven design,which succesfully operates as a steady state device /4,5/.

The basic design characteristics of the TETRASI-rig are:

- o Two independently monitored and controlled capsules, arranged vertically.
- o Asymmetrically positioned capsules in a protection tube to maximise outer diameter.
- o Each capsule contains one or two fuel elements.
- o Diameter of fuel elements is between 58 and 59 mm.
- o Spherical fuel element is enclosed by two graphite caps which are centered on middle line, eliminating influences of graphite shrinkage and differential thermal expansion on centering mechanism.
- o In excess of 95% the heat generated by gamma absorption and fissions is radially dissipated through the rig materials by conduction and through the gas gaps by conduction and radiation, to the outside can, which is cooled by the downflow of primary cooling water.
- o Adjustment of irradiation temperature in a wide range by changing thermal conductivity in the control gas gaps (helium/ neon) and by changing specimen position in the flux field (Fig.3).
- o Control range with constant gas gap width is about 350°C. Additional control range of up to about 650°C for temperature transients is achieved by varying the gas gap width between specimen and capsule wall, which is one of the special features of the TETRASI-rig.
- o Gas gap width variation is realized by a vertical movement of a thermal shield which surrounds the specimen zone. The movement is realized by a pressure controlled bellow. Two designs with different characteristics are presented.
  The thermal shield in capsule 1 (Fig.4) can be completely removed from the specimen zone, which results in a gas gap width enlargement by at least a factor 7 (e.g. 0.5 mm to 3.5 mm). The thermal shield of capsule 2 is designed as a cone of 4° (Fig.5). The effective enlargement of the basic gas gap width is 0.7 mm.
- o Further means for temperature or specimen power adjustment due to flux variations are provided by mechanisms which displace the complete in-pile section, continuously and remotely controlled out of the reactor fuel zone (stroke 630 mm) and/or independently displace capsule 1 over a stroke of 250 mm. These features provide a substantial operating mode flexibility.
  Flux variations by a factor 6 for capsule 2 and by a factor >10 for capsule 1 are possible.

The capsules are instrumented with thermocouples, fluence detector sets and SPN detectors. Thermocouples are located on the poles and on equator in bores of the spherical specimen. Each capsule is provided with two capillary tubes to sweep the capsule continuously with an appropriate gas mixture for temperature adjustment and fission gas release measurements.

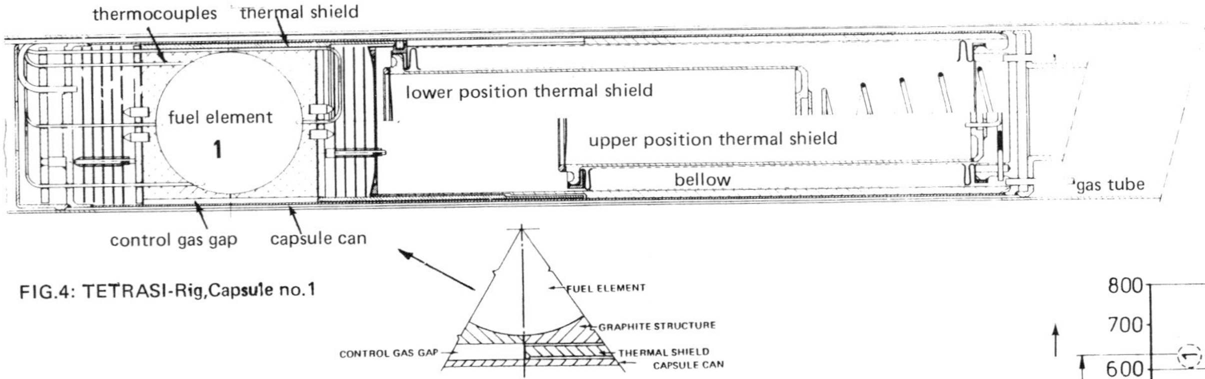

FIG.4: TETRASI-Rig,Capsule no.1

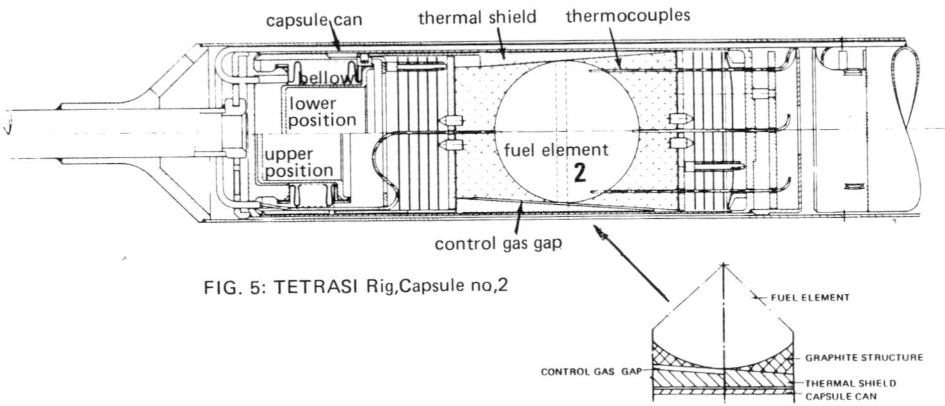

FIG. 5: TETRASI Rig,Capsule no,2

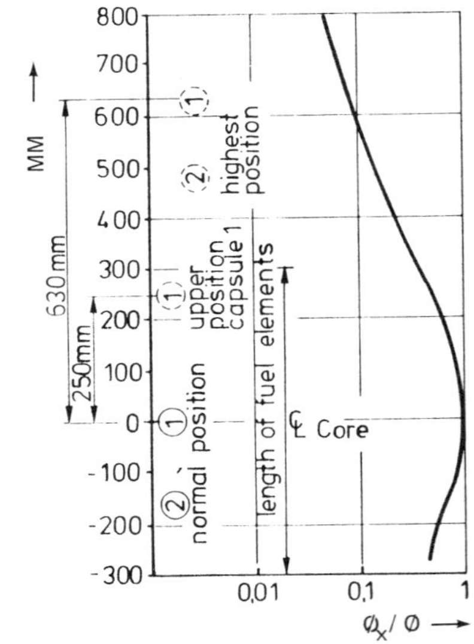

FIG.3: RELATIVE VERTICAL FLUX DISTRIBUTION

OUT-OF-PILE INSTALLATION

This experiment is planned to be operated and controlled by the Sweep Loops/6/which have been designed for HTR fuel irradiation requirements,e.g.:

- continuous transport, monitoring and analysis of gaseous fission products, released by HTR fuels under irradiation,
- temperature control of irradiation specimens by binary noble gas mixtures,
- data acquisition of irradiation and process parameters.

THERMAL ANALYSIS

The thermal analysis was performed by a two-dimensional stationary heat flow code WRZ in r-z geometry (14 x 81 structure) for BOL, EOL and transient conditions. The 2DT code calculates the temperature field with 'hot' radii. The parameters considered for different conditions are (Fig. 6):

- Maximum initial fission power of 3500 W
- Fission power depletion
- Adequate core position for burn-up/fluence correlation
- Material properties as a function of temperature and fluence
- Shrinkage phenomena of graphite structure.

The results of thermal analysis for different case studies are shown on figures 7 and 8 for capsule 1 and capsule 2 respectively. Initial irradiation conditions are achieved with a gas mixture of approximately 90% helium and 10% neon. The figures show the radial temperature distribution on centre line capsule for BOL and EOL conditions with and without a temperature transient. Temperatures are almost uniform on the pole and equator of the spherical specimen due to the conically stepped gas gap with the greatest width on the equator.

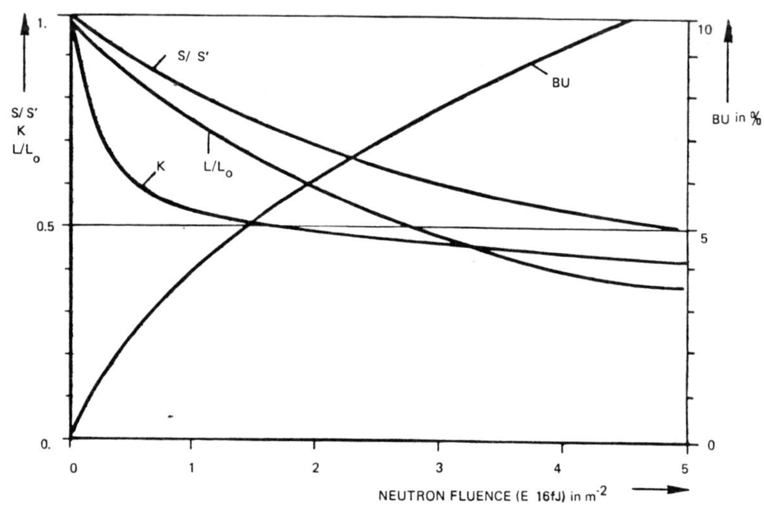

FIG. 6:Normalized thermal conductivity,k,graphite shrinkage, s/ s', power depletion,L/L_0,and burn-up,BU,vs.neutron fluence

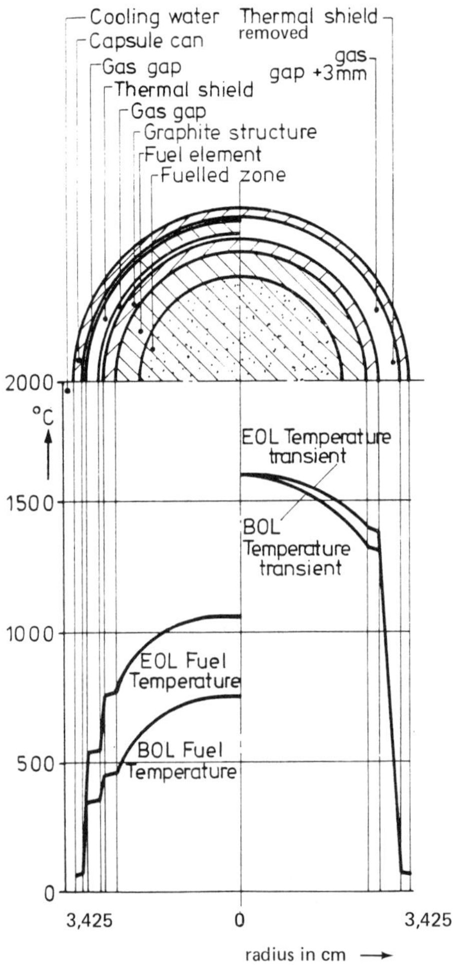

FIG.7: RADIAL TEMPERATURE DISTRIBUTION IN CAPSULE NO.1 AT NORMAL OPERATING, BOL-AND EOL-CONDITIONS

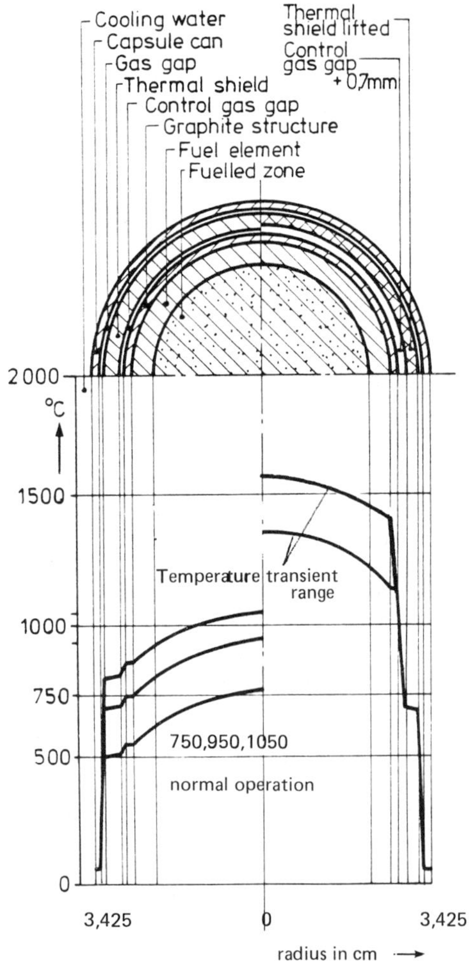

FiG.8: RADIAL TEMPERATURE DISTRIBUTION IN CAPSULE NO.2 AT NORMAL OPERATING, BOL-AND EOL-CONDITIONS

SUMMARY

A new irradiation facility for temperature transient and cycling in-pile experiments on HTR fuel elements under realistic operating conditions has been developed by JRC Petten.
Case studies, considering future HTR concepts with their specific temperature history and anticipated transients have been calculated. The main conclusions are as follows:

- o Full dose experiments with HTR specific temperature transients up to about 1600°C at BOL and EOL conditions are feasible.
- o Adaption to any transient course within 600 hours (duration of operating cycle) can be realized by a number of measures, i.e. variation of gas mixture, control gas gap width and specimen position.
- o Experiments can be conducted at real time temperature history and real burn-up/fluence correlation for single pass and multiple pass fuelling systems.
- o Constant nuclear conditions are assured throughout the testing period. No change of reactor core position is necessary.
- o Temperature cycling with superposed transient is feasible.
- o Experiments are automatically operated to specific operation mode.
- o Specimens are tested under controlled high purity inert gas atmosphere.
- o Doping of atmosphere with controlled impurities for safety related experiments is possible.
- o Fission gas release is under all operating modes continuously analysed.
- o Highly activating gases such as Argon are not applied for temperature control.
- o Safety containment of capsules operate below 100°C.
- o The rig is suitable for in-core power ramp tests.

After tentative trials in the past concerning cycled HTR fuel experiments with moderate temperature variations and numerous steady state experiments, it becomes evident that safety related experiments under realistic operating conditions is the next logical step in HTR fuel testing. The first JRC Petten temperature transient and temperature cycling experiment is planned for 1983, when well characterised experimental requirements are available.

REFERENCES

/1/ H. Reuther, G.H. Lohnert
Der Modulare HTR. Ein neues Konzept für den Kugelkaufenreaktor.
Atomwirtschaft, Januar 1982.

/2/ A.-W.Mehner, personal communication, 1982

/3/ E. Baust, J. Schöning, W. Wachholz
Status und Möglichkeiten des Hochtemperaturreaktors
Atomwirtschaft, Januar 1982

/4/ R. Conrad
Irradiation Facility for Large HTR Fuel Specimens in the HFR
Petten, EUR 5456, 1975

/5/ H. Röttger, A. Tas P. von der Hardt, W. Voorbraak
High Flux Materials Testing Reactor, HFR Petten Characteristics of
Facilities and Standard Irradiation Devices, EUR 5700e, 1981

/6/ D.W. Klage, R. Conrad
Application of a programmable logic controller for HTR fuel
irradiation facilities in the HFR Petten,
Atomkernenergie, Bd. 40, 1982

/7/ R. Conrad
Irradiation Device for HTR Fuel Testing under Abnormal Thermal
Conditions.
Atomkernenergie, Bd. 40, 1982.

EQUIPMENT FOR IRRADIATION EXPERIMENTS IN THE COOLANT CHEMISTRY OF THE AGR

H A Prior and A Harper

Atomic Energy Research Establishment, Harwell.

INTRODUCTION

The Advanced Gas Cooled Reactor (AGR) utilises carbon dioxide at 41 bar as its primary heat transfer medium. Radiolytic reactions in CO_2 produce species responsible for oxidation of the graphite from which the moderator and fuel sleeves are fabricated. In order to attain a satisfactory lifetime for these components, small quantities of carbon monoxide and methane (typically 1–2% CO and 100–200vpm CH_4) are maintained in the bulk coolant to inhibit radiolytic oxidation of the graphite. A consequence of this, however, can be the formation of carbonaceous deposit upon fuel cladding with resultant detrimental effects upon heat transfer from fuel to coolant leading to fuel pin overheating. It is therefore necessary to conduct controlled experiments, under conditions as close as possible to those pertaining in-reactor, to optimise the coolant composition and provide a satisfactory moderator life. At the same time the risk of carbonaceous deposition on the heat transfer surfaces of the fuel pins must be minimised.

The three high pressure gas loops in DIDO, a heavy water moderated materials testing reactor at AERE Harwell, provide facilities in which such studies may be made. Each facility comprises an out-of-reactor section where the composition of the bulk gas is maintained and from where the gas may be recirculated through the in-reactor rigs in which graphite or fuel can specimens may be mounted. These two sections of the experimental arrangement will be described separately.

P. von der Hardt and H. Röttger (eds.), Irradiation Technology, 509–518.

OUT-OF-REACTOR LOOPS.

Loop Design.

A schematic diagram of the main features of the out-of-reactor circuit is shown in Fig 1. The essential components of the loop are a pump and arrangements whereby the composition of the loop gas may be monitored and adjusted as necessary to satisfy the required experimental parameters.

The pump is of the diaphragm type and maintains flow around the in-reactor and out-of-reactor circuits. It is mounted in a pressure vessel filled with loop gas. This arrangement affords a maximum flow rate of *circa* 400 lh^{-1} around the loop. In general, however, flow rates much below this value are required (10 to 25 lh^{-1}) and may be achieved by suitable adjustment of the by-pass circuit. This allows a proportion of the gas to flow directly from pump outlet to inlet rather than around the loop. Reservoir or damper vessels, about 0.5 l in volume, are fitted between the main loop and both the inlet and outlet of the pump thus minimising any pressure variations in the main circuit due to pump action. Absolute values of flowrate are measured with calibrated ball-in-glass flowmeters. The direction of flow around the out-of-reactor circuit is fixed, as shown in Fig 1, but the flow reversal facility allows the direction of flow around the in-reactor rig to be reversed, even during the course of an experiment.

Both in– and out-of-reactor components of the equipment are designed to operate continuously at pressures up to 41 bar. This maximum value, being equal to the coolant pressure

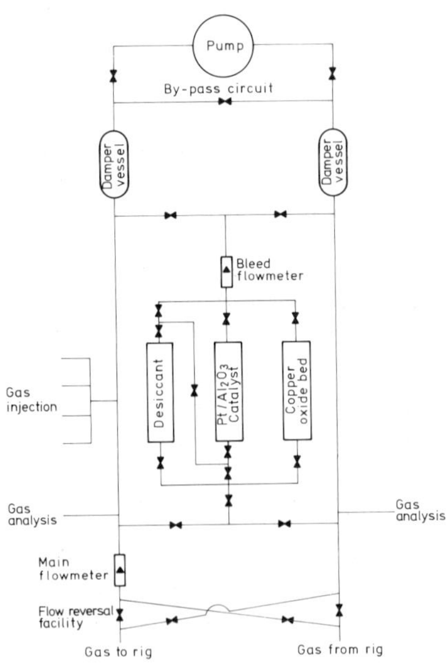

Fig. 1 Basic loop circuit

in an AGR, is the most usual for experimental work. The loop is protected against accidental overpressurisation by bursting discs, which yield at 50 bar. These are backed by relief valves operating at 43 bar. This arrangement provides a high integrity circuit through the absolute seal of bursting discs with the advantage that in the event of bursting disc failure only a small quantity of gas is lost from the loop. Loop pressure is measured using both Bourdon gauges and pressure transducers, the former possessing an accuracy of $\pm 1\%$ and the latter $\pm 0.2\%$. There is a continual loss of gas from the analytical bleeds which, if not compensated for, would result in the loss of 25% of the 30 litre volume of the system in 24 hours. It is therefore necessary to have a means of maintaining loop pressure. This is provided by an injection system operating on the voltage output of a pressure transducer. Carbon dioxide at 50 bar, produced by the vapourisation of solid CO_2, is purified by passing it over copper at 400°C and then fed to the arrangement shown in Fig 2(a). When the gas pressure in the loop falls below a preset value, the consequent change in the output from the pressure transducer causes the gas adder to operate solenoid valves S1 and S2 alternately, thus adding gas to the loop. The process is repeated until loop pressure has been restored. By suitable adjustment of the gas adder, it is possible to maintain loop pressure to better than $\pm 0.5\%$ of the required value.

Control of Coolant Additives.

In addition to controlling the pressure of CO_2 it is necessary to produce and maintain steady gas compositions and therefore to have facilities whereby the other major constituents of the gas may be continuously monitored and any necessary modifications made to their concentrations. Fig 2(b) shows, schematically, the arrangements used for gas analysis and the injection of methane and carbon monoxide. Gas from the inlet line to the in-reactor rig is bled off and reduced to a pressure of 2 bar. The gas is then passed, at a flowrate of 1.7×10^{-3} lh^{-1}, successively through a moisture meter, infra-red gas analysers (IRGAs) for determining carbon monoxide and methane, and a gas chromatograph for detailed analysis.

Loss of methane occurs primarily *via* the analytical bleeds and to a lesser extent by radiolytic destruction as the loop gas passes through the irradiation zone of the the in-reactor rig. The IRGA continuously monitors the transmittance of the loop gas to those wavelengths of infra-red radiation characteristically absorbed by methane. The resultant signal, related to the methane concentration, is used to drive a gas injection system similar to that described for carbon dioxide. In view of the much smaller quantities of gas involved, as concentrations in the range 50–1000 vpm are required, the injection volume used is 0.5 cm^3 and the feed gas is a 10% solution of high-purity methane in CO_2.

Carbon monoxide is always produced by reactions in-reactor but, in general, gas flow from the loop to the analytical system maintains a net loss of this component. In some experiments however, especially those involving graphite oxidation at low carbon monoxide concentrations, the rate of production may exceed the rate of loss. In order to control the concentration of this component it is therefore necessary to provide means for both its addition to, and removal from, the loop gas. The loop gas is continuously analysed for carbon monoxide with an IRGA. Addition of the gas is achieved by injection of CO at 43 bar from a 2cm^3 volume. Gaseous carbon monoxide is supplied from a cylinder of high purity gas and passed over a catalyst bed to remove free oxygen. In addition the carbonyls of nickel and iron are known catalysts for the carbon deposition process and must be rigorously excluded from the gas supply. To this end all carbon monoxide cylinders are analysed prior to use, and only those containing iron and nickel carbonyl

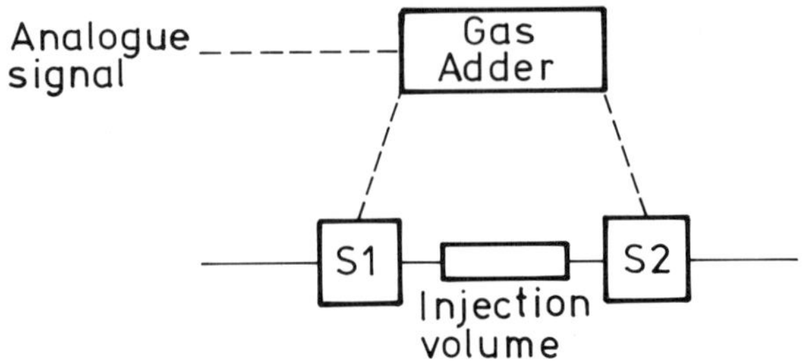

Fig. 2(a) Gas injector

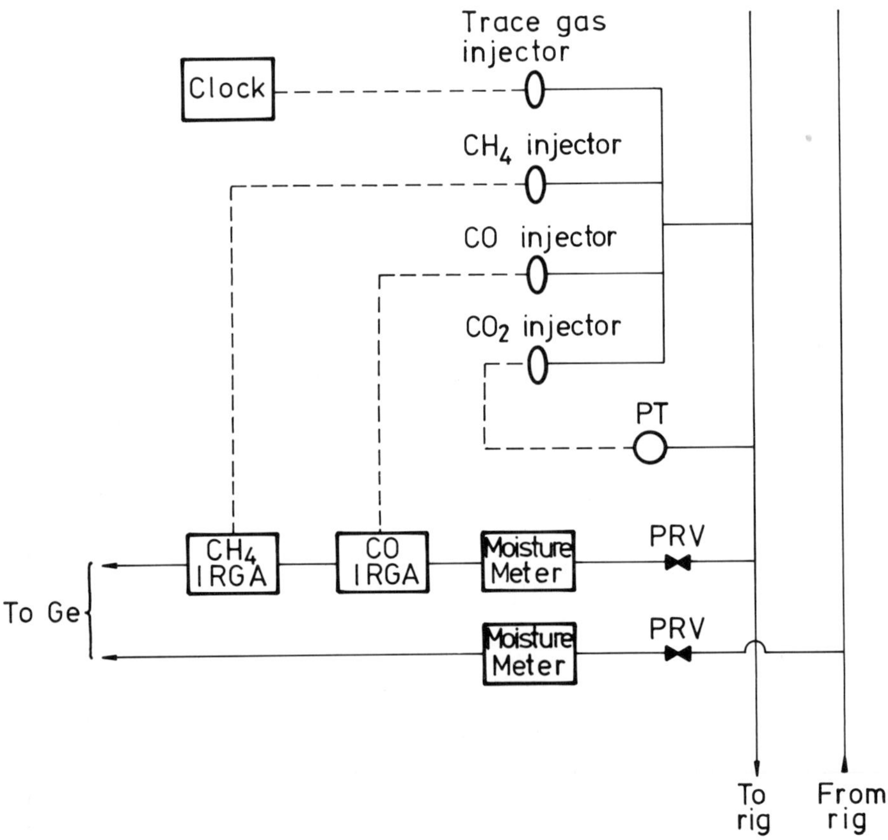

Fig. 2(b) Gas make-up system

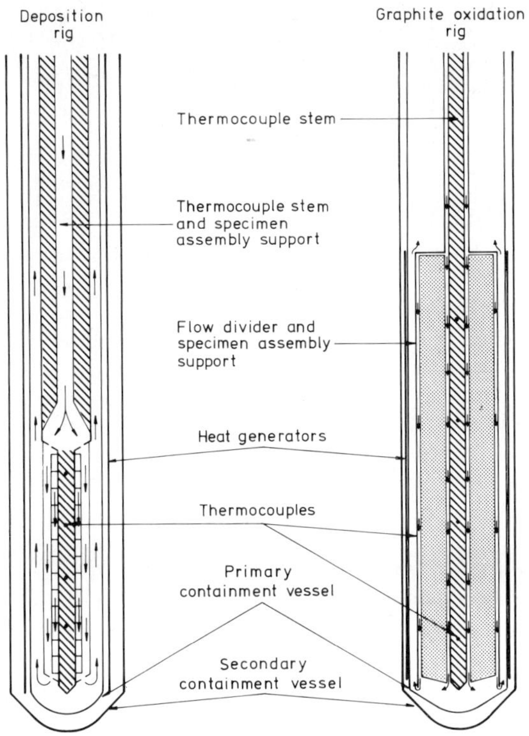

Fig. 3 General scheme of irradiation rigs

concentrations less than 5×10^{-7} gl^{-1} are employed. Carbon monoxide may be removed from the system by by-passing a proportion of the gas flow through a copper oxide bed at 400°C, where the carbon monoxide is converted to carbon dioxide.

Measurement of moisture levels in gases presents some difficulties. In the high pressure gas loops this is routinely achieved using electrolytic moisture monitors. These devices, which rely on desiccation of the gas and electrolysis of absorbed water, provide continuous analysis to an accuracy of $\pm 10\%$ at 300 vpm water. As a check upon the behaviour of this system facilities exist for passing samples of gas through a continuous reading dew-point meter which has an accuracy of $\pm 5\%$. Periodic checks upon the behaviour of the standard electrolytic monitors ensure that these devices are performing to specification.

The levels of moisture may be controlled in the loop system in the region 20–1000 vpm. Water is produced in the loop gas, to an equilibrium level of 200–300 vpm, as a result of radiolytic reactions in-reactor. In order to produce concentrations lower than the equilibrium value, a proportion of the loop gas may be by-passed from the direct circuit through a desiccant filled bed as shown in Fig 1. Control of the proportion of the flow by-passed allows various concentrations to be maintained. Concentrations above equilibrium levels are achieved by injection of carbon dioxide previously passed through a bubbler filled with demineralised water.

Suitable balancing of injection rate and flow through the drier bed allows the desired moisture level to be attained.

The methods described above allow the concentrations of methane and carbon monoxide in the loop to be controlled to better than $\pm 5\%$ of their required values and the concentration of water vapour at better than $\pm 10\%$ of the required value.

Free hydrogen is produced by reactions in-reactor, the resulting equilibrium levels being similar to those already noted for water. Although such levels are, in general, consistent with the desired experimental parameters, in some graphite oxidation experiments abnormal quantities of hydrogen are produced, probably within the graphite pores, and it is necessary to remove hydrogen from the system to maintain acceptable concentrations. Conversely it may be necessary to inject hydrogen in order to raise its concentration above equilibrium values to satisfy special experimental requirements. No continuous on-line monitoring of hydrogen concentration is carried out and these values must be obtained by gas chromatography as described below. One consequence of this is that, in the absence of an analogue signal related to hydrogen concentration, it is not possible to use an 'on-demand' injection system similar to that described for methane and carbon monoxide. The technique adopted is therefore to inject hydrogen at regular intervals using an electronic clock mechanism to drive solenoid valves, the period between injections determining the hydrogen concentration finally achieved. Free hydrogen may be removed from the system by passing a proportion of the gas flow through a bed containing an alumina-supported platinum catalyst at 400°C where hydrogen is converted to water. This facility may, if necessary, be operated in conjunction with a desiccant bed.

It is occasionally necessary to add small quantities (10–20 vpm) of organic gases such as ethylene in order to assess their significance in the carbon deposition process. This is effected using a timed injection system similar to that described for hydrogen although, like methane, these gases are usually injected in the form of dilute solutions in carbon dioxide to achieve and maintain the very low concentrations specified.

Gas Chromatographic Analysis

Accurate measurements of gas composition are performed by gas chromatography. Inlet and outlet lines from each of the rigs are sampled and the gas is passed to two gas chromatographs, one of which estimates carbon monoxide and methane and the other hydrogen. In addition to this, periodic analyses are made on gas sampled from a bottle of standard $CO/CH_4/H_2/CO_2$ mixture as a check upon the calibration of the instruments. Sampling sequence and frequency are defined and controlled by an electronic programmer driving an arrangement of solenoid valves. The signal from the gas chromatograph is passed to a programmable integrator which automatically assigns and integrates component peaks, thus producing output directly in terms of concentration. The inlet and outlet of each rig is sampled once every three hours providing a regular and fully automatic check upon gas composition. The accuracy of the determinations thus produced is $\pm 5\%$ of the true value for CO and CH_4 and $\pm 15\%$ of the true value for hydrogen.

The recirculating gas in the loop contains trace quantities of organic components derived primarily from radiolytic methane breakdown. The concentration spectrum of these components is determined by gas chromatographic analysis of samples of gas taken from the loop at high pressure and analysed remotely. The technique employed permits detection of these trace materials at concentrations greater than 0.2 vpm with an accuracy of $\pm 20\%$ at 10 vpm. Although this function currently requires taking and analysing gas samples manually it is hoped, in the near future, to instal equipment to allow automatic sampling and analysis in much the same manner as for CO, CH_4 and H_2.

IN-REACTOR RIGS.

Rig Design.

The irradiation rigs provide graphite and metal specimen irradiation in high pressure flowing carbon dioxide based coolants over a wide range of dose rates. The specimens are incorporated in assemblies that are readily retrieved from the rigs by simple remote handling techniques. No cutting of rig components is necessary and all items are reusable. A scheme showing the principal features of the irradiation rigs is shown in Fig 3. Two types are shown. One is for oxidation experiments using graphite specimens and the other is for the study of deposition on fuel can surfaces. Both types consist of concentrically arranged components comprising central thermocouple stem, specimen support and flow divider, heat generator stems and primary and secondary pressurised gas containment vessels. An additional, electrically heated, version of the deposition rig is installed in the out-of-reactor circuit and may be operated in series with an in-reactor rig in order to study thermal contributions to the reactions which produce carbon deposition.

The rig heating arises entirely from radiation absorption and temperature control depends upon the flow regime adopted for the particular experiment. Under normal conditions carbon dioxide based mixtures are used at 41 bar and at flow rates between 10 and 400 lh^{-1} at pressure. At 50 lh^{-1} or less the flow regime is wholly viscous and under these circumstances almost all the heat generated within the irradiation rig is conducted radially to the reactor D_2O in which the rig is immersed. Temperature control is achieved by limiting the rate at which heat is conducted to the D_2O. Coarse control is achieved by varying the radial dimensions of the heat generators, reduction of the gas gap between the primary and secondary containments increasing the rate of heat transfer. Fine control is brought about by varying the composition, and hence thermal conductivity, of the helium/neon/carbon dioxide mixture which is maintained in the secondary containment vessel. At flow rates in excess of 100 lh^{-1} turbulent flow conditions predominate, a significant fraction of the radiation-generated heat is carried from the rig by the flowing gas, and under these circumstances specimen temperatures are largely controlled by the flow rate.

Temperature Measurement.

The measurement of specimen temperatures has required considerable attention. They can not be determined directly because the irradiation of demountable assemblies precludes the possibility of attaching thermocouples directly to the specimens. In the graphite oxidation rig, Fig 3, thermocouples are positioned on the central stem and are complimented by others on the flow divider in the same axial planes. Measured temperatures are obtained, therefore, at the inner and outer diameters of the specimen carrier at four positions along its length and individual specimen temperatures are calculated, assuming that a linear temperature gradient exists radially between central stem and flow divider, by interpolating both radially and axially. Temperatures in deposition experiments are measured at four positions along the specimen support stem, Fig 3, by the built-in thermocouples and specimen temperatures are obtained by reference to the results of calibrations made with specially constructed calibration rigs. These are identical in construction and operation to the expermental rigs except that thermocouples are attached to fixed specimens.

Energy Deposition Rates.

Total energy deposition rates are measured with a graphite calorimeter of the steady state, temperature difference type. This consists of a cylindrical graphite sample supported centrally within a graphite tube. The calorimeter can be installed in the irradiation rigs in place of the specimen support stems. Thermocouples are installed in both components at the reactor core centre height and the rig is operated at low pressure in static gas. Heat is generated in the graphite sample by radiation absorption and is transferred across the intermediate gas gap to the outer graphite tube by radiation and conduction. The rate of heat loss from the central specimen, which is equal to the rate of radiation absorption at the rig centre, can be calculated from the measured temperature difference between graphite sample and tube and the heat transfer properties of the system. Neutron flux monitors are positioned at intervals along the specimen assemblies and the results are used in conjuction with the calorimetric data to estimate the dose rate delivered to individual specimens. An additional variation to this method is available for application to the graphite irradiation rigs. The precise positioning of the specimen assembly within the flow divider, Fig 3, and the measurement of temperatures within and without the annular specimens permits the calculation of dose rate at four positions along its length by a method analogous to that described above for the purpose-built calorimeter.

Specimen Assemblies.

A typical graphite specimen assembly is shown in Fig 4 where provision is made for the irradiation of 24 annular specimens over the full height of the reactor core. Considerable variation in the specimen inventory can be made by varying the specimen height and by incorporating complex assemblies of concentric specimens; a maximum of 48 specimens has been irradiated in a single specimen assembly by this means. These graphite assemblies can be irradiated at dose rates from 0.2 to 3.2 Wg^{-1}, depending upon the in-reactor position of the rig. The dose rate falls from the centre to each end of the assembly by about 40%. Specimen temperatures can be maintained between 300°C and 600°C with a maximum variation from centre to ends of about 30%. An important variation in specimen design is also employed where four solid specimens, 10cm long, are irradiated under closely controlled flow conditions. A description of its mode of operation appears elsewhere.[2]

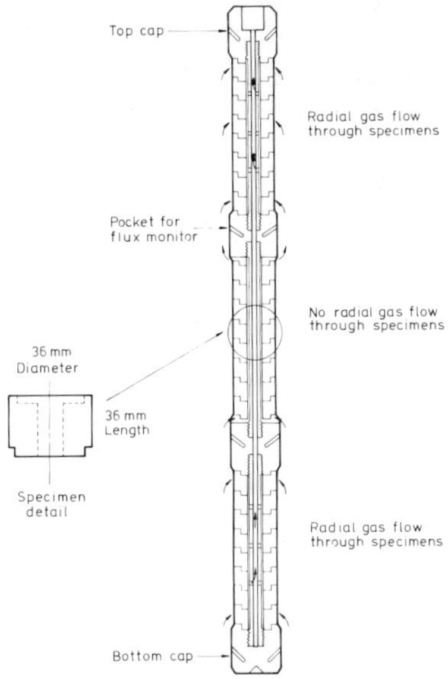

Fig. 4 Typical graphite irradiation assembly

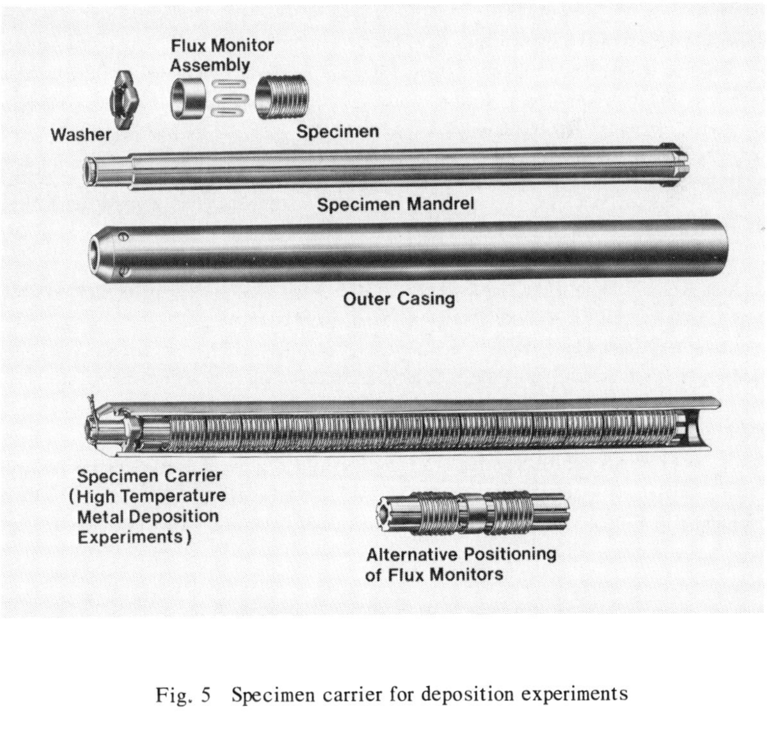

Fig. 5 Specimen carrier for deposition experiments

The deposition rig specimen assembly is shown in Fig 5 and permits the irradiation of 15 specimens of AGR fuel cladding steel, each 15mm high, with gas flow over both inner and outer cylindrical surfaces. The maximum dose rate plane passes through the middle of the assembly and the dose rate falls from the centre to each end by about 10%. The rig can be irradiated at dose rates from 0.7 to 1.5 Wg^{-1}, depending upon the in-reactor position adopted, and specimen temperatures can be maintained between 550°C and 800°C with a maximum variation from the centre to the ends of about 15%.

FUTURE DEVELOPMENTS AND LOOP UTILISATION

To date the bulk of the operational data generated during the operation of the gas loops has been gathered and recorded manually. A mini-computer has been installed, and is being commissioned, which will gather and store data on loop operational parameters automatically. In addition the system provides facilities for obtaining the current status of each loop at a site remote from the loops themselves. Ultimately the computer will perform not only these tasks but will also control loop operations such as the maintenance of gas composition, and will provide warnings and alarms should acceptable ranges for the value of any loop parameter be exceeded.

Over the past decade the facilities described above have been used to support a programme studying both graphite oxidation and carbon deposition. Some 28 experiments have been performed to date in studies of the radiolytic oxidation of moderator and sleeve graphites,[1,2] with gas compositions in the range 0.5–2.0% CO and 10–600 vpm CH_4, pressures of 20–41 bar and flowrates of up to 400 lh^{-1} at pressure. Data from these experiments have been used, in conjunction with similar information from other sources, to produce recommended oxidation rate data as a function of gas composition for moderator graphites.[3]

In excess of 100 experiments have been performed to study carbonaceous deposition on fuel cladding in gas compositions ranging from 0.2 to 6.0% CO and 4 to 698 vpm CH_4. Assessment of the data from such experiments has yielded, and continues to yield, considerable insight into the mechanism by which this process occurs.[4,5]

REFERENCES

1. Faircloth RL and Prior HA, The Oxidation of Moderator Graphites Irradiated in Carbon Dioxide Containing Carbon Monoxide, Methane and Water, Proceedings of a Conference held at the University of Salford, 21–24 April 1980, Heyden, 1980, 120–128

2. Faircloth RL and Prior HA, Experiments in DIDO to Measure Graphite Oxidation Rate Profiles in Thick Specimens, Nuclear Energy, February 1982

3. Shennan JV, Radiolytic Carbon Gasification, Proceedings of a Conference held at the University of Salford, 21–24 April 1980, Heyden, 1980, 98–110

4. Bennett MJ *et al*, Survey of the Principal Parameters Underlying Carbon Deposition on a 20% Cr/25% Ni/Niobium Stabilised Austenitic Stainless Steel from $CO_2/CO/CH_4/H_2O$ Gas Mixtures in in DIDO 6V3 Gas Loops, Proceedings of a Conference held at the University of Salford, 21–24 April 1980, Heyden, 1980, 81–89

5. Bennett MJ, Surface Coatings to Inhibit Carbonaceous Deposition on AGR Fuel Cladding, Nuclear Energy, February 1982

CHARACTERIZATION OF HTR FUEL ELEMENT SAMPLES BY ACTIVATION TESTING

A.-W. Mehner, J.J. Abassin, P. Sciers

NUKEM GmbH, Hanau FRG; CEA, CEN Grenoble France

ABSTRACT

Performance of HTR fuel particles is examined by irradiation experiments in Material Test Reactors. An important goal is the determination of particle failure fraction during irradiation. To do this a nondestructive method was developed to ensure zero failure at the begin of the irradiation test. The fueled samples are activated for half an hour in an electrically heated furnace in reactor Melusine. The resulting fission gases are trapped and quantitatively evaluated. From the resulting R/B-values the number of failed particles can be deduced. In two test series with a total number of 24 fuel compacts no defective particle could be found. Detection limit R/B Kr 85 m was 10^{-8}.

1. INTRODUCTION

The high temperature gas-cooled reactor (HTR) provides an alternative approach to the commercialization of nuclear power as compared to other fission power producing systems, such as LWR, HWR and LMFBR. The fuel for the HTR consists generally of uranium and thorium species in the form of microspheres encapsulated in layers of pyrolytic carbon and silicon carbide. The composition of the kernel material depends on the selected fuel cycle.

In 1979 the reference fuel cycle for HTR's in the Federal Republic of Germany was changed from high enriched uranium/thorium

P. von der Hardt and H. Röttger (eds.), Irradiation Technology, 519–528.

to low enriched uranium fuel. The reference HTR fuel element in the FRG is a spherical fuel element with an inner fueled zone and an outer fuel free zone.

All spherical fuel elements contain the fissile and fertile material as coated microspherical particles. The TRISO-coated particle consists of a fuel kernel of low enriched uranium as UO_2. The kernel is enclosed in a multilayered coating of porous buffer carbon, dense pyrolytic carbon, silicon carbide, and a final outer layer of dense pyrolytic carbon comprising a total coating thickness of about 200 µm. The porous buffer coating provides void space for gaseous fission products and prevents damage to outer coatings by recoil fission fragments. The silicon carbide and dense pyrolytic carbon layers serve as containment vessel for retention of gaseous and solid fission products.

After fabrication of the particles and after moulding them into spherical fuel elements, it is critically important to know the number with defective coatings that would allow the release of gaseous and solid fission products to the reactor cooling system.

2. OBJECTIVES

An important goal of the development of advanced concepts for HTR's such as generation of high temperature process heat or economic production of electric power with increased gas outlet temperatures is to improve the retention of fission products in the fuel, to limit the fission product release during normal operation and to satisfy regulatory requirements for public and plant personnel safety.

Three sources for volatile fission products circulating in the helium coolant have been identified so far:

- Fission of uranium present as contamination outside of the particle coating
- Missing or imperfect coatings caused by different manufacturing steps
- Total failure of the particle coating during irradiation so that the fuel kernel is directly exposed

The first two kinds of release are controlled by fuel specifications that limit the amount of uranium contamination in as manufactured fuel elements and by formation of a quality control programme for fuel manufacturing to ensure that fuel quality requirements are met in practice.
The third kind is controlled by specification of fuel kernel and coating parameters to ensure a high degree of coating integrity

throughout life and by establishing a core design which ensures that the limits of coating integrity are not exceeded.

The change from the Th/U fuel cycle into the U/Pu fuel cycle led to further development work to qualify the selected fuel element. Despite the similarity in fuel element design and manufacturing methods a complete qualification programme including process and product development as well as irradiation testing has to be performed. A major task in this field is the characterization of HTR fuel in terms of failure fractions of the coated fuel particles.

The exact knowledge of failure fraction of coated particles allows for

- prediction of fission product release with higher confidence
- verification of particle performance models
- detection of failure mechanisms and hence improvements in particle design

In our qualification programme for coated fuel particles with low enriched UO_2-kernels failure fractions are determined by irradiation experiments in Material Test Reactors (MTR).

A prerequisite for the optimized evaluation of an irradiation experiment is the knowledge of the number of defective as manufactured particles and the amount of contamination in coating and matrix material before irradiation in the test samples. Since core design asks for target values of less than 10^{-5} U/Uges in a fuel element large numbers of particles have to be evaluated to gain statistically significant data.

The determination of the defective particle fraction was originally done by several destructive methods. Wet chemical leaching of deconsolidated fuel elements, hot chlorine gas leaching of fuel elements or loose particles, mercury intrusion into loose particles, burn leach techniques may be mentioned as examples (ref. 1-4).

Though reliable data were gained with these methods, it was not satisfying to control only parallel samples and never the original test elements. To overcome this problem, nondestructive characterization methods for irradiation samples had to be developed.

3. EXPERIMENTAL

The fraction of coated fuel particles in HTR fuel elements can be determined by exposing the samples to a flux of neutrons. This creates volatile fission gases which are released liberately in the case of uncovered fissile material. The released fission gases can be trapped and determined by gammaspectrometry. From these data the amount of free uranium in the sample can be obtained. This method is already applied in the US to characterize fuel rods before being inserted into the blocksegments (rf.5) and proposed by CEA for the characterization of HTR fuel elements (ref. 6).

Two departments of the CEN Grenoble (Piles, DMG) have developed for the HBK project devices and codes to measure the release of gaseous fission products after short time activation and to determine from this data the initial fraction of free uranium. Since the irradiation samples have to be manipulated afterwards for inserting in the irradiation rig the neutron flux and the irradiation time has to be kept as low as possible.

3.1 DESCRIPTION OF TEST SAMPLES

To determine failure fractions during irradiation for HTR reference particles a special sample form was designed. The irradiation space in a Tumult-G rig is limited to samples with 35 mm diameter, therefore with a special moulding tool fueled spheres with an inner fuel zone of only 20 mm diameter were manufactured. These spheres were machined to compacts of about 30 mm height and 30 mm diameter (fig. 1).

Temperature in these irradiation sample is measured by thermocouples in channels along the fuel zone.

The fuel particles consisted in the first experiment of 500 µm (10 Th,U)O_2 LTI-TRISO-particles for the HEU fuel cycle and in the second experiment of 500 µm UO_2 LTI-TRISO-particles for the LEU fuel cycle.

3.2 DESCRIPTION OF THE IN PILE SECTION

The inpile section consists of a furnace and a sample holder. The furnace provides a volume 400 mm in length and 54 mm diameter. The heating is done by four independent heating elements each controlled by 2 thermocouples. Maximum power output is 6 KW for the furnace.

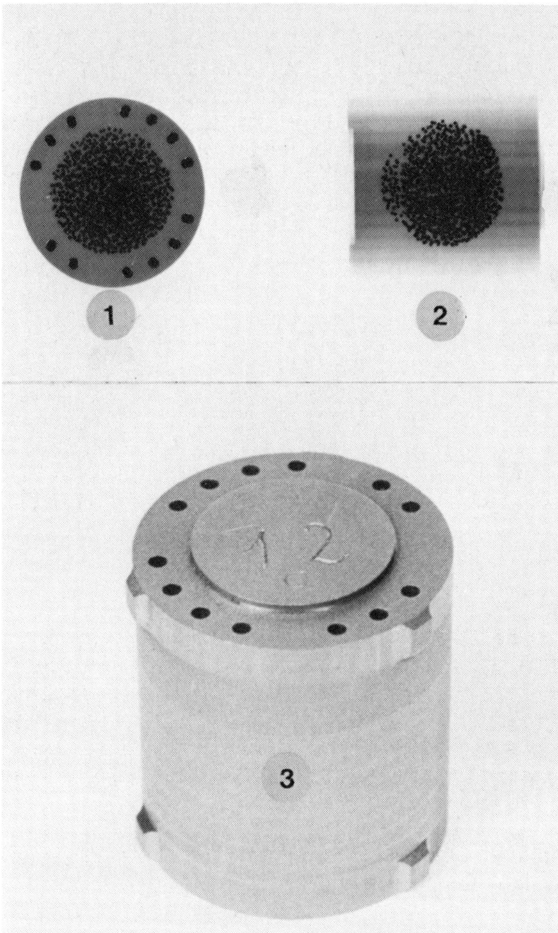

Fig. 1: Irradiation samples for activation tests
1 Topview, 2 Sideview, 3 Isometric view (1,5x)

The lower part of the furnace can be disconnected and then transported in a shielded cask to the Hot Cells for further disassembly. The sample carriers are individually designed and built for the different irradiation samples depending on number, geometry, temperature desired, and gamma heating available in the selected core position.

For our HTR-fuel compacts the sample holder is designed as follows (fig. 2).

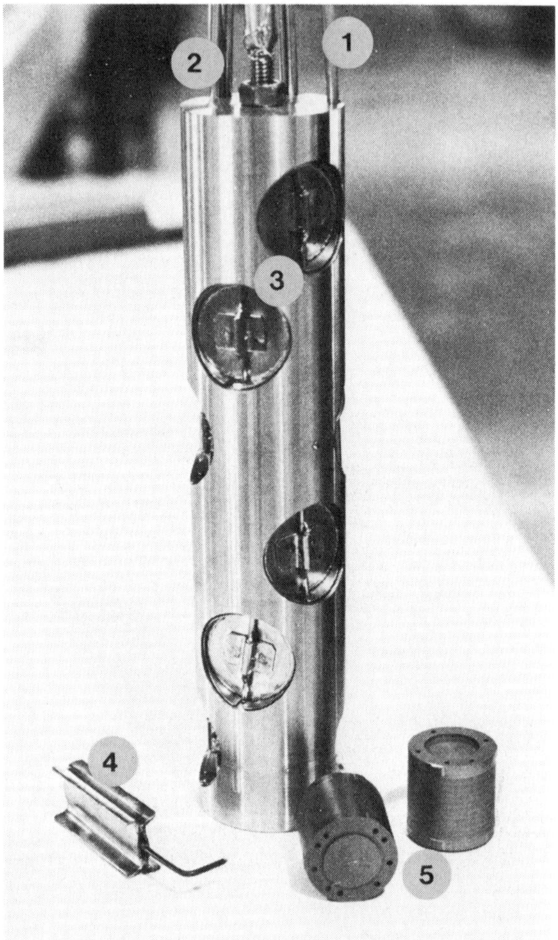

Fig. 2: Sample holder for activation tests (lower part)

1 Sweepgas tubes
2 Thermocouples
3 Holes with containers
4 Handling tool
5 Fuel compacts

- A stainless steel cylindrical block with 6 holes of 32,2 mm diameter. The distance between two holes is 35 mm. They are displaced by 60° on the envelope of the cylinder. These holes contain the different capsules

- Each capsule is provided with a separate sweep tube leading to the upper part of the furnace, three thermocouples to control the temperature and can be loaded with one compact.

- The upper part is linked to the lower part via a gasket. A tight duct is provided to connect the thermocouples and the sweepgas tubes to the out of pile section.

3.3 DESCRIPTION OF THE OUT OF PILE SECTION

The out of pile section consists of the stations to process and analyse the carrier gas with released fission gases and to adjust the desired temperature. The temperature is controlled by establishing a helium flux of about 60 cc/min and adjusting the heaters to a sample temperature of 600°C.

The carrier gas contains all emitted gases through out the activation experiment. By passing through char coal cooled with liquid nitrogen all fission gases are trapped. The first measurement is performed after a sampling time of 15 minutes, the second one after 20 minutes of sampling. The sampling time has to be measured precisely. Gamma spectrometry is performed on a 4000 channel analyser with Ge-Li detector. Absolut calibration is done with a source of Eu 152. All results are stored on discs for furhter processing.

3.4 DOSIMETRY

Single HTR-particles with well known U-235 content are used for dosimetry. They are activated in positions 2 and 5 of the sample holder under the same conditions as the fuel compacts. After cooling down the particle activity for several days, the fission product inventory is determined by gammaspectrometry. In parallel the particle inventory is calculated from the unperturbed flux in this reactor position, the irradiation time, and the cross section. From both values a correction factor is deduced which corrects the estimated flux at the positions of the test compacts.

4. RESULTS

Two series of activation tests have been performed in the Melusine reactor to characterize a total number of 24 fuel compacts.

4.1 HIGH ENRICHED $(TH,U)O_2$ PARTICLES

The characterization was performed in 3 steps with 4 fuel compacts each. The irradiation conditions can be taken from table 1.

	1. Irradiation	2. Irradiation	3. Irradiation
Date	03.06.1980	06.06.1980	11.06.1980
time (min.)	36	33	34
Flux $(n \times cm^{-2}s^{-1})$	$1,6 \times 10^{12}$	$1,96 \times 10^{12}$	$2,16 \times 10^{12}$
Temperatures (°C) TE 1,2,3	602, 613, 606	600, 605, 600	610, 616, 620
Compact Nr.	1, 2, 3, 4	5, 6, 7, 8	9, 10, 11, 12

Table 1: Irradiation data for fuel compacts with $(Th,U)O_2$ particles

The evaluation with the code JEROME gave R/B-values between 10^{-7} and 10^{-6} for all Kr- and Xe-isotopes.

4.2 LOW ENRICHED UO_2-PARTICLES

This characterization was performed in 4 steps with 3 compacts each. The irradiation conditions can be taken from tab. 2.

	1. Irradiation	2. Irradiation	3. Irradiation	4. Irradiation
Date	09.02.1982	12.02.1982	17.02.1982	23.02.1982
time (min.)	31	30	30	30
Flux $(n \times cm^{-2}s^{-1})$	$0,94 \times 10^{12}$	$1,08 \times 10^{12}$	$1,11 \times 10^{12}$	$0,84 \times 10^{12}$
Temperatures (°C) TE 1, 2, 3	661, 671, 663	678, 680, 667	675, 680, 671	669, 682, 672
Compact Nr.	1, 2, 3	4, 5, 6	7, 8, 9	10, 11, 12

Table 2: Irradiation data for fuel compacts with UO_2-particles

The examination of the fission gases revealed no significant release. After a counting time of five minutes for each gas sample no difference compared to the noise spectrum could be determined. In all cases less than 100 counts were detected during 5 minutes.

From both series one can conclude, that no particle is defect.

This is shown in table 3 for the case of the Kr 85 m when 1 % release is assumed for a defective particles.

Fuel	Particles per Compact	Atoms of Kr85m formed during activation	R/B for a defective particle
$(Th,U)O_2$	1490	$4,22 \times 10^9$	$6,7 \times 10^{-6}$
UO_2	1670	$1,82 \times 10^9$	$6,2 \times 10^{-6}$

Table 3: R/B values for a single defective particle

For a $(Th,U)O_2$ particle as well as for a UO_2-particle the gas release of one bare kernel should be clearly detectable since it is more than one order of magnitude beyond the detection limit.

5. CONCLUSIONS

Fission gas release measurements were performed on $(Th,U)O_2$- and on UO_2-TRISO particles. The results obtained show clearly that no failed particles are present in all irradiation samples. The contamination level can be determined separately for each compact and can thus be related to individual fission gas release measurements after irradiation. The applied method is sensitive enough to detect even a single particle with imperfect or missing coating layers. The computer codes involved in this work are applicable to a wide range of different fuel compositions ranging from 90 % thorium-containing particles to low enriched uranium particles.

6. REFERENCES

1. H. Bildstein, K. Knotik, Kerntechnik 15 (1973) 554
2. D.E. La Valle et al., Nucl. Techn. 33 (1977) 290
3. D.M. Hewette II and W.R. Laing, Nucl. Techn. 21 (1974) 149
4. GAC private communication
5. E.E. Anderson et al., Nucl. Techn. 11 (1971) 259
6. DE-OS 23 28 283, Verfahren zum Bestimmen des Anteils von Brennstoffteilchen mit defekter Umhüllung in einem Brennstoffelement für Kernreaktoren
7. M.L. Pointud, P. Chenebault, Nucl. Techn. 35 (1977)

IN-PILE HIGH TEMPERATURE IRRADIATION TESTING TECHNIQUES FOR THE MATERIALS AND FUELS OF HTGR IN JMTR

Isao Tanaka, Hidetake Takahashi, Hiroharu Itami,
Haruhiko Ito, Noriyoshi Tsuyuzaki, JAERI, Japan

Division of JMTR Project
Japan Atomic Energy Research Institute
Oarai Research Establishment, Ibaraki, Japan

ABSTRACT

The Japan Material Testing Reactor (JMTR) in Japan Atomic Energy Research Institute (JAERI) has been developing the many kinds of in-pile irradiation facilities for temperature control, power control and so on in order to serve adequate testing conditions according to the requested objects by researchers.

Recently we have developed the temperature ramp capsule for high speed temperature rising test of coated particle-fuels by using the thermal-bond moving system and vacuum temperature control system, and high temperature irradiation testing capsule for the graphite which is one of the main structural materials of HTGR.

INTRODUCTION

The JMTR in JAERI is a testing reactor, of which specifications are tank type, power of 50 MW moderated and cooled by light water, and highly enriched uraniums of 93%. The JMTR has various irradiation devices, such as loops, hydraulic rabbits, capsules and so on. Irradiation experiments in JMTR are mostly carried out by using capsules.

The vacuum control method is available for controlling the temperature in a capsule. The temperature control is made by means of regurating heat transfer through the gas gap by varying degree of vacuum in a capsule.

High temperature irradiation experiments for the fuels and materials in HTGR have been carried out using capsules.

P. von der Hardt and H. Röttger (eds.), Irradiation Technology, 529–537.

A temperature-ramp capsule has been developed in JMTR for investigating transient behaviors on the coated particle-fuels. The temperature-ramp experiment was carried out from normal irradiation temperature (about 1200°C) to higher temperature (above 1500°C) with high speed.

The paper describes irradiation experiences in the temperature-ramp capsule with including a graphite irradiation capsule for HTGR.

TEMPERATURE-RAMP CAPSULE

Objectives

A temperature-ramp capsule has been developed for the purpose of investigating in-reactor behaviors of compacted particle-fuels in graphite matrix in the accidental condition, when particle-fuels are exposed to high temperature of above 2000°C with fast temperature rising rate, due to reactivity change, loss of coolant, flow blockage and so on in the reactor. Main objectives of the first designed capsule are to verify irradiation techniques for rising temperature at fuels within a short time and to obtain technical data to fabricate sequentially designed temperature-ramp capsules at which temperature may attain up to above 2000°C from the base irradiation temperature (1200 - 1300°C).

Description of temperature-ramp capsule

The schematic drawing of the first designed capsule is shown in Fig. 1. Five compacted particles-fuels are held in a graphite holder which is placed in the inner capsule made of 316-stainless steel. Helium gas of about 1Kg/cm^{2}A is filled in the inner capsule. The fixation of graphite holder is hanged on the upper end plug of the inner capsule with three tie-rods, the guide tube and the pressurized chamber. The irradiation at the base irradiation temperature (1200 - 1300°C) is carried out by the graphite thermal bond (A) surrounding the graphite holder. The thermal bond (B), which forms a large gas gap against the inner capsule tube, is connected to the thermal bond (A). Irradiation at higher temperature by temperature-ramp can be executed with the thermal bond (B).

The temperature-ramp is performed by changing the thermal bond (A) to thermal bond (B). This change is made by moving the driving tube joined with two types of thermal bonds by means of pressurizing outer helium gas of the bellows, which is introduced in the pressurizing tube connected to the gas supplying system placed outside of the reactor. The displacement of the thermal bonds is detected as output of a linear variable differential transformer (LVDT) and also gas pressure measured in the gas supplying system.

The temperature at the specimens is monitored by a W-Re thermocouple placed in the center hole of the compacted particle-fuels. The inner capsule is enveloped by the outer capsule made of

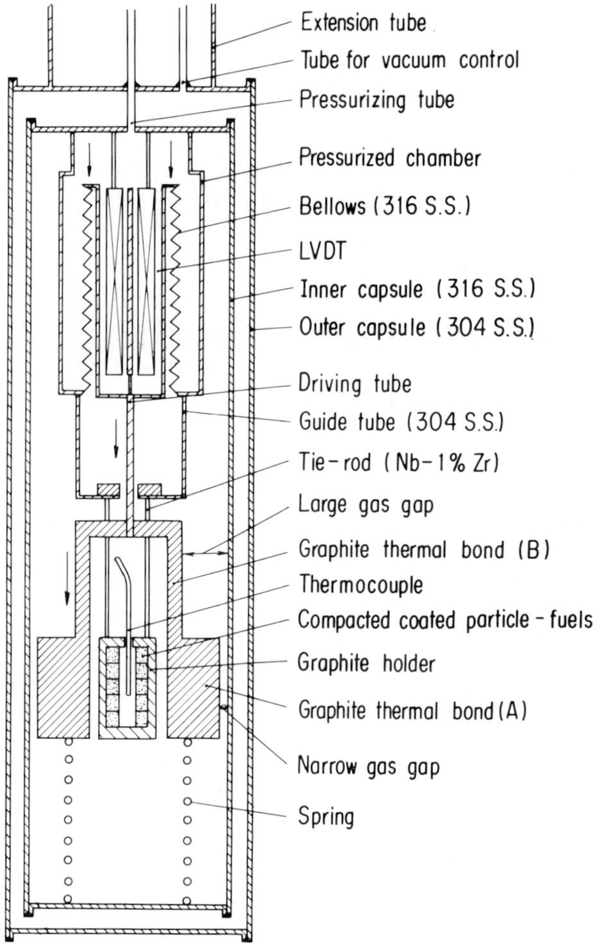

Fig. 1 Configuration of moving thermal bond type temperature-ramp capsule.

304-stainless steel cooled directly by the reactor primary coolant of about 50°C. The gas gap between the outer and inner capsule is utilized for temperature control by the vacuum method during base irradiation to maintain the fuel temperature of about 1200°C, and also used for a means of still more temperature rising after temperature-ramp with changing the thermal bonds.

Experiment and results

The irradiation experiment for the temperature-ramp capsule was carried out for 22 days during one reactor operation cycle of JMTR at 50 MW in 1981. The reactor power and center temperature of the fuels versus the irradiation time are shown in Fig. 2. The temperature is kept about 1200°C using the vacuum temperature controlling method until two hours before the end of experiment. The fuels in the capsule are irradiated at the power of about 130 W/cm.

Before the temperature-ramp test, the temperature at fuels was dropped from 1200 to 1110°C by dumping helium gas into the gas gap between inner and outer capsule. The temperature-ramp up was performed by pressurizing the bellows and changing the thermal bond (A) for (B) as mentioned above. The Fig. 3 shows position of thermal bonds and center temperature at the fuels vs. the pressure increase on the bellows. The thermal bonds moves totally about 56 mm and temperature-rise of 300°C is obtained by pressurizing the helium gas up to 7 kg/cm^{2}G, resulting that the center temperature of fuels attained 1410°C. A detailed temperature change during the temperature-ramp experiment is shown in Fig. 4. The temperature-ramp from 1110 to 1410°C is performed for about 20 minutes and thus the ramp rate is about 15°C/min. Still more temperature rising of about 100°C is carried out by the vacuum temperature controlling and thus maximum temperature is 1510°C during about one hour as shown in the figure.

A more detailed temperature change at the fuels, thermal bond, inside of the inner capsule-tube and so on are shown in Fig. 5 with time in minutes during temperature-ramp (between Time A and B in the figure). The temperature-rise at the inner capsule was within only about 30°C that was 10% of temperature-rise of 300°C at the fuels during temperature-ramp-up by moving the thermal bonds. Therefore, temperature-rise at the inner capsule is negligibly small enough. Temperature-rise at the fuels is about 100°C during temperature-rise by vacuum controlling (between Time B to C in the figure), corresponding temperature-rise of 100°C at the inner tube of capsule. It is confirmed in the experiment that temperature-rise by means of moving thermal bonds is by far greater than that with the vacuum controlling method.

The small temperature-rise at the inner capsule shows that the inner capsule should not be any damaged for a structual material during temperature-ramp-up test.

Results in the experiment are,

(a) it was demonstrated that the moving mechanism of thermal bonds

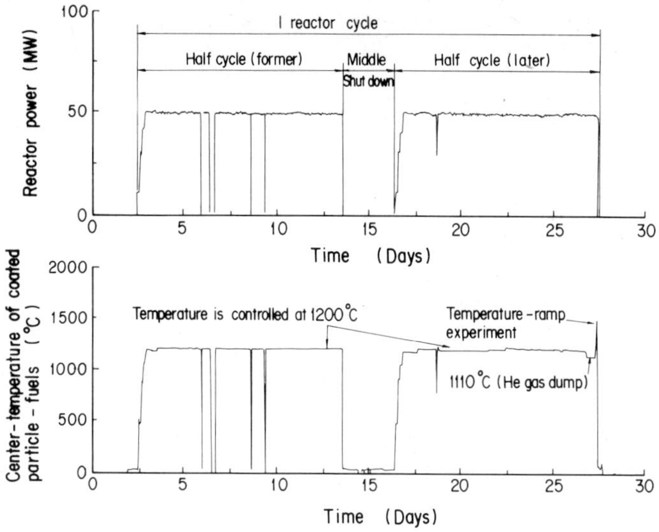

Fig. 2 Reactor power and obtained center-temperature at compacted coated particle-fuels in the temperature-ramp capsule during irradiation of one reactor operation cycle.

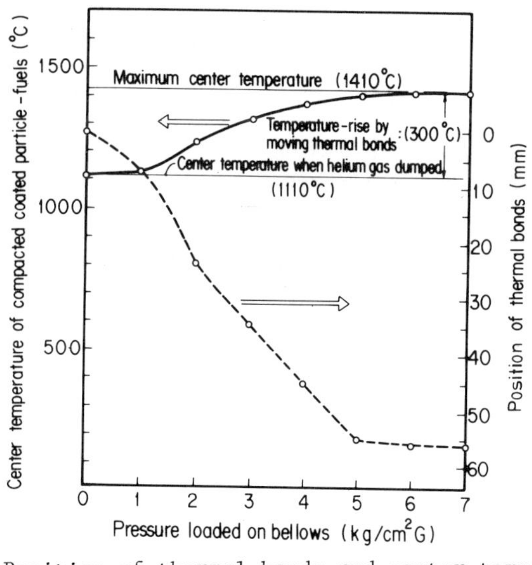

Fig. 3 Position of thermal bonds and center temperature of compacted coated particle-fuels versus loaded pressure on the bellows.

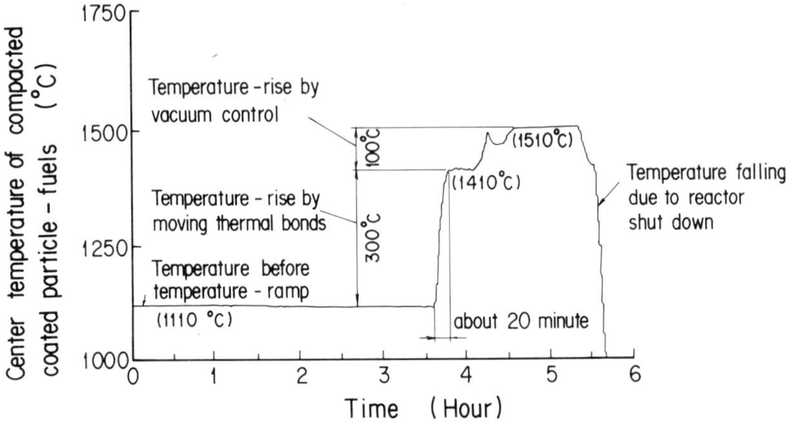

Fig. 4 Temperature history of center temperature at the compacted coated particle-fuels during the temperature-ramp experiment.

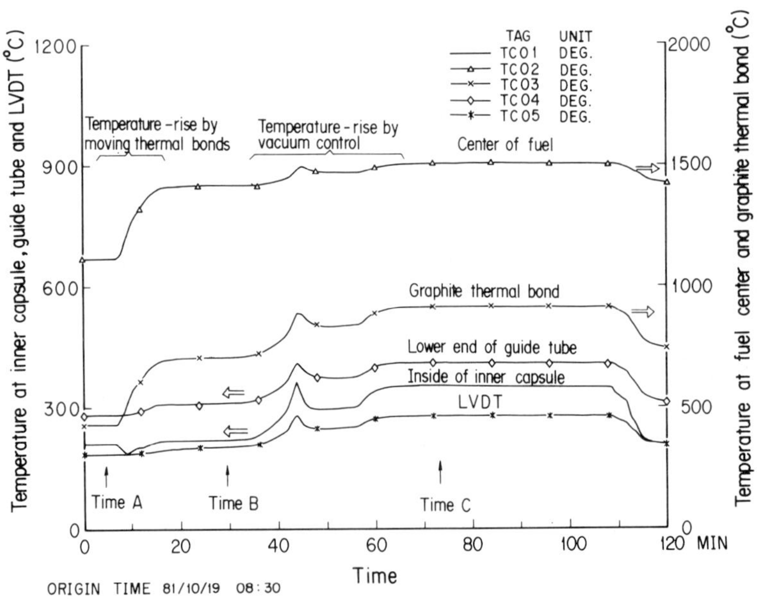

Fig. 5 Temperature history of temperature at fuel center, graphite thermal bond, guide tube, inner capsule and LVDT in the temperature-ramp capsule.

by actuating a bellows was able to be used for the temperature-ramp-up in the capsule,

(b) a large and fast temperature-rise at the fuels is obtained with little temperature-rise at the inner tube of capsule from moving thermal bonds,

(c) temperature-rise with moving thermal bonds is by far greater than that with the vacuum controlling method,

(d) the displacement of thermal bonds can be measured by the LVDT, and detected by a thermocouple in thermal bonds and also a pressure gage for measuring gas pressure at the bellows.

GRAPHITE IRRADIATION CAPSULE

The objective of a graphite irradiation capsule is to irradiate graphite specimens for in-pile physical testings i.e. tensile test, compression test, thermal expansion measurement and so on.

Description of capsule

The configuration of graphite irradiation capsule (80M-18A) is shown in Fig. 6. Graphite thermal bonds surrounded by rings, which hold graphite specimens for physical testings, are enclosed in the inner capsule. These rings are used as variable gamma heaters which consist of materials with various densities, such as graphite, molybdenum, niobium and tungsten. This is intending that gamma heat generated in the inner capsule along the axial direction in the capsule is about uniform, accordingly the temperature at all specimens in the capsule is expected to result in almost uniform (1100 ± 100°C). The inner tube with a co-extruded aluminum bond covers the rings and others, being placed in the outer capsule of stainless-steel cooled by the reactor coolant. The temperature of specimens is, additionally, regulated by adjusting thermal conduction through the gas gap between the outer and inner tube by means of the vacuum controlling method.

Five W-Re thermocouples are distributed in graphite holder along the axial direction of capsule to obtain temperature at graphite specimens. It is considered that measured temperatures show those at specimens which are held in graphite thermal bonds, because of good thermal conductivity of graphite.

Experiment and results

The 80M-18A capsule has been irradiated from 58th reactor cycle (March in 1982) in the core position G-6 of which thermal and fast neutron flux are about 2.0×10^{14} ($E < 0.625$ eV) and 1.0×10^{14} $n/cm^2 \cdot sec$ ($E > 1$ MeV), respectively.

Obtained results are shown in Fig. 7. The vertical axis in the figure shows positions of specimens placed along the axial direction in the capsule, with indication of position from the core

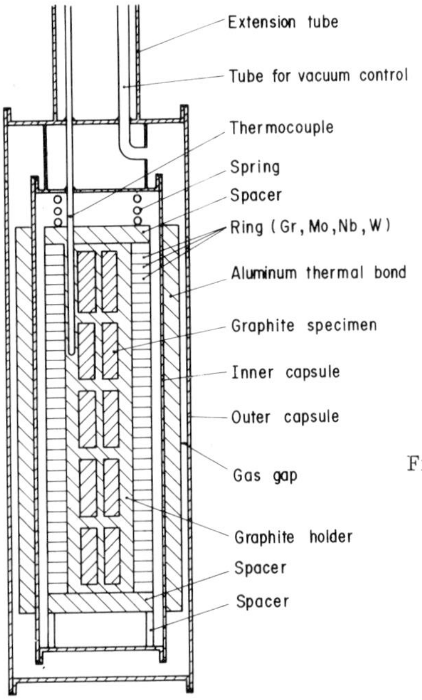

Fig. 6 Configuration of graphite irradiation capsule.

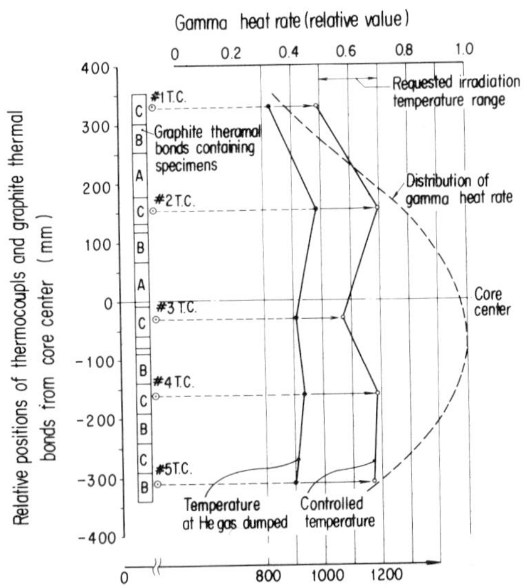

Fig.7

Obtained temperature at graphite thermal bonds ranged along axial direction and gamma heat rate distribution in the graphite irradiation capsule.

center.

Positions of five thermocouples in the graphite thermal bonds are also shown in the figure. The horizontal axis shows temperatures at specimens obtained from the thermocouples. The temperatures are ranged from about 827 to 984°C (about 905 ± 80°C) during helium gas dumped period. Temperatures are distributed from 995 to 1195°C (about 1100°C ± 100°C), when temperatures at specimens are raised by the vacuum controlling method, and thus the irradiation of graphite specimens at uniform temperature can be carried out. To irradiate graphite specimens at more uniform temperature, other capsules are being designed and fabricated with rearrangement of the rings based on the results in the experiment.

CONCLUSION

Various irradiation experiments on development the fuels and materials for HTGR have been carried out in JMTR.

The first temperature-ramp capsule was irradiated in JMTR in order to confirm the performance of the temperature ramping method that is based on a principle of changing heat transfer by displacing thermal bond with bellows. The test result shows that the temperature at specimens ramped from about 1100 to 1510°C, and the temperature-ramp capsule is thus verified by the irradiation experiments. The design and fabrication of second disigned capsule that is aimed above 500°C of temperature-ramp is now under way.

A graphite irradiation capsule with various densities was carried out to irradiate all specimens in the capsule at 1100 ± 100°C. The irradiation result shows that temperature at specimens is between 995 and 1195°C, and thus the irradiation was successfully performed.

The authors wish to thank to the people of Irradiation Section II who participated in the operation of the capsules.

SESSION IX
MISCELLANEOUS

DIE EXPERIMENTELLE NUTZUNG DES DEUTSCHEN FORSCHUNGSREAKTORS FR2 - EIN RÜCKBLICK

W. Steiger, G. Wittek

Kernforschungszentrum Karlsruhe GmbH

Weberstraße 5, D-7500 Karlsruhe 1

EINLEITUNG

Am 21. Dezember 1981 wurde der Forschungsreaktor FR2 im Kernforschungszentrum Karlsruhe letztmalig abgeschaltet. Damit ging ein zwanzigjähriger erfolgreicher Betrieb des ersten Eigenbaureaktors in der Bundesrepublik Deutschland zu Ende. Ursache für die Abschaltung waren nicht technische Unzulänglichkeiten des Reaktors und seiner Systeme, sondern die Aussage, daß die Möglichkeiten des Reaktors nicht mehr den forschungspolitischen Zielsetzungen des Kernforschungszentrums Karlsruhe genügten.

Mit dem Bau der Anlage war im Jahre 1957 begonnen worden, einer Bauzeit von nur 4 Jahren folgte im März 1961 das erste kritische Experiment. Der Leistungsbetrieb mit 12 MW wurde im Dezember 1962 aufgenommen, im Juni 1966 konnte die Leistung auf 44 MW erhöht werden. Beim FR2 handelte es sich um einen Reaktor vom Tank-Typ. Als Primärkühlmittel und als Moderator diente Schweres Wasser. Gesintertes Uranoxid mit einer Anfangsanreicherung von 2 % in Zircaloy-Umhüllung bildete den Brennstoff.

Für die Nutzung von Neutronen außerhalb des Reaktorkerns standen 12 Strahlkanäle und eine Thermische Säule zur Verfügung. Für Inpile-Experimente sind vor allem der Zentralkanal und zwei Nebenkanäle zu erwähnen. Daneben konnten Bestrahlungseinsätze auf Zwischengitterpositionen und auf Gitterpositionen der Brennelemente selbst eingebaut werden.

P. von der Hardt and H. Röttger (eds.), Irradiation Technology, 541–548.

NUTZUNG DES REAKTORS

Im Vordergrund der experimentellen Nutzung standen die Strahlrohrexperimente. Am letzten Betriebstag waren noch 20 von ehedem 22 Versuchsanordnungen aus dem Forschungsgebiet Nukleare Festkörperphysik und Strukturanalyse in Betrieb. Nutzer waren neben den entsprechenden Fachinstituten des Kernforschungszentrums Karlsruhe vor allem Universitäten und sonstige wissenschaftliche Institute vorwiegend aus dem süddeutschen Raum (Abb.1). Im Verlauf der Jahre waren Forschungsgruppen aus 16 Institutionen zum Teil über viele Jahre am FR2 tätig. An einem Strahlrohr in der Thermischen Säule war eine Neutronenradiographieanlage in Betrieb, in der hochradioaktive Prüflinge mit der Dimension von ganzen Brennelementen untersucht werden konnten.

Das nächstwichtigste Anwendungsgebiet war die Produktion von Radioisotopen (Abb.2). Auch in diesem Bereich gab es einen großen Kundenkreis (Abb.3).

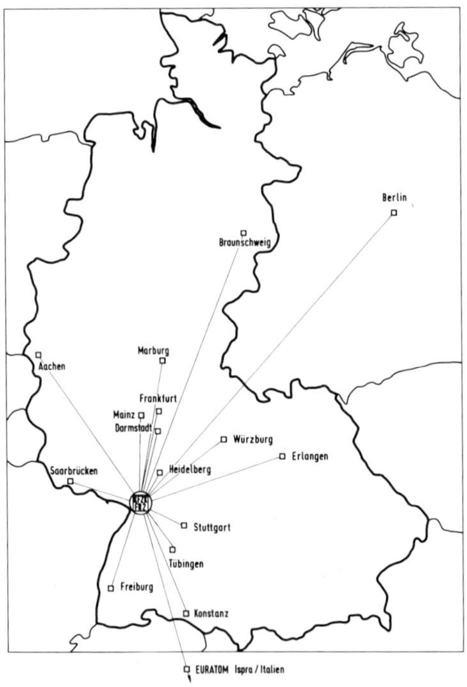

Abb.1 Einzugsbereich für Strahlrohrexperimente

Im Reaktorkern:	15 684 Kapseln
	283 Brennstoffplatten
	100 kg Seesand (markiert mit Scandium 45)
	1 606 kg Halbleitersilizium
Mit der Rohrpostanlage:	15 624 Kapseln
In der Thermischen-Säule:	2 418 Kapseln

Hauptauftraggeber waren Industrie und Forschungsinstitute.
Hauptanwendungsgebiete die Aktivierungsanalyse, die Nuklearmedizin, die (extra) terrestrische Altersbestimmung und die radiochemische Forschung.

Die interessantesten Proben: Mondgestein und Edelsteine.

Abb.2 Isotopenproduktion

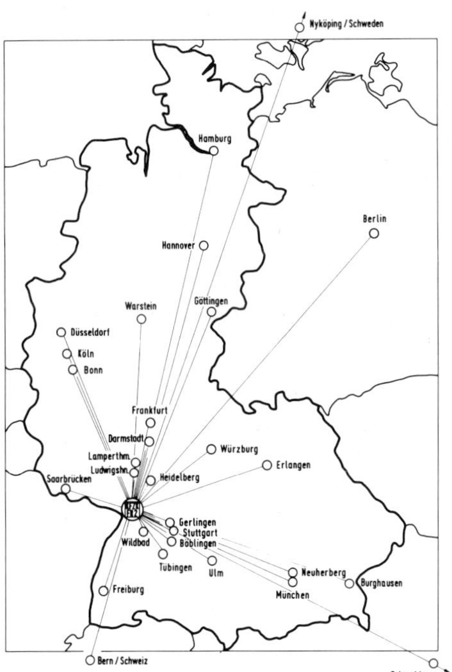

Abb.3 Einzugsbereich für Isotopenproduktion

Von ähnlicher Bedeutung war die Durchführung von Brennstoffkapselbestrahlungen vor allem für das in Karlsruhe im Vordergrund stehende Projekt Schneller Brüter (Abb.4). Es kamen 165 instrumentierte Kapseln aus 10 verschiedenen Kapseltypen zum Einsatz, unter anderem Blei-Wismut-Doppelkapseln, einwandige Natrium-Kapseln und Kriechkapseln unter Gasdruck. Im Ganzen ergaben sich 98 Kapselbestrahlungsjahre, ohne daß nennenswerte Störfälle zu verzeichnen waren. Bemerkenswert ist auch die lange Bestrahlungszeit - über 23.000 Stunden - eines Thermionischen Konverters, der eine elektrische Leistung von maximal 120 Watt abgab.

In der nächsten Gruppe von Experimentieranlagen sind die Kreislaufexperimente zusammenzufassen, die vom Aufbau und Betrieb her im Vergleich zu den bisher genannten Einrichtungen sehr komplex und anspruchsvoll waren. Im Laufe der Jahre standen 5 Anlagen zur Verfügung, wobei Temperaturen zwischen 20 und 800 Kelvin, Drucke von Hochvakuum bis zu 180 bar und Medien wie Helium, Wasserstoff, Heißdampf und Schwerwasser zu beherrschen waren (Abb.5). Auch hier kann von durchweg guten Betriebsbedingungen berichtet werden, wenn man von einzelnen Störungen an konventionellen Aggregaten absieht.

Auch dreiviertel Jahr nach Abschaltung des Reaktors ist immer noch eine Experimentiereinrichtung in Betrieb: Die γ-Bestrahlungseinrichtung im Lagerbecken für abgebrannte Brennelemente, die bis zum Abtransport der Elemente zur Wiederaufarbeitung ihren Dienst tun kann.

Erste Bestrahlung Exp.36 KVE Nr. 3 am 22.2.1965

Insgesamt wurden 165 Kapselbestrahlungen durchgeführt. Dabei kamen 10 verschiedene instrumentierte Kapseltypen zum Einsatz, u. a. Blei-Wismut-Doppelkapseln, einwandige Natrium Kapseln und Kriechkapseln unter Gasdruck.

Die Erfahrung aus 98 Kapselbestrahlungsjahren:

Keine Gefährdung von Mensch und Maschine.

Das interessanteste Experiment: Ein Thermionischer Wandler, der von 1972 bis 1976 in 23 240 Betriebsstunden 1637 kWh elektrische Energie erzeugte.

Abb.4 Brennstoffkapselbestrahlungen

Tieftemperatur-Bestrahlungsanlage:

In 75 420 Betriebsstunden wurden bei Temperaturen bis zu 20 Kelvin 445 Proben untersucht.

Kalte Neutronenquelle:

In 72 480 Betriebsstunden wurden mittels einer mit flüssigem Wasserstoff gefüllten Moderatorkammer subthermische Neutronen produziert.

Helium-Druckgaskreislauf:

In 58 075 Betriebsstunden wurden 146 Brennstoffprüflinge insgesamt 25 387 Stunden bei spezifischen Stableistungen bis 1100 W/cm bestrahlt.

Druckwasser-Kreislauf:

In 30 711 Betriebsstunden wurde an 57 Brennstoffplatten eine integrale Bestrahlungszeit von 34 810 Stunden erreicht.

Heißdampf-Hochdruckkreislauf:

In 18 965 Betriebsstunden wurde unter Störfallbedingungen (Hüllschadendefekte, Kühlmittelverlust) das Verhalten von insgesamt 75 Brennstäben untersucht.

Abb.5 Kreislaufexperimente

BESONDERHEITEN DES REAKTORKONZEPTS

Der Reaktor war ursprünglich ein schwerwassermoderierter Natururanreaktor. Bei der anfangs erwähnten Leistungssteigerung wurde auf leicht angereichertes Uranoxid übergegangen, was etwa dem üblichen Leichtwasserreaktorbrennstoff entspricht. Somit gab es zu keiner Zeit Schwierigkeiten mit der Beschaffung und der Wiederaufarbeitung der Kernbrennstoffe: Die abgebrannten Brennelemente werden in der auf dem Gelände des Kernforschungszentrums befindlichen Wiederaufarbeitungsanlage aufbereitet, es bestehen also auch keine Transportprobleme.

Aus dem ursprünglichen Reaktorkonzept folgte auch ein relativ großes Kernvolumen, was zusammen mit einer geeigneten Umladestrategie zu flachem Flußverlauf sowohl in axialer als auch in radialer Richtung führte. Diese Tatsache begünstigte eine homogene Bestrahlung großer Bestrahlungsproben. Als Beispiel sei hier die Dotierungsbestrahlung von Silizium genannt.

Andererseits sind gewisse Probleme mit dem Einsatz von schwerem Wasser verbunden. Das Primärkühlmittel hatte zu Betriebsende eine Tritiumkonzentration von etwa 6 Ci/l. Sie wäre noch höher gewesen, wenn nicht vor einigen Jahren etwa die Hälfte des Inventars ausgetauscht worden wäre. Dies bedingte ein sehr aufwendiges, aber auch gut funktionierendes Dichtheitskontrollsystem des Primärkreislaufs (Feuchtefühlerüberwachung aller Flanschverbindungen), einen großen

Aufwand bei Reparaturen an diesem System (Atemschutzanzüge) und eine nicht unbeträchtliche Belastung der Umgebung. So wurden im letzten Betriebsjahr etwa 150 Ci Tritium über den Abluftkamin abgegeben. Auch die 70.000 Ci Argon 41, die im gleichen Zeitraum ausgestoßen wurden, schlagen negativ zu Buche. Letztere waren bedingt durch die direkte Luftkühlung der Experimentierkanäle und der Tankspalte sowie der Thermischen Säule.

Bei den Bestrahlungsexperimenten und besonders bei der Isotopenproduktion führte das Schwerwasser-Tankreaktor-Konzept zu erheblichem konstruktivem und betrieblichem Aufwand. Dies war auch der Grund dafür, daß die Produktion von für Halbleiterzwecke neutroneninduziert dotiertem Silizium nach einem sehr guten Start bald wieder zu Ende ging. Der Konkurrenz von Swimmingpool-Reaktoren, die das Produkt billiger lieferten, konnte nicht standgehalten werden.

Das Ganze betrachtend muß gesagt werden, daß ein Mehrzweckforschungsreaktor, wie ihn der FR2 darstellte, wohl nicht die allerbeste Lösung darstellt. Besser ist die von Frankreich gewählte Philosophie, für das jeweilige Forschungsziel angepaßte spezielle Reaktoren zu erstellen, die verhältnismäßig unaufwendig in kürzerer Zeit die gewünschten Forschungsergebnisse erbringen können.

STILLEGUNGSMASSNAHMEN

Die Anlage ist derzeit in der Außerbetriebnahmephase. Die Brennelemente sind ausgeladen und werden noch dieses Jahr zur Wiederaufarbeitungsanlage Karlsruhe gebracht. Das schwere Wasser wird zur Zeit aufkonzentriert und soll später verkauft werden. Es ist beabsichtigt, die Hilfssysteme wie elektrische Energieversorgung und Lüftungsanlage weitgehend zu vereinfachen. Noch in diesem Jahr kann von einer Rund-um-die-Uhr-Überwachung der Anlage abgegangen werden, nachts und an Wochenenden wird sich kein Personal mehr darin aufhalten. Die Experimentiereinrichtungen werden, soweit sie sich nicht, wie einige Strahlrohrexperimente, an andere Reaktoren auslagern lassen, Zug um Zug verschrottet.

Nachdem eine Studie zur Untersuchung der Möglichkeit, den Reaktorblock abzureißen, ergeben hat, daß dies mehr kosten würde, als seinerzeit die Erstellung der gesamten Anlage, ist für die nächsten Jahre nicht mit einer solchen Maßnahme zu rechnen. Vielmehr wird das Konzept "Gesicherter Einschluß des Reaktorblocks" verfolgt. Dies ist eine Definition des Deutschen Atomgesetzes und damit genehmigungspflichtig. Nachdem sämtliche Öffnungen des Reaktorblocks dicht abgeschlossen sind, wird im Inneren Unterdruck gehalten, die abgesaugte Leckluft wird kontrolliert abgegeben. Alle anderen, nicht zur Infrastruktur gehörenden Systeme innerhalb des Reaktorgebäudes werden im Laufe der Zeit abgebaut.

KÜNFTIGE NUTZUNG

In der Zukunft soll das Gebäude mit seiner wertvollen Infrastruktur als Technikumshalle genutzt werden. Der stehenbleibende Reaktorblock nimmt nur etwa 2 % des gesamten Hallenvolumens ein, stellt also keine wesentliche Beeinträchtigung dar. Durch Einziehen einer Zwischenbühne auf Höhe der Reaktorplattform läßt sich eine ebene Experimentierfläche mit einem Durchmesser von 38 m gewinnen (Abb.6). Diese Experimentierfläche kann sehr vielseitig verwendet werden. In der Abbildung ist ein Teststand zur Hochleistungspulstechnik eingezeichnet, wie er im Rahmen eines neuen Forschungsprogramms des Kernforschungszentrums Karlsruhe aufgebaut werden könnte. Auch die darunterliegenden Stockwerke können voll genutzt werden. Das Bild ist jedoch nur beispielhaft aufzufassen, es stellt nur Realisierungsmöglichkeiten dar.

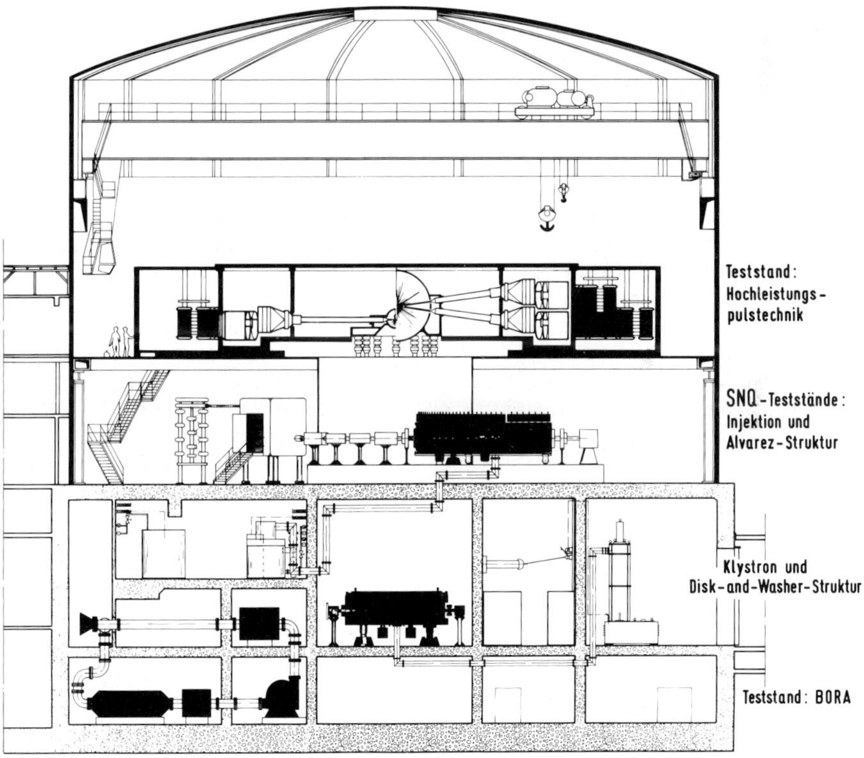

Abb.6 Nutzungsmöglichkeiten der Reaktorhalle

Das Interesse an der Aufnahme neuer Experimente ist nicht nur deshalb so groß, damit Gebäude und vorhandene technische Einrichtungen genutzt werden können. Auch für das bisherige Reaktorpersonal eröffnen sich dadurch Arbeitsmöglichkeiten, die der bisherigen Tätigkeit sehr ähnlich sind. Gerade für Mitarbeiter, die schon sehr lange, zum Teil seit Beginn des Reaktorbetriebs an der Anlage eingesetzt waren, ist eine solche Kontinuität in der Arbeitsweise äußerst wichtig.

BESTRAHLUNGSTESTS MIT NIEDRIGANGEREICHERTEN BRENNSTOFFEN UND BRENNELEMENTEN FÜR FORSCHUNGSREAKTOREN *

W. Krug, F. Schmidt, J. Seferiadis, G. Thamm

Kernforschungsanlage Jülich GmbH

Postfach 1913, 5170 Jülich

Zusammenfassung.

Zur Verminderung des Proliferationsrisikos läuft derzeit wie in vielen anderen Ländern auch in Deutschland ein Entwicklungsprogramm zur Verminderung der U-235-Anreicherung des Brennstoffs für Forschungsreaktoren. Eine Reduzierung der U-235-Anreicherung von derzeit etwa 90 % auf 20 % führt zu einer Erhöhung der Gesamturandichte um etwa den Faktor 5 bis 6. Diese hochdichten Brennstoffe werden zur Zeit entwicklt und umfangreichen Tests unter Bestrahlungsbedingungen unterzogen.

In Teil A werden die Testschritte, denen die neuentwickelten Brennstoffe in Deutschland unterzogen werden, beschrieben und der aktuelle Stand der Bestrahlungserprobung mitgeteilt.

Teil B beschäftigt sich mit der in Jülich entwickelten Bestrahlungseinrichtung, in der gleichzeitig 10 Testplatten (in Länge und Breite gegenüber den Standard-MTR-Brennstoffplatten reduziert) im FRJ-2 (DIDO) unter Bestrahlungsbedingungen getestet werden können. Die Einrichtung befindet sich zur Zeit im Bau, die wesentlichen Kalterprobungen sind erfolgreich abgeschlossen, so daß die Bestrahlungstests Anfang 1983 beginnen werden.

* Teil A: Überarbeitetes deutsches Testprogramm
Teil B: Bestrahlungseinrichtung für Brennstoffplatten, AF-Loop, FRJ-2

P. von der Hardt and H. Röttger (eds.), Irradiation Technology, 549–565.

Teil A

1. Einleitung.

Die weltweit durchgeführten INFCE-Untersuchungen kamen bekanntlich zu der Empfehlung, zur Verminderung des Proliferationsrisikos die U-235-Anreicherung des Brennstoffs in Forschungsreaktoren von derzeit ca. 90 % (HEU) auf < 20 % (LEU) zu vermindern.
Es wurde abgeschätzt, daß langfristig von den etwa 150 Reaktoren, die heute noch HEU-Brennstoff verwenden, weit über 90 % auf LEU-Brennstoff umgestellt werden können. Noch während der Laufzeit von INFCE bzw. kurz danach, wurden in mehreren Ländern (z.B. Amerika, Frankreich, Japan ...) Entwicklungsprogramme begonnen, die die Umstellung der Forschungsreaktoren von HEU- auf LEU-Brennstoffzyklen zum Ziel haben. Ende 1979 wurde auch in Deutschland im Zuge dieser Entwicklung das sogenannte AF-Programm (Anreicherungsreduzierung in Forschungsreaktoren) vom Bundesministerium für Forschung und Technologie in Gang gesetzt. In diesem Programm (mit einer Laufzeit von 5 Jahren) arbeiten Industrie (Brennstoffhersteller und Reaktorbauer) und Kernforschungszentren zusammen. Wesentliches Ziel dieses Programms ist die Entwicklung von Brennstoffen mit besonders hoher Urandichte, wie sie für die HEU-LEU-Umstellungen benötigt werden, bis zur Fertigungsreife hin. Diese Brennstoffe sind hinsichtlich ihrer Sicherheitsgrenzen und ihrer Zuverlässigkeit durch umfangreiche Bestrahlungstests mit anschließenden Nachuntersuchungen zu qualifizieren. Die Ergebnisse der Tests sind von wesentlicher Bedeutung für die ab Mitte der 80er Jahre notwendig werdenden Genehmigungsverfahren zur HEU-LEU-Konversion der bestehenden Forschungsreaktoren.

2. Arten von Bestrahlungstests.

Für die Untersuchungen sind grundsätzlich 4 Arten von Bestrahlungstests zu unterscheiden (Tabelle 1, Seite 3, gibt einen Überblick).

2.1 Das fundamentale Verhalten der neuentwickelten Brennstoffe unter Bestrahlungsbedingungen ist zu ermitteln. Hierzu werden Kleinplatten (Breite und Länge des Brennstoffmeats ~ 50 mm bzw. 120 - 150 mm, Platten- und Meatdicke überwiegend gemäß Standard 1,27/0,51 mm) in besonderen Bestrahlungseinrichtungen erprobt. Diese sind separat gekühlt und gestatten die genaue Erfassung aller wesentlichen Bestrahlungsdaten. Von einer Brennstoffart werden jeweils nur wenige Platten (3) getestet.

2.2 Mit einer wesentlich höheren Anzahl von Testplatten (Abmessungen z.T. wie unter 1. bis hin zur vollen Größe einer Standard-Brennstoffplatte) ist das „statistische" Bestrahlungsverhalten der einzelnen Brennstoffarten zu ermitteln. Dabei werden mindestens 10 Platten pro Brennstoffart in einem recht einfachen Einsatzelement getestet. Die Kühlung erfolgt durch Reaktor-Poolwasser, eine laufende Überwachung der wesentlichen Bestrahlungsparameter erfolgt nicht, Messungen vor allem zur Abbrandermittlung erfolgen zu Beginn der Tests.

Tabelle 1

AF-PROGRAMM BESTRAHLUNGSTESTS FÜR LEU-BRENNSTOFFE, ALLGEMEINE ÜBERSICHT

NR.	ART DES TESTS	TEST-OBJEKT	BESTRAHLUNGS-EINRICHTUNG, EINSATZORT	NACHUNTER-SUCHUNGEN	BEMERKKUNGEN
1	FUNDAMENTALES BESTRAHLUNGSVERHALTEN	KLEIN-PLATTEN (200x60x 1,27 cm^3)	LOOP, KAPSEL (GETRENNTE KÜHLUNG, INSTRUMENTIERUNG), INCORE, REFLEKTOR	AUFWENDIG ZERSTÖREND UND ZERSTÖRUNGSFREI	SEHR GENAUE ERFASSUNG DER BESTRAHLUNGSDATEN, Z.T. ÜBERLASTVERSUCHE ÄNDERUNG WICHTIGER SPEZ.-PARAMETER (KORNGRÖSSE, CLADDING ...)
2	STATISTISCHES BESTRAHLUNGSVERHALTEN	KLEIN-PLATTEN (Z. TEIL WIE 1)	EINSATZ-EL. (KÜHLUNG D. POOLWASSER, KEINE INSTRUMENTIERUNG), INCORE, REFLEKTOR	GERING ZERSTÖRUNGSFREI 1.)	GROBE ERFASSUNG DER BESTRAHLUNGSDATEN. BESTRAHLUNGSBEDINGUNGEN ABHÄNGIG VON TESTREAKTOR
3	TESTELEMENT-BESTRAHLUNGEN	KOMPLETTE BRENNELEMENTE (STAND.-ABMESSUNG) 2.)	INCORE	GERING ZERSTÖRUNGSFREI 1.)	GROBE ERFASSUNG DER BESTRAHLUNGSDATEN, MESSUNG REAKTORPHYSIKALISCHER PARAMETER, DEFINITION STANDARDISIERTER BE-SPEZIFIKATIONEN
4	PROTOTYP-ELEMENT-BESTRAHLUNGEN 3.)	KOMPLETTE BRENNELEMENTE 4.)	INCORE	SEHR GERING ZERSTÖRUNGSFREI 1.)	GROBE ERFASSUNG DER BESTRAHLUNGSDATEN, MESSUNG REAKTORPHYSIKALISCHER PARAMETER (FÜR EINZELELEMENTE, UNTERANORDNUNGEN, KOMPLETTKERN).

1.) ZERSTÖRENDE UNTERSUCHUNGEN NUR IN BESONDEREN FÄLLEN (Z.B. VERSAGEN).
2.) PLATTENDICKE ÜBERWIEGEND 1,27 MM (0,51), MEAT MAX. 1,5 MM. ELEMENTE NICHT PROTOTYPISCH FÜR DEN TESTREAKTOR.
3.) TESTS 1 - 3 = AF-PROGRAMM, TEST 4 = FORSCHUNGSREAKTORBETREIBER.
4.) ELEMENTE PROTOTYPISCH FÜR DEN TESTREAKTOR, TEST GEHÖRT ZUM HEU-LEU-KONVERSIONSVORGANG.

2.3 Komplette Brennelemente werden mit einer einzigen Brennstoffart ausgestattet und in besonderen Testreaktoren bestrahlt mit dem Ziel, für die spätere Fertigung von Brennelementen im industriellen Maßstab standardisierte Spezifikationen zu erarbeiten. Die Elemente entsprechen in der äußeren Form völlig den Brennelementen des jeweiligen Testreaktors. Allerdings kommen neben der Standardplattendicke von 1,27 mm (0,51 mm Meat) auch wesentlich dickere Platten (bis über 2,0 mm) zum Einsatz. Zum Teil erfolgen reaktorphysikalische Messungen und solche für die Abbrandermittlung.

2.4 In einem letzten Schritt werden sogenannte Prototyp-Brennelemente unter Bestrahlungsbedingungen erprobt. Sie sind prototypisch für den jeweiligen „Bestrahlungsreaktor" und liefern in erster Linie die wesentlichen reaktorphysikalischen und thermohydraulischen Daten, die für die Umstellung des jeweiligen Reaktors auf LEU-Brennstoffe benötigt werden.

Die Testschritte 1 bis 3 werden im Rahmen des AF-Programms abgewickelt, der Testschritt 4 ist in Zusammenarbeit mit dem AF-Programm Aufgabe des jeweiligen Reaktorbetreibers.

3. Nachuntersuchungen.

Art und Umfang der Nachuntersuchungen an den bestrahlten Testbrennstoffen hängt wesentlich von der Testart (siehe Tabelle 1) ab. Die Anforderungen sind für den Testschritt 1 am größten.

Tabelle 2 gibt einen Überblick über die Nachuntersuchungsmethoden und den Umfang ihrer Anwendung für die einzelnen Testschritte.

Tabelle 2

AF-Programm — Nachuntersuchungen für bestrahlte LEU-Brennstoffe

Methoden	Untersuchungsobjekt			
	Kleinplatten (Nr. 1)	Kleinplatten (Nr. 2)	Testelemente (Nr. 3)	Prototyp-Elemente (Nr.4)
Zerstörungsfrei				
Visuelle Inspektion (Fotodokumentation)	ja (100 %)	ja (100 %)	ja (100 %)	ja (100 %)
Dimensionsmessungen (Profilometrie)	ja (100 %)	ja (<100 %)	ja (<100 %)	ja 1.) 2.)
Dichtemessungen	ja (100 %)	ja (<100 %)	nein	nein
Gamma Scan	ja (100 %)	ja (<100 %)	nein	nein
Zerstörend	3.)	4.)		
Abbrandbestimmung (radiochemisch)	ja	teilweise		
Blistertests	ja	teilweise	zerstörende Untersuchungen nur in besonderen Fällen (Versagen)	zerstörende Untersuchungen nur in besonderen Fällen (Versagen)
Metallographie	ja	teilweise		
Phys. Eigenschaften	ja	nur in besonderen Fällen (Versagen)		
Mikrosonden-Analyse	ja (bei Bedarf)			
SEM-Untersuchungen	teilweise			

1.) einschliesslich Kühlkanalabmessungen
2.) Stichproben bei den ersten Elementen
3.) Untersuchungen jeweils an 1 Platte pro Brennstofftyp
4.) Stichprobenweise Untersuchungen nicht an allen Brennstofftypen

- An zerstörungsfreien Untersuchungen sind vorgesehen:
  - Visuelle Inspektionen mit Fotodokumentation.
  - Dimensionsmessungen.
  - Dichtmessungen (Immersions-Technik).
  - Gamma-Scan.

Die visuellen Inspektionen sollen vor allem Aufschluß über Oxydationsvorgänge während der Bestrahlung liefern. Die Ergebnisse werden z.T. fotografisch dokumentiert werden.

Dimensionsmessungen sollen Aufschluß über Verformungen und sonstige Veränderungen während der Bestrahlung liefern. Gemessen werden die Plattendicke, bei den Elementen auch die Kühlkanalabstände. Im Bereich von 0,01 mm wird mit einer Genauigkeit von $\pm$ 0,005 mm gemessen. Für Testschritt 1 erfolgen die Untersuchungen zu 100 %, für die übrigen nur zum Teil.

Dichtemessungen nach der Immersions-Technik sind eine weitere gute Methode zur Dickenmessung (nur für die Testschritte 1 (100 %) und 2 (< 100 %).

Mit Hilfe eines Gamma-Scan erfolgt die Bestimmung des relativen Abbrandes. Solche Messungen sind nur für die Testschritte 1 (100 %) und 2 (< 100 %) vorgesehen. Sie dienen zum Teil der Positionsbestimmung für die metallografischen Schliffe.

An zerstörenden Untersuchungen sind vorgesehen:

- Radiochemische Abbrandbestimmung.
- Blistertests.
- Metallographie.
- Physikalische Eigenschaften.
- Mikrosondenanalyse.
- SEM-Untersuchungen.

Zerstörende Untersuchungen sind nur für Testbrennstoffe gemäß Testschritt 1 (je 1 Platte für Metallographie und Blistertests) und stichprobenweise bei Testschritt 2 vorgesehen.
Bei Test- und Prototyp-Elementen (Schritte 3 und 4) erfolgen solche Untersuchungen nur in besonderen Fällen (z.B. beim Versagen unter Bestrahlung).

Die radiochemische Abbrandbestimmung benutzt das Neodym-Verfahren.

Die Blistertests erfolgen durch Glühen der Kleinplatte in Stufen von 100 °C (beginnend bei 200 °C). Die Temperaturschwellen, von denen ab sich deutlich Blasen bei etwa 20 bis 30minütiger Temperaturbehandlung bilden, sind zu ermitteln.

Mindestens 5 Schliffe werden an jeder Platte für die metallographischen Untersuchungen durchgeführt. Untersucht werden das Schwellverhalten der Brennstoffpartikel, chemische Reaktionen zwischen Brennstoff und Matrix, Änderungen der Cladding/ Meat-Bindung, Porositätsänderungen, Verteilung von Spaltgasblasen.

U.U. werden bei den Platten (Schritt 1) Zugversuche an herausgefrästen Proben unter Reaktorbetriebstemperatur durchgeführt.

Ergänzende Untersuchungen mit der Mikrosonde bzw. nach der SEM-Methode gestatten Aufschluß über die Verteilung der Spaltprodukte und Gasblasen (z.T. bei Schritt 1).

Die Kleinplatten werden vor und nach der Bestrahlung mit den selben Methoden untersucht.

4. Bestrahlungstests für die einzelnen Brennstoffe. Stand, Ausblick.

Tabelle 3 (Seite 7) gibt einen Überblick über die geplanten, laufenden, bzw. beendeten Testbestrahlungen für die einzelnen Brennstoffe. Getestet werden die herkömmlichen Brennstoffe UAl_x und U_3O_8, die Silizide U_3Si_2 und U_3Si und die neuartigen Brennstoffe U_xSi_y bzw. U_xNi_y in Al-Dispersion.

Bestrahlungstests mit im Rahmen des AF-Programms hergestellten Brennstoffen sind nicht nur auf deutsche Forschungsreaktoren beschränkt, sondern erfolgen zu einem recht großen Umfang im Ausland.

- UAl_x-Al.
  Hier sind die Tests mit Kleinplatten abgeschlossen. Dichten bis 2,2 g U/cm^3 wurden bis zu 77 % Abbrand erfolgreich getestet. Die Nachuntersuchungen laufen von Mitte bis Ende 1982.

  Insgesamt werden weit über 100 Elemente bis zu Dichten von maximal 1,8 g U/cm^3 in 10 Reaktoren und 6 Ländern getestet. Abbrände von mehr als 50 % wurden bis jetzt erfolgreich gefahren.

- U_3O_8-Al.
  Die Tests mit Kleinplatten sind beendet. Dichten bis 3,1 g U/cm^3 wurden bis zu einem Abbrand von 77 % erfolgreich bestrahlt. Nachuntersuchungen von Mitte bis Ende 1982.

  Mehr als 50 Testelemente sind gefertigt worden bzw. befinden sich noch in der Fertigung bis zu Dichten von 2,9 g U/cm^3. 7 Elemente sind im Test, 2 davon haben einen Abbrand von etwa 50 % erreicht. Bestrahlt wird in 7 Reaktoren und 6 Ländern.

  U_3Si_2-Al.
  Durch die Möglichkeit, vorab im ORR, USA, Bestrahlungstests durchführen zu können, hat hier der Testschritt 3 mit 2 Elementen (4,8 g U/cm^3) seit Ende Mai 1982 erfolgreich begonnen. Weitere Elemente werden nächstes Jahr im FRG-2 eingesetzt. Kleinplatten (11) werden ab 1983 gemäß Schritt 1 bestrahlt werden. Schwerpunktmäßig werden Brennstoffe mit Spezifikationsänderungen hinsichtlich Cladding und Feinkornanteil zum Ein-

Tabelle 3

AF-Programm — Übersicht der geplanten und laufenden Brennstofftest-Bestrahlungen

Brennstoff	Test-Nr.	Test-objekt	Zahl	Einsatzbeginn oder -ende, Stand, Dichte, Abmessungen, Abbrand, Reaktor, Land
UAl_x-Al	1, 2	Klein-platten	4	Bis Ende März 1982, beendet, 2,1-2,2 g U/cm^3, 0,51-0,77/1,27-1,53 mm Meat/Platte, 77 %, ORR, USA
	3, 4	Ele-mente	123	Seit Ende 1981, mehr als 50 % im Einsatz, 1,1-1,8 g U/cm^3, meist 0,51/1,27 mm Meat/Platte, z.T. > 50 %, 10 Reaktoren, 6 Länder
U_3O_8-Al	1, 2	Klein-platten	8	Bis Ende März 1982, beendet, 2,3-3,1 g U/cm^3, 0,51-0,77/1,27-1,53 mm Meat/Platte, 77 %, ORR, USA
	3 (4)	Ele-mente	59	Seit Anfang 1982, 5-7 im Einsatz, 2,0-2,9 g U/cm^3, 0,51-1,50/1,27-2,26 mm Meat/Platte, z.T. 50 %, 7 Reaktoren, 6 Länder
U_3Si_2-Al	1, 2	Klein-platten	11	Testbeginn Januar 1983, in Fertigung, 5,0 g U/cm^3, 0,51-0,65/1,27 mm Meat/Platte, FRG-2 und FRJ-2, BRD. Spezifikationsänderungen. Cladding ~ -15 % (von 0,38 mm), Feinkorn ~ verdoppelt
	3	Ele-mente	2 (2)	Testbeginn Ende Mai 1982, 4,8 g U/cm^3, 0,51/1,27 mm Meat/Platte, ORR, USA Testbeginn Frühjahr 1983, in Vorbereitung, 5,0 g U/cm^3, 0,51/1,27 mm Meat/Platte, FRG-2, BRD
U_3Si-Al	1, 2	Klein-platten	3 (10)	Testbeginn Januar 1983, in Fertigung, 7,0 g U/cm^3, 0,51/1,27 mm Meat/Platte, FRJ-2, BRD. Weitere Tests mit Spezifikationsänderungen ab Herbst 1983
	3	Ele-mente	2	Testbeginn 1984, in Planung, 5,0-6,0 g U/cm^3, 0,51/1,27 mm Meat/Platte, FRG-2, BRD
U_xF_y-Al	1 1)	Klein-platten	5	Testbeginn Herbst 1983, in Vorbereitung, 7,0 g U/cm^3, 0,51/1,27 mm Meat/Platte, FRG-2 und FRJ-2, BRD
U_xNi_y-Al	1 1)	Klein-platten	5	Testbeginn Mitte 1984, in Planung, 7,0 g U/cm^3, 0,51/1,27 mm Meat/Platte, FRG-2 und FRJ-2, BRD

1) Bestrahlungen mit Test- und Prototyp-Elementen nicht vor Ende 1985

satz kommen. Pro Brennstoffart werden jeweils 3 Platten in den Bestrahlungseinrichtungen im FRG-2, Geesthacht, und im FRJ-2, Jülich, bestrahlt werden. Tabelle 4 gibt einen Überblick über die Kleinplatten-Bestrahlungen in den nächsten 1,5 Jahren.

- U_3Si-Al.
  3 Kleinplatten sollen ab 1983 den Bestrahlungstest aufnehmen. Dichte 7,0 g U/cm^3, Standardspezifikation.

Der Test von Elementen (2) (Dichte 5 bis 6 g U/cm^3) ist nicht vor Anfang 1984 geplant. Als Testreaktor ist der FRG-2 in Geesthacht vorgesehen.

- U_xFe_y-Al und U_xNi_y-Al.
  Beide Brennstoffarten werden mit Kleinplatten gemäß Testschritt 1 getestet werden. Angestrebte Dichte 7,0 g U/cm^3. Nur bei U_xFe_y haben Vorbereitungen begonnen. Testbeginn Herbst 1983.

Tabelle 4 AF-Programm Kleinplatten für Bestrahlungstests (1. Teil)

Nr	Brenn-stofftyp	Korngrösse (μm)	U-Dichte (g/cm^3)	Platten/ Meatdicke (mm)	Hülle	Bemerkungen
1	U_3O_8-Al 1)	< 90 (<25% <40)	2,6	1,3/0,51	AlMg1	instrumentiert
2	U_3Si_2-Al	<150 (25% <40)	5,0	1,3/0,51	AlMg2	teilw. instrum
3	U_3Si_2-Al	<150 (50% <40)	5,0	1,3/0,51	AlMg2	teilw. instrum
4	U_3Si_2-Al	<150 (50% <40)	5,0	1,3/0,65	AlMg2	teilw. instrum
5	U_3Si-Al	<150 (<25% <40)	7,0	1,3/0,51	AlMg2	teilw. instrum
6	U_xFe_y-Al	<150 (<25% <40)	7,0	1,3/0,51	AlMg2	teilw. instrum
7	U_xNi_y-Al	<150 (<25% <40)	7,0	1,3/0,51	AlMg2	teilw. instrum

Standard-Beladung pro Bestrahlungseinrichtung:
min. 3 Platten pro Brennstoff-Typ und 1 Referenz-Platte

1) Referenz-Platte: 1. Bestrahlung

5. Schlußbemerkung.

Im vorstehenden Abschnitt wurde dargelegt, daß die sogenannten Performance-Bestrahlungstests zumindest im Fall des U_3Si_2-Al-Brennstoffs mit Testelementen bereits angelaufen sind, bevor die fundamentalen Bestrahlungsuntersuchungen gemäß Testschritt 1 begonnen haben.

Dadurch kann die begrenzte Bestrahlungskapazität (50 bis 60 Kleinplatten von 1983 - 1985) in erster Linie für neuartige („exotische") Brennstoffe besonders hoher Dichte und gezielte Spezifikationsänderungen genutzt werden. Die vom Testreaktor völlig getrennten, separat gekühlten und gut instrumentierten Bestrahlungseinrichtungen gestatten bei den Spezifikationen auch die Untersuchung sogenannter Grenzbereiche und außerdem in geringem Umfang (ca. 10 bis 15 %) Überlastuntersuchungen (erhöhte Temperaturen und Leistungsüberschläge).

Vor allem die Spezifikationsuntersuchungen sind für die Reaktorbetreiber sehr wichtig, weil sich dadurch die Möglichkeit eröffnet, solche Spezifikationsparameter schwerpunktmäßig zu untersuchen, die später einmal eine möglichst „billige" LEU-Brennstoffertigung gestatten.

Teil B

Bestrahlungseinrichtung

Das AF-Loop-Jülich dient als Bestrahlungseinrichtung zur Untersuchung von Brennstoffplatten mit niedrigangereichertem, hoch dichtem Brennstoff.
Das Loop besteht aus dem Bestrahlungseinsatz im 6V-Kanal des FRJ-2, dem Kreislaufsystem und der Instrumentierung.

Bestrahlungseinsatz

Der Bestrahlungseinsatz (Abb.1) enthält im Corebereich 10 Testplatten die in zwei Gruppen á 5 Platten übereinander angeordnet sind. Er besteht aus Aluminium und ist in ein 6V-Thimble eingesetzt. Der Zwischenraum im Thimble wird auf Feuchtigkeit überwacht.
Kühlwasserleitungen sowie Thermoelementleitungen werden durch den Abschirmstopfen gewendelt hindurchgeführt. Das Kühlwasser strömt zwischen Außenrohr und Innenrohr abwärts und durch das Zentralrohr wieder aufwärts, wobei es die Testplatten umströmt, diese kühlt und deren nukleare Wärme aufnimmt.

Daten für Testplatten:

Abmessungen:	200x60x1,27 mm
Meat:	180x50x0,51 mm
Max.Oberflächentemperatur:	130°C
Max.Flächenbelastung:	130 W/cm^2
Wassergeschwindigkeit:	3,2 m/s
Durchsatz:	11,5 m^3/h
Gesamtleistung:	220 KW

Kreislaufsystem

Das Kreislaufsystem besteht aus folgenden 4 Funktionsgruppen (Abb.2 Kreislaufschema)

1. Hauptkühlung
2. Hilfskühlung
3. Reinigung
4. Druckregelung

1. Hauptkühlung

Die Hauptkühlung ist an den Bestrahlungseinsatz angeschlossen und bildet einen geschlossenen Kreislauf. Als Kühlmittel wird Deionat bei einem Überdruck von 3 bar verwendet. Die von den Testplatten erzeugte Wärme wird durch die Hauptkühlung abgeführt und weiter über den Wärmeaustauscher WT1, einem Rohrbündelwärmeaustauscher, an den Experimentierkühlkreis des FRJ-2 abgegeben.
Der Durchsatz des Experimentekühlwassers wird durch das Regelventil RV76 auf 15m^3/h eingestellt und durch die Durchflußmesser Q5, Q6

angezeigt und überwacht.
Die Temperatur der Testplatten wird durch den Massenstrom der Hauptkühlung geregelt, was durch den Bypass zwischen Vor- und Rücklauf mit dem elektrisch gesteuerten Regelventil RV21 geschieht. Der Massenstrom wird durch die Normventuridüse Q_1 gemessen.
In diese sind die drei Meßstellen Q1a, Q1b und Q1c angeschlossen, welche den Massenstrom überwachen, anzeigen und registrieren.
Die zwei parallel geschalteten und hermetisch gekapselten Kreiselpumpen ZP1, ZP2 (Spaltrohrpumpen) erzeugen die erforderliche Druckdifferenz von ca. 5 bar. Eine dieser Pumpen ist für den Betrieb und die andere als Reserve vorgesehen.
Der Systemdruck wird durch die Drucktransmitter P_2, P_3, P_4 überwacht, angezeigt und geschrieben.
Die Hauptkühlung ist so ausgelegt, daß sie eine max. nukleare Wärmeleistung von 250 KW an den Experimentekühlkreis des FRJ2 abführen kann. Die Wassertemperatur der Hauptkühlung wird ebenfalls durch die Thermoelemente T10, T11, T12 überwacht, registriert und angezeigt. Die mittlere Temperatur des Kühlwassers liegt bei ca. 50°C. Um die Rücklaufleitung der Hauptkühlung nicht abschirmen zu müssen, ist der Verzögerungsbehälter FB4 mit einer 10 cm dicken Bleiabschirmung in der Rücklaufleitung eingebaut. Sein Volumen (120 cm lang, 43 cm breit, 50 cm hoch) beträgt ca. 240 l. Im Inneren des Behälters sind Schikane-Bleche eingebaut, die eine gute Durchmischung des einströmenden Kühlwassers gewährleisten. Der Verzögerungsbehälter ist in der Reaktor-Top-Nische untergebracht, dadurch ist nur die Verbindungsleitung (ca. 1,5 m) zwischen Verzögerungsbehälter und Bestrahlungseinsatz abgeschirmt.
Die drei Aktivitätsmeßköpfe X1, X2, X3 sind in einem Halter eingebaut, der in der Rücklaufleitung, direkt hinter dem Verzögerungsbehälter montiert ist.
Die Aktivität des Kühlmittels wird geschrieben und überwacht. Die Lecküberwachung des Kreislaufes geschieht durch die Niveaumeßsonde N1, die in dem Druckhaltetank FB1 eingebaut ist. Das Flüssigkeitsniveau im Druckhaltetank wird überwacht, angezeigt und registriert.
Das Rohrleitungssystem mit einer Nennweite von 34 mm und alle Bauteile der Hauptkühlung sind für einen Nenndruck von 10 bar ausgelegt, weil der Experimentekühlkreis des FRJ-2 auch für einen Nenndruck von 10 bar ausgelegt ist.

2. Hilfskühlung

Die Hilfskühlung ist, wie die Hauptkühlung an den Bestrahlungseinsatz angeschlossen und bildet einen geschlossenen Zwangsumlauf. Als Kühlmittel wird auch Deionat bei einem Überdruck von 3 bar ver wendet. Der Massenstrom von $2m^3/h$ und die Druckdifferenz von ca. 1,5 bar werden ebenfalls durch zwei parallel geschaltete Spaltrohrpumpen ZP3, ZP4 erzeugt, die jedoch aus Sicherheitsgründen immer beide gleichzeitig laufen, obwohl eine ausreichen würde. Da-

durch wird sichergestellt, daß die Hilfskühlung immer funktionsfähig ist.
Bei Ausfall der Hauptkühlung wird der Reaktor abgeschaltet (Scram) und die Hilfskühlung wird automatisch eingeschaltet. Gleichzeitig werden die Magnetventile MV1, MV2 geschlossen und die Magnetventile MV3 und MV4 geöffnet. Die von den Testplatten erzeugte Nachwärme (<7 % der Nennleistung) wird durch die Hilfskühlung über den Wärmeaustauscher WT2, ebenfalls ein Rohrbündelwärmeaustauscher, an den Sekundärkühlkreis des FRJ-2 abgeführt. Der Durchsatz des Sekundärkühlkreises wird durch das im Reaktor vorhandene Handventil AV4121 auf $2m^3/h$ eingestellt und durch den Durchflußmesser Q7 angezeigt und überwacht. Die Hilfskühlung ist so ausgelegt, daß sie eine max. Nachwärmeleistung von 25 KW an den Sekundärkühlkreis abführen kann. Der Durchsatz wird durch die Normventuridüse Q3 überwacht, registriert und angezeigt, der Druck durch das Kontaktmanometer P7 und die Temperatur durch das Thermoelement T16.
Die Lecküberwachung der Hilfskühlung geschieht durch die Niveaumeßsonde N2, die in dem Druckhaltetank FB2 eingebaut ist. Der Flüssigkeitsstand wird angezeigt, geschrieben und überwacht.
Das Rohrleitungssystem mit einer Nennweite von 18 mm und alle Bauteile der Hilfskühlung sind ebenfalls für einen Nenndruck von 10 bar ausgelegt.
Der Sammeltank FB3, 130 cm lang, 50 cm breit, 33 cm hoch, mit einem Volumen von ca. 200 l dient für die Entleerung der Kühlkreisläufe und ist so ausgelegt, daß er das Wasser des geplanten Kreislaufsystems aufnehmen kann.
Hauptkühlung und Hilfskühlung sind innerhalb des Einschubkopfes miteinander verbunden und die Magnetventile MV1, MV2, MV3 und MV4 sind direkt am Kopf befestigt. Dadurch wird verhindert, daß durch einen eventuellen Rohrbruch beide Kühlkreisläufe gleichzeitig ausfallen können.

3. Reinigung

Die Reinigung besteht im wesentlichen aus dem abgeschirmten Filter AF1 und hat die Aufgabe, die mechanischen und die im Wasser gelösten Verunreinigungen aus dem Kühlmittel zu entfernen, die unter Umständen aktiv sein können.
Sie arbeitet im Bypass zwischen Vor- und Rücklauf der Haupt- und Hilfskühlung. Beim normalen Betrieb sind die Ventile AV22, AV29 zur Hauptkühlung geöffnet und die Ventile AV23, AV30 zur Hilfskühlung geschlossen. Der Durchsatz wird durch das Regelventil RV24 eingestellt und beträgt 250 l/h. Er wird auch durch den Durchflußmesser Q2 angezeigt.
Bei diesem Durchsatz wird die gesamte Kühlwassermenge zweimal pro Stunde gereinigt.
Die Filterfüllung besteht aus 3 parallelgeschalteten Filterpatronen (Micro Clean II 5/µm) und eine Ionenaustauscherharzfüllung (4 Liter Levatit 112,8 Liter Levatit M500), die in einem abgeschirmten Behälter untergebracht ist.

Da eine Abschätzung der Aktivität der Reinigung sehr schwierig ist, wurde der Behälter nach Absprache mit dem Reaktorbetrieb und dem Strahlenschutz mit 10 cm Blei abgeschirmt. Nach bisherigen Erfahrungen im Reaktor dürfte dies mit Sicherheit ausreichend sein. Die Filterpatronen und die Harzfüllung werden ausgetauscht, wenn der Durchsatz um ca. 10 % geringer oder pH-Wert >5-6 wird.

4. Druckregelung

Der Druck im AF-Loop wird durch ein Druckhaltesystem erzeugt und auf einem Wert von 3 bar Überdruck konstant gehalten.
Die Druckregelung besteht aus den beiden Druckhaltetanks FB1 und FB2, der He-Versorgungsflasche GB1, der He-Vorratsflasche GB2 sowie einer Ventilstation.
Das als Druckgas verwendete Helium wird der Versorgungsflasche GB1 (Flaschendruck; 10 bar) entnommen und über das Druckminderventil DV53 konstant auf den Betriebsdruck von 3 bar reduziert (Anzeige an P10).
Über das Ventil AV56, das während des Betriebes geöffnet ist, wird der Behälter FB1 mit Druckgas beaufschlagt.
Der Behälter FB1 ist normalerweise zu 75 % mit Deionat und das Restvolumen mit Helium gefüllt. Gegen Überschreitung des Nenndrucks ist dieser Leitungsabschnitt und damit die Hauptkühlung durch das Sicherheitsventil SV 58 abgesichert.
Der Betriebsdruck in der Hilfskühlung beträgt ebenfalls 3 bar; er wird vor Betriebsbeginn eingestellt. Durch Öffnen des Ventils AV57 wird der Druckhaltetank FB2, der ebenfalls zu 75 % mit Deionat gefüllt ist, mit einem He-Überdruck von 3 bar beaufschlagt und das Ventil AV57 wieder geschlossen. Das Sicherheitsventil SV64 sichert die Hilfskühlung gegen eine unzulässige Überschreitung des Nenndruckes ab.
Zur Druckentlastung ist die Druckregelung mit dem Abluftsystem des Reaktors verbunden. Das abzulassende Helium strömt über den Verzögerungsbehälter der Abluftanlage in den Kamin. Hierfür sind 2 Möglichkeiten vorgesehen. Über die Ventile AV70, AV62 und AV61 kann Helium direkt in die Abluft abgeführt werden. Um eine definierte und kontrollierte Gasmenge abzuführen, wird das Helium über den Druchflußmesser Q4 mit dem Regelventil RV60 und die geöffneten Ventile AV70, AV59 und AV61 in die Abluft abgegeben.
Die Druckentlastung der Hilfskühlung kann einmal über die Ventile AV63, AV62 und AV61 vorgenommen werden. Zur Abführung einer dosierten Menge wird das Druckgas über die Ventile AV63, AV59, RV60, AV61 und den Durchflußmesser Q4 dem Abluftsystem zugeleitet.
Zur Sicherstellung der Gasversorgung für die Druckregelung wird der Versorgungsbehälter GB1 von der Meßstelle P5 drucküberwacht. Wird ein Mindestvordruck unterschritten, erfolgt über den Grenzwert der Meßstelle P5 eine Warnung. Aus dem Vorratsbehälter GB2 wird Helium von Hand in die Versorgungsflasche GB1 nachgespeist.

4. Elektrische Instrumentierung

Über die Instrumentierung des Experimentes werden alle wichtigen Betriebsgrößen überwacht. Sie ist so eingerichtet, daß das Experiment während des Betreibes keiner besonderen Aufsicht bedarf. Abweichungen vom Normalbetriebszustand werden von Überwachungseinrichtungen erfaßt. Sie lösen je nach der Bedeutung der Abweichung eine der folgenden Maßnahmen bzw. Kombinationen davon aus.

- Störungsmeldung, optisch und akustisch - Pultwarnung am Experiment und Reaktorwarnung bzw. Zusatzwarnung in der Reaktorwarte - als Aufmerksamkeitszeichen für Abweichungen vom Betriebszustand die zu Reaktorabschaltungen führen können.
  Reaktoreingriff (Reaktorabschaltung) RE wenn die spontane Energieerzeugung im Strahlrohreinschub unterbrochen werden muß.
  Notkühlung zur Sicherung von Versuchsergebnissen.

Konventionelle Sicherheitsgrößen

Kühlwasserdurchsatz

Der Kühlwasserdurchsatz wird durch die drei Meßstellen Q1a, Q1b, Q1c überwacht, registriert und angezeigt. Bei einem Ausfall des Kühlwassers der Hauptkühlung lösen zwei der drei obengenannten Meßstellen eine Reaktorabschaltung (Scram) aus. Gleichzeitig geht die Hilfskühlung in Betrieb, die Magnetventile MV1, MV2 schließen, die Magnetventile MV3, MV4 öffnen und die Hauptkühlpumpen ZP1 und ZP2 werden automatisch abgeschaltet.

Temperatur

Um eine Überhitzung der Testplatten zu vermeiden, wird die Temperatur des Kühlwassers durch die drei Meßstellen T10, T11, T12 überwacht, angezeigt und registriert. Bei einer Überschreitung der zulässigen Betriebstemperatur lösen zwei der drei obengenannten Meßstellen eine Reaktorabschaltung (Scram) aus.

Druck

Um ein Sieden des Kühlwassers im Bestrahlungseinsatz zu vermeiden, wird der eingestellte Betriebsdruck im Loop durch die drei Meßstellen P2, P3, P4 überwacht, registriert und angezeigt. Bei einer Unterschreitung des Betriebsdruckes lösen die drei obengenannten Meßstellen in 2 von 3-Schaltung eine Reaktorabschaltung (Scram) aus.

Aktivität

Durch eine Schädigung der Testplatten kann eine Freisetzung von Spaltprodukten in das Kühlwasser erfolgen, dessen Aktivität von den drei Meßstellen X1, X2, X3 gemessen und überwacht wird. Übersteigt die Aktivität die zulässigen Grenzwerte von $\mu Ci/cm^3$, so erfolgt in 2 von 3-Schaltung eine Reaktorabschaltung (Scram).

Strahlenschutz

Bestrahlungseinsatz

Der Bestrahlungseinsatz ist nach oben durch den Abschirmstopfen abgeschirmt. Der Innenraum des Stopfenrohres ist mit einer Masse aus Araldit/Stahlschrot im Verhältnis 50/50 ausgegossen. Die durch den Stopfen geführten Rohre sind aus Abschirmgründen je dreimal abgewinkelt.
Die Berechnung ergab folgende Werte:

Unter der Top-Platte

D_γ = 18 mrem/h (Normalbetrieb)
D_n = 270mrem/h (Normalbetrieb)
D_γ = 90 mrem/h (Störfall)

Oberhalb der Top-Platte

$D_\gamma \leq$ 1,0 mrem/h
$D_n \leq$ 1,0 mrem/h

Verzögerungsbehälter FB4

Das Volumen des Verzögerungsbehälter FB4 beträgt ca. 240 l. Im Inneren des Behälter sind Schikane-Bleche eingebaut, die eine gute Durchmischung des einströmenden Wassers gewährleisten.
Die Verzögerungszeit im Behälter von 76 s ist die Zeit, die ein Wasserpartikel mittlerer Geschwindigkeit im Behälter verweilt.
Die mittlere spezifische Aktivität des Wassers in dem Behälter ist 1,17 $\mu Ci/cm^3$.
In 100 cm Abstand vom unabgeschirmten Behälter entspricht das einer Dosisleistung von 125 mrem/h. Eine 10 cm dicke Bleiabschirmung rings um den Behälter vermindert die Dosisleistung auf 0,65 mrem/h. Der Abstand von 100 cm entspricht ungefähr dem möglichen Standort des Bedienungspersonals auf dem Reaktortop. In 30 cm Abstand wurden 1,63 mrem/h errechnet.
Treten bei einer eventuellen Schädigung der Testplatten Spaltprodukte in das Kühlwasser aus, so daß die spezifische Aktivität in der Rücklaufleitung der Hauptkühlung hinter dem Verzögerungsbehälter den Grenzwert von 30 $\mu Ci/cm^3$ erreicht, so erfolgt eine

Reaktorabschaltung (Scram).
Diese Aktivität und die sich daraus ergebende Dosisleistung setzen sich zusammen aus der Aktivität des Wassers (4,6 µCi/cm^3) und der Spaltproduktaktivität aus den Testplatten.
Die Dosisleistung an dem Verzögerungsbehälter (10 cm Bleiabschirmung) in 30 cm Abstand beträgt ≈7 mrem/h.
Die Dosisleistung an der Rücklaufleitung der Hauptkühlung (unabgeschirmt) hinter dem Verzögerungsbehälter in 30 cm Abstand beträgt ≈72 mrem/h.

Kühlwasser

Kreislauf-Volumen:	120,0 l
Aktivierbares Volumen:	3,1 l
Volumenstrom:	3,15l/s
Verweilzeit des aktivierbaren Volumens im "Strahlungsfeld":	0,97 S

Aktivierungs-Flußdichten im 6V-Kanal FRJ-2
(laut SUEL 1022-RX, Nov/73):

$$\phi_{th} = 1,6 \cdot 10^{14} cm^{-2} \cdot s^{-1}$$
$$\phi_{epi} = 1,6 \cdot 10^{14} cm^{-2} \cdot s^{-1}$$
$$\phi_{sn} = 1,2 \cdot 10^{12} cm^{-2} \cdot s^{-1}$$

Die spezifische Aktivität des Kühlwassers (Deionat) in der Rücklaufleitung der Hauptkühlung beim Austritt aus dem Stopfen wurde mit 4,6 µCi/cm^3 errechnet.
Diese Aktivität entspricht einer Dosisleistung von 26 mrem/h in einem Aufpunkt ca. 30 cm von der Leitung entfernt.
Durch den vorgebauten Verzögerungsbehälter ändert sich der ursprünglich errechnete Wert der Dosisleistung von 26 mrem/h in 30 cm Abstand von der unabgeschirmten Rücklaufleitung auf 0,444 mrem/h.

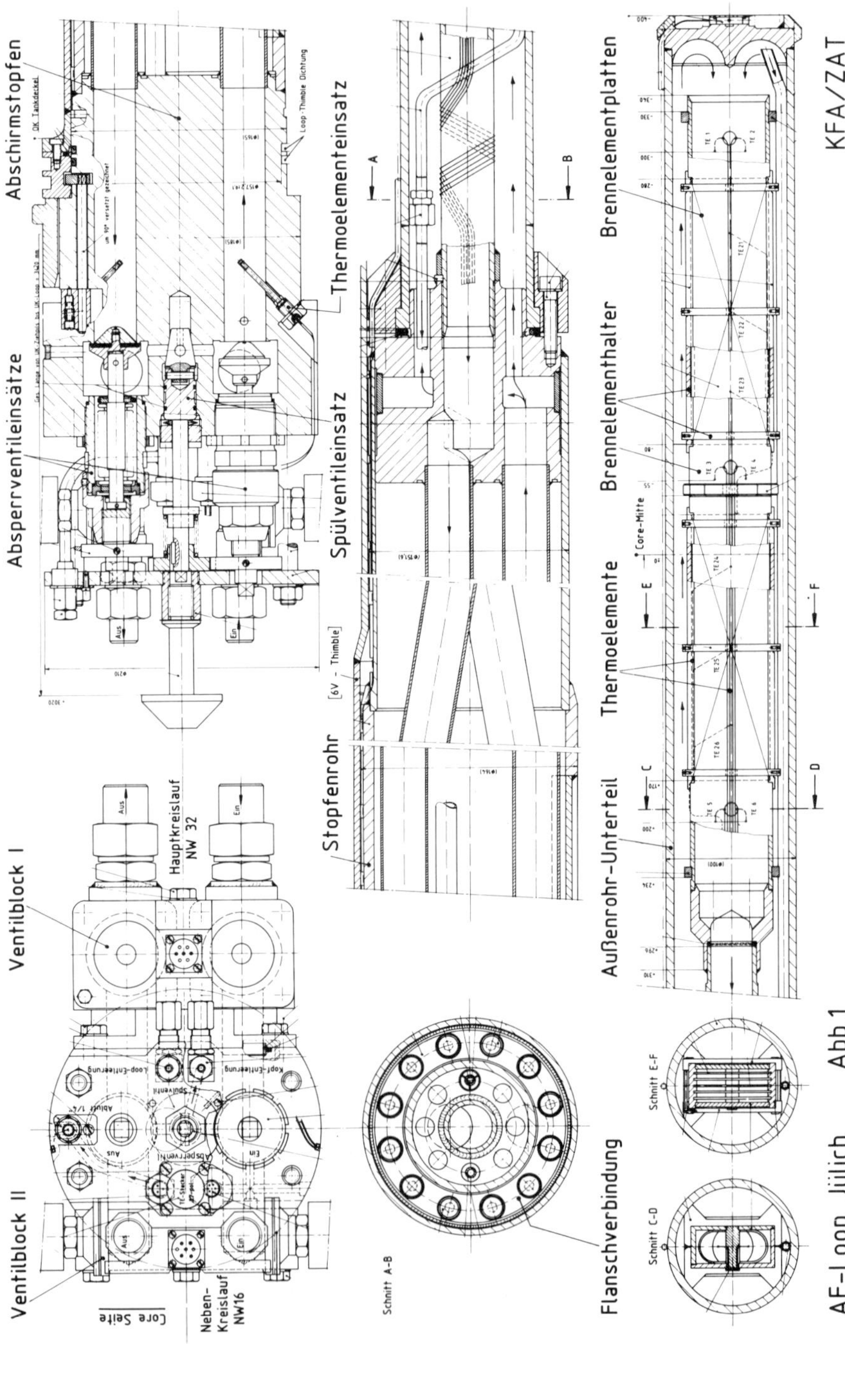
Ventilblock II
Ventilblock I
Core Seite
Neben-Kreislauf NW16
Hauptkreislauf NW 32
Absperrventileinsätze
Abschirmstopfen
Thermoelementeinsatz
Spülventileinsatz
Stopfenrohr
6V - Thimble
Schnitt A-B
Flanschverbindung
Außenrohr-Unterteil
Thermoelemente
Brennelementhalter
Brennelementplatten
Core-Mitte
Schnitt C-D
Schnitt E-F
AF-Loop Jülich
Abb.1
KFA/ZAT

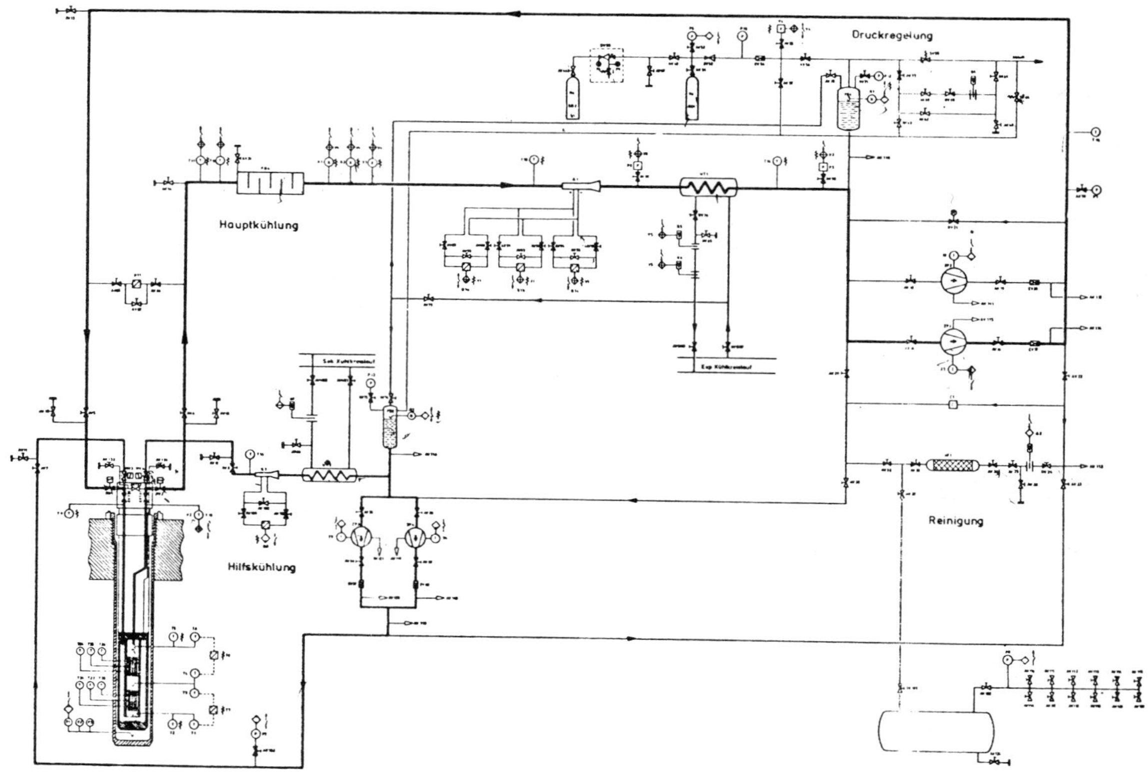

Abb. 2 AF-Loop Jülich (LV 50) - Kreislaufschema

AN AUTOMATIC CONTROLLED, HEAVY WATER COOLED FACILITY FOR IRRADIATION OF SILICON CRYSTALS IN THE DR 3 REACTOR AT RISØ NATIONAL LABORATORY, DENMARK

K. Hansen, K. Stendal, K. Andresen and K. Heydorn

Risø National Laboratory

4000 Roskilde, Denmark

ABSTRACT

Neutron transmutation doping of silicon crystals with phosphorus has been carried out in the DR 3 reactor since 1975.

The irradiation capacity has been enlarged concurrently with the growing demand for neutron transmutation doped crystals.

This report describes a recently developed facility for the irradiation of 4 inches silicon crystals. The irradiation rig is placed in one of the vertical experimental tubes in the heavy water moderator of the reactor.

The rig is filled with heavy water, and the nuclear heat absorbed in the crystals during irradiation is removed by convection.

During the irradiation the crystals are placed in round aluminum containers. Transport of the containers from the irradiation zone to the top of the rig is made by heavy water flow. The irradiation cycle is controlled automatically by instrumentation.

P. von der Hardt and H. Röttger (eds.), Irradiation Technology, 567–578.

INTRODUCTION

Pure silicon mono-crystals doped with phosphorous atoms for adjustment of the electrical resistivity to a specified value are used in the electronic industry for production of thyristors, integrated circuits, etc.

In the conventional method of doping phosphorous atoms are added during the crystal growth and temperature nonuniformities and dopant segration produce significant variations of the resistivity across the crystal, whereas doping of silicon crystals with phosphorous atoms by transmutation gives a small variation in the resistivity over the crystal.

Nonradioactive phosphorous atoms are formed from the ^{30}Si isotope by the following process:

$$^{30}Si + n \rightarrow {}^{31}Si \rightarrow {}^{31}P + e^-$$

The content of ^{30}Si isotope in natural silicon is 3.09% and the half-life of decay of ^{31}Si is 2.6 hours.

At Risø National Laboratory in Denmark, doping of silicon crystals by means of thermal neutrons was started in 1974 for the Topsil Co., in order to prove the validity of the method and to provide test samples. The first irradiations were made in the thermal column of the DR 2 reactor (MTR-tank type) without special equipment.

The promising results from the test samples resulted in a demand for silicon crystals doped by neutron irradiation and a rig designed for irradiation of 3 inches silicon crystals in the DR 3 reactor was then manufactured during 1975.

Due to demand for further irradiation the facility was enlarged with four 3 inches rigs, manufactured during 1977. Finally in 1981 a facility for irradiation of 4 inches crystals was ready for operation.

This report describes the design and operation experience of the 4 inches facility.

DESCRIPTION OF THE FACILITY

The Irradiation Position in the Reactor

The Danish DR 3 reactor is a 10 MW heavy water moderated and cooled research reactor. A vertical cross section of the reactor is shown in Figure 1. The 4 inches irradiation rig is placed in one of

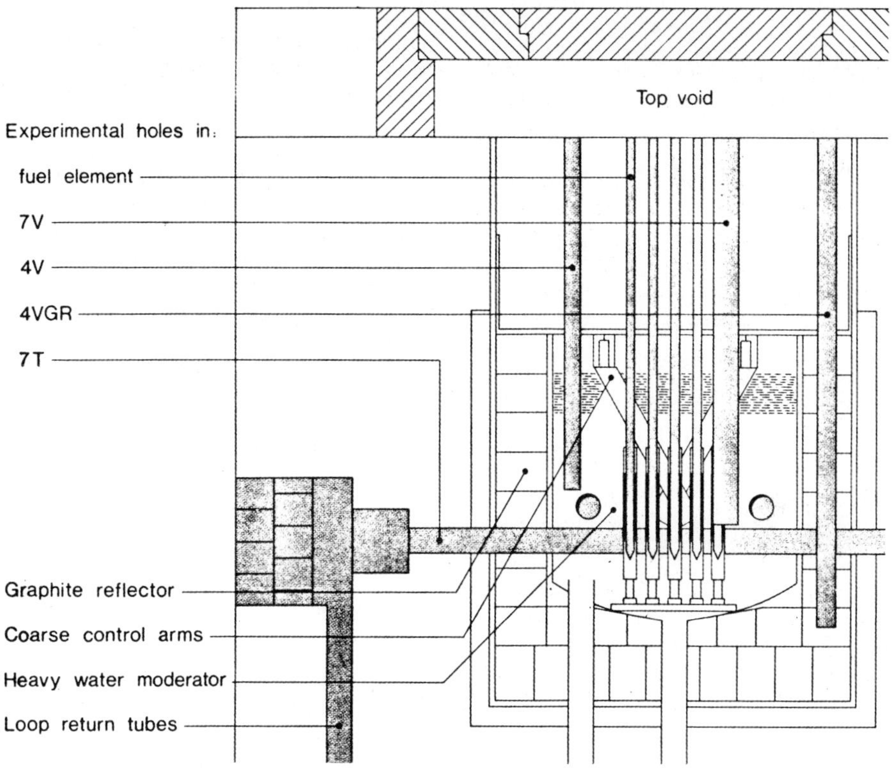

Fig. 1. Irradiation Positions in the DR 3 Reactor

the vertical experimental holes in the heavy water tank. The thermal neutron flux, smoothed by the absorber screen, is about 2 x 10^{13} n/cm^2 sec. and the γ-heat in the irradiation zone is about 250 mW/g. Due to the γ-heat, it is necessary to cool the silicon crystals during irradiation. The temperature of the crystals during the irradiation is about 50°C.

Specifications for the Rig

The rig is designed for irradiation of silicon crystals of a diameter up to 107 mm and length up to 400 mm.

Variation in the phosphorous doping produced in the crystal is designed to be less than +/- 5% and this is routinely achieved in practice.

Methods Used for Obtaining Uniform Doping

In the irradiation zone of the 7V hole the variation of the incident thermal neutron flux is about 20% in axial direction and about 25% in radial direction.

In order to reduce the variation in the received neutron dose the following methods are used:

- An absorber screen of stainless steel with varying thickness is used to equalize the axial flux profile.
- The crystal is rotated at a constant speed of 2 rpm around the vertical axis of the irradiation hole.
  The rotation eliminates the influence of the radial flux gradient.

Crystal Container

The unirradiated crystals are loaded in a thin walled (0.5 mm) round container made from Al-2S. The inside diameter of the container is 110 mm and the length is 405 mm.

The container is strongly radioactive after the irradiation; but most of the activity comes from the ^{28}Al isotope with a short half life (2.3 minutes). After a 30 minute decay under the top shield plug of the rig the container can be removed from the rig by the shielded handling flask. The container and crystal are dried in the flask for about 2 hours and then transferred to the storage facility for further decay for 4-5 days. The remaining activity is now acceptably low for handling without shielding in connection with control, packing and shipment of the irradiated crystal.

Irradiation Rig

During irradiation the container is placed at the bottom of the rotating aluminum tube, which is supported by a ball bearing at the top and a graphite bearing at the middle.

The tube is rotated by a geared electric motor placed at the top of the rig. The rotating tube is surrounded by a rig thimble, which acts as containment for the heavy water in the rig. The absorber screen is placed on the outside of the rig thimble. The rig has a top shield made from stainless steel with channels for the heavy water flow used to transport the container up from the rig. A lead filled stainless steel plug is placed in the top of the rotating tube. During loading operations the plug is removed and stored in inside the handling flask. Figure 2 shows the rig placed in the reactor.

Fig. 2. The Irradiation Rig Placed in the Reactor

Heavy Water Circuit

The heavy water circuit is used for transport and indication of containers in the rig.

The main components of the circuit are:

- The irradiation rig
- The levelling tank
- The pump unit

The pump unit consists of pumps, valves, ion exchanger, filter, drain tank and transducers for control of flow, pressure and conductivity.

Transport of the Container. When the irradiation of the crystal is finished one of the main pumps is started by a signal from the flux integrator in the instrumentation. The heavy water flow circulating up through the rotating tube and forces the container up to the top of the rig. The water flow is about 75 litres/minute and the gap between the container and the tube is 1.5 mm.

The container is stopped by the shield plug at the top of the rig. The transport time from bottom to top is 1-1.5 minutes. The container is then "cooling" for about 30 minutes at the top of the rig. The main pump is first stopped when the irradiated container is removed to the handling flask and an unirradiated container is placed in the top of the rig.

When the main pump is stopped, the container will sink down in the rotating tube and reach the bottom after 0.5-3 minutes depending on the crystal weight.

Indication for a Container in Irradiation Position. A nozzle is placed in the bottom of the rig. The bottom end plate of the container will close the nozzle, when the container is in irradiation position.

The indication pump (diaphragm type) forces a water flow (about 1.5 litres/minute) through the nozzle. The delivery pressure of the pump indicates when a container is in irradiation position. The indication pump is running continuously and is also acting as a cleaning pump by sending the water through a filter and an ion exchanger.

The heavy water from the rig and the circuit can be drained to the drain tank by means of the indication pump.

The pump unit is shown in Figure 3.

Handling Flask

The shielded handling flask (see Figure 4) is used for loading opations in the rig and in the storage facility and for transfer operations between the rig and the storage facility.

The flask is carried by a small wagon driven by a geared electric motor. The shield is formed as a cylindrical jacket with a thickness of 12 cm lead. A drum with a loading hole and positions for the rig plug, an irradiated container and an unirradiated container is placed inside the jacket.

The plug and the containers are lifted by a pneumatically activated grab during loading operations. The grab is supplied with compressed air through a reinforced rubber hose, which also is used

Fig. 3. The Pump Unit

Fig. 4. The Shielded Handling Flask

as rope when the grap is lifted by a winch placed on the top of the flask.

By means of an indication valve it is possible to control when the plug or a container are correctly placed in the grab. The order of the single operations are controlled by a special guide disc placed on the top of the flask.

The wet containers are dried in the flask for about 2 hours by means of heated air circulated through the container. The evaporated heavy water is recovered in a condenser. All inside surfaces in the flask are stainless steel in order to avoid rust contamination on the containers.

Storage Facility

After irradiation it is necessary to place the containers in a shielded storage facility 3-5 days before they can be handled without shielding. The duration is determined by the ^{72}Ga impurity in the aluminum of the containers.

The storage facility consists of 23 stainless steel tubes arranged in five rows and shielded by concrete containing lead shot and iron shot,

The top of the storage facility is in level with the reactor top (see Figure 5). Loading operations are made by means of the handling flask, which can be moved on the reactor top between the rig and the storage positions.

Instrumentation

Automatic Dose and Transport Control. The thermal neutron flux is not constant during the time required for the irradiation. It is therefore necessary to integrate the thermal flux over the irradiation time in order to obtain acceptable accuracy of the received neutron dose.

The thermal neutron flux is measured by three self-powered neutron detectors of the vanadium type placed in guide tubes in the irradiation zone of the rig.

The signal from one of the neutron detectors is connected to an analog to digital converter, which integrates by means of a frequency proportional to the signal from the neutron detector. Before irradiation the integrator is set at a predetermined value.

Fig. 5. The Storage Facility

When the container reaches the bottom of the rig a rise in the indication pressure starts the integrator and the rig rotation. The integrator counts downwards during the irradiation and at zero count the integrator starts the main pump and the container is lifted away from the irradiation zone by the heavy water flow.

Rotation Control. Control of the rig rotation is made by means of an opto-electrical detector activated by slits in a small aluminum disc.

Warnings. Warnings for abnormal conditions in the rig system during irradiation are announced by sound and by lamps on a display in the instrumentation rack.

Warnings during irradiation:

- Stop of rig rotation
- High and low level in the levelling tank
- Diaphragm fault in the indication pumps
- Low main flow
- Low pressure and flow in the indication system

Warnings at the end of irradiation:

- End of irradiation
- Low main flow
- Transport error (container not lifted from bottom of rig)

The warnings are connected to the "experiment warning guard" in the control room of the reactor.

The instrumentation panel is shown in Figure 6.

OPERATION EXPERIENCE

The facility was ready for operation in August 1981.

The planned irradiations have been performed without significant problems.

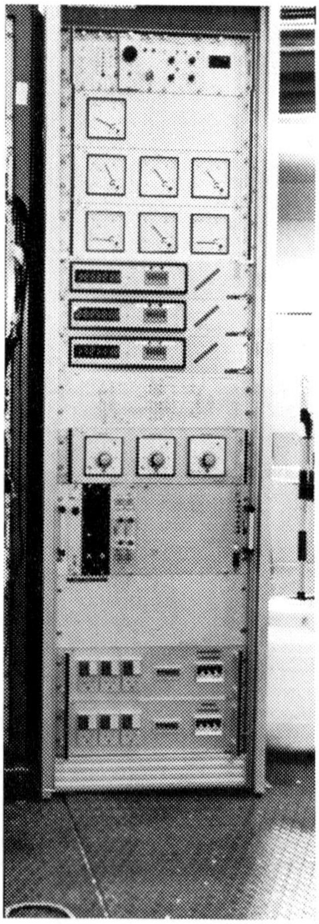

Fig. 6. The Instrumentation

Experience with Use of Heavy Water in the System

Heavy water is used in the system due to its low neutron absorption property. Measurements have shown that use of light water would reduce the neutron flux level more than 50% and give us difficulties in equalizing the axial flux profile.

The disadvantages of heavy water are high cost and tritium build-up during irradiation.

Losses of heavy water. Due to the high cost of heavy water it is important to minimize losses during operation.

Evaporation losses are prevented by having a tight system. Only during loading operations the rig is open to the reactor top.

The irradiated containers still contain a small amount of heavy water when they are removed from the rig. The containers are dried in the handling flask and the heavy water is recovered in the condenser, but it is degraded to about 80% by light water from the atmosphere.

Tritium build-up. The system was filled with "fresh" tritium free heavy water at the start in August 1981. A slow build-up of tritium takes place due to neutron irradiation.

Calculations have shown that the heavy water can be used for at least 2 years before the tritium concentration exceeds the tolerated level and exchange of the heavy water in the system is required.

Handling of Crystals

The crystals are protected by the aluminum containers during transport and irradiation. All operations in connection with loading and unloading of crystals in the containers are made in a special equipped room in the isotope laboratory, shown in Figure 7.

After opening of the irradiated containers the crystals are rinsed in demineralised water and, if required, ultrasonicly cleaned; no crystals have as yet been strongly contaminated.

Accuracy of the Neutron Dose

In order to obtain a quality control on the neutron dose two cobalt monitors are placed in each container, one at the bottom and one at the top.

Results have shown that the flux integrators are working precisely and are maintaining the neutron doping within the specified values.

Fig. 7. The Packing Room in the Isotope Laboratory

QUALITY ASSURANCE AND RELIABILITY ANALYSIS IN THE REALISATION OF IRRADIATION TESTING DEVICES

R. S. MATFIELD

JOINT RESEARCH CENTRE COMMISSION
OF THE EUROPEAN COMMUNITIES
21020 ISPRA, (VA) ITALY

INTRODUCTION

In-pile irradiation devices are usually expensive to design and manufacture and the experimental results which they produce are the culmination of many man-hours of technical studies. Such devices therefore represent a considerable financial investment and so they must be of adequate quality and they must be reliable.

This paper will therefore discuss the problems of producing such devices to a required level of reliability and the manner in which reliability prediction and quality assurance techniques can help to achieve this reliability.

RELIABILITY, QUALITY ASSURANCE AND THE ONE-OFF ITEM

The reliability of an item is usually expressed as a probability of failure; such probabilities can be calculated by using failure-data from data banks or from stress-strength calculations based upon a knowledge of the distribution of the stress and strength data. Later in this paper I will refer to some of the problems of these methods but at this point let us assume that the Reliability Analyst can provide a required failure rate for each component.

P. von der Hardt and H. Röttger (eds.), Irradiation Technology, 579–590.

Accepting that this data will also include assumptions on the frequency of maintenance; the quantitative manner in which maintenance restores the item to its original performance capability and the use of back-ups, or redundant, features in the design, then the results of the reliability study will be passed to the designer as the basis for the design.

Subsequently the Quality Assurance Engineer has the task of ensuring that these components can be manufactured and purchased to the required level of reliability. Clearly the ease with which this can be done will depend upon the manner in which the assumptions made by the Reliability Analyst and the Designer correspond to the realities of the world of manufacturing and procurement. Let us consider some of these realities for an in-pile irradiation device.

Most of the in-pile irradiation devices that I refer to are one-off, custom-built items and as such they present a particular quality/reliability problem. To illustrate this let me take the opposite case of the quantity produced item. Here we usually have the opportunity to measure reliability by field performance measured by operational reports (military equipment), maintenance reports (civil aircraft), customer complaints (cars and domestic products) etc., and whilst this performance is related to the way in which the item is used, the existance of a considerable body of data makes it possible to identify weaknesses in the design and thereby to eventually remove these faults and so increase the reliability of the item.

The one-off item cannot be the source of such data and the designer of a one-off in-pile irradiation device is therefore left with four possible choices if he wishes to control the reliability of his design.
He can:

a) build and test a number of prototypes,

b) build several operational items and test these,

c) use theoretical methods of reliability prediction,

or he could:

d) do neither (a), (b) or (c).

Clearly, choices (a) and (b) are expensive and, whilst the theoretical techniques for (c) are available, they are of limited use for mechanical equipment. As I suspect that many in-pile irradiation devices are produced to choice (d), the theme of the following discussion will be to offer an alternative to this choice which uses some of the features of (a), (b) and (c), and also to discuss the manner in which the use of quality assurance techniques can help to ensure that the equipment is designed, manufactured and tested with the minimum of faults and errors.

METHODS OF RELIABILITY PREDICTION

There are two approaches to reliability prediction that may be considered here:

a) the use of reliability analysis and

b) design analysis based upon the distribution of strength and stress.

In method (a) a "tree" is constructed of all the possible event-paths that could lead to a failure of the system. The probability of each event failure is taken from a data bank and these are summated along each event path to give the probability for each mode of failure.

As we are discussing the reliability of a one-off system then method (a) will only be applicable to those bought-out parts of the system which are produced in quantity. However, if failure rate data exists it must be used with care, for it is important that the circumstances of use to which this data relates is sufficiently close to the intended use for it to be valid.

On the other hand even if this data does not exist the construction of a fault tree, in which all the chains of events leading to a failure are written down, is a very useful tool for making a qualitative evaluation of a design and it can be used to highlight areas which require special design consideration to avoid failures.

If we now consider the use of design by analysis then this must take account of the variability of both the applied load due to adventitious influences and the variability of the strength of the material which must resist these loads, in terms of the distribution of physical properties, manufacturing variations that are allowed within the specification, etc. These distributions are described by Carter in references 1 and 2. He concludes that failure rates are heavily dependant upon the variations that occur in loading due to a lack of complete understanding of the environment. He states that ".... the control of the load distribution is a far more powerful method of achieving reliability than the control of strength through the design and manufacturing process ...." He also points to significant differences in the loading variations experienced by electrical and mechanical equipment.

But where does all this leave the designer of an in-pile irradiation device? Firstly, the environment of a testing reactor is very likely to produce smooth loading conditions, but it is incumbent upon the designer to ensure that this is so by being fully aware of all the loads that could be applied. Secondly, as Carter points out, he can use adequate safety margins, but these should be calculated from the standard deviation of the strength and load distributions and not established by simply applying a "factor of ignorance". However, such data is not always available and whilst this probalistic approach is highly desirable, it is not often possible to use it.

Nevertheless the designer should be aware of this fundamental approach because it provides a means of dealing with the uncertainties that often exist in the loading, the material properties, the design model etc, and shows the advantages to be gained from collecting data on such topics. It also provides an explanation of the difference between the statistical failure of electrical and mechanical equipment.

Now the above discussion is concerned with the calculation of reliability in terms of failure rates, let us now consider the contribution to reliability that can be made when these more theoretical concepts are being translated into hardware. Let us consider the contribution of quality assurance.

THE PRINCIPLES AND METHODS OF QUALITY ASSURANCE

The engineering process is concerned with taking a specification for a system and, by the use of design creativity and engineering knowledge, to transform this specification into working hardware in the minimum time and at the minimum cost. The hardware must fulfil all of the requirements of the specification without failure or breakdowns.

To achieve this it is necessary to have the right personnel, with the right knowledge and experience, working in an organisation in which there is the right production equipment and where there are no communication problems and, above all, nobody makes any mistakes.

That is what is required.

In practice, of course, all of these conditions are less than perfect and the resulting hardware is not 100% reliable. Quality assurance is an attempt to correct these imperfections.

In my view quality assurance is based upon four principles, three of which are organisational and one philosophical, namely:-

a) write it down (DOCUMENTATION)

b) check it systematically (VERIFICATION)

c) check it randomly (AUDITS)

d) quality is everybody's business (TRAINING).

I also believe that, for various reasons, the first three principles taken together represent a means of increasing individual competance, in that they help to:-

a) produce clearer thinking,

b) encourage more constructive criticism,

c) reduce undetected human error.

The fourth principle is undoubtedly the most important and often the least understood. It says that every person working on the quality assurance programme is responsible for achieving the required quality of the product and the Quality Assurance Engineer is responsible for verifying that everybody follows the rules of the quality assurance method.

I would define the essence of the quality assurance method as control and verification. These occur throughout every phase of the work, and because a full description of this method cannot be given here I will restrict further explanations to those areas that I consider bear most on reliability, i.e. design and testing.

DESIGN

Design is primarily a creative process, whilst quality assurance is a formal, stylised process and so it is not surprising that the use of quality assurance in design has lagged behind quality assurance in other areas, such as manufacturing. The quality assurance systems in design are therefore shaped to fit the existing pattern of work and to avoid the stifling of design creativity.

The main QA methods in design are:-

a) write it down

b) check it.

All features of the design are described in writing, either as reports or drawings, as this induces clearer thinking and it also allows constructive comments by specialists e.g., the welding engineer, the maintenance engineer. Such comments are made at periodic design review meetings at which the design is closely scrutinised by persons experienced in manufacture, operations etc... to ensure that all interface problems are identified, all tasks are allocated and that the design specification is satisfied.

The technical aspects of the design are verified at all stages by the checking of drawings and calculations by independent persons and by the independent checking and approval of all documents.

The various processes of design, e.g. from the analysis of the customer specification to the planning of installation activities. are described by written procedures in order that such procedures can be clearly defined and critically examined before they are put into practice.

Quality Assurance is also concerned with the use of good design practices, and the first of these is to ensure that the designer has adequate expertise for the tasks to be performed and that he is up to date in this expertise. Ensuring that all of the staff are adequately trained is a major item on the Quality Programme.

Pugh's [(3)] argument that "simplest is best" is also good design practice in that simplicity reduces the number of problems to be solved in all areas of design and manufacturing. It simplifies maintenance and generally leads to fewer human errors throughout all phases of the work.

Pugh gives another concept for judging complexity which he calls "load lines". This attempts to reduce the number of discontinuities in a system from the viewpoint that straight pipes are simpler to make than bent pipes, plain shafts are easier to machine than stepped ones, standard materials are easier to work than exotic materials, etc.

TESTING

A stipulated performance level has little meaning unless there is an associated demonstration of that performance by a well prepared test. Such tests are usually functional and overstress tests in which the weaker components are eliminated without damaging the stronger. These tests are very useful but they must be carefully designed to ensure that they are meaningful with regard to reproducing the environmental conditions that will exist in practice.

Testing is an essential method for demonstrating the quality of an item, indeed it can be the only way. The quality assurance method is therefore to verify that the tests are commensurate with the quality level required of the item.

AN ITERATION OF THE QUALITY/RELIABILITY PROBLEM

Let me now try to bring together the initial discussion on reliability and the subsequent description of quality assurance in design and testing by iterating and developing my earlier remarks.

The idealised reliability requirement is for components to be designed, manufactured maintained and used in a documented and uniform manner so that they have an identical performance and hence give rise to populations with well defined distributions.

In practice this is difficult to achieve with mechanical equipment because whilst it is possible to keep the design and manufacture under reasonable control when such equipment is made in large numbers it inevitably sells to many markets and it thereby experiences many forms of use.

A solution that is being adopted by manufacturers of nuclear instrumentation, which must be capable of withstanding earthquakes, loss of coolant accidents, etc, is to apply the type-tests described in IEEE 323. Such testing can take up to one year to qualify an instrument, it demands extensive and specialised testing equipment and is expensive. Unfortunately it is not an option for equipment that sells into general markets or for the one-off equipment.

In my view the present-day situation is that there is no reliable solution to the problem of one-off mechanical and electromechanical equipments to a high degree of reliability (10^{-2} failures per year or better) without a substantial increase in costs.

It is true that there are data banks which contain data on the failure rates of some of the components that may be used in such equipment, but close examination of the raw data reveals that it comes from a variety of sources and there is very little data on the manufacturing specifications that were followed, the standard of maintenance that was applied or the conditions of use to which the equipment was subjected. In many cases the reliability analyst uses his judgement and reduces the published failure rate by, say, an order of magnitude in order to allow for these uncertainties.

This method allows the analyst to ascribe a failure rate to say, a valve, and this is then passed to the QA Engineer as a requirement to obtain a valve with a particular failure rate. He also examines the raw data in the data bank and, of course, finds that it is insufficient to write a specification that will allow him to be confident that the failure rate can be achieved. Thus the circle is closed.

Fortunately, there are ways of breaking the circle. It may be possible to examine the customer-complaints/repair data held by the manufacturer of the valve in order to estimate the reliability that such a valve will have when it is used in certain markets. It is possible to request additional tests and inspections, over and above the standard manufacturing practise, in order to increase confidence in the performance of the valve.

THE USE OF ENGINEERING JUDGEMENT

The examples cited in the above paragraphs involve the use of engineering judgement; the analyst decides that an order of magnitude adjustment is adequate; the QA Engineer judges that the customer complaints data is from a market that is sufficiently similar to his own to allow the confident use of the data; the Testing Engineer judges that certain extra tests will increase confidence by some stated amount; etc. etc.

The necessity for this is that we are in a situation where there is a lack of hard data, but we must take decisions. The only recourse is to use judgement, but it is important that such judgement should be structured and, if possible, numerical.

One method of making a structured judgement on quality/reliability level is based upon a consideration of the documented attributes possessed by a component as shown in Table 1 (from ref.4). The method is to list the quality-evidence i.e. stress report, material certificate, inspection report etc. and then to rank these attributes in their order of importance to quality as seen by the engineers responsible for the work. Scores are given to each and then summated and multiplied by a weighting factor representing the effectiveness with which the contractor who built the component did the work.

In order to relate the quality-scores obtained by this method to a failure rate one must have at least one component-type which has a statistically determined failure rate. Such as pressure vessels designed and constructed to the ASME III codes. Components with the same quality-score as these reference components are judged to have the same order of failure rate.

Clearly this is a subjective method that can only give order of

TABLE 1 ATTRIBUTES SCORES

REF. TO QUAL. PLAN	ATTRIBUTES	COMPONENT / CONTRACTOR / QUALITY LEVEL / WEIGHTED SCORE	PRESSURE VESSEL		VALVE		PUMP	TEMP. TRANS	PRESS TRANS
			A		B		C	D	E
			1	3	1	3	1	1	3
A1	QUALIFICATION AUDIT	10	10	10	10	10	10	10	10
A2	CONTRACT AUDIT	10	10	10	10	10	10	10	10
B1	DESIGN STRESS REPORT	20	20		20		20		
B2	DESIGN CALCULATIONS	10		10		10			
B3	DESIGN-REVIEW REPORT	20	20	20					
C1	ASME MATERIAL	20	20	20	20		20		
C2	MATERIAL CERTIFICATE	10	10	10	10	10	10		
C3	MATERIAL INSP. REPORT	10	10	10			10		
D1	QUALITY PLAN	20	20	20	20	20	20		
D2	BURN-IN/FAULT FINDING	10							10
D3	WELD & WELDER QUALIFICATION	10	10	10	10		10		
E1	ROUTE CARDS	20	20	20	20	20	20	20	20
E2	INSPECTION REPORT	10	10	10	10	10	10	10	10
E3	INDEPENDANT INSPECTORS REPORT	10	10	10			10		
F1	TYPE TEST REPORT	30						30	
F2	PURCHASE EXPERIENCE	10					10	10	
F3	RELIABILITY ANALYSIS	10							
G1	DEVELOPMENT TEST PROGRAMME	20							
H1	SERIES PRODUCT	10				10		10	10
H2	IN-FIELD DEFECT REPORTING	20							20
	SUB-TOTAL		170	160	130	100	160	100	90
	RETROSPECTIVE JUDGEMENT OF CONTRACTOR		0.7	0.7	0.9	0.9	1.0	0.9	0.8
	TOTAL		119	112	117	90	160	90	72

magnitude answers but when there is no hard data available it is very useful in structuring a judgement on whether one item is likely to be more reliable than another. It also focuses attention on the role that information (documented evidence) plays in the recognition of quality/reliability.

If when using this method it is not possible to arrive at a sufficiently low failure rate then the list of attributes may be extended by extra tests and controls, i.e. by providing more of the type of evidence shown in Table 1.

In order to demonstrate the use of the QR matrix let us now consider its application to an in-pile irradiation device. I will assume that this device is basically an assembly that can apply test conditions to some samples; that it utilises thermocouples, flux monitors etc. and that there is an out of pile gas/water circuit which provides the required atmosphere within the device. There is also control equipment to maintain the test parameters and data logging equipment to record the test data.

The first step in the application is to decide upon the required reliability level for the complete system and express this as a failure rate. If we assume that the device must operate for several years then clearly the system failure rate must be better than one per year, so assume 10^{-2} or 10^{-3} failures per year for the system.

Next make an event/fault analysis in order to determine the required failure rate of the individual components of the system. Perhaps some 20% of these will be shown to require a reliability that is better than industrial level and hence require to have the fault rates calculated or estimated.

For items that are susceptible to calculation (e.g. the pressure retaining envelope, support structure etc.) determine the failure rates by the methods described by Carter, or use information from a data bank for valves, pumps etc., taking due care in the use of such data.

Where this is not possible select the appropriate codes, standards, additional tests, etc, in order to construct an attributes matrix in which the levels of quality are judged to be equal along each horizontal axis of the matrix (corresponding to each quality level), and are judged to be of an appropriate relationship along the vertical axes (eg. level 1 is clearly better than level 2, level 2 is clearly better than level 3, etc.).

Past experience is a valuable input to the required data. For example it may be possible to do a survey within your organisation of the failure rates of, say, heaters and thermocouples and, providing that the environment is known with sufficient certainty and the specified manufacturing processes have been faithfully carried out, this could be used to determine the failure rates.

CONCLUSION

The paper has proposed the use of reliability prediction methods, used in the nuclear and aerospace industries, for in-pile irradiation devices. Such devices do not have the public-safety requirements of nuclear reactors and aircraft but they do represent a considerable investment in experimental research, both financially and in research-programme time. It is therefore recommended that reliability and quality assurance methods should be applied to the production of these devices.

REFERENCES

1. Carter, A.D.C., The Interaction Between Some Simple Load and Strength Distributions. Qual. Ass. Vol. 5 No.1 March 1979.

2. Carter, A.D.S., Mechanical Reliability, Macmillan Press Ltd (1972)

3. Pugh S., Quality Assurance and Design: the Problem of Cost Versus Quality. Qual. Ass. Vol. 4 N° 1, March 1978.

4. Matfield R. A Method of Evaluating the Quality of a Component Based Upon its Attributes (to be published in the IQA Journal)

POSTER SESSION

OPERATING EXPERIENCE WITH CAPSULES FOR RPV STEEL IRRADIATIONS WITH LARGE SPECIMEN VOLUME

J. Ahlf, D. Bellmann, H. Martens

GKSS-Forschungszentrum Geesthacht GmbH

2054 Geesthacht, Bundesrepublik Deutschland

ABSTRACT

RPV steel specimens up to 120 mm in thickness are irradiated in rigs behind gamma shields of stainless steel and tungsten in reflector positions of FRG-2, a swimming pool type research reactor operating at 15 MW. To reduce fluence gradients in the specimens the rigs are rotated several times during the irradiation periods.

For fast neutron dosimetry the reactions Fe54(n,p)Mn54 and Ni58(n,p)Co58 are used. The evaluation of the monitor activation is based on two-dimensional DOT-IV calculations.

The temperature distribution in the specimen volume is traced by thermocouples and controlled by electrical heaters placed in front, behind and at the sides of the specimen volume.

By the measures taken temperature and fluence gradients in the specimen volume can be kept in tolerable limits.

1. INTRODUCTION

For a number of older nuclear power plants the embrittlement of the pressure vessel by fast neutrons may become the life limiting factor. For safety analysis based on fracture mechanics the fracture toughness of the RPV steel in the irradiated conditions must be known. According to the peculiarities of fracture toughness determination this necessitates the irradiation of large specimens. As the embrittlement is sensitive to irradiation temperature and neutron exposure (fluence, flux and spectrum), gradients of these influencing parameters must be kept as small as possible.

P. von der Hardt and H. Röttger (eds.), Irradiation Technology, 593–603.

For a large research program of the Federal Republic of Germany the irradiation techniques for this type of investigation has been developped and tested at GKSS in the past few years [1], [2]. The rig design is shown schematically in fig. 1. The specimen volume has a width of 260 mm and a height of 500 mm; the depth can be chosen between 50 mm and 120 mm according to the specimen cariage. The rigs are operated in FRG-2, a 15 MW swimming pool type research reactor. Three rigs can be arranged at the borders of the reactor behind gamma shields as shown in fig. 2.

Gamma shields of stainless steel (60 and 74 mm thick) and tungsten (46 mm thick) are used.

As yet the irradiation experiments listed in table 1 have been performed.

Table 1: List of Irradiation Experiments with Large Specimen Volume Capsules

Rig Designation	Depth of Spec. Vol. [mm]	Gamma Shield	Fluence E>1MeV [$10^{19} cm^{-2}$]	Status
CK50-1	50	SS,60 mm	2	A
CK50-2	50	SS,60 mm	7,5	A
CK60-1	60	SS,74 mm	7,5	B
CK60-2	60	SS,74 mm	2	C
CK120-1	120	SS,74 mm	7,5	B
CK120-2	120	SS,74 mm	2	C
CK100-1	100	Tungsten, 46 mm	2	C

A: irradiation finished, dosimetry evaluated, post irradiation examination in progress

B: irradiation finished, dosimetry in progress

C: under irradiation

2. FLUENCE DETERMINATION

The fluence distribution inside the rig is determined by evaluation of activation wires of stainless steel (X10CrNiNb 18 9), 0,8 mm in diameter. The Mn54 and Co58 activity measured by scanning the wire along a Ge(Li)-detector. In order to reduce fluence gradients in the irradiation direction the rig is revolved several times along its longitudinal axis during the irradiation period. This has to be taken into account for the evaluation, because each point of a horizontal plane in the rig had seen two different neutron spectra, except the vertical (u-w) middle plane.

The neutron field in the specimen volume is remarkably weakened by the gamma shields. Multigroup neutron transport calculations were performed for the different shield and rig configurations with DOT-IV code, using the 100 groups EURLIB-library with 31 groups > 1 MeV. The dosimetry cross sections were taken from ENDF/B V. Up to now complete results of neutron dosimetry are available only for the rigs CK50-1/2. As an example the distribution of the fast neutron fluence E > 1 MeV in the specimen volume of CK50-1 is shown in fig. 3. On the left hand the positions of the wires are marked. The fluence distribution as measured by the monitor wires was approximated in the w-direction by a modified cosine function; the fluence in u-direction was nearly constant, in v-direction described by a square function. The standard deviation of the approximation was small compared with the errors in the fluence determination. Using this distribution formula the fluence in the fracture plane of each specimen can be calculated [3].

3. TEMPERATURE CONTROL

As mentioned before embrittlement is very sensitive to irradiation temperature. Therefore the temperature inside the rig must be measured very exactly and must be constant over the whole irradiation period. Large gradients within the relevant specimen volume must be avoided.

The temperature inside the rigs is traced by a large number of thermocouples (up to 54) destributed in the specimen volume and can be controlled by electrical heaters. Three heaters are arranged at the front side of the specimen volume and three at the back side. By this arrangement it is possible to control the temperature profile in the vertical and in the irradiation direction. It is possible to assemble two additional electrical heaters at the sides of the specimen volume to control the temperature profile to the sides of the specimen volume.

All electrical heaters are controlled in bank by a regulator as shown in fig. 4. As a reference for the regulator one thermocouple is selected. The output signal of the regulator is given by amplifiers to the controller of each heater. The gain of the amplifiers can be adjusted by the experimentator and can be varied from 0 to 2. Fig. 5 shows a computer display of actual temperatures in the specimen volume. Such a computer display is generated every three minutes according to the actual temperature distribution and has proven to be a good tool for the experimentator in order to find the appropriate adjustments for the electrical heaters.

4. TEMPERATURE DISTRIBUTION

The measured temperature distribution in the specimen volume is approximated by the polynom:

$$T(u,v,w) = C_1w^4 + C_2w^3 + C_3u^2 + C_4v^2 + C_5w^2 + C_6uv + C_7uw$$

$$+ C_8vw + C_9u + C_{10}v + C_{11}w + C_{12}$$

wherein u, v, w are the coordinates of the specimen volume as indicated in fig. 6 to 9. The standard deviation of the measured temperatures from the approximated ones was better than ± 4 K for all rigs. The results for the u-w-plane in the middle of the irradiation direction v are shown in fig. 6 to 9 for the capsules CK50-1, CK60-1, CK120-1 and CK100-1. Within the specimen volume only that region is of interest where the crack planes are arranged. These regions are marked on the figures.

The irradiations of the rigs CK50-1, CK60-1 and CK120-1 are finished already. The shown temperature distributions are calculated from values averaged over the whole irradiation time, during which the rigs had been turned several times around the center lines. For the rig CK100-1 which is yet under irradiation the temperature distribution was calculated from instantaneous values.

The pronounced temperature gradient in u-direction in the rig CK50-1 could be avoided in the rig CK60-1 by the above mentioned auxiliary heaters at both sides of the specimen volume. As a consequence all rigs of type CK60 and the rig CK100-1 are now equipped with these side heaters.

The swings in w-direction originate from the three zone heaters before and behind the specimen volumes.

Fig. 10 shows the calculated and measured temperature profile in the irradiation direction v for different capsule types and different gamma-heating rates normalized to an average value of 290 ^{o}C. The calculated curves are obtained from one dimensional calculations for the infinitely extended plate [1].

The gamma-heating rate was measured before and behind a 60 mm steel shield. From this the gamma-heating rate behind the 74 mm steel and behind the 46 mm tungsten shield was estimated to 0.2 W/g and 0.1 W/g.

All rigs of types CK60 and CK120 are irradiated behind 74 mm steel shields and the rig CK100-1 behind the 46 mm tungsten shield. All measured temperature profiles are in the range of the calculated.

As a measure of the amplitude of the temperature profile the standard deviation from the mean value was calculated. In fig. 11 this standard deviation is plotted against the specific gamma-heating rate. The calculated profiles give straight lines in the diagram. For the measured profiles the values are marked at the appropriate lines. It can be seen that the 74 mm steel gamma-absorber shield gives a specific gamma-heating rate less than 0.2 W/g and the 46 mm tungsten gamma-absorber shield a specific gamma-heating rate of 0.1 W/g as expected.

REFERENCES

[1] J. Ahlf, D. Bellmann; Irradiation Techniques for Large Reactor Pressure Vessel Steel Specimens, p. 483 in Aspects of Nuclear Reactor Safety, P. von der Hardt, H. Röttger, Editor, EUR 6612, 1980

[2] J. Ahlf, D. Bellmann, H. Dittmer, H. Martens; An Irradiation Capsule for Reactor Pressure Vessel Steel with a Large Specimen Volume, GKSS 81/E/42

[3] D. Bellmann, J. Ahlf, P. Wille, G. Prillinger; Neutron Dosimetry in Irradiation Capsules for Large Reactor Pressure Vessel Steel Specimens, p. 579 in Radiation Metrology Techniques, Data Bases and Standardizations, FBK. Kam, Editor, NUREG/CP-0029 Vol. 1, CONF-820321/VI.

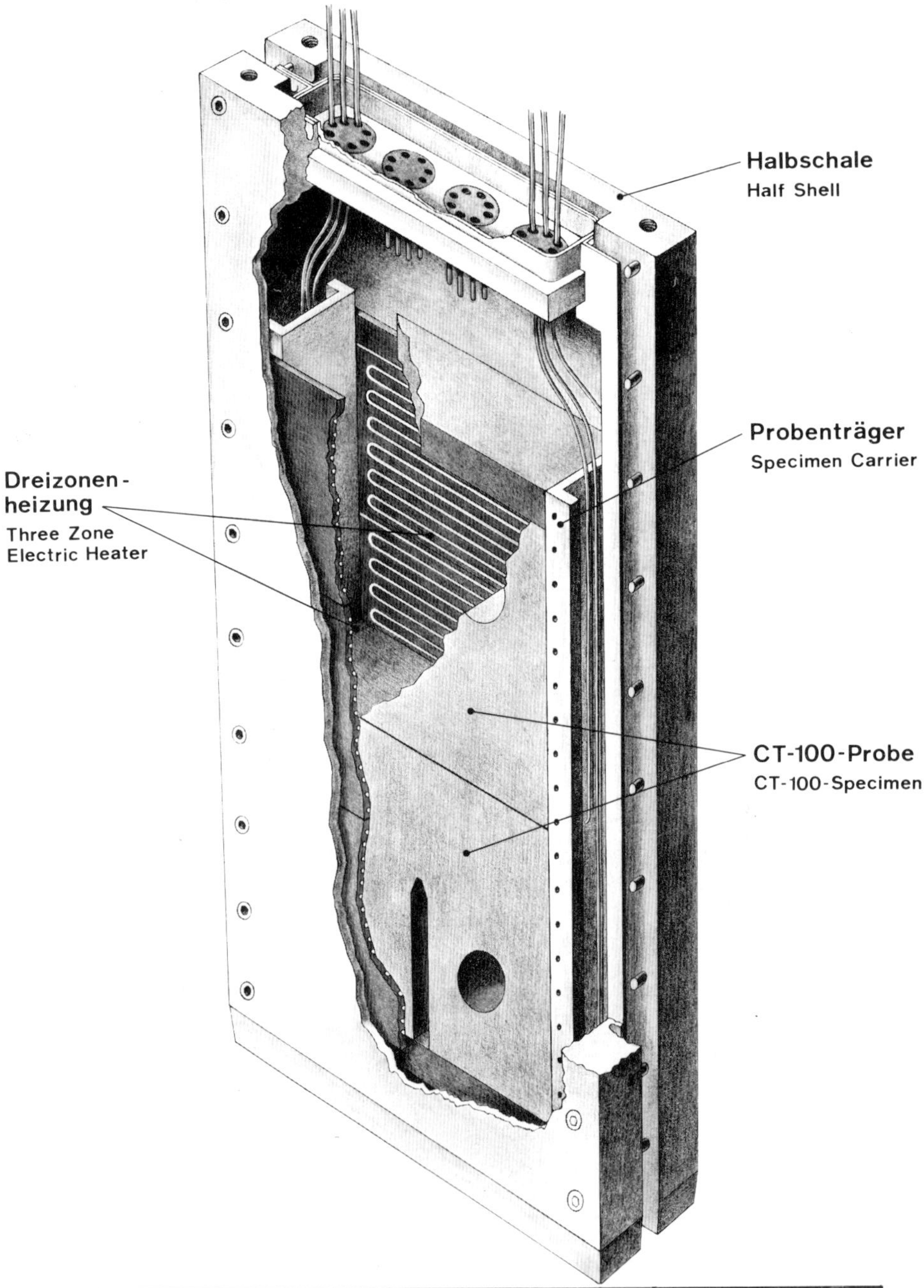

Design of Irradiation Rig	
Fig. 1	GKSS FORSCHUNGSZENTRUM GEESTHACHT GMBH

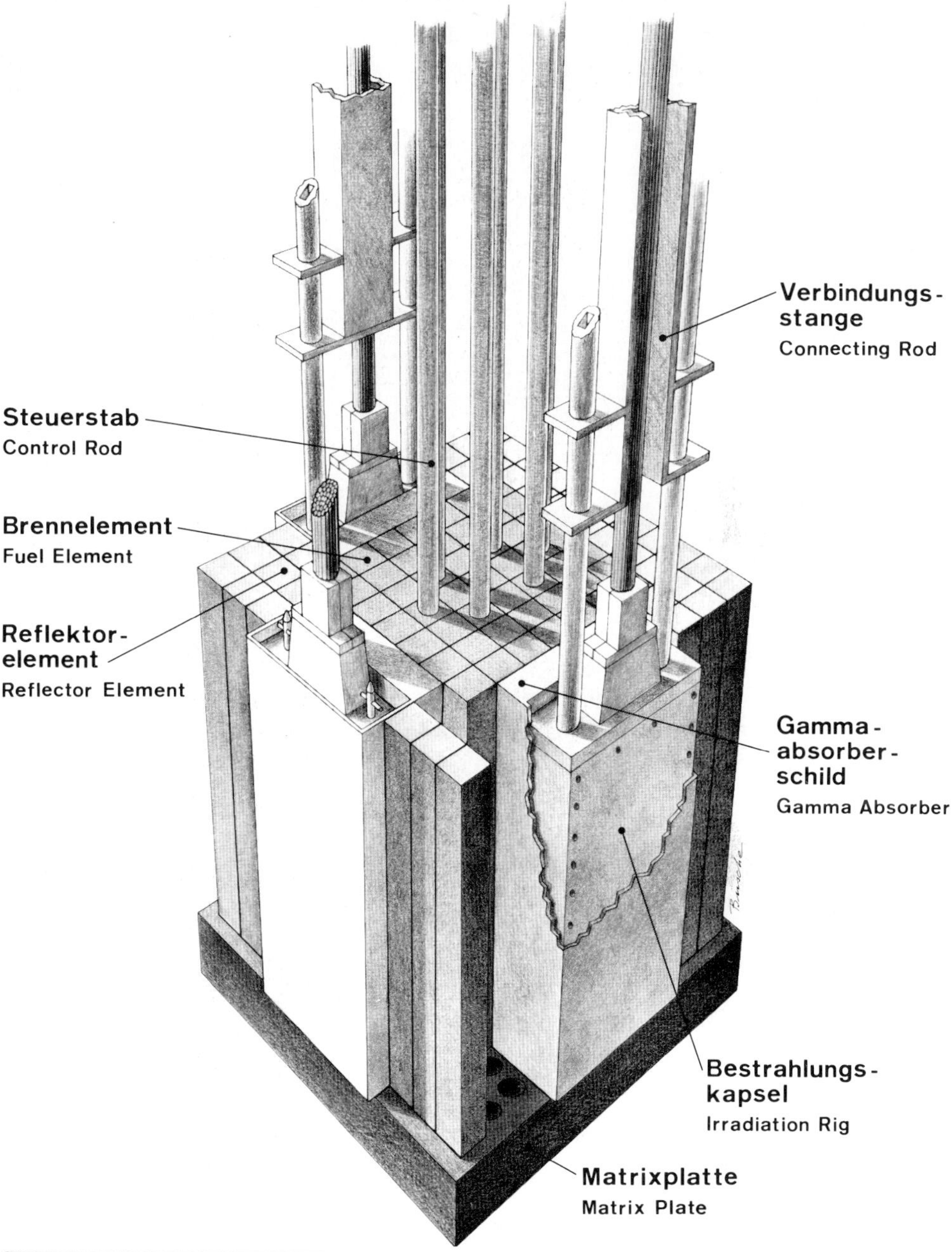

Position of Irradiation Rigs in FRG-2	
Fig.2	GKSS FORSCHUNGSZENTRUM GEESTHACHT GMBH

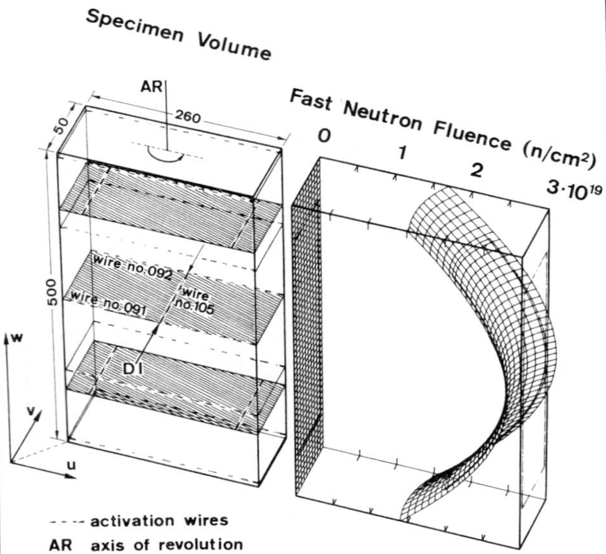

- - - - activation wires
AR axis of revolution
DI direction of irradiation

reaction: $^{54}Fe(n,p)^{54}Mn$

fluence distribution relation:

$\Phi(v,w) = (C_2 v^2 + C_1 v + C_0) \cdot \cos(C_3 \cdot (w - C_4))$, $\Phi(u) = \text{const.}$

standard deviation: $\sigma = \pm 0.07 \cdot 10^{19}\ n/cm^2$

Fast Neutron Fluence Distribution, E > 1 MeV
Rig: CK50-1, v-w-Plane

Fig. 3

GKSS FORSCHUNGSZENTRUM GEESTHACHT GMBH

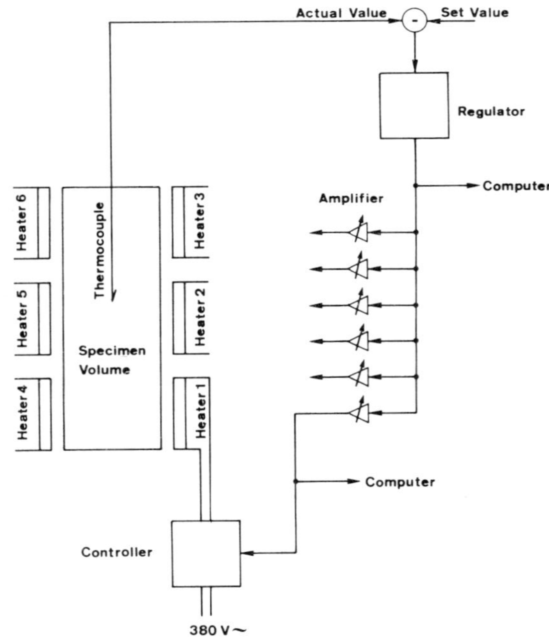

Temperature Control by Electrical Heaters (Principle)

Fig. 4

GKSS FORSCHUNGSZENTRUM GEESTHACHT GMBH

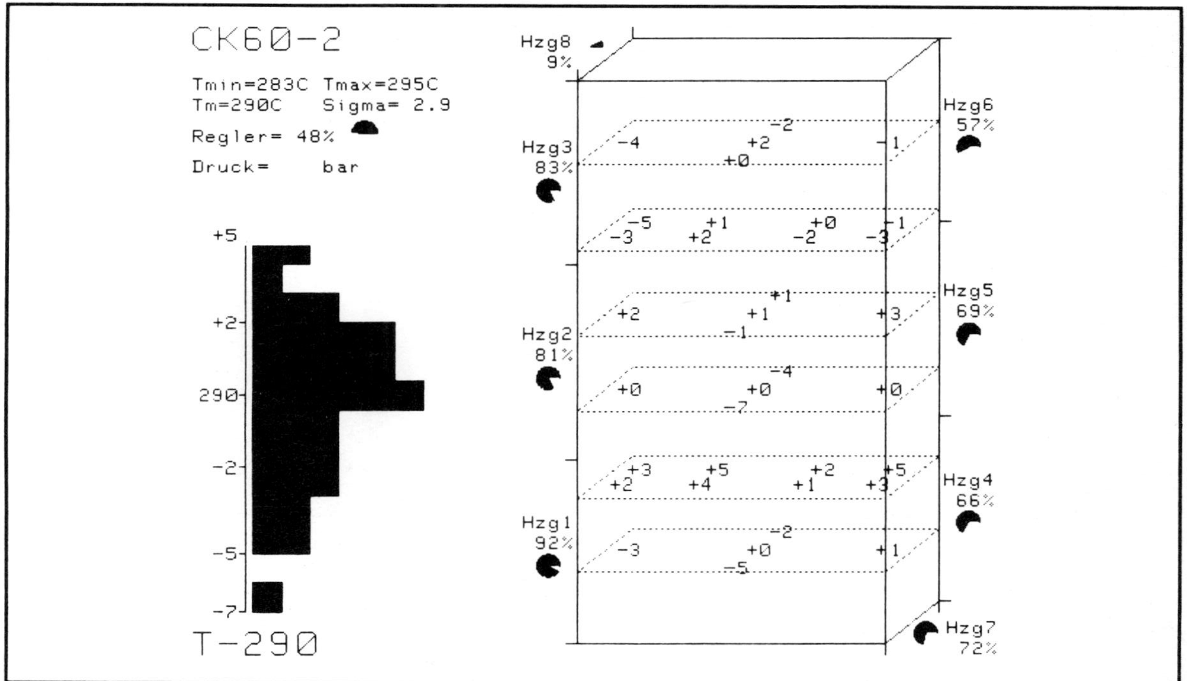

Fig. 5 Computer Display of Actual Temperature Distribution

GKSS FORSCHUNGSZENTRUM GEESTHACHT GMBH

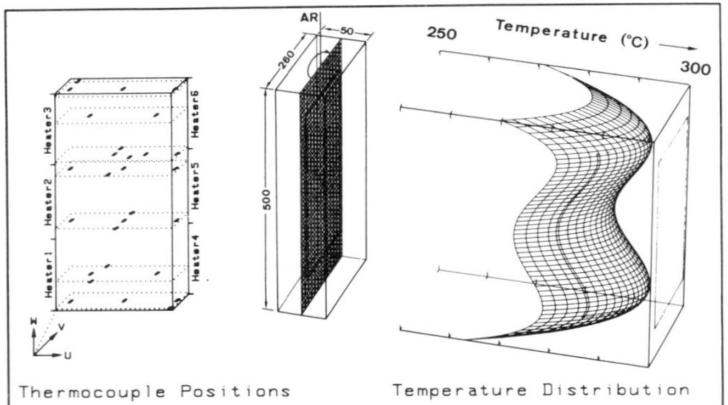

Fig. 6 Temperatures in Rig CK50-1 GKSS FORSCHUNGSZENTRUM GEESTHACHT GMBH

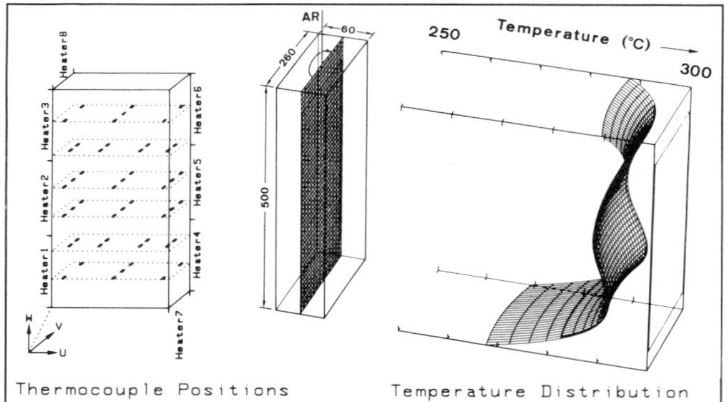

Fig. 7 Temperatures in Rig CK60-1 GKSS FORSCHUNGSZENTRUM GEESTHACHT GMBH

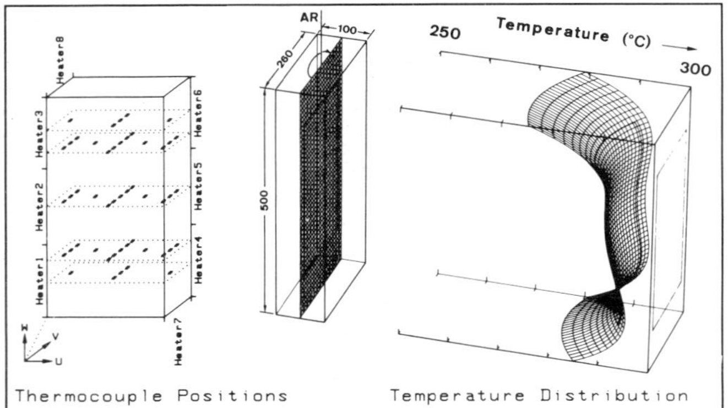

Fig. 8 Temperatures in Rig CK100-1 GKSS FORSCHUNGSZENTRUM GEESTHACHT GMBH

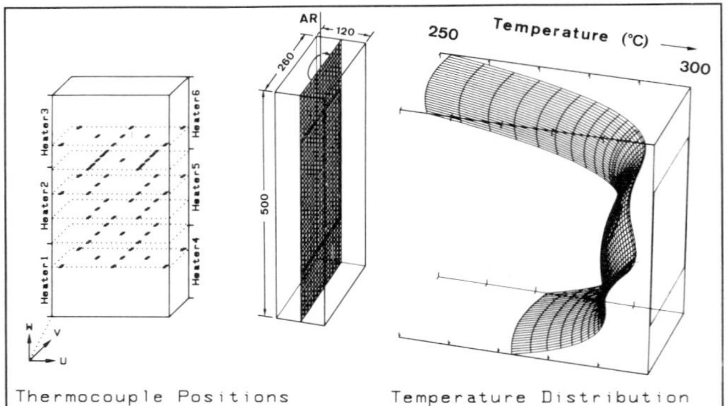

Fig. 9 Temperatures in Rig CK120-1 GKSS FORSCHUNGSZENTRUM GEESTHACHT GMBH

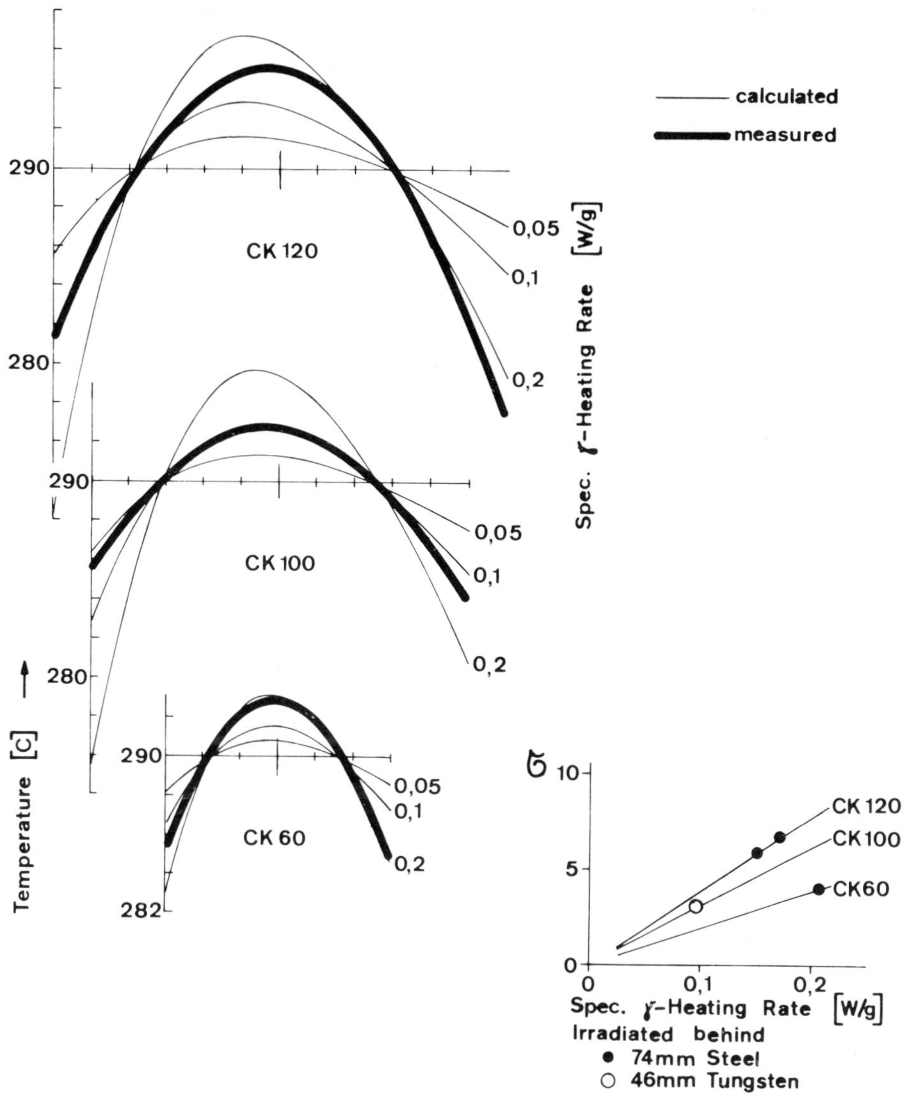

Temperature Profiles in Irradiation Direction

Fig. 10

GKSS FORSCHUNGSZENTRUM GEESTHACHT GMBH

DETERMINATION TOMOGRAPHIQUE DE LA DISTRIBUTION DES EMETTEURS GAMMA DANS UN CRAYON COMBUSTIBLE EN IRRADIATION (PROGRAMME ISARD)

G. DUCROS - F. MICHEL

CENTRE D'ETUDES NUCLAIRES DE GRENOBLE
SERVICE DES PILES/SEREG
85 X - 38041 GRENOBLE CEDEX (FRANCE)

1. INTRODUCTION

La méthode de scrutation des dispositifs d'irradiation par spectrométrie gamma en pile est l'un des moyens non destructifs classique de suivi des expériences en irradiation.

Lors d'une scrutation axiale (le long de l'axe principal de la colonne fissile), une variation d'activité gamma est directement reliée, aux problèmes d'auto-absorption près, à une variation de la quantité de l'isotope mesuré, puisque l'angle de visée délimite un volume de combustible constant selon la position de scrutation.

Lors d'une scrutation transversale, une variation d'activité gamma mesurée est la conséquence d'une variation éventuelle de la quantité de l'isotope mesuré, mais aussi de la variation simultanée du volume de combustible visé (puisque l'angle de visée délimite un volume de combustible variable selon la position de scrutation) ; l'interprétation des mesures est donc rendue beaucoup plus délicate, d'autant plus que le phénomène d'auto-absorption se retrouve ici au coeur même du problème.

L'information recherchée est la distribution de chaque isotope choisi dans une section de l'élément combustible au moyen de la scrutation transversale.

Un seul examen transversal, bien qu'insuffisant pour résoudre rigoureusement ce problème, permet toutefois de déterminer avec une bonne approximation dans un grand nombre de cas, la distribution radiale des concentrations [1], [2]. Deux projections ou plus,

P. von der Hardt and H. Röttger (eds.), Irradiation Technology, 605–614.

effectuées selon des angles d'incidences différents, permettent d'améliorer sensiblement la connaissance de cette distribution. La reconstruction de l'image isotopique à partir de plusieurs projections nécessite la mise en oeuvre d'un algorithme analogue à ceux utilisés en tomographie médicale.

Le code I.S.A.R.D., mis au point au Service des Piles du CEN/G (Groupe de Gammamétrie en pile) effectue cette reconstitution en utilisant une méthode itérative qui tient compte de l'auto-atténuation dans le combustible et de l'atténuation dans les structures du dispositif d'irradiation qui le contient.

2. LE PROGRAMME I.S.A.R.D.

2.1. Choix de la Méthode utilisée

La reconstruction tomographique de l'image isotopique se fait à partir de quelques projections effectuées sous différentes incidences (fig. 1).

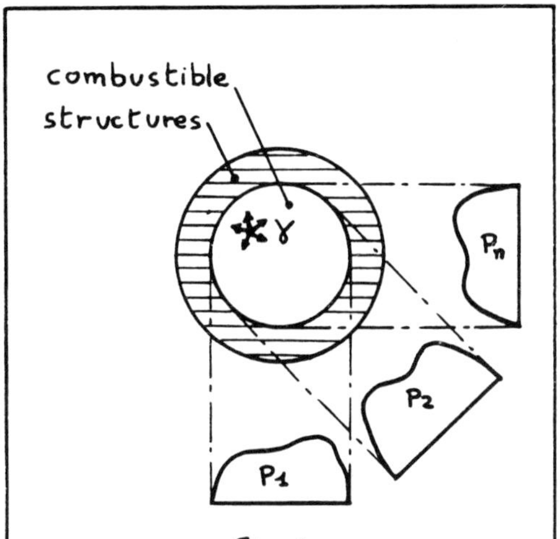

Fig 1

Les contraintes spécifiques imposées par la gammamétrie en pile nécessitent :

- de tenir compte de l'absorption du rayonnement gamma dans les structures du dispositif
- de réduire le nombre de projections, le temps total d'acquisition étant généralement limité par une reprise rapide de l'irradiation.

Plusieurs algorithmes de reconstruction ont déjà été mis en oeuvre par d'autres auteurs. A. SCHWARTZ et R. PINEIRA ont développé dans le code TOMOGAM [3] la méthode itérative des moindres carrés, décrite en détail par T.F. BUDINGER et G.T. GULLBERG ([4] et [5]). J.R. PHILLIPS et B.K. BARNES ([6] et [7]) ont développé quatre méthodes distinctes et étudiés la mieux adaptée à un faible nombre de projections. Le programme I.S.A.R.D. combine l'utilisation successive de 2 méthodes itératives : la méthode algébrique (A.R.T.) et la méthode des moindres carrés (I.L.S.T.) avec facteur d'amortissement.

La méthode algébrique assure une convergence rapide du système ; mais, sensible au bruit de fond de la mesure, elle n'est utilisée que pendant les premières itérations. La méthode des moindres carrés, qui est peu sensible au bruit, achève alors la convergence du système grâce à l'utilisation d'un facteur d'amortissement pour chaque itération.

2.2. Description de la méthode utilisée

L'objectif est de calculer les activités gamma en un point quelconque M (x, y) d'une section d'un crayon combustible, à partir de plusieurs projections de mesures effectuées sous des incidences différentes.

La reconstitution de l'activité gamma est élaborée sur une grille carrée composée de NxN cellules dont la largeur est égale à l'épaisseur de la fente de collimation (fig. 2).

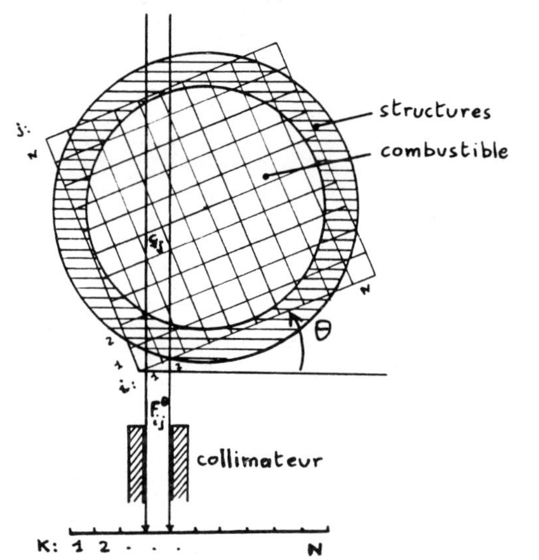

Fig 2

Les notations utilisées sont les suivantes :

C_{ij} : concentration locale du produit de fission émetteur dans la cellule (i,j)

f_{ij}^{θ} : facteur d'auto-atténuation (à travers le combustible et les structures) de la cellule (i,j), vue sous l'incidence θ.

(k,θ) : ensemble des couples (i,j) qui donnent une contribution non nulle au rayon k, d'incidence θ.

$P(k,\theta)$: valeur mesurée du rayon k, sous l'incidence θ ; une projection est constituée de l'ensemble des N rayons situés sous la même incidence.

$R(k,\theta)$: valeur calculée du rayon k, sous l'incidence θ.

$$\text{On a :} \quad R(k,\theta) = \sum_{(i,j)\varepsilon(k,\theta)} C_{ij} \cdot f_{ij}^{\theta}$$

La procédure itérative consiste, connaissant les $C_{ij}^{(n)}$, à la nième itération, à modifier ces valeurs de façon à minimiser l'écart entre les projections calculées et mesurées.

Organigramme - Il est le suivant :

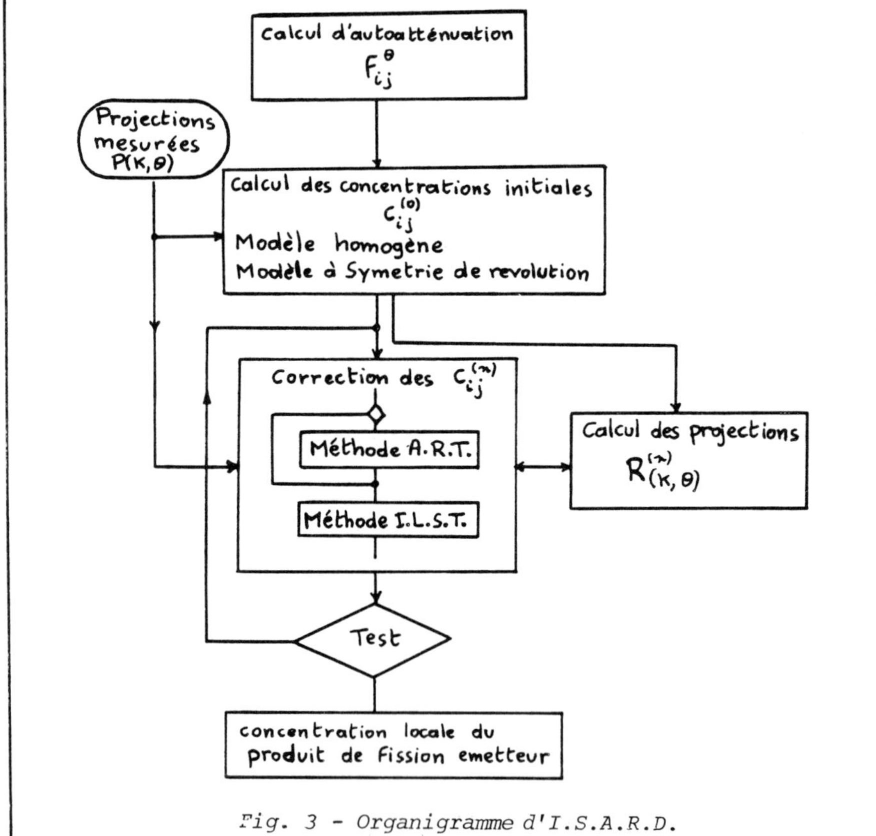

Fig. 3 - Organigramme d'I.S.A.R.D.

Initialisation du Calcul

Deux options peuvent être utilisées pour déterminer les concentrations initiales avant le lancement de la procédure itérative :

- *le modèle homogène consiste à attribuer la même valeur dans toutes les cellules*
- *le modèle à symétrie de révolution calcule les concentrations par couronnes concentriques depuis l'extrémité du crayon jusque vers son centre.*

Géométrie - La géométrie utilisée dans le programme pour calculer les facteurs d'auto-atténuation est composée de zones absorbantes circulaires et concentriques (15 zones possibles) et d'une zone semi-curviligne qui simule l'épaisseur d'eau entre le dispositif d'irradiation et le banc de gammamétrie lors de l'examen en pile (fig. 4).

La zone d'émission utilisée pour calculer les concentrations initiales est circulaire et indépendante des zones d'absorption.

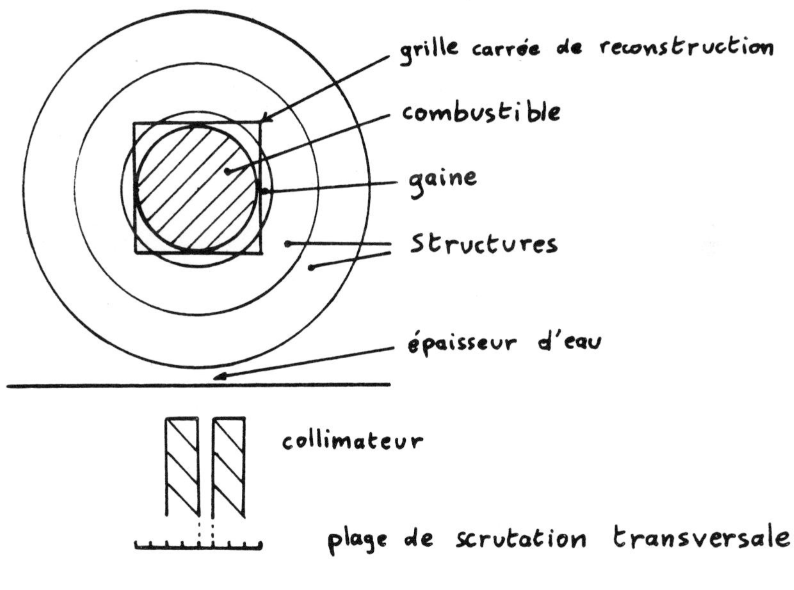

Fig. 4 - Géométrie de calcul

3. EXEMPLE DE RESULTATS

Les exemples suivants, qui se rapportent au cas réel d'un crayon UO2 faiblement enrichi de la filière à eau légère, irradié pendant plusieurs cycles dans SILOE, permettent d'esquisser les types d'interprétation possibles que permet la méthode.

Deux projections de 30 points chacune ont été effectuées selon deux génératrices à 90°.

Dans chacun des cas les concentrations initiales ont été calculées à l'aide du modèle à symétrie de révolution puis 4 itérations (1 ART suivie de 3 ILST) ont été effectuées.

3.1. Cas du Zirconium 95

Images mesurées : figure 5.

Résultats du calcul I.S.A.R.D. : figure 6.

La répartition du Zirconium 95, bon traceur de la matière fissile est uniforme dans la section du crayon combustible. La valeur mesurée de la dépression de flux est ici en bon accord avec le modèle théorique de Runnals utilisé pour ce type de crayon.

Ce résultat montre qu'il n'y a pas eu de mouvement radial de la matière fissile à ce niveau du crayon. Il renseigne également sur l'homogénéité du flux neutronique perturbé pour la position d'irradiation dans le coeur.

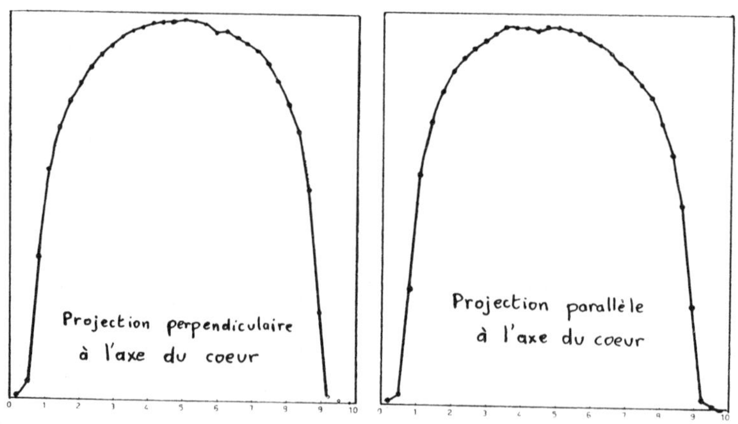

Fig. 5 - Zr 95 - Images transversales mesurées

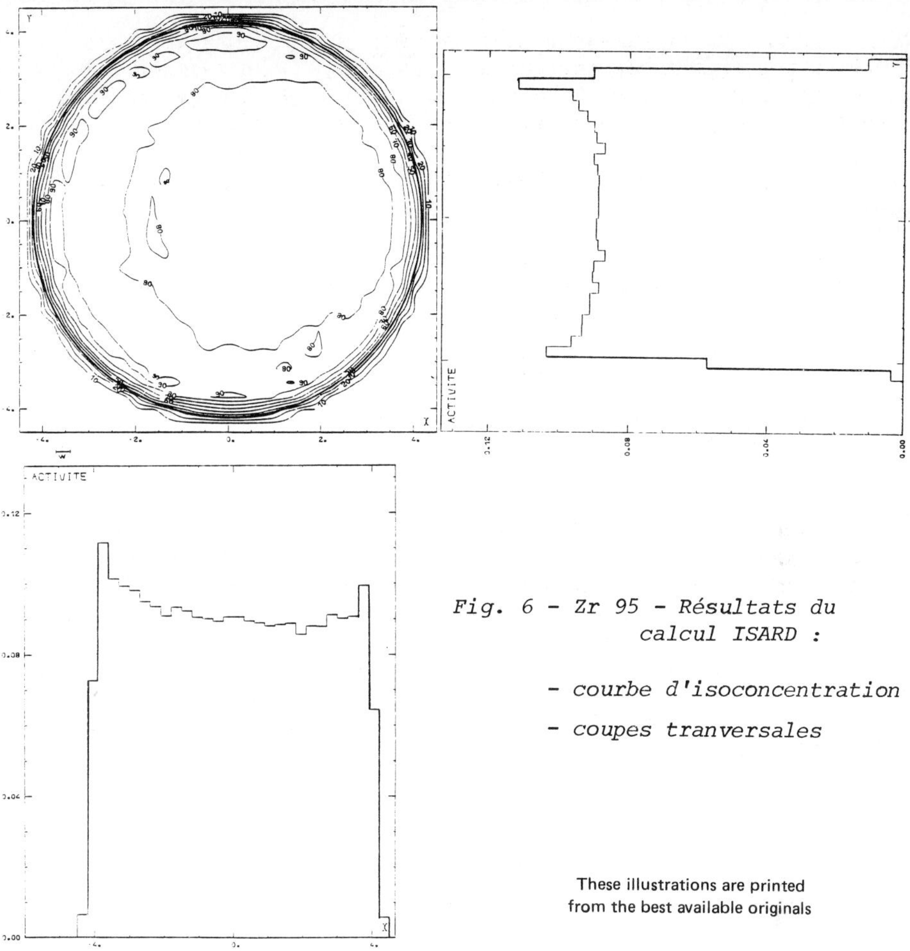

Fig. 6 - Zr 95 - Résultats du calcul ISARD :

- courbe d'isoconcentration

- coupes tranversales

These illustrations are printed from the best available originals

3.2. Cas du Baryum 140 (fig. 7)

La répartition obtenue montre que, pour le Baryum 140, à la distribution radiale des fissions se superpose un effet de migration conduisant à une accumulation vers le centre du crayon.

La comparaison de tels résultats pour divers nucléides, de comportements différents ou voisins, renseigne sur les conditions de température et de sévérité de l'irradiation.

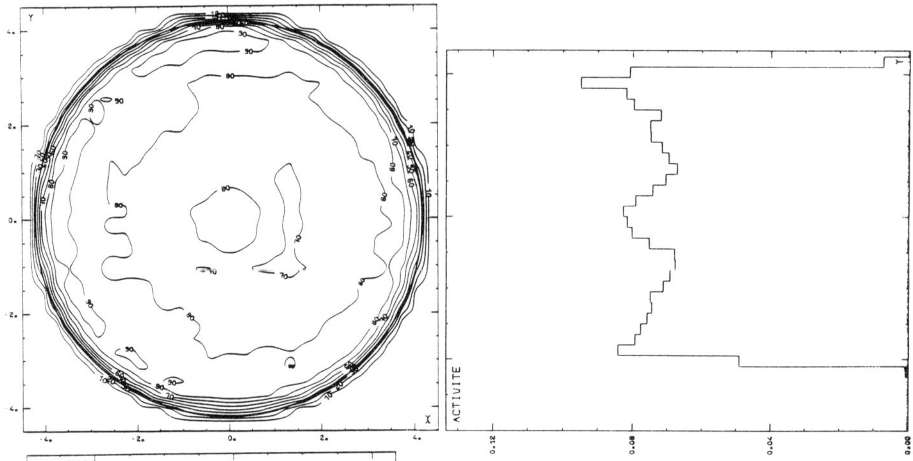

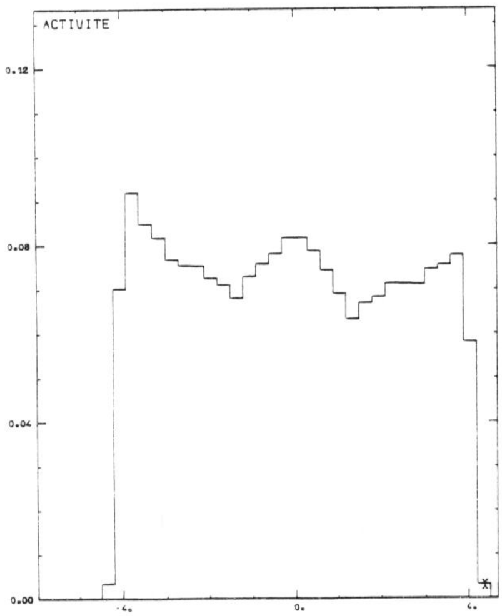

Fig. 7 - Ba 140 - Résultats du calcul ISARD :

- courbe d'isoconcentration
- coupes transversales

These illustrations are printed from the best available originals

3.3. <u>Cas du Césium 137</u> (fig. 8)

Le résultat plus classique observé ici montre une importante migration radiale du Césium, qui s'est réparti à la périphérie de la colonne fissile ; on note dans ce cas la légère dissymétrie diagonale de la distribution, mais aussi l'absence de singularité locale significative.

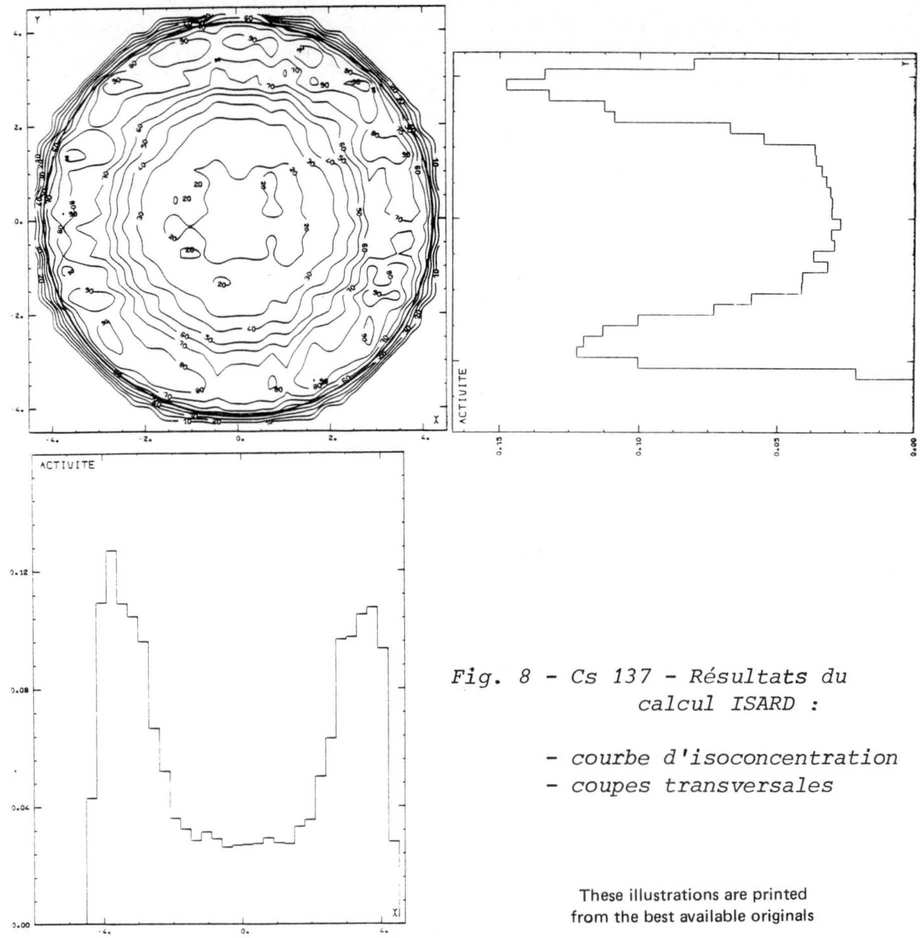

Fig. 8 - Cs 137 - Résultats du calcul ISARD :

- courbe d'isoconcentration
- coupes transversales

These illustrations are printed from the best available originals

4. CONCLUSION

Le contrôle non destructif par scrutation γ des combustibles irradiés tout au long de leur expérimentation en pile avait, dès le début, cherché à couvrir des résultats radiaux, notamment par l'interprétation d'un balayage tranversal sous certaines conditions de symétrie de révolution.

La méthode I.S.A.R.D. permet de s'affranchir de toute hypothèse de ce type et fournit une distribution assez précise des différents produits de fission dans une section du combustible, moyennant un petit nombre de projections mesurées (2 au minimum) compa-

tible avec les contraintes de temps de l'examen et de l'irradiation.

Ces résultats renseignent efficacement sur la distribution du flux neutronique et de la matière fissile, ainsi que sur l'état du combustible et les conditions de l'irradiation.

/ REFERENCES /

[1] - F. MICHEL
"Contribution à l'étude et à la réalisation d'une méthode d'examen par scrutation gamma, en réacteur de recherche, de dispositifs d'irradiation".
Thèse - Grenoble 1975./ CEA - R 4963

[2] - F. MICHEL et M.L. POINTUD
"Méthodes spécifiques de la gammamétrie en pile".
CEA-CEN/G - Pi/SEREG/920-188/79 (1979) - Communication CEA-CEN/G-SPi aux Journées de Spectrométrie γ du LMRI, Paris (mai 1979).

[3] - A. SCHARTZ et T. PINEIRA
"Une méthode de reconstruction bidimensionnelle de la répartition de nucléides dans un combustible nucléaire irradié à partir de scrutations gamma transversales à différentes incidences".
Rapport CEA - N - 2272 (1980).

[4] - T.F. BUDINGER et G.T. GULLBERG
"Three dimensional reconstruction in nuclear medicine emission imaging".
IEEE - Transactions on Nuclear Science - Vol. NS 21, P. 2-20 (1974).

[5] - T.F. BUDINGER et G.T. GULLBERG
"Three dimensional reconstruction in nuclear medicine by iterative Least-Squares and Fourier transform techniques".
UC-48 Biology and Medicine (1974).

[6] B.K. BARNES et J.R. PHILLIPS
"TWODIM, a computer code for unfolding diametral gamma-ray scans on reactor fuel elements".
LA-4676 - Los Alamos Scientific Laboratory (1971).

[7] - B.K. BARNES et J.R. PHILLIPS
"Reconstruction of radial fission product distributions in reactor fuels from a small number of projections".
Conference on characterization and quality control of Nuclear Fuels - Karlsruhe, FRG on 2-5 june 1981.

SUMMARY

Nicrosil-nisil thermocouples in fuel element irradiation experiments

J M Hartog
Springfields Nuclear Laboratories
United Kingdom Atomic Energy Authority

Part 1 - Introduction

In the final stages of operating the Windscale Advanced Gas-Cooled Reactor (WAGR) a series of fuel element experiments were planned to simulate certain transient faults such as loss-of-flow of coolant, or reactivity insertion accidents. The associated peak clad temperature excursions were expected to be in the order of 1350°C, and above the values where adequately reliable readings would be obtainable from the type K[(1)] couples normally used in WAGR to measure fuel cladding temperatures. Nicrosil-nisil thermocouples were specified and used successfully in two fuel element transient experiments. A pilot experiment was also operated for a few months in the WAGR core to investigate any irradiation-induced drift of the couples. They were also checked against Pt/Pt-Rh in a furnace.

(1) ANSI, 1964, C96.1.

Part 2 - Instrumentation for WAGR fuel elements

Instrumented fuel elements for WAGR have regularly been assembled at Springfields Nuclear Laboratories (SNL) and shipped to WAGR for building up into stringers and irradiation. The design mainly used is a special thick cladding tube with a partially penetrating chordal hole so that the sheathed hot junction tip of the thermocouple cable is located within the body of the cladding, Figs 2 & 3. For the normal type K thermocouple the diameter of the sheath of 1 mm (0.040 in.) is reduced to 0.5 mm (0.020 in.) prior to insertion in the cladding. The cable sheath alloy is identical to that used for the fuel cladding (20/25/Nb stainless steel). Depending on the axial location of the hot junction, the total length of the cable run to terminals on the reactor top plate can be up to 17 m (55 ft). This length of cable has a number of bends, Fig. 4, and is eventually twisted with other cables to form a rope, so that adequate ductility was required, together with ability to accept thermal expansion strains. This virtually ruled out the refractory metal W/Re alloys, while Pt/Pt-Rh (Type S) was not obtainable in the required length of sheath. The timescale on which the experimental programme was mounted and geometrical limitations did not allow development of an in-reactor junction between different thermocouple types.

P. von der Hardt and H. Röttger (eds.), Irradiation Technology, 615–622.

There is no previously published use of nicrosil/nisil under irradiation, but a quick appraisal of its likely behaviour suggested that its thermo-electrical characteristics, Fig. 1, in a neutron flux should be as stable as the extensively used Type K couples. In January 1981 a pilot experiment to check this was loaded into WAGR with 2 nicrosil/nisil and 2 type K thermocouples attached to fuel pins. The type K thermocouples were the standard type normally used to measure cladding temperatures, but for nicrosil/nisil specification the cable diameter was increased to 2.0 mm (instead of 1 mm) and the hot junction tip was reduced to 1.0 mm (instead of 0.5 mm). For applications with lengths of cable exposed to high temperature coolant this diameter increase has the benefit of allowing a thicker sheath (i.e. more oxidation corrosion acceptable) and increased diameter conductors (i.e. less loss of signal due to ohmic resistance).

In WAGR the interpretation of thermocouple signals is normally carried out automatically by a data logger and it was necessary to incorporate a sub-routine programme for calibration data for the nicrosil/nisil couples which has been published by the National Bureau of Standards, Washington DC and physically checked for each of the WAGR couples in the quality assurance schedules at several fixed temperatures up to the melting point of silver (960°). No problems were experienced in adjusting the data processing equipment although the apparent reluctance of individuals or organisations to make this change seems to be one reason why nicrosil/nisil has not been used more frequently in irradiation experiments and in other out-of-reactor applications in the nuclear industry. It is in the temperature range $1000-1370^{o}C$ where Type K couples experience drift due to oxidation of aluminium in one conductor and chromium migration in the other that nicrosil/nisil has its particular merit and for which its inventors intended it.

The WAGR pilot experiment (IE 1512) comparison of nicrosil/nisil with Type K showed minimal difference between the two thermocouple types in recording fuel pin temperatures over the 3 months operation before the reactor was finally shut down. In order to check any flux tilt effects on fuel pin temperatures the fuel element was rotated after 35 full power days. Allowing for flux tilt effects the measured temperatures were similar within 2° to $4^{o}C$, Fig. 5.

In the event the final specifications and mode of operating the two main transient experiments with coolant flow reduction and power increase did not call for extended operation at the peak cladding temperatures reached in the range $1250^{o}-1300^{o}C$ recorded by the thermocouples, Fig. 6. As there were Type K thermocouples on some fuel elements these experiments also allowed a comparison between the thermocouple types. The use of the two different types gives increased confidence in the accuracy of the cladding temperatures measured at the hot junction positions in the instrumented pins.

This is the first known use of nicrosil/nisil thermocouples in a reactor application and the experience has been wholly encouraging. All the couples installed operated satisfactorily throughout the experiments, and their emf readings appeared uninfluenced by neutron flux. Credit is due to the high grade manufacture and quality assurance by the supplier (SODERN, Paris); to skilled installation by staff of the Springfields and Windscale Nuclear Laboratories of the UK Atomic Energy Authority.

IE 1512: Flux level Fast (> 1.35 MeV) 1.33×10^{13} $n/cm^2/s$

Fluence 6.6×10^{19} n/cm^2

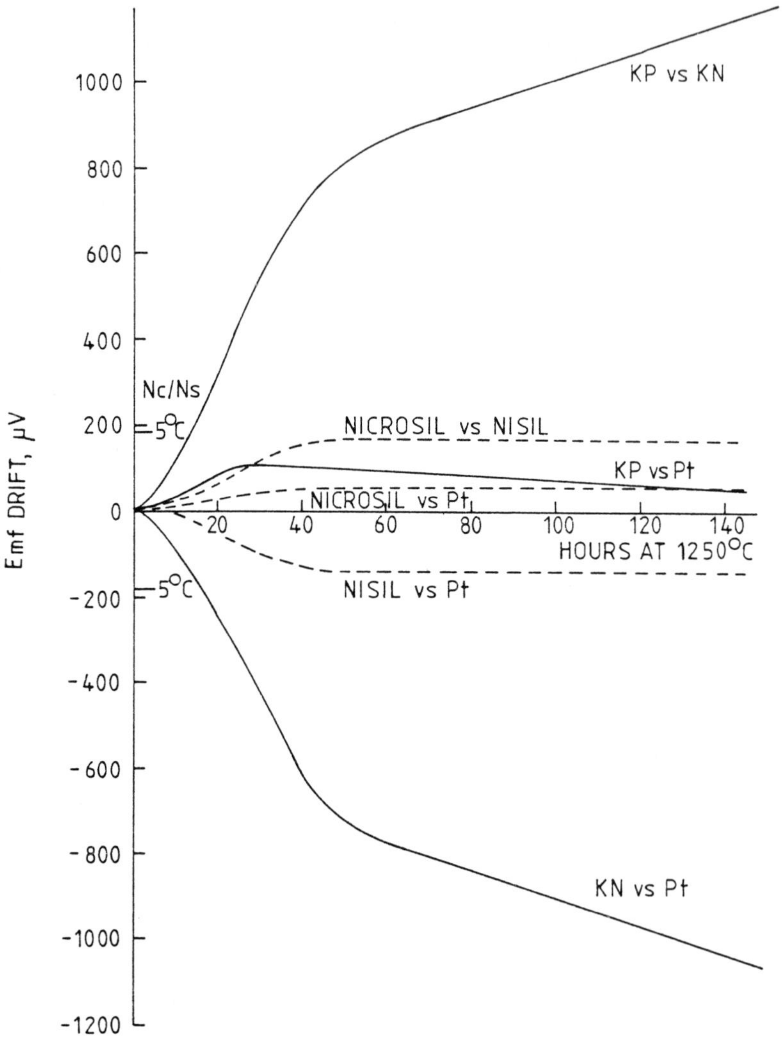

FIG.1. SHORT TERM CHANGES IN NICROSIL / NISIL AND TYPE K THERMO-ELEMENTS IN A CONSTANT TEMPERATURE OF 1250°C

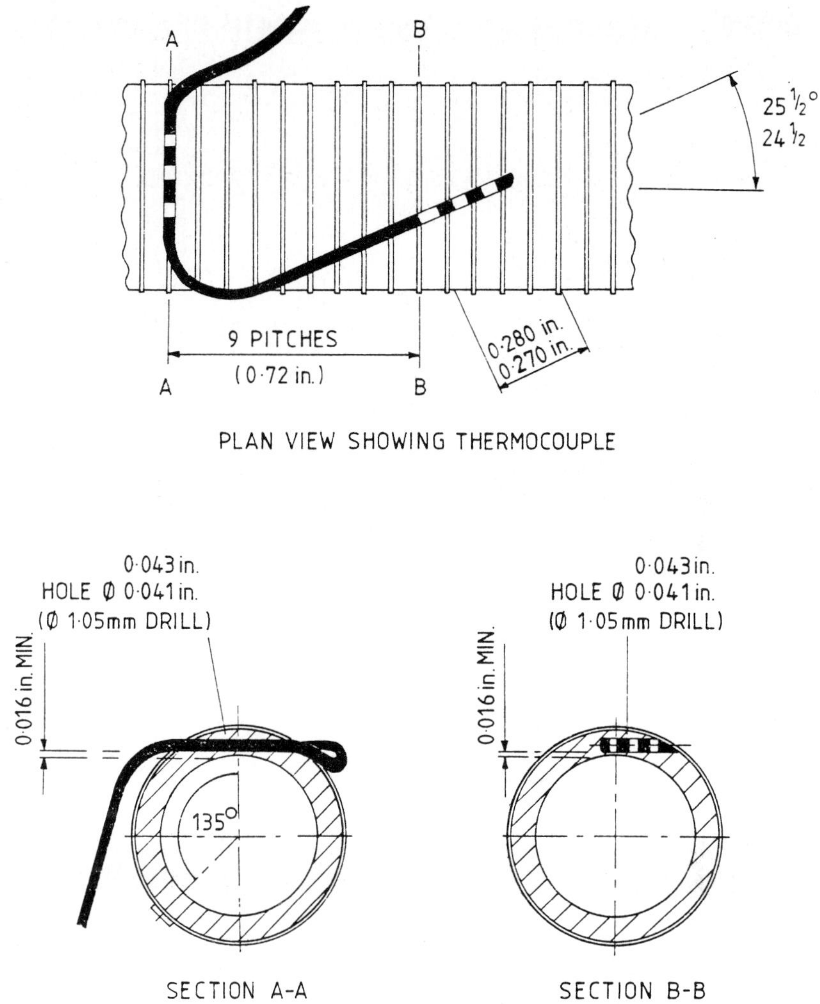

FIG. 2. PRINCIPAL DIMENSIONS FOR 1mm DIAMETER THERMOCOUPLE POCKET AND LOCKING HOLE WITH THERMOCOUPLE IN POSITION

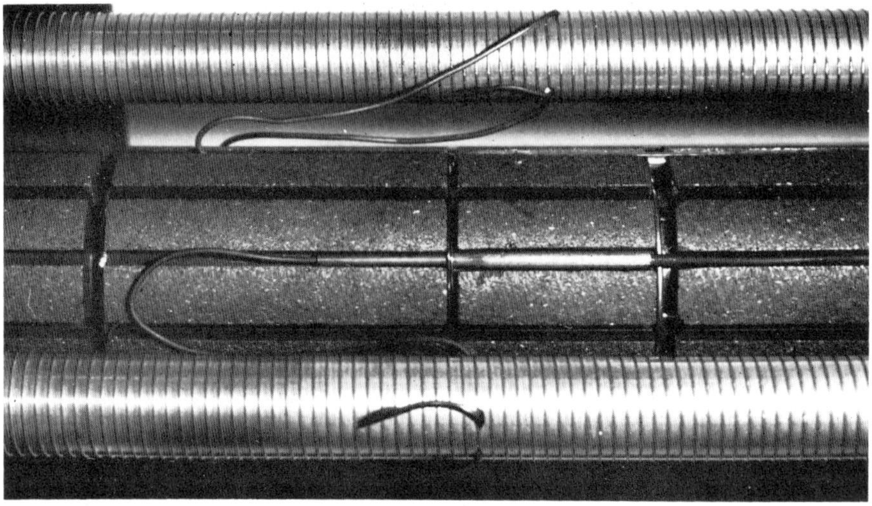

FIG.3. 1mm dia. THERMOCOUPLES MOUNTED IN FUEL PIN AND TIED TO CENTRAL GRAPHITE SPINE ALSO SHOWS S.S. SPLINT AT INCREASE FROM 1mm TO 2mm DIAMETER

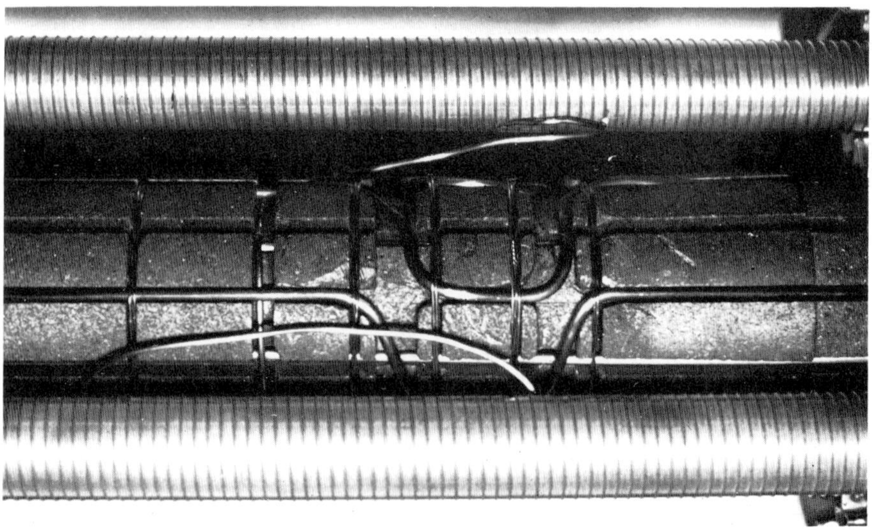

FIG.4. NICROSIL/NISIL THERMOCOUPLE EXPANSION LOOPS

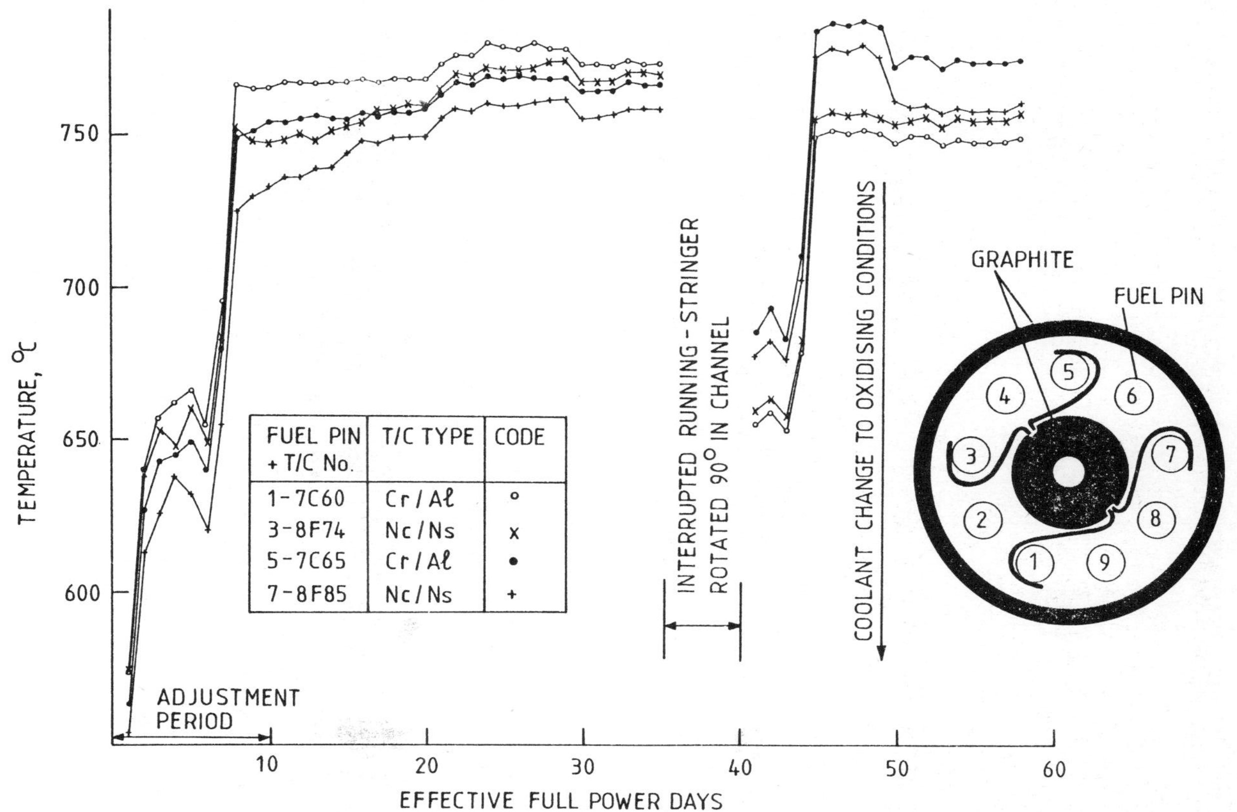

FIG.5. COMPARISON OF STANDARD WAGR 1mm ⟶ 1/2 mm CHROMEL/ALUMEL WITH 2mm ⟶ 1mm NICROSIL/NISIL THERMOCOUPLES

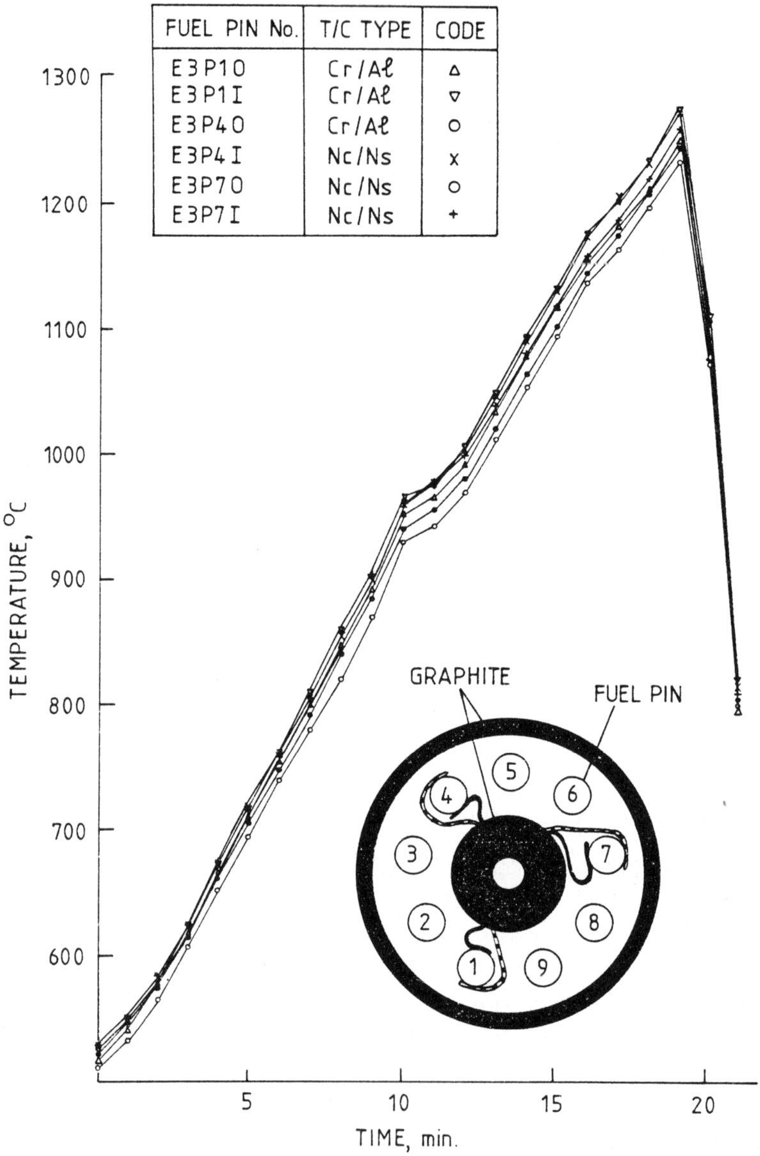

FIG.6. POWER TRANSIENT DURING CONCLUDING EXPERIMENTS IN WAGR

- D I A N E -

A SPALLATION NEUTRON SOURCE OF FUNDAMENTAL RESEARCH IN WEST GERMANY

(ADDENDUM TO THE POSTER SESSION CONTRIBUTION OF THE GERMAN SPALLATION NEUTRON SOURCE PROJECT) *

G.S. BAUER, H. STECHEMESSER, G. THAMM
KERNFORSCHUNGSANLAGE JÜLICH GmbH
D-5170 JÜLICH
FEDERAL REPUBLIC OF GERMANY

ABSTRACT

The spallation process, in which copious numbers of neutrons are freed from heavy nuclei by particles impinging with energies around 1 GeV has recently attracted considerable interest for neutron production for both research and industrial applications. The facts, that nuclear spallation is a driven reaction rather than a self-sustaining one and that the amount of heat dissipated per useful neutron is much lower than in nuclear fission have a number of important advantages, especially when designing a neutron source for beamhole research: The primary source of fast neutrons can be very compact, leading to high neutron densities and the particle beam driving the reaction can have almost any desired time structure. This enables one to design neutron moderators according to the experimental needs such as suppression of background, separation of orders from monochromators and time-of-flight experiments. The fact, that in the slowing-down regime, the width of the neutron pulse in time is small and inversely proportional to the neutron velocity, and that large numbers of these neutrons leak from a sufficiently small moderator, makes it possible to extend the useful neutron energies into the eV-range, are relatively low spallation neutron sources built or under construction so far, are relati-

* This is a preprint of a paper intended for presentation at a scientific meeting. Because of the provisional nature of its content and since changes of substance or detail may have to be made before publication, the preprint is made available on the understanding that it will not be cited in the literature or in any way be reproduced in its present form. The views expressed and the statements made remain the responsibility of the named author(s); the views do not necessarily reflect those of the government of the designating Member State(s) or of the designating organization(s).

P. von der Hardt and H. Röttger (eds.), Irradiation Technology, 623–645.

vely low time average flux facilities and are optimized for utilization of the time structure, a study performed for the German Ministry for Research and Technology (BMFT) by the two nuclear research centres at Jülich and at Karlsruhe was designed to provide a facility which can be competitive with a high flux reactor also with respect to time average neutron flux. This machine, driven by a 1.1 GeV proton beam with 100 Hz repetition rate and delivered in pulses of 100 mA peak current and0.5 ms duration would yield a time average thermal flux of 7.10^{14} and a peak flux of $1.3.10^{16}$ $cm^{-2}s^{-1}$, when using Pb or W as target material. When combined with a proton pulse compressor capable to deliver 0.8 us long pulses of 66 A, the peak flux is expected to rise as high as 7.10^{16} $cm^{-2}s^{-1}$. The target is designed in such a way that another important option has been retained: the use of uranium as target material. Depleted uranium would double the neutron flux relative to Pb and the question, to what extent a multiplying target based on LEU can be used with respect to heat generation has yet to be looked into in detail. By a suitable distribution inside the target to give an even heat load, a further substantial gain is expected.

INTRODUCTION

Many experiments at high flux research reactors using chopper techniques for high energy resolution are wasting a great protion of the available neutrons. Therefore pulsed neutron sources have been considered as an attractive alternative to even optimised reactor systems for many years. Pulsed neutron sources are capable to generate very high fluxes (pulses) at just the moment when they are required. In addition, DIANE is designed to deliver a high time average neutron flux. Among pulsed neutron systems a spallation neutron source comprising a particle (proton) accelerator and a neutron producing (heavy metal) target is superior to pulsed reactor systems for many reasons:

- Almost any desired time structure of the neutrons can be easily realised by particle beam
- Criticality problems are avoided
- The heat dissipation per spallation neutron is much lower than for nuclear fission neutrons
- High neutron fluxes are available also in the epithermal energy range
- Other experimental applications using protons, pions or other secondary particles are possible
- Irradiation facilities can be operated at various proton energies and with fast and thermal neutrons.

The rapid and impressive development of research reactors has reached a limit at a thermal neutron flux level of one or a few times 10^{15} $cm^{-2}s^{-1}$ as achieved in the High Flux Reactors at Oak Ridge, Brookhaven and Grenoble. There is general agreement that this limit will be difficult to overcome by a substantial amount because of the heat density in the reactor core. Because of this limitation and in view of the high standard of modern proton accelerator technology, when faced with the question of how to provide a successor for its presently operating research reactors, the German Ministry for Research and Technology asked the two research centres at Jülich and Karlsruhe to carry out a joint study to assess the feasibility and future prospects of a high power spallation neutron source. A short version of the report of this study /1/ was published in June 1981. Apart from a lower heat dissipation per neutron in the spallation target relative to a reactor core, this new type of neutron source offers an unusual degree of freedom in its choice of target and moderator design /12/.

THE SPALLATION PROCESS

Nuclear spallation occurs when a heavy target nucleus is hit by a medium energy particle such as a 1 GeV proton (Fig. 1). Due to its high energy and correspondingly small de Broglie wavelength this proton interacts with individual nuclei inside the nucleus, setting off a cascade of nucleon-nucleon collisions, the intranuclear cascade. In the course of this cascade, some energetic nucleons may escape from the nucleus and eventually hit another nucleus, thus giving rise to an internuclear cascade. The remaining nucleus is left behind in a highly excited state from which it returns to the ground state by "evaporating" off further nucleons. In the case of uranium, high energy fission may occur after evaporation of only a few nucleons, with further nucleons being evaporated from the fission fragments. On top of that, fast fission in U-238 (and thermal fission in natural uranium) may further enhance the neutron yield, but also the heat per neutron. The neutron yield per proton and the heat per neutron as a function of neutron energy are shown in Fig. 2.

A sophisticated system of computer programs to calculate the neutronics of spallation targets and moderators /3/ has been implemented at KFA Jülich, based on the work by Armstrong and Chandler /4/ and Atchison /5/ and including the MORSE transport code. In parallel, a large number of experiments have been performed to corroborate the theoretical assessments /13/.

THE CONCEPT OF THE NEUTRON SOURCE

Apart from opening up a new route to lead beyond the capabilities of the high flux reactors at the long run, the neutron source considered in the above study was to meet the following near-term design goals:

- Provide a time-average thermal neutron flux of more than $6 \cdot 10^{14}$ $cm^{-2}s^{-1}$,
- Have a foreseeable potential for further improvements,
- Serve thermal and cold neutron scattering as well as thermal and fast neutron irradiation facilities,
- Show an increased performance level for neutron scattering experiments by a pronounced time structure,
- Allow the option of exploiting epithermal neutron energies for neutron scattering experiments.

The accelerator designed to drive this source is a LINAC of 1.1 GeV final proton energy operating at 5 mA time average and 100 mA peak current and at a repetition rate of 100 Hz giving a pulse width of 500 μs /1/, /6/, /7/.

It is a unique feature of a spallation neutron source that the apparently conflicting above requirements can be reconciled by accommodating moderators of different characteristics around the target. This is, because no chain reaction needs to be maintained and even burnup is no concern. Nevertheless, the first of the above requirements means, that a proton beam power of 5.5 to 6 MW is needed. This leads to a power deposition of the order of 6 kW/cm^3

at the hot spot of a stationary non fissile target, as can be deduced from Fig. 3. If uncooled, a lead target would heat up by 4000 K/s at this position.

THE TARGET DESIGN

In order to be able to dissipate the high heat load in the target, a design as shown in Fig. 4 was selected. It consists of a large diameter (2.5 m) disc, 10 cm high, which is hit by the pulsed proton beam from the accelerator at its outer circumference and which is rotating at such a speed that consecutive pulses hit adjacent areas on the outer window. The resulting speed for a 4 cm diameter beam is 4 m/s or roughly 0.5 Hz. The internal structure of the disc consists of some 9000 individually clad target cylinders of 10 cm height and 18-24 mm diameter (see Table 1) which are cooled by water flowing between them. The water enters and leaves the target wheel through the hub. A few of the target elements are designed as tie-rods to avoid bulging of the structural material under the coolant pressure. Some relevant data are compiled in Tables 1 through 6 for Pb as target material /14,15/.

A calculated distribution of leakage neutrons for the target in operating position and for lead as target material is shown in Fig. 5. Obviously, the source is a very compact one with a FWHM along the proton beam with 25 cm and about 17 cm in the lateral direction.

THE MODERATOR ARRANGEMENT

As noted before, the various conflicting requirements to the performance of the source were reconciled by choosing two different moderator concepts above and below the target as shown in Fig. 6. The very compact source of primary neutrons is ideally suited for a small hydrogeneous moderator placed below the target. It yields a high instantaneous flux with good time structure. In order to improve the performance of this moderator, it is surrounded by a Pb-reflector /8/ and given a grooved surface with 6 cm deep and 1 cm high grooves /9/. This moderator serves 12 beam tubes.

The natural division of the space by the wheel into an upper and a lower half is taken advantage of to install a D_2O-tank above the target, which provides a large volume of high flux to accomodate irradiation positions and a cold neutron source. The neutron flux distribution expected in this tank is shown in Fig. 7. The thimble of the low temperature irradiation facility (not shown in Fig. 6) is designed such that it can either be located in a region of high thermal flux or lowered into the Pb-shield next to the target for fast neutron irradiation. The cold neutron source, which is located in the centre of the D_2O-tank, will be viewed by 12 neutron guides and 2 beam tubes.

The target-moderator arrangement will be surrounded by about 6 m of heavy shielding to stop the energetic cascade neutrons. Heavy rotatable shutters will allow to close off individual beams safely, while the source is running (see Fig. 8).

PERFORMANCE ASSESSMENT

Based on extensive theoretical and experimental investigations, the performance data listed in Table 1 have been deduced. With a time average thermal neutron flux of $7 \cdot 10^{14}$ $cm^{-2}s^{-1}$ the source is considered competitive to existing high flux reactors. On top of that, the time structure of the thermal neutron flux, which is shown in Fig. 9, will allow a much more efficient exploitation of the source than a continuous flux does. In fact, the peak flux is higher than in any other neutron source for beam hole research presently planned or operating (see Table 7). This constitutes a unique opportunity for time of flight experiments and white beam experiments. Also, it has been shown for a number of examples /10/, that by specially designed advanced instruments, this time structure can be exploited with great advantage.

POSSIBLE FUTURE IMPROVEMENTS

In Fig. 10, a comparison is shown between the peak flux of the spallation neutron source target station DIANE (Deutsche intensitätsmodulierte Anlage für Neutronen-Experimente) and the time average flux of the HFR Grenoble. The facility is referred to as intensity modulated bacause of the long pulse of 500 µs throughout the whole energy range as determined by the accelerator. This does not allow to exploit the slowing down region of neutron energies (0.1 to 1 eV). In this regime, intrinsically short pulses can be obtained for very short proton pulses. For this reason a proton pulse compressor /11/ has been studied which could be added to the source to yield proton pulses of only 0.7 µs duration. With this option, gain factors of the order of 300 would be obtained in the peak flux for epithermal energies, as shown in Fig. 10.

Another important option is the use of uranium as target material. Measurements and calculations have shown that by replacing Pb by depleted uranium in the target wheel, a factor of 2 in neutron flux can be gained, at the same time sharpening the neutron pulse in the D_2O-tank (Fig. 9b). Further improvements are possible, if some more fission is allowed. This will, of course, result in an increased heat deposition in the target but, as can be seen from Table 8, the peak energy deposition ratio between natural uranium and lead is only 3.2, whereas the total energy deposition increases by a factor of 6.1. In particular, there is only little difference in peak deposition between U-238 and natural uranium. This is, because most of the thermal fission occurs outside the proton beam area. Thermal fission also has a broadening effect on the leakage distribution from the target as can be seen from Fig. 11a and b. Since the heat deposition caused by the proton beam is very uneven (Fig. 3), there exists a possibility to distribute fissile material in the target volume in such a way as to level off the heat load within each target element and also among the individual target elements. No optimization has been carried out in this respect so far, but it is clear, that at most a low enrichment in fissible material would be needed to carry the time average neutron flux of the spallation source far beyond the HFR level, without harming the time structure. Even when run with such a target, the facility would always be non-critical and hence require no shut down or controll equipment. This is, why wee feel that, in a spallation neutron source, LEU is beautiful.

REFERENCES

/1/ G.S. Bauer, H. Sebening, J.E. Vetter and H. Willax (eds)
"Realisierungsstudie zur Spallations-Neutronenquelle"
report Jül-Spez 113 and KfK 3175 (1981)

/2/ G.A. Bartholomew and P.R. Tunnicliffe (eds)
"The AECL-Study for an Intense Neutron Generator"
report AECL 2600 (1966)

/3/ T.W. Armstrong, P. Cloth, D. Filges and R.D. Neef
"Theoretical Target Physics Studies for the SNQ Spallation Neutron Source"
report Jül-Spez 120 (1981)

/4/ T.W. Armstrong and K.C. Chandler
"HETC - A High Energy Transport Code"
Nucl. Sci. Eng. 43, 353 (1971)

/5/ F. Atchison
"A theoretical Study of a Target Reflector and Moderator Assembly for SNS"
report RL-81-006 (1981)

/6/ J.E. Vetter, ed.
"The Basic Concept of the SNQ Linear Accelerator"
report KfK 3180B (1981)

/7/ G.S. Bauer and J.E. Vetter
"The German Project for a High Power Spallation Neutron Source for Fundamental Research"
to be published in Proc. of an IIASA workshop on "A Perspective on Adaptive Nuclear Energy Evolutions: Towards a World of Neutron Abundance"

/8/ G.S.Bauer, H.M.Conrad, H.Spitzer, K.Friederich and G.Milleret
"Measurement of Time Structure and Thermal Neutron Spectra for Various Target-Moderator-Reflector Configurations of an Intensity-Modulated Spallation Neutron Source"
in Proc. of the 5th Meeting of the International Collaboration on Advanced Neutron Sources (ICANS-V)
to be published in the Jül-Conf series

/9/ G.S.Bauer
"Summary on a discussion on moderators with grooved surface"
in Proc. 4th Meeting of the International Collaboration on Advanced Neutron Sources (ICANS-IV)
report KENS II, p. 423 (1981)

/10/ B.Alefeld
"Some new Instruments Designed to Use the Time Structure of a Quasi Continuous Spallation Neutron Source"
in Proc. 4th Meeting of the International Collaboration on Advanced Neutron Sources (ICANS-IV)
eport KENS II, p.678 (1981)

/11/ G.Schaffer, ed.
IKOR, An Isochronous Pulse Compressor Ring for Proton Beams
Report Jül-Spez. 114 (1981)

/12/ G.S.Bauer
General concept for a spallation neutron source in der Fed.Rep.of Germany
Atomkernenergie 41/4 (1982)

/13/ P.Cloth, D.Filges, R.Hecker
Nuclear Assessment of the DINA target station
Atomkernenergie 41/4 (1982)

/14/ J.F.Stelzer
Behaviour of the spallation neutron source target wheel under thermal and mechanical loads
Atomkernenergie 41/4 (1982)

/15/ H.Stechemesser
Mechanical design of the target station DIANE of the German spallation neutron source project and its rotating target
Atomkernenergie 41/4 (1982)

Target concept	rotating solid target
Target material	Lead
Coolant	H_2O
Moderator concept	Hybride consisting of a D_2O tank and a reflected H_2O moderator
Mode of operation	100 Hz time structure
Time average neutron fluxes	
in D_2O tank	$7 \cdot 10^{14}$ cm^{-2} sec^{-1} (unperturbed)
at H_2O moderator	$7 \cdot 10^{14}$ cm^{-2} sec^{-1} (equivalent flux)
Peak neutron fluxes	
in D_2O tank	$5.5 \cdot 10^{15}$ cm^{-2} sec^{-1} (unperturbed)
at H_2O moderator	$1.3 \cdot 10^{16}$ cm^{-2} sec^{-1} (equivalent flux)
FWHM of thermal neutron pulses	
H_2O moderator	515 μs
D_2O tank	850 μs
Number of beam tubes	
thermal	12
subthermal	2
Diameter of beam tubes	10 cm
Number of neutron guides (subthermal)	12
Thickness of the biological shield	6 - 6.5 m
Experimental areas	main target hall (14 beam holes) 40 m Ø neutron guide hall (12 neutron guides) 40 x 50 m^2 target top hall (irradiation facilities) 14 m Ø 2 neutrino halls

Tab. 1 Main parameters of the target station

Diameter of the rotating target	2.5 m
Frequency of revolution	0.5 sec^{-1}
Target design	9304 individually canned target elements
Number of target elements serving as tie-bars	232
Depth of target	70 cm
Height of the target elements	10 cm
Diameter of the target elements	18 - 24 mm
Wall thickness of the Al canning	0.5 mm
Minimum coolant gaps	1 mm
Average materials composition in the target	Pb 76.5 %, H_2O 16.5 %, Al 7 %
Thickness of the upper and lower support plate	25 mm
Number of coolant channels per support plate	130
Thickness of the rotating beam window	5 mm
Structural material	Al containing 1 % Mg, 0.6 % Si, 0.25 % Cu, 0.2 % Cr
Total weight of target wheel	4600 kg
Total lead	3350 kg

Tab. 2 Mechanical layout data of the rotating target

Proton current	$3.12 \cdot 10^{16}$ p/sec
Proton energy	1100 MeV
Neutron production in the target	29.4 n/p
Neutron leakage from the target	
E > 15 MeV	1.8 n/p
E < 15 MeV	25.3 n/p
Maximum neutron current density at the target surface	$3 \cdot 10^{-2}$ n/cm^2 proton
Charged particle leakage from the target	
Protons	0.12 p/p
Pions	$3.5 \cdot 10^{-2}$ π/p
Muons	$1.0 \cdot 10^{-4}$ μ/p
Saturated activity of the target	2.3 MCi
Saturated activity of the coolant	34 kCi
Decay heat at shutdown (after infinitely long irradiation)	38 kW
Maxmimum decay heat density in the target	1 W/cm^3

Tab. 3 Nuclear and neutronic data of target

Total heat dissipation in target	2.9 MW
Coolant temperature at entry	50 ^oC
Coolant flow	180 m^3/h
Temperature rise in coolant	14 K
Coolant pressure	3.5 - 3 bar
Maximum heat density in the target during proton pulse	120 kW/cm^3
Maximum temperature rise per pulse in the target	42 ^oC (per revolution)
Maximum temperature in the target material	123 ^oC
Maximum heat density in the aluminum during the proton pulse	47 kW/cm^3
Maximum temperature of the target canning	91 ^oC
Maximum temperature rise per pulse in the canning	24 K (per revolution)
Maximum temperature of the rotating beam window	74 ^oC
Maximum temperature rise of the rotating beam window	11 K
Maximum heat load on the surface of the target elements	15 W/cm^2
Maximum reference stress in the canning	75 N/mm^2
Maximum reference stress in the rotating beam window	75 N/mm^2

Tab. 4 Thermomechanical data of the rotating target

Proton beam power	5.5 MW
Power dissipation in the target	2.9 MW
Maximum energy density in the target	120 kW/cm^3 during pulse
Maximum decay heat after shutdown	1 W/cm^3
Local power dissipation	
Structural material of the rotating target	308 kW
D_2O moderator	292 kW
Bottom of D_2O tank	72 kW
H_2O moderator	16 kW
Reflector (computed for Be)	143 kW
Support unit	∿ 9 kW
Shielding	407 kW

Tab. 5 Energy dissipation in the target and the moderators

Maximum damage rate	0.01 dpa/day
Maximum H production rate	1.6 appm/day
Maximum He production rate	0.6 appm/day

Tab. 6 Radiation effects in the rotating target (Al canning and Al window)

	HFR (ILL)	IBR II	DIANE
Peak thermal flux [$cm^{-2}s^{-1}$]	10^{15}	10^{16}	1.3×10^{16}
Average thermal flux [$cm^{-2}s^{-1}$]	10^{15}	7.5×10^{12}	7×10^{14}
Puls rep. rate [s^{-1}]	-	5	100
Pulse duration [µs]	-	150	500
Number of beam holes	14 + 11 N.G.	14	14 + 12 N.G.
Fuel resp. target material	U-235 (HEU)	Pu	Pb (W)
Mode of operation	critical	periodically super-critical	non-critical
Coolant	D_2O	Na	H_2O
Average thermal power	57 MW	4 MW	2.9 MW
Peak thermal power	57 MW	6000 MW	58 MW
Moving Part	-	Reflector	Target
Extensions	2nd cold neutron source	Electron Induction Linac → source pulse 7 µs	Compressor Ring → source pulse 0.7 µs U-Target → increase flux

Tab. 7 Comparison of some data for research neutron sources, the HFR at ILL Grenoble, the pulsed reactor IBR II at Dubna and the target station DIANE of the spallation source presented here.

	Pb	U-238	Natural U
1. Total Deposition in Target (MeV)			
- HET (a)	542	945	945
- MORSE (b)	29	1140	2540
- Total	571	2085	3485
2. Total Deposition/1100 MeV	0.52	1.90	3.17
3. Total Deposition Relative to Pb	1.0	3.7	6.1
4. Peak Deposition (MeV/cm^3-proton)			
- HET (a)	1.20	2.45	2.45
- MORSE (b)	≈0	0.55	0.76
- Total	1.20	3.00	3.21
5. Peak Deposition (kW/cm^3), for I = 100 mA	120	300	321
6. Peak Deposition Relative to Pb	1.0	2.5	2.7

(a) From the effects of high-energy particle transport (all charged particles and neutrons > 15 MeV)

(b) From the effects of low-energy ($\leq$ 15 MeV) neutron- and γ-transport

Tab. 8 Comparison of calculated heating data for different target materials in the SNQ reference design target.

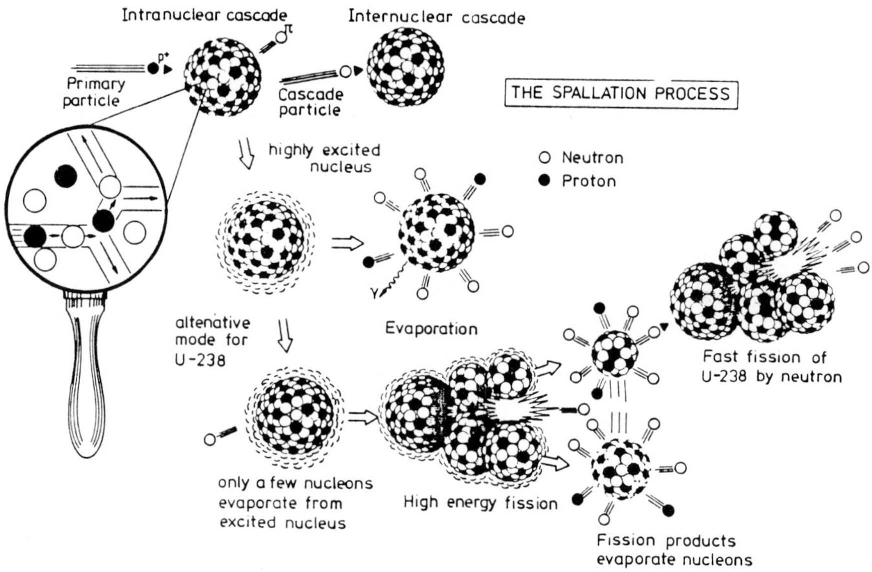

Fig. 1 Schematic representation of the spallation process as set off by a medium energy proton striking a heavy nucleus. About 5-6% of the neutrons released are cascade particles with energies up to that of the incident particle. The spectrum of the evaporation neutrons is similar to a fission spectrum.

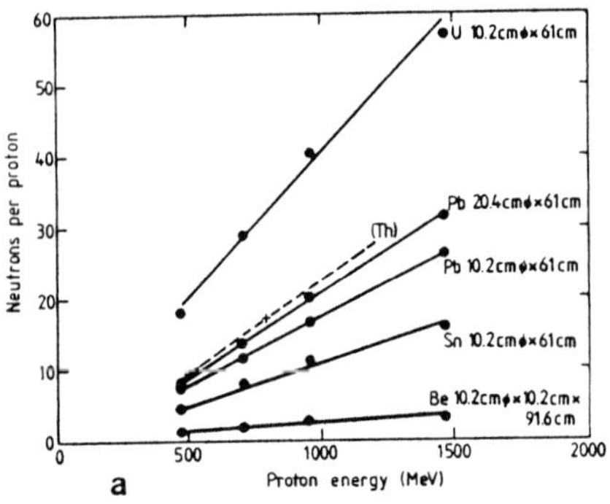

Fig. 2 a) Neutron yield per incident proton for various target materials as a function of proton energy (after /2/).

b) Heat deposition in a spallation target per neutron as a function of proton energy (after /2/).

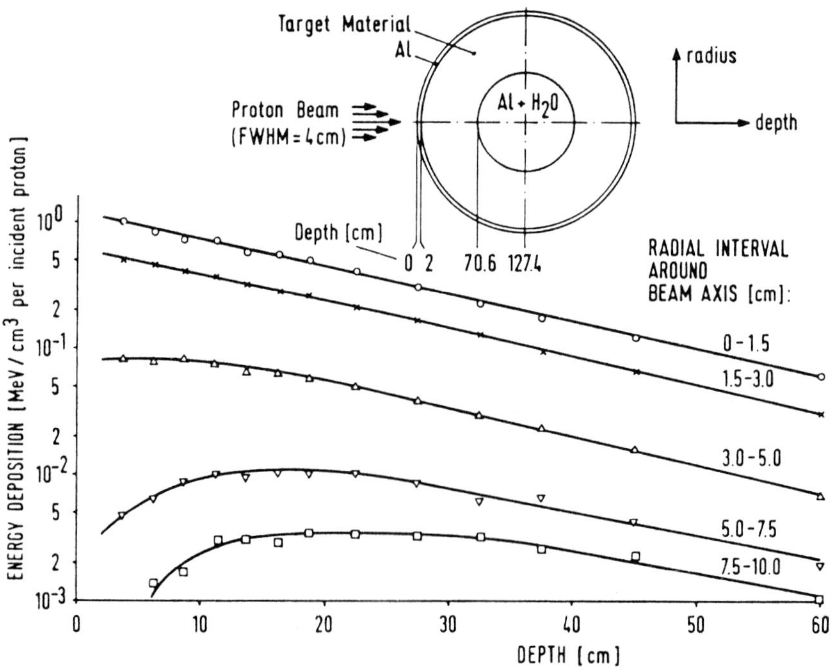

Fig. 3 Calculated power density in the target as designed for the SNQ, with lead as target material. The beam current is 5 mA, leading to peak power density of 6 kW/cm^3.

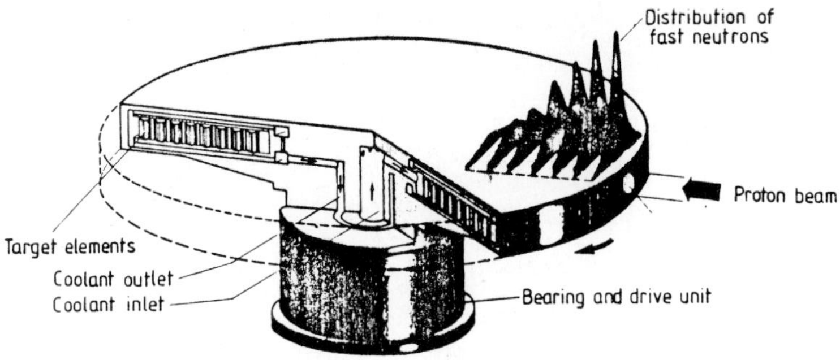

This illustration is printed from the best available original

Fig. 4 Schematic representation of the SNQ rotating target. 9304 individually clad cylindrical target elements are arranged over a depth to stop the proton beam completely (70 cm) around the circumference of a 2.5 m diameter wheel. The target is rotating at 0.5 Hz and the target elements are cooled by water flowing between them. The distribution of leakage neutrons (cf. Fig. 5) is indicated at the upper surface.

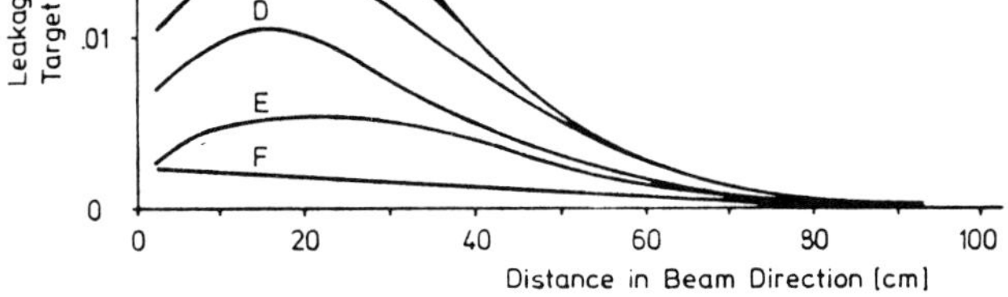

Fig. 5 Calculated distribution of neutrons leaking from the surface of the rotating target.

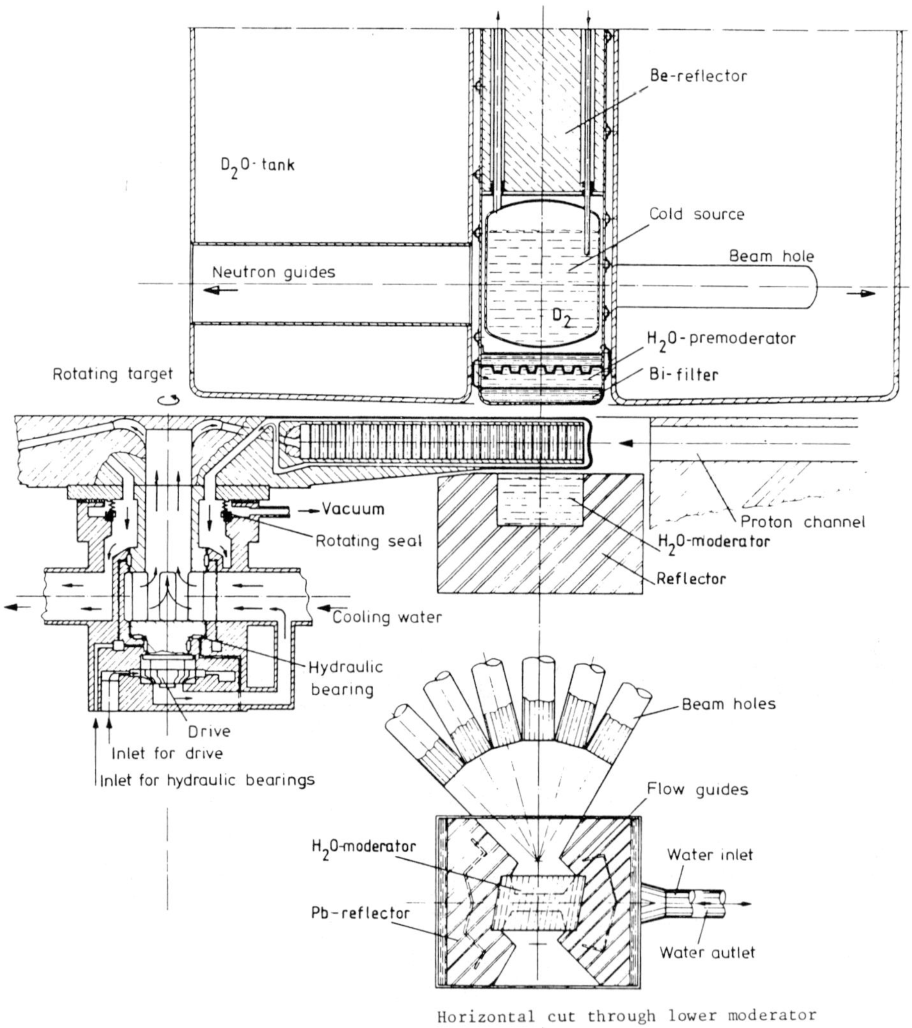

Fig. 6 Arrangement of moderators at the target in the SNQ-concept. A Pb-reflected H_2O-moderator with grooved surfaces below the target is designed for high thermal neutron leakage with good time characteristics. The D_2O-tank above the target provides a large volume of high flux to accomodate a cold neutron source and various irradiation thimbles.

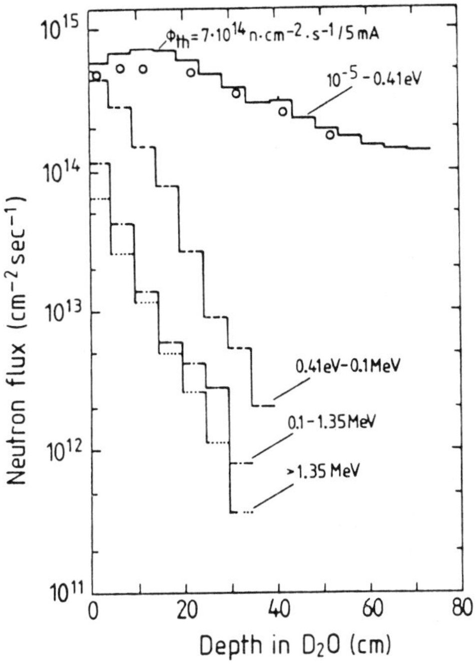

Fig. 7 Calculated and measured (o) neutron flux distributions in the D_2O-tank (Pb-target).

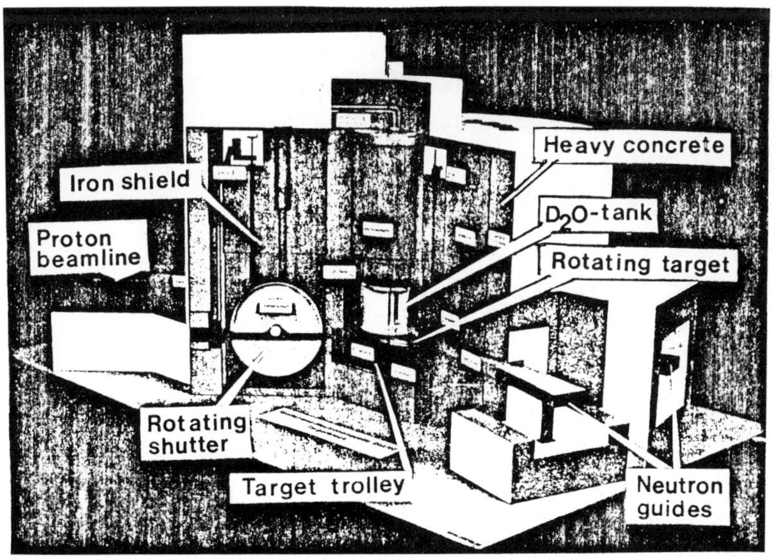

Fig.8 Cutaway view of the target block (model). This illustration is printed from the best available original

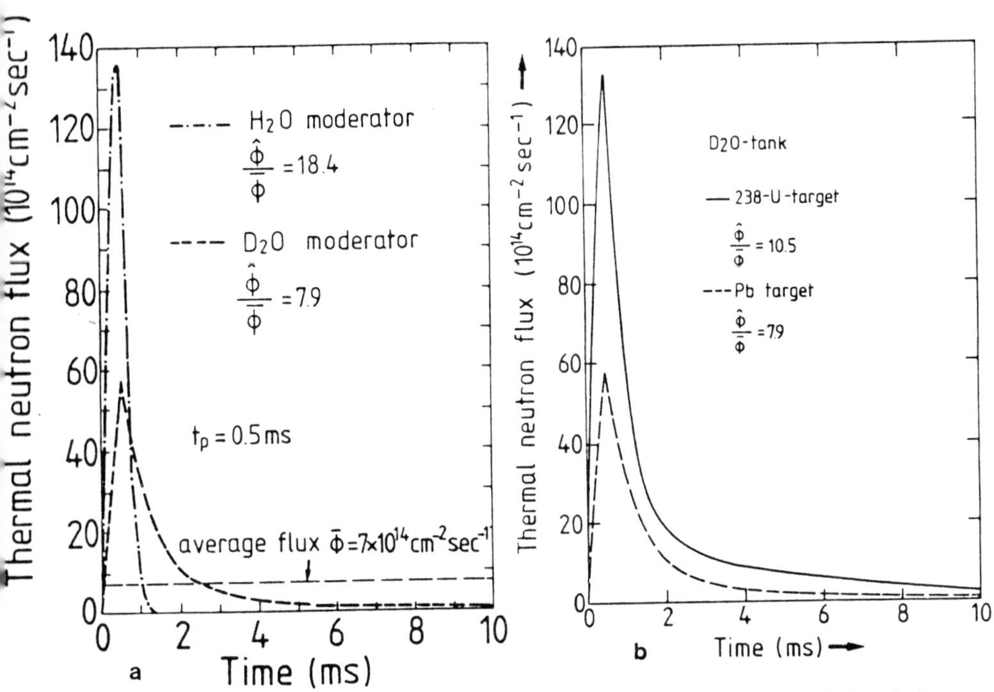

Fig. 9 Time-dependent thermal neutron flux (4π-equivalent flux deduced from leakage measurements in beam tubes) for the moderators of the SNQ-target station DIANE.
a) with lead target for both moderators
b) with target of depleted uranium for D_2O-tank.

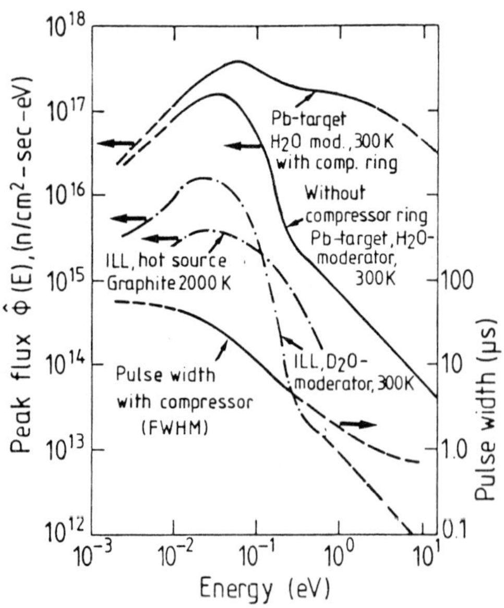

Fig. 10 Peak flux per unit energy as a function of neutron energy without and with a proton pulse compressor. For comparison, the time average flux for two moderators of the HFR Grenoble is also shown.

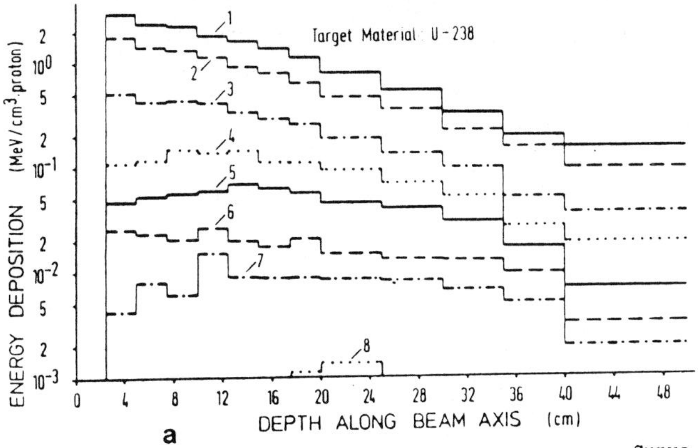

a

Curve	Radial Interval about Beam Axis (cm)
1	0 - 1.5
2	1.5 - 3.0
3	3.0 - 5.0
4	5.0 - 7.5
5	7.5 - 10.0
6	10.0 - 15.0
7	15.0 - 25.0
8	25.0 - 56.8

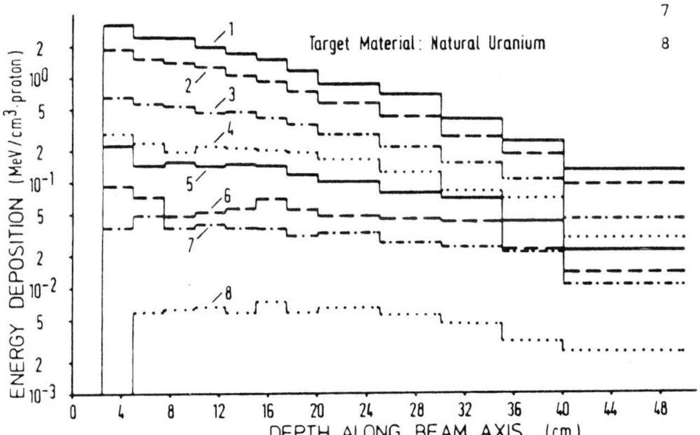

b

Fig. 11 Calculated heat distribution in the rotating target

a) target material U-238

b) target material natural Uranium

(for Pb see Fig. 3)

A SMALL, MULTIPURPOSE, LOW PRESSURE WATER LOOP AT BR2 REACTOR

F. MOONS

C.E.N./S.C.K. Nuclear Research Center

B - 2400 MOL - BELGIUM

ABSTRACT

The DGR (Doigt de Gant Refroidi) low pressure water loop is a multipurpose irradiation facility in use at BR2 for about two years. It allows the remote loading and unloading of irradiation specimens (also instrumented ones and fuel rods) during reactor operation.

A closed water circuit independent of the reactor primary coolant provides the specimens with an efficient cooling. The loop also offers the possibility of an axial shift of the specimens carrier over a distance of 900 mm in and out of the reactor core (e.g. for medium speed power cycling purposes).

The main operating characteristics are :

- cooling water pressure : 12 bar
- flow rate : 1,7 10^{-3} m^3/s (6 m^3/h)
- maximum specimen surface heat flux : 2,5 10^6 W/m^2 (250 W/cm^2)
- heat dissipation capacity of the loop : 100 kW

P. von der Hardt and H. Röttger (eds.), Irradiation Technology, 647–656.

CONTENTS

1. Introduction

2. Design specifications

3. Description

3.1. In pile section
3.2. Main coolant loop
3.3. Purification loop
3.4. Emergency cooling system

4. Data summary

5. Summarized operation manual

6. Application

7. Exploitation incidents

8. Conclusion

References

1. INTRODUCTION

The DGR (Doigt de Gant Refroidi - Cooled basket) low pressure water loop is a multipurpose irradiation facility in use at the BR2 reactor (CEN/SCK - MOL) for about two years.

It allows the remote loading and unloading of irradiation specimens (also instrumented ones and fuel rods) during reactor operation.

The in pile section is dimensionally suitable for accomodation into any standard BR2-reflector channel (Ø 85 mm) and, with minor modifications, into the central hole of a BR2-driver fuel element consisting of four concentric shells (BR2 type IVn).

Till now, the loop has been used for short term (4 days) irradiations for isotope production (^{133}Xe and ^{99}Mo).

2. DESIGN SPECIFICATIONS

The design specifications for the DGR are :

- Exploitation as much as possible independent of the BR2 reactor. It results in
  - possibility of loading and unloading of an experimental charge in the loop during reactor operation ;
  - cooling circuit separated from the primary reactor cooling water circuit (allowing irradiation of fuel pins till failure).
- A heat dissipation capacity of 100 kW.
- An admissible heat flux at the sample surface of 2,5 MW/m^2 (250 W/cm^2).

3. DESCRIPTION

The facility (fig. 1), is of a modular design and mainly consists of

- the in pile section
- the main cooling loop
- the purification loop
- and the emergency cooling system for delay heat dissipation.

3.1. In pile section

The in pile section covers
- an aluminium pressure tube (thimble) extending from the reactor cover into the reactor core
- a cooling water flow separating tube
- a sample carrier (basket) with suspension tube
- an elaborated stainless steel connecting head fixed to the thimble, locked to the irradiation channel sleeve and extending into the reactor pool.

3.1.1. Thimble

The total length of a sample holder and suspension tube with the rig head in its highest position amounts to about 6 m and the total length of the equipped "DGR" is about 7 m.

The cooling water entering into the thimble head flows downwards in the central guide tube which contains the sample holder. It flows in an annulus along the samples, passes at the lower end through a porous metal filter and mounts again to the thimble head by flowing upwards in an annular space between the central guide tube and the outside pressure tube. During upward flow the water dissipates its heat, taken up by cooling the samples, to the reactor cooling water flowing outside the pressure tube.

3.1.2. Sample holder (or basket)

The design of the basket is function of the irradiation specimen(s), taking into account mechanical rigidity, instrumentation and cooling requirements.

3.1.3. Head of the in pile section

The head of the in pile section fulfils the following tasks :
- closing the irradiation channel opening by a leaktight plug equipped with O-ring joints and screwed into the channel sleeve ;
- coupling the cooling water in- and outlet to the out-pile circulating system by means of quick couplings which shut-off automatically both ends when they are disconnected ;
- receiving the sample holder to be introduced into the thimble, assuring its leaktight penetration by gliding O-ring joints and locking the sample holder head in its proper position by a special slot device ;
- allowing, by means of a screwing nut, the 900 mm axial displacement of the samples. During this operation the sample holder suspension tube glides through O-ring joints which assure the leaktightness between the interior of the thimble (pressure tube) and the reactor pool ;
- allowing the remote insertion and withdrawal of the samples from the thimble whilst the latter stays in the reactor vessel.

The head is mainly characterized by a long sleeve (height of about 1.9 m above the channel sleeve) extending from the reactor cover into the water pool. The lower part of the connecting head allows for cooling water in- and outlet whereas the upper part houses the moving rig head, the gliding joints of the test section extension tube and, at the upper end, the locking device and the commanding screw for the axial displacement of the samples.

By means of a special manipulating tool connected to the rig head, the latter may be introduced in and locked to the "DGR" thimble. Then the tool becomes free of the rig head and may be withdrawn. Now the sample holder is safely locked to the "DGR" thimble and can only be unlocked by means of the tool. Unlocking the sample holder implies automatic locking to the tool so that there is no danger for dropping it. Loading and unloading of sample holders from the thimble may be handled remotely under the water level of the reactor pool.

3.1.4. Cross section at reactor core level

At core level (irradiation section) the "DGR" presents the following dimensions :

- Aluminium reflector channel plug : outside diameter 80mm
  inside diameter 50mm
- Aluminium pressure tube thimble: outside diameter 43mm
  inside diameter 38mm
- Central water flow guide tube (aluminium): outside diam. 30mm
  inside diam. 27mm
- Central experimental hole (for sample holder) 27mm
- Axial level of the lower end of the sample charge with respect to the core midplane,
  -completely inserted -350mm
  -completely withdrawn +550mm.

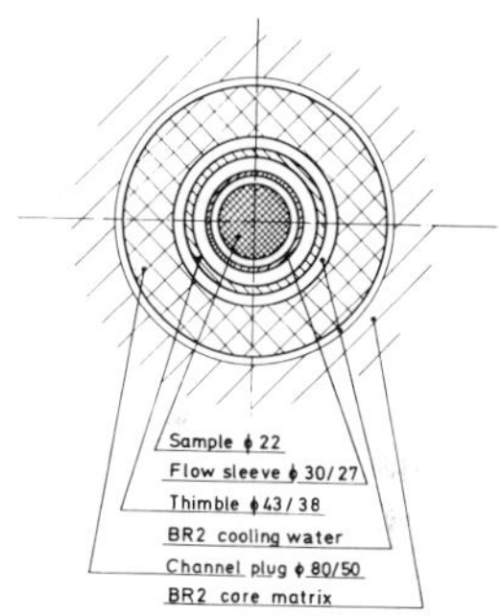

CROSS SECTION OF IN PILE PART AT MID PLANE

Above the reactor core, the outside diameter of the thimble increases to 70mm in order to reduce the cooling water velocity and hence its pressure drop.
The velocity of axial displacement of the samples can be chosen according to the rotating speed of the commanding srew ; by manual operation the axial movement may vary between 0 and about 250mm per minute, this allows for medium speed power transients.

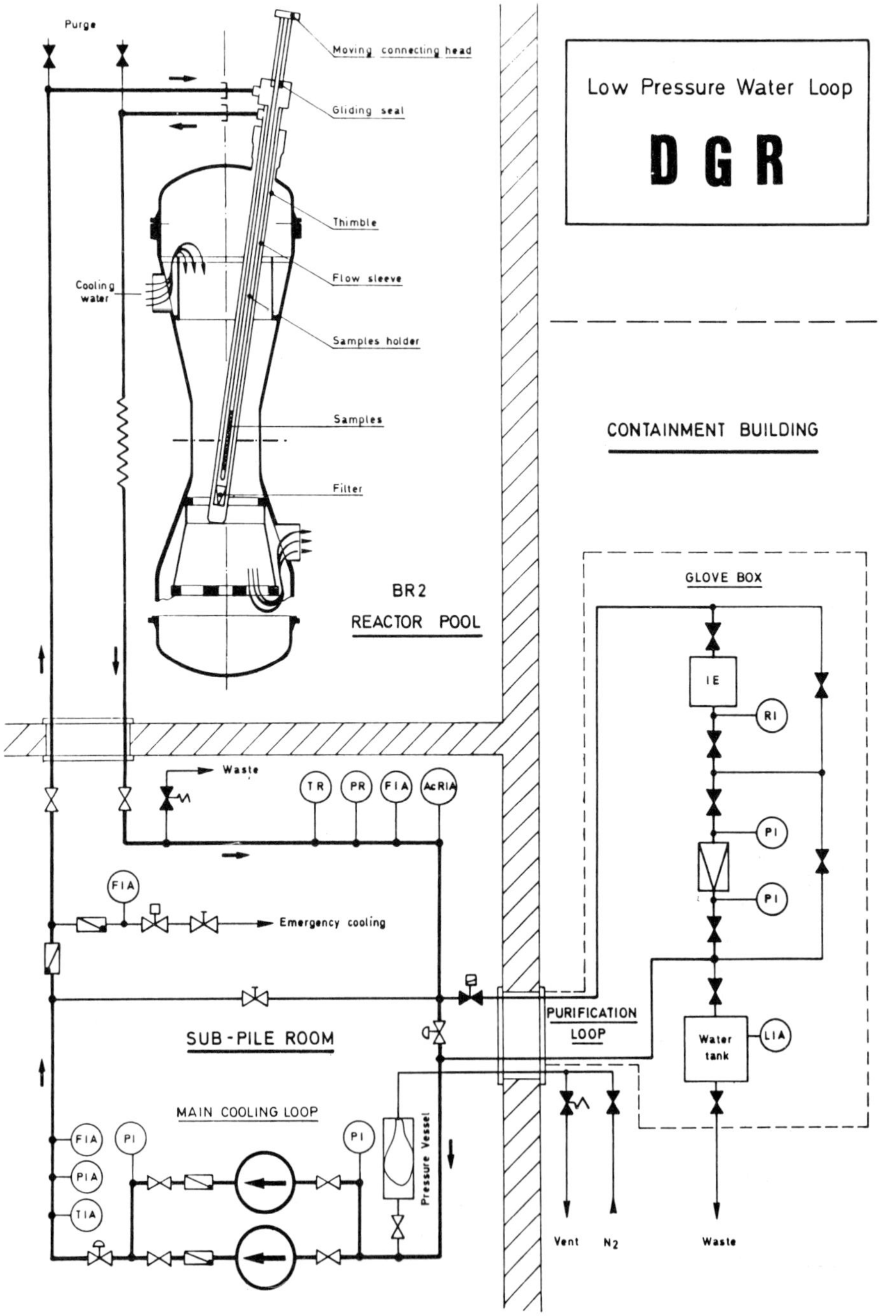
Low Pressure Water Loop
DGR
Purge
Moving connecting head
Gliding seal
Thimble
Flow sleeve
Cooling water
Samples holder
Samples
Filter
BR2
REACTOR POOL
CONTAINMENT BUILDING
GLOVE BOX
IE
RI
PI
PI
Waste
TR
PR
FIA
AcRIA
FIA
Emergency cooling
SUB-PILE ROOM
PURIFICATION LOOP
Water tank
LIA
MAIN COOLING LOOP
FIA
PI
PIA
TIA
PI
Pressure Vessel
Vent
N2
Waste

3.2. Main coolant loop

The main loop comprises the pump rack and the rigid and flexible connecting tubes (also heat exchangers) between pump rack and in pile section.

3.2.1. Pump rack

The pump rack is housed in the so-called "sub-pile room" which provides in shielding (concrete walls) and second containment (separate ventilation system) requirements.
A pressurizing and expansion vessel (20 dm^3), two equivalent circulating pumps in parallel and the proper manipulation, instrumentation and control equipment, for temperature, pressure, flow rate and activity of the coolant, are assembled in the pump rack.

3.2.2. Heat exchanger

About 100 kW can be dissipated. For nominal flow rate conditions about 45 kW are already transferred in the in pile section to the BR2-primary cooling water.
The remainder is transferred to the BR2-pool water by the connecting tubes (Ø $\sim$ 40 mm) between pump rack and in pile section.

3.3. Purification loop

The water purification annex is, for purification purposes, used as a by-pass (1% of nominal flow rate) of the main cooling loop. For draining, it is used as an open loop.

It is composed of a 5 dm^3 mixed bed ion-exchanger, a fine mesh (porosity grade 5 μm) particle filter, a 150 dm^3 storage tank and a sampling station.

The whole equipment is enclosed in a glove box. The major components (tank, filter and ion-exchanger) protected by individual biological shielding, can be removed and, if necessary, be treated in a cell.

3.4. Emergency cooling system

The emergency cooling system is fed by the BR2 auxiliary water cooling system (experimental cooling water circuit). The water flows through the in pile section into the purification loop and is collected in the storage tank (cfr. § 3.3.).

4. DATA SUMMARY

- nominal working pressure	12	bar
- maximum working pressure (pump outlet)	18	bar
- minimum working pressure (pump outlet)	7,5	bar
- opening pressure of safety valve (for nominal working pressure of 12 bar)	18	bar
- nominal flow rate of one pump	1,7	dm^3/s(6 m^3/h)
- maximum flow rate of one pump	2,5	dm^3/s(9 m^3/h)
- nominal pressure difference over the pump	6,5	bar
- cooling medium	demineralised water pH = 5,5 à 6,5 ρ = 5 à 10 cm	
- maximum working temperature (out-pile)	90	°C
- heat dissipation capacity about	100	kW
- maximum heat flux at the surface of the samples	2,5	MW/m^2(250 W/cm^2)
- nominal water velocity in the annulus between sample (22mm) and flow sleeve	8 to 9	m/s
- nominal water velocity in the annulus between flow sleeve and thimble tube (core section)	4	m/s
- central experimental hole	27	mm
- axial displacement of the sample holder	900	mm
- number of rotations for displacement of 900 mm	180	
- velocity of axial displacement of the samples (manual)	0 to 250	mm/min
- maximum sample diameter (depends on cooling requirements) about	22	mm
- maximum sample length about	600	mm
- contents of storage tank about	150	dm^3
- amount of resins in the ion-exchanger 3 dm^3 of cation exchanger 2 dm^3 of anion exchanger	5	dm^3
- porosity of filter in the purification unit	5	μm
- surface of filter in the purification unit	0,03	m^2

The above mentioned values may not be handled as absolute but must be considered in the right context, e.g., a hollow sample Ø 22/29 mm, length 600 mm delivering a heat flux of 2,5 MW/m^2 is not compatible with the DGR. On the other hand, it may be possible to handle samples up to about 27 mm diameter, if their heat production is of minor importance.

5. SUMMARIZED OPERATION MANUAL

Assuming the thimble is already in the reactor, the loading and unloading procedure for the sample holder goes as follows :

5.1. Loading

1. Load sample holder in thimble and lock
2. Vent the circuit
3. Pressurize the circuit
4. Start pump
5. Switch on the actions to the reactor (by-passed during DGR-start-up phase)
6. Turn the irradiation samples in the neutron flux.

5.2. Unloading

1. Turn the irradiation samples in the highest position, out of the neutron flux
2. Switch off the actions to the reactor (alarm actions in by-pass)
3. Stop pump
4. Depressurize the loop
5. Unlock and withdraw the sample holder.

6. APPLICATION

Up to now, the DGR irradiation facility has been used for the irradiation of highly enriched uranium samples for the production of ^{99}Mo and ^{133}Xe for medical purposes.

This application takes advantage of only one of the DGR-features i.e. capability for loading and unloading during reactor operation.

7. EXPLOITATION INCIDENTS

No major exploitation incidents can be reported for the DGR irradiation facility during two years of operation.

8. CONCLUSION

The DGR is a small, multipurpose, reliable low pressure water loop at BR2 with a lot of intrinsic safety features.

Its major advantages are :

- a separate cooling circuit and
- the possibility of sample loading and unloading during reactor operation.

REFERENCES

/1/ J.M. BAUGNET, W. BOEYKENS, F. LEONARD
The BR2 materials testing reactor. Past and future.
Seminar on research reactor operation and use.
Jülich (Germany) - September 14-18, 1981.
IAEA-SR-77/36.

/2/ F. MOONS
Project DGR, thimble with forced water cooling.
Descriptive report.
SCK/CEN NT. 81/0390/16/FM dd. 30.07.79 (internal document).

/3/ F. MOONS
Production of ^{133}Xe and ^{99}Mo in the DGR-facility.
SCK/CEN NT. 81/8956/09/FM dd. 11.07.80 (internal document).

AN ULTRASONIC GAUGE FOR MEASURING OF M.T.R. FUEL ELEMENT COOLING GAP DIMENSIONS

K. van Otterdijk

Reactor Department

Netherlands Energy Research Foundation E.C.N.

Abstract

As a contribution to the technological development of proliferation proof Research and Test Reactor Fuel (RERTR programme), fuel elements have been developed and designed for irradiation testing of research reactor fuel with 20% enrichment. The project, managed by E.C.N., is a collaboration project of E.C.N., A.N.L. and J.R.C.. In this framework some fuel assemblies have been manufactured by NUKEM and CERCA. As a part of the testing procedures, cooling channel widths and channel width profiles as a function of channel height have to be determined for different burn-up values. This has been achieved by the development and manufacturing of an installation for handling these fuel assemblies and a travelling probe based on ultrasonics for measuring the cooling channel dimensions.

1. Introduction

In the framework of an ANL/ECN/JRC collaboration project, cooling gap dimensions have been determined in low enriched fuel assemblies. The measurements were performed on several assemblies manufactured by both NUKEM and CERCA.
The aim of the experiment is to determine the absolute channel width and the channel width profile as a function of channel height for different burn-up values.
To meet these requirements an apparatus has been developed and constructed consisting of :

- poolside bench with fuel element holder, calibration block and manoeuvering mechanism
- travelling probe
- instrumentation for analogue and digital data registration and display.

The measuring method and the ultrasonic installation are briefly described in chapter 2 and 3, while signal and data processing is described in chapter 4.
Some measuring results and a survey of obtainable accuracies are presented in chapter 5 and 6.

P. von der Hardt and H. Röttger (eds.), Irradiation Technology, 657–667.

2. Measuring method

As a spacing sensor, an ultrasonic transducer has been used in a pulse-echo mode. The transmitted ultrasonic sound is reflected by a surface facing the crystal element. The propagation time of the ultrasonic wave is proportional to the physical distance between element and surface. The use of ultrasonics requires the fuel element being submerged.

The transducer has been developed and manufactured by E.C.N. and an outer firm (Sonimex). The latter manufactures and develops transducers under license of Sperry, a division of Automation Industries in the U.S.A..

The transducer has been depicted in Figure 1 and consists actually of a stainless steel ring mounted on a leaf spring; the ring contains the crystal element, the acoustic lens and the necessary acoustic damping material.

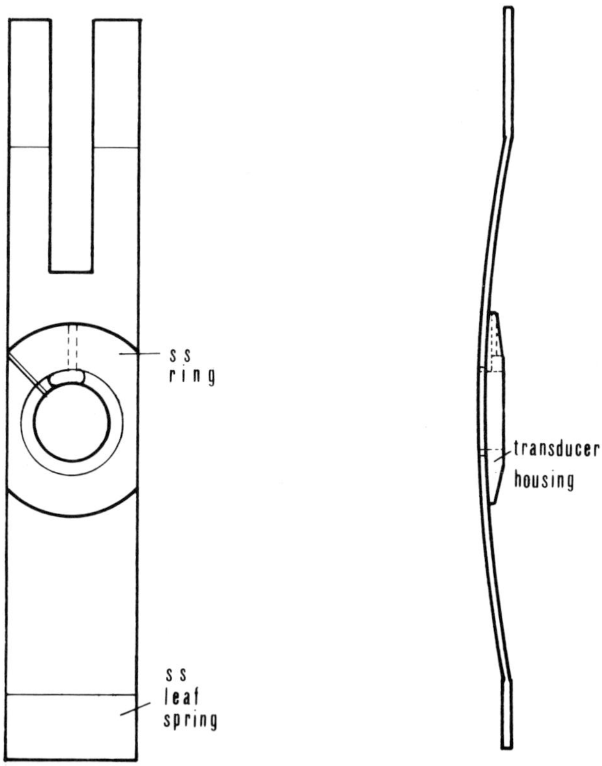

Figure 1. Ultrasonic probe

A shock pulse with an amplitude of 200 Volt and a risetime of 2 Nanoseconds is used to activate the crystal element which then produces a short burst of ultrasonic waves with a frequency of 8 Megacycles.

A lead-metaniobate crystal element has been chosen for obtaining optimum acoustic damping for best resolution. Special precautions have been taken in the acoustic interface material (acoustic lens) between element and water to guarantee a reflection from the surface facing the element, while the crystal element is used in the so called "near field".

3. Measuring installation

The installation actually consists of:

- poolside bench provided with fuel element holder, calibration block and operators console
- a travelling probe.

3.1 The poolside bench

The complete device is depicted in Figure 2 with some additional details shown in Figures 3 and 4.
The bench with a length of 4.5 meters and provided with a fuel element holder is attached to a wall of the reactor pool. From the operators desk the fuel element can be positioned very precisely with respect to the measuring sword as may be seen from Figure 3. Having chosen a specific channel, the probe can be lowered into that channel by a mechanism on the operators desk (Figure 4) until it reaches the bottom of the element. And again, by the same mechanism, the probe can be pulled up until it sticks in the calibration block, which provides a calibration facility for the ultrasonic transducers. Two calibration widths are available.
Provisions have been made for displaying probe depth, while pulling up the probe. The channel width and temperature can be displayed as a function of probe position within a channel.

3.2 Travelling probe

The travelling probe with three spring loaded ultrasonic sensors and an overall length of 1250 mm has been depicted in Figure 5. It actually consists of a stainless steel "sword" which contains three ultrasonic transducers. Figure 6 shows clearly the transducers mounting configuration. The transducer uses two leaf springs. A ring containing the crystal element, acoustic lens and the necessary sound damping material is mounted on one spring, while the opposite leaf spring contains a hole to transmit the sound, in order to obtain a reflection directly from the opposite channel wall. The connection of the transducer with the electronics is formed by a mineral insulated 50 Ohm transmission cable within the sword and outside by a radiation resistant soft coaxial cable. Besides a flexible transition is present between the transducer and the mineral insulated cable. By applying three sensors in one probe, simultaneous channel spacing measurements are possible in three positions within one channel.
Provisions have been made for displaying probe position and temperature.

Figure 2. Poolside bench

Figure 3. Lower part of the poolside bench

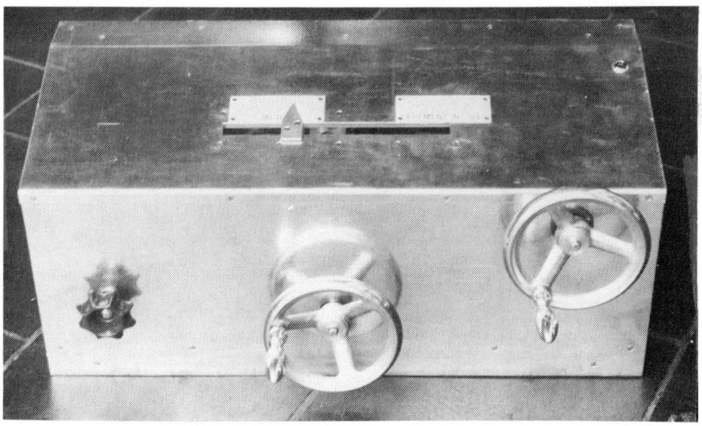

Figure 4. Upper part of the poolside bench

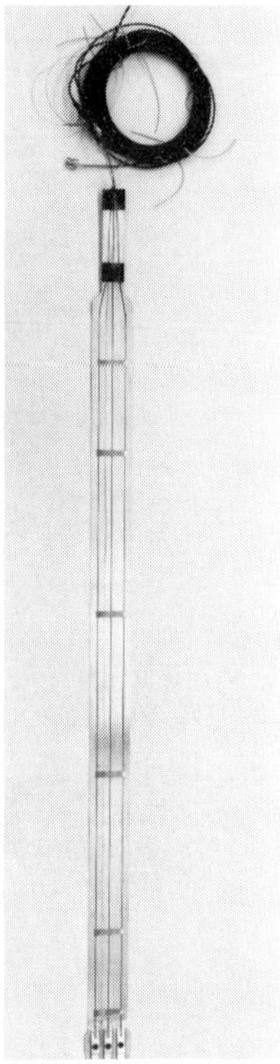

Figure 5. Travelling probe

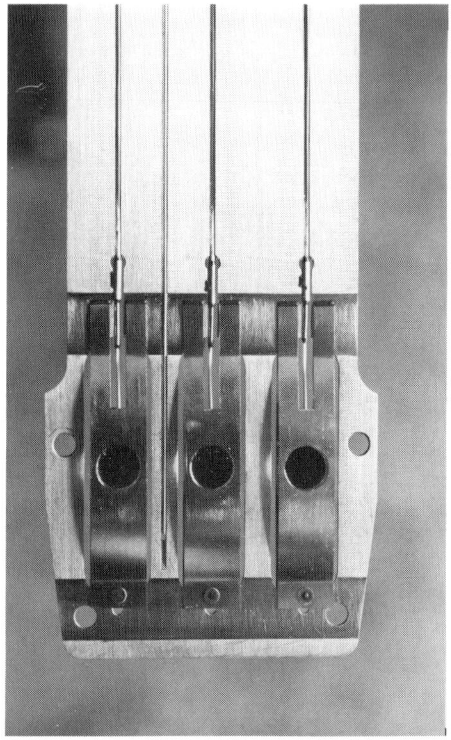

Figure 6. Mounting configuration of the ultrasonic sensors

3.3 Specification of the travelling probe

- Ultrasonic transducers :

range	: 1.70 mm to 2.60 mm
max. inaccuracy	: ± 0.05 mm
resolution	: 0.005 mm
max. non lineanity	: 0.05 mm
temperature range	: 20-40 Deg. C.

- temperature sensor
  max. inaccuracy : ± 1.1 Deg. C.
  temperature range : 20 to 100 Deg. C.
- probe position measurement
  range : 0 to 1000 mm
  max. inaccuracy : ± 3 mm
  max. non linearity : 0.25%

4. Signal and data processing

The electronic equipment, necessary to operate the travelling probe, consists of:

- the signal conditioning equipment of the ultrasonic transducers, being an instrument manufactured by Automation Industries, type NOVA 201S. This provides the excitation pulse, calibration control facilities and the necessary tools to translate a propagation time into a voltage representing a typical distance. An output for direct recording purposes on an X-Y recorder is present for a direct check of the measurement.
- thermocouple equipment, being an electronic cold junction reference unit and a D-C amplifier for providing the necessary levels for recording purposes.
- requirements for probe position indication, being a precise potentiometer and a power supply, providing the necessary output levels for the position of the ultrasonic gauge within a channel.

Five signals, representing three channel widths, the temperature within a channel and the probe position are recorded on an analogue tape recorder. Besides, voltages corresponding with calibration widths for each sensor, are recorded, so that a relationship between a recorded voltage and channel width is known. The temperature is recorded for having available a correction possibility because of the temperature dependance of the ultrasonic sound velocity in water. After having digitized the analogue tape, the measurement data are available as a permanent file. A plotting routine provides an output representing channel width as a function of height with all required parameters.

5. Measuring results

According to the aim of the experiment, measurements have been performed to determine the absolute channel width as a function of height and to detect a possible change in the channel width profile for increasing burn-up figures. As a typical example, the channel width as a function of height has been depicted in Figure 7 for different burn-up figures. An arbitrary channel of fuel element LN-01, manufactured by NUKEM, has been chosen and only the channel width profile in the middle of the channel is shown. The curves show clearly the reproducability of the measurements, the resolution of the sensors and the shape of the channel width profile.

Averaging all the average channel width values for one fuel element, yields an average coolant cross section.

As a typical example, the average coolant cross section of fuel assembly LN-01 as a function of burn-up is shown in Figure 8.

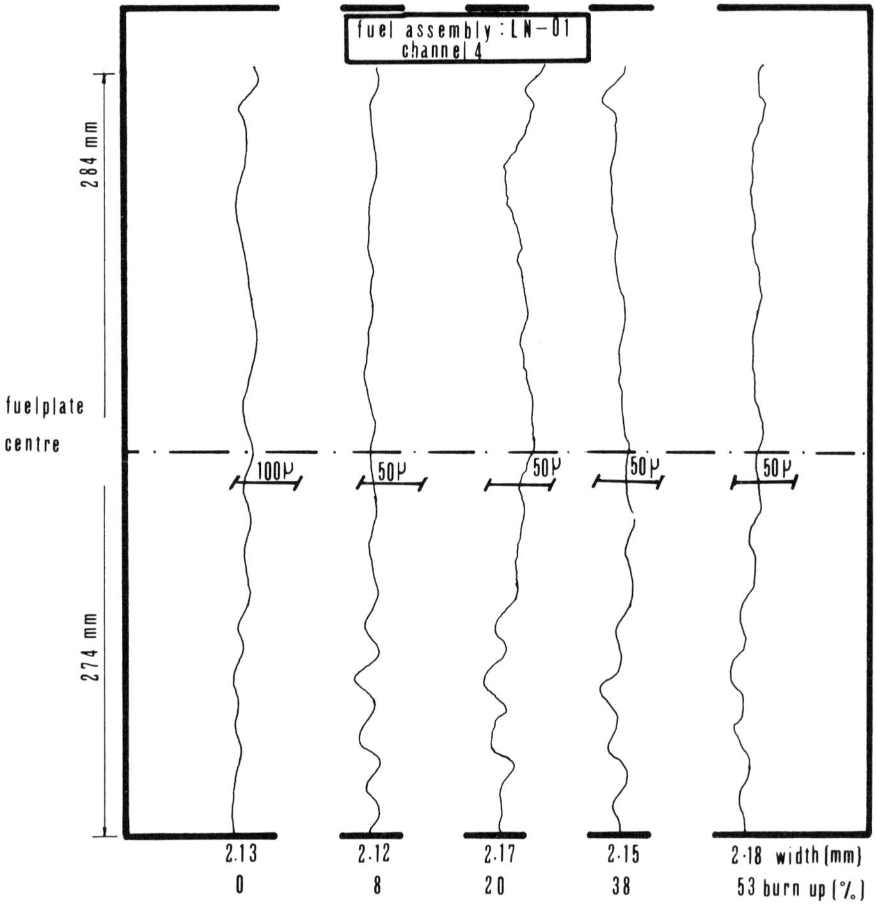

Figure 7. Channel width profile as a function of channel height for channel 4, fuel assembly LN-01

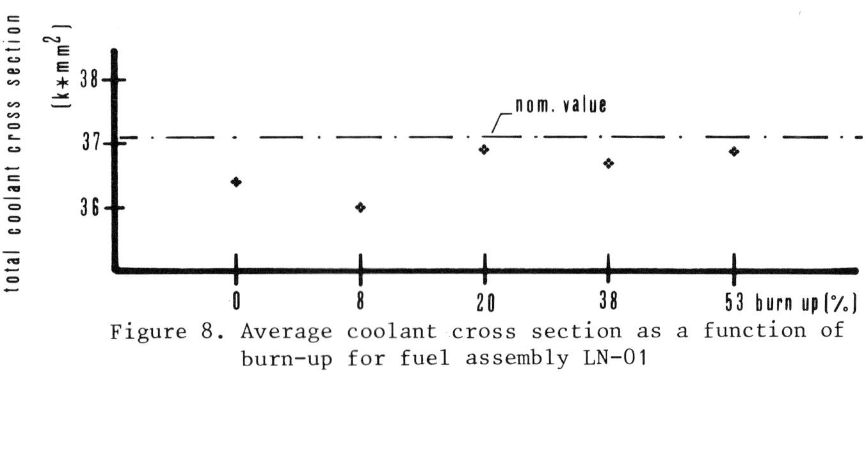

Figure 8. Average coolant cross section as a function of burn-up for fuel assembly LN-01

6. Error evaluation

Even for different burn-up figures, the typical width profile, as shown for instance in Figure 7, remains unchanged, though the absolute level of the profile can vary with a maximum of <u>± .04 mm</u>. This variation is mainly caused by calibration errors.

The resolution of the sensor and electronic equipment equals about 5 micron (.005 mm).

For the total uncertainty in the calculated average channel width a value of .05 mm is estimated.

MEASUREMENTS OF NUCLEAR HEATING
IN THE PETTEN HIGH FLUX REACTOR

A.J. Nolten

Reactor Department

Netherlands Energy Research Foundation E.C.N.

Abstract

The knowledge of the magnitude and distribution of the nuclear heating in all positions of the reactor core is an essential basis for the design of irradiation facilities. The principle for determining the nuclear heating is based on the caloric method. The nuclear heating is generated in a small sample suspended in the centre of a He-filled container. Since in steady state condition, the heat generation in the sample equals the heat removal over the gas annulus, the temperature difference is related to the nuclear heating in the sample. This relation is determined by means of an appropriate calibration. Two sets of three calorimeters with different sample materials, stainless steel, aluminium and graphite, are fitted in a measuring probe, which is placed in a thimble. The probe is vertically displacable in the thimble.

1. Introduction

For all lay-out and design work concerning irradiation equipment, the knowledge of the magnitude and distribution of the nuclear heating in a reactor core is of great importance.
For this reason measurements are performed in all experimental positions in the HFR-core at Petten, over more than ten years.
The measurements are done by means of calorimeters, which are described in this paper.
During the first eight years a number of calorimeters were vertically distributed in irradiation rigs. In the beginning the rigs were designed for one particular reactor position. Later rigs were designed, fitting in every reactor position. The calorimeters, however, were vertically distributed over the fuel length at fixed positions.
During the last years the calorimeters are placed in probes which can be moved vertically, so that each calorimeter covers either a vertical distance equal to half the fuel length, or the whole fuel length.
In this paper also a description is given of the last mentioned devices. In the last chapter some measuring results are given together with an accuracy consideration.

P. von der Hardt and H. Röttger (eds.), Irradiation Technology, 669–676.

2. Calorimeters

The calorimeter consists of a cylindrical piece of target material centrally placed in a stainless steel container. The sample is centered by means of a thermo-couple and a dummy couple. On the wall of the container an outer thermo-couple is welded. Both temperatures are measured and result in a temperature difference.
The target materials are graphite, aluminium and stainless steel, because these are the most frequently applied materials in the Petten reactor.
Forcing the heat flow, generated in the sample, over an annular gas gap implies slender sample dimensions. The heat losses through the thermo-couple wires are less than 3% of the heat produced. The gas gap width is limited to reduce the heat transfer by radiation.
The ratio between sample diameter and container diameter is an optimum of sample diameter and gas gap width for a given container diameter.

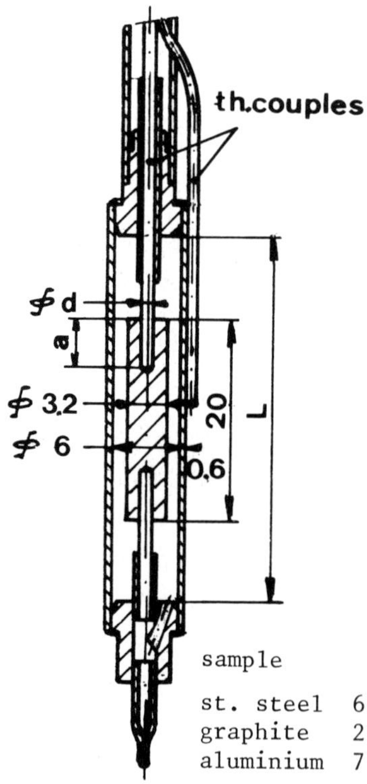

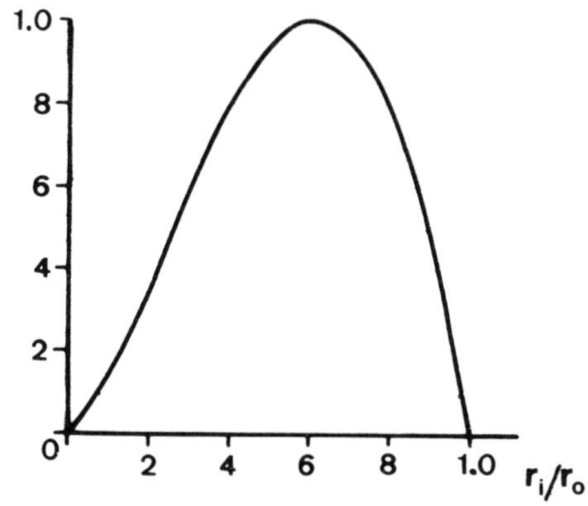

Sensitivity of the calorimeter versus the ratio of sample r_i, and container wall r_o

sample	a	d	L
st. steel	6.8	1	40
graphite	2.5	0.5	32
aluminium	7.5	0.5	32

TRAMP-calorimeter

3. Measuring rigs

3.1 General

At first the measurements were done with a so called TRAMP-rig (traveling and measuring probe) untill some of the thermo-couple failed. Next a "handstick" was used, a probe with calorimeters, fitting in the same extension tube.

3.2. Description of the TRAMP-rig

The rig consists of a thimble in which the measuring probe can be moved up and down, and a drive mechanism. The thimble consists of two concentric tubes. The outer tube is closed on the bottom side and fits in a standard HFR-filler element with a bore of 52 mm. On top of the thimble a cooling water inlet connection is placed. The water flows downwards through the space between the two tubes, rises through the inner tube and discharges via outlet holes into the reactor pool.
The measuring probe is provided with holes and grooves through which the cooling water enters and flows around six calorimeters, two groupes of three calorimeters each at a mutual distance of 35 cm.

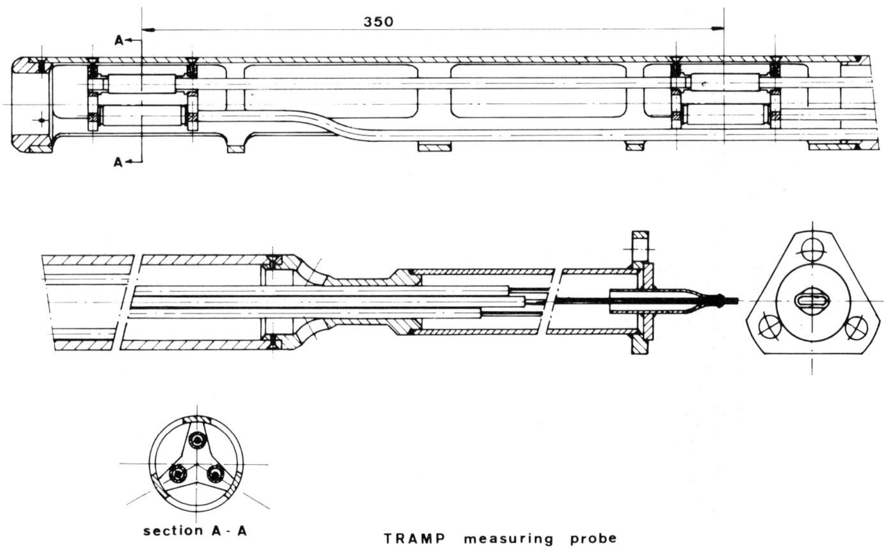

TRAMP measuring probe

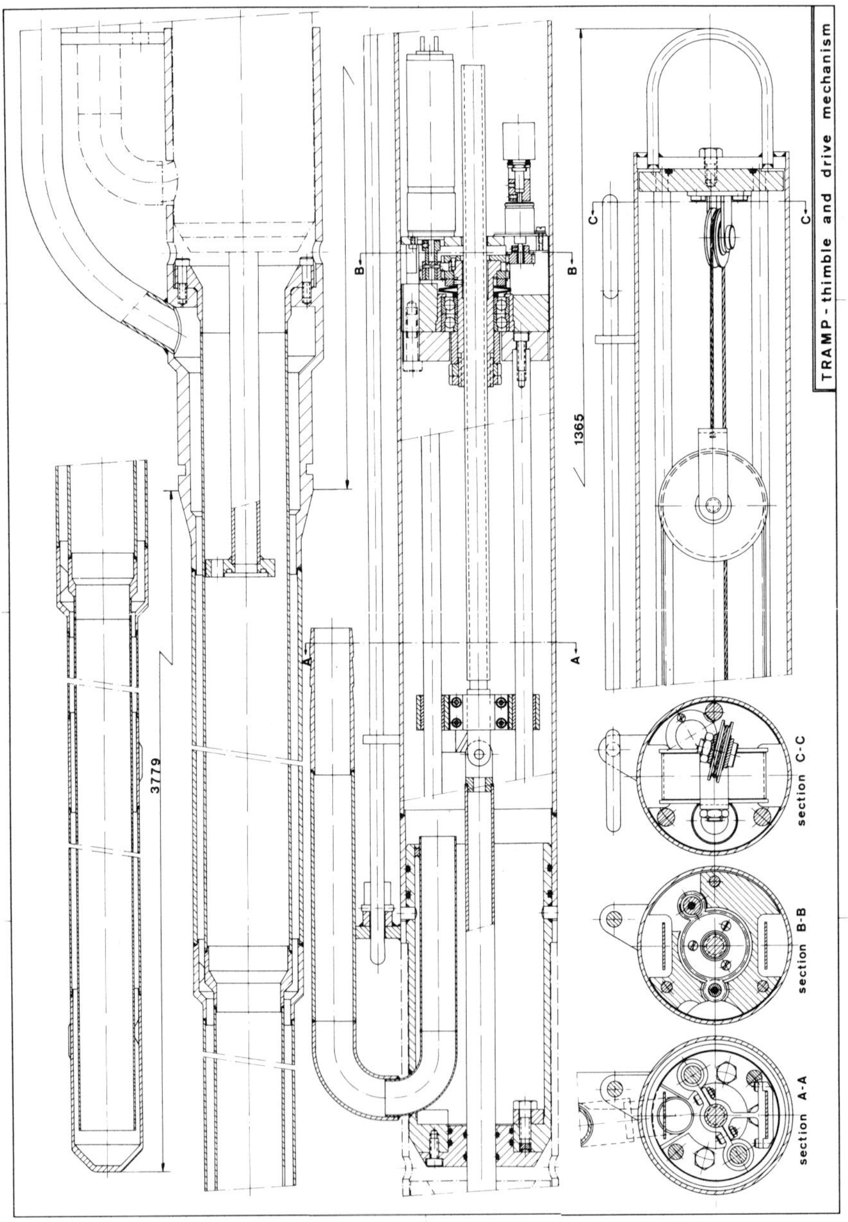

TRAMP - thimble and drive mechanism

The probe is connected to an extension tube with a watertight penetration to guide the thermo-coupe wires into the instrumentation box. In this box also a drive mechanism is installed, consisting of an electrical motor with a slip coupling, a gear drive and a threathed spindle. The displacement of the probe, which is 350 mm, is indicated by means of a potentiometer. The probe can be inserted in the thimble independent from the reactor operation.

3.3. Description of the "handstick"

The "handstick" is also a probe with holes and grooves in which six calorimeters are placed in two groups of three at a mutual distance of 100 mm. The thermo-couples are guided via an extension tube to an instrumentation box. On the box another extension tube is connected, which reaches the bridge over the reactor pool. This tube is provided with marking stripes. By hand the whole probe can be moved up- and downwards. The calorimeters can be displaced unrestrictedly, so each calorimeter covers the whole fuel length.

4. Determination of the nuclear heating and calibration

The basic principle for determination of the relation between temperature and nuclear heating is a heat balance. Assume that the nuclear heating in the sample is P watt per gram and that the mass of the sample is M gram, then the generated heat is P.M Watt.
The heat flows radially to the container wall and is proportional to the temperature difference $T_{sample} - T_{wall}$, in effect it is equal to $h.F.\Delta T$, where h is the overall heat transfer coefficient and F is the

balance :

$$P = \frac{h.F.\Delta T}{M} \qquad (1)$$

By means of calibration the exact value h.F has to be determined. Due to the small dimensions of the calorimeters, the calibration can only be done under dynamic conditions, where the cooling down curve is measured.
After some calculations equation (1) can be written as :

$$P = \frac{C_{po}}{\tau}(1 + \beta\Delta T)\,\Delta T \qquad (2)$$

where C_{po} is the specific heat of the sample at equilibrium conditions where $T_{sample} = T_{wall}$.
τ and β can be calculated from the calibration curve.
As an example two figures are given. The right figure gives the difference between the measured curve and the calculated curve belonging to the derived values of τ and β. In fact it is a reproduction of the accuracy of the calibration.

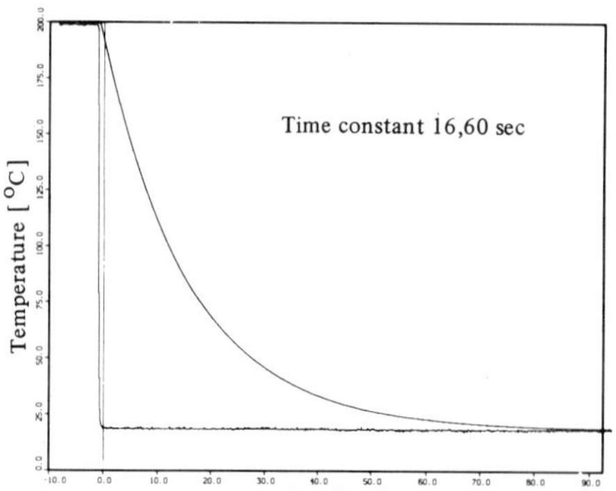

Experimental and fitted temperatures vs. time for sensor 9 (aluminium).

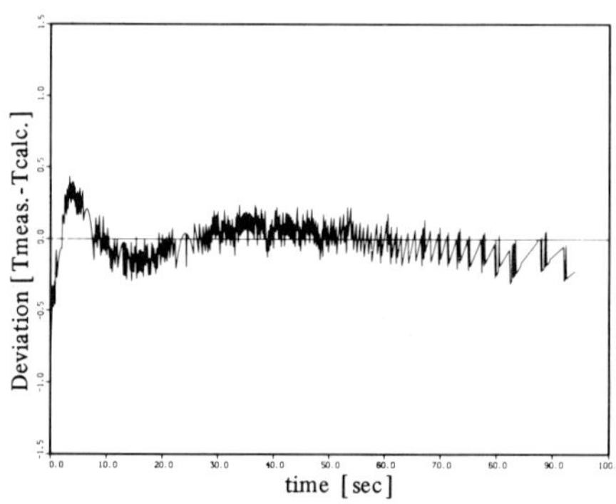

Error curve for sensor 9 (aluminium).

5. Measurements

5.1 Measurements in the 45 MW core

The measurements are carried out at a moment halfway the HFR cycles, which ensured average reactor conditions. Only measurements in one core position per cycle can be performed, because the thimble has to remain in the core position during the whole cycle.

5.2 Accuracy considerations

The error insolved in the determination of the nuclear heating from equation (2) can be estimated from its components, which are :

- end loss effects < 5%
- time constant < 1%
- specific heat sample < 5%
- measuring effects < 2%

The total statistical accuracy of the ensemble error derived from its components is then approximately 8%.

5.3 Measured values

The HFR core positions can be derived into three groups : central positions C5 and E5, intermediate positions C3, C7, E3, E7, G3, G5 and G7, and boundary positions D2, D8, F2, F8, H2, H4, H6 and H8. From each group the measuring values are given for stainless steel, aluminium and graphite.

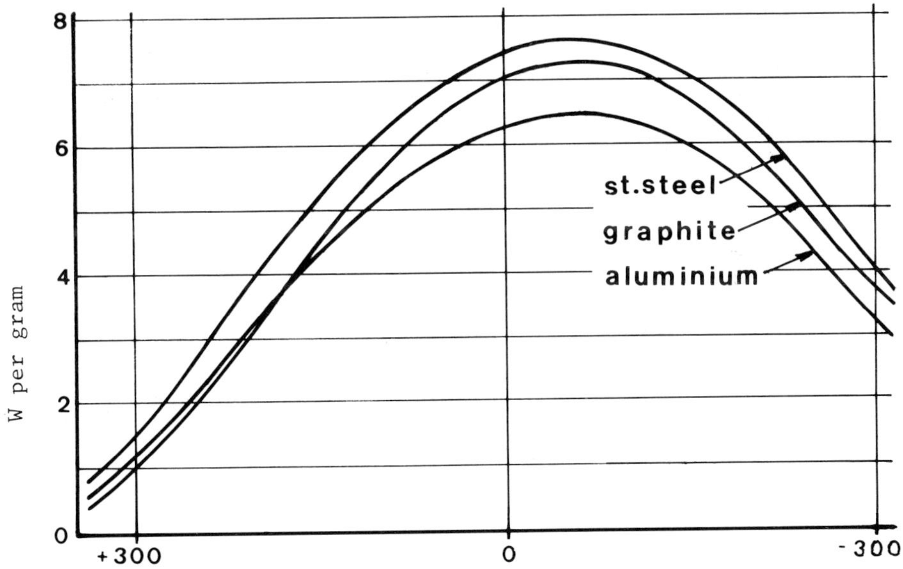

Nuclear heat in central positions

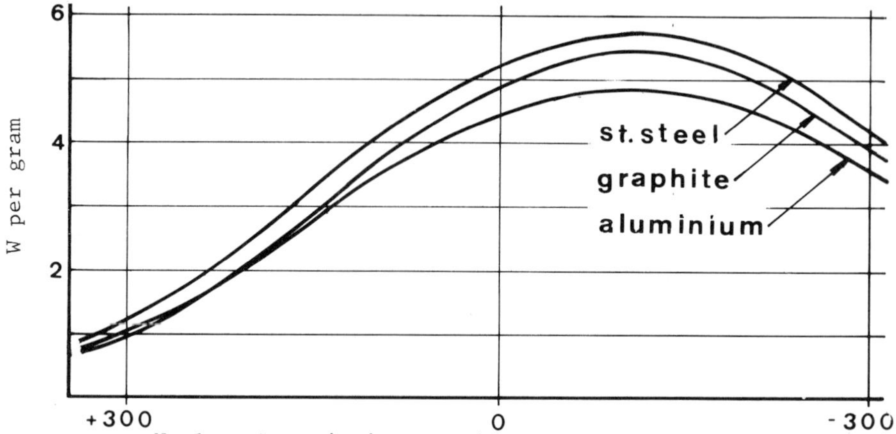

Nuclear heat in intermediate positions

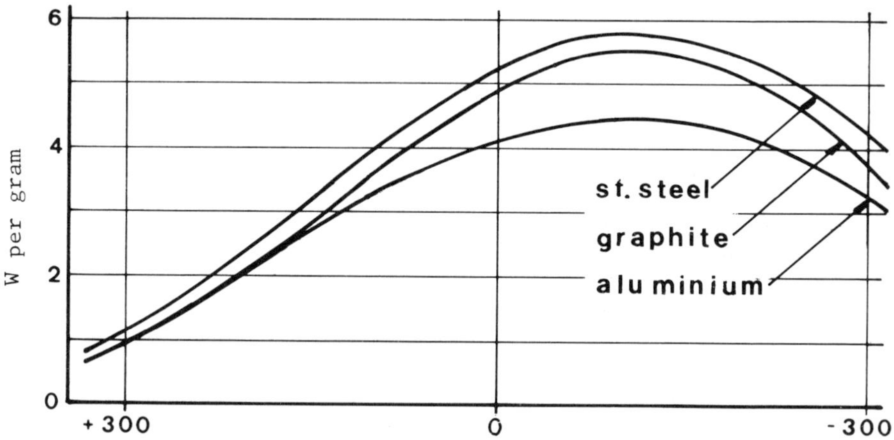

Nuclear heat in boundary positions

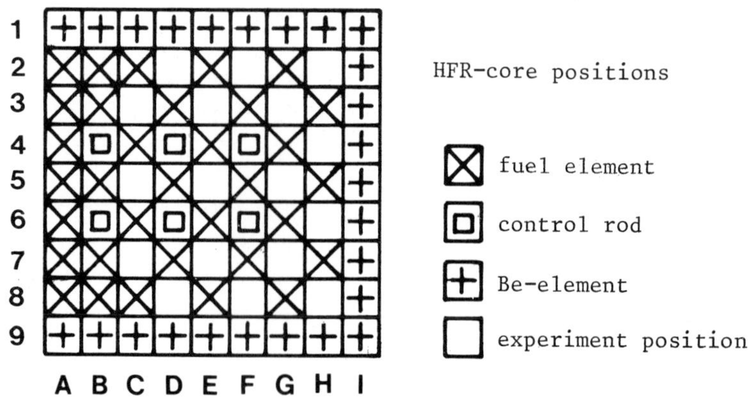

FACILITIES FOR THE IRRADIATION OF 235-U FOR THE PRODUCTION OF 99 Mo AT THE HFR PETTEN

J. Konrad

Commission of the European Communities
Joint Research Centre
Petten Establishment/The Netherlands

SUMMARY

One of the most applicated radioactive isotope in modern tumor diagnosis is technetium 99m, the daughter nuclide of molybdenum 99.
99-Mo can be obtained as a fission product of uranium-235. The 99-Mo in form of molybdate absorbed into an aluminium oxyde column, formed the generator, usually termed as radioactive "cow", where by "milking" 99m Tc with his short half-life of 6 hours can be obtained.
The basis product 235-U in form of plates or tubes, is similar to the plates used in MTR type assemblies. They are irradiated in special rigs for four to six days at a thermal neutron flux between 1×10^{14} and 2×10^{14} $cm^{-2}s^{-1}$, corresponding to a maximum specific heating rate of 130 W/cm^2.
The paper describes the used facilities and in special, a new in-core facility for the irradiation of 9 targets simultaneously. The design of the capsule allows the loading and unloading of the targets, making use of a special flooding chamber system during reactor operation, without interrupting the cooling circuit and the pressure of the reactor.

P. von der Hardt and H. Röttger (eds.), Irradiation Technology, 677–683.

1. INTRODUCTION

The High Flux Reactor (HFR) provides several irradiation devices for the radionuclide production. Seven devices are installed in the reflector inside the reactor vessel and three conveyor facilities, four other devices are positioned in the Pool Side Facility /Ref.1/ (Fig.1).
The object of this poster is to present two irradiation facilities for the irradiation of 235U for the production of 99Mo.
Diagnosis and therapy of cancer have made increasing use of the short-live radioisotope "Technetium 99m (^{99m}Tc) which can be produced in high purity by radioactive decay of molybdenum 99 (99 Mo).

99 Mo is produced in the HFR Petten by two different techniques:
1. Irradiation of molybdenum (activation route)
2. Irradiation of highly enriched uranium (fission route).
The production of 99 Mo starting with highly enriched uranium has been considerably increased during the last two years to fulfil the growing demand (Fig.2).

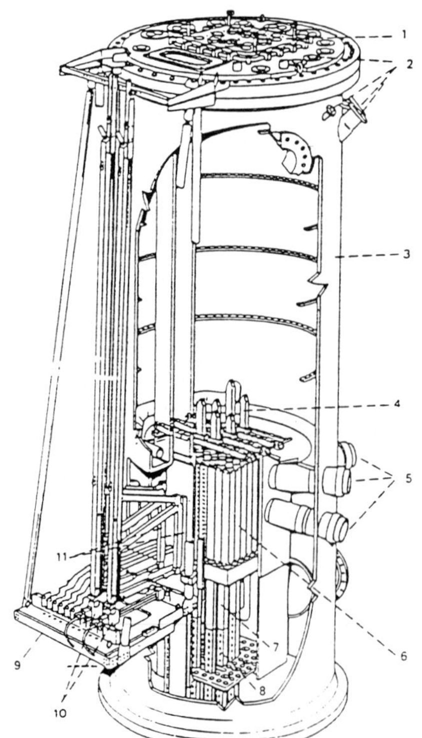

1. Cover
2. Penetrations
3. Reactor vessel
4. Control member absorber section
5. Horizontal beam tubes
6. Reactor core with fuel and reflector elements
7. Control member fuel section
8. Control member connection to drive mechanism
9. Pool side facility (PSF) with support table
10. PSF trolley
11. PSF rig

HFR core and pool side facility with a rig in the pool side facility.
a) Central reactor top lid (CRTL), used since August 1976.
b) General view of the HFR vessel.

Fig.1

This illustration is printed from the best available original

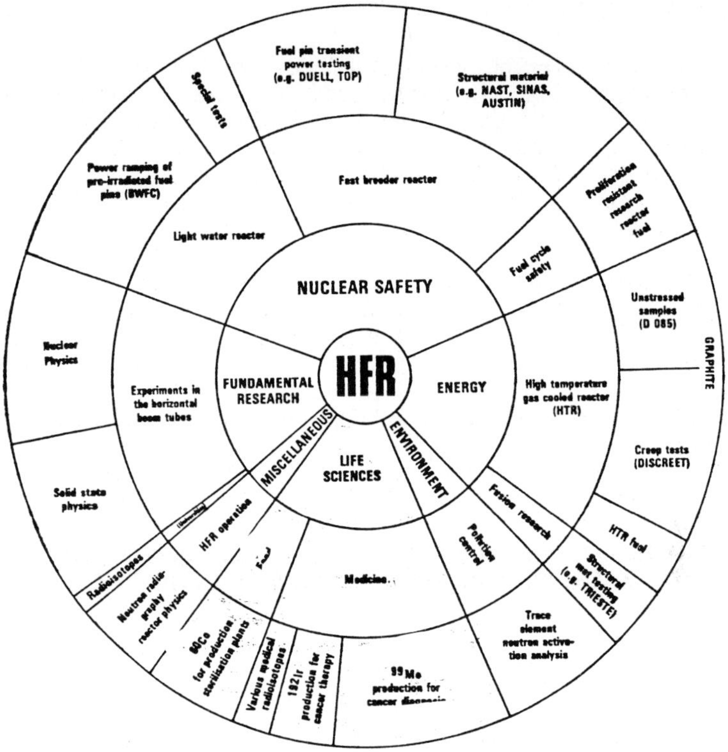

Fig. 2 HFR Petten 1981 utilization scheme

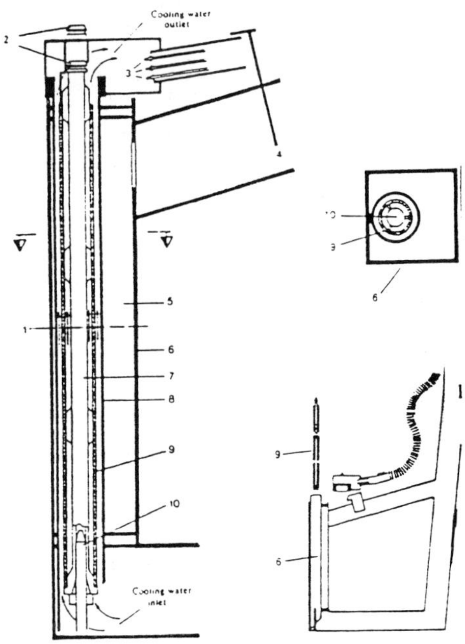

1. Core center line
2. Coupling for loading and unloading
3. Thermocouples
4. Locking device
5. Stagnant water
6. Standard carrier for PSF
7. Location pin
8. Guide tube
9. Targets
10. Centering pin

Fig. 3

Fissile Irradiation Target Device for a PSF position.

FIT

The two existing irradiation devices FIT (Ref.1) have been developed for the irradiation of fissile targets for the production of short-lived radioisotopes. One version is designed for a core position. The other one is used in the pool side facility (Fig.3)

IN-CORE FIT (Fig.4)

The targets for the In-Core FIT irradiation device consist of aluminium uranium sandwich fuel (MTR plate type), which is transformed into cylinders of 22/19.5 mm Ø and 160 mm height by bending and welding. Up to three cylinders are threaded onto a central location pin with centering plug and can be placed likewise centered via a modified head of a REFA-capsule (see Ref.1) in a tube of 3314 mm length and 27 mm inner diameter. The tube, which is manufactured from aluminium forms with its in-pile part the irradiation channel and is loaded in a standard filler element of a core position.

Cooling of the targets is provided by water entering the irradiation channel through two longitudinal slots which are machined in the upper stainless steel part of the tube above the targets. The water flows downwards through an outer and inner channel which are formed by the tube and targets, respectively by the targets and the central location pin. The cooling water which is supplied by the primary water system of the reactor leaves the channel through a hole in the lower end cup.

The construction of the capsule allows loading and unloading of the targets during the reactor operation, i.e. without interruption of the cooling circuit of the reactor. This was the main concept of the device. The maximum perturbed neutron flux density is depending on the core position in which the irradiation channel is loaded. At present the position H4 is used. For this position the maximum perturbed thermal neutron flux density, averaged over the radial fuel zone, amounts to $\phi_{th} = 1.6 \times 10^{18} m^{-2} s^{-1}$.

It is planned to use in 1983 an in-core FIT which follows the TRIO design and which allows to irradiate 9 targets in three parallel channels. For each channel of this TRIO-FIT the above description remains valid.

This illustration is printed from the best available original

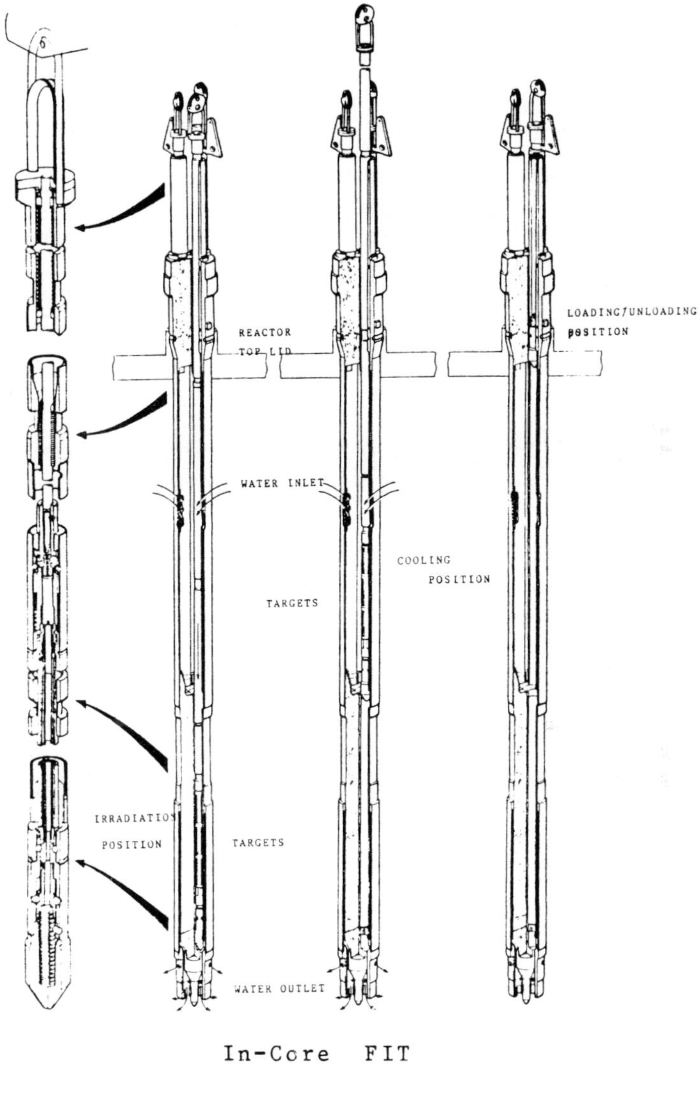

In-Core FIT

Fig.4

Fissile Irradiation Targets Device For A Core Position

3. PSF FACILITY (Fig.5)

Irradiation facility for fissile targets for the production of 99 MOLYbdene. MOLY

3.1 Fissile Targets

The targets are conform to the following specifications:

length : 220 mm
width : 40 mm
thickness : 1,3 mm
fuelled area: length: 220 mm
width : 30 mm
tot. U weight : 4,44 g
235U enrichment : 93%

The plates are fabricated and tested by the same technique as the HFR fuel plates.

3.2 Irradiation Facility - MOLY

The design of the Moly facility allows the irradiation in any PSF position. The facility consists of an aluminium carrier with on his front side the housing of the plates. The plates are charged in a plate holder (Fig.5). Cooling of the plates is provided by water entering on the bottom and flowing upwards. The cooling flow is guaranteed by a special cooling system. Coolant flow is 3 m^3/h. A maximum of eight plates can be irradiated simultaneously.

1. Core center line
2. Cooling water out
3. Cooling water in
4. Coupling
5. Coupling
6. Targets
7. Targets holder
8. Cover

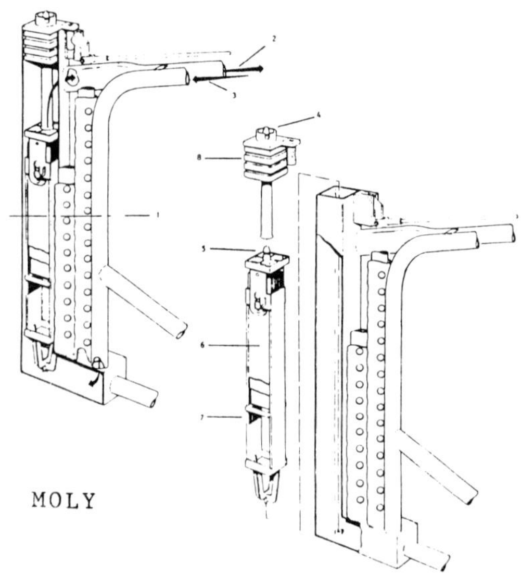

MOLY

Fig.5

Fissile Irradiation Targets Device For a PSF Position

REFERENCES

/1/ P. von der Hardt, H. Röttger
High Flux Materials Testing Reactor, HFR Petten
EUR 5700e, Rev. Edition 1981-1982

/2/ Patent N°. P/32/14628/0
In-core irradiation device for a reactor
Commission of the European Communities, DG XIII, Kirchberg, Luxembourg

/3/ A. Sameh, J. Konrad, H. Ache
Production of fission molybdenum on a technical scale at the Nuclear Research Centre Karlsruhe
Paper presented on the International Symposium on Application and Technology of Ionizing Radiations.
12-17 March 1982, Riyadh, Saudi Arabia.

/4/ P. von der Hardt, H. Röttger
Handbook of Materials Testing Reactors and associated Hot Laboratories in the European Community.
February 1981.

EVALUATION OF THE PRODUCED RADIOACTIVITY FROM RAPID IRRADIATION IN A NEUTRON FLUX DENSITY PROFILE.

Johnny Olesen Schmidt

Isotope Division, Chemistry Department

Risø National Laboratory, 4000 Roskilde, Denmark

INTRODUCTION

For rapid instrumental neutron activation analysis with the Mach-1 transfer facility at the heavy water moderated Danish DR-3 reactor (e.g. Schmidt et al. 1981), the neutron flux density decreases exponentially over a distance of 80 - 100 cm. In such a profile, the neutron flux density is time-dependent for the irradiated sample. For the evaluation of the radioactivity produced from rapid irradiations under these conditions, studies are reported on timing, sample transportation, and neutron density flux profile in the Mach-1 transfer facility.

THEORETICAL

For the description of the radioactivity produced from rapid irradiations in a neutron flux density profile, the transportation and the neutron flux density are combined to give the time-dependent neutron flux density for the sample.

Transportation

In the transport tube, the sample transportation is determined by the forces working on the capsule. In a simple model of the capsule transfer, the net driving force is supplied by the propellant gas pressure behind the capsule and is reduced by the pressure of the gas-compression in front of the capsule and the friction between the capsule and the wall of the transport tube.

P. von der Hardt and H. Röttger (eds.), Irradiation Technology, 685–695.

In addition, reflected expansion- and compression waves modify the pseudo-static pressure behind and in front of the capsule, just as the gravitation modifies the driving force for transportation into or out of the reactor, respectively. From the one-dimensional gas gun theory (e.g. Seigel 1965), the net pressure can be expressed

$$p = p_b(1 - u(g_b - 1)/2a_b) - p_f(1 + u(g_f - 1)/2a_f) \tag{1}$$

with u the velocity of the gas-expansion, g_b and g_f the specific heats, a_b and a_f the sound velocities, and p_b and p_f the pressures behind and in front of the capsule, respectively. For $g_b = g_f = 1.4$, $a_b = a_f = 346$ m/s, $p_b = 1$ MPa and $p_f = 0.1$ MPa, the net pressure is approximately linear in u for velocities less than 120 m/s. Under these conditions, Newton's second law yields the time-distance input-output functions in the eqs. (2) and (3) for the capsule transfer into and out of the irradiation zone, respectively (Schmidt 1982). With L the distance between the capsule start position and the irradiation position, the time-distance functions become

$$x_i(t) = L - b_i/c_i(t + (\exp(-c_i t) - 1)/c_i) \tag{2}$$

$$x_o(t) = b_o/c_o(t + (\exp(-c_o t) - 1)/c_o) \tag{3}$$

with the b's being the initial capsule accelerations and the c's being attenuation coefficients of the accelerations, respectively. In this simple model, it is further assumed that the frictional force is proportional to the capsule velocity and that the gravitation is constant.

Thermal neutron flux density Ø(t)

For an irradiation facility immerse into the moderator, the neutron flux density is determined for the sample by the type of moderator, the construction materials,and the cooling medium of the heat-exchanger. While the latter two factors cause a perturbation of the neutron flux density, the type of moderator determines the profile. Thus for heavy water moderation, the thermal neutron diffusion length is about 100 cm compared to about 2.9 cm for light water moderation (see e.g. Glasstone 1955). From measurement of the thermal neutron flux density profile Ø(x) along the length x of the transport tube, the time-dependent thermal neutron flux density Ø(t) can be found by means of the time-distance function x(t).

Produced radioactivity

For a given target nucleus and a constant neutron energy distribution, the activation probability p is determinded by the activation cross section σ, the neutron flux density $\emptyset(t)$ and the survival probability p for the radioactive nuclide. That is, p can be defined as the integral over the events that target nuclei become activated at time t and that they survive the rest irradiation time $t_i - t$ as radioactive nuclei without decaying or reacting further in the small time-intervals of length dt (Schmidt 1982).

For the transfer of the radioactive nuclei from the termination of the irradiation zone to the counting position, their survival probabilities depend both on the decay time t_d and on the decay constant λ. Provided n target nuclei are being irradiated, the number of radioactive nuclei n^+ becomes

$$n^+ = n \int_0^t \sigma\emptyset(t)\exp(-\lambda(t_i - t)dt \exp(-\lambda t_d) \quad (4)$$

where the life-times are assumed to be exponentially distributed for the radioactive nuclei (see e.g. Cannizzaro et al. 1978).

EXPERIMENTAL

Mach-I irradiation facility

In Figure 1, the rapid irradiation facility Mach-I is shown. In the irradiation position (1) of the in-pile rig, the transport tube terminates in a low-angle stopping cone (2) that is surrounded by a light water heat-exchanger (3). In the counting position (4), the transport tube terminates in a stopping cone (5) and is followed by a muffler (6) to attenuate the noise from the gas flow. In the transport tube, a revolving station (7) is inserted and the transport capsules are injected one by one from an inlet tube (8) that functions as capsule storage. The irradiation facility applies N_2 and is installed in the 4V3R test hole of the DR-3 reactor at Risø National Laboratory (Westermann et al. 1980).

Capsules

The sample material is contained in 1/2-dram low density polyethylene capsules that are transported in 2-dram low density polyethylene capsules with 1.6 cm diameter and 6 cm^3 volume. Both sample and transport capsules are sized by means of specially developed welding apparatus.

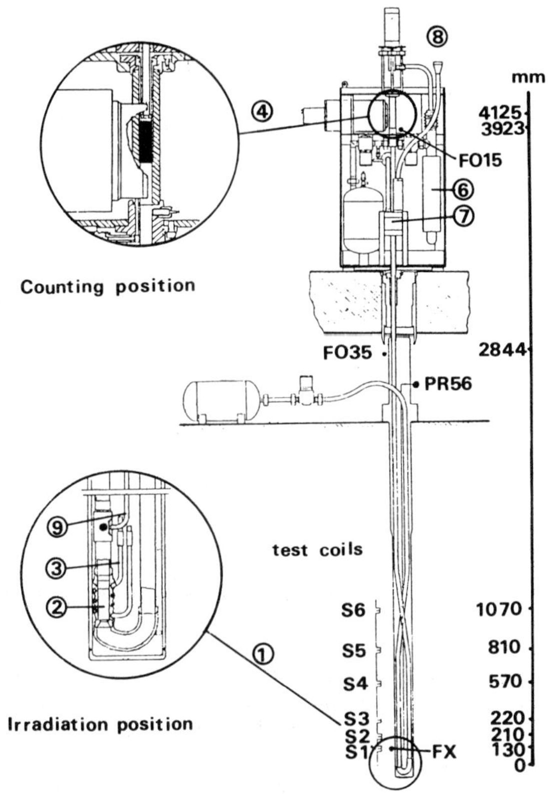

Figure 1: The rapid transfer facility Mach-I with blown-ups of the irradiation and counting positions.

For the transportation of solid magnets in the out-of-pile experiments, the sample capsules were packed between two nylon discs to prevent the magnets from penetrating the polyethylene capsules during the deceleration process.

Timing

For the timing of the capsule transfer into the reactor, the time of entrance is measured by the photosensor F035 and the arrival is monitored in the irradiation region by means of the pressure transducer PR56 that is placed outside the reactor. In the irradiation zone (1), a pressure monitoring tube (9) is mounted 13.0 cm above the irradiation position and extends to the pressure transducer over a distance of 248.0 cm. For the timing of the capsule transfer, the time intervals

Measured irradiation time	: T_i	= PR56 - F035
Measured transfer time out	: T_t	= F015 - PR56
Measured transfer time in	: F_i	= F035 - PR56
Sound transmission time	: F_o	= PR56 - FXo
True irradiation time	: t_i	= FXo - FXi
True transfer time	: t_t	= F015 - FXo

were defined between the various sensors and the pressure transducer.

During the out-of-pile investigations, a photosensor FX was mounted opposite the pressure monitoring tube, and 6 coils (S1-S6) were connceted in series and mounted on the transport tube as illustrated in Figure 1. For the transferred magnets, the inductive signals were recorded on a 4-channel Racal recorder and replayed on a Textronic 2623 Oscilloscope with storage mode. From known distances of the coils, the time-distance functions could then be constructed by means of the point of intersection of the coil signals and the photosensor times.

RESULTS AND DISCUSSION

In out-of-pile investigations (Schmidt 1982), the timing calibration and time-distance functions were studied of the capsule transfer under conditions of variable transfer parameters such as capsule weight [4 - 7 g], preset irradiation time [80 - 9900 ms], input pressure [0.75 - 0.81 MPa] with [1.5 - 4.0 turns] adjustment, output pressure [1.06 - 1.12 MPa], and rig/plug temperatures [20 - 60°C].

Timing calibration

From the time-interval definitions in Eqs. (3 - 8), the true irradiation time t_i and the true transfer time t_t becomes

$$t_i = T_i - F_i - F_o \quad (9)$$

$$t_t = T_t + F_o \quad (10)$$

For the conditions of a fixed input pressure of 0.70 MPa with 1.5 turns adjustment, 1.12 MPa output pressure, 40 and 70 ms opening times of the input- output magnetic valves, respectively, 20°C rig/plug temperatures, and preset irradiation times larger than 120 ms, i.e. the times between the opening of the input- and output magnetic valves, the following models are suggested:

$$t_i = T_i - 0.92\,w - 42.0 - 9.6, \quad s = 1.5 \text{ ms} \quad (11)$$

$$t_t = T_t + 9.6 \quad , \quad s = 1.0 \text{ ms} \quad (12)$$

with 0.92 w + 42.0 ms being the transfer time Fi into the irradiation position for a total capsule weight of w gram and 9.6 ms being the sound transmission time F_0. The standard deviations are 1.5 and 1.0 ms, respectively, for 153 and 63 degrees of freedom.

For the calibration models, the transfer time F_i and the sound transmission time F_0 were studied in a randomized block design with the rig/plug temperatures being the block-effects. In a five-way analysis of variance, the input pressure and capsule weight showed significant effects on F_i and the measured irradiation time T_i showed a significant effect on the sound transmission time F_0. For the latter, the significance indicates that the gas-system has not reestablished itself in the monitoring tube or in the transport system before the output gas-pulse arrives. For preset irradiation times shorter than 120 ms, the sound transmission F_0 can no longer be assumed constant, but must be assumed less than 9.6 ms.

Assuming that the thermal neutron flux can be approximated by a rectangular pulse shape, a first order evaluation gives a variation of 3.5% in the produced radioactivity for a nuclide with 20 ms half-life, when the 1.0 ms standard deviation is propagated for the transfer time.

Time distance functions

In the Figures 2 and 3, the observed and calculated time-distance functions are shown for two irradiations obtained under similar conditions of 7.5 g capsule weight, 80 ms preset irradiation time, 0.71 MPa input pressure with 4.0 turns adjustment, 1.12 MPa output pressure and 20°C rig/plug temperatures.

In Figure 2 it is seen that the two input functions are linear and that they coincide. From 16 transfer measurements it was observed that the capsule velocity depends on the input pressure conditions and that it can be estimated at the F035-photosensor by the length of the F035-signal and the length of the capsule.

In Figure 3 it is seen that the two output functions differ significantly from each other, although the transfer conditions are kept identical. By a non-linear optimization algorithm (Rothamsted 1977), the parameters have been estimated of the theoretical time-distance functions. The agreement is quite acceptable on the part of the transfer until the F035-photosensor. For the last part, the gas-compression becomes significant and the observed values differ from the expected since the gas-compression is not considered in the model for this part of the capsule transfer. From the transfer measurements, the strongly correlated parameters showed a large scatter but about 90% of their variation could be described by the measured transfer times in a weighted regression analysis with 12 degrees of freedom.

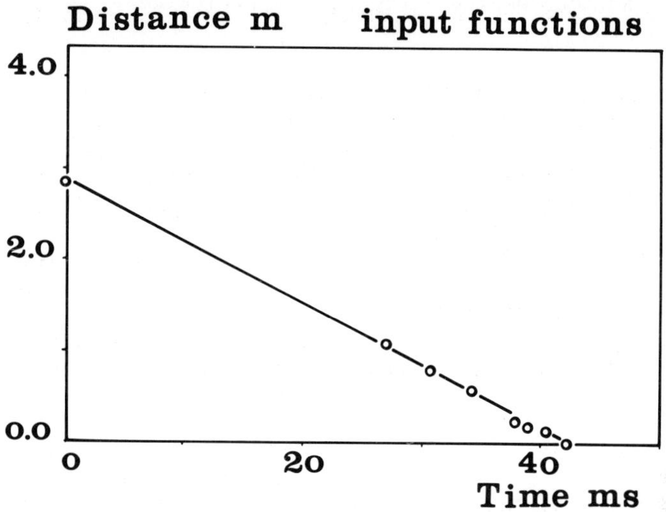

Figure 2: Time-distance input functions for two similar capsule transfers.

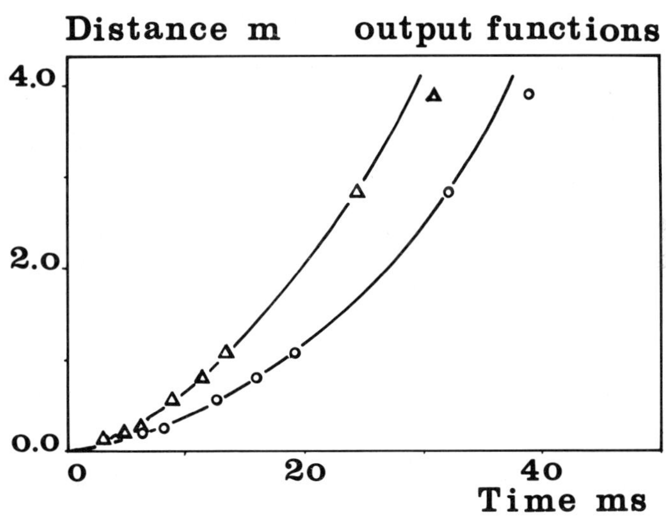

Figure 3: Time-distance output functions for two similar capsule transfers.

For the time-distance functions, the differences are caused by the different degrees of tightness for the capsule capture in the stopping cone at the irradiation position.

Thermal neutron flux density

For the 4RV3 test hole facility, the average thermal neutron flux density was measured with Co-monitors at approximately each 10 cm of the transport tube length in the Mach-I rapid transfer facility. At each point, the perturbed thermal neutron flux density was calculated by a simple method of neutron absorption (Wiig 1981). In Figure 4, the measured average thermal neutron flux density and the perturbed thermal neutron flux densities are shown.

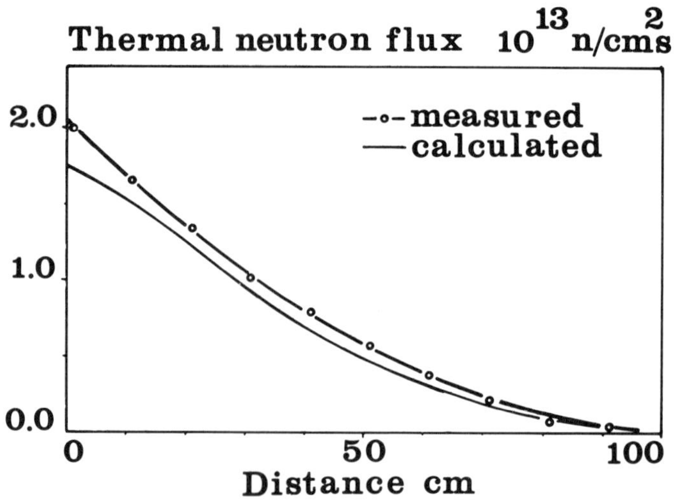

Figure 4: Measured and calculated (perturbed) thermal neutron flux densities in the Mach-I irradiation facility.

Combining the perturbed thermal neutron flux density profile and the time-distance functions of the capsule transfer, the perturbed time-dependent neutron flux density is obtained for the sample. For short irradiation times, it looks like the shape of a rectangular pulse with the edges rounded off. In Figure 5 two time-dependent thermal neutron flux densities are shown for two capsules transferred under similar conditions. It is seen that the two neutron flux density pulses do not coincide. They have the same front-edges, but both their length and their back-edges differ. As mentioned earlier, this is due to different captures of the capsules in the stopping cone at the irradiation position. Thus, depending on the size match between the cone and the capsule, the capture may vary. If the capsule is caught firmly, the irradiation time becomes relatively longer and the transfer time relatively shorter than otherwise.

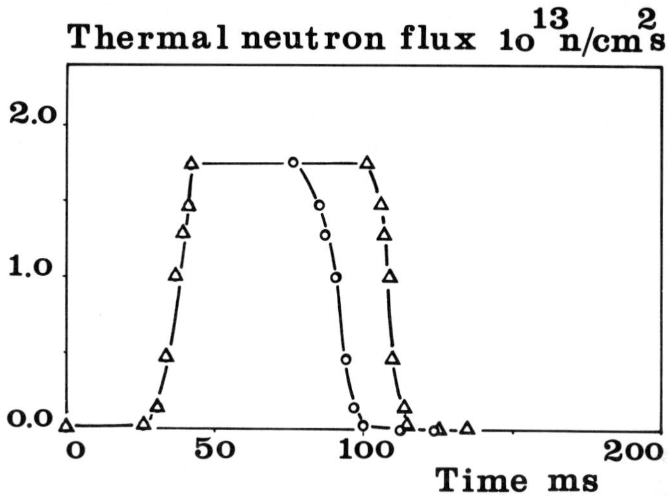

Figure 5: Time-dependent pertubed thermal neutron flux densities for the capsule transfers shown in Figure 2 and 3.

Variation of the produced radioactivity

For the observed variability of the thermal neutron flux density, the variation has been estimated of the produced radioactivity at the counting position for a radionuclide with 20 ms half-life. For the total radioactivity that is composed of a rectangular part and a back-edge part of the pulse, a total variation of 19.6% was obtained for the studied range of the transfer parameters with contributions of 45% and 55% respectively for the two regions. Extrapolating to an infinitely long neutron flux density pulse, the variation reduces to 4.3% for the total radioactivity.

In order to reduce this variation, regression analysis was performed both for the total radioactivity and for the radioactivity from the back-edge, respectively. For the total radioactivity produced, no transfer parameters could describe the variability, but for the radioactivity from the back-edge, the transfer time could explain 84% of the variation with a standard deviation of 1.6% for the extrapolated case of irradiation to saturation activity with 18 degrees of freedom. This shows that it is possible to reduce the variation of the produced radioactivity by means of an estimation of the perturbed time-dependent neutron flux density. The applicability of such a procedure, however, depends on the stability of the neutron flux density profile and on the quality of the estimation of the actual time-distance functions.

To avoid these difficulties, it was attempted to reduce the effect by omitting the sizing of the bottom-end of the capsule. Fortunately enough, in-pile irradiations showed that the standard deviation could be reduced from about 3 ms to 1 ms of the transfer time. The standard deviation could even be reduced to as little as 0.5 ms by using the same capsule 2-3 times. With such a small standard deviation, the capsules have almost the same time-distance functions and the variation is reduced to insignificance for the produced radioactivity at the counting position due to variations in the tansfer time.

CONCLUSION

For preset irradiation times exceeding 120 ms, the timing can be accurately estimated of the true irradiation and transfer times with standard deviations of 1.5 ms and 1.0 ms, respectively, by means of the capsule weight, the measured irradiation time and the measured transfer time. For preset irradiation times less than 120 ms, the in- and output gas pulses interact and yield sound transmission times less than 9.6 ms in the pressure monitoring tube. Due to different degrees of size mis-match between the capsule and the stopping cone, the output time-distance functions may differ between capsules transferred under similar conditions, whereas the input time-distance functions may be assumed constant. The variations may cause a standard deviation of about 3 ms in the transfer times, which can be reduced to about 0.5 ms by reducing the apparent size mis-match. For the produced radioactivity, the variation can be reduced to about 1.6% for a radionuclide with 20 ms half-life by assessment of the perturbed thermal neutron flux density. By removal of the size mis-match, the variation can be reduced to insignificance.

ACKNOWLEDGEMENT

I deeply wish to acknowledge the interest shown by J.Seir Olsen, L.Palgaard, J.Westermann, K.Heydorn and P.Wiig, just as I wish to thank the Danish Natural Research Council for supporting my work at the Atomic Institute of the Austrian University of Vienna.

REFERENCES

CANIZZARO F., GRECO G., RIZZO S., SINGRA E. (1978)
Results of the Measurement Carried out in Order to Varify the Validity of the Poisson-Exponential Distribution in Radioactive Decay Events.
Int.J.Appl.Rad. and Isotop. Vol 29, 649-652

GLASSTONE S. (1955)
Principle of Nuclear Reactor Engineering
Van Nostrand Company, INC, New York, pp 861

ROTHAMSTED EXPERIMENTAL STATION (1977)
GENSTAT. A general statistical program. Release 4.03.
Rothamsted, England

SCHMIDT J.O. (1981)
Statistical Treatment of Dual-Parameter Multiscaling Spectra from Fast Instrumental Neutron Activation Analysis.
Proceedings from Symposium i Anvendt Statistic Editor: A.Hüskuldsson, NEUCC, The Technical University of Denmark

SCHMIDT J.O., PALGAARD L., WESTERMANN J., MCLAIN M.E. (1981)
Determination of Boron by Fast Instrumental Neutron Activation Analysis.
International Conference on Modern Trends in Activation Analysis Toronto, Canada, June 15-19, published in J. Radioanal.Chem. Vol. 72, No. 1-2, 425-436

SCHMIDT J.O. (1982)
Aspects of Rapid Instrumental Neutron Activation Analysis Using Short-Lived Radionuclides. (Ph.D.Thesis, pp 285).
Isotope Division, Chemistry Department, Risø National Laboratory

SEIGEL A.E. (1965)
The Theory of High Speed Guns.
Agardograph 91, pp 313
United States Naval Ordance Laboratory
White Oak, Silver Spring, Maryland

WESTERMANN J., SCHMIDT J.O., PALGAARD L. (1980)
Beskrivelse af Mach-1 Ultrahurtigt Rørpostsystem til Neutronaktiveringsanalyse på DR3.
Internal report, Nr. 508, Forsøgssektionen, 81 pp
Risø National Laboratory

WIIG P. (1981)
Simple Model for Calculation of the Thermal Neutron Flux Pertubation.
Private communications, DR-3, Risø National Laboratory

CAPSULE TRANSPORT EVALUATION OF A RAPID He-TRANSFER SYSTEM WITH A ^{6}LiD-CONVERTER

A. Salahi, F. Grass, F. Bensch, G. Zugarek, J.O. Schmidt*

Atominstitut der Österreichischen Universitäten

Schuettelstrasse 115, A-1020 Wien, Austria

INTRODUCTION

For neutron activation analysis, Dams (1981) has shown that nearly all elements have potential short-lived radionuclides when different irradiation techniques are used. At the Atominstitut Wien we have been devoted to the field of short-lived radionuclides applying half lifes down to less than 20 ms (Brandstatter et al. 1970). To increase the range of analytical applications the activation can be performed with 14 MeV neutrons to measure or increase the sensitivities for a number of interesting elements like O, Si, P, K, As, Y, Ba, and Pb (see, e.g., Fanger et al. 1981). To meet the requirements of a high fast neutron flux density a rapid He-driven transportation facility was, therefore, suggested with a 14 MeV neutron converter (Salahi and Grass 1981) which enables both fast and thermal neutron activation analysis with short-lived radionuclides.

Following a description of the rapid He-driven irradiation facility and the preliminary test setup studies are reported of different parameters such as the gas pressure, transfer times, stopping effects in the irradiation position, and transfer velocities.

EXPERIMENTAL

A rapid He-driven irradiation facility with a 14 MeV neutron converter has been constructed for insertions in the light-water

*Isotope Division, Chemistry Department, Risø National Laboratory, DK-4000 Roskilde, Denmark

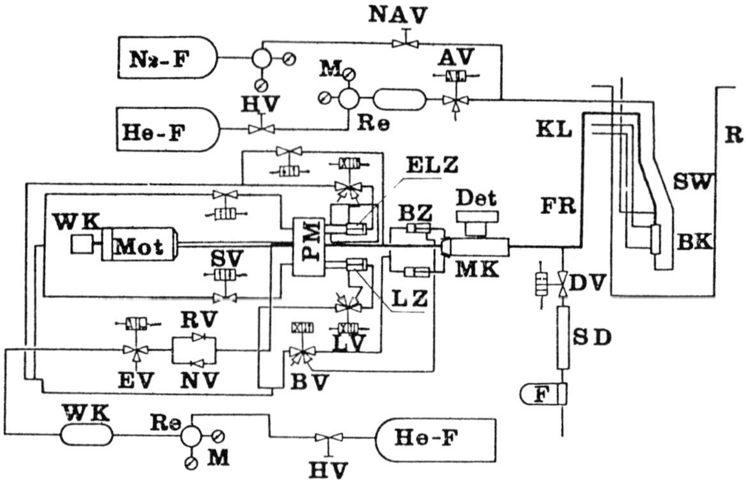

Figure 1 Sketch of the rapid He-driven irradiation facility

TRIGA Mark II reactor at the Atominstitut Wien. Sample management and irradiation processing are controlled by a microprocessor, photosensors, magnetic valves, optical encoders, and revolving stations.

System Description

In Figure 1 the rapid irradiation facility is sketched. By means of a loader (LZ) polyethylene capsules are injected into the sample changer (PM) for irradiation, discharge or storage controlled by a 8073 microprocessor.

The irradiation sequence is initialized by opening the valve (EV) upon which the capsule is transferred through the counting chamber (MK) into the irradiation position (BK) by means of the transport tube (FR). On its way the capsule triggers the Klixon K-X-4, the microswitch (SW; Texas Instruments) and is stopped by a narrowing of the transport tube. In the irradiation position a remotely controlled 14 MeV neutron converter surrounds the light-water cooled transport tube (KL).

After the preselected irradiation time the magnetic valve (AV) is opened and the capsule returns to the counting chamber (MK), where it is arrested by a specially developed stopping device. On its way the capsule triggers the microswitch (DA) and becomes decelerated by injecting He through the magnetic valve (EV) on the last part of the transfer. Behind the capsule the pressure is released through the valve (DV) and the gas is discharged into a filter (F) through a muffler (SD).

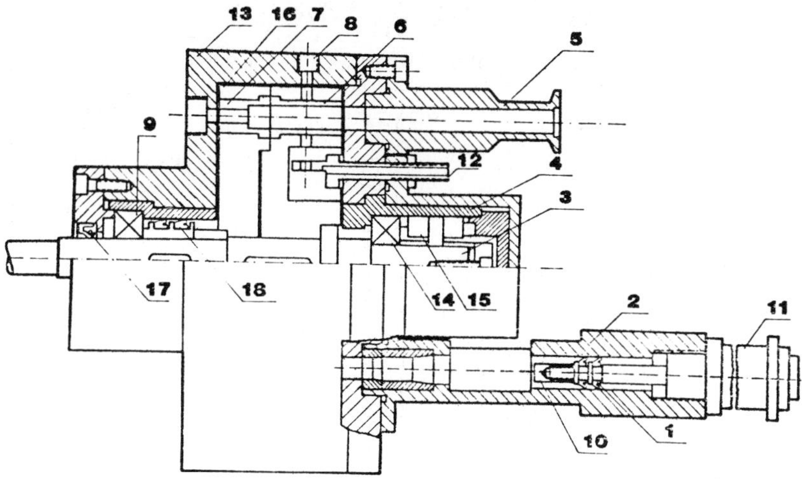

Figure 2 Sketch of the sample changer

Following the counting of the irradiated sample a new irradiation sequence can be initiated or the capsule can be returned to the sample changer for storage or discharge by an unloader (ELZ). In case of a return failure the capsule can be blown out manually by means of the valve (NAV). All magnetic valves are from Ross or Schrader.

Sample changer

In Figure 2 a vertical cross section is shown of the sample changer that consists of a revolving station (16) with storage capacity for 16 capsules in an aluminum housing (13). To ensure correct storage position an incremental optical encoder (Rotaswitch 882-400-1(b) LS-TTL, Disc Instrument, INC) is mounted on a shaft (3) with bearings (9, 14, 15) to resist the capsule impact. Each storage position consists of a polyamid stopping cone (7) and a perspex cladding that is transparent for the photosensor (8, 12) signals. The system is kept tight by sealing rings (17, 18).

Capsules are injected into the rotation station (16) by means of the loader (2) that consists of a pressure driven (10) piston (11). By means of the incremental optical encoder and a motor the chosen storage position is aligned (5) for either irradiation or discharge.

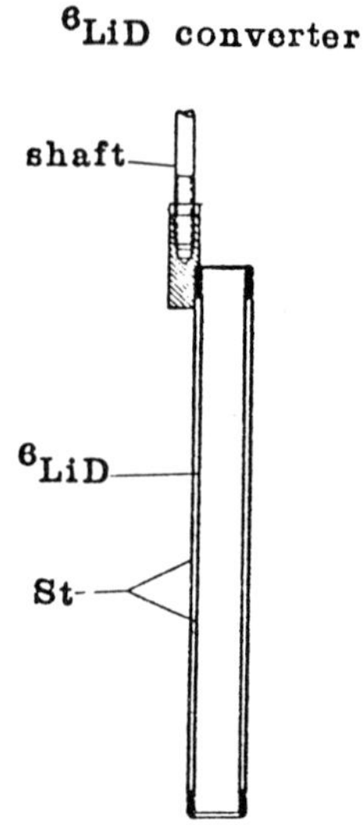

Figure 3 Sketch of the ^{6}LiD-converter

14 MeV-Converter

For the production of 14 MeV neutrons the thermal neutron reaction $^6Li(n_{th},\alpha)^3T(^2D,\alpha)n_{14}$ is used (see, e.g., Baumgärtner 1963). The dimensions of the converter are chosen based on the recommendations by Zillner and Bensch (1981). Accordingly the transport tube is surrounded by a 1 mm ^{6}LiD layer enclosed in two concentrical cylinders that can be moved up and down by remote control outside the reactor.

Counting Chamber

Figure 4 shows a vertical cross section of the counting chamber. In front of the Ge(Li)-detector (3) the capsule is fixed by means of a stopping device (6) made of spring steel and supported by two hydraulic cylinders (5) mounted to the support (4). The yellow brass supports (1, 8) are connected by a perspex block (2), which at the same time is applied as Cerenkov radiator.

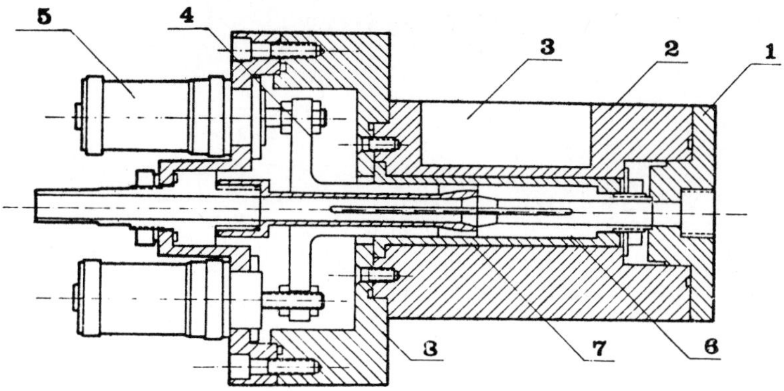

Figure 4 Sketch of the counting chamber

Irradiation Processing

In Figure 5 the flow chart of the irradiation processing is given. By means of a national semiconductor 8073 microprocessor (μP), signals are supplied to the step motor (SM) and the valve controls (V). From the sensors (S) and the incremental optical encoder (IOE), signals are received and information can be exchanged with the memory (M), the PDP 11/34 computer (PDP), the screen terminal (CRT) and the gamma-ray spectrometer (GS).

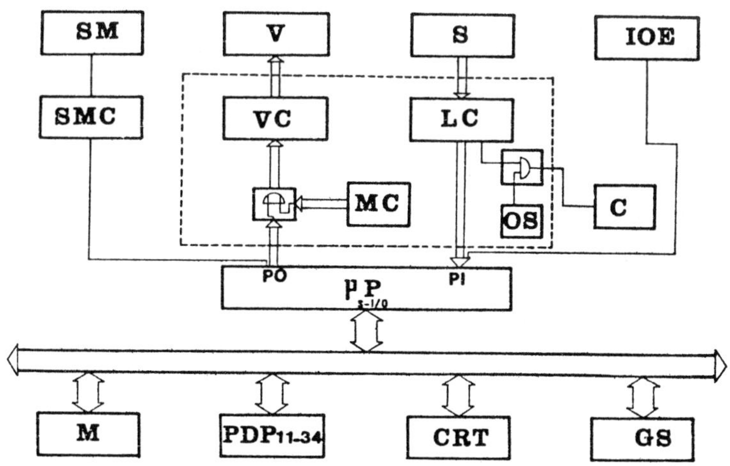

Figure 5 Flow chart of the irradiation processing

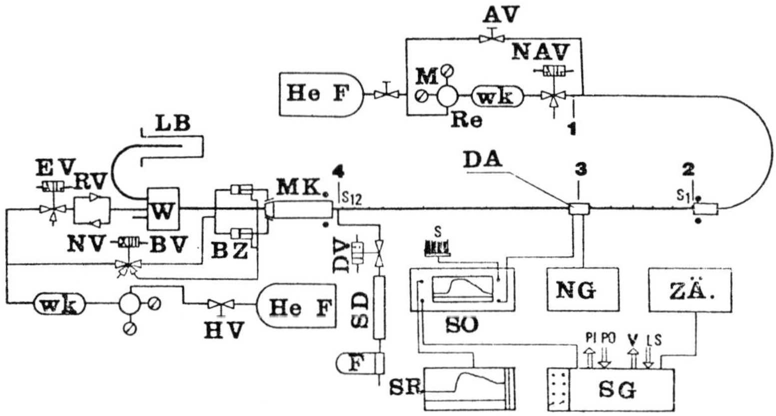

Figure 6 Sketch of the preliminary test setup

Preliminary test setup

To investigate the gas expansion that drives the capsule, the sample transportation is studied through measurements of the gas pressures, the time-distance functions and the total transfer times with the preliminary test setup shown in Figure 6.

The capsules are inserted at (4) and transferred to the irradiation position (2) by opening the magnetic valve (EV). The pressures are measured by means of an absolute pressure sensor (DA; national semiconductor LX1830A) which leads its signals directly to a Gould digital storage oscilloscope OS4100 (SO). For the study of time-distance functions, magnets are transferred which induce currents in the coils (S1 - S12) mounted on the transport tube. The total transfer time is measured by means of photosensors that are placed at the positions 2 and 4 over a length of 8 m.

RESULTS AND DISCUSSION

By means of the preliminary test setup a number of transfer characteristics have been studied without counter gas to decelerate the capsule before capture in the counting position. For real irradiations, however, the counter gas is an absolute requirement for a destruction-free arresting of the capsule.

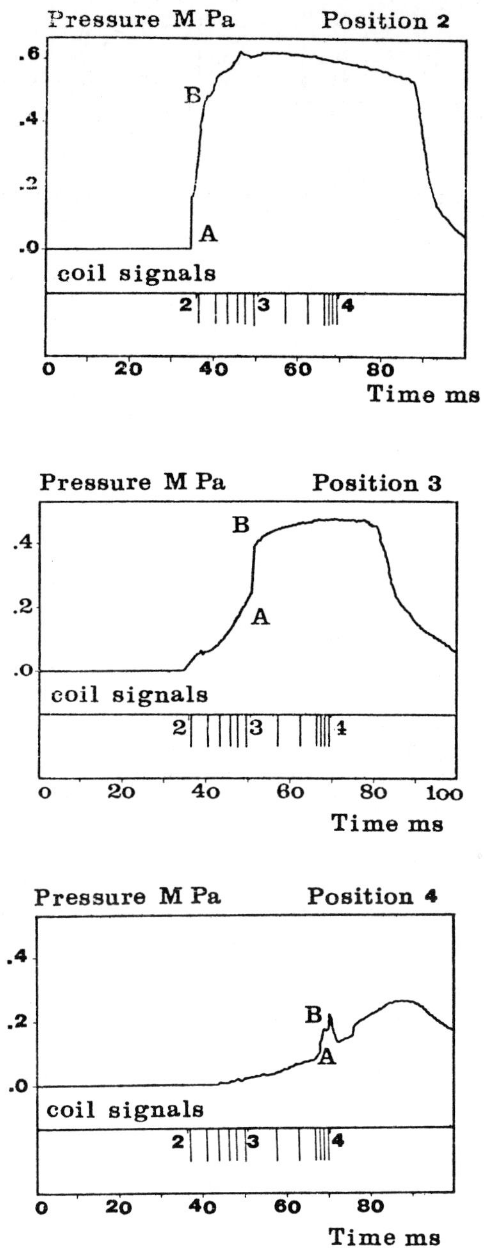

Figure 7 Pressures at the positions 2, 3 and 4 during capsule transfer

Capsule Driving Pressure

In Figure 7 the pressures and coil signals are shown for the preliminary test setup with the trigger pulse for the magnetic valve (NAV) defining the time origin. With the valve (NAV) set at an opening time of about 55 ms the pressure changes significantly to a chamber pressure of 1.23 MPa in position 1. As expected, both the maximum pressure as well as the shape of the pressure pulse decrease during the gas expansion (see, e.g., Seigel 1965 or Bratter et al. 1976).

From the shape of the pressure pulses in the positions 2 and 3 it is seen that the net pressure decreases from about 0.5 MPa to about 0.2 MPa over the capsule length (marked A and B). Further, it is seen that the He-gas partly passes the capsule and builds up a pressure together with compressions in the front gas. It is also seen that in front of the capsule, a fast moving shock wave is built up with a velocity of about 1000 m/s.

Pressure Characteristics

In Figure 8 a summary is given of the different pressure parameters at the various distances both with and without the capsule being transferred. When no capsule is being transferred it is seen that both the maximum dynamic plus static pressure (1) and the maxi-

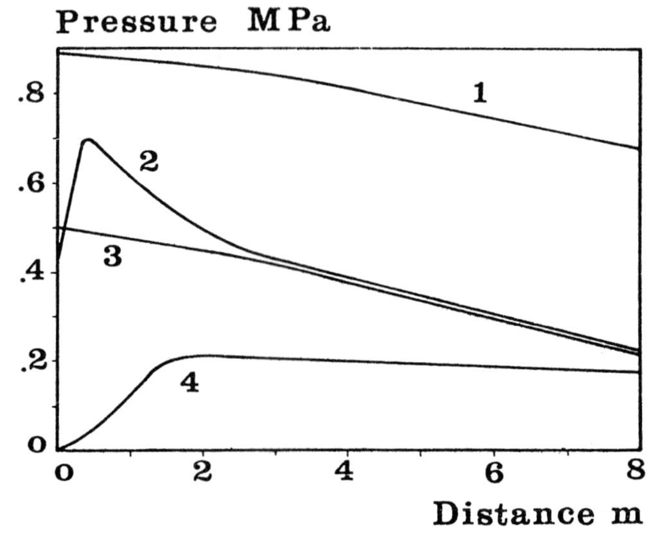

Figure 8 Dynamic and static pressure characteristics

mum static pressure (3) decrease significantly over the transport length of about 8 m. Also in the case of a capsule being transferred the pressure behind the capsule (2) decreases significantly, whereas the pressure in front of the capsule (4) remains stable having once reached its maximum.

Transfer Time

In Figure 9 the relation is given between the output chamber pressure (WK) and the transfer times. It is seen that the transfer times can only be obtained down to the 30 ms range for chamber pressures larger than 1.2 MPa.

Capsule Stopping Effect

Figure 10 shows the special effect of the different pressures behind the capsule when the capsule is caught more or less weakly in the irradiation position. The upper part shows the pressure pulse of a firmly caught capsule. It is seen that the pressure is much higher than for a weakly caught capsule. Further, the comparison shows that the transfer time is only increased about 1 ms, whereas the irradiation time is increased about 14 ms before the capsule take-off. The measurements thus indicated a bad size-match between the conical stopping cone and the capsule and a new cone was, therefore, constructed by which the effect could be reduced to insignificance.

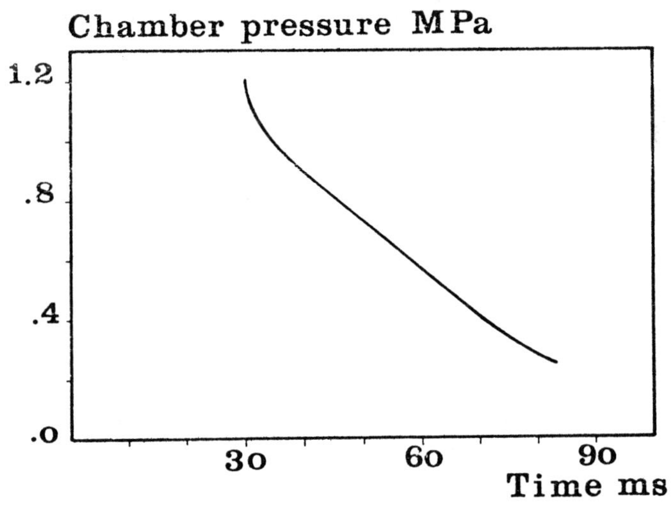

Figure 9 Output chamber pressure versus transfer time

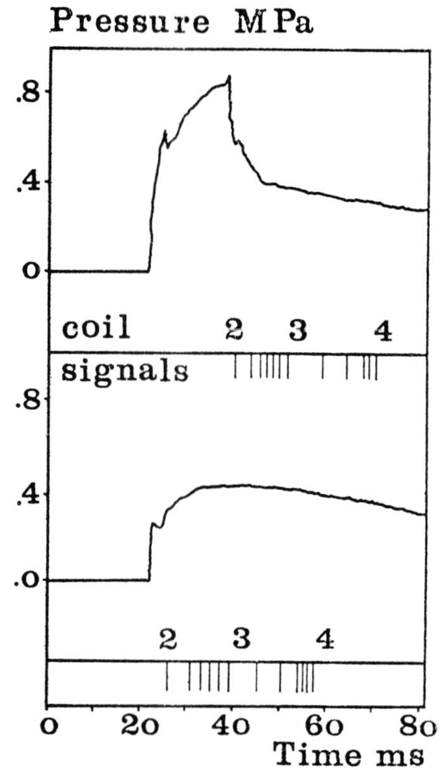

Figure 10 Capsule stopping effect in the irradiation position

Velocity Measurements

To illustrate the effect resulting from the pressure reduction of the expanding gas, the output valve (NAV) was moved to position (2) and the time-distance function was compared with the one obtained from the original location at (1). In Figure 11 two time-distance functions are shown. It is seen that the limiting velocity is significantly larger for the valve placed at (2) than at (1).

Vertical Thermal Neutron Flux Profile

From the obtained reproducibility of transfer conditions the variations can be kept to a minimum for the radioactivity produced due to variations in the time-distance functions for the individual irradiation. Further, in comparison with heavy-water moderation the thermal neutron flux density profile is limited to a short region, which reduces the influence from differences in time-distance func-

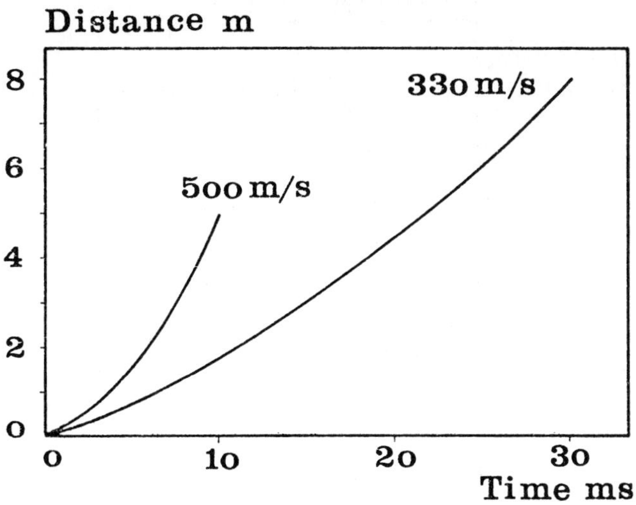

Figure 11 Time-distance functions for two different pressures behind the capsule

tions. The changes of sample activity are more sensitive to rebound distances of the sample capsule than to variations of the neutron flux density profile. For irradiation times long enough to allow stabilization of the capsule in the irradiation position and leading to saturation activity of the radionuclide, the effect is reduced to significance.

CONCLUSION

For application in neutron activation analysis with thermal and fast neutrons a rapid He-driven irradiation facility with a 14 MeV neutron converter has been constructed and its sample transportation has been evaluated. The studies showed that transfer times less than 30 ms require pressures larger than 1.2 MPa for a distance of 8 m and that the net pressure changes significantly for the sample transportation. Furthermore, the studies shown that the capsule stopping effect can be reduced to insignificance in the irradiation position by means of a correct size-match between the capsule and the stopping cone.

ACKNOWLEDGEMENTS

We wish to thank the Danish National Research Council, Mr. G.Kutschera and Mr. P.Schröder for constructing assistance as well as the Fonds zur Förderung der wissenschaftlichen Forschung for financial support.

REFERENCES

BAUMGÄRTNER, F (1963)
Der Umwandlungsfaktor für thermische Neutronen in schnelle von 14 MeV durch ^{6}LiD und die Herstellung einer ^{6}LiD-Bestrahlungsröhre.
Radiochemica Acta, Bd. 1, Heft 4, 222 - 225

BERWANGER, H G N (1964)
Der Neutronenfluß im Austria 100 TRIGA Mark II Reactor.
Thesis, Atominstitut der Österr. Universitäten

BRANDSTÄTTER, O, GIRSIG, F, GRASS, F, KLENK, R, BAUER, R (1970)
Zerstörungsfreie Analysen mit kurzlebigen Kernzuständen
Atomkernenergie 15, 285 - 287

BRÄTTER, R, GATSCHKE, W, GAWLIK, D, MÜLLER, J, WENKEL, W (1976)
Meßsystem für schnelle Neutronen - Aktivierungsanalyse an BER II.
HMI-B 188, 125

DAMS, R (1981)
Selection of Short-Lived Isotopes for Activation Analysis with Respect to Sensitivity.
J.Radioanal.Chem. Vol. 61, 1 - 2, 13 - 36

FANGER, H, PEPELNIK, R, MICHAELIS, W (1981)
Fast-Neutron Activation Analysis with Short-Lived Nuclides.
J.Radioanal.Chem. Vol. 61, 1 - 2, 147 - 163

ZILLNER, K, BENSCH, F (1981)
Neutron-Physical Aspects of an Incore Generator for 14 MeV Neutrons.
J.Radioanal.Chem. Vol. 61, 1 - 2, 191 - 194

SALAHI, A, GRASS, F (1981)
A Rapid Transportation Facility for Irradiation with Thermal and Fast Neutrons.
J.Radioanal.Chem. Vol. 61, 1 - 2, 63 - 68

SEIGEL, A E (1965)
The Theory of High Speed Guns.
Agardograph 91, 313, U.S. Naval Ordance Lab., White Oak, Silver Springs, Maryland

Experimental facilities for PEC Reactor Design

Central Channel Test Loop: CPC-1 - Thermal Shocks Loop: CEDI

C.CALVARESI - L.F. MORESCHI

E.N.E.A. Centro del Brasimone -

Casella postale, 1 - Castiglione dei Pepoli (BO)

1. Introduction

PEC (Prova Elementi di Combustibile: Fuel Elements Test) is an experimental fast sodium-cooled reactor with a power of 120 MWt. This reactor aims at studying the behaviour of feul elements under thermal and neutron conditions comparable with those existing in fast power nuclear facilities.

In the core of the PEC reactor the elements are positioned so as to form groups of seven. In order to ensure the stability of their position, the six outermost elements are elastically forced against the central element.

Thanks to a central test channel, installed in the core, it is possible to submit an exprimental fuel element to a radiation by a neutron flux of $4.3 \cdot 10^{15}$ n/cm^2 . sec.), under thermal condition which are more critical than those in the rest of the core.

The central test channel, with a length of about 16 m. and an external diameter of about 185 mm., is placed vertically on the axis of the reactor thank; the connections to its cooling circuit and the acces for the insertion or extraction of the experimental element are located in the transfer cell;

The indepentent cooling of the central channel allows us to carry out some particular test, such as thermal cycles and the simula-

von der Hardt and H. Röttger (eds.), Irradiation Technology, 709–717.

tion of thermo-hydraulic transients in the fuel elements. Therefore, it will be possible to enlarge the existing knowledge about the fuel to be used in fast power reactors, as far as both the behaviour of materials and the diagnostics of accidents are concerned, thus raising the safety standars of this kind of reactors.
Gives the particular structure of the core, the complex operations to be performed in the transfer cell and the strict operating conditions of the central channel, two experimental facilities, CPC-1 and CEDI, have been designed as a support to the construction of the reactor.
CPC-1 is a 1:1 scale model of the channel, transfer-celle and loop unit of the channel, whereas CEDI is a sodium-cooled loop which allows us to carry out test of isothermal endurance and thermal shocks on the group of seven forced elements, by simulating the thermo-hydraulic and mechanical conditions exiusting in the reactor.

2. CPC-1 experimental tests

These experiments aim to carry out an overall check on the test channel, trying to reproduce on a prototype of the component all the working, constraint and interface conditions that will occur in the component to be installed on the reactor.
In this way is possible to check the channel on an endurance basis, by reproducing on its structure all those stress distributions that, if associated with high temperatures, can bring about phenomena of permanent deformation (creep, ratcheting); such phenomena are capable of impairing, in the long run, the functional features and the structural integrity of the component.
As a matter of fact, such stress distributions (particularly the secondary and localized ones) are brought about by not completely predictable causes. In fact, as they can be foressen just on a qualitative basis, these causes cannot be reduced to fixed patterns by means of calculation, nor can they be reproduced on semplified models.
Besides, owing to the particular functions of the test channel wich need frequent and complex handling operations (for instance the disassembly of the channel itself) that, in turn, can be impaired by structural deformations caused by the hot-functioning and/or by sodium, a check of all the handling procedures concerning the channel proves necessary.
Thus, it can be stated that all those aspects of the test channel behaviour which are not linked to neutron radiation can be studied. It is so possible to provide useful evidence for the construction

of the channel to guarantee both the good functioning of the component and its handling at the end of the running period.

Description of the CPC-1 facilities

The sodium-cooled CPC-1 facilities (Fig. 1) consist of two independent loops called "CS1" and "CS2" and of a mock-up of the transfer celle with its own loop "CS3". The main loop "CS1" reproduces in 1:1 scale the primary loop of the test channel with its main components. In particular, it is formed by a components tank "S1" which contains two prototypical centrifugal pumps, by the channel and by the connection pipes. By means of on-off valves it is possible to obtain an "8 shaped" course and by using an electric heater, an aereotherm and a heat exchanger, we con obtain an high temperature (600°-650°C) in the channel, keeping loow ($<$450°C) the Components tank temperature.

The in reactor-temperature distribution along the channel is obtained, in CPC-1 facility, by means of an electrically heated fuel element and reduced flow.

Outside the channel, the CS2 loop reproduces environemntal conditions similar to those existing in the reactor. It consists of a tank "S2" that simulates the reactor tank of an electromagnetic pump, a heater and a cooler.

The transfer cell is a 1:1 scale mock-up of the corresponding PEC cell (Fig. 2).

In the Cell there are two areas, both reproducing the PEC central experimental area.

The first area is bound to contain a channel prototype (CP) for thermo-fluid-mechanical tests connected to the CS1 loop.

The second area is bound to contain another channel prototype (CM) for handling tests, with or without sodium; when the circulation of sodium takes place, this prototype is connected to the CS3 loop.

Besides, other connections have been placed on the floor of the Cell: they are for the transit channel prototype (CT), the mock-up of the decay pool and the equiments storage.

On the roof of the Cell there are two openings, endowed with sealing valves, placed on the axis of channel prototype (CP) and of the handling channel (CM).

The Cell is gas-tight and is endowed with a conditioning equipment to reproduce the PEC transfer-cell environmental conditions.

TABLE I contains the CPC-1 main parameters.

- nominal flow-rate	70 m^3 h^{-1}
- channel pressure drop	12.5 Kg cm^{-2}
- channel max temperature	650°C
- components tank max. temp.	450°C
- transfer cell loop (CS3) max. temperature	250°C
- thermal shocks: max. Δ T	175°C (625° ÷ 450°)

4. CEDI experimental test

The experimental programme supporting the desing of the PEC reactor core, envisages a series of endurance and thermal shock test to be carried out in the CEDI sodium-cooled loop for different groups of seven elements forced among them.
The models of the elements are exposed, during a given period of time, related to their life in the reactor, to flow rate, temperature and thermal shock conditions that simulate as much as possible the real operating conditions. This aims at:

a) studyng the behaviour of the group as well as the effects and the evolution of mechanical interactions among the various elements and between the foot and its housing after a long permanence in sodium, in order to discover possible phenomena of seizure, sticking, deposits, plugging, cavitation and vibration brought about by the cooling sodium;

b) checking the effects of the thermal shocks caused by the thermo-hydraulic differences that exist between the various elements of the group;

c) checking the contact force of the elements near the plates and its possible variation at the end of the test;

d) checking the dimensional stability and the good thermo-mechanical behaviour of the elements while the core is operating. In particular, we want to study the effects of thermal gradients, stresses, release of internal stresses, thermal fatigue, etc., considering the deformation of the welded joints on the element and its components;

e) establishing the quality and reliablity standards of the manufacturing process, particularly as to the welded joints and the deposits of stellite, by checking:

- the dimensional stability of the welded components
- the dimensional stability of the stellited areas
- the possible appearance of superficial flaws or cracks in the welded joints and in the deposits of stellite
- the loss of adhesion in the deposit, as far as the stellited areas are concerned.

5. Description of the CEDI facilities

The CEDI facility (Fig. 3) consists of a sodium-cooled loop which allows to carry out tests of isothermal endurance and thermal shocks on the group of seven forced elements, by simulating the conditions that exist in the PEC reactor.
The loop, built in AISI 316 S.S. consists of an expansione tank (VE), which contains a vertical-axis centrifugal pump (PC) for the circulation of sodium. A 6" pipe, equipped with a flow-rate meter (FT1) connects the centrifugal pump to the test section, where the models to be tested are contained.
The 10" return pipe is equipped with, a heater (R 1), and with an unit heater (AE), acting as a cooler. These two components allows us to control the temperature of the circulating sodium.
The loop also consists of a 3" pipe of by-pass to the test section equipped with a flow-rate meter (FT2), which connects the delivery pipe (6") to the return pipe (10") and which displays a 2" motor-driven register (VR) used for the prearrangement of the thermal shocks; a 3" pipe of by-pass to the 6" manual on-off valve (V1), equipped with a 2" rapid-opening pneumatic valve (V2) just used to carry out the thermal shocks.
The test section is composed of a cylindrical tank measuring about 8000 mm. in length and having an internal diameter of 411 mm.
The sodium is let in throungh a 6" gate situated in the lower section of the component, close to the distribution chamber, which contains the grid with the models to be tested.
From the distribution chamber, the sodium passes through a 10" gate.
At the end of the upper section of the well there is a flange, coupled with a blind flange, closing the well and crossed by the penetrations of the equipment.

The equipment of the well is made up of:

- a barrel displaying a series of 8 thermocouples placed at different levels, for the measurement of the axial outline of pre-shock temperature
- a barrel, at the end of which there is a "spider" supporting 7 thermocouples, used to measure the sodium temperature at the head of the test elements during thermal shocks
- a continuos-type indicator to measure the sodium level
- two flow-rate meters FT1 and FT3

The test well is also provided with a heating system, situated outside the bulb screw base to keep hot the loop pipes and to realize the pre-shock temperature distribution along the test section.

TABLE II contains the CEDI main parameters.

- nominal flow-rate	320 m^3 h^{-1}
- pressure drop	12.5 Kg. cm^{-2}
- max temperature	600°C
- thermal shock max ΔT	200°C (600°-400°C)

6. Experimental measurements during operation

During operation it is possible to obtain the following data:

a) Endurance test

- overall sodium flow-rate at the entrance of the well;
- rate of the sodium downward leaking through the clearances between fuel element fast and its housing
- temperature of the sodium at the head of the fuel elements
- pressure drop in the well
- oxygen concentration in the sodium

b) Thermal shock test

- temperature distribution along the test section before carryng out thermal shock
- temperature of the sodium at the leed of the fuel elements to Know $\Delta T/\Delta t$
- overall flow-rate in the test well

CENTRAL CHANNEL TEST LOOP : C P C 1

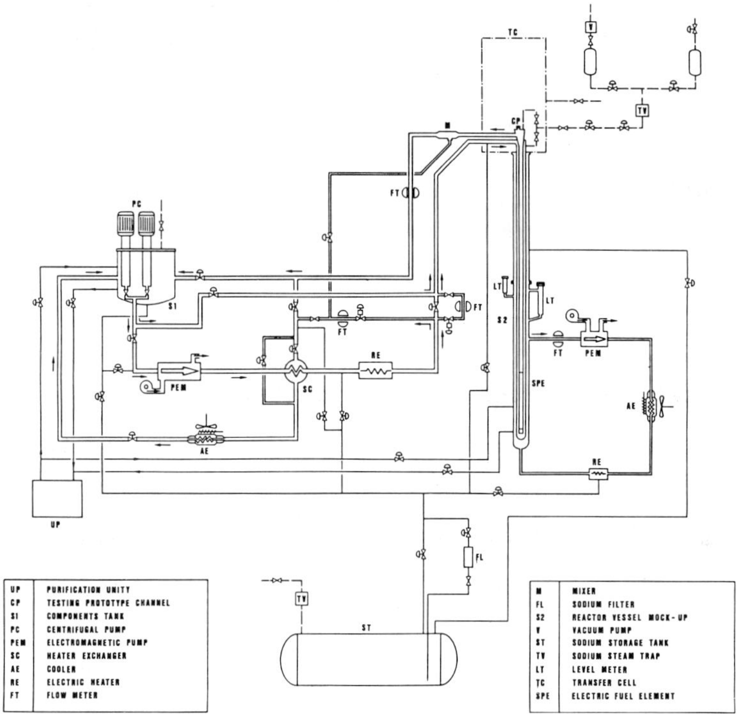

FIG. 1

CPC 1 - TRANSFER CELL MOCK-UP

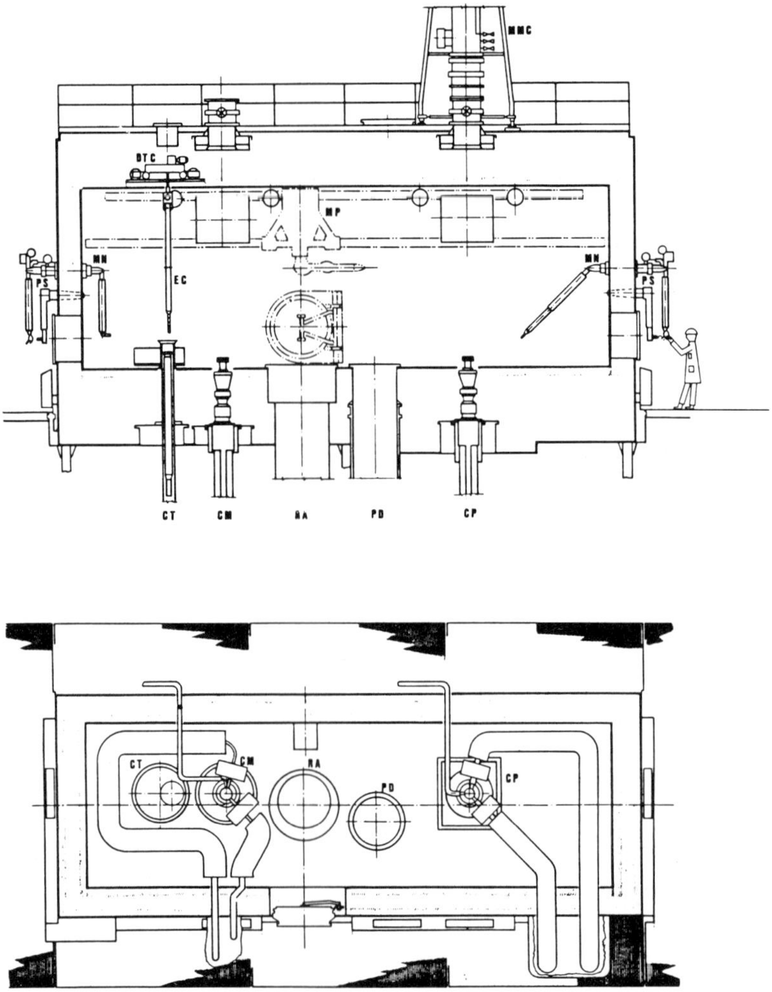

CP	CHANNEL PROTOTYPE
CM	HANDLING CHANNEL
CT	TRANSIT CHANNEL
MMC	HANDLING CHANNEL MACHINE
RA	EQUIPMENTS STORAGE
PD	DECAY POOL

DTC	TRANSFER FUEL ELEMENT MACHINE
MN	MANIPULATOR
MP	POWER MANIPULATOR
EC	FUEL ELEMENT
PS	PERISCOPE

FIG. 2

THERMAL SHOCKS LOOP : CEDI

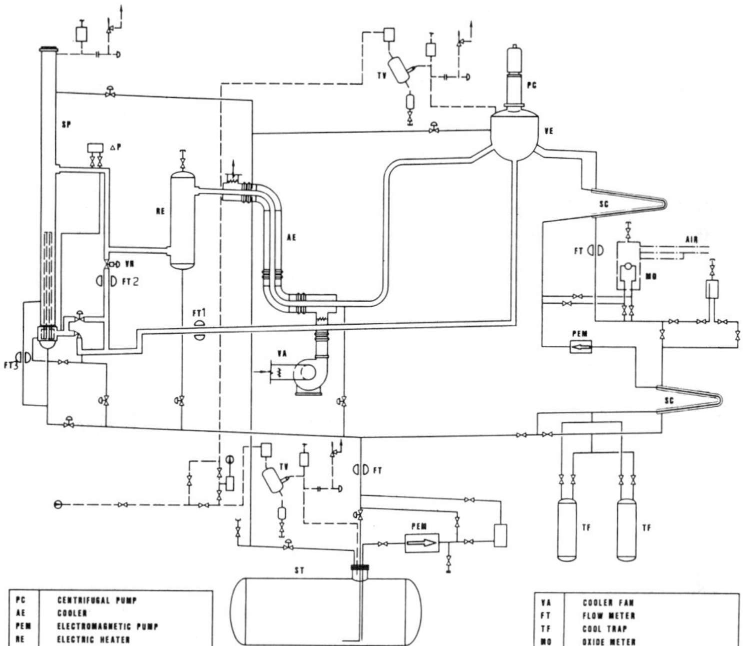

PC	CENTRIFUGAL PUMP
AE	COOLER
PEM	ELECTROMAGNETIC PUMP
RE	ELECTRIC HEATER
SP	TEST SECTION
VE	EXPANSION TANK
SC	HEAT EXCHANGER

VA	COOLER FAN
FT	FLOW METER
TF	COOL TRAP
MO	OXIDE METER
TV	SODIUM STEAM TRAP
ΔP	DIFFERENTIAL PRESSURE METER
ST	SODIUM STORAGE TANK

FIG. 3

PUISSANCE DEPOSEE DANS UN DISPOSITIF COMPLEXE PAR LES GAMMAS DE REACTEUR D'ESSAIS EXPERIENCES ET CALCULS REALISES A SILOE

Collaboration ENEA-CEA

G. Cosoli, R. Caracchini - ENEA Casaccia
H. Petitcolas, A. Besson, A.Bevilacqua, S. Bloch - CEN Grenoble

NATURE DU PROBLEME

Dans une géométrie simple, on sait déterminer la puissance déposée par les gammas de réacteur dans un échantillon, si on connait sa nature sa densité et ses dimensions - si toutefois elles sont petites - à partir d'une mesure de la puissance dans un calorimètre moniteur.
Par contre, cette méthode est insuffisante si les dimensions sont grandes et surtout si la géométrie est complexe, avec juxtaposition de matériaux différents.
Or la puissance déposée par les gammas dans un dispositif d'irradiation est souvent considérable. Seul un calcul performant peut la prévoir, en raison
- de la complexité de la géométrie
- de la multiplicité des sources gamma à spectres différents,
- de la diversité des matériaux plus ou moins absorbants

OBJECTIF GENERAL

Qualifier, par un programme expérimental, le code MERCURE 4, qui calcule, dans une géométrie complexe, la distribution spatiale et spectrale du flux gamma et les dépôts de puissance.

OBJECTIFS PARTICULIERS

Déterminer les valeurs relatives des puissances déposées dans chaque matériau, en fonction du numéro atomique et de la densité.
Corréler la puissance déposée dans les matériaux du dispositif à celle mesurée par le calorimètre moniteur au graphite.

P. von der Hardt and H. Röttger (eds.), Irradiation Technology, 719–737.

SCHEMA D'ENSEMBLE DU DISPOSITIF EXPERIMENTAL

Le prgramme expérimental a été éffectué dans un FOUR CYRANO, dispositif que peut, en particulier, mesurer la puissance totale.

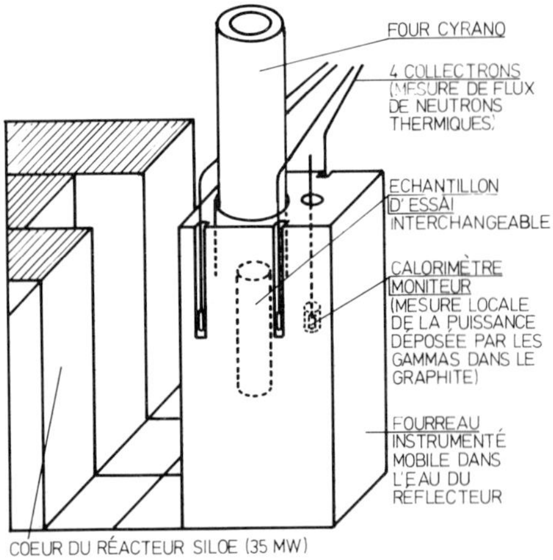

SCHEMA DU FOUR CYRANO

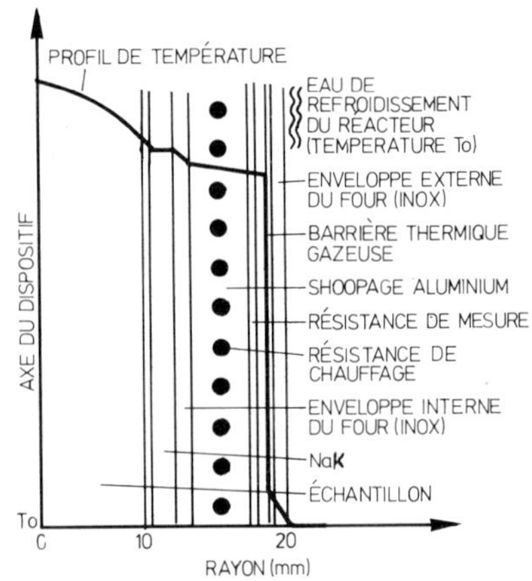

PRINCIPE DE FONCTIONNEMENT

Le four Cyrano est un dispositif d'irradiation servant à tester les matériaux qui permet de connaitre à tout instant la puissance nucléaire totale dissipée.

Pour mesurer la puissance sortant du dispositif on utilise une résistance de mesure qui fonctionne en thermomètre à résistance.

En effet, sa température dépend de la puissance traversant le dispositif, que celle-ci soit nucléaire ou électrique.

L'étalonnage préalable est effectué en injectant une puissance électrique connue dans le résistance de chauffage.

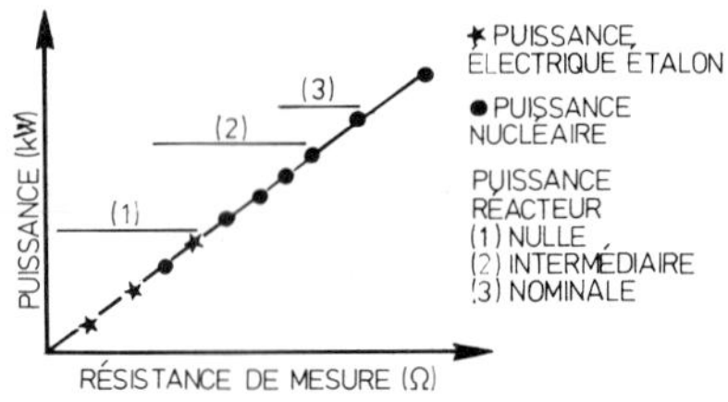

ECHANTILLONS IRRADIES

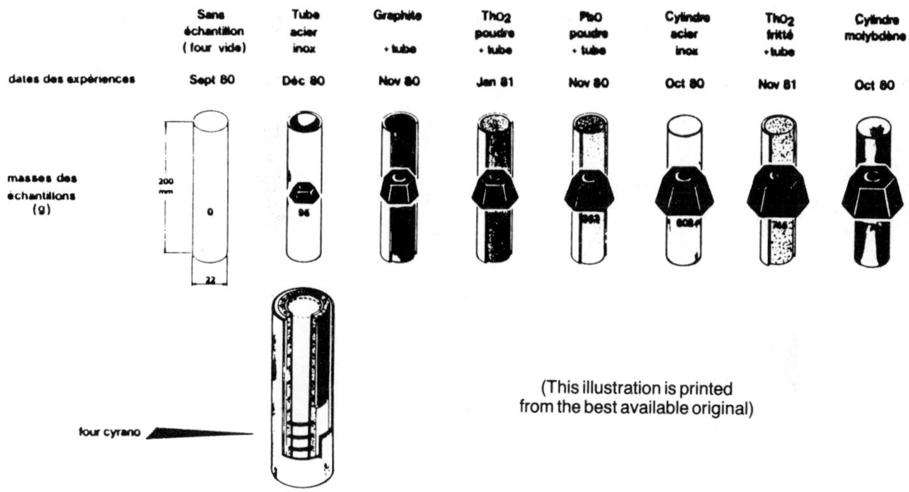

(This illustration is printed from the best available original)

CRITERES DE CHOIX DES ECHANTILLONS:

- représentatifs des matériaux utilisés dans les réacteurs d'essais
  ex: molybdène - zirconium et niobium
  ThO_2- UO_2 et PuO_2
- couvrant l'échelle des numéros atomiques (z de 6 à 90)
- densités variables (de 0 à 10,2 g/cm^3)
- faible sensibilité aux neutrons thermiques
- non fusibles à la température d'irradiation

DEROULEMENT DU PROGRAMME EXPERIMENTAL

Avant l'irradiation de chaque échantillon

- mesure des champs de rayonnements incidents gamma et neutronique par déplacement du fourreau vide instrumenté.
- étalonnage du systéme Cyrano

Irradiation des échantillons en déplaçant le dispositif dans tout le champs de rayonnement entre 34 et 179 mm du cœur

et mesures des grandeurs caractéristiques:

- "puissance Cyrano", déposée par les divers rayonnements dans l'échantillon et dans le four
- puissance massique deposée dans le calorimètre
- flux de neutrons thermiques dans les 4 collectrons
- température à cœur de certains échantillons
- température dans le four Cyrano

MOYENS DE CALCUL: CODE MERCURE 4
A trois dimensions pour l'intégration de noyaux ponctuels d'atténuation en ligne droite.

PRINCIPE DU CALCUL

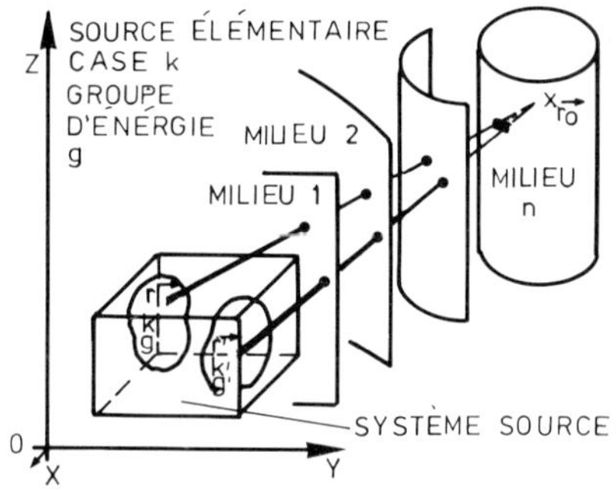

de la puissance gamma en un point $\vec{r_0}$

$$J = \sum_k \sum_g J(k,g)$$

$$J(k,g) = \iiiint_{\Delta V_k \, \Delta E_g} S(\vec{r},E)\, G(\vec{r},\vec{r_0},g)\, dE\, d\vec{r}$$

V : DOMAINE SPATIAL DE LA SOURCE
$S(\vec{r},E)$: SOURCE DISTRIBUEE DANS L'ESPACE EN 18 GROUPES D'ENERGIE
$G(\vec{r},\vec{r_0},g)$: NOYAU, OU CONTRIBUTION EN $\vec{r_0}$ D'UNE SOURCE UNITE PLACEE EN $\vec{r}$ A L'ENERGIE E_g

SYNOPTIQUE DU CODE

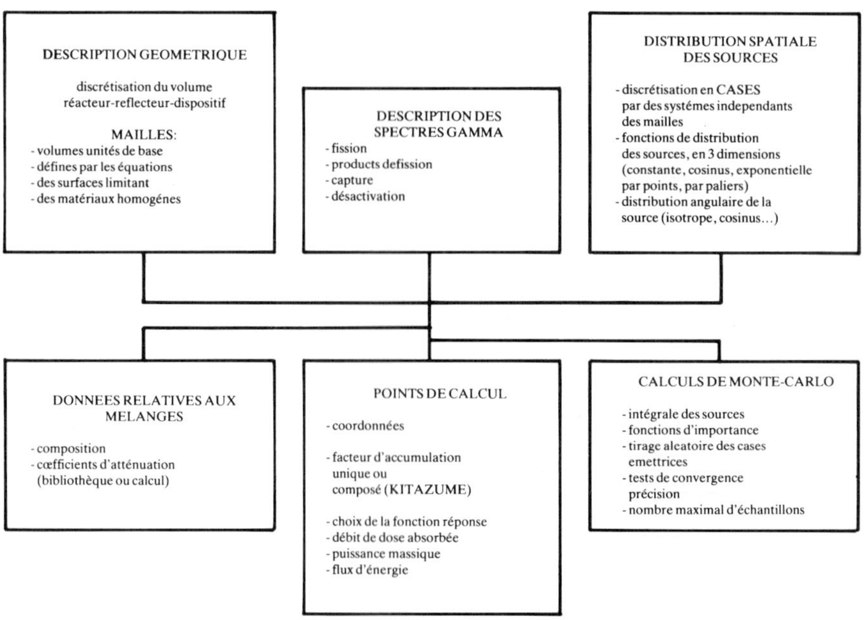

APPLICATIONS: PUISSANCES GAMMA DEPOSEES DANS LE FOUR ET DANS LES ECHANTILLONS

COUPE HORIZONTALE DU DOMAINE DE CALCUL

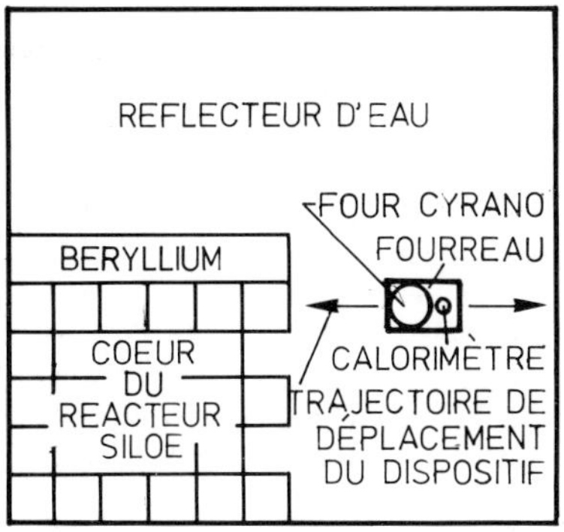

POSITION DES POINTS DE CALCUL
dans un plan horizontale

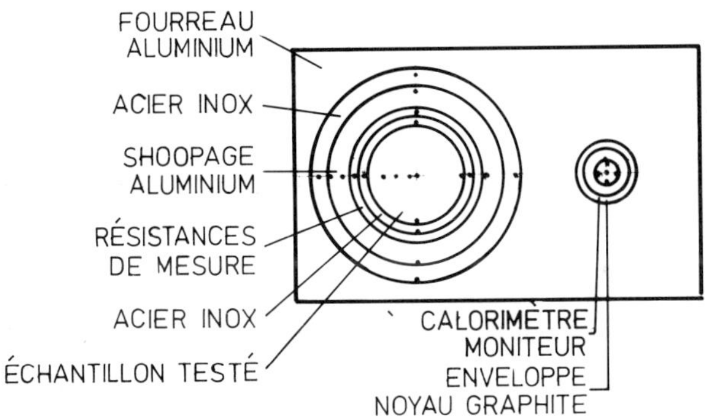

- 21 cas traités
- 3 positions par rapport à la face du cœur 34-99 et 179 mm.

échantillons testés: MOLYBDENE
OXYDE DE THORIUM
CYLINDRE ACIER
OXYDE DE PLOMB
GRAPHITE
TUBE ACIER
FOUR VIDE

Les nombreux points de calculs permettent de construire les fonctions de répartition de puissance gamma dans les différents milieux.

RESULTATS

Les valeurs expérimentales obternues avec les différents échantillons ont été normalisées entre elles à l'aide du calorimètre moniteur.

PUISSANCE EN FONCTION DE LA MASSE

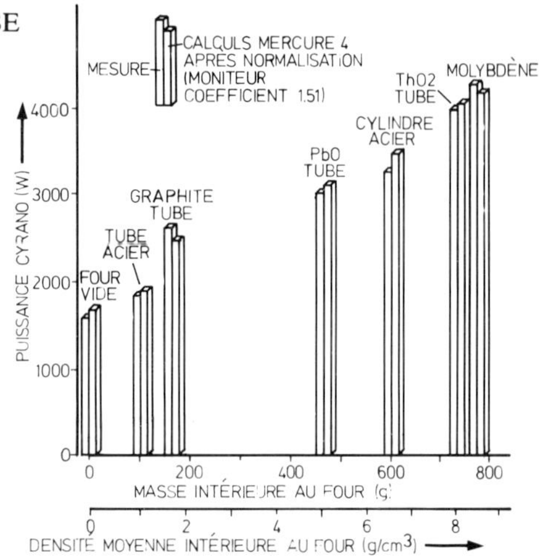

EN RELATIF

D'un échantillon à l'autre, les puissances relatives mesurées et calculées sont en assez bon accord, quelle que soit la nature de l'échantillon et la distance considérée.

COMPARAISON, CALCULS, MESURES

De la puissance Cyrano pour différentes distances au cœur et pour les divers échantillons après normalisations (moniteur-cœfficient 1,51).

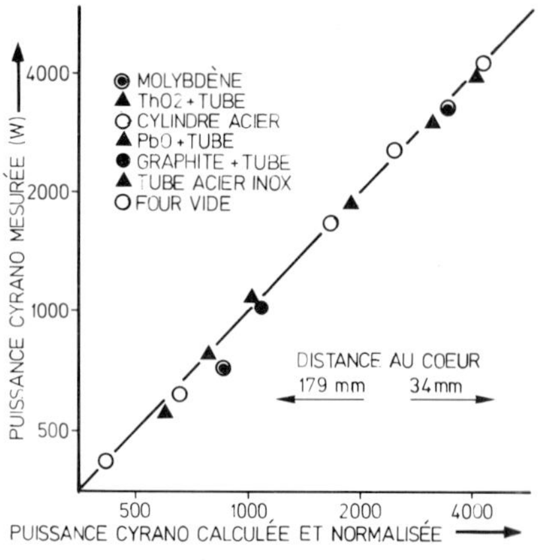

EN ABSOLU

Pour passer des valeurs calculées aux valeurs mesurées dans le four, il faut tenir compte des valeurs calculées et mesurées dans le calorimètre moniteur, cependant, un cœfficient de normalisation complémentaire 1,51 est nécessaire; il a été trouvé constant à 6% prés, quelque soit le cas considéré.

CONCLUSION

Les calculs et les mesures des puissances déposées par les gammas dans le four et ses échantillons sont en assez bon accord relatif lorsque varient le distance, le numéro atomique et la densite de l'échantillon.

Ce resultat permet de penser que les effets d'autoabsorbtion et le spectre gamma sont correctement calculés à l'interieur du four.

Le code mercure 4 permet donc de calculer en relatif les puissances déposées par les gammas dans une géométrie complexe.

Les valeurs absolues sont atteintes à l'aide d'un facteur de normalisation complémentaire qu'il est nécessaire de mesurer.

OUT-OF-PILE EXPERIMENTS IN SUPPORT OF THE HFR-TOP IRRADIATION

J.M. Plaum, C. Croon, I. Monster, H. Manuel, J.E. de Vries

Netherlands Energy Research Foundation ECN,

P.O. Box 1, 1755 ZG Petten (N.-H.), The Netherlands

SUMMARY

This paper describes the design and testing of different components of a loop-type irradiation facility for transient over-power experiments (HFR-TOP), and the experimental results of an out-of-pile test with a real scale simulation of this facility. The experimental results demonstrate a good thermal-hydraulic response during a TOP and a safe behaviour under extreme situation.

1. INTRODUCTION

The experimental programme on LMFBR's in progress at the Netherlands Energy Research Foundation ECN at Petten comprises a study on the transient fuel pin behaviour, the so-called "HFR-TOP Programme". A loop-type irradiation facility has been developed in which single fuel pins are being exposed to mild overpower transients under forced-convection cooling conditions with liquid sodium. |1| Different components of this irradiation facility, like a sodium pump, a capacity controlled sodium heat exchanger, both cooled by water from the reactor pool and two types of sodium flowmeters have been developed and truly tested in a component test facility. Besides, a real scale simulation of the irradiation facility has been constructed (in which the fuel pin has been replaced by an electrically heated fuel pin simulator) in order to investigate its transient thermal-hydraulic behaviour and to determine the performance characteristics of its components.

This paper deals with the experimental results as

P. von der Hardt and H. Röttger (eds.), Irradiation Technology, 725–737.

- the specific characteristics of the components and measuring devices,
- the behaviour of the facility under extreme conditions as sodium pump failure, failure of the cooling water system to the sodium-water heat exchanger and the behaviour under sodium boiling conditions, in order to demonstrate the safety of the irradiation device and how to handle under these circumstances,
- the thermal-hydraulic response of the out-of-pile facility under power transients up to conditions representative for the HFR-TOP experiment,
- the verification of a computer model which simulates the thermal-hydraulic system behaviour during a power transient.

2. COMPONENT DEVELOPMENT AND TEST RESULTS

2.1. Loop Description of HFR-TOP Component Test Facility (CTF)

This loop consists of a sodium pump, a seven-rod electrically heated bundle with a power output of 45 kW at inlet sodium temperature between 150 and 500°C and an outlet sodium temperature up to 950°C, a calibrated electromagnet flowmeter, a sodium-air heat exchanger which consists of three parallel units and a dummy pipe in which the components have been placed and tested.

2.2. Development and Test Results of Sodium Flowmeters

To measure the sodium flow around the fuel pin before and during a TOP, two different flowmeters have been developed and tested. The first one is the well-known probetype |2|, the second one is an external type (the magnets are placed outside the sodium). Both types have been built and tested at the ECN laboratories mainly because calibration of the instruments is needed under the same circumstance as expected in the HFR-TOP irradiation facility. The flowmeters have been calibrated against the calibrated CTF flowmeter.

The construction of the external type flowmeters, in which two magnet pairs have been mounted, is presented in Fig. 1. This flowmeter is placed downstream the fuel pin. Consequently, during a power transient the coolant temperature will increase from 520 to 700 - 1000°C. To keep the magnet temperature under the Curie point the magnets have been cooled with air during the out-of-pile experiments. In this way the conditions in the HFR-TOP irradiation facility are simulated. In that case the magnets are mounted against the wall of the capsule housing which is in contact with the water of the pool of the HFR.

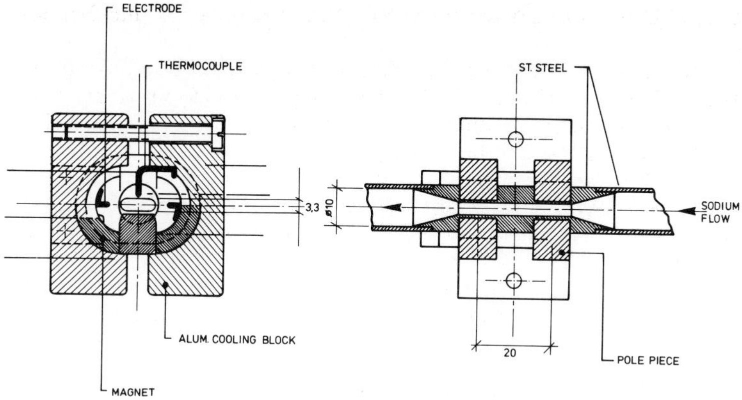

FIG. 1 Construction external flowmeter

The calibration of this flowmeter is presented in Fig. 2.

Within the temperature range of interest the temperature dependency is negligible in case of magnet cooling, without cooling the effect was comparable with temperature dependency of the probe-type. Both types of flowmeters gave a reproducible signal during longterm operation and a second set of flowmeters had equal characteristics. So, it was possible to give a once-time out-of-pile calibration for these measuring devices.

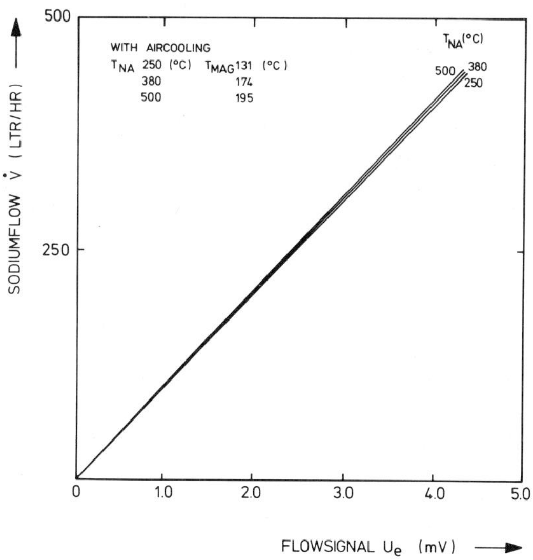

FIG. 2 Calibration external flowmeter

2.3. Development and Test Results of an Asynchrone Linear Induction Pump (ALIP)

An axial cross-sectional view of the sodium pump is given in Fig. 3.

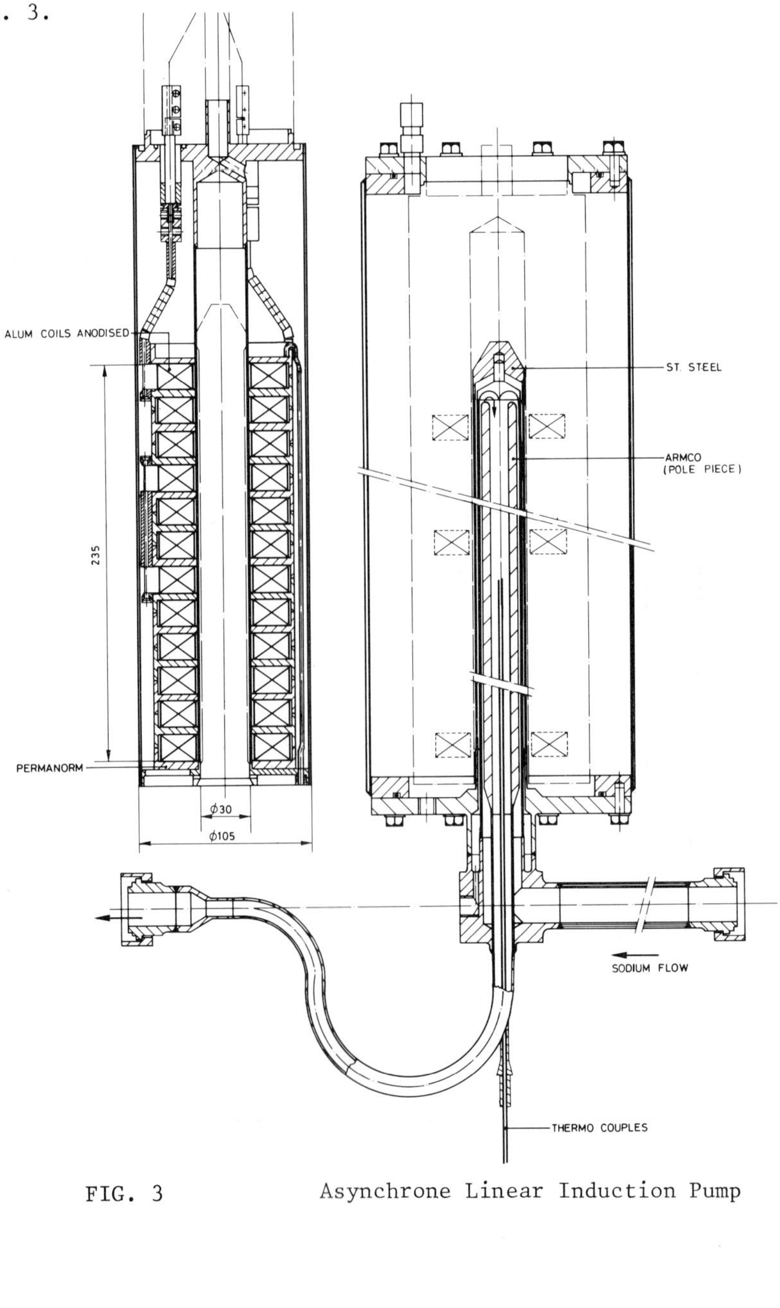

FIG. 3 Asynchrone Linear Induction Pump

This sodium pump consists of two main parts:

- a removable stator containing the coils which produce an axially moving magnetic field like in a three-phase induction motor,
- a sodium housing containing an annulus inlet part and a central pipe through which the sodium returns. Between the annulus and central pipe a pole piece has been placed for conductance of the travelling magnetic field which induces eddy currents in all conductors it passes, hence, also in the sodium. These currents undergo Lorentz-forces and the sodium moves in the direction of the travelling magnetic field.

The stator part is separated from the sodium housing by a gas gap filled with N_2 for thermal insulation. The coils of the stator are cooled by demineralized water.

The test results for 60 Amps are presented in Fig. 4 and it is shown how the pressure head reduces for increasing mass flow and temperature of the coolant.

Due to the large gas gap the efficiency of the pump is very low (∿ 1%) but the advantage of the desired low heat losses and the absence of moving parts is of a greater importance.

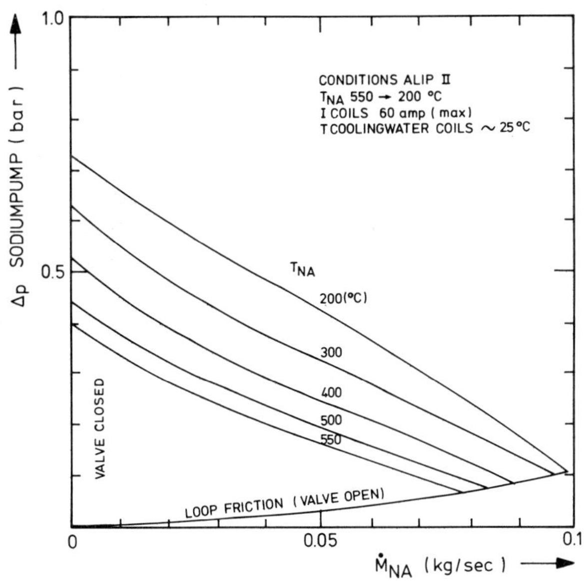

FIG. 4 Thermal-hydraulic characteristics sodium pump

2.4. Development and Test Results of the Sodium-Water Heat Exchanger

A schematic cross-sectional view of the sodium-water heat exchanger is given in Fig. 6 of the out-of-pile test facility. This component principally consists of three coaxially arranged sodium pipes. Sodium enters at the top, flows downwards through the innermost pipe and is deflected upwards through the annulus formed by the two outer pipes.

The outermost pipe forms the first containment. It is surrounded by a second containment and the narrow gap between them is filled with an adjustable mixture of He and N_2. The outer wall of the gas gap is cooled by forced-convection with water. The heat transfer during steady-state conditions is controlled by changing the composition of the He-N_2 mixture in the gap by means of an automatic controlling system . During the transient additional cooling capacity is obtained by closure of the gas gap due to thermal expansion of the inner wall. This explains the choice of rather large diameters for the tubes forming the gap. The cooling characteristic of the heat exchanger is given in Fig. 5. The dotted line gives the situation for a selected outlet sodium temperature of 380°C at a heat input of 11 kW and $T_{NA_{inlet}}$ at 530°C (a typical HFR-TOP condition during pre-irradiation). It shows a range of about 7 to 50 kW at this sodium outlet condition, which is in good agreement with the design value.

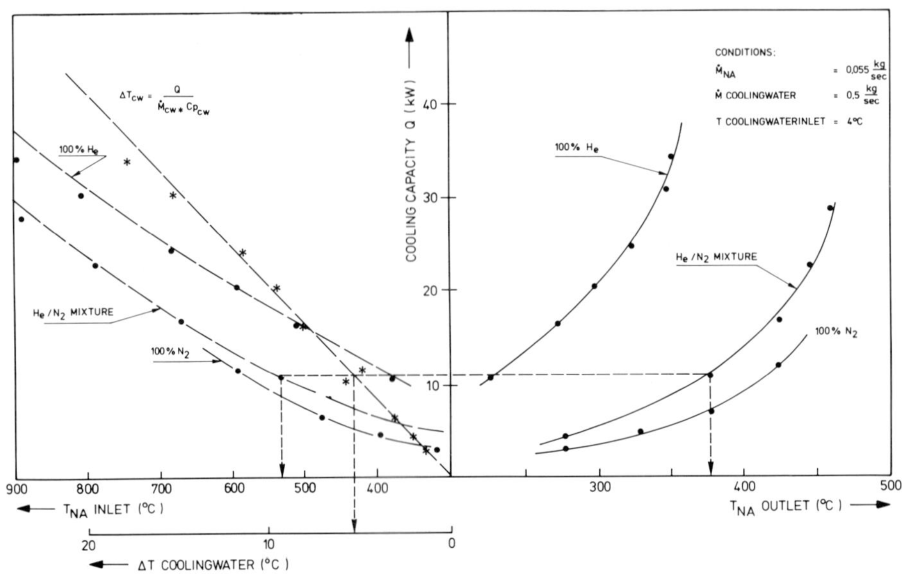

FIG. 5 Cooling characteristic sodium-water heat exchanger

3. EXPERIMENTAL RESULTS OUT-OF-PILE TESTS WITH REAL SCALE HFR-TOP FACILITY

3.1. Description of the Facility

Figure 6 gives a schematic diagram of the out-of-pile HFR-TOP capsule simulating the HFR-TOP irradiation facility. In this loop the fuel pin is simulated by an electrically heated pin of which the electrode has been made of nichrome with a temperature resistance up to 1050°C.

3.2. The Behaviour of the Facility under Extreme Conditions

3.2.1. Simulation of the Sodium Pump Failure. To investigate the system behaviour during the occurrence of sodium pump failure (with traceheating or some power left on the heater pin to simulate decay heat after a reactor scram) several experiments have been done. One of them has been selected as being representative.

Figure 7 shows the behaviour of the system after a sodium pump coast-down at maximum traceheating (5.5 kW). Under these conditions it takes about 10 minutes to reach a new stationary condition with some natural convection.

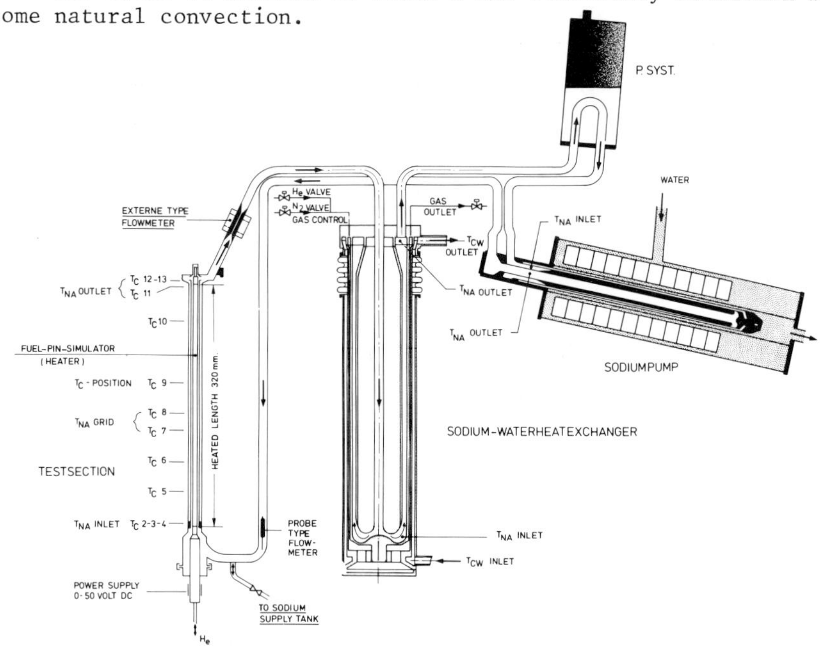

FIG. 6 HFR-TOP out-of-pile test facility

The maximum coolant temperature of $\sim$ 700°C has been observed at position 9 in the annulus surrounding the pin.

We learned that a sodium pump failure does not lead to a critical situation in case of maximum traceheating on the system or about 4 kW left on the heater pin.

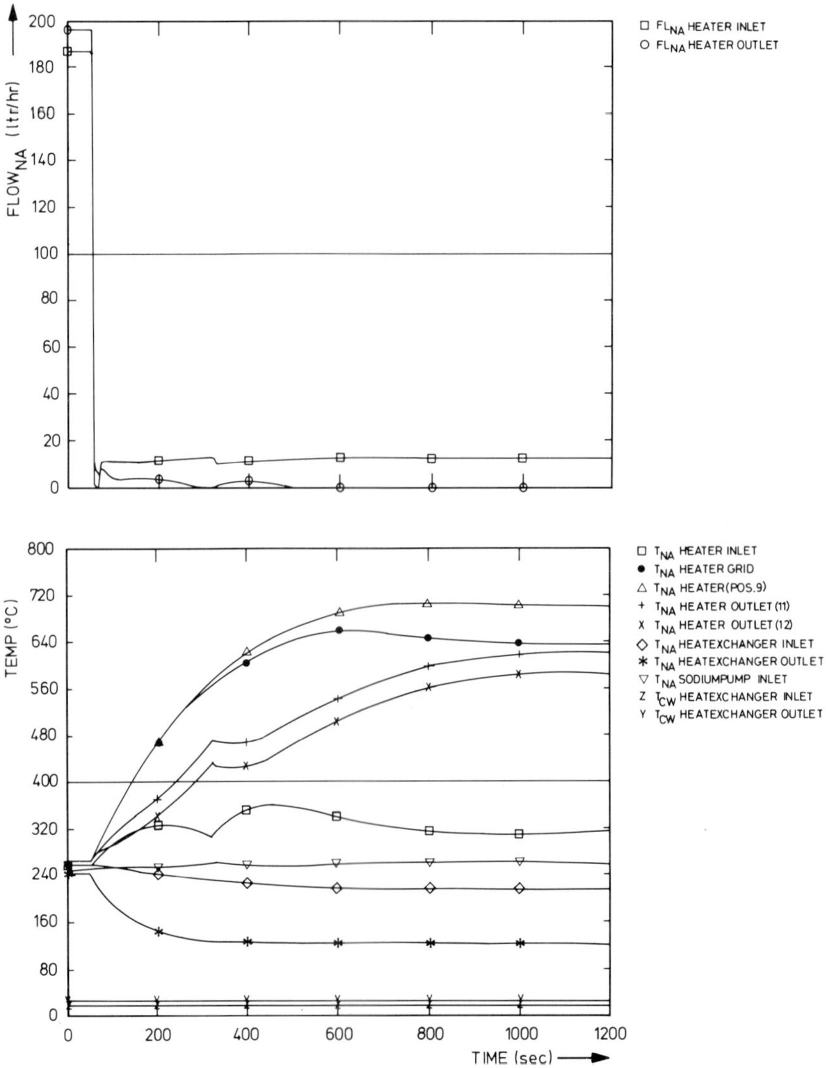

FIG. 7 Simulation sodium pump failure (5.5 kW power traceheating)

3.2.2. Simulation of Cooling Water Failure of the Sodium-Water Heat Exchanger. This experiment has been done with maximum traceheating (5.5 kW) on the system to simulate enough power input. (The heater pin did not function at that time.)

Figure 8 shows the sodium temperature through the system and the cooling water inlet/outlet temperature. The cooling water failure starts 43 seconds on the time scale. The outlet cooling water temperature reaches the saturation temperature 1.5 minutes later. It takes about 3 minutes to evaporate all the water in the heat exchanger. Till this moment the sodium temperature is practically constant after which the temperature rises slowly.

We learned that a failure of the cooling water does not lead at once to a critical situation. There is much time left to switch off the power and to control the temperature by traceheating.

3.2.3. Behaviour of the System under Moderate Boiling Condition. During present proposed HFR-TOP experiments the sodium coolant temperatures do not exceed the boiling point. However, boiling experiments have been done to establish the system capabilities under all circumstances, with in mind possible future work.

We have done two boiling experiments, one at 4 kW heater power with the sodium flow nearly zero, and one with 3 kW heater power during which the sodium pump has been switched off for a few seconds. In both cases boiling occurs around the upper part of the heater pin and the system behaves practically the same.

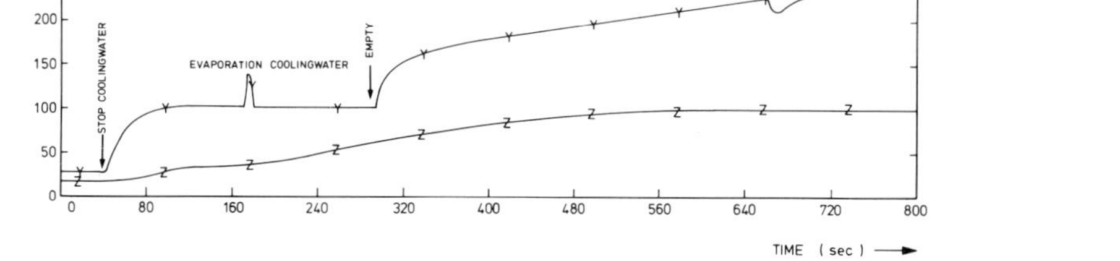

FIG. 8 Simulation cooling water failure of sodium-water heat exchanger

Figure 9 presents the data during the last test. At t = 470 seconds, the coolant in the upper part starts boiling after a short boiling delay. Fifty seconds later nearly 4 cm of the upper part is boiling. The sodium pump power is on its lowest value and there still is some flow. At 640 seconds the sodium pump is switched off. As shown in the lower figure, boiling is limited to the upper part of the coolant channel under these conditions.

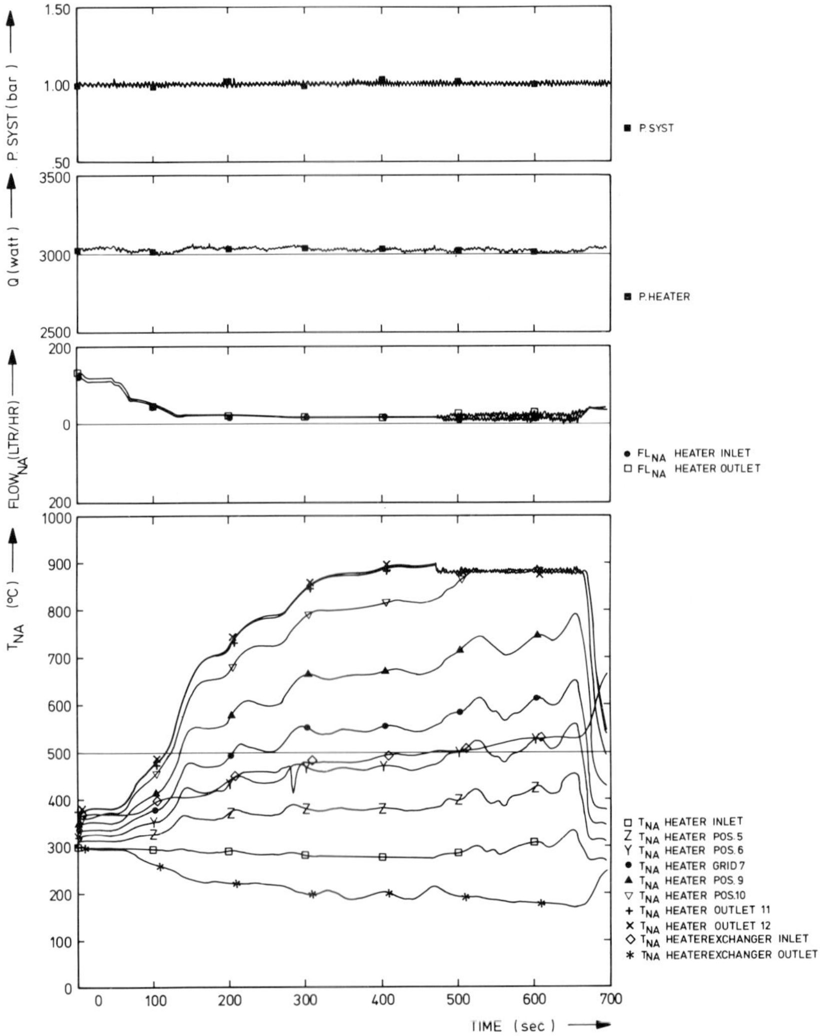

FIG. 9 Behaviour facility under boiling conditions

From the axial temperature gradient in the non-boiling part of the coolant channel and from the flowmeter signals it is concluded that natural convection occurs also under these moderate boiling conditions. The two-phase heat transfer is not interrupted by dryout of the heater pin.

To save the heater pin (at these conditions the electrode temperature is nearly 1050°C) these two-phase tests have not been repeated at higher power levels.

3.3. The Thermal-Hydraulic Response of the Out-of-Pile Facility under Power Transients up to Conditions of the HFR-TOP Experiment

Many power transients have been done under different conditions to learn the transient behaviour of the system without pin failure, especially the dynamic behaviour of the sodium-water heat exchanger.

As mentioned in chapter 2.4, during a power transient, additional cooling capacity is obtained by closure of the gas gap in the heat exchanger due to thermal expansion of the inner wall. In addition an automatic gas mixture control can be applied to keep the coolant temperature constant at the test section inlet. To investigate the contribution of both mechanisms experiments with and without N_2/He gas control have been carried out.

During a power transient from 4 kW to 16 kW and a total energy input of about 250 kJ, the increase of the coolant temperature of the heat exchanger outlet is very small and the temperature of the coolant at the test section inlet remains practically constant with gas mixture control. Without gas mixture control the temperature rise at the heat exchanger outlet is about 70°C and at the inlet of the test section about 40°C, which still is less compared to the coolant temperature rise at the test section outlet. This last demonstrates the automatic capacity control by the closure of the gas gap.

Last but not least, a simulation of a typical HFR-TOP transient has been done.

The conditions of a typical HFR-TOP transient are:

$T_{NA\,inlet\ test\ section}$	:	380 °C
$\dot{M}_{NA}$	:	0.055 kg/s
$P_{H\,stationary}$	:	12 kW
$P_{H(transient)}$	:	12 → 24 kW in 8 s.

To simulate the HFR-TOP transient without the consequences of a direct pin failure we have started at a test section inlet temperature of 210°C. The next test with 250°C at the test section inlet is given in Fig. 10.

The temperature of the coolant at the inlet of the test section is almost constant owing to the automatic temperature control by the heat exchanger. The outlet coolant temperature increases up to 600°C. The power plot gives the results of the heater power, the power transferred by the sodium flow and the cooling power of the heat exchanger as a function of time. The plot above gives the integrated value of these three signals. It shows that at the moment of power scram, the heat generated in the pin has been transferred largely by the sodium coolant in case of this transient.

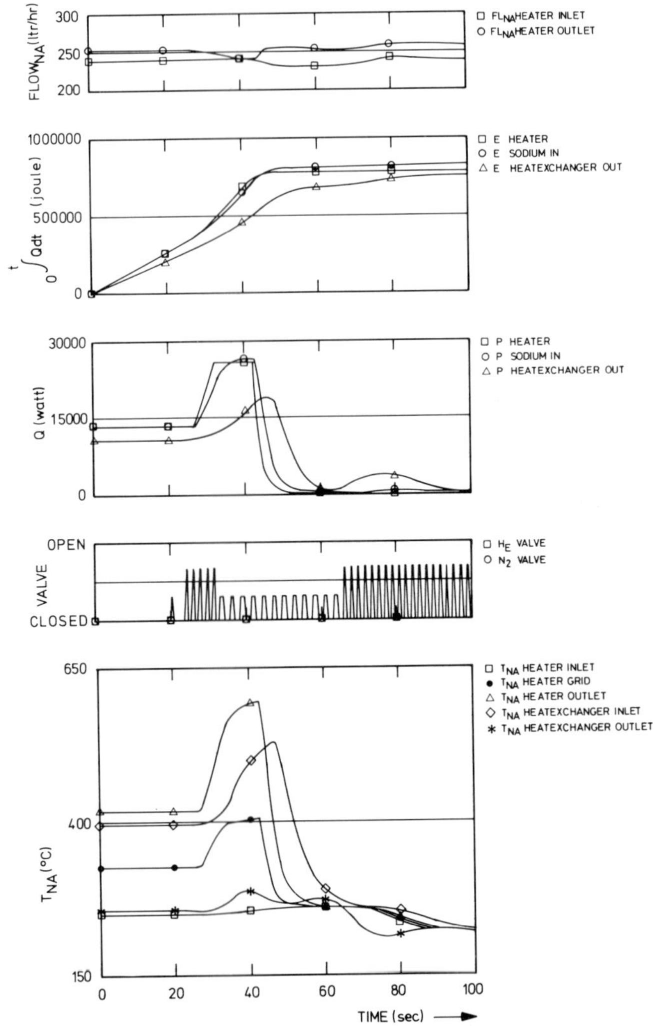

FIG. 10 Simulation HFR-TOP transient

The upper plot gives the M_{NA}, which is nearly constant. This was the highest transient simulated in the out-of-pile facility. During this test the electrode has been damaged due to overheating (maximum calculated electrode temperature was 1100°C).

4. CONCLUSIONS

The tests have demonstrated that the planned HFR-TOP experiments can be executed with the developed irradiation facility. The different components meet the requirements. The facility behaved well even under conditions more severe than the conditions of the planned HFR-TOP tests. The verification of a computer model which simulates the thermal-hydraulic behaviour during a power transient was in good agreement with the experimental results.

5. REFERENCES

|1| Van der Koogh, G., et al.,
Technical Aspects of the HFR-TOP Irradiation,
Paper presented at this meeting.

|2| Baker, R.C.,
Electromagnetic Flowmeters for Fast Reactors,
Progress in Nuclear Energy, Vol. 1, pp. 41 - 61, 1977.

|3| Müller, St., Thun, G.,
Permanentmagnetische Durchflussmesser Sonde für flüssige Metalle,
KFK-2479, 1979.

ACKNOWLEDGMENT

This work would not have been possible without the effort of many people of our institute, particularly those who did the design, mechanical and experimental work.

PIRAMID, A LOOP FOR EUROPEAN PAHR IRRADIATIONS ACCORDING TO A MOL INTEGRATED DEVICE

C. JOLY, M. SOENEN, W. DETAVERNIER, CH. DE RAEDT, L. FOUARGE,
S.C.K./C.E.N., MOL, BELGIUM

H. BACKS,
J.R.C., ISPRA, ITALY

ABSTRACT

On behalf of the Commission of the European Communities, the Nuclear Research Centre of Mol, S.C.K./C.E.N., is in charge of the execution in the BR2 reactor of three in-pile experiments aiming at the study of the post-accident heat removal (PAHR) phenomenology. The purpose of these experiments is to validate the physical models and the computer codes used to predict the temperature fields in a sodium saturated bed and in debris retention devices.
The European desiderata are extremely severe as they assume that the loop design will allow for experiments with stainless steel melting (second experiment) and with UO_2 melting (third experiment).

From the large experience gained with in-pile Na-loops, the S.C.K./C.E.N. has followed the approach of an integrated design, which presents numerous advantages.

The paper describes the lay-out of the first experiment - PIRAMID 1 -, which is foreseen for temperatures up to 1400 °C in a UO_2 debris bed with a bed loading of 1000 kg UO_2/m^2.
It emphasizes also the importance of the back-up programme related to the in-pile PAHR programme.

INTRODUCTION

As a consequence of a hypothetical accident in a Liquid Metal Fast Breeder Reactor (LMFBR), part of the core - mainly oxide fuel and stainless steel - may melt.

P. von der Hardt and H. Röttger (eds.), Irradiation Technology, 739–748.

Upon contact with the subcooled sodium, the molten materials may quench and fragment : the generated particles will settle under gravity on various parts in the reactor vessel where they form particulate debris beds whose geometries and material compositions depend on the accident scenario and local in-pile conditions.

Hence the size of fuel and stainless steel particles, the concentration of steel and UO_2, the possibility of particle size and/or material stratification in the bed are parameters to be considered in debris bed studies. Boundary conditions as bed height or the bulk temperature of the surrounding sodium and the burn-up of the fuel have an influence on the coolability of the debris beds.

DESCRIPTION OF THE LAYOUT. [1] [2]

From the large experience obtained with in-pile Na-loops irradiated in BR2, the S.C.K./C.E.N. has followed the approach of an integrated facility containing the debris bed in a controlled capsule (fig. 1). The advantages are :

- compact integrated design with well known basic principles,
- easy loading and unloading operations in the reactor,
- no special mounting or connections to be performed at the reactor site,
- possible transport under water as one piece by means of the existing reactor transfer chute towards either the storage channel or the high activity hot cells for dismantling,
- easy performance of neutron radiographies.

The design is conceived with the aim of only modifying the central part of the test section for the second and third experiments.

Disposing the successive barriers - the Na containment tube, the pressure tube and the safety tube - each with its well-defined thermal and mechanical functions and the needed passages for the instrumentation, a cross-section at the reactor mid-plane is obtained, a typical example of which - with a bed diameter of 135 mm - is given by figure 2. This cross-section will perhaps be modified by the introduction of a neutron- and/or a gamma absorbing screen with a view to the realization of flatter power and temperature profiles. The closed pressure tube must take up the pressure shock waves in case of accidental MFCI while the safety tube encircles all circuits to guarantee absolute integrity of the test facility in the reactor. Finally a jacket tube surrounds the safety tube to distribute correctly the primary reactor water.

The proposed <u>primary circuit</u> contains the following items :

- a one meter long crucible for which a niobium alloy is retained as the best candidate material,

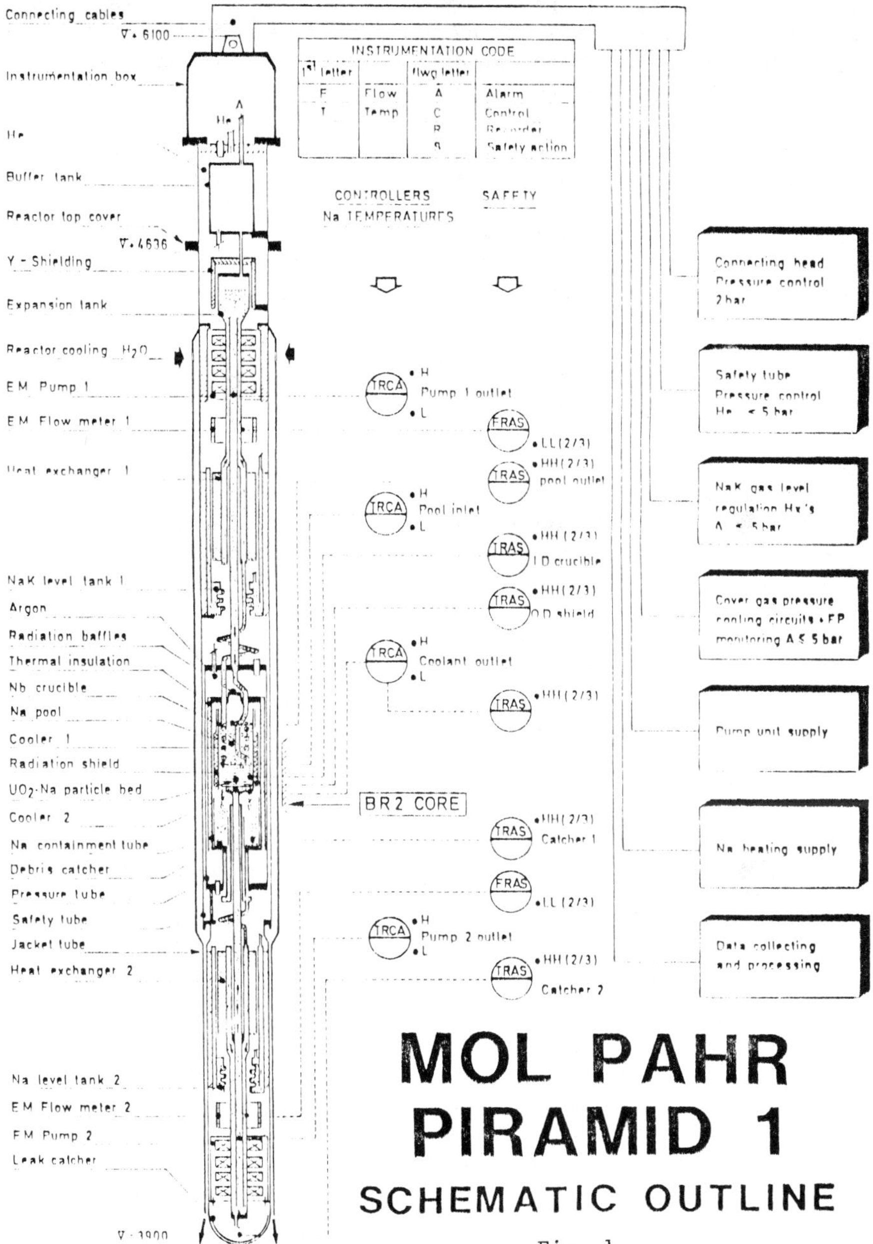

MOL PAHR PIRAMID 1

SCHEMATIC OUTLINE

Fig. 1

This illustration is printed
from the best available original

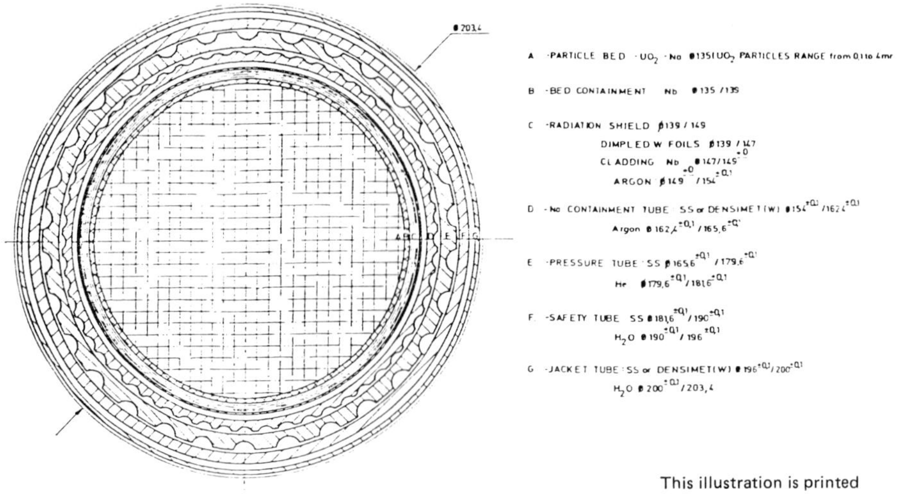

This illustration is printed from the best available original

Fig.2. PIRAMID 1. CROSS SECTION AT REACTOR MID-PLANE

- a radiation shield consisting of concentric W sheets surrounding the crucible to limit the radial heat losses to a very low level,
- the upper primary heat exchanger inside the crucible consisting of the Na pool above the UO_2 debris bed in which are immersed the closed coils of the upper secondary circuit,
- the argon cover gas common to both the upper and lower Na coolants : the gas pressure can be regulated between 1 and 5 bar passing over a buffer tank, manual valves and active coal absorption,
- the Na containment tube which encloses all the above items and which is equipped with a debris catcher.

A Ti-plate will be added to the bottom of the crucible to realize the simulation of the heat transfer through a 2 cm thick SS core catcher.

The upper and lower Na secondary circuits contain each :

- a heat exchanging area, i.e. the upper cooling coils immersed in the Na pool and the lower open cooling circuit in which the Na is pumped up against the bottom of the crucible,
- a Na-Ni-NaK-Cu-H_2O heat exchanger, 2 annular linear induction pumps in series, a permanent electromagnetic flowmeter, all with a concentric lay-out; the unit of pumps and flowmeter is installed in the coldest part (max. 650° C) of the circuit, the hot part of which is composed of ducts separated by an insulating gas gap,
- an expansion tank.

Such a lay-out permits the maximum flexibility to regulate the Na temperature in the upper and lower heat exchanging areas in the range 400° C - 800° C. There are three ways of regulating the Na-temperature : by changing the NaK level along the heat exchanger length, by changing the Na flow, and by manual control of the different heaters. These heaters (4 heaters and 5 preheaters on each circuit) are disposed such that the complete Na circuit can be liquified in steps starting from the expansion tank.

The primary reactor water cooling circuit forms the tertiary circuit.

BACK-UP PROGRAMME [1]

In order to guarantee the successful execution of the test and its interpretation a so-called back-up programme is being performed in the following main areas. This back-up programme is partly supported by the Commission of the European Community.

Materials research. To validate the choice of Nb, both for the crucible and for the thermocouple sheaths, compatibility tests are now being performed at Mol in presence of Na. The main parameters are test temperature and duration. The first results confirm the good behaviour of Nb and new tests characterized by more severe conditions are now under way and will be repeated for Nb-alloy.

In parallel JRC-Ispra is developing a kind of tissue composed of refractory material compatible with the debris. This tissue could be inserted as protecting liner in a given crucible. First compatibility experiments at 1800° C with molten steel in a Mo crucible protected by such a tissue showed no interaction.

Radiation shield. One of the best ways of realizing an efficient radiation shield would be the use of concentrically Mo or/and W dimpled foils. JRC-Ispra has undertaken a series of tests to qualify such a thermal shield. A second possibility would be the use of ZrO_2 foam for the fabrication of which a specialized firm has been contacted.

Instrumentation. Behind the choice of adequate high temperature thermocouples, the main problems in the field of instrumentation concern the ultrasonic thermometers and the sodium level detectors, both being part of the back-up programme supported by JRC-Ispra.

The potential of ultrasonic thermometers was demonstrated years ago [3] but now this system has to be refined and adapted for specific in-pile conditions. The sensor should have more (~ 10) and shorter (~ 10 mm) sensing sections to provide representative axial temperature profiles of the bed. So a certain number of judiciously distributed ultrasonic thermometers may be sufficient to reveal the temperature profile and thereby limit bed disturbance.

Although they are not absolutely indispensable, it is foreseen to load and test the ultrasonic thermometers during the first irradiation.

Sodium level detectors which allow to measure relative surface variations of 1 mm at temperatures of 800° C with small enough diameters to house the detectors inside the capsule are being tested.

Phenomenological studies. To get a better understanding of what could happen in debris beds and also to do parametric phenomenan studies, a comprehensive out-of-pile programme has been conceived under the responsability of JRC-Ispra concerning the following main aspects :

- densification of UO_2 and UO_2+stainless steel particle beds in sodium atmosphere. The degree of densification and eventual crust formation are studied as a function of time and temperature.
- steel migration. An orientative experiment performed at 2500° C has shown that steel particulates may agglomerate and that possibly vaporising steel condenses in the cold part of the experimental set-up. A study contract has been concluded to study these phenomena under well defined conditions. The top of the bed will be heated to 1600, 2000 and > 2200° C by an arc lamp whereas the bed bottom is maintained at 900° C.
- study of the physico-chemical reactions of sodium with UO_2 and $(UPu)O_2$. The aim of this study is to determine the composition and the physical state of the products formed after the interaction of liquid Na or of Na vapour with fast reactor fuel material. The oxide reactants that are being studied are : urania, urania-plutonia and urania-plutonia containing fission product simulants. The tests will be conducted at 850, 1050 and even 1600° C.

Development of models and codes. One of the main objectives of the in-pile experiments is to put in evidence and to understand the important basic phenomena characterizing the behaviour of a particulate debris bed in order to allow the physical modelling of these phenomena and to validate computer codes.

A collaboration began 3 years ago between S.C.K./C.E.N. and the von KARMAN Institute with a view to the development of a computer tool - called TORPEDO - for the prediction and interpretation of the in-pile PAHR experiments. [4] [5]
TORPEDO is a 2D code based on a finite difference solution of the relevant transport equations and on the modelling of convective effects and of the PAHR phenomena. It allows the prediction and interpretation of the Mol in-pile experiments as well as the calculation of the temperature field in all components of the device. JRC-Ispra will join S.C.K./C.E.N. and VKI to participate in the further developments of the code.

INSTRUMENTATION AND DATA ACQUISITION SYSTEM

In addition to the instrumentation necessary for the control and operation, the test section is equipped with an appropriate number of thermocouples, 30 type W/Re, 5 Cr/Al plus 5 ultrasonic thermometers to measure and follow the temperature evolution in the bed. A sodium level detector should help to define the extension of boiling layers and of dry zones. Calibrated flowmeters and thermocouples in the secondary circuits allow the evaluation of upward and downward heat fluxes. The radial heat losses of the bed through the thermal barrier will be evaluated by calculations, relaying on measured temperature gradients. Collectrons arranged outside the bed will complete the power balance. 2 pressure transducers will indicate the argon cover gas pressure. Furthermore, in a first approach for the whole PIRAMID 1 loop, a total of 160 Cr/Al thermocouples will be installed.

The data acquisition system for result analysis was looked at in more detail with a view to a continuous recording during the irradiation. The system is now envisioned with three ways of recording : 1 analog recording, 3 digital recordings - low (200 records/ minute), medium (3000 records/second) and high (50000 records/ second) speed - and the BR2 safety measurement system. More than 200 channels are foreseen and visualization of selected data will also be possible, 28 channels being duplicated for the analog recording and 50 channels for the high speed digital recording. The principal data will also be directed to the BR2 measurement system, a scram being only possible with the occurrence of 2 out of 3 predefined actions.

CHARACTERIZATION OF THE LAY-OUT

A large number of neutron and gamma calculations were performed, with the S.C.K./C.E.N. version of the neutron transport code DTF-IV and adequate neutron and gamma libraries, for the following four loop concepts :

- irradiation devices without neutron-absorbing screens, a typical example being given by figure 2,
- irradiation devices with a Cd neutron-absorbing screen cooled on both sides, which leads to a re-thermalization of the incident neutrons,
- irradiation devices with a B_4C neutron-absorbing screen placed nearer towards the crucible, for which case no re-thermalization occurs. This loop concept, which leads to the best characteristics, requires highly enriched fuel in the bed and appeared to be impossible to execute in practice because of the high heat production in the B_4C screen.

- irradiation devices with increased total W thickness in the radiation shield, leading to a larger absorption of the gamma rays.

In all cases, the absolute heating power level in the bed due to the neutron induced fission and the gamma ray captures is such that the BR2 reactor would operate at a reduced power level. Figure 3 gives typical curves of the power ratio p_{axis}/p_{border} as a function of the ^{235}U enrichment for the three first cases. The study of the fourth case is still under way.

As to the thermal calculations, TORPEDO indicates that a mean specific bed power of about 5.10^6 W/m³ bed will be necessary to reach 1400° C in the debris bed. Figure 4 gives typical temperature fields in the 1000 kg UO_2/m^2 loaded debris bed as foreseen in the first experiment. These temperature fields correspond to two stable situations obtained with non-uniform power profiles, the first one for a mean specific bed power of about 2.4 10^6 W/m³ bed and the second one for about 5.6 10^6 W/m³ bed. In the first case the maximum temperature is 707° C and in the second case a stable dryout situation is reached with a maximum temperature of 1180° C. The results of the TORPEDO code emphasize also the great importance of the natural convection both in the upper Na layer and in the upper part of the debris bed.

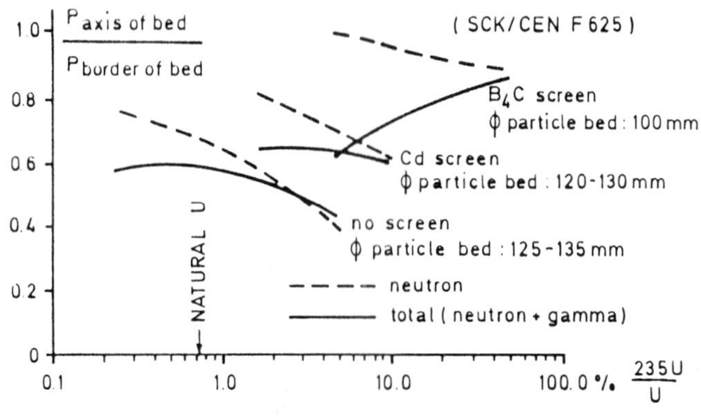

Fig. 3 Calculated radial depression in the fuel debris bed.

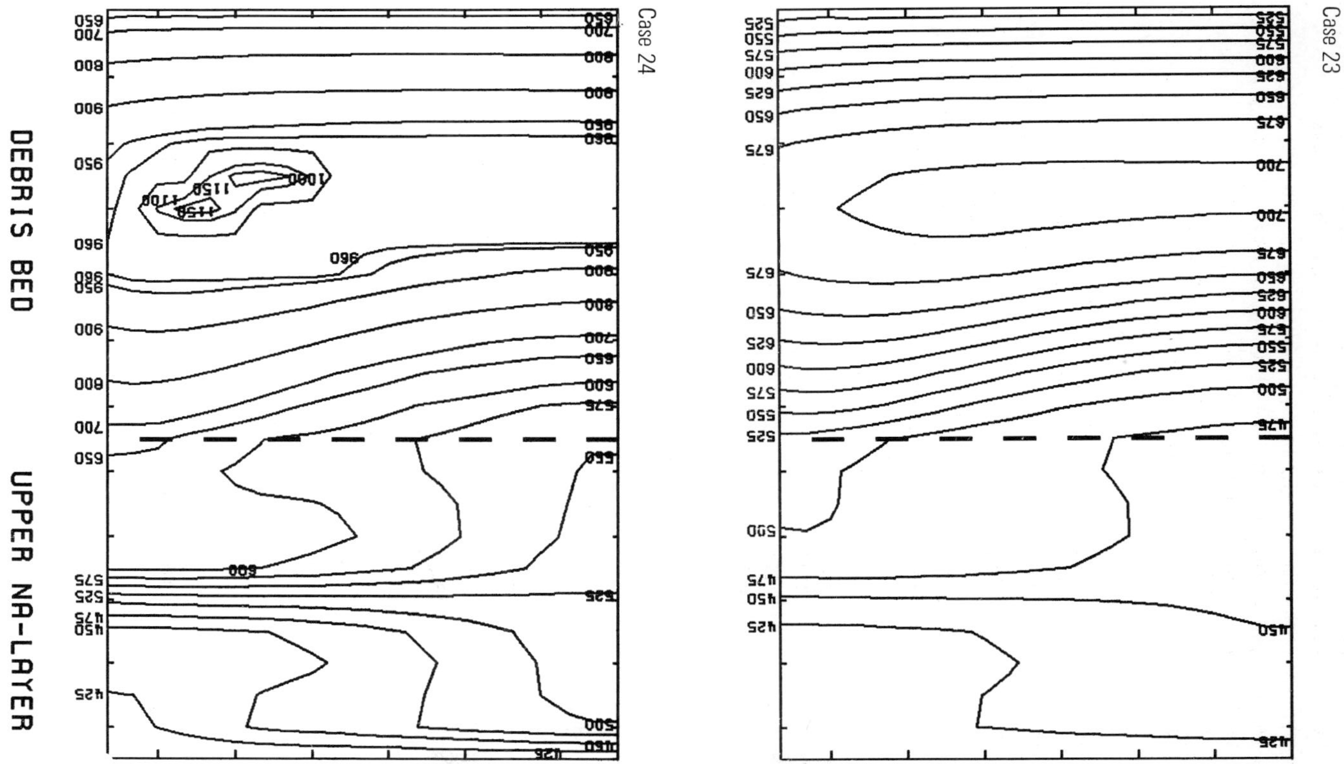

UO_2 debris bed : Bed diameter : 135 mm, Bed loading : 1000 kg UO_2/m^2, Measured porosity : 0.35
Non uniform power profile : p_{axis}/p_{border} = 0.750, Mean specific power : 2.4 10^6 W/m^3 bed(case 23) and 5.6 10^6 W/m^3 bed (case 24). Bottom and top cooled - Heat evacuated by reactor cooling water.
FIG. 4 : TYPICAL TEMPERATURE FIELDS IN THE DEBRIS BED AND IN THE Na POOL.

CONCLUSIONS

This short paper stresses the main characteristics of the European PAHR programme, of the irradiation device PIRAMID conceived by S.C.K./C.E.N.-Mol and the solution given to the main problems.

REFERENCES

[1] BACKS, H., HOLTBECKER, H., JOLY, C., SOENEN, M., ROUSSEAU, D., VEYRAT, J., "The European PAHR In-Pile Programme : Objectives and General Lay-out", International Topical Meeting on LMFBR. Safety and related design and operational aspects, Lyon, July 1982, Poster Session 2.

[2] JOLY, C., SOENEN, M., DETAVERNIER, W., DE RAEDT, C., FOUARGE, L., BACKS, H., "European PAHR Experiments in the BR2 reactor. Lay-out and Characterization of the First Experiment". Information Exchange Meeting on Post Accident Debris Cooling, K.f.K., July 1982.

[3] TASMAN, H.A., CAMPANA, M., PEL, D., RICHTER, J., "Ultrasonic thinwire thermometry for nuclear applications", The 6th Temperature Symposium, March 15-18, 1982, Washington D.C.

[4] BUCHLIN, J.M., BENOCCI, C., JOLY, C., SIEBERTZ, A., "Theoretical Modelling of Thermal Hydraulics in PAHR debris bed up to extended Dry-out", International Topical Meeting on LMFBR. Safety and related design and operational aspects, Lyon, July 1982, Poster Session 2.

[5] BENOCCI, C., BUCHLIN, J.M., JOLY, C., "A two-dimensional finite difference modelling of the thermohydraulic behaviour of the PAHR debris bed up to extended dry-out", Information Exchange Meeting on Post Accident Debris Cooling, K.f.K., July 1982.

EXPOSE POUR LA CLOTURE DE LA CONFERENCE

(F. MERCHIE / CEN de GRENOBLE)

Au moment de clore cette Conférence sur la Technologie des Irradiations, je pense pouvoir dire sans risque d'être contredit qu'un travail important a été accompli pendant les différentes sessions plénières et pendant la session "posters". Mais un premier bilan serait sans doute incomplet s'il ne soulignait pas, non seulement le volume de travail effectué, mais également la variété et la qualité des présentations et des échanges et enfin leur utilité car il est intéressant à tous égards que périodiquement chacun fasse une synthèse de ses travaux pour les présenter, les comparer et les confronter aux travaux de ses collègues au sein de la communauté scientifique et technique à laquelle il appartient.

Peut-on dire que, dans cet esprit, la deuxième partie du pari que j'évoquai lors de l'ouverture de la Conférence a été gagnée? C'est à chacun de vous qu'il appartient de répondre. Pour ma part, je crois effectivement que ce brassage et cet échange d'idées et d'informations porteront leurs fruits et marqueront de façon directe ou indirecte l'évolution future des techniques d'irradiation et donc, dans une moindre mesure, l'avenir des réacteurs de recherche et d'essais de matériaux.

Dans le monde, plusieurs centaines d'ingénieurs, de scientifiques et de techniciens consacrent l'essentiel de leurs activités aux réacteurs expérimentaux, à leur exploitation et à leur utilisation dans les domaines les plus variés.

Notre Conférence aura démontré la vitalité et le dynamisme de toutes ces équipes, ainsi que le rôle primordial qu'elles jouent pour participer, à leur niveau, au développement et à la maturité de l'énergie nucléaire, énergie nucléaire parfois contestée mais dont dépend chaque jour davantage l'approvisionnement du monde en énergie, lequel détermine à son tour les conditions de vie et le mode de vie de millions d'hommes dans les pays industrialisés et dans les pays en voie de développement.

Véritable carrefour des hommes et des techniques de toutes spécialités, je suis convaincu que les réacteurs expérimentaux, dans leur spécificité et leur diversité, contribueront longtemps encore à l'acquisition, au développement et à la diffusion de la technologie et du savoir faire nucléaires qu'ils ont d'abord initié et auxquels ils ont tant apporté depuis.

Mesdames, Messieurs, Chers Amis, la Conférence Internationale sur la Technologie des Irradiations est terminée.

Il me reste à vous souhaiter des visites agréables et intéressantes au Centre d'Etudes Nucléaires de Grenoble cet après-midi et demain à SUPER PHENIX, à l'usine d'enrichissement d'EURODIF et à la Centrale du BUGEY.

P. von der Hardt and H. Röttger (eds.), Irradiation Technology, 749.

PROGRAMME SUMMARY

Tuesday, September 28, 1982

Welcome and Opening

Session I .... Light Water Reactors
Session II .... Light Water Reactors
Session III .... Light Water Reactors
Session IV .... Light Water Reactors

Wednesday, September 29, 1982

Session V .... Light Water Reactors. Instrumentation.
Session VI .... Fast Breeder Reactors
Session "Poster"

Thursday, September 30, 1982

Session VII .... Instrumentation
Session VIII .... Gas Cooled Reactors and Fast Breeder Reactors

Session IX .... Miscellaneous

Closure of the Meeting
Visit of CEN Grenoble

Friday, October 1, 1982

Visits to:

- "Superphenix" Fast breeder nuclear power station (prototype under construction)
- "Bugey" Nuclear power station
- "EURODIF" Enrichment plant

LIST OF CHAIRMEN

J. AHLF
J.M. BAUGNET
E. BOJARSKY
A. CALZA-BINI
P. VON DER HARDT
K.J. HENRY
M. LEHRHEUER
J.A. LETH
J.P. PAGÈS
J.F. VEYRAT
H. VIDAL

LIST OF PARTICIPANTS

ABASSIN, Jean-Jack	CEN Grenoble, France
AGAISSE, Roland	CEN Cadarache, France
AHLF, Jürgen	GKSS Geesthacht, Germany
ALI Abdul A.K.	IRAQI Atomic Energy Commission Baghdad, Iraq
BALOURDET, Marcel	CEN Grenoble, France
BARON, Daniel	FRAGEMA, France
BAUER, Ekkerhard	Laue-Langevin Institute Grenoble, France
BAUGNET, Jean-Marie	CEN/SCK Mol, Belgium
BAUMONT, Geneviève	CEN Cadarache, France
BERTIN, Jean-Luc	CEN Saclay, France
BLANCHARD, Robert	CEN Cadarache, France
BOJARSKY, Eberhard	KfK Karlsruhe, Germany
Bruet, Michel	CEN Grenoble, France
CADALBERT, Robert	CEN Saclay, France
CALVARESI, Camillo	ENEA Brasimone, Italy
CALZA-BINI, Alberto	ENEA Casaccia, Italy
CAMPANA, Maurice	JRC Karlsruhe, Germany
CARACCHINI, Rino	ENEA Casaccia, Italy
CARTERET, Yvette	CEN Saclay, France
CHALONY, André	CEN Cadarache, France
CHENEBAULT, Pierre	CEN Grenoble, France
COLOMEZ, Gérard	CEN Saclay, France
CONRAD, Rainer	JRC Petten, The Netherlands
COQUERELLE, Michel	JRC Karlsruhe, Germany
COSOLI, Giorgio	ENEA Casaccia, Italy
DE RAEDT, Charles	CEN/SCK Mol, Belgium
DE ZAAIJER, Dik	ECN Petten, The Netherlands
DEKEYSER, Jean	CEN/SCK Mol, Belgium
DELLA CASA, Ennio	ENEA Casaccia, Italy
DEWES, Peter	KWU Erlangen, Germany
DUCROS, Gérard	CEN Grenoble, France
DUNCAN, Thomas	NUCOR Pretoria, Rep. of South-Africa
DUPONT, Georges	CEN Grenoble, France
FALLA, André	CEN/SCK Mol, Belgium
FARNY, Gérard	CEN Saclay, France
FEHRENBACH, Paul	A.E.C.L./Ontario, Canada
FERRY, Roger	CEN Cadarache, France
FRANCARD, Jean-Louis	CEN Saclay
FRIBOULET, Claude	CEN Grenoble, France
FUJISHIRO, Toshio	JAERI, Japan

GÄRTNER, Michel	KWU Erlangen, Germany
GENTHON, Jean-Pierre	CEN Saclay, France
GILBERT, Jean	CEN Saclay, France
GOLINELLI	CEN Cadarache, France
GOUTAGNY, Luc	CEN Cadarache, France
GOUX, Dominique	CEN Cadarache, France
GRASS, Friedrich	Atominstitut Wien, Austria
HAESSLER, Maurice	CEN Cadarche, France
HAGEN, Siegfried	KfK Karlsruhe, Germany
HAIRION, Jean-Pierre	CEN Grenoble, France
HAMOODI, Skaker K.	IRAQI Atomic Energy Commission Baghdad, Iraq
HANSEN, Kai	RISØ National Lab., Denmark
HARBOTTLE, John	CEGB Berkeley, United Kingdom
HARTOG, J.M.	UKAEA/AERE Harwell, United Kingdom
HENRY, Kenneth James	UKAEA/AERE Harwell, United Kingdom
HURAND, Jean-Pierre	ÇERCA Romans, France
JACOBI, Siegfried	KfK Karlsruhe, Germany
JANVIER, Jean-Claude	CEN Grenoble, France
JOHANSSON, Juhani	IMATRAN VOIMA Helsinki, Finland
KELLER, Roman	JRC Petten, The Netherlands
KENNEDY, Kenneth James	UKAEA/AERE Harwell, United Kingdom
KERMARREC, Jacques	CEN Cadarache, France
KLAGE, Dieter	JRC Petten, The Netherlands
KONNO, Koichi	Power Reactor and Nuclear Fuels Dpt. Tokyo, Japan
KONRAD, Joseph	JRC Petten, The Netherlands
LANCHI, Massimo	ENEA Casaccia, Italy
LANGUILLE, Alain	CEN Cadarache, France
LEFORT, Alain	CEN Grenoble, France
LEHRHEUER, Werner	KFA Jülich, Germany
LETH, Jorgen	RISØ National Lab., Denmark
LUCOT, Michel	CEN Saclay, France
MACCARDELL, Richard K.	EG&G Idaho Falls, U.S.A.
MAILLARD, Arlette	CEN Saclay, France
MARIN, François	CEN Cadarache, France
MARINUCCI, Gianfranco	ENEA Casaccia, Italy
MARKGRAF, Joachim	JRC Petten, The Netherlands
MARTENS, Herning	GKSS Geesthacht, Germany

MAS, Pierre	CEN Grenoble, France
MASSON, Maurice	CEN Grenoble, France
MATFIELD, Ronald Sydney	JRC Ispra, Italy
MEHNER, Alfred	KFA Jülich, Germany
MERCHIE, Francis	CEN Grenoble, France
MESNAGE, Bernhard	CEN Cadarache, France
MITCHELL, Eric James	UKAEA/AERE Harwell, United Kingdom
MONAHAN, James	UKAEA/AERE Harwell, United Kingdom
MOONS, Frans	CEN/SCK Mol, Belgium
MORESCHI, Luigi F.	ENEA Brasimone, Italy
NACKAERTS, Herman	CEN/SCK Mol, Belgium
NATHANSON, Paul	NUCOR Pretoria, Rep. of South Africa
NEGRINI, Alberto	AMN Genova, Italy
NOBILI, Antonio	ENEA Casaccia, Italy
NOLTEN, Antonius	ECN Petten, The Netherlands
NOVAK, Jiri	IRNE Rez, Czechoslovak
OGATA, Keizo	Nippon Nuclear Fuel Devel. Ibaraki, Japan
OJANEN, Matti	Institute of Radiation Protection Helsinki, Finland
OLSEN, Jorgen Seir	RISØ National Lab., Denmark
OMARINI, Sergio	ENEA Rome, Italy
PAGÈS, Jean-Paul	CEN Cadarache, France
PINEIRA, Thomas	CEN Saclay, France
PLAUM, Josephus	ECN Petten, The Netherlands
PLUCHERY, Max	CEN Grenoble, France
PRIOR, Harry	UKAEA/AERE Harwell, United Kingdom
QUIST, Jens	RISØ National Lab., Denmark
REYMANN, Alfred	GKSS Geesthacht, Germany
RÖTTGER, Heinz	JRC Petten, The Netherlands
ROUSSEAU, Denys	CEN Grenoble, France
ROUSSET, Pierre	CEN Saclay, France
SAINFORT, Gérard	CEN Grenoble, France
SALAHI, Abolghassem	Atominstitut Wien, Austria
SALLE, Pierre	CEN Saclay, France
SCANDOLA, Gianni	ENEA Bologne, Italy
SCHLEY, Robert	CEN Saclay, France
SCHMIDT, Johny Olesen	RISØ National Lab., Denmark
SEFERIADIS, Johann	KFA Jülich, Germany

SIMONI, Oscar	JRC Ispra, Italy
SNELSON, W.K.	A.E.C.L. Ontario, Canada
SOEKARNO, Suyudi	BATAN Jakarta, Indonesia
SPITZ, Jean-Pierre	CEN Grenoble, France
STECHEMESSER, Horst	KFA Jülich, Germany
STENDAL, Karsten	RISØ National Lab., Denmark
STOLZ, Erich	TüV e. V. Essen, Germany
SUHARNO	BATAN Jarkarta, Indonesia
SWANENBURG DE VEYE, Robert	ECN Petten, The Netherlands
THAMM, Gerd	KFA Jülich, Germany
TIIHONEN, Olli	Technical Research Centre of Finland Helsinki, Finland
TROTABAS, Marie	CEN Saclay, France
TSCHAMPER, Pierre	GE Co. Sunnyvale, U.S.A.
VAN DER BOORN	CEN/SCK Mol, Belgium
VAN DER KOOGH, Gerbrand	ECN Petten, The Netherlands
VAN OTTERDYK, Karel	ECN Petten, The Netherlands
VERNAY, Michel	CEN Saclay, France
VEYRAT, Jean-François	CEN Grenoble, France
VIDAL, Henry	CEN Saclay, France
VON DER HARDT, Peter	JRC Petten, The Netherlands
VULLIEZ, Pierre	CEN Grenoble, France
WARLOP, Raymond	CEN Grenoble, France
WITTEK, Gerd	KfK Karlsruhe, Germany
YUKIO, Ikawa	Mitsubishi Atomic Power Industries Saitam-Ken, Japan
ZEISSER, Peter	JRC Petten, The Netherlands

AUTHOR INDEX

A

ABASSIN, J.J. 519
AHLF, J. 101, 133, 593
ALTER, D. 77
ANDERS, D. 133
ATABEK, R. 267
ANDRESEN, K. 567

B

BACKS, H. 405, 739
BAUER, G.S. 623
BELMANN, D. 101, 593
BENSCH, F. 697
BESSON, A. 719
BEVILACQUA, A. 719
BLOCH, S. 719
BLOM, K.H. 101
BOURBOUSSON, P. 93

C

CADALBERT, R. 451
CALVARESI, G. 709
CAMPANA, M. 379
CARACCHINI, R. 369, 719
CAUVIN, J.C. 299
CHAMRÁD, B. 161
CONRAD, R. 499
COQUERELLE, M. 461
COSOLI, G. 719
CROON, C. 725
CUTFORD, D.C. 391

D

DEBRUE, J. 475
DEKEYSER, J. 355
DE RAEDT, Ch. 475, 739
DETA VERNIER, W. 739
DEREYMEZ, P. 405
DE VRIES, J.E. 725
DEWES, P. 77
DUCROS, G. 605
DUPONT, G. 369

E

ERBEN, O. 161

F

FALLA, A. 439
FARNY, G. 117
FEHRENBACH, P.J. 3
FOUARGE, L. 739
FRIBOULET, C. 93
FRIČ, Z. 161
FUJISHIRO, T. 191

G

GACHE, G. 335
GÄRTNER, M. 77
GALTIER, J. 369
GOUTAGNY, L. 335
GRASS, F. 697
GUYON, H. 405

H

HANSEN, K. 567
HARPER, A. 509
HARTOG, J.M. 615
HEYDORN, K. 567
HOSHI, T. 231
HYODO, H. 309

I

IKAWA, Y. 279
ISHIDA, Y. 309
ITAMI, H. 529
ITO, H. 529

J

JACOBI, S. 429
JEGAT, A. 93
JENSEN, A.M. 175
JOLY, C. 739
JUNOD, E. 405

K

KATHOL, W. 345
KELLER, R. 499
KENNEDY, T.D.A. 143
KERMARREC, J. 299
KLAGE, D. 255
KNAPPIK, P. 461
KONRAD, J. 672
KOMINE, I. 279
KŘÍŽ, A. 161
KRUG, W. 549
KUBOTA, T. 279

L

LAMBERT, J.D.B. 391
LANCHI, M. 369
LECARDINAL, G. 335
LEIPOLD, M. 241
LENTZ, G.L. 391
LUCOT, M. 35

M

MANVEL, H. 725
MCCARDELL, R.K. 175, 213
MACDONALD, P.E. 213
MAENE, N. 475
MARIN, F. 291
MARINUCCI, G. 369
MARKGRAF, J. 241, 255
MARTENS, H. 593
MATFIELD, R.S. 579
MEHNER, A.W. 519
MENIL, R. 475
MICHEL, F. 605
MITCHELL, E.J. 49
MONAHAN, J. 49
MONSTER, I. 725
MOONS, F. 489, 647
MORESCHI, L.F. 709

N

NACKAERTS, H. 23
NOBILI, A. 369
NOLTEN, A.J. 669

NOVÁK, J. 13

O

OKADA, K. 309
OGATA, K. 309

P

PAGES, J.P. 299
PERRIER, C. 461
PERRY, D. 229, 243
PETITCOLAS, H. 719
PÍNA, J. 161
PLAUM, J.M. 725
PLOUYOUX, M. 405
PLUCHERY, M. 417
PRIOR, H.A. 509

R

REYMANN, A. 101
ROUSSEAU, D. 405
ROUSSET, P. 451

S

SAITO, S. 191
SALAMON, J. 161
SALAHI, A. 697
SALLE, P. 267
SCHLEY, R. 267
SCHMIDT, F. 549
SCHMIDT, J.O. 685, 697
SCHREIBER, R. 461
SCIERS, P. 519
SEFERIADIS, J. 549
SEIDEL, B.R. 391
ŠMÍD, J. 13
SNELSON, W.K. 201
SOENEN, M. 739
SPENCER, W.A. 175
STECHEMESSER, H. 623

STEIGER, W. 541
STENDAL, K. 567

T

TAKAHASHI, H. 529
TANAKA, I. 529
TANZAWA, S. 191
THAMM, G. 549, 623
THE, H.L. 101
TOURNEBIZE, F. 405
TSUYOZAKI, N. 529

V

VAN DER KOOGH, G. 319
VAN OTTERDIJK, K. 657
VULLIEZ, P. 93, 417

W

WITTEK, G. 541

Z

ZBĚHLIK, J. 161
ZUGAREK, G. 697